The HCS12/9S12:

An Introduction to Software and Hardware Interfacing

Second Edition

Han-Way Huang
Minnesota State University · Mankato

DELMAR
CENGAGE Learning

Australia · Brazil · Japan · Korea · Mexico · Singapore · Spain · United Kingdom · United States

DELMAR
CENGAGE Learning™

The HCS12 / 9S12: An Introduction to Software and Hardware Interfacing, 2nd Edition
Han-Way Huang

Vice President, Career and Professional Editorial: Dave Garza

Director of Learning Solutions: Sandy Clark

Acquisitions Editor: Stacy Masucci

Managing Editor: Larry Main

Senior Product Manager: John Fisher

Senior Editorial Assistant: Dawn Daugherty

Vice President, Career and Professional Marketing: Jennifer McAvey

Executive Marketing Manager: Deborah S. Yarnell

Senior Marketing Manager: Erin Coffin

Marketing Coordinator: Shanna Gibbs

Production Director: Wendy Troeger

Production Manager: Mark Bernard

Art Director: David Arsenault

Technology Project Manager: Christopher Catalina

Production Technology Analyst: Thomas Stover

For product information and technology assistance, contact us at
Professional Group Cengage Learning Customer & Sales Support, 1-800-354-9706

For permission to use material from this text or product, submit all requests online at **cengage.com/permissions**
Further permissions questions can be emailed to **permissionrequest@cengage.com**

Library of Congress Control Number: 2009920892

ISBN-13: 978-1-4354-2742-6

ISBN-10: 1-4354-2742-4

Delmar
5 Maxwell Drive
Clifton Park, NY 12065-2919
USA

Cengage Learning is a leading provider of customized learning solutions with office locations around the globe, including Singapore, the United Kingdom, Australia, Mexico, Brazil, and Japan. Locate your local office at: **international.cengage.com/region**

Cengage Learning products are represented in Canada by Nelson Education, Ltd.

For your lifelong learning solutions, visit **delmar.cengage.com**

Visit our corporate website at **cengage.com.**

Notice to the Reader
Publisher does not warrant or guarantee any of the products described herein or perform any independent analysis in connection with any of the product information contained herein. Publisher does not assume, and expressly disclaims, any obligation to obtain and include information other than that provided to it by the manufacturer. The reader is expressly warned to consider and adopt all safety precautions that might be indicated by the activities described herein and to avoid all potential hazards. By following the instructions contained herein, the reader willingly assumes all risks in connection with such instructions. The publisher makes no representations or warranties of any kind, including but not limited to, the warranties of fitness for particular purpose or merchantability, nor are any such representations implied with respect to the material set forth herein, and the publisher takes no responsibility with respect to such material. The publisher shall not be liable for any special, consequential, or exemplary damages resulting, in whole or part, from the readers' use of, or reliance upon, this material.

Printed in the United States of America
1 2 3 4 5 6 7 12 11 10 09

Contents

Chapter 2   HCS12 Assembly Programming   39

Chapter 3 Hardware and Software Development Tools for the HCS12 87

Chapter 4 Advanced Assembly Programming 145

Chapter 5 C Language Programming 205

Chapter 6 Interrupts, Clock Generation, Resets, and Operation Modes 261

Chapter 7   Advanced Parallel I/O   303

Chapter 8   Timer Functions   363

Chapter 9 Serial Communication Interface (SCI) 443

Chapter 10   The SPI Function   473

Chapter 11 Inter-Integrated Circuit (I²C) Interface   529

Chapter 12 Analog-to-Digital Converter   591

Chapter 13 Controller Area Network (CAN) 633

Chapter 14   Internal Memory Configuration and External Memory Expansion   693

Appendices   755

References   827

Glossary   829

Index   839

Preface to Second Edition

The Freescale HCS12 (also known as 9S12) microcontroller family was initially designed for automotive applications. The design of the HCS12 combines most features common in major 8-bit and 16-bit microcontrollers.

1. *Full-feature timer system*. The HCS12 timer system provides input-capture, output-compare, pulse-width modulation, pulse accumulator, modulus down counter, real-time interrupt, and computer-operate-properly systems.

2. *Background debug mode (BDM)*. The BDM circuit provides a single-wire interface for accessing the internal resources of the HCS12 and hence allows a low-cost debug adapter to be constructed.

3. *Multiple serial interfaces*. The HCS12 supports industrial-standard UART, SPI, I^2C, BDLC, and the CAN bus. The UART allows the HCS12 to interface with the PC using the popular RS232 protocol. The SPI and I^2C allow the HCS12 to interface with numerous peripheral devices (e.g., LED drivers, LCDs, matrix displays, A/D converters, D/A converters, real-time clocks, EEPROMs, Ethernet controllers, phase-locked-loops, and so on).

4. *In-system programming (ISP) capability*. Most HCS12 members provide on-chip flash memory and allow the software to be upgraded in the system.

5. *Fuzzy-logic support*. The HCS12 provides a group of instructions to support fuzzy-logic operations. These instructions should facilitate the programming of fuzzy-logic applications in assembly language.

With these features, the HCS12 is very suitable for those who want to learn modern microcontroller interfacing and applications.

Intended Audience

This book is written for three groups of readers.

1. Students in electrical and computer engineering and technology who are taking an introductory course in microprocessor interfacing and applications. For this group of readers, this book provides a broad and systematic introduction to microprocessors and microcontrollers.

2. Students in electrical and computer engineering and technology who are taking an embedded-system design course. This book pays attention to design methodology, programming style, and debug strategy in addition to discussing the general HCS12 programming and interfacing and hence should be suitable for an embedded system design course.

3. Senior electrical engineering and computer engineering students and working engineers who want to learn the HCS12 and use it in design projects. For this group of readers, this book provides numerous more complicated examples to explore the functions and applications of the HCS12.

Prerequisites

The writing of this book has assumed that the reader has taken a course on digital logic design and has been exposed to high-level language programming. Knowledge of digital logic will greatly facilitate learning the HSC12. Knowledge of assembly language programming is not required because one of the goals of this book is to teach the HCS12 assembly language programming.

Approach

Both assembly and C languages are used to illustrate the programming of the HCS12 microcontroller. Learning about the microcontroller using assembly language may produce an intimate feel for the functioning of the hardware. However, the programming productivity of assembly language is low because the programmer needs to implement the program logic at a very low level. C language has the edge in programming productivity. However, the code generated by a C compiler is still much larger than its equivalent in assembly language. Many time-critical applications are still written in assembly language, and many applications mix the use of assembly and C languages.

Organization of the Book

Chapter 1 starts with a discussion of the number system issue. It then presents the hardware structure of the computer, explains how the computer is started and executes instructions, elaborates the addressing modes, and discusses the operations of a subset of the HCS12 instructions. Chapter 2 starts with the format of the HCS12 assembly program. It then progresses to discuss assembler directives, software development methodology, arithmetic programming, program loops, and the HCS12 instructions.

Chapter 3 gives an overview of the hardware and software development tools and then goes on to give tutorials on the use of MiniIDE and CodeWarrior. A few tips on debugging assembly programs are also given in this chapter. Chapter 4 discusses the stack data structure, subroutine mechanism, software reuse, parallel I/O ports, and simple I/O devices. Many subroutine examples, including bubble sort, binary search, 32-bit division, square root, and prime test, are given. An example is given to illustrate the top-down design and hierarchical refinement system design methodology. Chapter 5 starts with a summary of the syntax of C language. Examples are then used to illustrate how to write single-function, multiple-function, and multiple-file C programs. Tutorials on the use of CodeWarrior IDE and ImageCraft ICC12 IDE in entering, compiling, and debugging C programs conclude Chapter 5. The tutorial on using the Embedded GNU (EGNU) IDE is given in Appendix E.

Chapter 6 introduces the concepts of interrupt and reset. Examples are then used to illustrate how interrupt programming is done in CodeWarrior, ICC12, and EGNU. Topics such as clock generation, real-time interrupt, computer operating properly (COP), and low-power modes are also elaborated in this chapter. Chapter 7 discusses I/O ports in detail and spells out the consideration for electrical compatibility. It then continues to elaborate on the topics of liquid crystal display (LCD), keypad, stepper motor, and D/A converter. Chapter 8 explores the operation and applications of the timer system, including input-capture, output-compare, real-time interrupt, pulse accumulator, and pulse-width modulation. Chapter 9 deals with serial communication interface (SCI). Chapter 10 examines the SPI interface and the applications of the SPI-compatible peripheral chips.

Chapter 11 introduces the I^{2}C protocol and several peripheral chips with I^{2}C interface. Chapter 12 discusses the A/D converter and its applications in temperature, humidity, and barometric pressure measurement. Chapter 13 presents the CAN 2.0 protocol and the HCS12 CAN module. Several examples of the programming of the CAN module are provided. Chapter 14 describes the HCS12 internal SRAM, EEPROM, and flash memory. This chapter also explores issues related to external memory expansion: address space assignment, address decoder design, and timing anaylsis.

Pedagogical Features

Each chapter starts with a list of objectives. Every subject is presented in a step-by-step manner. Background issues are presented before the specifics related to each HCS12 function are discussed. Numerous examples are then presented to demonstrate the use of each HCS12 I/O function. Procedural steps and flowcharts are used to help the reader to understand the program logic in most examples. Each chapter concludes with a summary, numerous exercises, and lab assignments.

Software Development Tools

MiniIDE, AsmIDE, and CodeWarrior are recommended for the development of assembly programs. MiniIDE and AsmIDE require the user to use a demo board programmed with a resident monitor, for example, D-Bug12. Users can only download their programs onto SRAM for execution. CodeWarrior allows users to perform source-level debugging to their programs, in which they can quickly identify and locate the program bugs and be able to resolve the problems. CodeWarrior allows users to download their programs onto both SRAM and flash memory for execution. CodeWarrior can work with a demo board programmed with the serial monitor (from Freescale) or with a BDM debug adapter (P&E's BDM adapters, Abatron's BDI adapter, Softec's inDART, or TBDML).

When using C language to program the HCS12, the CodeWarrior IDE is recommended for program entering, compiling, and debugging. The source-level debugging capability provided by CodeWarrior is a great help to pinpoint and locate program errors. The demo version of the ICC12 IDE from ImageCraft and the freeware Embedded GNU (EGNU) IDE can also be used. However, the user will need to use a demo board programmed with a resident debug monitor program such as D-Bug12. Neither of them can program the on-chip flash memory of the HCS12.

The tutorials for using MiniIDE, AsmIDE, CodeWarrior, ICC12, and EGNU IDEs are provided in this textbook.

Demo Boards

Demo boards are the most important hardware tool for learning a microcontroller. There are a few demo boards available to the reader for experimenting with the HCS12 hardware. The combination of the student project board and an HCS12 MCU module (e.g., 9S12C32, 9S12C128, or 9S12DT256) provided by Freescale is a viable option. More information can be found at www .freescale.com. The user needs to use CodeWarrior to enter, compile (or assemble), and debug his or her programs with this option.

Another viable demo board for experimenting with the HCS12 is the Dragon12-Plus made by Wytec. This demo board packs a lot of features and is very popular in universities. The user can choose a Dragon12-Plus demo board programmed with the D-Bug12 monitor or the serial monitor. When programmed with the serial monitor, the user will be able to use the source-level debugging capability provided by CodeWarrior. Most programs in this book are tested using the Dragon12-Plus demo board. Information about the Dragon12-Plus demo board can be found at www.evbplus.com.

To Instructors

It is unnecessary for instructors to follow strictly the order of chapters of this book in their teaching. If only assembly language programming is to be taught, then the following order is recommended:

- Chapters 1 through 4 in that order
- Chapters 6 and 7 in either order
- Chapter 8
- Chapters 9 through 14 in any order

If your microprocessor (or microcontroller) course only covers C language, then the following order is recommended:

- Sections 1.7 and 1.8
- Subjects related to I/O ports in Chapter 4
- Chapter 5
- Chapters 6 and 7 in either order
- Chapter 8
- Chapters 9 through 14 in any order

If both assembly and C languages are to be taught, then the following order is recommended:

- Chapters 1 through 4 in that order
- Chapter 5
- Chapters 6 and 7 in either order
- Chapter 8
- Chapters 9 through 14 in any order

Complementary Material

The following materials are useful to the learning of the HCS12 and are provided in the complementary CD included with this text:

- Source code of all example programs in the text
- The PDF files of datasheets of the HCS12 and peripheral chips
- The software (including CodeWarrior, demo version of ICC12, and EGNU)
- Utility programs (in assembly and C languages) for time delays, LCD, UART, and SPI

Supplements

A CD dedicated to instructors who adopt this text is also available from the publisher. This CD contains solutions to all exercise problems and the lecture notes in PowerPoint format. Professors are encouraged to modify the PowerPoint lecture notes to suit their teaching needs. ISBN: 1-4354-2743-2

Feedback and Update

The author has tried his best to eliminate errors from this text. However, it is impossible to eliminate all errors. The solutions in the examples of this book may not be the best either. Error reports and suggestions are welcomed. Please send them to hanwayh@yahoo.com or han-way .huang@mnsu.edu.

Acknowledgements

This book would not be possible without the help of a number of people. I would like to thank the following reviewers for their valuable opinions on how to improve the quality of this book: Shu-Jen Chen, DeVry University; Norm Grossman, DeVry University; and Abhiman Hande, Lake Superior State University. I would also like to thank Stacy Masucci, acquisition editor, and John Fisher, product manager, of Cengage Learning for their enthusiastic support during the preparation of this book and the Cengage staff for their outstanding production work. I would like to express my heart-felt appreciation to my students and colleagues at the Department of ECET at Minnesota State University, Mankato, who allowed me to test out the manuscript.

Finally, I would like to express my thanks to my wife, Su-Jane, and my sons, Craig and Derek, for their encouragement and support during the entire preparation of this book.

Han-Way Huang
Mankato, Minnesota
July 31, 2008

1

Introduction to the HCS12 Microcontroller

1.1 Objectives

On the successful completion of this chapter, you should be able to

- Use the appropriate prefix to represent numbers in different number bases
- Design the arithmetic logic units (ALUs) that implement multiple functions
- Design the circuit of the program counter
- Explain the memory technology
- Explain the operation of a memory system
- Explain the instruction execution process
- Use addressing modes to specify operands for the HCS12 instructions
- Use data transfer, add, and subtract instructions to perform simple operations

1.2 Number System Issue

Computers were initially designed as a number crunching machine. Due to the on-and-off nature of electricity, numbers were represented in binary base from the beginning of the electronic computer age. However, we are more used to the decimal number system due to the fact that it has been used for thousands of years. Therefore, entering numbers to be processed by the computer and outputting numbers to be viewed by the user would be done in decimal format. Computers need to perform conversion between binary and decimal representations. The unit used to represent the on or off state is referred to as a *bit*.

The number of bits used by a computer to represent a number is usually a multiple of 8. The computation capacity of a computer is also often expressed using the number of bits that it can operate on in one operation. For example, there are 8-bit, 16-bit, 32-bit, and 64-bit computers. Computers need to deal with both signed and unsigned numbers. Two's complement method is used to represent negative numbers. A number with its most significant bit set to 1 is negative. Otherwise it is a nonnegative number. Using two's complement representation allows the computer to use the same circuit to perform addition and subtraction.

In the microprocessor and microcontroller development tool's environment, we often see the mixed use of different number bases. Binary, decimal, octal, and hexadecimal have all been used. The binary number system has 2 digits: 0 and 1. The octal number system uses 8 different digits: 0 to 7. The decimal number system uses 10 different digits: 0 to 9. The hexadecimal number system uses 16 different digits: 0 to 9 and A through F. Since these four different number systems share some common digits, ambiguity is unavoidable. To clarify the ambiguity, a prefix is added to each number to indicate its base. The prefixes used in the Freescale microcontrollers are listed in Table 1.1. Microprocessors and microcontrollers from other vendors may use a different method to differentiate number bases.

A brief introduction to the conversion among different number bases is given in Appendix B.

Base	Prefix	Example
Binary	%	%10001010
Octal	@	@123467
Decimal		12345678
Hexadecimal (shorthand hex)	$	$392

Table 1.1 ■ Prefixes for number bases

1.3 Computer Hardware Organization

A computer consists of hardware and software. The hardware of a computer consists of the processor, input devices, output devices, and memory.

- *Processor.* The processor is responsible for performing all of the computational operations and the coordination of the usage of resources of a computer. A computer system may consist of one or multiple processors. A processor may perform general-purpose computations or special-purpose computations, such as graphics rendering, printing, or network processing.

- *Input devices.* A computer is designed to execute programs that manipulate certain data. Input devices are needed to enter the program to be executed and the data to be processed into the computer. There are a wide variety of input devices: keyboards, keypads, scanners, bar code readers, sensors, and so on.

- *Output devices.* Whether the user uses the computer to perform computations or to find information from the Internet or a database, the end results must be displayed or printed on certain media so that the user can see them. There are many media and devices that can be used to display information: cathode ray tube (CRT) displays, flat-panel liquid crystal displays (LCDs), seven-segment displays, printers, light-emitting diodes (LEDs), and so on.

- *Memory devices.* Users write *programs* to tell the computer what to do with the data at hand. Programs to be executed and data to be processed must be stored in memory devices so that the processor can readily access them.

1.3.1 The Processor

A processor is also referred to as the central processing unit (CPU). A processor consists of three major components: *arithmetic logic unit*, *control unit*, and *registers*.

THE ARITHMETIC LOGIC UNIT

The Arithmetic Logic Unit (ALU) performs arithmetic and logic operations requested by the user's program. The complexity of the ALU varies from one computer to another. If the processor designer wants to implement more operations directly in the hardware, then the ALU will get more complicated. An ALU that implements addition, subtraction, AND, and OR operations is illustrated in Figure 1.1. In Figure 1.1, all four operations are

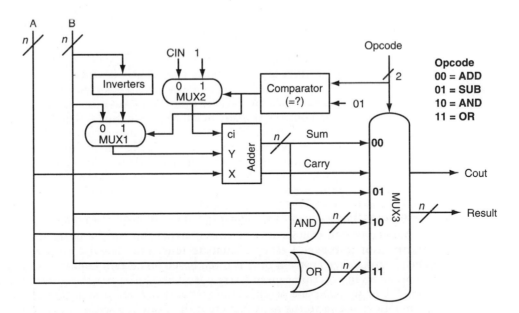

Figure 1.1 ■ An ALU that implements ADD, SUB, AND, and OR operations

performed simultaneously by different circuits whereas the opcode tells the multiplexer to select one of the four units' outputs as the result. The adder is used to perform addition and subtraction operations.

The four-operation ALU operates in the following manner:

When opcode = 00, the adder selects the n-bit A as its X input, the n-bit B as its Y input, and CIN as its Carry input (ci) and generates SUM and Carry. MUX3 selects SUM to become Result whereas Carry is connected to Cout directly. For this opcode, the ALU performs the ADD operation.

When opcode = 01, the comparator output is 1, the inversion of the B input is selected as the Y input, and 1 is selected as the ci to the adder. The adder essentially adds the two's complement of B to A, which is equivalent to performing the SUB operation. Since opcode is 01 and SUM is connected to both the 00 and 01 inputs, it will be selected and sent to Result. For this opcode, the ALU performs the SUB operation.

When opcode = 10, the MUX3 multiplexer selects the value connected to the 10 input and sends it to Result. Therefore, the ALU performs the AND operation.

When opcode = 11, the MUX3 multiplexer selects the value connected to the 11 input and sends it to Result. Therefore, the ALU performs the OR operation.

An ALU that performs more operations can be implemented by expanding the circuit shown in Figure 1.1.

CONTROL UNIT

From the beginning, the electronic digital computer is designed to execute *machine instructions* only. A machine instruction is a combination of 0s and 1s. To simplify the computer hardware design, most computers limit the instruction length to a few choices that are a multiple of 8 bits. For example, the HCS12 microcontroller from Freescale has instructions that are 8 bits, 16 bits, 24 bits, 32 bits, 40 bits, and 48 bits. You don't see instructions that are 13 bits, 29 bits, and so on.

A machine instruction has several fields. A mandatory field for every instruction is *opcode*, which tells the ALU what operation to perform. Other fields are optional; when they exist, they specify the operand(s) to be operated on.

To make the instruction execution time predictable, a *clock* signal is used to synchronize and set the pace of instruction execution. A clock signal is also needed to control the access of registers in the processor and external memory. The clock frequencies of the 8-bit and 16-bit microcontrollers range from a few MHz to over 100 MHz.

Since a program consists of many machine instructions, there is a need to keep track of what instruction to execute next. The control unit has a register called *program counter* (PC) that serves this function. Whenever the processor fetches an instruction from memory, the program counter will be incremented by the length of that instruction so that it points to the next instruction. The fetched instruction will be placed in the *instruction register* (IR), decoded, and executed. During this process, appropriate control signals will be generated to control the hardware circuit operation.

A program is normally not sequential. The execution order of machine instructions may be changed due to the need to execute instructions on the basis of the value of a certain condition or to repeat a group of instructions. This is called *program flow control*. The decision to change program flow is often based on certain conditions, for example, whether the previous instruction caused Carry out to be 1, whether the result of the previous operation is 0, or whether the result of the previous operation is negative. These conditions are often collected in a *status register* so that they can be used to make a decision. This type of program flow change is implemented by a *conditional branch* instruction (may also be called *conditional jump* instruction). There is a limit to the distance that the processor can branch (or jump) conditionally. The *branch distance* (referred to as *branch offset*) is from −128 byte to 127 bytes for most 8-bit and 16-bit microcontrollers because they use 8 bits to specify branch offset. In other situations, the programmer wants to force the processor to execute the instruction in any location within the available memory space. A jump instruction is used for this purpose. The *target of jump* may be specified in 16 bits, 32 bits, or 64 bits depending on the width of the program counter.

It is easy to figure out that writing programs in machine instructions is extremely difficult. Over the years, assembly language and high-level languages such as FORTRAN, COBOL, BASIC, C, C++, JAVA, and so on have been invented. Assembly language uses a mnemonic symbol to represent each machine instruction. The result is that each machine instruction is represented by an *assembly instruction*. The programmer can see the assembly instruction and figure out what operation is going to be performed quickly for most instructions. The assembly language makes programming much easier than in machine language. However, assembly language is still at a very low level. It is not very productive to write large and complicated programs in assembly language. Moreover, it needs a translator, called an *assembler*, to translate assembly instructions into machine instructions so that they can be executed by the computer.

High-level language is at a much higher level. Therefore, one statement written in high-level language may be translated into tens or even hundreds of machine instructions. A program written in high-level language also needs a translator to translate it into machine instructions so that it can be executed by the computer. The translator of a high-level programming language is called a *compiler*. The translation from high-level language to machine language is often not optimal. Therefore, there are some applications that require very tight performance controls that are still written in assembly language. It is not unusual to find large programming projects that are written in both assembly and high-level languages.

Programs written in assembly language or high-level languages are referred to as *source code* whereas the outputs of assembler and compiler are called *object code*.

REGISTERS

A register is a storage location inside the CPU. It is used to hold data and/or a memory address during the execution of an instruction. Because the register is very close to the CPU, it can provide fast access to operands for program execution. The number of registers varies greatly from processor to processor.

A processor may add a special register called an *accumulator* and include it as one of the operands for most instructions. The Intel 8051 variants, the Microchip PIC18, and the Freescale HCS12 microcontroller use this approach. Using the dedicated accumulator as one of the operands can shorten the instruction length. Other processors, for example, Atmel AVR and Microchip PIC32, may include many general-purpose data registers (16 or 32) in the CPU and allow any data register to be used as any operand of most instructions with two or three operands. This provides great freedom to the compiler during the program translation process. A processor designed using this approach is considered to be *orthogonal*.

1.3.2 Microprocessor

The earlier processors may be implemented in one or multiple printed circuit boards. With the advancement of integrated circuit technology, a complete processor can be implemented in one integrated circuit (an integrated circuit is often called a *chip*). A microprocessor is a processor implemented in a single integrated circuit.

In 1968, the first microprocessors Intel 4004 and TI TMS 1000 were introduced. Both the Intel 4004 and TI TMS 1000 are 4-bit microprocessors. In 1972, Intel introduced the Intel 8008, which was the first 8-bit microprocessor in the world. Several other 8-bit microprocessors were introduced after the Intel 8008 including the Intel 8080, Zilog Z80, Motorola 6800, Rockwell 6502, and so on. Microprocessors were quickly used as the controller of many products. Because of their small size (compared to discrete logic), programmability, ease of use, and low cost, microprocessors were well received and quickly replaced discrete logic devices.

However, the microprocessor still has a few disadvantages.

1. The microprocessor does not have on-chip memory. The designer needs to add external memory chips and other glue logic circuit such as decoder and buffer chips to provide program and data storage.

2. The microprocessor cannot drive the input/output (I/O) devices directly due to the fact that the microprocessor may not have enough current to drive the I/O devices or the voltage levels between the microprocessor and I/O devices may be incompatible. This problem is solved by adding peripheral chips as a buffer between the microcontroller and the I/O devices. The Intel 8255 parallel interface chip is one of the earliest interface chips.

3. The microprocessor does not have peripheral functions such as parallel I/O ports, timers, analog-to-digital (A/D) converter, communication interface, and so on. These functions must be implemented using external chips.

Because of these limitations, a product designed with microprocessors cannot be made as compact as might be desired. One of the design goals of microcontrollers is to eliminate these problems.

1.3.3 Microcontroller

A microcontroller (MCU) incorporates the processor and one or more of the following modules in one very large-scale integrated circuit (VLSI):

- Memory
- Timer functions
- Serial communication interfaces such as the Universal Synchronous Asynchronous Receiver Transmitter (USART), serial peripheral interface (SPI), interintegrated circuit (I²C), and controller area network (CAN)
- A/D converter
- Digital-to-analog (D/A) converter
- Direct memory access (DMA) controller
- Parallel I/O interface (equivalent to the function of Intel 8255)
- Memory component interface circuitry
- Software debug support hardware

The discussion of these functions is the subject of this textbook. Since their introduction, MCUs have been used in almost every application that requires a certain amount of intelligence. They are used as controllers for displays, printers, keyboards, modems, charge card phones, palm-top computers, and home appliances such as refrigerators, washing machines, and microwave ovens. They are also used to control the operations of engines and machines in factories. One of the most important applications of MCUs is probably the automobile control. Today, a luxurious car may use more than 100 MCUs. Today, most homes have one or more MCU-controlled consumer electronics appliances.

1.3.4 Embedded Systems

An embedded system is a special-purpose computer system designed to perform a dedicated function. Unlike a general-purpose computer, such as a personal computer, an embedded system performs one or a few predefined tasks, usually with very specific requirements, and often includes task-specific hardware and mechanical parts not usually found in a general-purpose computer. Since the system is dedicated to specific tasks, design engineers can optimize it, reducing the size and cost of the product. Embedded systems are often mass produced, benefiting from economy of scale.

Physically, embedded systems range from portable devices such as digital watches and MP3 players, to large stationary installations like traffic lights, factory controllers, or the systems that control power plants. In terms of complexity, embedded systems run from simple, with a single microcontroller chip, to very complex with multiple units, peripherals, and networks mounted inside a large chassis or enclosure.

Mobile phones or handheld computers share some elements with embedded systems, such as the operating systems and microprocessors that power them, but are not truly embedded systems themselves because they tend to be more general purpose, allowing different applications to be loaded and peripherals to be connected.

CHARACTERISTICS OF EMBEDDED SYSTEMS

Embedded systems have the following characteristics:

- Embedded systems are designed to perform some specific task, rather than being a general-purpose computer for multiple tasks. Some also have real-time performance constraints that must be met, for reasons such as safety and usability; others may have low or no performance requirements, allowing the system hardware to be simplified to reduce costs.

- An embedded system is not always a separate block; very often it is physically built into the device it is controlling.

- The software written for embedded systems is often called firmware and is stored in read-only memory or flash memory chips rather than a disk drive. It often runs with limited computer hardware resources: small or no keyboard or screen and little memory.

USER INTERFACES

Embedded systems range from having no interface at all—dedicated to only one task—to full user interfaces similar to desktop operating systems in devices such as personal digital assistants (PDAs).

A simple embedded system may use buttons for input and use *LEDs* or small character-only display for output. A simple *menu system* may be provided for users to interface with.

A more complex system may use a full graphical screen that has *touch* sensing or screen-edge buttons to provide flexibility while at the same time minimize space. The meaning of the buttons can change with the screen.

Handheld systems often have a screen with a "joystick button" for a pointing device. The rise of the World Wide Web has given embedded designers another quite different option: providing a webpage interface over a network connection. This avoids the cost of a sophisticated display, yet provides complex input and display capabilities when needed, on another computer. This is successful for remote, permanently installed equipment. In particular, routers take advantage of this ability.

1.4 Memory

There are three major memory technologies in use today: magnetic, optical, and semiconductor.

1.4.1 Magnetic Memory

Magnetic drum, magnetic tape, and magnetic hard disk are three major magnetic memory devices that have been invented. Magnetic drum has long been obsolete, and magnetic tape is only used for data archival. Currently only magnetic hard disk is still being used in almost every PC, workstation, *server*, and *mainframe* computer. Hard-drive vendors are still vigorously improving the hard-disk density. It doesn't seem possible that any memory technology can totally replace the hard disk yet.

1.4.2 Optical Memory

There are two major optical memory technologies in use today: compact disc (CD) and digital videodisc (DVD). The CD was introduced to the market in 1982 and has several variations. The most popular single-sided CD has a 12-cm diameter and can hold about 700 MB of data. The CD-R version of the compact disc can be recorded once whereas the CD-RW disc can be re-recorded many times. A single-sided DVD with 12-cm diameter can hold 4.7 GB of data. There are several versions of the DVD; among them, DVD-R can be recorded only once whereas DVD-RW can be re-recorded many times by the end user. There are several possible competing successors to the current DVD technology. They have single-sided capacities ranging from 15 to 25 GB.

1.4.3 Semiconductor Memory

Semiconductor memory is the dominant memory technology used in embedded systems. Memory technologies can be classified according to several criteria. Two common criteria are volatility and read-writability. On the basis of volatility, semiconductor memories are divided into *volatile* and *nonvolatile* memories. On the basis of read-writability, semiconductor memories are divided into *random-access memory* (RAM) and *read-only memory* (ROM).

1.4.4 Nonvolatile and Volatile Memory

A memory device is *nonvolatile* if it does not lose the information stored in it even without the presence of power. If a memory device cannot retain its stored information in the absence of power, then it is *volatile*.

1.4.5 Random-Access Memory

Random-access memory allows the CPU to read from or write to any location within the chip for roughly the same amount of time. RAM can be *volatile* or *nonvolatile*. RAM is also called *read/write memory* because it allows the processor to read from and write to it. As long as the power is on, the microprocessor can write data to a location in the RAM chip and later read back the same contents. Reading memory is nondestructive. Writing memory is destructive. When the microprocessor writes data to memory, the old data is written over and destroyed.

There are four types of commercially available RAM technology: *dynamic* RAM (DRAM), *static* RAM (SRAM), *magnetoresistive* RAM (MRAM), and *ferroelectric* RAM (FRAM).

DRAMs are memory devices that require periodic refreshes of the stored information. *Refresh* is the process of restoring binary data stored in a particular memory location. The dynamic RAM uses one transistor and one capacitor to store 1 bit of information. The information is stored in the capacitor in the form of electric charges. The charges stored in the capacitor will leak away over time, so periodic refresh operations are needed to maintain the contents in the DRAM. The time interval over which each memory location of a DRAM chip must be refreshed at least once in order to maintain its contents is called its *refresh period*. Refresh periods typically range from a few milliseconds to over a hundred milliseconds for today's high-density DRAMs.

SRAMs are designed to store binary information without needing periodic refreshes and require the use of more complicated circuitry for each bit. Four to six transistors are needed to store 1 bit of information. As long as power is stable, the information stored in the SRAM will not be degraded.

MRAMs were first developed by IBM. Several other companies were also involved in the research, development, and marketing of this technology. MRAMs use a magnetic moment to store data. A MRAM chip combines a magnetic device with standard silicon-based microelectronics to achieve the combined attributes of nonvolatility, high-speed operation, and unlimited read and write endurance. The first MRAM device from Freescale is the 4-Mbit MR2A16A. This device is a parallel memory (8 or 16 bits can be accessed in one operation) and has a 35 ns access time, reported in 2007.

FRAMs use the property of ferroelectric crystal to store data bits. Much of the present FRAM technology was developed by Ramtron International. The FRAM technology has already achieved high maturity. Both the serial and parallel versions of FRAM chips are available. Ramtron even incorporates FRAM in some of its 8051 microcontroller products. The fastest access time of FRAM from Ramtron is 55 ns, reported in 2007. However, the access time of FRAM may improve in the future.

RAM is mainly used to store *dynamic* programs or data. A computer user often wants to run different programs on the same computer, and these programs usually operate on different sets of data. The programs and data must therefore be loaded into RAM from the hard disk or other secondary storage, and for this reason they are called *dynamic*.

1.4.6 Read-Only Memory

ROM is nonvolatile. When power is removed from ROM and then reapplied, the original data will still be there. As its name implies, ROM data can only be read. If the processor attempts to write data to a ROM location, ROM will not accept the data, and the data in the addressed ROM memory location will not be changed. However, this statement is not completely true. For some ROM technologies (EEPROM and flash memory), the user program can still write data into the memory by following a special procedure prescribed by the manufacturer. However, it would take a much longer time to write than to read from the flash memory.

Mask-programmed read-only memory (MROM) is a type of ROM that is programmed when it is manufactured. The semiconductor manufacturer places binary data in the memory according to the request of the customer. To be cost-effective, many thousands of MROM memory units, each consisting of a copy of the same data (or program), must be sold. MROM is the major memory technology used to hold microcontroller application programs and constant data. Most people simply refer to MROM as ROM. The design of MROM prevents it from being written into.

Programmable read-only memory (PROM) was invented in 1956 by Wen Tsing Chow. It is a form of memory where the setting of each bit is locked by a fuse or antifuse. The memory can be programmed just once after manufacturing by blowing the fuses (using a *PROM blower*), which is an irreversible process. Blowing a fuse opens a connection whereas blowing an antifuse closes a connection (hence the name). Programming is done by applying high-voltage pulses that are not encountered during normal operation (typically 12 to 21 volts). Fused-based PROM technology is no longer in use today.

Erasable programmable read-only memory (EPROM) was invented by the Israeli engineer Dov Frohman in 1971. It is an array of floating-gate transistors individually programmed by an electronic device that supplies higher voltages than those normally used in electronic circuits. Programming is achieved via *hot carrier injection* onto the floating gate. Once programmed, an EPROM can be erased only by exposing it to strong ultraviolet (UV) light. That UV light usually has a wavelength of 235 nm for optimum erasure time. EPROMs are easily recognizable by the transparent fused quartz window in the top of the package, through which the silicon chip can be seen and which permits UV light to go through during erasing.

As the quartz window is expensive to make, one-time programmable (OTP) chips were introduced; the only difference is that the EPROM chip is packed in an opaque package, so it cannot be erased after programming. OTP versions are manufactured for both EPROMs themselves and EPROM-based microcontrollers. However, OTP EPROM (whether separate or part of a larger chip) is being increasingly replaced by EEPROM for small amounts where the cell cost isn't too important and flash memory is used for larger amounts.

A programmed EPROM retains its data for about 10 to 20 years and can be read an unlimited number of times. The erasing window must be kept covered with a foil label to prevent accidental erasure by sunlight. Old PC *basic input/output system* (BIOS) chips were often EPROMs, and the erasing window was often covered with a label containing the BIOS publisher's name, the BIOS revision, and a copyright notice.

Electrically erasable programmable read-only memory (EEPROM) was developed in 1983 by George Perlegos at Intel. It was built on earlier EPROM technology, but used a thin gate oxide layer so that the chip could erase its own bits without requiring a UV source. EEPROM is programmed and erased using the process called field emission (more commonly known in the industry as *Fowler-Nordheim tunneling*). EEPROM allows the user to selectively erase a single location, a row, or the whole chip. This feature requires a complicated programming circuitry. Because of this, the EEPROM cannot achieve the density of the EPROM technology.

Flash memory was invented by Fujio Masuoka while working for Toshiba in 1984. Flash memory incorporates the advantages and avoids the drawbacks of EPROM and EEPROM technologies. The flash memory can be erased and reprogrammed in the system without using a dedicated programmer. It achieves the density of EPROM, but it does not require a window for erasure. Like EEPROM, flash memory can be programmed and erased electrically. However, it does not allow individual locations to be erased; the user can only erase a block or the whole chip. Today, the BIOS programs of many high-performance PCs are stored in flash memory. Most microcontrollers introduced today use on-chip flash memory as their program memory.

Flash memory chips have also been used in flash disk memory, personal digital assistants, digital cameras, cell phones, and so on.

1.5 Memory System Operation

A simplified memory system block diagram is shown in Figure 1.2. A memory system may consist of one or multiple memory chips. Both memory chips and memory systems are organized as an array of memory locations. A memory location may hold any number of bits (most common numbers are 4 bits, 8 bits, 16 bits, 32 bits, and 64 bits). The memory organization of a memory chip or a memory system is often indicated by **m × n;** where m specifies the number of memory locations in the memory chip or memory system and n specifies the number of bits in each location. Every memory location has two components: *contents* and *address*.

A memory location can be used to store data, instruction, and the status of peripheral devices. The size of memory is measured in bytes; a byte consists of 8 bits. A 4-bit quantity is

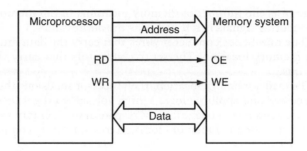

Figure 1.2 ■ Block diagram of a simplified memory system

called a *nibble*. A 16-bit quantity is called a *word*. To simplify the quantification of memory, the units *kilobyte* (kB), *megabyte* (MB), and *gigabyte* (GB) are often used. The value of k is given by the following formula:

k = 2^{10} = 1024

M is given by the following formula:

M = k^2 = 2^{20} = 1024 × 1024 = 1,048,576

G is given by the following formula:

G = k^3 = 2^{30} = 1024 × 1024 × 1024 = 1,073,741,824

In this book, we will use the notation [addr] and [reg] to refer to the contents of a memory location at addr and the contents of register reg, respectively. We will use the notation m[addr] to refer to the memory location at addr. For example,

[$40]

refers to the contents of the memory location at $40 and [A] refers to the contents of accumulator A. The notation **m[$40] ← [A]** refers to saving the contents of accumulator A in the memory location at $40. Registers are referred to by their names whereas memory locations are referred to by their addresses.

The memory chip or memory system can only be accessed (read or written) one location at a time. This is enforced by implementing a decoder on the memory chip to select one and only one location to be accessed. There are two types of memory accesses: *read* and *write*.

1.5.1 Read Operation

Whenever the processor wants to read a memory, it sends out the address of the location it intends to read. Since the memory access can be a read or a write, the processor needs to use a control signal to inform the memory of the type of access. In Figure 1.2, the RD signal from the processor indicates a read access whereas the WR signal indicates a write operation. The memory chip also has control signals to control the read or write operation. The OE signal in Figure 1.2 means *output enable* and is connected to the RD signal from the processor whereas the WE signal means *write enable* and is connected to the WR signal from the processor. For digital systems, there are three logic states for each signal: high, low, and high impedance (no current flows). When the OE input to the memory chip is low, the data pins are in a high-impedance state.

The processor uses a set of signals, referred to as *address signals*, to specify a memory location to access. The number of address signals needed for selecting a memory location is $\log_2 m$, where m is the number of memory locations in the memory. The set of conductor wires that carry address signals is referred to as the *address bus*.

The number of conductor wires that carry the data must be equal to the number of bits in each memory location. The set of conductors that carry the data to be accessed is called the *data bus*.

To read a memory location, the processor sends out the address of the memory location to be accessed and applies a logic 1 (high voltage) to the RD signal and a logic 0 to the WR signal (this specifies a read operation). In response, the memory system decodes the address input and enables the specified memory location to send out its contents to the data bus to be read by the processor.

1.5.2 Write Operation

To write a value to a location of the memory system in Figure 1.2, the processor places the data to be written on the data bus and places the address of the memory location on the address bus and applies a logic 1 (high voltage) to the WR signal and a logic 0 to the RD signal (this specifies a write operation). In response, the memory system uses its address decoder to select a location and writes the value on the data bus to that location.

The actual memory system design and the signals involved may be different from those in Figure 1.2 but the concept would be the same. The semiconductor vendors may use $\overline{\text{RD}}$ instead of RD and $\overline{\text{WR}}$ instead of WR to refer to read and write signals. These types of signals are active low; that is, when they are low, they are considered to be at logic 1.

1.6 Program Execution

In order to allow the computer to execute the program immediately after the power is turned on, part of the program must be stored in nonvolatile memory. Some computers placed the startup program in the nonvolatile memory, which will perform the system initialization. After the system initialization is completed, it loads additional programs from secondary storage such as hard disk or optical storage into the semiconductor memory (often called *main memory*) for execution. Mainframe computers, workstations, and personal computers follow this approach. After power is turned on, the processor starts to execute the program from the BIOS, which performs the system initialization. After system initialization is completed, the processor loads additional programs such as Windows operating system into the main memory for execution. Other computers, including most embedded systems, place all their programs in the nonvolatile memory. After power-up, the processor starts to execute the program from the nonvolatile memory.

The following sections deal with several important issues related to program execution.

1.6.1 The Circuit of the Program Counter

The program counter consists of flip-flops and other additional logic gates. There are several types of flip-flops in use. Among them, the D-type flip-flop is the most popular one. The circuit of a D-type flip-flop with set and reset capability is shown in Figure 1.3.

In Figure 1.3,

- Depending on the design, the D value may be transferred to Q on either the rising or the falling edge (but not both edges) of the CLK input.
- The CLK signal is the clock input signal of the D flip-flop.

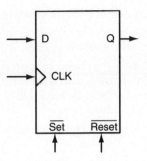

Figure 1.3 ■ Block diagram of a D flip-flop with a set and reset

- The Q signals of all the flip-flops of the program counter determine the address of the next instruction to be fetched.
- The $\overline{set}$ and $\overline{reset}$ inputs are active low (low voltage means logic 1) and cannot be low at the same time. When $\overline{set}$ is low, the Q signal is forced to 1. When $\overline{reset}$ is low, the Q signal is forced to 0.

As described in Section 1.3.1, a microprocessor or microcontroller has instructions to change the program flow. The design of the program counter circuit must take this into account. Figure 1.4 shows the block diagram of a program counter of an 8-bit microcontroller that allows the program counter to be

- Forced to 0
- Incremented by 1
- Incremented by a field in the IR
- Loaded with a jump target

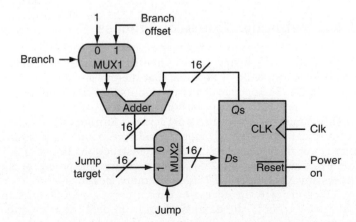

Figure 1.4 ■ A simplified block diagram of the program counter (PC) of an 8-bit microcontroller

For the program counter circuit shown in Figure 1.4,

- Whenever power is turned on to the microcontroller, the program counter is forced to 0 and the instruction fetch will start from address 0.
- If the instruction being executed is a conditional branch instruction and the branch condition is true, then the branch signal will be 1, the sum of the current PC and *branch offset* will be loaded into the PC, and instruction execution will continue from that address.
- If the instruction being executed is a jump instruction, then the value *Jump target* will be loaded into the PC.
- If the instruction being executed is not a program flow control instruction, then the PC is simply incremented by 1 after each instruction byte is fetched.

Other microprocessors or microcontrollers may have a different program flow control scheme and may fetch more instruction bytes in one fetch. In that case the program counter shown in Figure 1.4 will need to be modified accordingly.

1.6.2 Where Does the Processor Start to Execute the Program?

As discussed earlier in this chapter, the program counter holds the address of the next instruction to be fetched, so the value of the program counter must be known when power is turned on. One approach is to force the PC to a fixed value when power is turned on. The circuit shown in Figure 1.4 forces the PC to 0 whenever power is turned on. Many 8-bit micro-controllers including Microchip PIC, all Intel 8051 variants, and Atmel AVR use this approach because it is easy to implement.

Another approach is to fetch the program starting address from a fixed (known) memory location whenever the power is turned on. The Freescale microcontrollers use this approach. The HCS12 microcontroller from Freescale fetches the program starting address from memory locations at 0xFFFE and 0xFFFF into PC and then start program execution from there. This approach is slightly more complicated.

Another way to restart program execution is to apply a reset signal to the processor. All microprocessors and microcontrollers have a reset pin that allows the user to force the processor to start from scratch. The effect is identical to turning on the power.

1.6.3 Instruction Execution Process

The instruction sets of most commercial processors are irregular and complicated. The complexity of the instruction set makes it difficult to explain the instruction execution process. In the following, we assume that there is an 8-bit processor X with instruction set shown in Table 1.2. The opcode of any instruction is 1 byte and is always the first byte of the instruction. The processor X has an 8-bit accumulator A and a 16-bit pointer register ptr. The data memory and program memory are separate and are each 64 kB in size. The register ptr is used to point to data memory and supports indirect memory addressing for data memory. The instruction set of the processor X allows the instructions to use an 8-bit address to access the lowest 256 bytes (addresses 0 to 255) of data memory. The processor X can use the 16-bit ptr register to access any location of the 2^{16} data memory locations.

To facilitate the access of data memory, processor X includes the memory data register (MDR) to hold the data received from data memory and data to be written to the data memory.

Assembly Instruction Mnemonic		Machine Code	Meaning
ld	**addr, #val**	**75 aa xx**	Load the 8-bit value (val) into memory location at addr.
ld	**ptr, #data16**	**90 yyyy**	Load the 16-bit value (data16) into the register ptr.
ld	**A, @ptr**	**E0**	Load the contents of memory location pointed to by ptr into A.
and	**A, #val**	**54 xx**	And the 8-bit value (val) with A and leave the result in A.
bnz	**addr, offset**	**70 zz**	Branch to a location that is offset from the next instruction if the value at addr is not zero.
inc	**addr**	**05 aa**	Increment the contents of memory location at addr.
dbnz	**addr, offset**	**D5 aa zz**	Decrement the contents of memory location at addr and branch if the result is not zero. The branch distance is offset.

Note: aa: an 8-bit value that represent an 8-bit address.

xx: an 8-bit value.

yyyy: a 16-bit value.

zz: distance of branch from the first byte of the instruction after the branch instruction.

Machine codes are expressed in hex format.

Table 1.2 ■ The instruction set of the processor X

1.6.4 Instruction Sequence Example

Assume that the following instruction sequence is stored in the program memory starting from address 0 so that it will be executed immediately after a power-on or reset:

```
         ld      0x20,#0        ; place 0 in data memory located at address 0x20
         ld      0x21,#20       ; place 20 in data memory located at address 0x21
         ld      ptr,#0x2000    ; load 0x2000 into the register ptr
loop:    ld      A,@ptr         ; load the memory contents pointed to by ptr into A
         and     A,#0x03        ; and the value 0x03 with A and leave the result in A
         bnz     next           ; branch if the result in A is not 0
         inc     0x20           ; increment the memory location at 0x20 by 1
next:    dbnz    0x21,loop      ; decrement the memory location at 0x21 and branch if
                                ; the result is not 0
```

The corresponding machine code of the given instruction sequence is shown in Table 1.3. The next section explains the process of instruction execution.

1.6.5 Instruction Execution Process

Processor X executes the instruction sequence given in Table 1.3 as follows:

Instruction   ld 0x20,#0 (machine code 75 20 00)

When the processor comes out of a reset or power-on process, the program counter is forced to 0 and this instruction will be fetched and executed. The execution of this instruction involves the following steps:

Step 1
The value in the PC (0x0000) is placed on the address bus of the program memory with a request to read the contents of that location.

Assembly Instruction Mnemonic			Address	Machine Code	Comment
	ld	0x20,#0	0x0000	75 20 00	
	ld	0x21,#20	0x0003	75 21 14	
	ld	ptr,#0x2000	0x0006	90 20 00	
loop:	ld	A,@ptr	0x0009	E0	
	and	A,#0x03	0x000A	54 03	
	bnz	next	0x000C	70 02	02 is the branch offset.
	inc	0x20	0x000E	05 20	
next:	dbnz	0x21,loop	0x0010	D5 21 0A	0A is the branch offset.

Note: 1. The user uses a label to specify the instruction to branch to and the assembler needs to figure out the branch offset.
2. The assembler figures out that the label next is 2 bytes away from the inc 0x20 instruction.
3. The assembler figures out that the label *loop* is 10 (0A) bytes away from the first byte after the **"next: dbnz 0x21,loop"** instruction.

Table 1.3 ■ The processor X instruction sequence to be executed

Step 2

The 8-bit value at the location 0x0000 is the instruction opcode 0x75. At the end of this read cycle, the PC is incremented to 0x0001. The opcode byte 0x75 is fetched. Figure 1.5 shows the opcode read cycle.

Step 3

Control unit recognizes that this version of the ld instruction requires one read cycle to fetch the direct address and another cycle to read the data operand. These 2 bytes are stored immediately after the opcode byte. Two more read cycles to program memory are performed to access the data memory address 0x20 (held in IR) and the value 0x00 (held in IR). After these two read cycles, the PC is incremented to 0x0003.

Step 4

Control unit places 0x0020 on the data memory address bus and the value 0x00 on the data memory data bus to perform a write operation. The value 0x00 is to be stored at data memory location 0x0020, as shown in Figure 1.6.

Instruction   ld 0x21,#20 (machine code 75 21 14)

The execution of this instruction is identical to that of the previous instruction. After the execution of this instruction, the PC is incremented to 0x0006 and the data memory location 0x21 receives the value of 20.

Instruction   ld ptr,#0x2000 (machine code 90 20 00)

Step 1

The value in the PC (0x0006) is placed on the program memory address bus with a request to read the contents of that location.

Step 2

The 8-bit value at the location 0x0006 is the instruction opcode 0x90. At the end of this read cycle, the PC is incremented to 0x0007. The opcode byte 0x90 is fetched. Figure 1.7 shows the opcode read cycle.

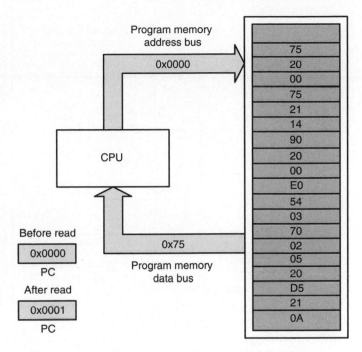

Figure 1.5 ■ Instruction 1–opcode read cycle

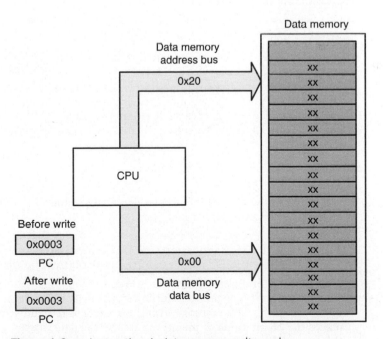

Figure 1.6 ■ Instruction 1–data memory write cycle

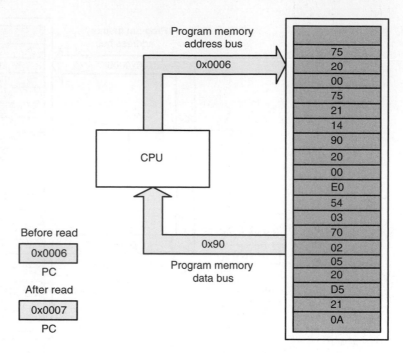

Figure 1.7 ■ Instruction 3–opcode read cycle

Step 3

The control unit recognizes that this instruction requires two more read cycles to the program memory to fetch the 16-bit value to be placed in the ptr register. These 2 bytes are stored immediately after the opcode byte. The control unit continues to perform two more read cycles to the program memory. At the end of each read cycle, the processor X stores the received byte in the ptr register upper and lower bytes, respectively. After these two read operations, the PC is incremented to 0x0009.

Instruction   ld A,@ptr (machine code E0)

Step 1

The value in the PC (0x0009) is placed on the program memory address bus with a request to read the contents of that location.

Step 2

The 8-bit value at the location 0x0009 is the instruction opcode 0xE0. At the end of this read cycle, the PC is incremented to 0x000A. The opcode byte 0xE0 is fetched.

Step 3

The control unit recognizes that the current instruction requires performing a read operation to the data memory with the address specified by the ptr register. The processor places the 16-bit value of the ptr register on the data memory address bus and indicates this is a read operation.

Step 4

The data memory returns the contents to the processor and the processor places it in accumulator A. The process is shown in Figure 1.8.

Instruction and A,#0x03 (machine code 54 03)

Step 1

The value in the PC (0x000A) is placed on the program memory address bus with a request to read the contents of that location.

Step 2

The 8-bit value at the location 0x000A is the instruction opcode 0x54. At the end of this read cycle, the PC is incremented to 0x000B. The program memory returns the opcode byte 0x54 to the CPU.

Step 3

The control unit recognizes that the current instruction requires performing a read operation on the program memory to fetch the operand for the AND operation. It then places the PC value on the program memory address bus again with a read request.

Step 4

The program memory returns the value 0x03 to the CPU. The PC is incremented to 0x000C.

Step 5

The CPU then performs an AND operation on the contents of accumulator A and the value 0x03 and places the result in A.

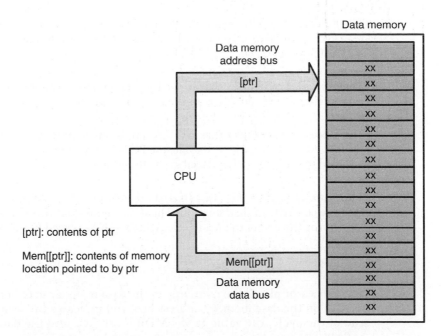

Figure 1.8 ■ Instruction 4–data memory and read cycle

Instruction   bnz next (machine code 70 02)

Step 1
The value in the PC (0x000C) is placed on the program memory address bus with a request to read the contents of that location.

Step 2
The 8-bit value at the location 0x000C is the instruction opcode 0x70. At the end of this read cycle, the PC is incremented to 0x000D. The program memory returns the opcode byte 0x70 to the CPU.

Step 3
The processor recognizes that this is a conditional branch instruction and it needs to fetch the branch offset from the program. So it places the PC value (0x000D) on the program memory address bus with a read request. At the end of this read cycle, the processor increments the PC to 0x000E.

Step 4
The program memory returns the branch offset 0x02 to the CPU (held in IR). The CPU checks the contents of accumulator A to determine whether the branch should be taken. Let's assume that A contains zero and the branch is not taken. The PC remains at 0x000E. If A contains a nonzero value, the next instruction will be skipped.

Instruction   inc 0x20 (machine code 05 20)

Step 1
The value in the PC (0x000E) is placed on the program memory address bus with a request to read the contents of that location.

Step 2
The 8-bit value at the location 0x000E is the instruction opcode 0x05. At the end of this read cycle, the PC is incremented to 0x000F. The program memory returns the opcode byte 0x05 to the CPU.

Step 3
The processor recognizes that it needs to increment a data memory location; this requires it to fetch an 8-bit address from the program memory.

Step 4
The processor places the value in the PC on the program memory address bus with a read request. At the end of the read cycle, the PC is incremented to 0x0010 and the value 0x20 is returned to the CPU and is placed in the IR register.

Step 5
The processor places the value 0x20 on the data memory address bus with a request to read the contents of that location. The data memory returns the value of that memory location at the end of the read cycle, which will be placed in the MDR.

Step 6
The processor adds 1 to the MDR.

Step 7
The processor places the contents on the data memory data bus and places the value 0x0020 on the data memory address bus, and indicates this is a write cycle. At the end of the cycle, the value in the MDR is written into the data memory location at 0x20.

Instruction   dbnz 0x21,loop (machine code = D5 21 0A)

Step 1

The value in the PC (0x0010) is placed on the program memory address bus with a request to read the contents of that location.

Step 2

The 8-bit value at the location 0x0010 is the instruction opcode 0xD5. At the end of this read cycle, the PC is incremented to 0x0011. The program memory returns the opcode byte 0xD5 to the CPU.

Step 3

The CPU recognizes that it needs to read a data memory address and a branch offset from the program memory.

Step 4

Processor X performs two more read operations to the program memory. The program memory returns 0x21 and 0x0A (both are held in IR). At the end of these two read cycles, the PC is incremented to 0x0013.

Step 5

Processor X places 0x21 on the data memory address bus with a read request. At the end of the read cycle, the value of the data memory location at 0x21 is returned to the CPU which will be held in the MDR.

Step 6

Processor X decrements the contents of the MDR. The contents of the MDR are then placed on the data memory data bus. Processor X also places the address 0x21 on the data memory address bus with a write request to store the contents of the MDR in data memory.

Step 7

If the value stored in the MDR is not zero, processor X adds 0x0A to the PC and places the result in the PC (this causes a branch behavior). Otherwise, the PC is not changed.

This section demonstrates the activities that may occur during the execution of a program. Overall, the operations performed by the processor are dictated by the opcode.

1.7 Overview of the HCS12 Microcontroller

Freescale designed the 68HC12 as an upgrade to the 8-bit 68HC11 microcontroller. However, Motorola discovered that the performance of the 68HC12 microcontroller was not satisfactory after it was introduced to the market. The 68HC12 has the highest bus clock speed of 8 MHz. To be competitive, Freescale revised the design to achieve a bus clock rate of 25 MHz (a few microcontrollers can run at 33 MHz). The revised 68HC12 was referred to as the Star12 family. It was also named the HCS12 family. The HCS12 MCU has the same instruction set and addressing modes as does the 68HC12. However, many of the internal designs have been changed.

Automotive and process control applications are the two major target markets of the HCS12. This is evidenced by the inclusion of such peripheral functions as *input capture* (IC), *output compare* (OC), *pulse-width modulation* (PWM), *controller area network* (CAN), and *byte data link control* (BDLC). Other peripheral functions such as *serial peripheral interface* (SPI), *serial communication interface* (SCI), and *interintegrated circuit* (I²C) are also included to facilitate interconnection with a wide variety of peripheral chips.

Using flash memory to hold application programs has become the trend of microcontroller design. All HCS12 members incorporate on-chip flash memory to hold programs. Most HCS12

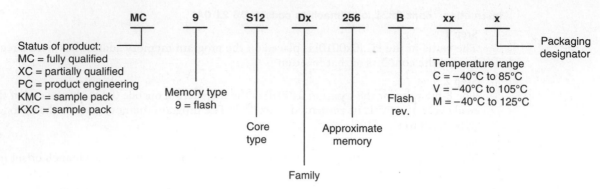

Figure 1.9 ■ Freescale product numbering system for the HCS12

devices also include a certain amount of on-chip SRAM and EEPROM to hold data and/or programs needed in different applications.

Most HCS12 devices have many I/O pins to interface with I/O devices. When on-chip memory is not adequate, external memory can be added. All HCS12 devices adopt the same design for the same peripheral function to facilitate the migration from one device (with less memory or fewer peripheral functions) to another.

The features of all HCS12 devices are shown in Appendix C. All devices with the CAN module are for automotive applications. The numbering system for the HCS12 is shown in Figure 1.9.

In addition to automotive and control applications, Freescale is also trying to attract users from other application areas. For example, the MC9S12NE64 was designed for applications that need to access the Internet whereas the MC9S12UF32 was designed for interfacing with the USB bus.

Software debugging support is an important issue for embedded applications. Freescale has implemented the Background Debug Mode (BDM) in each HCS12 member. With this BDM module, the tool developers can design inexpensive software debugging tools for the HCS12. Most of the HCS12 features and peripheral functions will be discussed in this text.

1.8 The HCS12 CPU Registers

The HCS12 microcontroller has registers for supporting general-purpose operations and controlling the functioning of peripheral modules. These registers are divided into two categories: CPU registers and I/O registers. CPU registers are used solely to perform general-purpose operations such as arithmetic, logic, and program flow control. I/O registers are mainly used to configure the operations of peripheral functions, to hold data transferred in and out of the peripheral subsystem, and to record the status of I/O operations. The I/O registers in a microcontroller can further be classified into *data*, *data direction*, *control*, and *status registers*. These registers are treated as memory locations when they are accessed. CPU registers do not occupy the HCS12 memory space.

The CPU registers of the HCS12 are shown in Figure 1.10 and are listed next. Some of the registers are 8 bit and others are 16 bit.

General-purpose accumulators A and B. Both A and B are 8-bit registers. Most arithmetic functions are performed on these two registers. These two accumulators can also be concatenated to form a single, 16-bit accumulator referred to as the D accumulator.

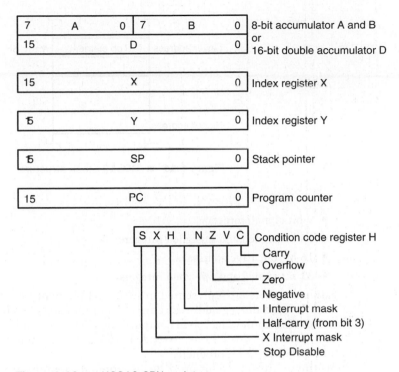

Figure 1.10 ■ HCS12 CPU registers

Index registers X and Y. These two registers are used mainly in forming operand addresses during the instruction execution process. However, they are also used in several arithmetic operations.

Stack pointer (SP). A stack is a *last-in-first-out* data structure. The HCS12 has a 16-bit stack pointer which points to the top byte of the stack (shown in Figure 1.11). The stack grows toward lower addresses. The use of the stack will be discussed in Chapter 4.

Program counter. The 16-bit PC holds the address of the next instruction to be executed. After the execution of an instruction, the PC is incremented by the number of bytes of the executed instruction.

Condition code register (CCR). This 8-bit register is used to keep track of the program execution status, control the execution of conditional instructions, and enable/disable the interrupt handling. The contents of the CCR register are shown in Figure 1.10. The function of each condition code bit will be explained in later sections and chapters.

The HCS12 supports the following types of data:

- Bits
- 5-bit signed integers
- 8-bit signed and unsigned integers
- 8-bit, two-digit binary-coded-decimal (BCD) numbers

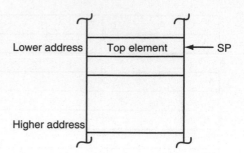

Figure 1.11 ■ HCS12 stack structure

- 9-bit signed integers
- 16-bit signed and unsigned integers
- 16-bit effective addresses
- 32-bit signed and unsigned integers

Negative numbers are represented in two's complement format. Five-bit and 9-bit signed integers are formed during addressing mode computations. Sixteen-bit *effective addresses* are formed during addressing mode computations. Thirty-two-bit integer dividends are used by extended division instructions. Extended multiply and extended multiply-and-accumulate instructions produce 32-bit products.

A multibyte integer (16 bit or 32 bit) is stored in memory from most significant to least significant bytes, starting from low to higher addresses.

1.9 HCS12 Addressing Modes

An HCS12 instruction consists of 1 or 2 bytes of opcode and 0 to 5 bytes of operand addressing information. The opcode specifies the operation to be performed and the addressing mode(s) used to access the operand(s). The addressing mode determines how the CPU accesses registers or memory locations to be operated on. The HCS12 supports the following address modes:

1.9.1 Inherent Mode

In this mode, the instruction does not use extra bytes to specify the operand. Instructions using this mode either have no operands or all operands are in internal CPU registers. Therefore the CPU does not need to access memory in order to fetch operands. For example, the following two instructions use inherent mode:

```
nop      ; this instruction has no operand
inx      ; operand is a CPU register
```

1.9.2 Immediate Mode

In this mode, the value to be operated on has been included in the instruction itself. An immediate value can be 8 bit or 16 bit depending on the context of the instruction. An immediate value is preceded by a # character in the assembly instruction. For example,

```
ldaa     #$55          ; A ← $55
```

places the hex value $55 in accumulator A when this instruction is executed.

    ldx   #$2000   ; X ← $2000

places the hex value $2000 in index register X when this instruction is executed.

    movw   #$10,$100   ; m[$100] ← $00; m[$101] ← $10

stores the hex value $00 and $10 in memory locations at $100 and $101, respectively, when this instruction is executed. Only an 8-bit value was supplied in this instruction. However, the assembler will generate the 16-bit value $0010 because the CPU expects a 16-bit value when this instruction is executed.

1.9.3 Direct Mode

This addressing mode is sometimes called *zero-page* addressing because it is used to access operands in the address range of $0000 to $00FF. Since these addresses begin with $00, only the 8 low-order bits of the address need to be included in the instruction; this saves program space and execution time. For example,

    ldaa $20   ; A ← [$20]

fetches the contents of the memory location at $0020 and puts it in accumulator A.

    ldx $20   ; X_H ← [$20], X_L ← [$21]

fetches the contents of memory locations at $0020 and $0021 and places them in the upper and lower bytes (X_H and X_L), respectively, of the index register X.

1.9.4 Extended Mode

In this addressing mode, the full 16-bit address of memory location to be operated on is provided in the instruction. This addressing mode can be used to access any location in the 64-kB memory map. For example,

    ldaa $2000   ; A ← [$2000]

copies the contents of the memory location at $2000 into accumulator A.

1.9.5 Relative Mode

The relative addressing mode is used only by branch instructions that may change the program flow. The distance of the branch (or jump) is referred to as *branch offset*. Short and long conditional branch instructions use the relative addressing mode exclusively. Branching versions of bit manipulation instructions (BRSET and BRCLR) may also use the relative addressing mode to specify the branch target. A short branch instruction consists of an 8-bit opcode and a signed 8-bit offset contained in the byte that follows the opcode. Long branch instructions consist of an 8-bit prebyte, an 8-bit opcode, and a signed 16-bit offset contained in 2 bytes that follow the opcode.

Each conditional branch instruction tests certain status bits in the condition code register. If the bits are in a specified state, the offset is added to the address of the next instruction to form an effective address, and execution continues at that address; if the bits are not in the specified state, execution continues with the instruction next to the branch instruction.

Both 8-bit and 16-bit offsets are signed two's complement numbers to support branching forward and backward in memory. The numeric range of the short branch offset values is $80 (−128) to $7F (127). The numeric range of the long branch offset values is $8000 (−32768) to $7FFF (32767). If the offset is zero, the CPU executes the instruction immediately following the branch instruction, regardless of the test result.

Branch offset is often specified using a label rather than a numeric value due to the difficulty of calculating the exact value of the offset. For example, in the following instruction segment:

```
minus  .
       .              ; if N (of CCR) = 1
       .              ; PC ← PC + branch offset
bmi    minus          ; else
       ...            ; PC ← PC
```

The instruction **bmi minus** causes the HCS12 to execute the instruction with the label **minus** if the N flag of the CCR register is set to 1.

The assembler will calculate the appropriate branch offset when the symbol that represents the branch target is encountered. Using a symbol to specify the branch target makes the programming task easier and the resultant program more readable.

1.9.6 Indexed Addressing Modes

The indexed addressing mode uses two components to compute the effective address of an operand or the target of a jump instruction. The first component is called the *base address,* which is stored in a base register. The base register can be X, Y, SP, or PC. The second component is called the *offset,* which is the distance of the target from the base address. The effective address of the operand or jump target is the sum of these two components.

The offset may be a constant (5 bits, 9 bits, or 16 bits) or the contents of accumulator A, B, or D. The base register may be pre- or postincremented or pre- or postdecremented. The size of increment or decrement may be specified by the user and can be from −8 to +8. In addition, the HCS12 also provides the user one level of indirection. That is, the sum of the contents of the base register and the offset does not point to the actual operand or the jump target. Instead, it points to the memory location that holds the address of the actual operand or jump target.

The variations of the indexed addressing mode are described in the following subsections.

1.9.7 Indexed Addressing Modes with Constant Offsets

The syntax of the indexed addressing mode with constant offset is as follows:

n, r

where

n is a 5-bit, 9-bit, or 16-bit constant
r is the base register and can be X, Y, SP, or PC

For example,

```
ldaa   4,X       ; A ← [4 + [X]]
```

loads the contents of the memory location with the address equal to the sum and 4 and X into A.

The HCS12 performs the following two operations for the ldd 100, Y instruction:

```
A ← [100 + [Y]];
B ← [101 + [Y]];
```

1.9.8 Indexed Addressing Mode with Offset in an Accumulator

The syntax of this form of indexed address mode is as follows:

acc, r

where

acc can be A, B, or D
r is the base register and can be X, Y, SP, or PC

For example,

staa   B, X   ; m[[B] + [X]] ← [A]

stores the contents of A in the memory location of which the address equals the sum of the contents of B and X.

For the instruction

ldx   D, SP

the HCS12 performs the following operations:

X ← [[D] + [SP]]:[1 + [D] + [SP]]   ; 2 bytes are loaded into X

1.9.9 Auto Pre-/Postdecrement/-Increment Indexed Addressing Modes

The syntax and resultant effective address of this mode are shown in Table 1.4. For the predecrement/preincrement version of this addressing mode, the HCS12 decrements/increments the specified base register by the specified amount (n in Table 1.4) before using the contents of the base register as an effective address to access memory. For the postdecrement/postincrement version of this address mode, the HCS12 uses the contents of the specified base register as the effective address to access memory and then decrements/increments the specified base register. For example, if index register X contains $1000, then

staa   2, −X   ; predecrement X

stores the contents of accumulator A in the memory location at $9FE and the new value in X becomes $9FE.

ldaa   2, +X   ; preincrement X

loads the contents of memory location at $1002 into A and the new value of X is $1002.

sty   2, X−

stores the high and low bytes of Y in memory locations at $1000 and $1001, respectively. After that, index register X receives the new value of $9FE.

ldaa   4, X+

loads the contents of the memory location at $1000 into A. After that, index register X receives the new value of $1004.

Syntax	Effective Address	New Value of Base Register r	Example	Comment
n, −r	[r] − n	[r] − n	std 2, −SP	Predecrement
n, +r	[r] + n	[r] + n	ldd 2, +SP	Preincrement
n, r−	[r]	[r] − n	std 2, X−	Postdecrement
n, r+	[r]	[r] + n	std 2, Y+	Postincrement

Note: n = amount of decrement or increment.
        r = base register (may be X, Y, or SP).

Table 1.4 ■ Auto predecrement/increment or auto postdecrement/increment indexed modes

1.9.10 16-Bit Offset Indexed Indirect Mode

The syntax of this addressing mode is as follows:

[n, r]

where

n is the 16-bit offset

r is the base register and can be X, Y, SP, or PC

In this mode, the HCS12 fetches the actual effective address from the memory location with address equal to the sum of the 16-bit offset and the contents of the base register and then uses that effective address to access the operand. The square brackets distinguish this addressing mode from 16-bit constant offset indexing; for example,

ldaa   [10, X]

In this example, index register X holds the base address of a table of pointers. Assume that X has an initial value of $1000, and that $2000 is stored at addresses $100A and $100B. The instruction first adds the value 10 to the value in X to form the address $100A. Next, an address pointer ($2000) is fetched from memory locations at $100A and $100B. Then, the value stored in $2000 is read and loaded into accumulator A.

1.9.11 Accumulator D Indirect Indexed Addressing

The syntax of this addressing mode is as follows:

[D, r]

where

r is the base register and can be X, Y, SP, or PC

This indexed addressing mode adds the value in accumulator D to the value in the base index register to form the address of a memory location that contains a pointer to the memory location affected by the instruction. The instruction operand points not to the memory location to be acted on but rather to the location of a pointer to the location to be acted on. The square brackets distinguish this addressing mode from accumulator D offset indexing. For example, the following instruction sequence implements a computed GOTO statement:

```
        jmp     [D, PC]
GO1     dc.w    target1     ; the keyword dc.w reserves 2 bytes to hold the
GO2     dc.w    target2     ; value of the symbol that follows
GO3     dc.w    target3     ;                  "
        . . .
target1 . . .
        .
        .
target2 . . .
        .
        .
target3 . . .
        .
        .
```

In this instruction segment, the names (also called labels) *target1*, *target2*, and *target3* are labels that represent the addresses of the memory locations that the jmp instruction may jump to. The names GO1, GO2, and GO3 are also labels. They represent the memory locations that hold the values of the labels *target1*, *target2*, and *target3*, respectively.

The values beginning at GO1 are addresses of potential destinations of the jump instructions. At the time the jmp [D, PC] instruction is executed, the PC points to the address GO1 and D holds one of the values $0000, $0002, or $0004 (determined by the program some time before the jmp). Assume that the value in D is $0002. The jmp instruction adds the values in D and PC to form the address of GO2 and jumps to target2. The locations of target1 through target3 are known at the time of program assembly, but the destination of the jmp depends on the value in D computed during program execution.

1.10 Addressing More than 64 kB

The HCS12 devices incorporate hardware that supports addressing a larger memory space than the standard 64 kB. The expanded memory system is accessed by using the bank-switching scheme. The HCS12 treats the 16 kB of memory space from $8000 to $BFFF as a program memory window. The HCS12 has an 8-bit program page register (PPAGE), which allows up to 256 16-kB program memory pages to be switched into and out of the program memory window. This provides up to 4 MB of paged program memory space.

1.11 A Sample of HCS12 Instructions

It would be very helpful to learn a small set of HCS12 instructions that are used most often before we formally learn HCS12 assembly language programming. In the following, we will examine data movement, addition, and subtraction instructions. The HCS12 provides a large group of data movement instructions. Some of them may transfer data between a CPU register and a memory location. Some of them may transfer or exchange data between two registers. Others may transfer data from one memory location to another memory location.

1.11.1 The Load and Store Instructions

The load instruction copies the contents of a memory location or places an immediate value into an accumulator or a register. Memory contents are not changed. Store instructions copy the contents of a CPU register into a memory location. The contents of the accumulator or CPU register are not changed. Store instructions automatically update the N and Z flags in the condition code register (CCR). Table 1.5 is a summary of load and store instructions.

There are restrictions on the addressing modes that can be used in a load and a store instruction:

- For the load instruction, all except for the relative addressing mode can be used to select the memory location or value to be loaded into an accumulator or a CPU register.
- For the store instruction, all except for the relative and immediate addressing modes can be used to select the memory location to store the contents of a CPU register.

For example, the following instruction loads the contents of the memory location pointed to by index register X into accumulator A:

```
ldaa    0,X
```

The following instruction loads the contents of the memory location at $1004 into accumulator B:

```
ldab    $1004
```

The following instruction stores the contents of accumulator A in the memory location at $20:

```
staa    $20
```

Load Instructions		
Mnemonic	Function	Operation
LDAA <opr>	Load A	A ← [opr]
LDAB <opr>	Load B	B ← [opr]
LDD <opr>	Load D	A:B ← [opr]:[opr+1]
LDS <opr>	Load SP	SP ← [opr]:[opr+1]
LDX <opr>	Load index register X	X ← [opr]:[opr+1]
LDY <opr>	Load index register Y	Y ← [opr]:[opr+1]
LEAS <opr>	Load effective address into SP	SP ← effective address
LEAX <opr>	Load effective address into X	X ← effective address
LEAY <opr>	Load efective address into Y	Y ← effective address
Store Instructions		
Mnemonic	Function	Operation
STAA <opr>	Store A in a memory location	m[opr] ← [A]
STAB <opr>	Store B in a memory location	m[opr] ← [B]
STD <opr>	Store D in a memory location	m[opr]:m[opr+1] ← [A]:[B]
STS <opr>	Store SP in a memory location	m[opr]:m[opr+1] ← [SP]
STX <opr>	Store X in a memory location	m[opr]:m[opr+1] ← [X]
STY <opr>	Store Y in a memory location	m[opr]:m[opr+1] ← [Y]

Table 1.5 ■ Load and store instructions

The following instruction stores the contents of index register X in memory locations at $8000 and $8001:

```
stx      $8000
```

When dealing with a complex data structure such as a record, we often use an index register or the stack pointer to point to the beginning of the data structure and use the indexed addressing mode to access the elements of the data structure. For example, a record contains the following four fields:

- ID number (unit *none*, size 4 bytes)
- Height (unit *inch*, size 1 byte)
- Weight (unit *pound*, size 2 bytes)
- Age (unit *year*, size 1 byte)

Suppose this record is stored in memory starting at $6000. Then we can use the following instruction sequence to access the weight field:

```
ldx      #$6000      ; set X to point to the beginning of data structure
ldd      5, X        ; copy weight into D
```

1.11.2 Transfer and Exchange Instructions

A summary of transfer and exchange instructions is displayed in Table 1.6. Transfer instructions copy the contents of a register or accumulator into another register or accumulator. Source content is not changed by the operation. TFR is a universal transfer instruction, but other mnemonics are accepted for compatibility with the MC68HC11. The TAB and TBA instructions affect the N, Z, and V condition code bits. The TFR instruction does not affect the condition code bits.

It is possible to transfer from a smaller register to a larger one or vice versa. When transferring from a smaller register to a larger one, the smaller register is signed-extended to 16-bit and then assigned to the larger register. When transferring from a larger register to a smaller one, the smaller register receives the value of the lower half of the larger register. For example,

tfr	A,X	; A is signed-extended to 16 bits and then assigned to X
tfr	X,B	; B← X[7:0], B receives bits 7 to 0 of X

Exchange instructions (exg r1, r2) exchange the contents of pairs of registers or accumulators. For example,

    exg   A, B

exchanges the contents of accumulator A and B.

    exg   D,X

exchanges the contents of double accumulator D and index register X.

The r1 register does not need to have the same size as r2. If r1 has a larger size than r2 does, then r2 will be 0-extended to 16 bits and loaded into r1 whereas r2 will receive the lower half of r1. For example,

    exg   X,A   ; X ← $00:[A], A ← X[7:0]

Transfer Instructions		
Mnemonic	Function	Operation
TAB	Transfer A to B	B ← [A]
TAP	Transfer A to CCR	CCR ← [A]
TBA	Transfer B to A	A ← [B]
TFR	Transfer register to register	A, B, CCR, D, X, Y, or SP ← [A, B, CCR, D, X, Y, or SP]
TPA	Transfer CCR to A	A ← [CCR]
TSX	Transfer SP to X	X ← [SP]
TSY	Transfer SP to Y	Y ← [SP]
TXS	Transfer X to SP	SP ← [X]
TYS	Transfer Y to SP	SP ← [Y]
Exchange Instructions		
Mnemonic	Function	Operation
EXG	Exchange register to register	[A, B, CCR, D, X, Y, or SP] ⇔ [A, B, CCR, D, X, Y, or SP]
XGDX	Exchange D with X	[D] ⇔ [X]
XGDY	Exchange D with Y	[D] ⇔ [Y]
Sign Extension Instructions		
Mnemonic	Function	Operation
SEX	Sign extend 8-bit operand	X, Y, or SP ← [A, B, CCR]

Table 1.6 ■ Transfer and exchange instructions

If the r2 register is larger in size than r1, then r1 will be 0-extended to 16 bits and loaded into r2 whereas r1 will receive the lower half of r2. For example,

```
exg      A,Y           ; A ← Y[7:0], Y ← $00:[A]
```

The sex instruction is a special case of the universal transfer instruction that is used to sign-extend 8-bit two's complement numbers so that they can be used in 16-bit operations. The 8-bit number is copied from accumulator A, accumulator B, or the condition code register to accumulator D, index register X, index register Y, or the stack pointer. All the bits in the upper byte of the 16-bit result are given the value of the most significant bit of the 8-bit number. For example,

```
sex      A,X
```

copies the contents of accumulator A to the lower byte of X and duplicates bit 7 of A to every bit of the upper byte of X.

```
sex      B,Y
```

copies the contents of accumulator B to the lower byte of Y and duplicates bit 7 of B to every bit of the upper byte of Y.

Transfer instructions allow operands to be placed in the right register so that the desired operation can be performed. For example, if we want to compute the squared value of accumulator A, we can use the following instruction sequence:

```
tab              ; B ← [A]
mul              ; A:B ← [A] × [B]
```

Applications of other transfer and exchange instructions will be discussed in Chapters 2 and 4.

1.11.3 Move Instructions

A summary of move instructions is listed in Table 1.7. These instructions move data bytes or words from a source $(M\sim M+1_1)$ to a destination $(M\sim M+1_2)$ in memory. Six combinations of immediate, extended, and indexed addressing are allowed to specify source and destination addresses (IMM $\Rightarrow$ EXT, IMM $\Rightarrow$ IDX, EXT $\Rightarrow$ EXT, EXT $\Rightarrow$ IDX, IDX $\Rightarrow$ EXT, IDX $\Rightarrow$ IDX). Move instructions allow the user to transfer data from memory to memory or from I/O registers to memory and vice versa.

For example, the following instruction copies the contents of the memory location at $1000 to the memory location at $2000:

```
movb     $1000, $2000
```

The following instruction copies the 16-bit word pointed to by X to the memory location pointed to by Y:

```
movw     0,X, 0,Y
```

Transfer Instructions		
Mnemonic	Function	Operation
MOVB <src>, <dest>	Move byte (8-bit)	dest ← [src]
MOVW <src>, <dest>	Move word (16-bit)	dest ← [src]

Table 1.7 ■ Move instructions

Add Instructions		
Mnemonic	Function	Operation
ABA	Add B to A	$A \leftarrow [A] + [B]$
ABX	Add B to X	$X \leftarrow [X] + [B]$
ABY	Add B to Y	$Y \leftarrow [Y] + [B]$
ADCA <opr>	Add with carry to A	$A \leftarrow [A] + [opr] + C$
ADCB <opr>	Add with carry to B	$B \leftarrow [B] + [opr] + C$
ADDA <opr>	Add without carry to A	$A \leftarrow [A] + [opr]$
ADDB <opr>	Add without carry to B	$B \leftarrow [B] + [opr]$
ADDD <opr>	Add without carry to D	$D \leftarrow [D] + [opr]$
Subtract Instructions		
Mnemonic	Function	Operation
SBA	Subtract B from A	$A \leftarrow [A] - [B]$
SBCA <opr>	Subtract with borrow from A	$A \leftarrow [A] - [opr] - C$
SBCB	Subtract with borrow from B	$B \leftarrow [B] - [opr] - C$
SUBA <opr>	Subtract memory from A	$A \leftarrow [A] - [opr]$
SUBB <opr>	Subtract memory from B	$B \leftarrow [B] - [opr]$
SUBD <opr>	Subtract memory from D	$D \leftarrow [D] - [opr]$

Table 1.8 ■ Add and subtract instructions

1.11.4 Add and Subtract Instructions

Add and subtract instructions allow the HCS12 to perform fundamental arithmetic operations. A summary of add and subtract instructions is in Table 1.8. The <opr> field in Table 1.8 is specified using one of the legal addressing modes. All except inherent and relative modes are legal addressing modes for these two groups of instructions.

Example 1.1
▼

Write an instruction sequence to add 3 to the memory locations at $10 and $15.

Solution: A memory location cannot be the destination of an ADD instruction. Therefore, we need to copy the memory content into an accumulator, add 3 to it, and then store the sum back to the same memory location.

```
ldaa    $10    ; copy the contents of memory location at $10 to A
adda    #3     ; add 3 to A
staa    $10    ; store the sum back to memory location at $10
ldaa    $15    ; copy the contents of memory location at $15 to A
adda    #3     ; add 3 to A
staa    $15    ; store the sum back to memory location at $15
```

▲

Example 1.2

Write an instruction sequence to add the byte pointed to by index register X and the following byte and place the sum at the memory location pointed to by index register Y.

Solution: The byte pointed to by index register X and the following byte can be accessed by using the indexed addressing mode.

```
ldaa    0,X     ; put the byte pointed to by X in A
adda    1,X     ; add the following byte to A
staa    0,Y     ; store the sum at the location pointed to by Y
```

Example 1.3

Write an instruction sequence to add the numbers stored at $1000 and $1001 and store the sum at $1004.

Solution: To add these two numbers, we need to put one of them in an accumulator.

```
ldaa    $1000   ; copy the number stored in memory location at $1000 to A
adda    $1001   ; add the second number to A
staa    $1004   ; save the sum at memory location at $1004
```

Example 1.4

Write an instruction sequence to swap the 2 bytes at $100 and $200.

Solution: To swap the 2 bytes, we need to make a copy of one of the 2 bytes and then the swapping can proceed.

```
ldaa    $100        ; make a copy of m[$100] in A
movb    $200,$100   ; store [$200] in m[$100]
staa    $200        ; store the original [$100] in m[$200]
```

1.12 Instruction Queue

The HCS12 uses a three-stage instruction queue to facilitate instruction fetching and increase execution speed. Queue logic prefetches program information and positions it for sequential execution, one instruction at a time. The relationship between bus cycles and execution cycles is straightforward and facilitates tracking and debugging.

There are three 16-bit stages in the instruction queue. Instructions enter the queue at stage 1 and roll out after stage 3. Each byte in the queue is selectable. An opcode-prediction algorithm determines the location of the next opcode in the instruction queue.

Each instruction refills the queue by fetching the same number of bytes that the instruction uses. Program information is fetched in aligned 16-bit words. Each program fetch indicates that 2 bytes need to be replaced in the instruction queue. Each optional fetch indicates that only 1 byte needs to be replaced. For example, an instruction composed of 5 bytes does two program fetches and one optional fetch. If the first byte of the 5-byte instruction was even aligned, the optional fetch is converted into a free cycle. If the first byte was odd aligned, the optional fetch is executed as a program fetch.

Two external pins, IPIPE[1:0], provide time-multiplexed information about instruction execution and data movement in the queue. Decoding and using the IPIPE[1:0] signals are discussed in Chapter 14.

The content of queue stage 1 advances to stage 2, stage 2 advances to stage 3, and stage 1 is loaded with a word of program information from the data bus.

1.13 Summary

The invention of the microprocessor in 1968 resulted in a revolution in the electronics industry. The first microprocessor, the Intel 4004, incorporated a simplified CPU into an integrated circuit. Following the introduction of the 4-bit 4004, Intel introduced the 8-bit 8008, 8080, and 8085 over three years. The 8085 was a big success because of its programmability. Through this programmability, many products could be designed and constructed. Other companies joined in the design and manufacturing of microprocessors, Zilog, Motorola, and Rockwell being among the more successful.

The earliest microprocessors still needed peripheral chips to interface with I/O devices such as seven-segment displays, printers, timers, and so on. Memory chips were also needed to hold the application program and dynamic data. Because of this, the products designed with microprocessors could not be made as small as desired. Then came the introduction of microcontrollers, which incorporated the CPU, some amount of memory, and peripheral functions such as parallel I/O ports, timer, and serial interface functions onto one chip. The development of microcontrollers has had the following impacts:

- I/O interfacing is greatly simplified.
- External memory is no longer needed for many applications.
- System design time is greatly shortened.

A microcontroller is not designed to build a desktop computer. Instead, it is used as the controller of many products. End users of these products do not care what microcontrollers are used in their appliances; they only care about the functionality of the product. A product that uses a certain microcontroller as a controller and has this characteristic is called an *embedded system*. Cell phones, automobiles, cable modems, HDTVs, and home security systems are well-known embedded systems.

Over the last 20 years, we have clearly seen that a microcontroller needs to incorporate some or all of the following peripheral functions in order to be useful:

- Timer module that incorporates input capture, output compare, real-time interrupt, and counting capability
- Pulse-width modulation function for easy waveform generation
- Analog-to-digital converter
- Digital-to-analog converter
- Temperature sensor
- Direct memory access controller

- Parallel I/O interface
- Serial I/O interface such as UART, SPI, I²C, and CAN
- Memory component interface circuitry

The HCS12 from Freescale incorporates most of these peripheral modules and the CPU in one VLSI chip.

Memory is where software programs and data are stored. Semiconductor memory chips can be classified into two major categories: random-access memory and read-only memory. RAM technology includes DRAM, FRAM, MRAM, and SRAM. MROM, PROM, EPROM, EEPROM, and flash memory are read-only memories.

Programs are known as software. A program is a set of instructions that the computer hardware can execute. In the past, system designers mainly used assembly language to write microcontroller application software. The nature of assembly language forces an assembly programmer to work on the program logic at a relatively low level. This hampers programming productivity. In the last 15 years, more and more people have turned to high-level language to improve their programming productivity. C is the most widely used language for embedded system programming.

Although system designers use assembly or high-level language to write their programs, the microcontroller can only execute machine instructions. Programs written in assembly or high-level language must be translated into machine instructions before they can be executed. The program that performs the translation work is called an *assembler* or *compiler* depending on the language to be translated.

A machine instruction consists of *opcode* and *addressing information* that specifies the operands. Addressing information is also called *addressing mode*. The HCS12 implements a rich instruction set along with many addressing modes for specifying operands. This chapter examines the functions of a few groups of instructions. Examples were used to explore the implementation of simple operations using these instructions.

The execution of an instruction may take several clock cycles. Because the HCS12 does not access memory in every clock cycle, it performs instruction prefetch to speed up the instruction execution. A two-word (16-bit word) instruction prefetch queue and a 16-bit buffer are added to hold the prefetched instructions.

1.14 Exercises

E1.1 What is a processor?

E1.2 What is a microprocessor? What is a microcomputer?

E1.3 What makes a microcontroller different from a microprocessor?

E1.4 How many bits can the HCS12 CPU manipulate in one operation?

E1.5 How many different memory locations can the HCS12 access without the expanded memory?

E1.6 Why must every computer have some amount of nonvolatile memory?

E1.7 Why must every computer have some amount of volatile memory?

E1.8 What is source code? What is object code?

E1.9 Convert 5K, 8K, and 13K to decimal representation.

E1.10 Write an instruction sequence to swap the contents of memory locations at $1000 and $1001.

E1.11 Write an instruction sequence to add 10 to memory locations at $1000 and $1001, respectively.

E1.12 Write an instruction sequence to set the contents of memory locations at $1000, $1010, and $1020 to 10, 11, and 12, respectively.

E1.13 Write an instruction sequence to perform the operations equivalent to those performed by the following high-level language statements:

```
I = 11;
J = 33;
K = I + J − 5;
```

Assume that variables *I*, *J*, and *K* are located at $1000, $1005, and $1010, respectively.

E1.14 Write an instruction sequence to subtract the number stored at $1010 from that stored at $1000 and store the difference at $1005.

E1.15 Write an instruction sequence to add the contents of accumulator B to the 16-bit word stored at memory locations $1000 and $1001. Treat the value stored in B as a signed number.

E1.16 Write an instruction sequence to copy 4 bytes starting from $1000 to $1100~$1103.

E1.17 Write an instruction sequence to subtract the contents of accumulator B from the 16-bit word at $1000~$1001 and store the difference at $1100~$1101. Treat the value stored in B as a signed value.

E1.18 Write an instruction sequence to swap the 16-bit word stored at $1000~$1001 with the 16-bit word stored at $1100~$1101.

E1.19 Give an instruction that can store the contents of accumulator D at the memory location with an address larger than the contents of X by 8.

E1.20 Give an instruction that can store the contents of index register Y at the memory location with an address smaller than the contents of X by 10.

2

HCS12 Assembly
Programming

2.1 Objectives

After completing this chapter, you should be
able to

- Use assembler directives to allocate memory
  blocks, define constants, and create a message
  to be output

- Write assembly programs to perform simple
  arithmetic operations

- Write program loops to perform repetitive
  operations

- Use program loops to create time delays

- Use Boolean and bit manipulation instructions
  to perform bit field manipulation

2.2 Assembly Language Program Structure

An assembly language program consists of a sequence of statements that tells the computer to perform the desired operations. From a global point of view, an HCS12 assembly program consists of three sections. In some cases, these sections can be mixed to provide better algorithm design.

1. *Assembler directives.* Assembler directives instruct the assembler how to process subsequent assembly language instructions. Directives also provide a way to define program constants and reserve space for dynamic variables. Some directives may also set a location counter.

2. *Assembly language instructions.* These instructions are HCS12 instructions. Some instructions are defined with labels.

3. *Comments.* There are two types of comments in an assembly program. The first type is used to explain the function of a single instruction or directive. The second type explains the function of a group of instructions or directives or a whole routine. Adding comments improves the readability of a program.

Each line of an HCS12 assembly program, excluding certain special constructs, is made up of four distinct fields. Some of the fields may be empty. The order of these fields is

1. Label
2. Operation
3. Operand
4. Comment

2.2.1 The Label Field

Labels are symbols defined by the user to identify memory locations in the programs and data areas of the assembly module. For most instructions and assembler directives, the label is optional. The rules for forming a label are as follows:

- The label field can begin in column one or in any column if it is terminated by a colon (:). It must begin with a letter (A–Z, a–z), and the letter can be followed by letters, digits, or special symbols. Some assemblers permit special symbols to be used. For example, the assembler from IAR Inc. allows a symbol to start with a question mark (?), a character (@), and an underscore (_) in addition to letters. Digits and the dollar ($) character can also be used after the first character in the IAR assembler.

- Most assemblers restrict the number of characters in a label name. The **as12** assembler reference manual does not mention the limit. The IAR assembler allows a user-defined symbol to have up to 255 characters.

- The as12 assembler from several companies allows a label to be terminated by a colon (:).

Example 2.1 Examples of Valid and Invalid Labels

▼

The following instructions contain valid labels:

```
begin   ldaa    #10             ; label begins in column 1
print:  jsr     hexout          ; label is terminated by a colon
        jmp     begin           ; instruction references the label begin
```

The following instructions contain invalid labels:

```
here is    adda    #5        ; a space is included in the label
   loop    deca              ; label begins at column 2
```

▲

2.2.2 The Operation Field

This field specifies an assembler instruction, a directive, or a macro call. Assembler instructions or directives are case insensitive. The operation field must not start in column one. If a label is present, the opcode or directive must be separated from the label field by at least one space. If there is no label, the operation field must be at least one space from the left margin.

Example 2.2 Examples of Operation Fields

▼

```
           adda    #$02      ; adda is the instruction mnemonic
   true    equ     1         ; the equ directive occupies the operation field
```

▲

2.2.3 The Operand Field

If an *operand field* is present, it follows the *operation field* and is separated from the operation field by at least one space. The operand field is composed of one or more operands separated by commas, followed by optional space or tab characters. The operand field is used to supply arguments to the assembler instruction, directive, or macro that has been used in the operation field. The following instructions include the operand field:

```
TCNT    equ    $0084     ; the value $0084 is the operand field
TC0     equ    $0090     ; the value $0090 is the operand field
```

2.2.4 The Comment Field

The comment field is optional and is added mainly for documentation purposes. The comment field is ignored by the assembler. Here are the rules for comments.

- Any line beginning with an * is a comment.
- Any line beginning with a semicolon (;) is a comment. In this book, we use a ";" to start a comment.
- You must have a ";" prefixing any comment on a line with mnemonics.

Examples of comments are shown in the following instructions:

```
; this program computes the square root of N 8-bit integers.
      org    $1000     ; set the location counter to $1000
      dec    lp_cnt    ; decrement the loop count
```

In this chapter, we use the Freescale Freeware cross assembler as12 as the standard to explain every aspect of assembly programming. The as12 assembler contained in the MiniIDE from Mgtek is compatible with the as12 freeware from Freescale.

2.3 Assembler Directives

Assembler directives look just like instructions in an assembly language program, but they tell the assembler to do something other than create the machine code for an instruction. The available assembler directives vary with the assembler. Interested readers should refer to the user's manual of the specific assembler for details.

We discuss assembler directives supported by the as12 in detail here. In the following discussion, statements enclosed in square brackets [ ] are optional. All directives and assembly instructions can be in either upper- or lowercase.

end

The **end** directive is used to end a program to be processed by the assembler. In general, an assembly program looks like this:

        (your program)
        end

The end directive indicates the logical end of the source program. Any statement following the end directive is ignored. A warning message will occur if the end directive is missing from the source code; however, the program will still be assembled correctly.

org (origin)

The assembler uses a *location counter* to keep track of the memory location where the next machine code byte should be placed. If the programmer wants to force the program or data array to start from a certain memory location, then the **org** directive can be used. For example, the statement

        org $1000

forces the location counter to be set to $1000.

The org directive is mainly used to force a data table or a segment of instructions to start with a certain address. As a general rule, this directive should be used as infrequently as possible. Using too many orgs will make your program less reusable.

db (define byte), dc.b (define constant byte), fcb (form constant byte)

These three directives define the value of a byte or bytes that will be placed at a given memory location. The **db** (or **dc.b** or **fcb**) directive assigns the value of the expression to the memory location pointed to by the location counter. Then the location counter is incremented. Multiple bytes can be defined at a time by using commas to separate the arguments. For example, the statement

    array   db $11,$22,$33,$44,$55

initializes 5 bytes in memory to

        $11
        $22
        $33
        $44
        $55

and the assembler will use *array* as the symbolic address of the first byte whose initial value is $11. The program can also force these 5 bytes to a certain address by adding the org directive. For example, the sequence

        org $1000
    array   db $11,$22,$33,$44,$55

initializes the contents of memory locations at $1000, $1001, $1002, $1003, and $1004 to $11, $22, $33, $44, and $55, respectively.

dw (define word), dc.w (define constant word), fdb (form double bytes)

These three directives define the value of a word or words that will be placed at a given address. The value can be specified by an integer or an expression. For example, the statement

```
vect_tab      dw       $1234, $5678
```

initializes the two words starting from the current location counter to $1234 and $5678, respectively. After this statement, the location counter will be incremented by 4.

fcc (form constant character)

This directive allows us to define a string of characters (a message). The first character in the string is used as the delimiter. The last character must be the same as the first character because it will be used as the delimiter. The delimiter must not appear in the string. The space character cannot be used as the delimiter. Each character is encoded by its corresponding American Standard Code for Information Interchange (ASCII) code. For example, the statement

```
alpha         fcc      "def"
```

will generate the following values in memory:

```
           $64
           $65
           $66
```

and the assembler will use the label *alpha* to refer to the address of the first letter, which is stored as the byte $64. A character string to be output to the LCD display is often defined using this directive.

fill (fill memory)

This directive allows a user to fill a certain number of memory locations with a given value. The syntax of this directive is as follows:

```
              fill      value, count
```

where the number of bytes to be filled is indicated by *count* and the value to be filled is indicated by *value*. For example, the statement

```
spaceLine     fill     $20, 40
```

will fill 40 bytes with the value of $20 starting from the memory location referred to by the label *spaceLine*.

ds (define storage), rmb (reserve memory byte), ds.b (define storage bytes)

Each of these three directives reserves a number of bytes given as the arguments to the directive. The location counter will be incremented by the number that follows the directive mnemonic. For example, the statement

```
buffer        ds       100
```

reserves 100 bytes in memory starting from the location represented by the label *buffer*. After this directive, the location counter will be incremented by 100. The content(s) of the reserved memory location(s) is(are) not defined.

ds.w (define storage word), rmw (reserve memory word)

Each of these directives increments the location counter by the value indicated in the number-of-words argument multiplied by 2. In other words, if *ds.w* evaluates to k, then the location counter is advanced by $2k$. These directives are often used with a label. For example, the statement

```
dbuf          ds.w      20
```

reserves 40 bytes starting from the memory location represented by the label *dbuf*. None of these 40 bytes is initialized.

equ (equate)

This directive assigns a value to a label. Using **equ** to define constants will make a program more readable. For example, the statement

```
loop_cnt      equ       40
```

informs the assembler that whenever the symbol *loop_cnt* is encountered, it should be replaced with the value of 40.

loc

This directive increments and produces an internal counter used in conjunction with the backward tick mark (`). By using the loc directive and the ` mark, you can write a program segment like the following example without thinking up new labels:

```
              loc
              ldaa      #2
loop`         deca
              bne       loop`
              loc
loop`         brclr     0,x $55 loop`
```

This code segment will work perfectly fine because the first loop label will be seen as loop001, whereas the second loop label will be seen as loop002. The assembler really sees this:

```
              loc
              ldaa      #2
loop001       deca
              bne       loop001
              loc
loop002       brclr     0,x $55 loop002
```

You can also set the loc directive with a valid expression or number by putting that expression or number in the operand field. The resultant number will be used to increment the suffix to the label.

macro, endm

Macro is a name assigned to a group of instructions or directives. There are situations in which the same sequence of instructions needs to be included in several places. This sequence of instructions may operate on different parameters. By placing this sequence of instructions in a macro, the sequence need be typed only once. The macro capability not only makes the programmer more productive but also makes the program more readable. The freeware MiniIDE supports macro directives. However, the freeware AsmIDE does not.

The keyword *macro* starts a new macro definition, whereas the keyword *endm* ends the macro definition. For example, a macro may be defined for the HCS12 as follows:

```
sumOf3        macro     arg1,arg2,arg3
              ldaa      arg1
              adda      arg2
              adda      arg3
              endm
```

If you want to add the values of three memory locations at $1000, $1001, and $1002 and leave the sum in accumulator A, you can use the following statement to invoke the previously defined macro:

sumOf3 $1000,$1001,$1002

When processing this macro call, the assembler will insert the following instructions in the user program:

```
ldaa     $1000
adda     $1001
adda     $1002
```

2.4 Software Development Issues

A complete discussion of issues involved in software development is beyond the scope of this text. However, we do need to take a serious look at some software development issues because embedded system designers must spend a significant amount of time on software development.

Software development starts with *problem definition*. The problem presented by the application must be fully understood before any program can be written. At the problem definition stage, the most critical thing is to get the programmer and the end user to agree on what needs to be done. To achieve this, asking questions is very important. For complex and expensive applications, a formal, written definition of the problem is formulated and agreed on by all parties.

Once the problem is known, the programmer can begin to lay out an overall plan of how to solve the problem. The plan is also called an *algorithm*. Informally, an algorithm is any well-defined computational procedure that takes some value, or a set of values, as input and produces some value, or a set of values, as output. An algorithm is thus a sequence of computational steps that transforms the input into the output. We can also view an algorithm as a tool for solving a well-specified computational problem. The statement of the problem specifies in general terms the desired input/output relationship. The algorithm describes a specific computational procedure for achieving that input/output relationship.

An algorithm is expressed in *pseudocode* which is very much like C or PASCAL. What separates pseudocode from "real" code is that in pseudocode, we employ whatever expressive method is most clear and concise to specify a given algorithm. Sometimes, the clearest method is English, so do not be surprised if you come across an English phrase or sentence embedded within a section of "real" code.

An algorithm provides not only the overall plan for solving the problem but also documentation to the software to be developed. In the rest of this book, all algorithms will be presented in the following format:

Step 1

. . .

Step 2

. . .

An earlier alternative for providing the overall plan for solving software problems was the use of flowcharts. A flowchart shows the way a program operates. It illustrates the logic flow of the program. Therefore, flowcharts can be a valuable aid in visualizing programs. Flowcharts are not only used in computer programming; they are also used in many other fields, such as business and construction planning.

The flowchart symbols used in this book are shown in Figure 2.1. The *terminal symbol* is used at the beginning and the end of each program. When it is used at the beginning of a program, the word *Start* is written inside it. When it is used at the end of a program, it contains the word *Stop*.

The *process box* indicates what must be done at this point in the program execution. The operation specified by the process box could be shifting the contents of one general-purpose register to a peripheral register, decrementing a loop count, and so on.

The *input/output box* is used to represent data that are either read or displayed by the computer.

The *decision box* contains a question that can be answered either yes or no. A decision box has two exits, also marked yes or no. The computer will take one action if the answer is yes and will take a different action if the answer is no.

The *on-page connector* indicates that the flowchart continues elsewhere on the same page. The place where it is continued will have the same label as the on-page connector.

The *off-page connector* indicates that the flowchart continues on another page. To determine where the flowchart continues, one needs to look at the following pages of the flowchart to find the matching off-page connector.

Normal flow on a flowchart is from top to bottom and from left to right. Any line that does not follow this normal flow should have an arrowhead on it. When the program gets complicated, the flowchart that documents the logic flow of the program also becomes difficult to follow. This is the limitation of the flowchart. In this book, we will mix both the flowchart and the algorithm procedure to describe the solution to a problem.

After the programmer is satisfied with the algorithm or the flowchart, it is converted to source code in one of the assembly or high-level languages. Each statement in the algorithm (or each block of the flowchart) will be converted into one or multiple assembly instructions

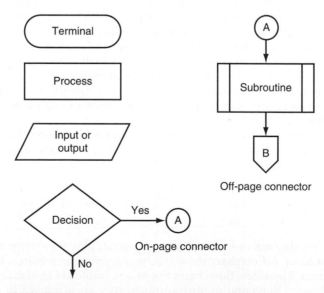

Figure 2.1 ■ Flowchart symbols used in this book

or high-level language statements. If an algorithmic step (or a block in the flowchart) requires many assembly instructions or high-level language statements to implement, then it might be beneficial either to convert this step (or block) into a subroutine and just call the subroutine or to further divide the algorithmic step (or flowchart block) into smaller steps (or blocks) so that it can be coded with just a few assembly instructions or high-level language statements.

The next major step is *testing the program*. Testing a program means testing for anomalies. The first test is for normal inputs that are always expected. If the result is what is expected, then the borderline inputs are tested. The maximum and minimum values of the input are tested. When the program passes this test, then illegal input values are tested. If the algorithm includes several branches, then enough values must be used to exercise all the possible branches. This is to make sure that the program will operate correctly under all possible circumstances.

In the rest of this book, most of the problems are well defined. Therefore, our focus is on how to design the algorithm that solves the specified problem as well as how to convert the algorithm into source code.

2.5 Writing Programs to Do Arithmetic

In this section, we use small programs that perform simple computations to demonstrate how a program is written.

Example 2.3

Write a program to add the numbers stored at memory locations $1000, $1001, and $1002 and store the sum at memory location $1010.

Solution: This problem can be solved by the following steps:

Step 1
Load the contents of the memory location at $1000 into accumulator A.

Step 2
Add the contents of the memory location at $1001 into accumulator A.

Step 3
Add the contents of the memory location at $1002 into accumulator A.

Step 4
Store the contents of accumulator A at memory location $1010.

These steps can be translated into the as12 assembly program as follows:

```
org    $1500        ; starting address of the program
ldaa   $1000        ; A ⇐ [$1000]
adda   $1001        ; A ⇐ [A] + [$1001]
adda   $1002        ; A ⇐ [A] + [$1002]
staa   $1010        ; $1010 ⇐ [A]
end
```

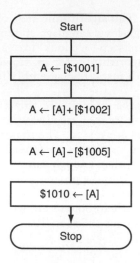

Figure 2.2 ■ Logic flow of program 2.4

Example 2.4

Write a program to subtract the contents of the memory location at $1005 from the sum of the memory locations at $1000 and $1002 and store the result at the memory location $1010.

Solution: The logic flow of this program is illustrated in Figure 2.2. The assembly program is as follows:

```
org     $1500       ; starting address of the program
ldaa    $1000       ; A ⇐ [$1000]
adda    $1002       ; A ⇐ [A] + [$1002]
suba    $1005       ; A ⇐ [A] − [$1005]
staa    $1010       ; $1010 ⇐ [A]
end
```

Example 2.5

Write a program to subtract 5 from four memory locations at $1000, $1001, $1002, and $1003.

Solution: In the HCS12, a memory location cannot be the destination of an ADD or SUB instruction. Therefore, three steps must be followed to add or subtract a number to or from a memory location.

Step 1
Load the memory contents into an accumulator.

Step 2
Add (or subtract) the number to (from) the accumulator.

Step 3

Store the result at the specified memory location.
The program is as follows:

```
org     $1500
ldaa    $1000        ; A ⇐ [$1000]
suba    #5           ; A ⇐ [A] − 5
staa    $1000        ; $1000 ⇐ [A]
ldaa    $1001        ; A ⇐ [$1001]
suba    #5           ; A ⇐ [A] − 5
staa    $1001        ; $1001 ⇐ [A]
ldaa    $1002        ; A ⇐ [$1002]
suba    #5           ; A ⇐ [A] − 5
staa    $1002        ; $1002 ⇐ [A]
ldaa    $1003        ; A ⇐ [$1003]
suba    #5           ; A ⇐ [A] − 5
staa    $1003        ; $1003 ⇐ [A]
end
```

Example 2.6

▼

Write a program to add two 16-bit numbers that are stored at $1000~$1001 and $1002~$1003 and store the sum at $1010~$1011.

Solution: This program is very straightforward.

```
org     $1500
ldd     $1000        ; place the 16-bit number at $1000~$1001 in D
addd    $1002        ; add the 16-bit number at $1002~$1003 to D
std     $1010        ; save the sum at $1010~$1011
end
```

2.5.1 Carry/Borrow Flag

The HCS12 can add and subtract either 8-bit or 16-bit numbers and place the result in either 8-bit accumulators, A or B, or the double accumulator D. The 8-bit number stored in accumulator B can also be added to index register X or Y. However, programs can also be written to add numbers larger than 16 bits. Arithmetic performed in a 16-bit microprocessor/microcontroller on numbers that are larger than 16 bits is called *multiprecision arithmetic*. Multiprecision arithmetic makes use of the carry flag (C flag) of the condition code register (CCR).

Bit 0 of the CCR register is the C flag. It can be thought of as a temporary 9th bit that is appended to any 8-bit register or 17th bit that is appended to any 16-bit register. The C flag allows us to write programs to add and subtract hex numbers that are larger than 16 bits. For example, consider the following two instructions:

```
ldd     #$8645
addd    #$9978
```

These two instructions add the numbers $8645 and $9978.

```
    $ 8 6 4 5
  + $ 9 9 7 8
    $ 1 1 F B D
```

The result is $11FBD, a 17-bit number, which is too large to fit into the 16-bit double accumulator D. When the HCS12 executes these two instructions, the lower 16 bits of the answer, $1FBD, are placed in double accumulator D. This part of the answer is called *sum*. The leftmost bit is called a *carry*. A carry of 1 following an addition instruction sets the C flag of the CCR register to 1. A carry of 0 following an addition clears the C flag to 0. This applies to both 8-bit and 16-bit additions for the HCS12. For example, execution of the following two instructions

```
    ldd      #$1245
    addd     #$4581
```

will clear the C flag to 0 because the carry resulting from this addition is 0. In summary,

- If the addition produces a carry of 1, the carry flag is set to 1.
- If the addition produces a carry of 0, the carry flag is cleared to 0.

2.5.2 Multiprecision Addition

For a 16-bit microcontroller like the HCS12, multiprecision addition is the addition of numbers that are larger than 16 bits. To add the hex number $1A598183 to $76548290, the HCS12 has to perform multiprecision addition.

```
      1     1 1
    $ 1 A 5 9 8 1 8 3
  + $ 7 6 5 4 8 2 9 0
    $ 9 0 A E 0 4 1 3
```

Multiprecision addition is performed 1 byte at a time, beginning with the least significant byte. The HCS12 does allow us to add 16-bit numbers at a time because it has the addd instruction. The following two instructions can be used to add the least significant 16-bit numbers together:

```
    ldd      #$8183
    addd     #$8290
```

Since the sum of the most significant digit is greater than 16, it generates a carry that must be added to the next more significant digit, causing the C flag to be set to 1. The contents of double accumulator D must be saved before the higher bytes are added. Here the 2 bytes are saved at $1002~$1003.

```
    std      $1002
```

When the second most significant bytes are added, the carry from the lower byte must be added in order to obtain the correct sum. In other words, we need an "add with carry" instruction. There are two versions of this instruction: the adca instruction for accumulator A and the adcb instruction for accumulator B. The instructions for adding the second significant bytes are

```
    ldaa     #$59
    adca     #$54
```

We need also to save the *second most significant* byte of the result at $1001 with the following instruction:

```
    staa     $1001
```

The most significant bytes can be added using similar instructions, and the complete program with comments appears as follows:

```
ldd     #$8183      ; D ⇐ $8183
addd    #$8290      ; D ⇐ [D] + $8290
std     $1002       ; $1002−$1003 ⇐ [D]
ldaa    #$59        ; A ⇐ $59
adca    #$54        ; A ⇐ [A] + $54 + C
staa    $1001       ; $1001 ⇐ [A]
ldaa    #$1A        ; A ⇐ $1A
adca    #$76        ; A ⇐ [A] + $76 + C
staa    $1000       ; $1000 ⇐ [A]
end
```

Note that the load and store instructions do not affect the value of the C flag (otherwise, the program would not work). The HCS12 does not have a 16-bit instruction with the carry flag as an operand. Whenever the carry needs to be added, we must use the 8-bit instruction adca or adcb. This is shown in the previous program.

Example 2.7

▼

Write a program to add two 4-byte numbers that are stored at $1000~$1003 and $1004~$1007 and store the sum at $1010~$1013.

Solution: The addition should start from the least significant byte and proceed to the most significant byte. The program is as follows:

```
org     $1500       ; starting address of the program
ldd     $1002       ; D ⇐ [$1002~$1003]
addd    $1006       ; D ⇐ [D] + [$1006~$1007]
std     $1012       ; $1012~$1013 ⇐ [D]
ldaa    $1001       ; A ⇐ [$1001]
adca    $1005       ; A ⇐ [A] + [$1005] + C
staa    $1011       ; $1011 ⇐ [A]
ldaa    $1000       ; A ⇐ [$1000]
adca    $1004       ; A ⇐ [A] + [$1004] + C
staa    $1010       ; $1010 ⇐ [A]
end
```

▲

2.5.3 Subtraction and the C Flag

The C flag also enables the HCS12 to borrow from the high byte to the low byte during a multiprecision subtraction. Consider the following subtraction problem:

$$\begin{array}{r} \$39 \\ -\$74 \\ \hline \end{array}$$

We are attempting to subtract a larger number from a smaller one. Subtracting $4 from $9 is not a problem.

$$\begin{array}{r} \$39 \\ -\$74 \\ \hline 5 \end{array}$$

Now we need to subtract $7 from $3. To do this, we need to borrow from somewhere. The HCS12 borrows from the C flag, thus setting the C flag. When we borrow from the next higher digit of a hex number, the borrow has a value of decimal 16. After the borrow from the C flag, the problem can be completed.

$$\begin{array}{r} \$ \, 3 \, 9 \\ - \, \$ \, 7 \, 4 \\ \hline \$ \, C \, 5 \end{array}$$

When the HCS12 executes a subtract instruction, it always borrows from the C flag. The borrow is either 1 or 0. The C flag operates as follows during a subtraction:

- If the HCS12 borrows a 1 from the C flag during a subtraction, the C flag is set to 1.
- If the HCS12 borrows a 0 from the C flag during a subtraction, the C flag is set to 0.

2.5.4 Multiprecision Subtraction

For a 16-bit microcontroller, multiprecision subtraction is the subtraction of numbers that are larger than 16 bits. To subtract the hex number $16753284 from $98765432, the HCS12 has to perform a multiprecision subtraction.

$$\begin{array}{r} \$ \, 9 \, 8 \, 7 \, 6 \, 5 \, 4 \, 3 \, 2 \\ - \, \$ \, 1 \, 6 \, 7 \, 5 \, 7 \, 2 \, 8 \, 4 \end{array}$$

Like multiprecision addition, multiprecision subtraction is performed 1 byte at a time, beginning with the least significant byte. The HCS12 does allow us to subtract 2 bytes at a time because it has the subd instruction. The following two instructions can be used to subtract the least significant 2 bytes of the subtrahend from the minuend:

```
ldd     #$5432
subd    #$7284
```

Since a larger number is subtracted from a smaller one, there is a need to borrow from the higher byte, causing the C flag to be set to 1. The contents of double accumulator D should be saved before the higher bytes are subtracted. Let's save these 2 bytes at $1002~$1003.

```
std     $1002
```

When the second most significant bytes are subtracted, the borrow 1 has to be subtracted from the second most significant byte of the result. In other words, we need a "subtract with borrow" instruction. There is such an instruction, but it is called *subtract with carry*. There are two versions: the sbca instruction for accumulator A and the sbcb instruction for accumulator B. The instructions to subtract the second most significant bytes are

```
ldaa    #$76
sbca    #$75
```

We also need to save the second most significant byte of the result at $1001 with the following instruction:

```
staa    $1001
```

The most significant bytes can be subtracted using similar instructions, and the complete program with comments is as follows:

```
org     $1500        ; starting address of the program
ldd     #$5432       ; D ⟸ $5432
subd    #$7284       ; D ⟸ [D] − $7284
std     $1002        ; $1002~$1003 ⟸ [D]
```

```
ldaa    #$76         ; A ⇐ $76
sbca    #$75         ; A ⇐ [A] − $75 − C
staa    $1001        ; $1001 ⇐ [A]
ldaa    #$98         ; A ⇐ $98
sbca    #$16         ; A ⇐ [A] − $16 − C
staa    $1000        ; $1000 ⇐ [A]
end
```

Example 2.8

▼

Write a program to subtract the hex number stored at $1004~$1007 from the hex number stored at $1000~$1003 and save the difference at $1010~$1013.

Solution: We will perform the subtraction from the least significant byte toward the most significant byte as follows:

```
org     $1500        ; starting address of the program
ldd     $1002        ; D ⇐ [$1002 − $1003]
subd    $1006        ; D ⇐ [D] − [$1006 − $1007]
std     $1012        ; $1012 − $1013 ⇐ [D]
ldaa    $1001        ; A ⇐ [$1001]
sbca    $1005        ; A ⇐ [A] − [$1005] − C
staa    $1011        ; $1011 ⇐ [A]
ldaa    $1000        ; A ⇐ [$1000]
sbca    $1004        ; A ⇐ [A] − [$1004] − C
staa    $1010        ; $1010 ⇐ [A]
end
```

▲

2.5.5 Binary-Coded-Decimal (BCD) Addition

Although virtually all computers work internally with binary numbers, the input and output equipment generally uses decimal numbers. Since most logic circuits only accept two-valued signals, the decimal numbers must be coded in terms of binary signals. In the simplest form of binary code, each decimal digit is represented by its binary equivalent. For example, 2538 is represented by

0010 0101 0011 1000

This representation is called a *binary coded decimal* (BCD). If the BCD format is used, it must be preserved during arithmetic processing.

The principal advantage of the BCD encoding method is the simplicity of input/output conversion; its major disadvantage is the complexity of arithmetic processing. The choice between binary and BCD depends on the type of problems the system will be handling.

The HCS12 microcontroller can add only binary numbers, not decimal numbers. The following instruction sequence appears to cause the HCS12 to add the decimal numbers 25 and 31 and store the sum at the memory location $1000:

```
ldaa    #$25
adda    #$31
staa    $1000
```

This instruction sequence performs the following addition:

$25
+ $31
+ $56

When the HCS12 executes this instruction sequence, it adds the numbers according to the rules of binary addition and produces the sum $56. This is the correct BCD answer, because the result represents the decimal sum of 25 + 31 = 56. In this example, the HCS12 gives the appearance of performing decimal addition. However, a problem occurs when the HCS12 adds two BCD digits and generates a sum greater than 9. Then the sum is incorrect in the decimal number system, as the following three examples illustrate:

$18 $35 $19
+ $47   + $47   + $47
$5F $7C $60

The answers to the first two problems are obviously wrong in the decimal number system because the hex digits F and C are not between 0 and 9. The answer to the third example appears to contain valid BCD digits, but in the decimal system 19 plus 47 equals 66, not 60; this example involves a carry from the lower nibble to the higher nibble.

In summary, a sum in the BCD format is incorrect if it is greater than $9 or if there is a carry to the next-higher nibble. Incorrect BCD sums can be adjusted by adding $6 to them. To correct the examples,

1. Add $6 to every sum digit greater than 9.

2. Add $6 to every sum digit that had a carry of 1 to the next higher digit.

Here are the problems with their sums adjusted.

$18 $35 $19
+ $47   + $47   + $47
$5F $7C $60
+ $ 6   + $ 6   + $ 6
$65 $82 $66

The fifth bit of the condition code register is the *half-carry*, or H flag. A carry from the lower nibble to the higher nibble during the addition operation is a half-carry. A half-carry of 1 during addition sets the H flag to 1, and a half-carry of 0 during addition clears it to 0. If there is a carry from the high nibble during addition, the C flag is set to 1; this indicates that the high nibble is incorrect. $6 must be added to the high nibble to adjust it to the correct BCD sum.

Fortunately, we don't need to write instructions to detect the illegal BCD sum following a BCD addition. The HCS12 provides a *decimal adjust accumulator A* instruction, daa, which takes care of all these detailed detection and correction operations. The daa instruction monitors the sums of BCD additions and the C and H flags and automatically adds $6 to any nibble that requires it. The rules for using the daa instruction are

1. The daa instruction can only be used for BCD addition. It does not work for subtraction or hex arithmetic.

2. The daa instruction must be used immediately after one of the three instructions that leave their sum in accumulator A. (These three instructions are adda, adca, and aba.)

3. The numbers added must be legal BCD numbers to begin with.

Example 2.9

▼

Write an instruction sequence to add the BCD numbers stored at memory locations $1000 and $1001 and store the sum at $1010.

Solution:

ldaa	$1000	; load the first BCD number in A
adda	$1001	; perform addition
daa		; decimal adjust the sum in A
staa	$1010	; save the sum

Multiple-byte BCD numbers can be added and the correct result can be obtained by executing the daa instruction immediately after the addition of each byte.

▲

2.5.6 Multiplication and Division

The HCS12 provides three multiply and five divide instructions. A brief description of these instructions is shown in Table 2.1.

The emul instruction multiplies the 16-bit unsigned integers stored in accumulator D and index register Y and leaves the product in these two registers. The upper 16 bits of the product are in Y, whereas the lower 16 bits are in D.

The mul instruction multiplies the 8-bit unsigned integer in accumulator A by the 8-bit unsigned integer in accumulator B to obtain a 16-bit unsigned result in double accumulator D. The upper byte of the product is in accumulator A, whereas the lower byte of the product is in B.

Mnemonic	Function	Operation
emul	Unsigned 16 by 16 multiply	$(D) \times (Y) \rightarrow Y{:}D$
emuls	Signed 16 by 16 multiply	$(D) \times (Y) \rightarrow Y{:}D$
mul	Unsigned 8 by 8 multiply	$(A) \times (B) \rightarrow A{:}B$
ediv	Unsigned 32 by 16 divide	$(Y{:}D) \div (X)$ Quotient $\rightarrow$ Y Remainder $\rightarrow$ D
edivs	Signed 32 by 16 divide	$(Y{:}D) \div (X)$ Quotient $\rightarrow$ Y Remainder $\rightarrow$ D
fdiv	16 by 16 fractional divide	$(D) \div (X) \rightarrow X$ Remainder $\rightarrow$ D
idiv	Unsigned 16 by 16 integer divide	$(D) \div (X) \rightarrow X$ Remainder $\rightarrow$ D
idivs	Signed 16 by 16 integer divide	$(D) \div (X) \rightarrow X$ Remainder $\rightarrow$ D

Table 2.1 ■ Summary of 68HC12 multiply and divide instructions

The ediv instruction performs an unsigned 32-bit by 16-bit division. The dividend is the register pair Y and D with Y as the upper 16 bits of the dividend. Index register X is the divisor. After division, the quotient and the remainder are placed in Y and D, respectively.

The edivs instruction performs a signed 32-bit by 16-bit division using the same operands as the ediv instruction. After division, the quotient and the remainder are placed in Y and D, respectively.

The fdiv instruction divides an unsigned 16-bit dividend in double accumulator D by an unsigned 16-bit divisor in index register X, producing an unsigned 16-bit quotient in X, and an unsigned 16-bit remainder in D. The dividend must be less than the divisor. The radix point of the quotient is to the left of bit 15. In the case of overflow (the denominator is less than or equal to the nominator) or division by zero, the quotient is set to $FFFF and the remainder is indeterminate.

The idiv instruction divides an unsigned 16-bit dividend in double accumulator D by the unsigned 16-bit divisor in index register X, producing an unsigned 16-bit quotient in X and an unsigned 16-bit remainder in D. If both the divisor and the dividend are assumed to have radix points in the same positions (to the right of bit 0), the radix point of the quotient is to the right of bit 0. In the case of division by zero, the quotient is set to $FFFF and the remainder is indeterminate.

The idivs instruction divides the signed 16-bit dividend in double accumulator D by the signed 16-bit divisor in index register X, producing a signed 16-bit quotient in X and a signed 16-bit remainder in D. If division by zero is attempted, the values in D and X are not changed, but the values of the N, Z, and V status bits are undefined.

Example 2.10

▼

Write an instruction sequence to multiply the contents of index register X and double accumulator D and store the product at memory locations $1000~$1003.

Solution: There is no instruction to multiply the contents of double accumulator D and index register X. However, we can transfer the contents of index register X to index register Y and execute the emul instruction. If index register Y holds useful information, then we need to save it before the data transfer.

```
sty     $1010           ; save Y in a temporary location
tfr     x,y             ; transfer the contents of X to Y
emul                    ; perform the multiplication
sty     $1000           ; save the upper 16 bits of the product
std     $1002           ; save the lower 16 bits of the product
ldy     $1010           ; restore the value of Y
```

▲

Example 2.11

▼

Write an instruction sequence to divide the signed 16-bit number stored at memory locations $1005~$1006 by the 16-bit signed number stored at memory locations $1020~$1021 and store the quotient and remainder at $1030~$1031 and $1032~$1033, respectively.

Solution: Before we can perform the division, we need to place the dividend and divisor in D and X, respectively.

```
ldd     $1005    ; place the dividend in D
ldx     $1020    ; place the divisor in X
idivs            ; perform the signed division
stx     $1030    ; save the quotient
std     $1032    ; save the remainder
```

Because most arithmetic operations can be performed only on accumulators, we need to transfer the contents of index register X to D so that further division on the quotient can be performed. The HCS12 provides two exchange instructions in addition to the TFR instruction for this purpose.

- The xgdx instruction exchanges the contents of accumulator D and index register X.
- The xgdy instruction exchanges the contents of accumulator D and index register Y.

The HCS12 provides instructions for performing unsigned 8-bit by 8-bit and both signed and unsigned 16-bit by 16-bit multiplications. Since the HCS12 is a 16-bit microcontroller, we expect that it will be used to perform complicated operations in many sophisticated applications. Performing 32-bit by 32-bit multiplication will be one of them.

Since there is no 32-bit by 32-bit multiplication instruction, we have to break a 32-bit number into two 16-bit halves and use the 16-bit by 16-bit multiply instruction to synthesize the operation. Assume that M and N are the multiplicand and the multiplier, respectively. These two numbers can be broken down as follows:

$M = M_H M_L$

$N = N_H N_L$

where M_H and N_H are the upper 16 bits and M_L and N_L are the lower 16 bits of M and N, respectively. Four 16-bit by 16-bit multiplications are performed, and then their partial products are added together, as shown in Figure 2.3.

The procedure is as follows:

Step 1
Allocate 8 bytes to hold the product. Assume these 8 bytes are located at P, P+1, . . . , and P+7.

Step 2
Compute the partial product $M_L N_L$ (in Y:D) and save it at locations P+4~P+7.

Step 3
Compute the partial product $M_H N_H$ (in Y:D) and save it at locations P~P+3.

Step 4
Compute the partial product $M_H N_L$ (in Y:D) and add it to memory locations P+2~P+5. The C flag may be set to 1 after this addition.

Step 5
Add the C flag to memory location P+1 using the adca (or adcb) instruction. This addition may also set the C flag to 1. So, again, add the C flag to memory location P.

Step 6
Generate the partial product $M_L N_H$ (in Y:D) and add it to memory locations P+2~P+5. The carry flag may be set to 1, so add the C flag to memory location P+1 and then add it to memory location P.

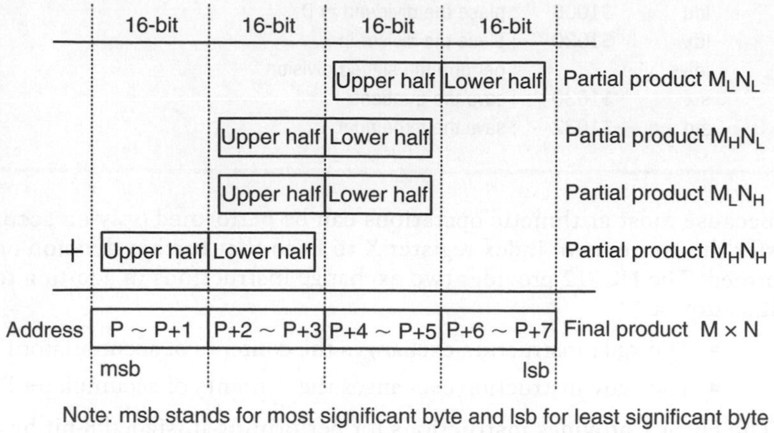

Note: msb stands for most significant byte and lsb for least significant byte

Figure 2.3 ■ Unsigned 32-bit by 32-bit multiplication

Example 2.12

Write a program to multiply the 32-bit unsigned integers stored at M~M+3 and N~N+3, respectively, and store the product at memory locations P~P+7.

Solution: The following program is a direct translation of the previous multiplication algorithm:

```
            org     $1000
M           rmb     4                    ; multiplicand
N           rmb     4                    ; multiplier
P           rmb     8                    ; product
            org     $1500
            ....................         ; some other instructions
            ldd     M+2                  ; place M_L in D
            ldy     N+2                  ; place N_L in Y
            emul                         ; compute M_L N_L
            sty     P+4                  ; save the upper 16 bits of the partial product M_L N_L
            std     P+6                  ; save the lower 16 bits of the partial product M_L N_L
            ldd     M                    ; place M_H in D
            ldy     N                    ; place N_H in Y
            emul                         ; compute M_H N_H
            sty     P                    ; save the upper 16 bits of the partial product M_H N_H
            std     P+2                  ; save the lower 16 bits of the partial product M_H N_H
            ldd     M                    ; place M_H in D
            ldy     N+2                  ; place N_L in Y
            emul                         ; compute M_H N_L
; the following seven instructions add M_H N_L to memory locations P+2~P+5
            addd    P+4                  ; add the lower half of M_H N_L to P+4~P+5
            std     P+4                  ;       "
```

```
            tfr        y,d             ; transfer Y to D
            adcb       P+3
            stab       P+3
            adca       P+2
            staa       P+2
; the following six instructions propagate carry to the most significant byte
            ldaa       P+1
            adca       #0              ; add C flag to location P+1
            staa       P+1             ;      "
            ldaa       P
            adca       #0              ; add C flag to location P
            staa       P
; the following three instructions compute M_L N_H
            ldd        M+2             ; place M_L in D
            ldy        N               ; place N_H in Y
            emul                       ; compute M_L N_H
; the following seven instructions add M_L N_H to memory locations P+2~P+5
            addd       P+4             ; add the lower half of M_L N_H to P+4~P+5
            std        P+4             ;      "
            tfr        y,d             ; transfer Y to D
            adcb       P+3
            stab       P+3
            adca       P+2
            staa       P+2
; the following six instructions propagate carry to the most significant byte
            ldaa       P+1
            adca       #0              ; add C flag to location P+1
            staa       P+1
            ldaa       P
            adca       #0              ; add C flag to location P
            staa       P
            end
```

Example 2.13

Write a program to convert the 16-bit binary number stored at $1000~$1001 to BCD format and store the result at $1010~$1014. Convert each BCD digit into its ASCII code and store it in 1 byte.

Solution: A binary number can be converted to BCD format using repeated division by 10. The largest 16-bit binary number corresponds to the 5-digit decimal number 65,535. The first division by 10 computes the least significant digit and should be stored in the memory location $1014, the second division-by-10 operation computes the 10s digit, and so on. The ASCII code of a BCD digit can be obtained by adding $30 to each BCD digit. The program is as follows:

```
            org        $1000
data        fdb        12345           ; place a number for testing
            org        $1010
```

```
result      rmb     5                   ; reserve 5 bytes to store the result
            org     $1500
            ldd     data                ; make a copy of the number to be converted
            ldy     #result
            ldx     #10                 ; divide the number by 10
            idiv                        ;         "
            addb    #$30                ; convert to ASCII code
            stab    4,Y                 ; save the least significant digit
            xgdx                        ; swap the quotient to D
            ldx     #10
            idiv
            addb    #$30                ; convert to ASCII code
            stab    3,Y                 ; save the second-least significant digit
            xgdx
            ldx     #10
            idiv
            addb    #$30
            stab    2,Y                 ; save the middle digit
            xgdx
            ldx     #10
            idiv                        ; separate the most significant and second-most
                                        ; significant digits
            addb    #$30
            stab    1,Y                 ; save the second-most significant digit
            xgdx                        ; swap the most significant digit to B
            addb    #$30                ; convert to ASCII code
            stab    0,Y                 ; save the most significant digit
            end
```

▲

2.6 Program Loops

Many applications require repetitive operations. We can write programs to tell the computer to perform the same operation over and over. A *finite loop* is a sequence of instructions that will be executed by the computer for a finite number of times; an *endless loop* is a sequence of instructions that the computer will execute forever.

There are four major loop constructs.

DO STATEMENT S FOREVER

This is an endless loop in which statement S is repeated forever. In some applications, we might add the statement "If C then exit" to leave the infinite loop. An infinite loop is shown in Figure 2.4.

FOR *i* = *N1* TO *N2* DO S OR FOR *i* = *N2* DOWNTO *N1* DO S

Here, the variable *i* is the *loop counter*, which keeps track of the current iteration of the loop. The loop counter can be incremented (the first case) or decremented (the second case). Statement S is repeated $n_2 - n_1 + 1$ times. The value of n_2 is assumed to be no smaller than that of n_1.

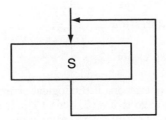

Figure 2.4 ■ An infinite loop

If there is concern that the relationship $n_1 \leq n_2$ may not hold, then it must be checked at the beginning of the loop. Four steps are required to implement a FOR loop.

Step 1
Initialize the loop counter and other variables.

Step 2
Compare the loop counter with the limit to see if it is within bounds. If it is, then perform the specified operations. Otherwise, exit the loop.

Step 3
Increment (or decrement) the loop counter.

Step 4
Go to step 2.

A **For loop** is illustrated in Figure 2.5.

WHILE C DO S

Whenever a While construct is executed, the logical expression C is evaluated first. If it yields a false value, statement S will not be executed. The action of a While construct is illustrated in Figure 2.6. Four steps are required to implement a While loop.

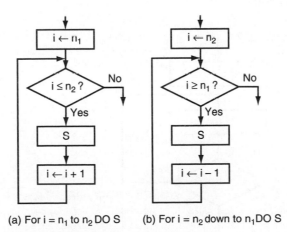

(a) For i = n_1 to n_2 DO S   (b) For i = n_2 down to n_1 DO S

Figure 2.5 ■ For looping construct

Step 1
Initialize the logical expression C.

Step 2
Evaluate the logical expression C.

Step 3
Perform the specified operations if the logical expression C evaluates to true. Update the logical expression C and go to step 2. (*Note:* The logical expression C may be updated by external conditions or by an interrupt service routine.)

Step 4
Exit the loop.

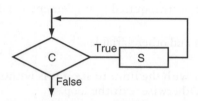

Figure 2.6 ■ The While ... Do looping construct

REPEAT S UNTIL C

Statement S is first executed then the logical expression C is evaluated. If C is false, the next statement will be executed. Otherwise, statement S will be executed again. The action of this construct is illustrated in Figure 2.7. Statement S will be executed at least once. Three steps are required to implement this construct.

Step 1
Initialize the logical expression C.

Step 2
Execute statement S.

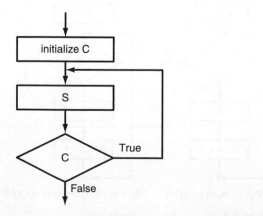

Figure 2.7 ■ The Repeat ... Until looping construct

Step 3

Go to step 2 if the logical expression C evaluates to true. Otherwise, exit.

To implement one of the looping constructs, we need to use the unconditional branch or one of the conditional instructions. When executing conditional branch instructions, the HCS12 checks the condition flags in the CCR register.

2.6.1 Condition Code Register

The contents of the condition code register are shown in Figure 2.8. The shaded characters are condition flags that reflect the status of an operation. The meanings of these condition flags are as follows:

Figure 2.8 ■ Condition code register

- *C: the carry flag.* Whenever a carry is generated as the result of an operation, this flag will be set to 1. Otherwise, it will be cleared to 0.
- *V: the overflow flag.* Whenever the result of a two's complement arithmetic operation is out of range, this flag will be set to 1. Otherwise, it will be set to 0. The V flag is set to 1 when the carry from the most significant bit and the second most significant bit differ as the result of an arithmetic operation.
- *Z: the zero flag.* Whenever the result of an operation is zero, this flag will be set to 1. Otherwise, it will be set to 0.
- *N: the negative flag.* Whenever the most significant bit of the result of an operation is 1, this flag will be set to 1. Otherwise, it will be set to 0. This flag indicates that the result of an operation is negative.
- *H: the half-carry flag.* Whenever there is a carry from the lower four bits to the upper four bits as the result of an operation, this flag will be set to 1. Otherwise, it will be set to 0.

2.6.2 Branch Instructions

Branch instructions cause program flow to change when specific conditions exist. The HCS12 has three kinds of branch instructions, *short branches*, *long branches*, and *bit-conditional branches*.

Branch instructions can also be classified by the type of condition that must be satisfied in order for a branch to be taken. Some instructions belong to more than one category.

- ***Unary (unconditional) branch*** instructions always execute.
- ***Simple branches*** are taken when a specific bit in the CCR register is in a specific state as a result of a previous operation.
- ***Unsigned branches*** are taken when a comparison or a test of unsigned quantities results in a specific combination of condition code register bits.
- ***Signed branches*** are taken when a comparison or a test of signed quantities results in a specific combination of condition code register bits.

When a short-branch instruction is executed, a signed 8-bit offset is added to the value in the program counter when a specified condition is met. Program execution continues at the new address. The numeric range of the short-branch offset value is $80 (-128) to $7F (127) from the address of the instruction immediately following the branch instruction. A summary of the short-branch instructions is in Table 2.2.

When a long-branch instruction is executed, a signed 16-bit offset is added to the value in the program counter when a specified condition is met. Program execution continues at the

Unary Branches		
Mnemonic	Function	Equation or Operation
bra rel8 or lbra rel16	Branch always	1 = 1
brn rel8 or lbrn rel16	Branch never	1 = 0
Simple Branches		
Mnemonic	Function	Equation or Operation
bcc rel8 or lbcc rel16	Branch if carry clear	$C = 0$
bcs rel8 or lbcs rel16	Branch if carry set	$C = 1$
beq rel8 or lbeq rel16	Branch if equal	$Z = 1$
bmi rel8 or lbmi rel16	Branch if minus	$N = 1$
bne rel8 or lbne rel16	Branch if not equal	$Z = 0$
bpl rel8 or lbpl rel16	Branch if plus	$N = 0$
bvc rel8 or lbvc rel16	Branch if overflow clear	$V = 0$
bvs rel8 or lbvs rel16	Branch if overflow set	$V = 1$
Unsigned Branches		
Mnemonic	Function	Equation or Operation
bhi rel8 or lbhi rel16	Branch if higher	$C + Z = 0$
bhs rel8 or lbhs rel16	Branch if higher or same	$C = 0$
blo rel8 or lblo rel16	Branch if lower	$C = 1$
bls rel8 or lbls rel16	Branch if lower or same	$C + Z = 1$
Signed Branches		
Mnemonic	Function	Equation or Operation
bge rel8 or lbge rel16	Branch if greater than or equal	$N \oplus V = 0$
bgt rel8 or lbgt rel16	Branch if greater than	$Z + (N \oplus V) = 0$
ble rel8 or lble rel16	Branch if less than or equal	$Z + (N \oplus V) = 1$
blt rel8 or lblt rel16	Branch if less than	$N \oplus V = 1$

Note: 1. Each row contains two branch instructions that are separated by the word or.

2. The instruction to the left of or is a short branch with 8-bit offset.

3. The instruction to the right of or is a long branch with 16-bit offset.

Table 2.2 ■ Summary of short and long branch instructions

new address. Long-branch instructions are used when large displacements between decision-making steps are necessary.

The numeric range of long-branch offset values is $8000 ($-32,768$) to $7FFF ($32,767$) from the instruction immediately after the branch instruction. This permits branching from any location in the standard 64-kB address map to any other location in the map. A summary of the long-branch instructions is in Table 2.2.

Although there are many possibilities in writing a program loop, the following one is a common format:

```
loop:           .

                .

                .

        Bcc (or LBcc) loop
```

where **cc** is one of the condition codes (CC, CS, EQ, MI, NE, PL, VC, VS, HI, HS, LO, LS, GE, GT, LS, and LT). Usually, there will be a comparison or arithmetic instruction to set up the condition code for use by the conditional branch instruction. Unsigned branch instructions treat the numbers compared previously as nonnegative numbers. Signed branch instructions treat the numbers compared previously as signed numbers.

2.6.3 Compare and Test Instructions

The HCS12 has a set of compare instructions that are dedicated to the setting of condition flags. The compare and test instructions perform subtraction between a pair of registers or between a register and a memory location. The result is not stored, but condition codes are set by the operation. In the HCS12, most instructions update condition code flags automatically, so it is often unnecessary to include a separate test or compare instruction. Table 2.3 is a summary of compare and test instructions.

Compare Instructions		
Mnemonic	Function	Operation
cba	Compare A to B	$(A) - (B)$
cmpa <opr>	Compare A to memory	$(A) - (M)$
cmpb <opr>	Compare B to memory	$(B) - (M)$
cpd <opr>	Compare D to memory	$(D) - (M:M+1)$
cps <opr>	Compare SP to memory	$(SP) - (M:M+1)$
cpx <opr>	Compare X to memory	$(X) - (M:M+1)$
cpy <opr>	Compare Y to memory	$(Y) - (M:M+1)$
Test Instructions		
Mnemonic	Function	Operation
tst <opr>	Test memory for zero or minus	$(M) - \$00$
tsta	Test A for zero or minus	$(A) - \$00$
tstb	Test B for zero or minus	$(B) - \$00$

Note: <opr> represents an immediate value or a memory location and can be specified by using the immediate, direct, extended, and indexed addressing modes.

Table 2.3 ■ Summary of compare and test instructions

2.6.4 Loop Primitive Instructions

A lot of the program loops are implemented by incrementing or decrementing a loop count. The branch is taken when either the loop count is equal to zero or not equal to zero, depending on the applications. The HCS12 provides a set of loop primitive instructions for implementing this type of looping mechanism. These instructions test a counter value in a register or accumulator (A, B, D, X, Y, or SP) for zero or nonzero value as a branch condition. There are predecrement, preincrement, and test-only versions of these instructions.

The range of the branch is from $80 (−128) to $7F (127) from the instruction immediately following the loop primitive instruction. Table 2.4 shows a summary of the loop primitive instructions.

Mnemonic	Function	Equation or Operation
dbeq cntr, rel	Decrement counter and branch if = 0 (counter = A, B, D, X, Y, or SP)	counter ← (counter) − 1 If (counter) = 0, then branch; else continue to next instruction.
dbne cntr, rel	Decrement counter and branch if ≠ 0 (counter = A, B, D, X, Y, or SP)	counter ← (counter) − 1 If (counter) ≠ 0, then branch; else continue to next instruction.
ibeq cntr, rel	Increment counter and branch if = 0 (counter = A, B, D, X, Y, or SP)	counter ← (counter) + 1 If (counter) = 0, then branch; else continue to next instruction.
ibne cntr, rel	Increment counter and branch if ≠ 0 (counter = A, B, D, X, Y, or SP)	counter ← (counter) + 1 If (counter) ≠ 0, then branch; else continue to next instruction.
tbeq cntr, rel	Test counter and branch if = 0 (counter = A, B, D, X, Y, or SP)	If (counter) = 0, then branch; else continue to next instruction.
tbne cntr, rel	Test counter and branch if ≠ 0 (counter = A, B, D, X, Y, or SP)	If (counter) ≠ 0, then branch; else continue to next instruction.

Note: 1. **cntr** is the loop counter and can be accumulator A, B, or D and register X, Y, or SP.
2. **rel** is the relative branch offset and is usually a label.

Table 2.4 ■ Summary of loop primitive instructions

2.6.5 Implementation of Looping Constructs

The statement "for i = n1 to n2 do S" can be implemented as follows:

```
n1      equ     xx              ; start index (a nonnegative number)
n2      equ     yy              ; end index (a positive number)
        ...
i       ds.b    1
        ...
        movb    #n1,i           ; initialize loop index i
loop    ldaa    i               ; check loop index
        cmpa    #n2             ;       "
        bgt     next            ; if all iterations have been performed, then exit
        ...                     ; perform the loop operation
        ...                     ;       "
        inc     i               ; increment loop index
```

```
                    bra        loop                ; go back to the loop body
next                ...
```

The implementation of that statement "for i = n2 to n1 do S" can be modified from the previous instruction sequence as follows:

```
n1                  equ        xx                  ; start index (a nonnegative number)
n2                  equ        yy                  ; end index (a positive number)
                    ...
i                   ds.b       1
                    ...
                    movb       #n2,i               ; initialize loop index i
loop                ldaa       i                   ; check loop index
                    cmpa       #n1                 ;        "
                    blt        next                ; if all iterations have been performed, then exit
                    ...                            ; perform the loop operation
                    ...                            ;        "
                    dec        i                   ; decrement loop index
                    bra        loop                ; go back to the loop body
next                ...
```

Like the for loop, the while-loop construct also checks the condition at the start of the loop, and its implementation is similar with the following exceptions:

- The condition to be checked may be an external event instead of a variable.

- Updating the condition may be done by an external event such as an interrupt or the change of an input signal.

Assume that the CPU will keep performing a certain operation as long as the variable *icount* is not zero and *icount* is decremented by the interrupt service routine (discussed in Chapter 6). Then the following instruction sequence implements a loop using the while-loop construct:

```
N                   equ        xx
                    ...
icount              ds.b       1
                    ...
                    movb       #N, icount
wloop               ldaa       #0
                    cmpa       icount
                    beq        next
                    ...                            ; performed loop operation
                    ...
                    bra        wloop
next                ...
```

The "repeat S until C" looping construction is used more often to perform some operation a certain number of times. The following instruction sequence will perform a certain operation *N* times:

```
N                   equ        xx          ; define the constant N
                    ...
                    ...
                    ldy        #N          ; use Y to hold loop count
loop                ...                    ; perform operations
                    ...
                    dbeq       Y,loop      ; is loop count decremented to 0 yet?
```

Example 2.14

Write a program to add an array of N 8-bit numbers and store the sum at memory locations $1000~$1001. Use the "for i = n1 to n2 do" looping construct.

Solution: We will use variable i as the array index. This variable can also be used to keep track of the current iteration being performed. We will use a 2-byte variable sum to hold the sum of array elements. The logic flow of the program is illustrated in Figure 2.9. The program is a direct translation of the flowchart in Figure 2.9.

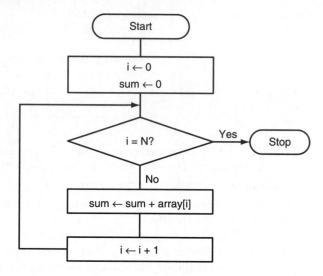

Figure 2.9 ■ Logic flow of example 2.14

```
N       equ     20              ; array count
        org     $1000           ; starting address of on-chip SRAM
sum     rmb     2               ; array sum
i       rmb     1               ; array index
        org     $1500           ; starting address of the program
        ldaa    #0
        staa    i               ; initialize loop (array) index to 0
        staa    sum             ; initialize sum to 0
        staa    sum+1           ;      "
loop    ldab    i
        cmpb    #N              ; is i = N?
        beq     done            ; if done, then branch
        ldx     #array          ; use index register X as a pointer to the array
        abx                     ; compute the address of array[i]
        ldab    0,x             ; place array[i] in B
        ldy     sum             ; place sum in Y
        aby                     ; compute sum <- sum + array[i]
        sty     sum             ; update sum
        inc     i               ; increment the loop count by 1
        bra     loop
```

```
done        swi                             ; return to D-Bug12 monitor
; the array is defined in the following statement
array       db          1,2,3,4,5,6,7,8,9,10,11,12,13,14,15,16,17,18,19,20
            end
```

It is a common mistake for an assembly language programmer to forget about updating the variable in memory. For example, we will not get the correct value for **sum** if we did not add the instruction **sty sum** in the program of Example 2.14.

Loop primitive instructions are especially suitable for implementing the "repeat S until C" looping construct, as demonstrated in the following example:

Example 2.15

Write a program to find the maximum element from an array of N 8-bit elements using the "repeat S until C" looping construct.

Solution: We will use the variable i as the array index and also as the loop count. The variable max_val will be used to hold the array maximum. The logic flow of the program is shown in Figure 2.10. The program is as follows:

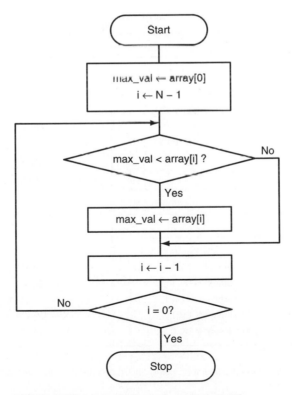

Figure 2.10 ■ Logic flow of example 2.15

```
N               equ     20                      ; array count
                org     $1000                   ; starting address of on-chip SRAM
arrmax          ds.b    1                       ; memory location to hold array max
                org     $1500                   ; starting address of program
                movb    array,arrmax            ; set array[0] as the temporary array max
                ldx     #array+N−1              ; start from the end of the array
                ldab    #N−1                    ; use B to hold variable i and initialize it to N−1
loop            ldaa    arrmax
                cmpa    0,x                     ; compare arrmax with array[i]
                bge     chk_end                 ; no update if max_val is larger
                movb    0,x,arrmax              ; update arrmax
chk_end         dex                             ; move the array pointer
                dbne    b,loop                  ; decrement the loop count, branch if not zero yet
forever         bra     forever
array           db      1,3,5,6,19,41,53,28,13,42,76,14,20,54,64,74,29,33,41,45
                end
```

2.6.6 Decrementing and Incrementing Instructions

We often need to add 1 to or subtract 1 from a variable in our program. Although we can use one of the add or sub instructions to achieve this, it would be more efficient to use a single instruction. The HCS12 has a few instructions for us to increment or decrement a variable by 1. A summary of decrement and increment instructions is listed in Table 2.5.

Decrement Instructions		
Mnemonic	Function	Operation
dec <opr>	Decrement memory by 1	$M \leftarrow [M] - 1$
deca	Decrement A by 1	$A \leftarrow [A] - 1$
decb	Decrement B by 1	$B \leftarrow [B] - 1$
des	Decrement SP by 1	$SP \leftarrow [SP] - 1$
dex	Decrement X by 1	$X \leftarrow [X] - 1$
dey	Decrement Y by 1	$Y \leftarrow [Y] - 1$
Increment Instructions		
Mnemonic	Function	Operation
inc <opr>	Increment memory by 1	$M \leftarrow [M] + 1$
inca	Increment A by 1	$A \leftarrow [A] + 1$
incb	Increment B by 1	$B \leftarrow [B] + 1$
ins	Increment SP by 1	$SP \leftarrow [SP] + 1$
inx	Increment X by 1	$X \leftarrow [X] + 1$
iny	Increment Y by 1	$Y \leftarrow [Y] + 1$

Note: <opr> can be specified using direct, extended, or indexed addressing modes.

Table 2.5 ■ Summary of decrement and increment instructions

Example 2.16

Use an appropriate increment or decrement instruction to replace the following instruction sequence:

```
ldaa      i
adda      #1
staa      i
```

Solution: These three instructions can be replaced by the following instruction:

```
inc       i
```

2.6.7 Bit Condition Branch Instructions

In certain applications, one needs to make branch decisions on the basis of the value of a few bits. The HCS12 provides two special conditional branch instructions for this purpose. The syntax of the first special conditional branch instruction is

```
brclr opr, msk, rel
```

where

opr   specifies the memory location to be checked and can be specified using direct, extended, and all indexed addressing modes.

msk   is an 8-bit mask that specifies the bits of the memory location to be checked. The bits to be checked correspond to those bit positions that are 1s in the mask.

rel   is the branch offset and is specified in 8-bit relative mode.

This instruction tells the HCS12 to perform bitwise logical AND on the contents of the specified memory location and the mask supplied with the instruction, then branch if the result is 0. For example, for the instruction sequence

```
here      brclr     $66,$80,here
          ldd       $70
```

the HCS12 will continue to execute the first instruction if the most significant bit of the memory location at $66 is 0. Otherwise, the next instruction will be executed.

The syntax of the second special conditional branch instruction is

```
brset opr, msk, rel
```

where

opr   specifies the memory location to be checked and can be specified using direct, extended, and all indexed addressing modes.

msk   is an 8-bit mask that specifies the bits of the memory location to be checked. The bits to be checked correspond to those bit positions that are 1s in the mask.

rel   is the branch offset and is specified in 8-bit relative mode.

This instruction tells the HCS12 to perform the logical AND of the contents of the specified memory location inverted and the mask supplied with the instruction, then branch if the result is 0 (this occurs only when all bits corresponding to 1s in the mask byte are 1s in the tested byte). For example, for the instruction sequence

```
loop    inc     count
                ...
        brset   $66,$E0,loop
                ...
```

the branch will be taken if the most significant 3 bits of the memory location at $66 are all 1s.

Example 2.17

▼

Write a program to count the number of elements that are divisible by 4 in an array of N 8-bit numbers. Use the "repeat S until C" looping construct.

Solution: The lowest 2 bits of a number divisible by 4 are 00. By checking the lowest 2 bits of a number, we can determine if a number is divisible by 4. The program is as follows:

```
N               equ     20
                org     $1000
total           rmb     1

                org     $1500           ; starting address of the program
                clr     total           ; initialize total to 0
                ldx     #array          ; use index register X as the array pointer
                ldab    #N              ; use accumulator B as the loop count
loop            brclr   0,x,$03,yes
                bra     chkend
yes             inc     total           ; add 1 to the total
chkend          inx                     ; move the array pointer
                dbne    b,loop
forever         bra     forever
array           db      2,3,4,8,12,13,19,24,33,32,20,18,53,52,80,82,90,94,100,102
                end
```

▲

2.6.8 Instructions for Variable Initialization

We often need to initialize a variable to 0 when writing a program. The HCS12 has three instructions for this purpose. They are

```
clr             opr
```

where opr is a memory location specified using the extended mode and all indexed addressing (direct and indirect) modes. The memory location is initialized to 0 by this instruction.

```
clra
```

Accumulator A is cleared to 0 by this instruction.

```
clrb
```

Accumulator B is cleared to 0 by this instruction.

2.7 Shift and Rotate Instructions

Shift and *rotate instructions* are useful for bit field manipulation. They can be used to speed up the integer multiply and divide operations if one of the operands is a power of 2.

A shift/rotate instruction shifts/rotates the operand by 1 bit. The HCS12 has shift instructions that can operate on accumulators A, B, and D or a memory location. A memory operand must be specified using the extended or indexed (direct or indirect) addressing modes. A summary of shift and rotate instructions is shown in Table 2.6.

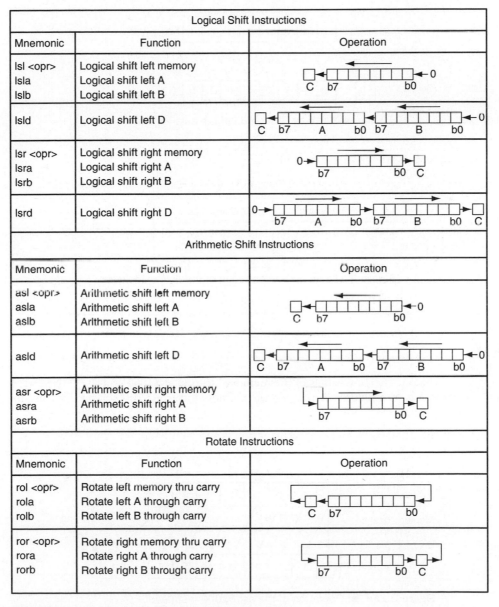

Table 2.6 ■ Summary of shift and rotate instructions

Example 2.18

What are the values of accumulator A and the C flag after executing the **asla** instruction assuming that originally A contains $95 and the C flag is 1?

Solution: The operation of this instruction is shown in Figure 2.11a.

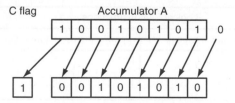

Figure 2.11a ▪ Operation of the asla instruction

The result is shown in Figure 2.11b.

Original Value	New Value
[A] = 10010101 C = 1	[A] = 00101010 C = 1

Figure 2.11b ▪ Execution result of the asla instruction

Example 2.19

What are the new values of the memory location at $1000 and the C flag after executing the instruction **asr $1000**? Assume that the memory location $1000 originally contains the value of $ED and the C flag is 0.

Solution: The operation of this instruction is shown in Figure 2.12a.

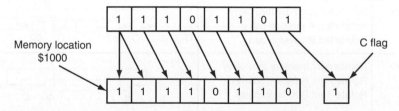

Figure 2.12a ▪ Operation of the **asr $1000** instruction

The result is shown in Figure 2.12b.

Original value	New value
[$1000] = 11101101 C = 0	[$1000] = 11110110 C = 1

Figure 2.12b ■ Result of the **asr $1000** instruction

Example 2.20

What are the new values of the memory location at $1000 and the C flag after executing the instruction **lsr $1000** assuming that the memory location $1000 originally contains $E7 and the C flag is 1?

Solution: The operation of this instruction is illustrated in Figure 2.13a.

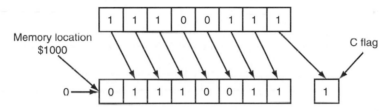

Figure 2.13a ■ Operation of the **lsr $1000** instruction

The result is shown in Figure 2.13b.

Original value	New value
[$1000] = 11100111 C = 1	[$1000] = 01110011 C = 1

Figure 2.13b ■ Execution result of **lsr $1000**

Example 2.21

What are the new values of accumulator B and the C flag after executing the **rolb** instruction assuming that the original value of B is $BD and the C flag is 1?

Solution: The operation of this instruction is illustrated in Figure 2.14a.

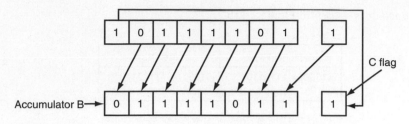

Figure 2.14a ■ Operation of the **rolb** instruction

The result is shown in Figure 2.14b.

Original value	New value
[B] = 10111101 C = 1	[B] = 01111011 C = 1

Figure 2.14b ■ Execution result of **rolb**

Example 2.22

What are the values of accumulator A and the C flag after executing the instruction **rora** assuming that the original value of A is $BE and C = 1?

Solution: The operation of this instruction is illustrated in Figure 2.15a.

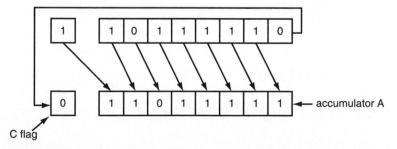

Figure 2.15a ■ Operation of the **rora** instruction

The result is shown in Figure 2.15b.

Original value	New value
[A] = 10111110 C = 1	[A] = 11011111 C = 1

Figure 2.15b ■ Execution result of **rora**

Example 2.23

Write a program to count the number of 0s contained in memory locations $1000~$1001 and save the result at memory location $1005.

Solution: The logical shift-right instruction is available for double accumulator D. We can load this 16-bit value into D and shift it to the right 16 times or until it becomes 0. The algorithm of this program is as follows:

Step 1
Initialize the loop count to 16 and the zero count to 0.

Step 2
Place the 16-bit value in D.

Step 3
Shift D to the right one place.

Step 4
If the C flag is 0, increment the zero count by 1.

Step 5
Decrement the loop count by 1.

Step 6
If the loop count is 0, then stop. Otherwise, go to step 3.

The program is as follows:

```
          org      $1000
          dc.w     $2355
zero_cnt  rmb      1
lp_cnt    rmb      1
          org      $1500
          clr      zero_cnt      ; initialize the zero count to 0
          movb     #16,lp_cnt    ; initialize the loop count to 16
          ldd      $1000         ; place the 16-bit number in D
again     lsrd
          bcs      chk_end       ; branch if the lsb is 1
          inc      zero_cnt
```

```
chk_end    dec    lp_cnt
           bne    again          ; have we tested all 16 bits yet?
forever    bra    forever
           end
```

▲

Sometimes we need to shift a number larger than 16 bits. However, the HCS12 does not have an instruction that does this. Suppose the number has k bytes and the most significant byte is located at **loc**. The remaining $k - 1$ bytes are located at loc + 1, loc + 2, ..., loc + $k - 1$, as shown in Figure 2.16.

Figure 2.16 ■ k bytes to be shifted

The logical shift-one-bit-to-the-right operation is shown in Figure 2.17.

Figure 2.17 ■ Shift-one-bit-to-the-right operation

As shown in Figure 2.17,

- Bit 7 of each byte will receive bit 0 of the byte on its immediate left with the exception of the most significant byte, which will receive a 0.
- Each byte will be shifted to the right by one bit. Bit 0 of the least significant byte will be shifted out and lost.

The operation can therefore be implemented as follows:

Step 1
Shift the byte at loc to the right one place (using the lsr <opr> instruction).

Step 2
Rotate the byte at loc + 1 to the right one place (using the ror <opr> instruction).

Step 3
Repeat step 2 for the remaining bytes.

By repeating this procedure, the given k-byte number can be shifted to the right as many bits as desired. The operation to shift a multibyte number to the left should start from the least significant byte and rotate the remaining bytes toward the most significant byte.

Example 2.24

▼

Write a program to shift the 32-bit number stored at $1000~$1003 to the right four places.

Solution: The most significant to the least significant bytes are stored at $1000~$1003. The following instruction sequence implements the algorithm that we just described:

```
            ldab     #4              ; set up the loop count
            ldx      #$1000
again       lsr      0,x
            ror      1,x
            ror      2,x
            ror      3,x
            dbne     b,again
```

▲

2.8 Boolean Logic Instructions

When dealing with input and output port pins, we often need to change the values of a few bits. For these types of applications, the Boolean logic instructions come in handy. A summary of the HCS12 Boolean logic instructions is given in Table 2.7.

The operand **opr** can be specified using all except the relative addressing modes. Usually, we would use the **and** instruction to clear one or a few bits and use the **or** instruction to set one

Mnemonic	Function	Operation
anda <opr>	AND A with memory	$A \leftarrow (A) \bullet (M)$
andb <opr>	AND B with memory	$B \leftarrow (B) \bullet (M)$
andcc <opr>	AND CCR with memory (clear CCR bits)	$CCR \leftarrow (CCR) \bullet (M)$
eora <opr>	Exclusive OR A with memory	$A \leftarrow (A) \oplus (M)$
eorb <opr>	Exclusive OR B with memory	$B \leftarrow (B) \oplus (M)$
oraa <opr>	OR A with memory	$A \leftarrow (A) + (M)$
orab <opr>	OR B with memory	$B \leftarrow (B) + (M)$
orcc <opr>	OR CCR with memory	$CCR \leftarrow (CCR) + (M)$
clc	Clear C bit in CCR	$C \leftarrow 0$
cli	Clear I bit in CCR	$I \leftarrow 0$
clv	Clear V bit in CCR	$V \leftarrow 0$
com <opr>	One's complement memory	$M \leftarrow \$FF - (M)$
coma	One's complement A	$A \leftarrow \$FF - (A)$
comb	One's complement B	$B \leftarrow \$FF - (B)$
neg <opr>	Two's complement memory	$M \leftarrow \$00 - (M)$
nega	Two's complement A	$A \leftarrow \$00 - (A)$
negb	Two's complement B	$B \leftarrow \$00 - (B)$

Table 2.7 ■ Summary of Boolean logic instructions

or a few bits. The **exclusive or** instruction can be used to toggle (change from 0 to 1 and from 1 to 0) one or a few bits. For example, the instruction sequence

```
ldaa           $56
anda           #$0F
staa           $56
```

clears the upper four pins of the I/O port located at $56. The instruction sequence

```
ldaa           $56
oraa           #$01
staa           $56
```

sets the bit 0 of the I/O port at $56. The instruction sequence

```
ldaa           $56
eora           #$0F
staa           $56
```

toggles the lower 4 bits of the I/O port at $56. The instructions (**coma** and **comb**) that perform one's complementing can be used if all of the port pins need to be toggled.

2.9 Bit Test and Manipulate Instruction

These instructions use a mask value to test or change the value of individual bits in an accumulator or in a memory location. The instructions **bita** and **bitb** provide a convenient means of testing bits without altering the value of either operand. Table 2.8 shows a summary of bit test and manipulation instructions.

Mnemonic	Function	Operation
bclr <opr>², msk8	Clear bits in memory	M ← (M) • $\overline{(mm)}$
bita <opr>¹	Bit test A	(A) • (M)
bitb <opr>¹	Bit test B	(B) • (M)
bset <opr>², msk8²	Set bits in memory	M ← (M) + (mm)

Note: 1. <opr> can be specified using all except relative addressing modes
            for bita and bitb.
        2. <opr> can be specified using direct, extended, and indexed
            (exclude indirect) addressing modes.
        3. msk8 is an 8-bit value.

Table 2.8 ■ Summary of bit test and manipulation instructions

For example, the instruction

```
bclr       0,x,$81
```

clears the most significant and the least significant bits of the memory location pointed to by index register X. The instruction

```
bita       #$44
```

tests bit six and bit two of accumulator A and updates the Z and N flags of the CCR register accordingly. The V flag in CCR register is cleared. The instruction

```
bitb    #$22
```

tests bit five and bit one of accumulator B and updates the Z and N flags of CCR register accordingly. The V flag in the CCR register is cleared. The instruction

```
bset    0,y,$33
```

sets bits five, four, one, and zero of the memory location pointed to by index register Y.

2.10 Program Execution Time

The HCS12 uses the bus clock (we will call it the E-clock from now on) signal as a timing reference. The generation of the E-clock is described in Chapter 6. The execution times of instructions are also measured in E cycles. The execution time of each instruction can be found in the column "Access Detail" in Appendix A. The number of letters in that column indicates the number of E cycles that a specific instruction takes to complete the execution. For example, the Access Detail column of the **pula** instruction contains three letters, ufo, which indicates that the pula instruction takes three E cycles to complete.

There are many applications that require the generation of time delays. Program loops are often used to create a certain amount of delay unless the time delay needs to be very accurate. The creation of a time delay involves two steps.

1. Select a sequence of instructions that takes a certain amount of time to execute.

2. Repeat the instruction sequence for the appropriate number of times.

For example, the following instruction sequence takes 40 E-clock cycles to execute:

```
loop    psha                    ; 2 E cycles
        pula                    ; 3 E cycles
        psha
        pula
        psha
        pula
        psha
        pula
        psha
        pula
        psha
        pula
        psha
        pula
        nop                     ; 1 E cycle
        nop                     ; 1 E cycle
        dbne    x,loop          ; 3 E cycles
```

Example 2.25

▼

Write an instruction sequence to create a 100-ms time delay for a demo board with a 24-MHz bus clock.

Solution: In order to create a 100-ms time delay, we need to repeat the preceding instruction sequence 60,000 times [100 ms ÷ (40 ÷ 24,000,000) μs = 60,000]. The following instruction sequence will create the desired delay:

```
        ldx     #60000        ; 2 E cycles
loop    psha                  ; 2 E cycles
        pula                  ; 3 E cycles
        psha                  ; 2 E cycles
        pula                  ; 3 E cycles
        psha                  ; 2 E cycles
        pula                  ; 3 E cycles
        psha                  ; 2 E cycles
        pula                  ; 3 E cycles
        psha                  ; 2 E cycles
        pula                  ; 3 E cycles
        psha                  ; 2 E cycles
        pula                  ; 3 E cycles
        psha                  ; 2 E cycles
        pula                  ; 3 E cycles
        nop                   ; 1 E cycle
        nop                   ; 1 E cycle
        dbne    x,loop        ; 3 E cycles
```

▲

Example 2.26

▼

Write an instruction sequence to create a delay of 10 sec.

Solution: The instruction sequence in Example 2.25 can only create a delay slightly longer than 100 ms. In order to create a longer time delay, we need to use a two-layer loop. For example, the following instruction sequence will create a 10-sec delay:

```
            ldab    #100      ; 1 E cycle
out_loop    ldx     #60000    ; 2 E cycles
inner_loop  psha
            pula
            psha
            pula
            psha
            pula
            psha
            pula
            psha
            pula
            psha
            pula
```

```
        psha
        pula
        psha
        pula
        psha
        pula
        nop
        nop
        dbne    x,inner_loop
        dbne    b,out_loop      ; 3 E cycles
```

▲

The time delay created by using program loops is not accurate. Some overhead is required to set up the loop count. For example, the one-layer loop has a 2-E-cycle overhead while the two-layer loop has much more overhead.

overhead =   1 E cycle (caused by the ldab #100 instruction)
            + 100 × 2 E cycles (caused by the out_loop   ldx #60000 instruction)
            + 100 × 3 E cycles (caused by the dbne b, out_loop instruction)
            = 501 E cycles = 20.875 μs (at 24-MHz E-clock)

To reduce the overhead, one can adjust the number to be placed in index register X. For example, by placing 59999 in X, one can create a delay of 9.99985 sec with the previous program, which is closer to 10 sec.

2.11 The Multiply-and-Accumulate (emacs) Instruction

The multiply-and-accumulate (**emacs**) instruction multiplies two 16-bit operands stored in memory and adds the 32-bit result to a third memory location (32 bit). EMACS can be used to implement simple digital filters, defuzzification routines, and any operation that involves the evaluation of linear polynomial functions.

When the emacs instruction is executed, the first source operand is fetched from a location pointed to by index register X, and the second source operand is fetched from a location pointed to by register Y. Before the instruction is executed, the X and Y index registers must contain values that point to the most significant bytes of the source operands. The most significant byte of the 32-bit result is specified by an extended address supplied with the instruction.

Example 2.27
▼

Write an assembly program to compute the following expression using the emacs instruction:

$a \times y + b$

where a, y, and b are in memory locations $1000, $1002, and $1004, respectively. Leave the result in memory location $1010~1013.

Solution: The value of b should be stored in memory locations $1012~$1013, and index registers X and Y should be set up to point to memory locations $1000 and $1002. We also need to store the constant in the destination. The following program performs the desired computation:

```
        org     $1000
aa      dc.w    10
xx      dc.w    7
```

```
bb       dc.w     6
         org      $1010
result   ds.w     2
         org      $1500
         ldx      #aa            ; set X to point to constant aa
         ldy      #xx            ; set Y to point to variable xx
         movw     bb,result+2    ; store b in the result
         movw     #0,result      ;     "
         emacs    result         ; perform the multiplication and accumulation
         swi
         end
```

Many applications (for example, *digital filtering*) involve the evaluation of polynomials. The emacs instruction can be used to compute the value of a polynomial. Before utilizing this instruction, we must transform the polynomial to a form that can take advantage of this instruction. The polynomial

$$a_n x^n + a_{n-1} x^{n-1} + \cdots + a_1 x + a_0$$

can be transformed into the format

$$(x(x(\cdots(a_n x + a_{n-1}) + a_{n-2}) + \cdots a_1) + a_0)$$

The expression enclosed by each pair of parentheses in the polynomial requires the computation of a multiplication followed by an addition and hence can be evaluated using the emacs instruction. The actual computation is quite straightforward and is left as an exercise problem.

2.12 Summary

An assembly language program consists of three major parts: *assembler directives, assembly language instructions*, and *comments*. A statement of an assembly language program consists of four fields: *label, operation code, operand*, and *comment*. Assembly directives supported by the freeware **as12** were all discussed in this chapter.

The HCS12 instructions were explained category by category. Simple program examples were used to demonstrate the applications of different instructions. The HCS12 is a 16-bit microcontroller. Therefore, it can perform 16-bit arithmetic. Numbers greater than 16 bits must be manipulated using multiprecision arithmetic.

Microcontrollers are designed to perform repetitive operations. Repetitive operations are implemented by program loops. There are two types of program loops: *infinite loops* and *finite loops*. There are four major variants of the looping constructs.

- **Do** statement S **forever**
- **For** $i = n_1$ **to** n_2 **do** S or **For** $i = n_2$ **downto** n_1 **do** S
- **While** C **do** S
- Repeat S until C

In general, the implementation of program loops requires

- The initialization of loop counter (or condition)
- Performing the specified operation

- Comparing the loop count with the loop limit (or evaluating the condition)
- Making a decision regarding whether the program loop should be continued

The HCS12 provides instructions to support the initialization of the loop counter, decrementing (or incrementing) the loop counter, and making a decision whether looping should be continued.

The shifting and rotating instructions are useful for bit field operations. Integer multiplication by a power of 2 and division by a power of 2 can be sped up by using the shifting instructions.

The HCS12 also provides many Boolean logical instructions that can be very useful for setting, clearing, and toggling the I/O port pins.

2.13 Exercises

E2.1 Find the valid and invalid labels in the following statements, and explain why the invalid labels are invalid:

```
column 1

    |
    ↓

a.  ABC         decb
b.  lp+:        adda    #1
c.  too:        mul
d.  not_true    nega
e.  star+=      ldaa    #10
f.  too_big     dec     count
```

E2.2 Identify the four fields of the following instructions:

```
a.              bne         not_done
b.  loop        brclr       0,x,$01,loop    ; wait until the least significant bit is set
c.  here:       dec         lp_cnt          ; decrement the variable lp_cnt
```

E2.3 Write a sequence of assembler directives to reserve 10 bytes starting from $1000.

E2.4 Write a sequence of assembler directives to build a table of ASCII codes of lowercase letters a–z. The table should start from memory location $2000.

E2.5 Write a sequence of assembler directives to store the message "Welcome to the robot demonstration!" starting from the memory location at $2050.

E2.6 Write an instruction sequence to add the two 24-bit numbers stored at $1010~$1012 and $1013~$1015 and save the sum at $1100~$1102.

E2.7 Write an instruction sequence to subtract the 6-byte number stored at $1000~$1005 from the 6-byte number stored at $1010~$1015 and save the result at $1020~$1025.

E2.8 Write a sequence of instructions to add the BCD numbers stored at $1000 and $1001 and store the sum at $1003.

E2.9 Write an instruction sequence to add the 4-digit BCD numbers stored at $1000~$1001 and $1002~$1003 and store the sum at $1010~1011.

E2.10 Write a program to compute the average of an array of N 8-bit numbers and store the result at $1000. The array is stored at memory locations starting from $1010. N is no larger than 255.

E2.11 Write a program to multiply two 3-byte numbers that are stored at $1000~$1002 and $1003~$1005 and save the product at $1010~$1015.

E2.12 Write a program to compute the average of the square of all elements of an array with 32 8-bit unsigned numbers. The array is stored at $1000~$101F. Store the result at $1020~$1021.

E2.13 Write a program to count the number of even elements of an array of N 16-bit elements. The array is stored at memory locations starting from $1010. N is no larger than 255.

E2.14 Write an instruction sequence to shift the 32-bit number to the left four places. The 32-bit number is located at $1000~$1003.

E2.15 Write a program to count the number of elements in an array that are smaller than 16. The array is stored at memory locations starting from $1010. The array has N 8-bit unsigned elements.

E2.16 Write an instruction sequence to swap the upper four bits and the lower four bits of accumulator A (swap bit 7 with bit 3, bit 6 with bit 2, and so on).

E2.17 Write a program to count the number of elements in an array whose bits 3, 4, and 7 are 0s. The array has N 8-bit elements and is stored in memory locations starting from $1000.

E2.18 Write an instruction sequence to set bits 3, 2, 1, and 0 of memory location at $1000 to 1 and leave the upper 4 bits of the same location unchanged.

E2.19 Find the values of condition flags N, Z, V, and C in the CCR register after the execution of each of the following instructions, given that [A] = $50 and the condition flags are N = 0, Z = 1, V = 0, and C = 1.

    a. suba #40

    b. tsta

    c. adda #$50

    d. lsra

    e. rola

    f. lsla

E2.20 Find the values of condition flags N, Z, V, and C in the CCR register after executing each of the following instructions independently, given that [A] = $00 and the initial condition codes are N = 0, C = 0, Z = 1, and V = 0.

    a. tsta

    b. adda #$40

    c. suba #$78

    d. lsla

    e. rola

    f. adda #$CF

E2.21 Write an instruction sequence to toggle the odd-number bits and clear the even-number bits of memory location at $66.

E2.22 Write a program to shift the 8-byte number located at $1000~$1007 to the left four places.

E2.23 Write a program to shift the 6-byte number located at $1010~$1015 to the right three places.

E2.24 Write a program to create a time delay of 100 sec by using program loops, assuming that the frequency of the bus clock is 24 MHz.

E2.25 Write a program to create a time delay of 5 sec using program loops, assuming that the frequency of the bus clock is 24 MHz.

E2.26 Write a program to evaluate

$$a_2x^2 + a_1x + a_0$$

Plug in values to verify that your program is correct.

E2.27 Write a program to evaluate

$$a_3x^3 + a_2x^2 + a_1x + a_0$$

Plug in values to verify that your program is correct.

3

Hardware and Software Development Tools for the HCS12

3.1 Objectives

After completing this chapter, you should be able to

- Explain the differences among different HCS12 members

- Know the peripheral functions available at different HCS12 members

- Understand the types of hardware and software development tools available

- Explain the functions of a source-level debugger

- Use the D-Bug12 commands to view and change the contents of memory locations and CPU registers

- Use the D-Bug12 commands to set breakpoints and trace program execution on the demo board

- Use the MiniIDE program to enter, assemble, and download programs onto the demo board for execution

- Use the D-Bug12 bootloader mode to upgrade D-Bug12 monitor and program application code into flash memory

- Use CodeWarrior to enter, assemble, and debug HCS12 assembly programs

- Understand the function of the background debug module

3.2 Development Tools for the HCS12

Microcontroller development tools can be divided into two categories: hardware and software.

3.2.1 Software Development Tools

A user needs a good *text editor* to enter his or her program. After the program is entered, the user needs an *assembler* or a *compiler* to convert the program into machine code for execution. A complicated program may consist of multiple computer files. A *linker* will be needed to resolve the variable cross-reference and memory allocation issue.

Before the user has the hardware to execute the program, a software *simulator* can be used to verify whether the program is logically correct. A *simulator* allows the user to execute microcontroller programs without having the actual hardware. It uses the computer memory to represent microcontroller registers and memory locations. The simulator interprets each microcontroller instruction by performing the operation required by the instruction and then saves the execution results in the computer memory. The simulator also allows the user to set the contents of memory locations and registers before the simulation run starts.

After the software has been assembled or compiled without syntax errors, the user may want to transfer the resultant machine code to the target hardware for execution. This will require a *communication* program (for example, the HyperTerminal bundled with Windows operating system) or even a hardware programmer.

When the software becomes complicated, the *program management* issue becomes complicated. The common approach is to use *project* as a unit to manage the software development process.

The *source-level debugger* is a program that allows the user to find problems in her or his code at the high-level-language (such as C) or assembly-language level. A debugger may have the option to run the program on the target hardware or using a simulator. Like a simulator, a debugger can display the contents of registers and memory (internal and external) and program code in separate windows. With a debugger, all debugging activities are done at the source level. The user can see the value change of a variable after a statement has been executed. The user can also set a breakpoint at a statement in a high-level language. However, a source-level debugger requires a lot of computation. A source-level debugger needs to communicate with the monitor program on the demo board in order to display the contents of CPU registers and memory locations, set or delete breakpoints, trace program execution, and so on. Since the monitor programs on different evaluation boards may not be the same, a source-level debugger may be used only with one type of demo board. The BDM mode of the HCS12 (and other microcontrollers) offers an alternative for implementing the source-level debugger.

The software vendors often integrate a text editor, an assembler and/or compiler, a linker, a simulator, a source-level debugger, and a project manager into a package so that the user can switch from one tool to another without exiting any of these programs. This integrated software environment is called an *integrated development environment* (IDE). The complexity of IDEs varies significantly. For example, the freeware MiniIDE consists of an assembler, a linker, a project manager, and a terminal program for the PC to communicate with the demo board. Using the MiniIDE, the user can enter and assemble his or her assembly program and download it onto a HCS12 demo board (this HCS12 demo board must have the D-Bug12 monitor program) for execution. After downloading the machine code onto the demo board, the user can then use commands supported by the D-Bug12 monitor to perform debug activities.

The CodeWarrior from Freescale is an IDE that consists of a text editor, an assembler, a linker, a C compiler, a project manager, a simulator, device drivers, and a source-level debugger. Both the MiniIDE and CodeWarrior will be used in this text for developing and testing programs. The AsmIDE written by Eric Engler is similar to MiniIDE in function.

3.2.2 Hardware Development Tools

For learning a microcontroller, the most important hardware tool is the microcontroller demo board. Tools such as function generators and oscilloscopes will be helpful for testing a user's program. Hardware tools such as in-circuit emulators and logic analyzers will be very useful in helping users in debugging their software. However, these are expensive debugging tools and hence we will not discuss them in this text.

The background debug module (BDM) of the HCS12 microcontroller allows the user to trace instruction execution on the target hardware (a HCS12 microcontroller) from a PC or another HCS12-based demo board running appropriate software. When performing debug activities in the BDM mode, the PC or the host HCS12 communicates with the target HCS12 via the BDM serial interface. This approach allows debug activities to be performed less intrusively.

3.2.3 Types of HCS12 Demo Boards

On the basis of the supporting debug environment, a demo board can be classified into the following three categories:

DEMO BOARD WITH THE D-BUG12 MONITOR

An HCS12 demo board may have its microcontroller programmed with the D-Bug12 monitor to support debug activities. The D-Bug12 monitor occupies the flash memory and supports a set of debug commands. The user can use these commands to display and modify register and memory contents, set breakpoints at certain memory locations (breakpoints can only be set in SRAM), step through the program, and download the program onto the demo board for execution. Using this approach, the user can only download his or her program onto the on-chip SRAM of the HCS12. For learning assembly language programs, this type of demo board provides a simple environment for testing and debugging user programs. Both the MiniIDE and the AsmIDE may work with this type of demo board.

DEMO BOARD WITH THE SERIAL MONITOR

An HCS12 demo board may also choose to have its microcontroller programmed with the serial monitor from Freescale. The serial monitor occupies 2 kB of flash memory and allows the user program to employ the remaining memory space. The CodeWarrior IDE from Freescale may communicate with the serial monitor and allow the user to carry out all debug activities.

DEMO BOARD WITH THE BDM ADAPTOR

This type of demo board relies on the HCS12 background debug module to carry out debug activities. The user runs an appropriate program on the PC to communicate with the HCS12 via a BDM adaptor to perform debug activities. The BDM adaptor may be a separate hardware kit or a part of the demo board. There is a public domain BDM-based debugging interface called Turbo BDM Light (*TBDML*) that every user can add to her or his demo board or make into a separate debug adaptor. The TBDML is supported by the CodeWarrior IDE and provides a nice debugging environment. CodeWarrior also supports a few other BDM adaptors including P&E micro's BDM adaptor. To use this approach to debug programs, the demo board must have a BDM interface circuit.

A demo board can be switched from one category to another. For example, a demo board programmed with the D-Bug12 monitor can be reprogrammed with the serial monitor and allow the user to employ CodeWarrior to develop software. If a demo board has the BDM interface circuit (a 6-pin connector), then the user can also use a BDM adaptor and CodeWarrior (or other IDE) to develop the software. If the user prefers to employ the D-Bug12 monitor, he or she can use a BDM adaptor to program the D-Bug12 monitor into the microcontroller of the demo board and start to use the D-Bug12 commands to debug the software.

It is obvious that different people will have different opinions on the choice of which HCS demo board is best for learning the HCS12 microcontroller. Some educators prefer a demo board to be a bare kit but allow the user to connect a wide variety of chips and I/O devices to the demo board. Other educators may prefer a demo board to have a lot of features and also allow the user to connect many other I/O devices and peripheral chips to the demo board. However, all educators would agree that the demo board must be affordable. For the demo board to have many features and still be affordable, in this text we use the Dragon12-Plus demo board from Wytec to test and debug all the programs. The Dragon12-Plus and its predecessors have been very popular in universities.

3.3 The Dragon12-Plus Demo Board

A photo of the Dragon12-Plus demo board is shown in Figure 3.1. In addition to the 9S12DG256B MCU, the Dragon12-Plus demo board has the following features:

- Dual RS232 communication ports
- DS1307 real-time clock (RTC) with backup battery included for testing I^2C interface
- CAN port
- A 10-bit dual-channel DAC LTC1661
- Four robot servo controllers with terminal block for external 5 V
- Four 7-segment displays

Figure 3.1 ▪ Dragon12-Plus demo board

- Eight LEDs
- Eight-position dual inline package (DIP) switch
- Four push button switches
- 5-V regulator with DC jack and terminal block for external 9-V battery input
- A speaker to be driven by timer, or DAC or PWM signal for alarm or music applications
- A dual H-bridge motor driver with motor feedback or rotary encoder interface for controlling two DC motors or one Stepper motor
- An IR transceiver with an on-board 38-kHz oscillator
- A BDM-in connector to be connected with a BDM from multiple vendors for debugging
- An opto-coupler output
- A logic probe with LED indicator
- An abort switch for stopping program when program is hung in a dead loop
- A mode switch for selecting four operating modes: EVB, Jump-to-EEPROM, BDM POD, and Bootloader
- A 4 × 4 keypad
- A form-C relay output rated at 3 A/30 V or 1 A/125 V
- An X-Y-Z accelerometer interface or GP2-D12 distance-measuring sensor interface
- A potentiometer trimmer pot for analog input
- A temperature sensor
- A communication port for a video graphics array (VGA) camera with built-in Joint Photographic Experts Group (JPEG) compression (camera is optional.)
- A light sensor
- A low-battery detection circuit
- Female and male headers provide access to all I/O pins of the MC9S12DG256

Additional product information about the Dragon12-Plus demo board can be found on the website at http://www.evbplus.com. Wytec also provides the miniDragon-Plus2 demo board, which is a bare kit that allows the user to add her or his circuit. This demo board has fewer features than the Dragon12-Plus demo board. The photo of this demo board is shown in Figure 3.2.

3.4 The D-Bug12 Monitor

The D-Bug12 is a monitor program designed for the HCS12 microcontrollers. Version 4 of the D-Bug12 supports the following devices:

- MC9S12Dx256 (x = G, P, or T)
- MC9S12A256
- MC9S12Dx128 (x = G, P, or T)
- MC9S12H256
- MC9S12A128

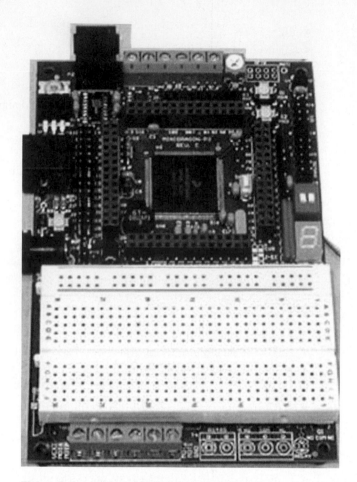

Figure 3.2 ■ Wytec MiniDragon demo board

This monitor is used in several HCS12 demo boards from several companies. It facilitates the writing, evaluation, and debugging of user programs.

Version 4.x.x of the D-Bug12 requires a host terminal program that supports XON/XOFF software handshaking for proper operation. The HyperTerminal program that comes with Windows 2000 and Windows XP can work with the D-Bug12. In addition, the terminal programs bundled with MiniIDE, AsmIDE, and EmbeddedGNU IDE can all work with the D-Bug12 monitor. The default baud rate out of reset is 9600. However, the baud rate can be set to a higher value if a higher communication speed is desired.

3.4.1 The D-Bug12 Operating Modes

The D-Bug12 monitor has four operating modes. When the D-Bug12 monitor first starts (at power-up or reset), it reads the logic levels on the PAD0 and PAD1 pins to enter different operating modes. The operating modes for the four logic level combinations are listed in Table 3.1.

PAD1	PAD0	Operating Mode
0	0	D-Bug12; EVB
0	1	Jump to internal EEPROM
1	0	D-Bug12; POD
1	1	SerialBootloader

Table 3.1 ■ D-Bug12 operating modes

3.4.2 EVB Mode

The EVB mode is the most important mode for beginners to learn the HCS12 microcontroller using a demo board. In this mode, the monitor operates as a ROM resident monitor/debugger executing from the on-chip flash memory. Although this mode provides a good environment for testing new algorithms or conducting performance benchmarks, it does have a few limitations. In this mode, the flash memory, 1024 bytes of on-chip SRAM, and one of the SCI serial ports are not available to the developer. The portion of the SRAM that may be used by the user program begins at $1000 and ends at $3BFF. D-Bug12 uses the remainder of the SRAM that begins at $3C00 and ends at $3FFF. User programs can also be loaded onto the on-chip EEPROM for execution. The conceptual operating model for this mode is illustrated in Figure 3.3.

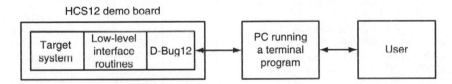

Figure 3.3 ■ EVB mode conceptual model

Suppose we have run a terminal program on the PC and selected the EVB mode; then either powering on the demo board or pressing the reset button brings out the sign-on message shown in Figure 3.4 on the terminal screen. The D-Bug12 monitor displays the ASCII greater than character (>), indicating that it is ready to accept a command. When issuing a command that causes a program to run from the internal RAM or EEPROM, D-Bug12 will place the terminal cursor on a blank line, where it will remain until control is returned to D-Bug12. If a running program fails to return to D-Bug12, pressing the EVB's reset button causes the running program to halt execution and initiate the D-Bug12 initialization sequence. Using this method to regain control of an executing program fails to report any information to the programmer on why or how the program may have failed.

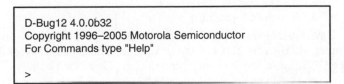

Figure 3.4 ■ D-Bug12 EVB mode sign on message

Alternatively, if an optional, normally open switch is wired to the $\overline{\text{XIRQ}}$ pin, pressing it generates an $\overline{\text{XIRQ}}$ interrupt and causes the running program to halt execution and return control back to D-Bug12 where the CPU register contents are displayed. The Dragon12-Plus demo board implements this option by providing the **abort** switch.

As with all ROM-based monitors, D-Bug12 utilizes some of the on-chip resources to perform its debugging functions. The D-Bug12's memory maps when running on 256-kB and 128-kB devices are detailed in Table 3.2. For the HCS12 devices with 256-kB flash memory, there are 11 kB of SRAM available for the development of application programs. For devices with 128-kB flash memory, only 7 kB is available for program development. The current implementation of D-Bug12 does not allow any of the on-chip flash memory to be utilized by application code when running in EVB mode.

Note that even though the HCS12Dx256 parts contain 4 kB of EEPROM, only the upper 3 kB are visible, as the lower 1 kB is overlaid with the I/O registers. Table 3.2 shows only the 64-kB memory map. Most of the D-Bug12 code occupies the on-chip paged flash memory beginning on page $38.

Address Range	Description
$0000~$03FF	I/O registers
$0400~$0FFF	On-chip EEPROM
$1000~$3BFF	On-chip SRAM (available to user)
$3C00~$3FFF	On-chip SRAM (D-Bug12)
$4000~$EE7F	D-Bug12 code
$EE80~$EEBF	User-accessible function table
$EEC0~$EEFF	Customization data
$EF00~$EF8B	D-Bug12 startup code
$EF8C~$EFFF	Secondary reset/interrupt table
$F000~$FFFF	Bootloader

Table 3.2a ■ D-Bug12 memory map for HCS12Dx256

Address Range	Description
$0000~$03FF	I/O registers
$0800~$0FFF	On-chip EEPROM
$2000~$3BFF	On-chip SRAM (available to user)
$3C00~$3FFF	On-chip SRAM (D-Bug12)
$4000~$EE7F	D-Bug12 code
$EE80~$EEBF	User accessible function table
$EEC0~$EEFF	Customization data
$EF00~$EF8B	D-Bug12 startup code
$EF8C~$EFFF	Secondary reset/interrupt table
$F000~$FFFF	Bootloader

Table 3.2b ■ D-Bug12 memory map for HCS12Dx128

D-Bug12 supports a set of commands that can be used for program development on the demo board. A summary of the command set is given in Table 3.3. One can request the D-Bug12 monitor to show the complete command set by typing **help** at the D-Bug12 monitor prompt.

3.5 Using a Demo Board with the D-Bug12 Monitor

We need at least the following software programs in order to develop assembly programs to be downloaded onto a demo board for execution:

1. A text editor
2. An HCS12 cross assembler
3. A terminal program

We prefer using an IDE program in developing assembly programs. Both the AsmIDE by Eric Engler and the MiniIDE from Mgtek are well-designed IDEs for developing assembly programs for the HCS12 and 68HC11 microcontrollers. These two freeware IDEs allow us to enter, assemble,

Command	Description
ALTCLK	Specify an alternate BDM communication rate
ASM	Single line assembler/disassembler
BAUD <baudrate> [;t]	Set communications rate for the terminal
BDMBD	Enter the BDM command debugger
BF <startAddress><EndAddress> [<data>]	Fill memory with data
BR [<Address>]	Set/Display breakpoints
BULK	Erase entire on-chip EEPROM contents
CALL [<address>]	Execute a user subroutine; return to D-Bug12 when finished
DEVICE [see description]	Select/define a new target MCU device
EEBASE <Address>	Inform D-Bug12 of the Target's EEPROM base address
FBULK	Erase the target processor's on-chip flash EEPROM
FLOAD <AddressOffset>	Program the target processor's on-chip Flash EEPROM from S-records
FSERASE	Erase one or more sectors of target flash EEPROM
G [<Address>]	Go-begin execution of user program
GT <Address>	Go-Till—set a temporary breakpoint and begin execution of user program
HELP	Display D-Bug12 command set and command syntax
LOAD [<AddressOffset>]	Load user program in S-record
MD <StartAddress> [<EndAddress>]	Memory display—display memory contents in hex bytes/ASCII format
MDW <StartAddress> [<EndAddress>]	Memory display word—display memory contents in hex words/ASCII format
MM <Address> [<data>]	Memory modify—interactively examine/change memory contents
MMW <address> [<data>]	Memory modify word—interactively examine/change memory contents
MOVE <StartAddress> <EndAddress> <DestAddress>	Move a block of memory
NOBR [<Address> <Address>...]	Remove individual user breakpoints
PCALL	Execute a user subroutine in expanded memory, return to D-Bug12 when finished
RD	Register display—display the CPU registers
REGBASE	Inform D-Bug12 of the target I/O register's base address
RESET	Reset the target CPU
RM	Register modify—interactively examine/change CPU register contents
SO	Step over subroutine calls
STOP	Stop execution of user code on the target processor and place it in background mode
T [<count>]	Trace <count> instructions
TCONFIG	Configure target before erasing or programming
UPLOAD <StartAddress><EndAddress>	S-record memory display
USEHBR	Use hardware breakpoints
VER	Display the running version of D-Bug12
VERF [<AddressOffset>]	Verify S-records against memory contents
<Register Name><Register Value>	Set register contents
Register Names:	PC, SP, X, Y, A, B, D, PP
CCR Status Bits:	S, XM, H, IM, N, Z, V, C

Table 3.3 ■ D-Bug12 command-set summary

and download the S-record file onto a demo board for execution without quitting any one of them. AsmIDE can be downloaded from the website at http://www.geocities.com/englere_geo; MiniIDE can be downloaded from the website at http://www.mgtek.com/miniide. The tutorial for using the MiniIDE is given in the next section; the tutorial for using the AsmIDE is given in Appendix E.

3.5.1 Starting the MiniIDE

The MiniIDE can be started by clicking on its icon. The startup screen should be similar to that in Figure 3.5.

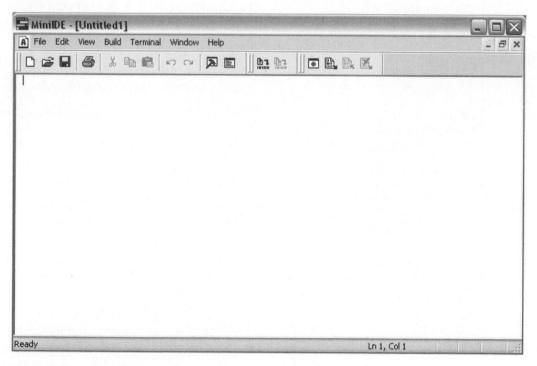

Figure 3.5 ■ MiniIDE startup screen

3.5.2 Communicating with the Demo Board

To communicate with the demo board using the D-Bug12 monitor, press **Terminal** menu and select **Show Terminal Window** and **Connected**. After this, the lower half becomes the terminal window as shown in Figure 3.6. Press the reset button of the demo board and the screen will change to Figure 3.7.

One can then enter D-Bug12 commands in the terminal window to display registers and memory contents, modify memory and register contents, set breakpoints, trace instruction execution, and download the program onto the demo board for execution.

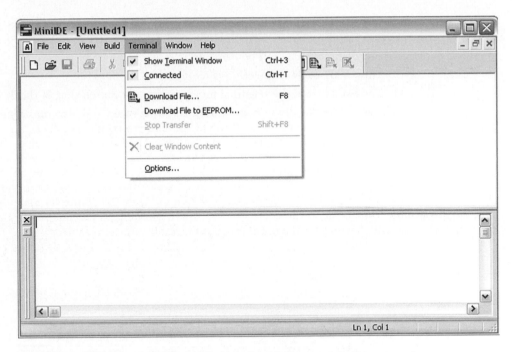

Figure 3.6 ■ Screen to show terminal window and connect to demo board

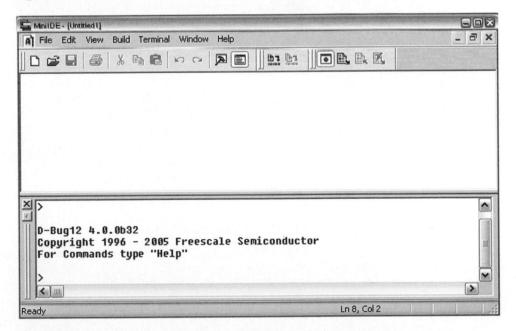

Figure 3.7 ■ MiniIDE terminal window with D-Big12 command prompt

If the D-Bug12 command prompt does not appear on the screen, the most likely three causes are

1. The demo board is not powered up.
2. The RS232 cable connection is wrong.
3. The baud rate of the terminal program does not match that of the demo board.

Make sure that the demo board has been powered up and also make sure that the RS232 cable is connected to the right connector (if the demo board has two RS232 connectors). If the D-Bug12 prompt still does not appear, then check the baud rate setting. Select the **Terminal** menu and unselected **Connected** and click on **Options.** The Options dialog will appear as shown in Figure 3.8. The settings shown in Figure 3.8 are acceptable. Another option that the user needs to set is **Tools.** Click on **Tools** in the dialog box and make sure that asm12.exe is selected as the tool for assembling HCS12 programs as shown in Figure 3.9. The user may also want to set the options in the General category. The settings in the General category are shown in Figure 3.10. The default settings for the options of the assembler are acceptable.

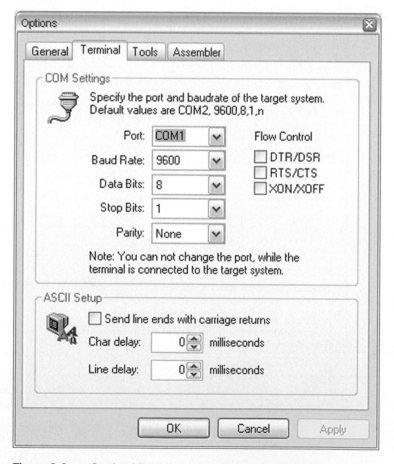

Figure 3.8 ■ Setting MiniIDE communication parameters

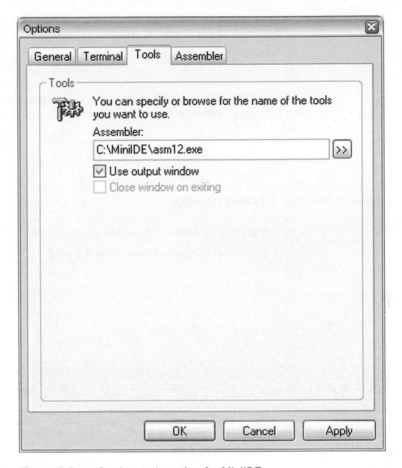

Figure 3.9 ■ Setting tools option for MiniIDE

3.5.3 Using the D-Bug12 Commands

The D-Bug12 commands are provided to help the program-debugging process. Some of these commands can be used to set and display the contents of memory locations. In this section, character strings entered by the user are in boldface and optional fields are enclosed in brackets [ ]. In the following section, the syntax of a command is presented and then examples are given.

bf <StartAddress> <EndAddress> [<Data>]

The **bf** command is used to fill a block of memory locations with the same value. For example, the following command clears the internal memory locations from $1000 to $10FF to zero:

>bf 1000 1FFF 0

The data field is optional. If we did not specify the data to be filled in, then zero would be entered in the specified memory locations.

md <StartAddress> [<EndAddress>]

The **md** (*memory display*) command is used to display memory contents. This command displays memory contents as both hexadecimal bytes and ASCII characters, 16 bytes on each line. The <StartAddress> parameter must be supplied; the <EndAddress> parameter is optional. When the <EndAddress> parameter is not specified, a single line is displayed.

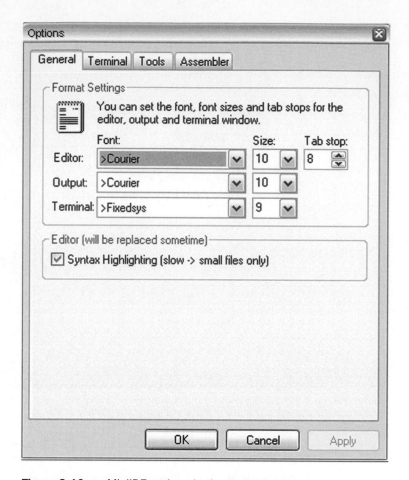

Figure 3.10 ■ MiniIDE options in the general category

The number supplied as the <StartAddress> parameter is rounded down to the next-lower multiple of 16; the number supplied as the <EndAddress> parameter is rounded up to the next-higher multiple of 16 minus 1. This causes each line to display memory in the range of $xxx0 to $xxxF. For example, if **$1005** is entered as the start address and **$1020** as the ending address, then the actual memory range displayed would be $1000 through $102F. This command allows the user to examine the program execution result.

Example 3.1

```
>md 1000
1000   00 00 00 00 - 00 00 00 00 - 00 00 00 00 - 00 00 00 00      ............
>md 1005 1020
1000   00 00 00 00 - 00 00 00 00 - 00 00 00 00 - 00 00 00 00      ............
1010   00 00 00 00 - 00 00 00 00 - 00 00 00 00 - 00 00 00 00      ............
1020   00 00 00 00 - 00 00 00 00 - 00 00 00 00 - 00 00 00 00      ............
```

mdw <StartAddress> [<EndAddress>]

The **mdw** (*memory display words*) command displays the contents of memory locations as hexadecimal words and ASCII characters, 16 bytes on each line. The *<StartAddress>* parameter must be supplied; the *<EndAddress>* parameter is optional. When the <EndAddress> parameter is not supplied, a single line is displayed.

The number supplied as the <StartAddress> parameter is rounded down to the next lower multiple of 16, while the number supplied as the <EndAddress> parameter is rounded up to the next higher multiple of 16 minus 1. This causes each line to display memory in the range of $xxx0 through $xxxF.

Example 3.2

▼

```
>mdw 1000
1000   FC08 00CD - 0900 CE00 - 0A18 10CB - 306B 44B7        . . . . . . . . . . . . OkD.
>mdw 1000 1020
1000   FC08 00CD - 0900 CE00 - 0A18 10CB - 306B 44B7        . . . . . . . . . . . . OkD.
1010   C5CE 000A - 1810 CB30 - 6B43 B7C5 - CE00 0A18        . . . . . . . OkC . . . . . .
1020   10CB 306B - 42B7 C5CE - 000A 1810 - CB30 6B41        . . OkB . . . . . . . . OkA
>
```

▲

mm <Address> [<Data>]

The **mm** (*memory modify*) command allows us to examine and modify the contents of memory locations 1 byte at a time. If the 8-bit data parameter is present on the command line, the byte at memory location *<Address>* is replaced with *<Data>* and the command is terminated. If no optional data is provided, then D-Bug12 enters the *interactive memory modify mode*. In the interactive mode, each byte is displayed on a separate line following the address of data. Once the memory modify command has been entered, single-character subcommands are used for the modification and verification of memory contents. These subcommands have the following format:

[<Data>]	<CR>	Optionally update current location and display the next location.
[<Data>]	</> or <=>	Optionally update current location and redisplay the same location.
[<Data>]	<^> or <->	Optionally update current location and display the previous location.
[<Data>]	<.>	Optionally update current location and exit Memory Modify.

With the exception of the carriage return (CR), the subcommand must be separated from any entered data with at least a one-space character. If an invalid subcommand character is entered, an appropriate error message is issued and the contents of the current memory location are redisplayed.

Example 3.3

▼

In this example, each line is terminated with a carriage-return character. However, the carriage return character is nondisplayable and hence is not shown.

```
>mm 1000
1000 00
1001 00 FF
1002 00 ^
1001 FF
```

```
1002 00
1003 00 55 /
1003 55 .
>
```

mmw <Address> [<Data>]

The **mmw** (*memory modify, word*) command allows the contents of memory to be examined and/or modified as 16-bit hex data. If the 16-bit data is present on the command line, the word at memory location <*Address*> is replaced with <*Data*> and the command is terminated. If not, D-Bug12 enters the interactive memory modify mode. In the interactive mode, each word is displayed on a separate line following the address of data. Once the memory modify command has been entered, single-character subcommands are used for the modification and verification of memory contents. These subcommands have the following format:

[<Data>]	<CR>	Optionally update current location and display the next location.
[<Data>]	</> or <=>	Optionally update current location and redisplay the current location.
[<Data>]	<^> or <−>	Optionally update current location and display the previous location.
[<Data>]	<.>	Optionally update current location and exit memory modify.

With the exception of the carriage return (CR), the subcommand must be separated from any entered data with at least a one-space character. If an invalid subcommand character is entered, an appropriate error message is issued and the contents of the current memory location are redisplayed.

Example 3.4

In this example, each line is terminated with a carriage-return character. However, the carriage-return character is nondisplayable and hence is not shown.

```
>mmw 1100
1100 00F0
1102 AA55 0008
1104 0000 ^
1102 0008 aabb
1104 0000
1106 0000 .
>
```

move <StartAddress> <EndAddress> <DestAddress>

The **move** (*move memory block*) command is used to move a block of memory from one location to another, 1 byte at a time. Addresses are specified in 16-bit hex values. The number of bytes moved is one more than <*EndAddress*> − <*StartAddress*>. The block of memory beginning at the destination address may overlap the memory block defined by <StartAddress> and <EndAddress>. One of the uses of the move command might be to copy a program from RAM into the on-chip EEPROM.

Example 3.5

```
>move 1000 10ff 1100
>
```

rd
The **rd** (*register display*) command is used to display the HCS12 CPU registers, including the *ppage* register.

Example 3.6

```
>rd
PP   PC        SP        X        Y        D = A:B      CCR = SXHI      NZVC
38 1521      3C00     2014     0000       6E:14               1001      0100
xx: 1521     9C42              CPD        $0042
>
```

rm
The **rm** (*register modify*) command is used to examine and/or modify the contents of the CPU12 registers interactively. As each register and its contents are displayed, D-Bug12 allows the user to enter a new value for the register in hex. If modification of the displayed register is not desired, entering a carriage return will cause the next CPU register and its contents to be displayed on the next line. When the last of the CPU registers has been examined and/or modified, the rm command displays the first register, giving the user an opportunity to make additional modifications to the CPU register contents. Typing a period as the first non-space character on the line will exit the interactive mode of the register modify command and return to the D-Bug12 command prompt. The registers are displayed in the following order, one register per line: PC, SP, X, Y, A, B, CCR. The PPAGE register (represented by PP) is not displayed.

Example 3.7

```
>rm
PC=0000 1500
SP=0A00
IX=0000 0100
IY=0000
A=00
B=00 ff
CCR=90 d1
PC=1500 .
>
```

pc <RegisterName> <RegisterValue>

The **pc** command allows one to change the value of any CPU register (PC, SP, X, Y, A, B, D, CCR). Each of the fields in the CCR may be modified by using the bit names shown in Table 3.4.

CCR Bit Name	Description	Legal Values
S	STOP enable	0 or 1
H	Half carry	0 or 1
N	Negative flag	0 or 1
Z	Zero flag	0 or 1
V	Two's complement overflow flag	0 or 1
C	Carry flag	0 or 1
IM	IRQ interrupt mask	0 or 1
XM	XIRQ interrupt mask	0 or 1

Table 3.4 ■ Condition code register bits

Example 3.8

```
>pc 2000
PC      SP      X      Y      D = A:B    CCR = SXHI    NZVC
2000    0A00    0100   0000     00:FF          1101    0001
>x 800
PC      SP      X      Y      D = A:B    CCR = SXHI    NZVC
2000    0A00    0800   0000     00:FF          1101    0001
>c 0
PC      SP      X      Y      D = A:B    CCR = SXHI    NZVC
2000    0A00    0800   0000     00:FF          1101    0000
>z 1
PC      SP      X      Y      D = A:B    CCR = SXHI    NZVC
2000    0A00    0800   0000     00:FF          1101    0100
>d 2010
PC      SP      X      Y      D = A:B    CCR = SXHI    NZVC
2000    0A00    0800   0000     20:10          1101    0100
>
```

asm <Address>

The **asm** command invokes the one-line assembler/disassembler. It allows memory contents to be viewed and altered using assembly-language mnemonics. Each entered source line is translated into object code and placed into memory at the time of entry. When displaying memory contents, each instruction is disassembled into its source mnemonic form and displayed along with the hex object code and any instruction operands.

Assembly mnemonics and operands may be entered in any mix of upper- and lowercase letters. Any number of spaces may appear between the assembler prompt and the instruction mnemonic or between the instruction mnemonic and the operand. Numeric values appearing in the operand field are interpreted as *signed* decimal numbers. Placing a $ in front of any number will cause the number to be interpreted as a hex number.

When an instruction is disassembled and displayed, the D-Bug12 prompt is displayed following the disassembled instruction. If a carriage return is the first nonspace character entered following the prompt, the next instruction in memory is disassembled and displayed on the next line. If an HCS12 instruction is entered following the prompt, the entered instruction is assembled and placed into memory. The line containing the new entry is erased and the new instruction is disassembled and displayed on the same line. The next instruction location is then disassembled and displayed on the screen.

When entering branch instructions, the number placed in the operand field should be the absolute destination address of the instruction. The assembler calculates the two's complement offset of the branch and places the offset in memory with the instruction.

The assembly/disassembly process may be terminated by entering a period as the first nonspace character following the assembler prompt.

The following example displays the assembly instructions from memory location $2000 until $2011. The carriage-return character is entered at the > prompt of each line and the period character is entered at the last line.

```
>asm 2000
2000    FC0800    LDD     $0800       >
2003    CD0900    LDY     #$0900      >
2006    CE000A    LDX     #$000A      >
2009    1810      IDIV                >
200B    CB30      ADDB    #$30        >
200D    6B44      STAB    4,Y         >
200F    B7C5      XGDX                >
2011    CE000A    LDX     #$000A      >.
>
```

The following example enters a short program that consists of three instructions starting from the memory location at $1500:

```
>asm 1500
1500    FC0800    LDD     $0800
1503    F30802    ADDD    $0802
1506    7C0900    STD     $0900
1509    E78C      TST     12,SP       >.
>
```

br [<Address> ...]

The **br** (*breakpoint set*) command sets a breakpoint at a specified address to display any previously set breakpoints. The function of a breakpoint is to halt user program execution when the program reaches the breakpoint address. When a breakpoint address is encountered, D-Bug12 disassembles the instruction at the breakpoint address, prints the CPU register contents, and waits for a D-Bug12 command to be entered by the user.

Breakpoints are set by typing the breakpoint command followed by one or more breakpoint addresses. Entering the breakpoint command without any breakpoint addresses will display all the currently set breakpoints. A maximum of two user breakpoints may be set at one time. Whenever the user program is not working correctly and the user suspects that the instruction at a certain memory location is incorrect, he or she can set a breakpoint at that location and check the execution result by looking at the contents of CPU registers or memory locations.

Example 3.9

```
>br            1020      1040      1050
Breakpoints:   1020      1040
Breakpoint Table Full
>
```

nobr [<Address> <Address>]
The **nobr** command removes one or more previously entered breakpoints. If the nobr command is entered without any argument, all user breakpoints are removed from the breakpoint table.

Example 3.10

```
>br            2000      2010      2020      2040      1090
Breakpoints:   2000      2010
Breakpoint Table Full
>nobr          2000
Breakpoints:   2010
>
```

g [<Address>]
The **g** command is used to begin execution of user code in real time. Before beginning the execution of user code, any breakpoints that were set with the br command are placed in memory. Execution of the user program continues until a user breakpoint is encountered, a CPU exception occurs, the stop or reset command is entered, or the EVB's reset switch is pressed. When the user code halts for any of these reasons and control is returned to D-Bug12, a message is displayed explaining the reason for user program termination. In addition, D-Bug12 disassembles the instruction at the current PC address, prints the CPU register contents, and waits for the next D-Bug12 command to be entered by the user. If the starting address is not supplied in the command line parameter, program execution will begin at the address defined by the current value of the program counter.

Example 3.11

```
>g    1500
User Bkpt Encountered
PP  PC      SP     X      Y         D = A:B    CCR = SXHI   NZVC
38 150C    3C00   7B48   0000          03:E8          1001   0001
xx: 150C   911E           CMPA       $001E
>
```

gt <Address>
The **gt** (*go till*) command is similar to the g command except that a temporary breakpoint is placed at the address supplied at the command line. Any breakpoints that were set by the use

of the br command are **not** placed in the user code before program execution begins. Program execution begins at the address defined by the current value of the program counter. When user code reaches the temporary breakpoint and control is returned to D-Bug12, a message is displayed explaining the reason for user program termination. In addition, D-Bug12 disassembles the instruction at the current PC address, prints the CPU register contents, and waits for a command to be entered by the user.

Example 3.12

▼

```
>pc 1500
PP  PC     SP      X      Y        D = A:B    CCR = SXHI   NZVC
381500  3C00   1000   1002      00:00          1001       0101
xx: 1500  CF1500          LDS     #$1500
>gt 1540
Temporary Breakpoint Encountered
PP  PC     SP      X      Y        D = A:B    CCR = SXHI   NZVC
381510  1500   1000   1002      1E:00          1001       0000
xx: 1510  3B              PSHD
>
```

▲

t [<count>]

The **t** (*trace*) command is used to execute one or several user program instructions beginning at the current program counter location. As each instruction is executed, the CPU register contents and the next instruction to be executed are displayed. A single instruction may be executed by entering the trace command immediately followed by a carriage return.

Because of the method used to execute a single instruction, branch instructions (Bcc, LBcc, BRSET, BRCLR, DBEQ/NE, IBEQ/NE, and TBEQ/NE) that contain an offset that branches back to the instruction opcode do not execute. D-Bug12 appears to become stuck at the branch instruction and does not execute the instruction even if the condition for the branch instruction is satisfied. This limitation can be overcome by using the gt command to set a temporary breakpoint at the instruction following the branch instruction.

Example 3.13

▼

```
>pc 1500
PP  PC     SP      X      Y        D = A:B    CCR = SXHI   NZVC
381500  1500   1000   1002      1E:00          1001       0000
xx: 1500  CF1500          LDS     #$1500
>t
PP  PC     SP      X      Y        D = A:B    CCR = SXHI   NZVC
381503  1500   1000   1002      1E:00          1001       0000
xx: 1503  CE1000          LDX     #$1000
>t 2
PP  PC     SP      X      Y        D = A:B    CCR = SXHI   NZVC
381506  1500   1000   1002      1E:00          1001       0000
xx: 1506  34              PSHX
```

PP PC	SP	X	Y	D = A:B	CCR = SXHI	NZVC
38 1507	14FE	1000	1002	1E:00	1001	0000
xx: 1507	861E		LDAA	#$1E		
>						

The first command in this example sets the program counter to $1500 so that the user knows where the program execution starts. This command is normally needed when tracing a program.

call [<Address>]

The **call** command is used to execute a subroutine and returns to the D-Bug12 monitor program when the final **rts** instruction of the subroutine is executed. All CPU registers contain the values at the time the final rts instruction was executed, with the exception of the program counter. The program counter contains the starting address of the subroutine. If a subroutine address is not supplied on the command line, the current value of the program counter is used as the starting address.

No user breakpoints are placed in memory before execution is transferred to the subroutine. If the called subroutine modifies the value of the stack pointer during its execution, it must restore the stack pointer's original value before executing the rts instruction. This restriction is required because a return address is placed on the user's stack that returns to D-Bug12 when the final rts of the subroutine is executed. Obviously, any subroutine must obey this restriction in order to execute properly.

Example 3.14

```
>call 1600
Subroutine Call Returned
```

pp PC	SP	X	Y	D = A:B	CCR = SXHI	NZVC
38 1600	0A00	0032	0900	00:31	1001	0000
xx: 1600	FC1000		LDD	$1000		
>						

This command is useful for testing a subroutine without writing a testing program.

3.5.4 Entering an Assembly Program

When a program is first created, one needs to open a new file to hold it. To do that, one presses the **File** menu from the MiniIDE window and selects **New,** as shown in Figure 3.11. After one selects the New command from the menu, an empty screen will appear allowing the user to enter a new program. The user can now start to enter a new program.

The editing window (upper half of the IDE window) may be too small and should be adjusted. After adjusting the window size, the user enters the program that converts a hex number into BCD digits as in Example 2.13. This program performs the repeated divide-by-10 operation to the given number and adds hex number $30 to each remainder to convert to its corresponding ASCII code. The result is shown in Figure 3.12.

3.5.5 Assembling the Program

To assemble an assembly program, press the **Build** menu and select **Build eg02_13.asm,** as shown in Figure 3.13. If the program is assembled successfully, the status window will display the corresponding message, as shown in Figure 3.14.

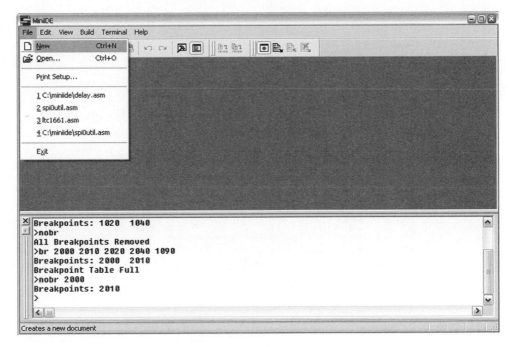

Figure 3.11 ■ Select New from the File menu to create a new file

After the program is assembled successfully, we are ready to run and debug the program. The output of the assembly process is an S-record file *eg2_13.s19*. The file name has a suffix **s19.** The S-record format is a common file format defined by Freescale to allow tools from different vendors to work on the same project.

3.5.6 Downloading the S-Record File onto the Demo Board for Execution

The D-Bug12 provides the following command for the user to download the S-record file onto the demo board:

load [<AddressOffset>]

The **load** command is used to load S-record objects into memory from an external device. The *<AddressOffset>*, if supplied, is added to the load address of each S-record before its data bytes are placed in memory. Providing an address offset other than zero allows an object code or data to be loaded into memory at a location other than that for which it was assembled. During the loading process, the S-record data is not echoed to the control console. However, for each 10 S-records that are successfully loaded, an ASCII asterisk character (*) is sent to the control console. When an S-record file has been successfully loaded, the D-Bug12 prompt reappears on the screen.

The **load** command is terminated when D-Bug12 receives an S9 end-of-file record. If the object being loaded does not contain an S9 record, D-Bug12 does not output its prompt and continues to wait for the end-of-file record. Pressing the reset switch returns D-Bug12 to its command line prompt.

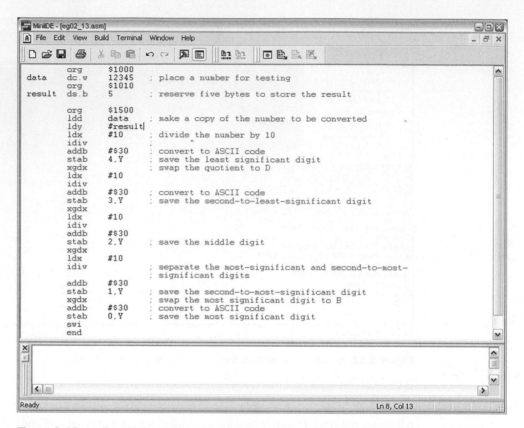

Figure 3.12 ■ Enter the program to convert hex number to BCD ASCII digits

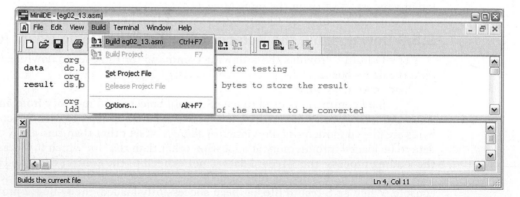

Figure 3.13 ■ Prepare to assemble the program eg213.asm

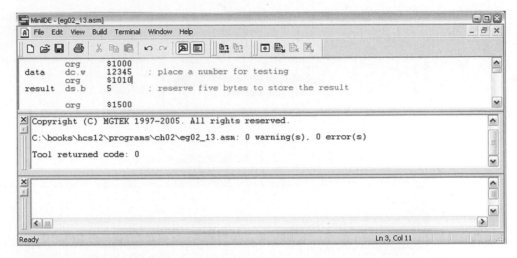

Figure 3.14 ■ Message window (in the middle) shows that previous assembling is successful

Example 3.15

>**load**
*
>

This example downloads a file that is equal to or shorter than 10 S-records and hence only one * character is displayed on the screen. In addition to entering the load command followed by a carriage return, one also needs to go back to the IDE window to specify the file to be downloaded. An example of selecting the **Download File** command in MiniIDE is shown in Figure 3.15.

After the Download File command is selected, a popup dialog box, as shown in Figure 3.16, will appear. This dialog box allows one to specify the file to be downloaded. Click on the **Open** button on the popup window and the file will be transferred to the demo board.

3.5.7 Running and Debugging the Program

We need to go back to the terminal window to run and debug the program. The screen after the program download would look like Figure 3.17.

Before running the program, we should verify that the test data is downloaded correctly into the memory. In this program, the test data is the decimal number 12,345, which corresponds to the hex number $3039. One can use the command **md 1000** to verify it. The contents of the memory locations $1000 and $1001 displayed by the D-Bug12 monitor should be $3039. Otherwise, some error might have occurred.

Also make sure that the program has been downloaded onto the right memory area. This can be verified by using the **asm 1500** command (the eg2_13 program starts at $1500) and pressing the **Enter** key several times. The first few lines should look as follows:

>**asm 1500**
xx:1500   FC1000   LDD $1000   >

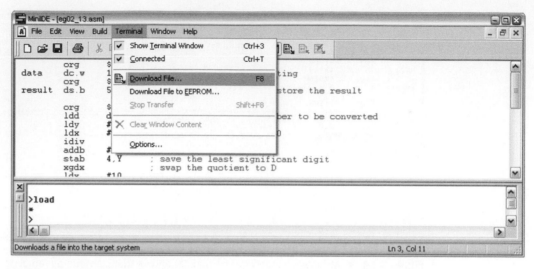

Figure 3.15 ■ Prepare to download a .s19 file

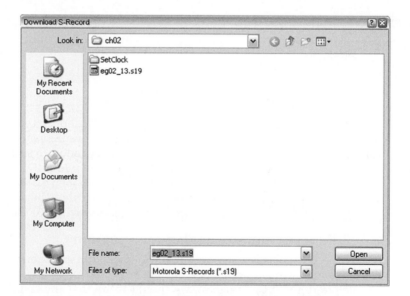

Figure 3.16 ■ Select an S-record file to be downloaded

xx:1503	CD1010	LDY	#$1010	>
xx:1506	CE000A	LDX	#$000A	>
xx:1509	1810	IDIV		>
xx:150B	CB30	ADDB	#$30	>
xx:150D	6B44	STAB	4,Y	>

After making sure that the program has been downloaded correctly, execute the program without setting any breakpoint and check the result to see if it is correct. The screen should look like Figure 3.18.

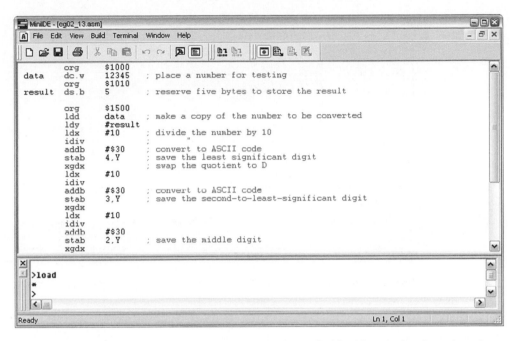

```
          org     $1000
data      dc.w    12345      ; place a number for testing
          org     $1010
result    ds.b    5          ; reserve five bytes to store the result

          org     $1500
          ldd     data       ; make a copy of the number to be converted
          ldy     #result
          ldx     #10        ; divide the number by 10
          idiv               ;            "
          addb    #$30       ; convert to ASCII code
          stab    4,Y        ; save the least significant digit
          xgdx               ; swap the quotient to D
          ldx     #10
          idiv
          addb    #$30       ; convert to ASCII code
          stab    3,Y        ; save the second-to-least-significant digit
          xgdx
          ldx     #10
          idiv
          addb    #$30
          stab    2,Y        ; save the middle digit
          xgdx
```

```
>load
*
>
```

Figure 3.17 ■ Screen after downloading the program eg2_13.s19 onto the demo board

The second line of the terminal window that contains the statement **User Bkpt Encountered** is caused by the **swi** (software interrupt) instruction. The swi instruction caused the program control to be returned to the D-Bug12 monitor. This instruction is often used as the last instruction of a program to be executed on a demo board with the D-Bug12 monitor. The last line of the terminal window displays the contents of the memory locations $1010 to $101F. The first 5 bytes represent the ASCII codes of 1, 2, 3, 4, and 5. Therefore, the program executes correctly.

If the program does not work correctly, we can set breakpoints at locations we suspect or we can even trace the execution of some instructions to find out the error. Suppose the execution result (we take out the first **xgdx** instruction) looks as follows:

>md 1010

1010   30 30 35 33 - 35 2D 2D 2D - 2D 2D 2D 2D - 2D 2D 2D 2D   00535---------------
>

The first four digits were found to be incorrect. Since this program is short, we can trace through it. One approach is the following:

Step 1
Trace through the first six instructions and check to see if the quotient (in index register X) and the memory contents at $1014 are correct. The last seven lines of the terminal window are

```
PP  PC     SP     X      Y         D = A:B    CCR = SXHI   NZVC
38 150F    3C00   04D2   1010      00:35           1001    0000
xx: 150F   CE000A         LDX      #$000A
```

>md 1010

1010   30 30 35 33 - 35 2D 2D 2D - 2D 2D 2D 2D - 2D 2D 2D 2D   00535---------------
>

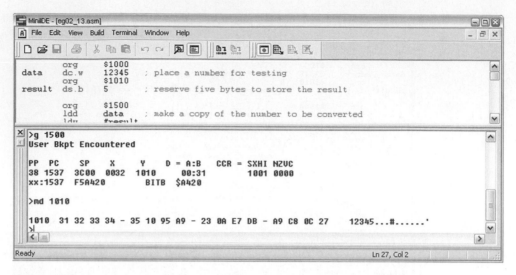

Figure 3.18 ■ Screen after running the eg2_13 program and displaying the result

These six lines tell us that

- The index register contains hex value $04D2 (equal to decimal 1234) and is the correct quotient.
- The memory location at $1014 contains hex value $35 and is the ASCII code of 5.
- In the next division, the number to be divided by 10 will be 1234.
- The next instruction to be executed is **ldx #$0A.**

Step 2
Trace the next two instructions. Oops! We find one error. The double accumulator D does not contain the value 1234 (it contains $35 instead) before the **idiv** instruction is executed. We forgot to swap the value in index register X with double accumulator D before performing the second division!

```
>t
PP  PC     SP     X      Y        D = A:B    CCR = SXHI   NZVC
381512   3C00   000A   1010      00:35           1001    0000
xx: 1512  1810                    IDIV
>t
PP  PC     SP     X      Y        D = A:B    CCR = SXHI   NZVC
381514   3C00   0005   1010      00:03           1001    0000
xx: 1514  CB30   ADDB   #$30

>
```

Step 3
Fix the error by inserting the **xgdx** instruction before the **ldx #$0a** instruction and rerun the program. After rerunning the program, check the contents at $1000 to $1004 again. The last few lines on the screen of the terminal window would be as follows:

```
>load
*
>g 1500
User Bkpt Encountered
PP  PC     SP      X        Y         D = A:B    CCR = SXHI   NZVC
38 1537    3C00    0032     1010        00:31           1001   0000
xx: 1537   2D2D             BLT       $1566
>md 1010
1010   31 32 33 34 - 35 2D 2D 2D - 2D 2D 2D 2D - 2D 2D 2D 2D   12345--------------
>
```

The memory locations from $1000 to $1004 contain correct values. So, we have fixed the error.

Most debug sessions are similar to this. Of course, longer programs will take a longer time and require more commands to try different things before we can fix the bugs.

3.6 Other D-Bug12 Modes

The other three D-Bug12 modes are not used as often as the EVB mode.

3.6.1 The Pod Mode

The pod mode is intended to use the demo board as a BDM host to control a target board. The arrangement is shown in Figure 3.19. This mode will be discussed in Section 3.9.5.

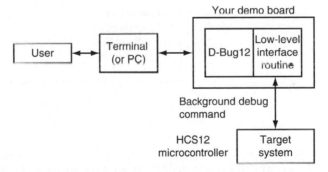

Figure 3.19 ■ D-Bug's POD mode conceptual model

3.6.2 The Jump-to-EEPROM Mode

The **Jump-to-EEPROM** mode allows a small program to be executed from the on-chip EEPROM whenever the EVB is powered up or reset. When running on a demo board with an MC9S12Dx256, MC9S12H256, or MC9S12A256 MCU, the bootloader startup code jumps directly to address $0400 without performing any initialization of the CPU registers or peripherals. When running on a demo board with an MC9S12Dx128 or MC9S12A128 MCU, the EEPROM is not visible at its default address of $0000 because it is overlaid by the on-chip SRAM. In this case, the SRAM is relocated to $2000 and the bootloader start code jumps directly to address $0400 without performing any additional initialization of the CPU registers or peripherals. This mode provides a convenient way to execute a program in a stand-alone manner without having to erase and program the on-chip flash using the bootloader. Program code and data can be programmed into the EEPROM using the D-Bug12's load command.

3.6.3 The Bootloader Mode

The on-chip flash memory includes a boot block area from $F000 to $FFFF containing an S-record bootloader program. The bootloader can be used to erase and reprogram the remainder of on-chip flash memory or erase the on-chip byte-erasable EEPROM. The bootloader utilizes the on-chip SCI for communications and does not require any special programming software on the host computer. The only host software required is a simple terminal program that is capable of communicating at 9600 to 115,200 baud and supports XON/XOFF handshaking.

The bootloader mode is mainly used to program the D-Bug12 monitor into the flash memory or download a user's fully debugged code into the D-Bug12 portion of flash memory. The latter allows the board to be operated in EVB mode or restart user code every time the board is turned on or reset.

When the user code is programmed into the D-Bug12 portion of flash memory, it wipes out the D-Bug12 monitor. We can restore it anytime because bootloader is not erased. We cannot erase the bootloader in bootloader mode. The bootloader can only be erased by a BDM kit via the BDM port.

Invoking the bootloader causes the prompt shown in Figure 3.20 to be displayed on the host terminal's screen.

```
D-Bug12 Bootloader v1.0.0

a) Erase Flash
b) Program Flash
c) Set Baud Rate
d) Erase EEPROM
?
```

Figure 3.20 ■ Serial bootloader prompt

ERASE FLASH COMMAND

This command is selected by typing **a** on the terminal after the **?** character in Figure 3.20. This command causes a bulk erase of all four 64-kB flash memory arrays except for the 4-kB boot block in the upper 64-kB array where the S-record bootloader resides. After the erase operation is completed, a verify operation is performed to ensure that all locations were properly erased. If the erase operation is successful, the bootloader's prompt is redisplayed. If any locations were found to contain a value other than $FF, an error message is displayed on the screen and the bootloader's prompt is redisplayed. If the HCS12 MCU will not erase after one or two attempts, the device may be damaged.

PROGRAM FLASH COMMAND

The bootloader uses interrupt-driven, buffered serial I/O in conjunction with XON/XOFF software handshaking to control the flow of S-record data from the host computer during the programming process. This allows the bootloader to continue receiving S-record data from the host computer while the data from the previously received S-record is programmed into the flash memory. The terminal program must support XON/XOFF handshaking to properly reprogram the HCS12's flash memory.

Typing a lowercase **b** on the keyboard causes the bootloader to enter the programming mode and wait for S-records to be sent from the host computer. The bootloader will continue to receive and process S-records until it receives an S8 or S9 end-of-file record. If the object file being sent to the bootloader does not contain an S8 or S9 record, the bootloader will not return its prompt and will continue to wait for the end-of-file record. Pressing the reset button on the demo board will cause the bootloader to return to its prompt.

If a flash memory location does not program properly, an error message is displayed on the terminal screen and the bootloader's prompt is redisplayed. If the device does not program correctly after one or two attempts, the device may be damaged or an S-record with a load address outside the range of the available on-chip flash may have been received. The S-record data must have load addresses ranging from \$C0000 to \$FFFFF. This address represents the upper 256 kB of the 1-MB address space of the MC9S12DP256.

SET-BAUD-RATE COMMAND

The default baud rate of 9600 is too slow for programming a large S-record file. The user has the option to select a higher baud rate in programming the on-chip flash memory. The **Set-Baud-Rate** command allows the bootloader communication rate to be set to one of four standard baud rates.

Typing a lowercase **c** on the terminal keyboard causes the prompt shown in Figure 3.21 to be displayed on the monitor screen. Entering **1** through **4** on the keyboard will select the associated baud rate and issue a secondary prompt indicating that the terminal baud rate should also be changed. After changing the terminal baud rate, pressing the **Enter** key will return to the main bootloader prompt (in Figure 3.20).

```
1) 9600
2) 38400
3) 57600
4) 115200
? 3
Change Terminal BR, Press Return
```

Figure 3.21 ■ Change baud rate (both the demo board and terminal program)

RELOADING D-BUG12

Freescale adds new features and enhancements to the D-Bug12 monitor frequently. It may be desirable to update to the latest version. A .zip distribution file containing the current version of D-Bug12, including the bootloader and the reference manual, can be downloaded from Freescale's website.

LOADING USER PROGRAMS INTO FLASH MEMORY

A demo board preloaded with the D-Bug12 monitor can also be reprogrammed with the user application to prototype an embedded application. When using the board in this manner, the user-supplied code may occupy all of the on-chip flash memory except the address range from $F000 to $FFFF in the fixed flash memory page that begins at $C000. To begin execution of the user's application program, both the PAD0 and PAD1 pins must have jumpers placed in the 0, or off, position. This will cause the bootloader startup code to jump to the address in the alternate reset vector at $EFFE. When the user code is programmed into flash, an address must be placed in the reset vector position ($EFFE) of the alternate interrupt vector table.

3.7 Tips for Assembly Program Debugging

Assembly program errors can be classified into two categories:

- Syntax errors
- Logical errors

3.7.1 Syntax Errors

Syntax errors are common for beginners. Syntax errors can be divided into following categories:

- Misspelling of instruction mnemonics. This type of error will be highlighted by the assembler and can be fixed easily.
- Starting the assembly program from column 1. If a line of the assembly program has a label, then this does not cause any error. However, if a line of the assembly program does not have a label, then the instruction mnemonic is treated as a label by the assembler, whereas the operand (represented in one of the addressing modes) is treated as the instruction mnemonics. This will always cause the undefined mnemonic error!
- Missing operands. Depending on what is missing, the error message will vary. For example, for the instruction

```
brclr   PORTA,$04
```

the resultant error message is "Missing target address." For the instruction

```
ldaa
```

the assembler outputs the message "Inherent addressing illegal."

As time goes on and you gain experience and also memorize the instruction mnemonics better, this type of error will reduce and even disappear.

3.7.2 Logical Errors

Beginners make many logical errors. The most common ones are as follows:

USING EXTENDED (OR DIRECT) MODE INSTEAD OF IMMEDIATE MODE

This error is very common for beginners. The following assembly program is written to compute the sum of an array of N 8-bit elements:

```
N        equ      20                                       ; array count
         org      $1000
array    dc.b     2,4,6,8,10,12,14,16,18,20
         dc.b     22,24,26,28,30,32,34,36,38,40
sum      ds.w     1
         org      $1500
```

```
                ldx      array         ; place the starting address of array in X
                movw     0,sum         ; initialize sum to 0
                ldy      N             ; initialize loop count to N
        loop    ldab     1,x |         ; place one number in B and move array pointer
                sex      B,D           ; sign-extend the 8-bit number to 16-bit
                addd     sum           ; add to sum
                std      sum           ; update the sum
                dbne     y,loop        ; add all numbers to sum yet?
                swi                    ; return to monitor
                end
```

At first look, this program appears to be fine and should work. After assembling the program, we download it onto the demo board and get the following screen output:

```
>load
....
done
>
```

It is a good idea to make sure that the program has been downloaded onto the demo board by using the **asm** command.

```
>asm 1500
xx:1500    FE1000          LDX      $1000          >
xx:1503    180400001014    MOVW     $0000,$1014    >
xx:1509    DD14            LDY      $0014          >
xx:150B    E630            LDAB     1,X+           >
xx:150D    B714            SEX      B,D            >
xx:150F    F31014          ADDD     $1014          >
xx:1512    7C1014          STD      $1014          >
xx:1515    0436F3          DBNE     Y,$150B        >
xx:1518    3F              SWI                     >
```

Indeed, the program is downloaded correctly. The next thing to check is to make sure that the array data is downloaded correctly by using the md command.

```
>md 1000 1010
1000   02 04 06 08 - 0A 0C 0E 10 - 12 14 16 18 - 1A 1C 1E 20   ................
1010   22 24 26 28 - 00 00 B9 A9 - 2A CA FA DB - AC DA 18 97   "$&(....*.......
>
```

Again, we are convinced that the data array has been downloaded correctly. The array is stored at locations from $1000 to $1013. The variable *sum* occupies the memory locations from $1014 to $1015. Since the program has not been run yet, these 2 bytes contain 0s.

The next thing to do is to run the program. This program should run to the swi instruction and stop. The screen should look as follows:

```
>g 1500
User Bkpt Encountered
PP  PC     SP      X       Y       D = A:B    CCR = SXHI   NZVC
38 1519    3C00    0213    0000    FF:07            1001    1000
xx: 1519   88F4            EORA    #$F4

    >
```

Checking the resultant sum, we can see that it ($FF07) is incorrect. The correct answer should be $1A4 (or 420). The sum can be found by using the md command as follows:

```
>md 1010
1010   22 24 26 28 - FF 07 B9 A9 - 2A CA FA DB - AC DA 18 97   "$&(....*.......
>
```

Since this program is short, we can trace through it. To trace this program, we set the program counter value to the start of the program ($1500) as follows:

```
>pc 1500
PP PC      SP      X       Y         D = A:B    CCR = SXHI   NZVC
38 1500  3C00    0213    0000       FF:07            1001    1000
xx: 1500  FE1000            LDX      $1000
>
```

Trace one instruction at a time.

```
>t 1
PP PC      SP      X       Y         D = A:B    CCR = SXHI   NZVC
38 1503  3C00    0204    0000       FF:07            1001    0000
xx: 1503  180400001014     MOVW   $0000,$1014
>
```

The executed instruction is **ldx array.** The purpose of this instruction is to place the starting address of the array into X. After the execution of this instruction, the value of X should change to $1000. However, the instruction trace shows otherwise. This is because of the incorrect use of the addressing mode. Change the instruction to **ldx #array**, rerun the program, and the sum is still incorrect.

```
>md 1010
1010   22 24 26 28 - FF F0 B9 A9 - 2A CA FA DB - AC DA 18 97   "$&(....*.......
>
```

Trace the program again. This time we trace up to the second instruction and examine the contents of *sum*.

```
>t 2
PP PC      SP      X       Y         D = A:B    CCR = SXHI   NZVC
38 1503  3C00    1000    0000       FF:F0            1001    0000
xx:1503  180400001014     MOVW  $0000,$1014

PP PC      SP      X       Y         D = A:B    CCR = SXHI   NZVC
38 1509  3C00    1000    0000       FF:F0            1001    0000
xx:1509  DD14             LDY      $0014
>md 1010
1010   22 24 26 28 - FF 00 B9 A9 - 2A CA FA DB - AC DA 18 97   "$&(....*.......
>
```

At this point, the value of *sum* should be changed to 0 because the second instruction intends to initialize it to 0. Again, the addressing mode is wrong. The correct instruction should be **movw #0,sum.** This is not easy to figure out by looking at the program. Rerun the program and display the contents of *sum*.

```
>load
*
>g 1500
```

```
User Bkpt Encountered
PP PC    SP     X      Y         D = A:B    CCR = SXHI   NZVC
381519   3C00   100F   0000      00:F0            1001   0000
xx: 1519 88F4          EORA      #$F4
>md 1010
1010   22 24 26 28 - 00 F0 B9 A9 - 2A CA FA DB - AC DA 18 97   "$&(....*.......
>
```

The value of *sum* is still incorrect. Again trace the program up to the third instruction.

```
>pc  1500
PP PC    SP     X      Y         D = A:B    CCR = SXHI   NZVC
381500   3C00   100F   0000      00:F0            1001   0000
xx: 1500 CE1000         LDX      #$1000
>t 3
PP PC    SP     X      Y         D = A:B    CCR = SXHI   NZVC
381503   3C00   1000   0000      00:F0            1001   0000
xx: 1503 180300001014   MOVW     #$0000,$1014

PP PC    SP     X      Y         D = A:B    CCR = SXHI   NZVC
381509   3C00   1000   0000      00:F0            1001   0000
xx: 1509 DD14           LDY      $0014

PP PC    SP     X      Y         D = A:B    CCR = SXHI   NZVC
38150B   3C00   1000   000F      00:F0            1001   0000
xx: 150B E630           LDAB     1,X+
>
```

The third instruction intends to load the array count *N* into index register Y. Y should receive the value of *N*, which is 20. The instruction trace shows that Y receives 15 ($F) instead of 20. Again, this is due to the incorrect addressing mode. Change the instruction to *ldy #N*, assemble, download, rerun the program, and display the contents of *sum*.

```
>g 1500
User Bkpt Encountered
PP PC    SP     X      Y         D = A:B    CCR = SXHI   NZVC
38151A   3C00   1014   0000      01:A4            1001   0000
xx: 151A F421BD         ANDB     $21BD
>md 1010
1010   22 24 26 28 - 01 A4 B9 A9 - 2A CA FA DB - AC DA 18 97   "$&(....*.......
>
```

This time the value of *sum* is $1A4 (420) and is correct.

MISMATCH OF OPERAND SIZE

This is another common mistake made by beginners. For example, some people might write the previous program as follows:

```
N       equ     20                                        ; array count
        org     $1000
array   dc.b    2,4,6,8,10,12,14,16,18,20
        dc.b    22,24,26,28,30,32,34,36,38,40
```

```
sum        ds.w     1
           org      $1500
           ldx      #array          ; place the starting address of array in X
           movw     #0,sum          ; initialize sum to 0
           ldy      #N              ; initialize loop count to N
loop       ldd      1,x+            ; place one number in D and move array pointer
           addd     sum             ; add to sum
           std      sum             ; update the sum
           dbne     y,loop          ; add all numbers to sum yet?
           swi                      ; return to monitor
           end
```

At the first run of this program, *sum* ($A61F) is also incorrect.

>md 1010

1010   22 24 26 28 - **A6 1F** B9 A9 - 2A CA FA DB - AC DA 18 97   "$&(....*.......
>

Again, the error can be found by tracing. This time the error will not be found until the instruction **loop ldd 1,x+** is traced.

```
>pc 1500
PP  PC      SP      X       Y       D = A:B     CCR = SXHI   NZVC
38 1500   3C00    1014    0000      A6:1F            1001   1000
xx: 1500  CE1000          LDX     #$1000
>t
PP  PC      SP      X       Y       D = A:B     CCR = SXHI   NZVC
38 1503   3C00    1000    0000      A6:1F            1001   0000
xx: 1503  180300001014    MOVW    #$0000,$1014
>t
PP  PC      SP      X       Y       D = A:B     CCR = SXHI   NZVC
38 1509   3C00    1000    0000      A6:1F            1001   0000
xx: 1509  CD0014          LDY     #$0014
>t
PP  PC      SP      X       Y       D = A:B     CCR = SXHI   NZVC
38 150C   3C00    1000    0014      A6:1F            1001   0000
xx: 150C  EC30            LDD      1,X+
>t
PP  PC      SP      X       Y       D = A:B     CCR = SXHI   NZVC
38 150E   3C00    1001    0014      02:04            1001   0000
xx: 150E  F31014          ADDD     $1014
>
```

We expect this instruction to place the first array element ($02) in D. Instead, it places $0204 in D. This is obviously wrong! The error is due to the fact that the **ldd** instruction loads a word (2 bytes) instead of 1 byte into D. At this point, we should be able to figure out the right instruction to use and fix the problem.

INAPPROPRIATE USE OF INDEX ADDRESSING MODE

When the index addressing mode is used to step through an array, one needs to increment or decrement the index register to reach the next element. Depending on the size of the element, one needs to increment or decrement the index register by 1, 2, or some other value. When such errors occur, the program execution result will be incorrect. This type of error can be discovered by checking the computation result of the first two elements.

STACK FRAME ERRORS

This type of error will be discussed in Chapter 4.

INCORRECT ALGORITHM

An incorrect algorithm can never result in a correct program. This type of problem cannot be fixed by tracing the program. After program tracing fails to fix the problem, we must reexamine the algorithm to see if it is incorrect.

3.8 Using CodeWarrior

CodeWarrior is an IDE designed to support the software development for all microcontroller products manufactured by Freescale. CodeWarrior allows the user to debug his or her software using the following three approaches:

1. Running the program using the simulator
2. Running the program on the target hardware programmed with the serial monitor (The HCS12 MCU is programmed with the serial monitor.)
3. Running the program on the target hardware connected to a BDM-based debug adapter

CodeWarrior has a built-in simulator that can be used by the user to debug her or his software. CodeWarrior can support software debugging via the **serial monitor**. Using this approach, the user needs to connect the demo board to the COM port of the PC using a serial cable. At the time of this writing, CodeWarrior can work with the following BDM-based adaptors to debug users' software:

1. P&E Multilink/CyclonePro BDM adaptor
2. TBDML adaptor
3. Abatron BDI adaptor
4. Softec's inDART debugger

A BDM adaptor can also be part of the demo board. For example, the Freescale student project board includes the P&E Multilink BDM adaptor. By plugging in different HCS12 microcontroller kits, the user can perform software debugging for several different HCS12 microcontrollers. Freescale publishes the public domain TBDML interface to be used by anyone interested in building her or his BDM debugger kit.

3.8.1 Building a Software Project Using CodeWarrior

CodeWarrior can be started by clicking on its icon. The startup screen is shown in Figure 3.22. CodeWarrior displays **Tip of the Day** whenever it starts up. The user can get rid of Tip of the Day by clicking on the **Close** button.

CodeWarrior uses *project* as the unit for managing the software development task. The software development process under CodeWarrior involves four stages.

Stage 1
Project setup

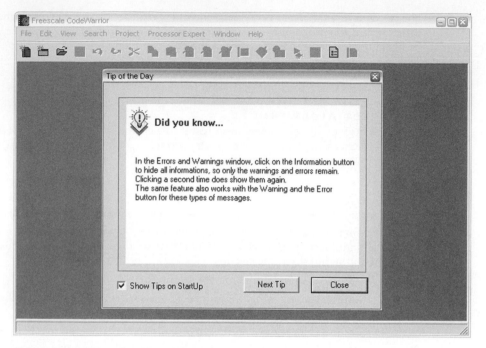

Figure 3.22 ■ CodeWarrior startup screen with Tip of the Day

Stage 2
Source code entering

Stage 3
Code compiling or assembling

Stage 4
Code debugging

3.8.2 Project Setup

Project setup involves the following steps:

Step 1
Create a new project by pressing the **File** menu and selecting **New**. A popup dialog box will appear to allow you to enter the project name as shown in Figure 3.23.

Step 2
Enter the project name and the project directory and then click on **OK.** A project wizard as shown in Figure 3.24 will appear.

Step 3
Select the HCS12 device (select MC9S12DG256B for Dragon12-Plus demo board) and then click on Next.

Step 4
Select the set of languages to be supported initially as shown in Figure 3.25. Choose Assembly for this tutorial and then click on **Next**.

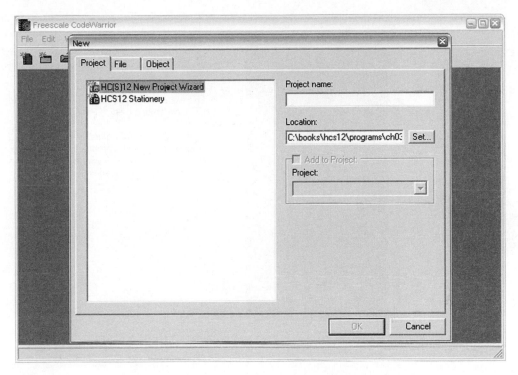

Figure 3.23 ■ Screen for creating a new project under CodeWarrior IDE

Step 5
Select absolute or relocatable assembly as shown in Figure 3.26. Absolute assembly fixes the memory location for the assembly program. It does not need linker to make memory assignment. Choose absolute assembly for this tutorial and then click on **Next.**

Step 6
Choose the connection method to the target hardware. You can choose only one method or multiple methods. Choosing multiple methods allows you to switch from one method to another. The screen in Figure 3.27 selects three connection methods. Click **Finish** to complete the project setup. The resultant screen is shown in Figure 3.28.

3.8.3 Source Code Entering

During the project creation process, CodeWarrior also creates the required files for the project and puts them under different directory names (Sources, Prm, Libraries, Debugger Project File, and Debugger Cmd Files). The user can display the file names under these directories by clicking on the + character to their left as shown in Figure 3.29. Figure 3.30 shows the files created under these directories.

It would be helpful for the first-time user to browse through the files readme.txt and tips.txt. A file can be opened by double-clicking on it using the left mouse button.

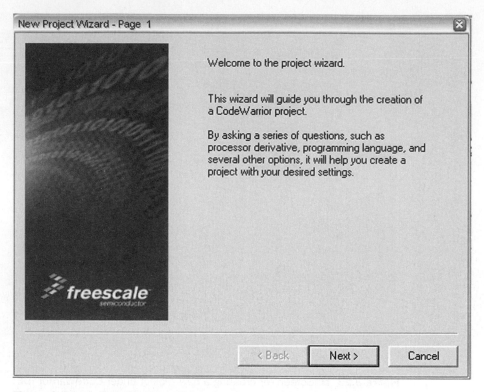

Figure 3.24 ■ CodeWarrior project wizard

To enter a program source file into the project, the user can modify the main.asm program or enter his or her program with a different name and add it to the project. Using the second approach, the user must remove the main.asm file from the project.

In this section, we use the following assembly program to illustrate the project build and debug process:

```
        include   "c:\miniide\hcs12.inc"
ARCNT   equ       20
        org       $1000
sum     ds.w      1
average ds.w      1
icnt    ds.b      1
        org       $1500
start   movw      #0,sum
        movb      #ARCNT,icnt      ; initialize loop count
        ldx       #array           ; use X as a pointer to the array
addloop ldd       sum              ; add array[i] to sum and move pointer
        addd      2,x+             ;   "
        std       sum              ;   "
        dec       icnt             ; decrement loop count
        bne       addloop          ; not done yet?
```

```
        ldd      sum                                  ; compute array average by dividing array count into
        ldx      #ARCNT                               ; array sum
        idiv                                          ;          "
        stx      average                              ; save the average
        bra      $                                    ; stay here forever
array   dc.w     11,12,13,14,15,16,17,18,19,20
        dc.w     21,22,23,24,25,26,27,28,29,30
        org      $FFFE                                ; set up reset vector
        dc.w     start
        end
```

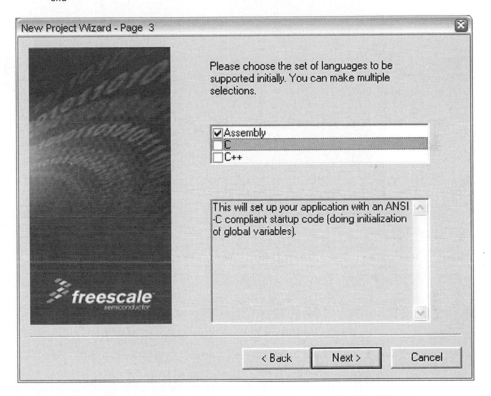

Figure 3.25 ■ Select language to be supported under CodeWarrior Wizard

The user can open main.asm by double clicking on it and then selecting the whole contents and replacing them with the previous program.

There are a few things that the user needs to know about when entering a program to be run by CodeWarrior.

- *The need to set up reset vector.* The reset vector is the address of the first instruction to be executed after the power is turned on or a reset. The reset vector must be stored at $FFFE. The following instruction sequence sets up the reset vector accordingly:

```
        org      $FFFE
        dc.w     start                                ; start is the label of the first instruction of the program
```

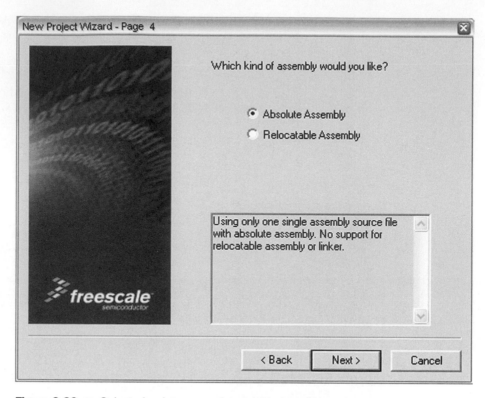

Figure 3.26 ■ Select absolute or a relocatable assembly

■ *The need to set up system clock.* The HCS12 includes an on-chip phase-locked-loop (PLL) circuit that allows the user to use a low-frequency crystal oscillator (e.g., 4 MHz or 8 MHz) to generate a higher-frequency E-clock signal. To achieve this, the user will need to write and call a subroutine similar to the following to multiply the frequency (this has been done in the D-Bug12 monitor):

```
; **************************************************************************
; The following subroutine enables PLL and uses an external oscillator to generate the system
; clock. Set the system clock (E-clock) to 24 MHz from a 4-MHz external crystal oscillator.
; **************************************************************************
;
SetClk     movb     #0,REFDV            ; set SYSCLK to 24 MHz from a 4 MHz oscillator
           movb     #$05,SYNR          ; "
;          movb     #$02,SYNR          ; use this value if external oscillator is 8 MHz
           movb     #$60,PLLCTL        ; turn on PLL, select high-bandwidth filter
           brclr    CRGFLG,LOCK,*       ; wait until PLL lock into target frequency
           bset     CLKSEL,PLLSEL       ; use PLL to derive system clock
           rts
```

The functioning and programming of the HCS12 on-chip phase-locked-loop are discussed in Chapter 6. To call this subroutine to set up the system clock, add the following instruction to your program:

```
jsr     SetClk
```

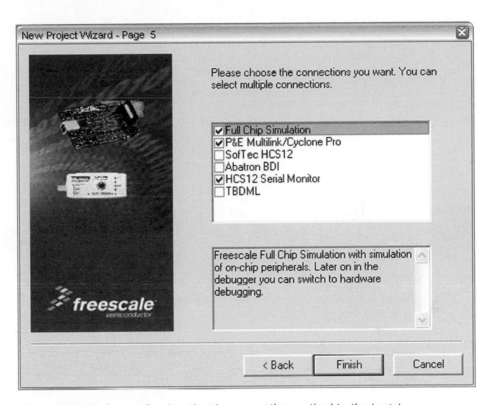

Figure 3.27 ■ Screen for choosing the connection method to the target

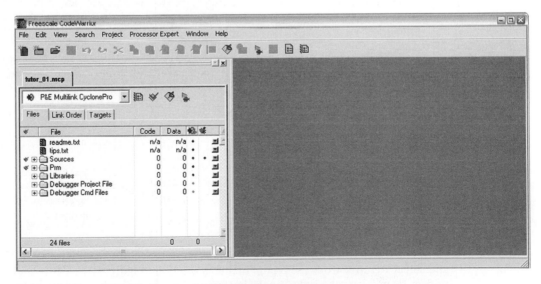

Figure 3.28 ■ CodeWarrior screen after project setup

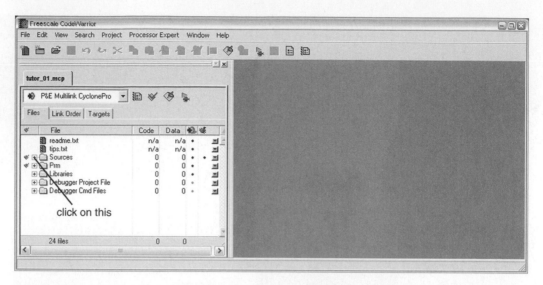

Figure 3.29 ■ Display files under each directory by clicking the '+' character to the left of the directory name

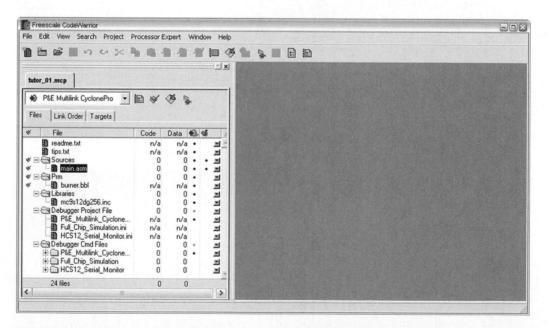

Figure 3.30 ■ Files created by CodeWarrior Wizard during the project-creating phase

To delete a file from the project, press the right mouse button on the file to be removed and select **Remove** from the popup menu as shown in Figure 3.31. A file can be added into the project by pressing the right mouse button on **Sources** and selecting the file name from the directory that contains the file.

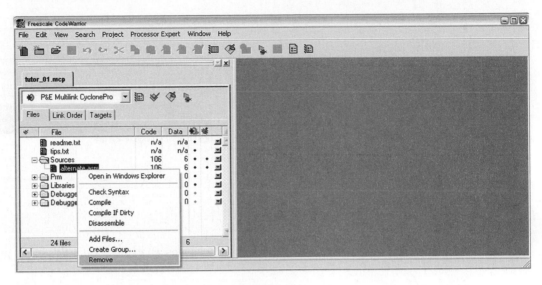

Figure 3.31 ■ Remove a file from the project

3.8.4 Project Build

A project can be built by pressing the **Project** menu and selecting **Make** or simply by pressing the F7 function key on the keyboard. This is shown in Figure 3.32. If there is no syntax error in the program, CodeWarrior won't display anything. However, if there are some errors in the program, CodeWarrior will display the error messages. One example is shown in Figure 3.33. The user can double-click on the message, and CodeWarrior will guide the user to the error location to fix it. There is no syntax error in this example program.

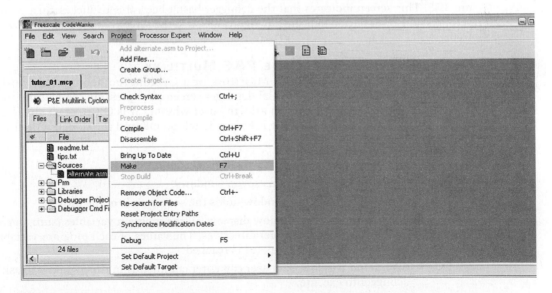

Figure 3.32 ■ Action to build a project

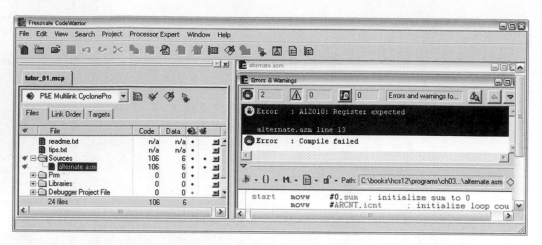

Figure 3.33 ■ CodeWarrior display error messages when there are errors in the program

3.8.5 Program Debugging

This phase is entered by selecting **Debug** from the **Project** menu or pressing the F5 function key of the keyboard. The initial response of CodeWarrior varies with the connection methods.

INITIAL RESPONSE UNDER SERIAL MONITOR

Suppose the user has a Dragon12-Plus demo board (or other demo board) programmed with the serial monitor; by pressing the function key F5, the debugger screen will look like that in Figure 3.34. The user needs to select the COM port before she or he can proceed. After selecting the correct COM port and clicking on OK, the debugger screen will change to that shown in Figure 3.35. This screen indicates that the debugger hasn't been able to successfully reset the MCU yet. Press the reset button indicated in Figure 3.35, and the screen should change to that shown in Figure 3.37.

INITIAL RESPONSE UNDER THE P&E MULTILINK

Suppose the user uses a P&E Multilink or a CyclonePro debugger to work with the Dragon12-Plus demo board; the initial debugger screen after pressing the F5 function will look like that in Figure 3.36. This screen asks the user whether it is OK to erase the flash memory and download the program onto the memory. Click on OK and the screen will change to that shown in Figure 3.37.

DEBUGGER SCREEN LAYOUT

The CodeWarrior screen contains seven windows.

- *Source window.* This window holds the source code of the project.
- *Data window.* This window displays all the program variables (*sum*, *average*, *icnt*) declared by assembler directives. The values of each program variable will be updated after each breakpoint is reached.
- *Command window.* The user can enter commands in this window to ask the debugger to execute.
- *Assembly window.* This window contains the program as shown in the Source window but it also displays the line number of each instruction.

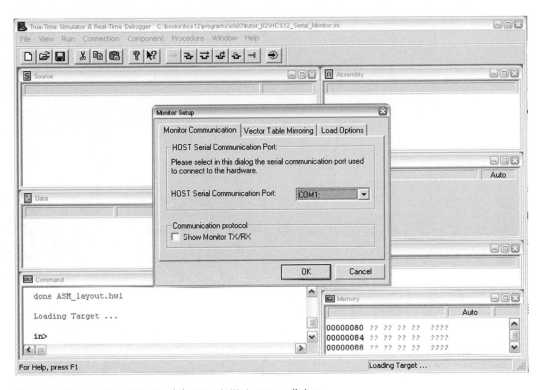

Figure 3.34 ■ CodeWarrior debugger initial popup dialog

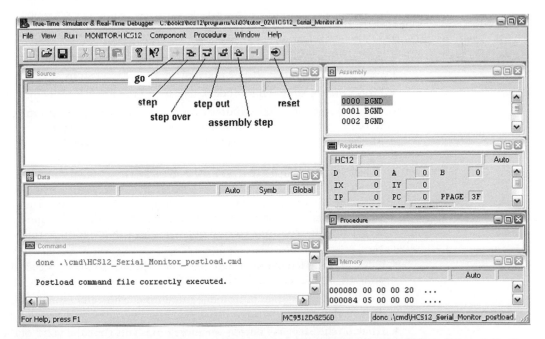

Figure 3.35 ■ CodeWarrior debugger startup screen and commonly used debug command buttons

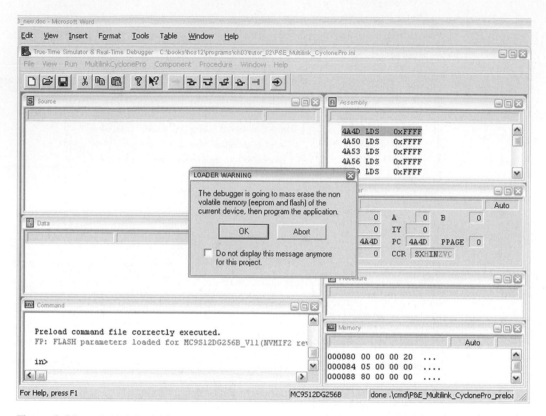

Figure 3.36 ■ Initial CodeWarrior debugger response screen under P&E Multilink debugger

- *Register window.* This window displays the value of each CPU register. The value of each CPU register will be updated after the MCU stops at each breakpoint.
- *Procedure window.* This window displays all the procedures defined in the program.
- *Memory window.* This window displays the contents of every memory location in the hardware. When the program contains one or multiple arrays, the user will need to use this window to examine the values of arrays.

EXAMPLE OF A CODEWARRIOR DEBUG SESSION

The user can resize the seven windows within the CodeWarrior debugger screen or even close some of them. We will get rid of the command window and resize the source window to display the whole program. A debug session may be similar to the following description:

Single step a few instructions to find out whether the program variables are initialized correctly.

- Click on the **Step over** button once. The value of *sum* is changed to 0 in the data window. This is correct.
- Click on the **Step over** button one more time. The value of *icnt* is changed to 20. This is also correct.
- Click on the **Step over** button one more time. This time the value of the index register is changed to 1529 (hex). This is also correct if we examine the assembly window.

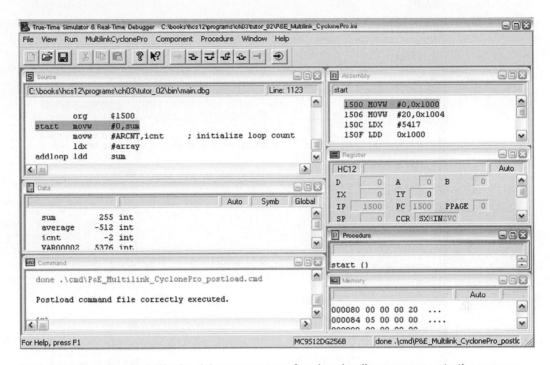

Figure 3.37 ■ The CodeWarrior debugger screen after downloading program onto the memory

After executing the first three instructions, the debugger window is changed as shown in Figure 3.38.

Use the **Run to Cursor** command to find out the program execution result. For example, we can move the mouse cursor on the **bne addloop** instruction and press the right mouse button. This will bring up a popup window (shown in Figure 3.39) that displays all the debug commands that the user can apply. Select the **Run to Cursor** command by clicking on it. After this, the value of *sum* is changed to 11 whereas the value of *icnt* is changed to 19. Both are correct. Now move the mouse cursor to the next instruction (ldd sum) and then press the right mouse button and select the Run to Cursor command. This time, the value of *sum* is changed to 410 whereas the value of *icnt* is changed to 0. Both are also correct. Now move the mouse cursor to the **bra $** instruction and then execute the Run to Cursor command again. This time, the value of the variable *average* is changed to 20 which is the quotient of 410 divided by 20 (the fraction is truncated because the CPU is performing integer divide). This is also correct. We have verified that the program is correct (shown in Figure 3.40).

OTHER COMMONLY USED DEBUGGER COMMANDS

Set Program Counter

Sometimes the user may need to execute an instruction sequence starting from a certain instruction. This command allows the user to do just that. The user can set the program counter to the address of any instruction by taking the following actions:

1. Place the mouse cursor at the instruction at which the program counter is to be set in the Source window.
2. Press the right mouse button to bring up the debugger command menu.
3. Select the **Set Program Counter** command.

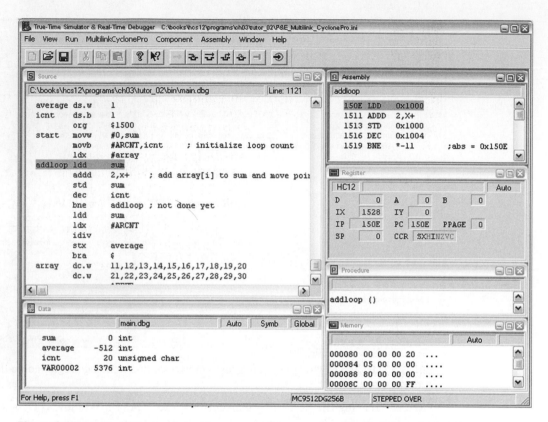

Figure 3.38 ■ CodeWarrior debugger screen after executing the first three instructions

The instruction to which the program counter is pointing at is darkened as shown in Figure 3.41. Executing this command will also bring the instruction pointed to by the program counter to the first line in the assembly window.

Set Breakpoint

To set a breakpoint at an instruction in the Source window, take the following actions:

1. Place the mouse cursor at the instruction in the source window.
2. Press the right mouse button to bring up the debugger command menu.
3. Select the **Set Breakpoint** command.

After these, a right-pointing red arrow will appear to the left of the instruction at the breakpoint as shown in Figure 3.41. A breakpoint can be disabled. A disabled breakpoint will not cause the program to stop. A disabled breakpoint can be re-enabled.

Show Location

Sometimes the user wants to find out the address of an instruction or a memory location that contains data; she or he can use this command to move that location to the first line in

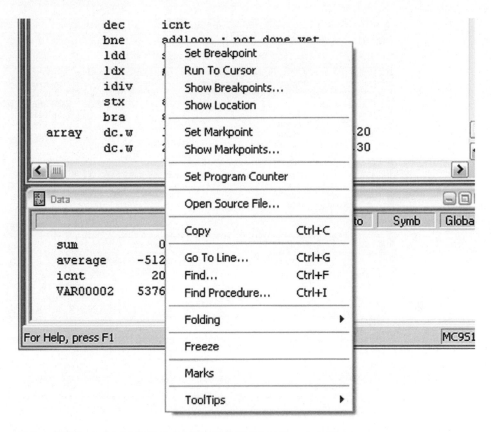

Figure 3.39 ■ CodeWarrior debugger debugging commands

the assembly window. To show the location of a line in the source window, take the following actions:

1. Place the mouse cursor at the instruction in the source window.
2. Press the right mouse button to bring up the debugger command menu.
3. Select the **Show Location** command.

3.9 BDM Serial Interface

The BDM module communicates with external devices via the BKGD pin. During reset, this pin is a mode-select input, which selects between the normal and special modes of operation. After reset, this pin becomes the dedicated serial interface pin for the BDM.

The timing of the BDM serial interface is controlled by the target clock. The target clock is programmable and can be either the bus clock or the oscillator clock. The BDM serial interface

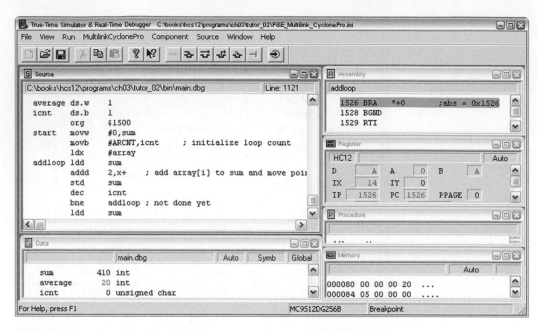

Figure 3.40 ■ CodeWarrior debugger screen after the whole program executed

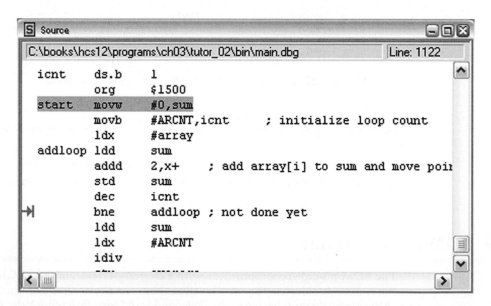

Figure 3.41 ■ CodeWarrior debugger source window after setting program counter and a breakpoint

uses a clocking scheme in which the external host generates a falling edge on the BKGD pin to indicate the start of each bit time. This falling edge is sent for every bit whether data is transmitted or received. Data is transferred most significant bit first at 16 target clock cycles per bit. The interface times out if 512 clock cycles occur between falling edges from the development host.

The BDM module implements a set of hardware and software commands that allows the external debug adapter to access CPU registers and memory locations to facilitate software debug activities. For hardware **Read** commands, the development host must wait for 150 clock cycles after sending the address before attempting to obtain the read data. For hardware **Write** commands, the host must wait for 150 clock cycles after sending data to be written before attempting to send a new command. For firmware Read commands, the external host should wait for 44 clock cycles before attempting to obtain the read data. For firmware Write commands, the external host must wait for 32 clock cycles after sending data to be written before attempting to send a new command.

The detailed timing of the BDM serial interface and the BDM commands are beyond the scope of this text. Interested readers should refer to the *Background Debug Module Guide* published by Freescale.

3.10 The BDM-Based Debugger

The block diagram of a BDM-based development system is shown in Figure 3.42. The BDM pod is connected to the target board via a BDM cable. To promote the portability of tools that use BDM mode, Freescale defines a 6-pin connector to be installed on the target board, shown in Figure 3.43.

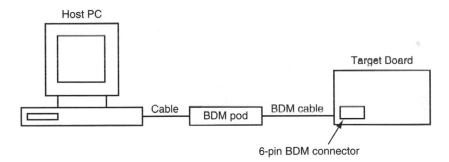

Figure 3.42 ■ BDM development system

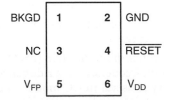

Figure 3.43 ■ BDM tool connector

The BDM pod can be connected to the PC or workstation via the serial, parallel, or USB port. There must be a software program running on the PC or workstation to send commands to the POD to perform the appropriate debug operations. This PC-based software should have a good user interface and allow the user to perform at least the following operations:

- Do basic configuration
- Download the program to the target board for execution
- Set breakpoints
- Display and change the values of the CPU registers, I/O registers, and memory locations
- Display the source code
- Set up a watch list to display the values of selected program variables
- Trace the program execution to the breakpoint

The PC sends the intended operation of the user to the POD and the POD sends the corresponding BDM commands to the target board to perform the desired operations. A source-level debugger using this approach need not be expensive. A BDM-based source-level debugger can be integrated with an IDE, which allows the user to perform program editing, assembling or compiling, and source-level debugging in the same environment. CodeWarrior supports the BDM-based debug adapters from several vendors. The BDM pod can also be part of the target board.

3.11 Summary

The HCS12 microcontroller was developed as an upgrade for the 8-bit 68HC11 microcontroller family. All members implement the same architecture; that is, they have the same instruction set and addressing modes, but differ in the number of peripheral functions that they implement.

All HCS12 members use the same design in each peripheral function; this simplifies the migration from one device to another. All of the HCS12 members implement a 10-bit A/D converter. The HCS12 microcontroller implements a complicated timer system.

- The *input-capture* (IC) function latches the arrival time of a signal edge. This capability allows the user to measure the frequency, the period, and the duty cycle of an unknown signal.

- The *output compare* (OC) function allows the user to make a copy of the main timer, add a delay to this copy, and store the sum to an output compare register. This capability can be used to create a delay, generate a digital waveform, trigger an action on a signal pin, and so on.

- The *pulse accumulation* function allows one to count the events that arrived within an interval.

- The *pulse-width modulation* (PWM) module is used to generate a digital waveform of a certain frequency with a duty cycle ranging from 0 to 100 percent. This capability is useful in DC motor control.

The HCS12 microcontroller also provides a wide variety of serial interface functions to appeal to different applications.

- The *serial communication interface* (SCI) function supports the EIA232 standard. A microcontroller demo board would use this interface to communicate with a PC.

- The *serial peripheral interface* (SPI) function is a synchronous interface that requires a clock signal to synchronize the data transfer between two devices. This interface is mainly used to interface with peripheral chips such as shift registers, seven-segment displays and LCD drivers, A/D and D/A converters, SRAM and EEPROM, phase-locked-loop chips, and so on. All of these devices must have the SPI interface.

- The *inter-integrated circuit* (I²C) is a serial interface standard proposed by Phillips. This interface standard allows microcontrollers and peripheral devices to exchange data.

- The *byte data link communication* (BDLC) module was proposed for low-speed data communication in automotive applications. It provides access to an external serial communication multiplexed bus that operates according to the SAE J1850 protocol.

- The *controller area network* (CAN) was proposed to be used mainly as a vehicle serial data bus to provide reliable operation in the EMI environment and achieve the high bandwidth required in that environment.

The MC9S12DG256-based demo board Dragon12-Plus from Wytec is recommended for learning the HCS12 microcontrollers.

Software development tools include the text editor, the terminal program, the cross assembler, the cross compiler, the simulator, the source-level debugger, and the integrated development environment (IDE). A sophisticated IDE should contain a text editor, a terminal program, a cross compiler, a cross assembler, and a source-level debugger. It allows the user to perform all the development work without leaving any program.

In this text, we use the freeware from Mgtek and CodeWarrior from Freescale to enter and test all the assembly programs. Tutorials on how to use MiniIDE and CodeWarrior IDE are provided. The debugging activities on a demo board are facilitated by the command set provided by the D-Bug12 monitor.

The D-Bug12 monitor has four modes.

- *EVB mode.* Most users use this mode to perform development and evaluation work.

- *POD mode.* This mode allows one to use the demo board to debug another HCS12 target board.

- *Jump-to-EEPROM mode.* This mode allows the user to run the application programmed into the on-chip EEPROM after power-up or reset.

- *Bootloader mode.* This mode allows the user to update the D-Bug12 monitor or program the application code into the flash memory out of reset or after being powered on.

The only freeware source-level debugger is the limited version of CodeWarrior from Freescale. CodeWarrior requires a demo board to be programmed with the 2-kB serial monitor from Freescale or connected to a BDM-based debug adapter.

3.12 Lab Exercises and Assignments

L3.1 Turn on the PC and start the MiniIDE (or AsmIDE) program to connect to a demo board with the D-Bug12 monitor. Perform the following operations:

   a. Enter a command to set the contents of the memory locations from $1000 to $10FF to 0.

   b. Display the contents of the memory locations from $1000 to $10FF.

   c. Set the contents of the memory locations $1000~$1003 to 1, 2, 3, and 4, respectively.

   d. Verify that the contents of memory locations $1000~$1003 have been set correctly.

L3.2 Enter monitor commands to display the breakpoints. Set new breakpoints at memory locations $1520 and $1550. Delete breakpoints at $1550, and redisplay the breakpoints.

L3.3 Enter commands to place $10 and $0 in accumulators A and B.

L3.4 Use appropriate D-Bug12 commands to perform the following operations:

   a. Set the contents of the memory location at $1000 to 3.

   b. Set the contents of the memory location at $1001 to 4, and redisplay the location $1000.

   c. Set the contents of the memory location at $1002 to 5, and redisplay the same location.

   d. Set the contents of the memory location at $1003 to 6, and return to the D-Bug12 command prompt.

L3.5 Invoke the one-line assembler to enter the following instructions to the demo board: starting from address $1500, trace through the program, and examine the contents of the memory locations at $1000 and $1001.

```
ldd     #$0000
std     $1000
ldab    #$00
incb
ldx     $1000
abx
stx     $1000
cmpb    #$14
bne     $1508
swi
```

L3.6 Use the text editor of MiniIDE (or AsmIDE) to enter the following assembly program as a file with the filename **lab03_6.asm:**

```
        org     $1000
sum     rmb     1
arcnt   rmb     1
        org     $1500
        ldaa    #20
        staa    arcnt
        ldx     #array
        clr     sum
```

```
again    ldaa    0,X
         lsra
         bcs     next
         ldaa    sum
         adda    0,X
         staa    sum
next     inx
         dec     arcnt
         bne     again
         swi
         org     $2000
array    fcb     1,3,5,7,2,4,6,8,9,11,13,10,12,14,15,17,19,16,18,20
         end
```

After entering the program, perform the following operations:

   a. Assemble the program.

   b. Download the S-record file (filename **lab03_6.s19**) to the demo board.

   c. Display the contents of memory locations from $1000 to $100F and $2000 to $202F.

   d. Execute the program.

   e. Display the contents of the memory location at $1000.

    This program adds all even numbers in the given array and stores the sum at $1000.

L3.7 Write a program to count the number of elements in an array that are divisible by 8. The array has 30 8-bit elements and is stored immediately after your program. Use the Repeat S until C looping construct. Leave the result at $1000. If you have a demo board programmed with the serial monitor, then use CodeWarrior to enter and debug the your program.

L3.8 Write a program to swap the last element of an array with the first element, second element with the second-to-last element, and so on. The array has 30 8-bit elements.

L3.9 Write a program to find the greatest common divisor (gcd) of two 16-bit numbers stored at $1000~$1001 and $1002~$1003, and store the result at $1010~$1011. When testing the program, enter two 16-bit numbers manually before running the program. The most efficient method for finding the greatest common divisor is the Euclidean method. This algorithm can be found on the website at www.fact-index.com/e/eu/euclidean_algorithm.html.

4

Advanced Assembly Programming

4.1 Objectives

After completing this chapter, you should be able to

- Explain the reason why the subroutine is useful

- Explain the issues involved in a subroutine call

- Write subroutines to perform arithmetic and string processing operations

- Access parameters stored in the stack and manipulate stack data structure

- Perform binary-to-ASCII string and ASCII string-to-binary conversion

- Write subroutines to perform certain functions

- Make subroutine calls

- Perform terminal I/O by calling D-Bug12 library functions

- Configure an I/O pin for input or for output

- Use HCS12 I/O ports to interface with simple I/O devices such as LEDs, seven-segment displays, and DIP switches

- Use an HCS12 I/O pin to generate waveforms and make sounds

- Write subroutines to create time delays that are a multiple of certain base values

4.2 Introduction

It is common that the same sequence of instructions need to be executed in several places of the program. The user may define macros as a solution to this issue. However, using the macro call will duplicate the same sequence of instructions in places where the macro is invoked and cause the program size to be bloated. Fortunately, all microprocessors and microcontrollers provide the *subroutine mechanism* in which the program can invoke the same sequence of instructions in many places of the program without duplicating it.

A subroutine is often written to perform operations on the basis of inputs provided by the caller. The inputs provided by the caller are called *incoming parameters*. The caller of the subroutine often expects the subroutine to return certain results to it.

A subroutine call instruction causes the program control flow to change. When calling a subroutine, the processor loads the starting address of the subroutine into the program counter and then the CPU starts to execute the subroutine instructions. When the processor finishes execution of the subroutine, it should return to the instruction immediately after the instruction that makes the subroutine call. This is achieved by executing a *return* instruction.

The subroutine call and return instructions work together to make the subroutine mechanism work. The subroutine call instruction saves the *return address* in the stack data structure while at the same time changing the program control flow to the start of the subroutine. The return instruction fetches the return address from the stack and places it in the program counter and hence returns the program control back to the instruction immediately after the subroutine call instruction.

The subroutine mechanism has great impact on the program development methodology. It is mentioned in Chapter 2 that the most popular software development methodology is *top-down design with hierarchical refinement*. The subroutine mechanism makes this approach possible.

Reusable macros and subroutines should be made into files and included in programs that need them. This approach can increase programmers' productivity. Subroutines are especially convenient for this approach. One of the objectives of this text is to promote software reuse and writing reusable software.

4.3 Stack

A *stack* is a data structure from which elements can be accessed from only its top. The processor can add a new element to the stack by performing a *push* operation and remove an element by performing a *pull* (or *pop*) operation. Physically, a stack can grow from a high address toward lower addresses or from a low address toward higher addresses. Depending on the processor, the *stack pointer* can point to the top element of the stack or to the byte immediately above the top element of the stack. As shown in Figure 4.1, the HCS12 stack grows from a high address toward lower addresses and has a 16-bit stack pointer (SP) that points to the top byte of the stack. The memory space available for use by the stack is limited in a computer system. There is always a danger of *stack overflow* and *stack underflow*. Stack overflow is a situation in which the processor pushes data into the stack too many times so that the SP points to a location outside the area allocated to the stack. Stack underflow is a situation in which the processor pulls data from the stack too many times so that the SP points to an area below the stack bottom. We must check the stack overflow and underflow in order to make sure that the program won't crash. Of course, this checking adds overhead to the stack access.

The HCS12 provides instructions for pushing and pulling all CPU registers except the stack pointer. A push instruction writes data from the source to the stack after decrementing

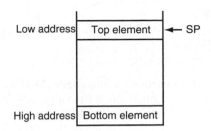

Figure 4.1 ■ Diagram of the HCS12 stack

the stack pointer. There are six push instructions: **psha, pshb, pshc, pshd, pshx,** and **pshy.** A pull instruction loads data from the top of the stack to a register and then increments the stack pointer. There are also six pull instructions: **pula, pulb, pulc, puld, pulx,** and **puly.** These instructions have equivalent store and load instructions combined with predecrement and postincrement index addressing modes, as shown in Table 4.1.

Mnemonic	Function	Equivalent Instruction
psha	push A into the stack	staa 1, −SP
pshb	push B into the stack	stab 1, −SP
pshc	push CCR into the stack	none
pshd	push D into stack	std 2, −SP
pshx	push X into the stack	stx 2, −SP
pshy	push Y into the stack	sty 2, −SP
pula	pull A from the stack	ldaa 1, SP+
pulb	pull B from the stack	ldab 1, SP+
pulc	pull CCR from the stack	none
puld	pull D from the stack	ldd 2, SP+
pulx	pull X from the stack	ldx 2, SP+
puly	pull Y from the stack	ldy 2, SP+

Table 4.1 ■ HCS12 push and pull instructions and their equivalent load and store instructions

Sometimes the programmer may need to push a value into the stack directly. He or she may use one of the following instructions to do that:

```
movb    #val1, 1, −SP    ; push the 8-bit val1 into the stack
movw    #val2, 2, −SP    ; push the 16-bit val2 into the stack
```

Example 4.1

▼

Assuming that we have the following instruction sequence to be executed by the HCS12, what would be the contents of the stack after the execution of these instructions?

```
lds     #$1500
ldaa    #$20
psha
```

```
ldab     #40
pshb
ldx      #0
pshx
```

Solution: The first instruction initializes the stack pointer to $1500. The second and the third instructions together push the 8-bit value $20 into the stack. The fourth and fifth instructions push the 8-bit value 40 (hex $28) into the stack. The sixth and seventh instructions push the value 0 to the top 2 bytes of the stack. The contents of the HCS12 stack are shown in Figure 4.2.

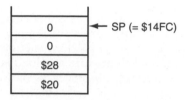

Figure 4.2 ■ The contents of the HCS12 stack

One of the main uses of the stack is saving the return address for a subroutine call. Before the stack can be used, we need to set up the stack pointer. Since the stack can only be implemented in the RAM, we must make sure not to violate this requirement. Because the stack is a last-in-first-out data structure, it can be used to reverse a data structure.

4.4 What Is a Subroutine?

A *subroutine* is a sequence of instructions that can be called from many different places in a program. A key issue in a subroutine call is to make sure that the program execution returns to the point immediately after the subroutine call (this address is called the *return address*) when the subroutine completes its execution. This is normally achieved by saving and retrieving the return address in and from the stack. The program flow change involved in a subroutine call is illustrated in Figure 4.3.

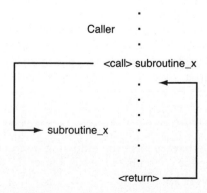

Figure 4.3 ■ Program flow during a subroutine call

The HCS12 has dedicated instructions for making the subroutine call. The subroutine call instruction saves the return address, which is the address of the instruction immediately following the subroutine call instruction, in the system stack. When completing the computation task, the subroutine will return to the instruction immediately following the instruction that makes the subroutine call. This is achieved by executing a return instruction, which will retrieve the return address from the stack and transfer the CPU control to it.

The HCS12 provides three instructions (**bsr, jsr**, and **call**) for making subroutine calls and two instructions for returning from a subroutine. The syntax and operations performed by these instructions are as follows:

- **bsr <opr>:** bsr stands for branch subroutine. This instruction requires the user to use the *relative addressing mode* to specify the subroutine to be called. When this instruction is executed, the stack pointer is decremented by 2 and the return address is saved in the stack; then the offset in the instruction is added to the current PC value and instruction execution is continued from there. In assembly language, the relative address is specified by using a label, and the assembler will figure out the relative offset and place it in the program memory. For example, the following instruction calls the subroutine *bubble:*

```
bsr     bubble
```

- **jsr <opr>:** jsr stands for jump subroutine. The user can use *direct, extended, indexed,* and *indexed indirect* addressing modes to specify the subroutine to be called. The subroutine can be located anywhere within 64 kB. As with the bsr instruction, the HCS12 first saves the return address in the stack and then jumps to execute the subroutine. Examples of the jsr instruction are as follows:

```
jsr     $ff         ; call the subroutine located at $ff
jsr     sq_root     ; call the subroutine sq_root
jsr     0,x         ; call a subroutine pointed to by index register X
```

- **call <opr>:** This instruction is designed to work with expanded memory (larger than 64 kB) supported by some HCS12 members. Members with expanded memory treat the 16-kB memory space from $8000 to $BFFF as a program memory window. An 8-bit program page register (PPAGE) is added to select one of the 256 16-kB program memory pages to be accessed. To support subroutine calls in expanded memory, the *call* instruction pushes the current value of the PPAGE register along with the return address onto the stack and then transfers program control to the subroutine (3 bytes are pushed into the stack). The **<opr>** field in the call instruction specifies the page number and the starting address of the subroutine within that page. The new page number will be loaded into the PPAGE register when the call instruction is executed. Extended, indexed, and indexed indirect addressing modes can be used to specify the subroutine address within a page. Writing assembly programs to be run in expanded memory requires an assembler that supports this feature.

- **rts:** rts stands for *return from subroutine*. This instruction pops the 16-bit value from the stack onto the program counter and increments the stack pointer by 2. Program execution continues at the address restored from the stack. This instruction is used with the jsr and bsr instructions.

- **rtc:** rtc stands for *return from call*. This instruction terminates subroutines in expanded memory invoked by the call instruction. The program page register and the return address are restored from the stack; program execution continues at the restored address. For code compatibility, call and rtc are also executed correctly by the HCS12 members that do not have expanded memory capability, that is, with memory capacity no larger than 64 kB.

Example 4.2

Write a subroutine to convert an unsigned 16-bit binary number into an ASCII string that represents the decimal value of the original binary number. The resultant string must be terminated by a NULL character (the ASCII code of a NULL character is 0). For example, the binary value $0101,1011,1010,0000_2$ (or 23456_{10}) will be converted to $32 $33 $34 $35 $36 $00. The 16-bit number to be converted is held in D, and the pointer (address) to the buffer to hold the resultant string is stored in Y.

Solution: The individual decimal digit can be separated by repeatedly performing divide by 10 to the given number and adding $30 to each remainder. Let *ptr*, *quo*, and *rem* represent the pointer to the buffer to hold the resultant string, the quotient after the divide operation, and the remainder of the divide operation, respectively. The repeated divide-by-10 operation will separate the least significant digit first and the more significant digits later; we cannot save them directly in the buffer pointed to by ptr. The solution is to push these digits into the stack in the order that they are separated and then pop them out and save them in the buffer. The following algorithm implements this idea:

Step 1
Push a 0 into the stack; quo ← number to be converted; the zero (NULL character) pushed into the stack is used to tell the program to stop popping from the stack.

Step 2
rem ← quo % 10; quo ← quo ÷ 10.

Step 3
Push the sum of $30 and rem into the stack.

Step 4
If (quo == 0) go to step 6.

Step 5
Go to step 2.

Step 6
Pull a byte from the stack. Save the byte in the memory location pointed to by ptr.

Step 7
ptr ← ptr + 1;

Step 8
If the byte pulled out in step 6 is NULL, then stop; else go to Step 6.

The subroutine that implements this algorithm is as follows:

```
bin2dec   pshx                      ; save X in stack
          movb    #0, 1, –SP        ; push NULL character into stack
divloop   ldx     #10               ; divide the number by 10
          idiv                      ;   "
          addb    #$30              ; convert the remainder to ASCII code
          pshb                      ; push it into the stack
```

```
              xgdx                      ; swap quotient to D
              cpd       #0              ; if quotient is 0, then prepare to pull out
              beq       revloop         ; the decimal digit characters
              bra       divloop         ; quotient is not 0, continue to perform divide-by-10
     revloop  pula
              staa      1, y+           ; save ASCII string in the buffer
              cmpa      #0              ; reach the NULL pushed previously?
              beq       done            ; if yes, then done
              bra       revloop         ; continue to pup
     done     pulx                      ; restore the index register X
              rts
```

Before calling this subroutine, the caller must place the number to be converted in D and place the starting address of the buffer to hold the resultant string in Y. The following instruction sequence is needed to call the above subroutine to convert the value 23,456 to an ASCII string:

```
     ldd       #23456
     ldy       #buffer       ; use Y as the pointer to the buffer that holds the string
     jsr       bin2dec
```

This subroutine can be modified to convert a signed 16-bit number into the ASCII string that represents its decimal value. A negative value would have a minus sign (−) in the front after the conversion. The modification to the subroutine to convert signed numbers is minor and hence will be left as an exercise problem.

This example points out several issues involved in the subroutine call.

- Passing parameters
- Returning results
- Saving registers used in the subroutine

A more complicated subroutine may have variables that are local to the subroutine but invisible to the caller. These variables are referred to as *local variables*. Examples of local variables include loop indices and temporal results. These variables do not exist if the subroutine is not entered. The next section deals with all of these issues.

4.5 Issues Related to Subroutine Calls

The program unit that makes the subroutine call is referred to as a *caller*, and the subroutine called by other program units is referred to as a *callee*. As was pointed out in the previous section, there are four major issues in a subroutine call. These issues are explored in the following subsections:

4.5.1 Parameter Passing

The caller usually wants the subroutine to perform a computation using the parameters passed to it. The caller may use the following methods to pass parameters to the subroutine:

- *Use registers.* In this method, parameters are placed in CPU registers before the subroutine is called. This method is very convenient when there are only a few parameters to be passed.

- *Use the stack.* In this method, parameters are pushed into the stack before the subroutine is called. The stack must be cleaned up after the computation is completed. This can be done by either the caller or the callee.

- *Use the global memory.* Global memory is accessible to both the caller and the callee. As long as the caller places parameters in global memory before it calls the subroutine, the callee will be able to access them.

4.5.2 Result Returning

The result of a computation performed by the subroutine can be returned to the caller using three methods:

- *Use registers.* This method is most convenient when there are only a few bytes to be returned to the caller.

- *Use the stack.* The caller creates a hole of a certain size in the stack before making the subroutine call. The callee places the computation result in the hole before returning to the caller.

- *Use global memory.* The callee simply places the value in the global memory and the caller will be able to access them.

4.5.3 Allocation of Local Variables

In addition to the parameters passed to it, a subroutine may need memory locations to hold temporary variables and results. Temporary variables are called *local variables* because they only exist when the subroutine is entered. Local variables are always allocated in the stack so that they are not accessible to any other program units.

Although there are several methods for allocating local variables, the most efficient one is using the **leas** instruction. This instruction loads the stack pointer with an effective address specified by the program. The effective address can be any indexed addressing mode except an indirect address. For example, to allocate 10 bytes, we can use the indexed addressing mode with SP as the base register and −10 as the offset:

```
leas    −10, sp    ; allocate 10 bytes in the stack
```

This instruction simply subtracts 10 from the SP and puts the difference back to the SP. The general format for allocating space to local variables is

```
leas    −n,sp      ; allocate n bytes in the stack
```

where *n* is the number of bytes to be allocated.

Before the subroutine returns to the caller, the space allocated to local variables must be deallocated. Deallocation is the reverse of allocation and can be achieved by the following instruction:

```
leas    n,sp       ; deallocate n bytes from the stack
```

4.5.4 Saving the CPU Registers

A subroutine may use CPU registers to hold local variables to perform certain operations. Sometimes, the subroutine needs to perform some operations that involve certain CPU registers. For example, the emul and emuls instructions use register D and Y whereas the ediv and the edivs instructions use registers D, X, and Y. However, these CPU registers may also be used by the caller of the subroutine. If the values of these CPU registers were not preserved during a subroutine call, the execution result of the caller cannot be correct. Either the caller or the callee may be responsible for saving these registers. However, it is more efficient for the subroutine to do the saving because the same subroutine may be called by many callers. CPU registers are usually saved at the entrance of the subroutine before they are modified.

Only those registers that are not used for passing incoming parameters and returning results need to be saved. The subroutine should restore the values of the saved registers immediately before returning to the caller. The saving and restoring must be done in the reverse order. For example, if the subroutine saves D, X, and Y in the following order:

```
pshd
pshx
pshy
```

then it should have the following instruction sequence to restore them before returning to the caller:

```
puly
pulx
puld
```

4.6 The Stack Frame

The stack is used heavily during a subroutine call: The caller may pass parameters to the callee, and the callee may need to save registers and allocate local variables in the stack. The region in the stack that holds incoming parameters, return addresses, saved registers, and local variables is referred to as the *stack frame*. Some microprocessors have a dedicated register for managing the stack frame; the register is referred to as the *frame pointer*. The HCS12, however, does not have a register dedicated to the function of the frame pointer. Since the stack frame is created during a subroutine call, it is also called the *activation record* of the subroutine. The stack frame exists as long as the subroutine is not exited. The structure of a stack frame is shown in Figure 4.4.

The reason for having a dedicated frame pointer is that the stack pointer may change during the lifetime of a subroutine. Once the stack pointer changes value, there can be problems in accessing the variables stored in the stack frame. The frame pointer is added to point to a fixed

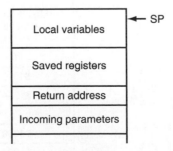

Figure 4.4 ■ Structure of the HCS12 stack frame

location in the stack and can avoid this problem. Since the HCS12 does not have a dedicated frame pointer, we will not use the term *frame pointer* in the following discussion.

Example 4.3

▼

Draw the stack frame for the following program segment after the last **leas −10,sp** instruction is executed:

```
            ldd    #$1234
            pshd
            ldx    #$4000
            pshx
            jsr    sub_xyz
            ...
sub_xyz     pshd
            pshx
            psy
            leas   −10, sp
            ...
```

Solution: The caller pushes two 16-bit words into the stack. The subroutine **sub_xyz** saves three 16-bit registers in the stack and allocates 10 bytes in the stack. The resultant stack frame is shown in Figure 4.5.

▲

4.6.1 Subroutines with Local Variables in Stack

When a subroutine has very few local variables, the subroutine can use CPU registers to hold them (need to save these registers in the stack). However, when there are not enough CPU registers to hold local registers, the subroutine will need to assign local variables to the stack.

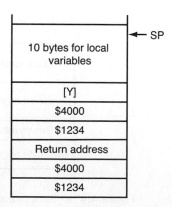

Figure 4.5 ■ Stack frame of Example 4.5

Example 4.4

Write a subroutine that can convert a BCD ASCII string to a binary number and leave the result in double accumulator D. The ASCII string represents a number in the range of $-2^{15} \sim 2^{15} - 1$. A pointer to the string is passed to this subroutine in X.

Solution: The subroutine needs to perform error checking. If there is any illegal character, the error flag will be set to 1. The subroutine will return the error flag to the caller using the CY flag in the CCR register. Let *in_ptr*, *sign*, *error*, and *number* represent the pointer to the BCD string, sign flag, error flag, and the number represented by the BCD string. The algorithm of the subroutine is as follows:

Step 1

sign ← 0; error ← 0; number ← 0;

Step 2
If the character pointed to by in_ptr is the minus sign, then

sign ← 1; in_ptr ← in_ptr + 1

Step 3
If the character pointed to by in_ptr is the NULL character,
then go to step 4.
else if the character is not a BCD digit (i.e., m[in_ptr] > $39 or m[in_ptr] < $30), then:

error ← 1;
go to step 4;

else:

number ← number × 10 + m[in_ptr] − $30;
in_ptr ← in_ptr + 1;
go to step 3;

Step 4
If sign = 1 and error = 0, then

number ← two's complement of number;

else

stop;

In this example, the subroutine allocates local variables in the stack as shown in Figure 4.6.

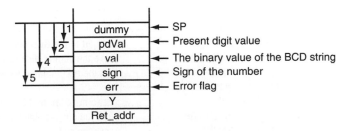

Figure 4.6 ■ Stack frame of Example 4.4

The conversion subroutine and its test program are as follows:

```
              #include "c:\miniide\hcs12.inc"
minus   equ   $2D             ; ASCII code of minus sign
dummy   equ   0               ; offset of dummy to hold a 0
pdVal   equ   1               ; offset of present digit value from SP in stack
val     equ   2               ; offset of the 2-byte binary value of the BCD string
                              ; from SP in stack
sign    equ   4               ; offset of the sign from SP in stack
err     equ   5               ; offset of error flag from SP in stack
locVar  equ   6
        org   $1000
StrBuf  dc.b  "−9889",0       ; input ASCII to be converted
result  ds.w  1
        org   $1500
start   lds   #$1500

        ldx   #StrBuf

        jsr   bcd2bin

        std   result
        swi
;***************************************************************************************
; This subroutine converts a BCD string into its equivalent binary value and also uses a CY
; flag to indicate error condition. The CY flag is set to 1 to indicate error.
;***************************************************************************************
bcd2bin pshy
        leas  −locVar,SP      ; allocate 4 bytes for local variables
        movw  #0,val,SP       ; initialize accumulated value to 0
        movb  #0,dummy,SP
        movb  #0,sign,SP      ; initialize sign to positive
        movb  #0,err,SP       ; clear error flag initially
        ldaa  0,x             ; check the first character
        cmpa  #minus          ; is the first character a minus sign?
        bne   GEZero          ; branch if not minus
        movb  #1,sign,SP      ; set the sign to 1
        inx                   ; move the pointer
GEZero  ldab  1,x+            ; is the current character a NULL character?
        lbeq  done            ; yes, we reach the end of the string
        cmpb  #$30            ; is the character not between 0 to 9?
        blo   inErr           ; "
        cmpb  #$39            ; "
        bhi   inErr           ; "
        subb  #$30            ; convert to the BCD digit value
        stab  pdVal,SP        ; save the current digit value
        ldd   val,SP          ; get the accumulated value
        ldy   #10
        emul                  ; perform 16-bit by 16-bit multiplication
        addd  dummy,SP        ; add the current digit value
        std   val,SP          ; save the sum
        bra   GEZero
inErr   movb  #1,err,SP       ; set the error flag to indicate error
        bra   chkout
```

```
done      ldaa     sign,SP        ; check to see if the original number is negative
          beq      chkout
          ldd      #$FFFF         ; convert to two's complement format
          subd     val,SP         ; if the number is negative
          addd     #1             ;    "
          std      val,SP         ;    "
chkout    ldaa     err,SP         ; check the error flag
          beq      clrErr         ; go to clear CY flag before return
          sec                     ; clear the C flag
          bra      dealloc
clrErr    clc                     ; set the C flag
dealloc   ldd      val,SP
          leas     locVar,SP      ; deallocate local variables
          puly                    ; restore Y
          rts
;         org      $FFFE          ; needed under CodeWarrior
;         dc.w     start          ;    "
          end
```

4.6.2 Bubble Sort

Sorting is among the most common ingredients of programming and many sorting methods are being used. Sorting makes many efficient search methods possible. The *bubble sort* is a simple, widely known, but inefficient, sorting method. Many other more efficient sorting methods require the use of recursive subroutine calls, which fall outside the scope of this book.

The basic idea underlying the bubble sort is to go through the array or file sequentially several times. Each iteration consists of comparing each element in the array or file with its successor (x[i] with x[i + 1]) and interchanging the two elements if they are not in proper order (either ascending or descending). Consider the following array:

157 13 35 9 98 810 120 54 10 30

Suppose we want to sort this array in ascending order. The following comparisons are made in the first iteration:

x[0] with x[1] (157 and 13) interchange

x[1] with x[2] (157 with 35) interchange

x[2] with x[3] (157 with 9) interchange

x[3] with x[4] (157 with 98) interchange

x[4] with x[5] (157 with 810) no interchange

x[5] with x[6] (810 with 120) interchange

x[6] with x[7] (810 with 54) interchange

x[7] with x[8] (810 with 10) interchange

x[8] with x[9] (810 with 30) interchange

Thus, after the first iteration, the array is in the order

13 35 9 98 157 120 54 10 30 810

Notice that after this first iteration, the largest element (in this case 810) is in its proper position within the array. In general, $x[n - i]$ will be in its proper position after iteration i. The method is called the *bubble sort* because each number slowly bubbles up to its proper position. After the second iteration the array is in the order

13 9 35 98 120 54 10 30 157 810

Notice that 157 is now in the second-highest position. Since each iteration places a new element into its proper position, an array or a file of *n* elements requires no more than $n - 1$ iterations. The complete set of iterations is as follows:

Iteration	
	0 (original array) 157 13 35 9 98 810 120 54 10 30
1	13 35 9 98 157 120 54 10 30 810
2	13 9 35 98 120 54 10 30 157 810
3	9 13 35 98 54 10 30 120 157 810
4	9 13 35 54 10 30 98 120 157 810
5	9 13 35 10 30 54 98 120 157 810
6	9 13 10 30 35 54 98 120 157 810
7	9 10 13 30 35 54 98 120 157 810
8	9 10 13 30 35 54 98 120 157 810
9	9 10 13 30 35 54 98 120 157 810

There are some obvious improvements to the foregoing method.

- First, since all elements in positions greater than or equal to $n - i$ are already in proper position after iteration *i*, they need not be considered in succeeding iterations. Thus, in the first iteration $n - 1$ comparisons are made, on the second iteration $n - 2$ comparisons are made, and on the $(n - 1)$th iteration, only one comparison is made (between x[0] and x[1]). Therefore, the process is sped up as it proceeds through successive iterations.

- Second, although we have shown that $n - 1$ iterations are sufficient to sort an array or a file of size *n*, in the preceding sample array of 10 elements, the array was sorted after the seventh iteration, making the last two iterations unnecessary. To eliminate unnecessary iterations, we must be able to detect the fact that the array is already sorted. An array is sorted if no swaps are made in an iteration. By keeping a record of whether any swaps are made in a given iteration, it can be determined whether any further iteration is necessary. The logic flow of the bubble sort algorithm is illustrated in Figure 4.7. The following example implements the bubble sort as a subroutine:

Example 4.5

Write a subroutine to implement the bubble sort algorithm and a sequence of instructions along with a set of test data for testing this subroutine. Use an array that consists of *n* 8-bit unsigned integers for testing purposes.

Solution: The bubble sort subroutine has several local variables.

- *buf*. buffer space for swapping adjacent elements
- *inOrder*. flag to indicate whether the array is in order after an iteration
- *inner*. number of comparisons remaining to be performed in the current iteration
- *iteration*. number of iterations remaining to be performed

The stack frame of the bubble sort subroutine is shown in Figure 4.8.

```
arr      equ      13       ; distance of the variable arrayX from stack top
arcnt    equ      12       ; distance of the variable arcnt from stack top
buf      equ      3        ; distance of local variable buf from stack top
```

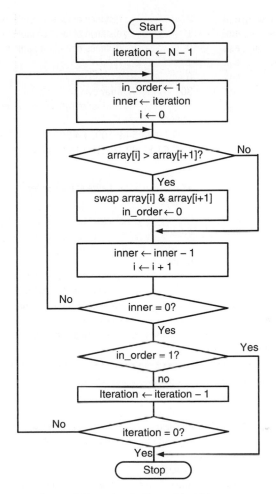

Figure 4.7 ■ Logic flow of bubble sort

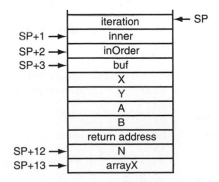

Figure 4.8 ■ Stack frame for bubble sort

```
        inOrder   equ      2                ; distance of local variable inOrder from stack top
        inner     equ      1                ; distance of local variable inner from stack top
        iteration equ      0                ; distance of local variable iteration from stack top
        true      equ      1
        false     equ      0
        n         equ      30               ; array count
        local     equ      4                ; number of bytes used by local variables
                  org      $1000
        arrayX    dc.b     3,29,10,98,54,9,100,104,200,92,87,48,27,22,71
                  dc.b     1,62,67,83,89,101,190,187,167,134,121,20,31,34,54

                  org      $1500
        start     lds      #$1500           ; initialize stack pointer
                  ldx      #arrayX
                  pshx
                  ldaa     #n
                  psha
                  jsr      bubble
                  leas     3,sp             ; deallocate space used by outgoing parameters
        ;         bra      $                ; uncomment this instruction under CodeWarrior
                  swi                       ; break to D-Bug12 monitor
; ********************************************************************************
; The following subroutine implements the bubble sort with the array address and array count
; pushed into the stack.
; ********************************************************************************
        bubble    pshd
                  pshy
                  pshx
                  leas     -local,sp        ; allocate space for local variables
                  ldaa     arcnt,sp         ; compute the number of iterations to be performed
                  deca                      ; "
                  staa     iteration,sp     ; "
        ploop     ldaa     #true            ; set array inOrder flag to true before any iteration
                  staa     inOrder,sp       ; "
                  ldx      arr,sp           ; use index register X as the array pointer
                  ldaa     iteration,sp     ; initialize inner loop count for each iteration
                  staa     inner,sp         ; "
        cloop     ldaa     0,x              ; compare two adjacent elements
                  cmpa     1,x              ; "
                  bls      looptst
; the following five instructions swap the two adjacent elements
                  staa     buf,sp           ; swap two adjacent elements
                  ldaa     1,x              ; "
                  staa     0,x              ; "
                  ldaa     buf,sp           ; "
                  staa     1,x              ; "
                  ldaa     #false           ; reset the inOrder flag
                  staa     inOrder,sp       ; "
        looptst   inx
                  dec      inner,sp
```

```
            bne       cloop
            tst       inOrder,sp      ; test array inOrder flag after each iteration
            bne       done
            dec       iteration,sp
            bne       ploop
; the following instruction deallocates local variables
done        leas      local,sp        ; deallocate local variables
            pulx
            puly
            puld
            rts
;           org       $FFFE           ; uncomment for CodeWarrior
;           dc.w      start           ; uncomment for CodeWarrior
            end
```

4.6.3 Binary Search Subroutine

Searching is another frequently performed operation. When an array is not sorted, the search program may need to compare every element of the array. When an array or a file needs to be searched frequently, it would be more efficient for the programmer to sort the array and apply a more efficient search method to find the desired element from the array or file.

The binary search algorithm is one of the most efficient search algorithms in use. Suppose the sorted array (in *ascending* order) has n elements and is stored at memory locations starting at the label **arr**. Let *max* and *min* represent the highest and lowest range of array indices to be searched, and the variable *mean* represent the average of *max* and *min*. The idea of a binary search algorithm is to divide the sorted array into three parts.

- *The upper half.* the portion of the array with indices ranging from *mean* + 1 to *max*
- *The middle element.* the element with index equal to *mean*
- *The lower half.* the portion of the array with indices ranging from *min* to *mean* − 1

The binary search algorithm compares the key with the middle element and takes one of the following actions on the basis of the comparison result:

- If the key equals the middle element, then stop.
- If the key is larger than the middle element, then the key can be found only in the upper half of the array. The search will be continued in the upper half.
- If the key is smaller than the middle element, then the key can be found only in the lower half of the array. The search will be continued in the lower half.

The binary search algorithm can be formulated as follows:

Step 1
Initialize variables *max* and *min* to $n - 1$ and 0, respectively.

Step 2
If *max* < *min*, then stop. No element matches the key.

Step 3
Let *mean* = (*max* + *min*)/2.

Step 4
If *key* equals **arr**[mean], then key is found in the array; exit.

Step 5

If *key* < arr[mean], then set *max* to *mean* − 1 and go to step 2.

Step 6

If *key* > arr[mean], then set *min* to *mean* + 1 and go to step 2.

The algorithm works for arrays sorted in ascending order, but can be modified to work for arrays sorted in descending order.

Example 4.6

Write a subroutine to implement the binary search algorithm and a sequence of instructions to test it. Use an array of *n* 8-bit elements for implementation.

Solution: In this example, we will use the stack to pass parameters (base address of the array, array count, and search key) and also allocate local variables in the stack. The stack frame is shown in Figure 4.9.

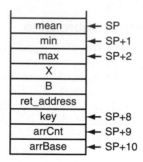

Figure 4.9 ■ Stack frame for binary search

```
n        equ    30          ; array count
srch     equ    69          ; key to be searched
mean     equ    0           ; stack offset for local variable mean
min      equ    1           ; stack offset for local variable min
max      equ    2           ; stack offset for local variable max
key      equ    8           ; stack offset for local variable key
arrCnt   equ    9           ; stack offset for local variable arrCnt
arrBas   equ    10          ; stack offset for local variable arrBas
locvar   equ    3           ; number of bytes for local variables
         org    $1000
result   ds.b   1           ; search result
         org    $1500
         lds    #$1500
         movw   #arr,2,-SP   ; pass array base address
         movb   #n,1,-SP     ; pass array count
         movb   #srch,1,-SP  ; pass key for search
         jsr    binsearch
         leas   4,SP         ; deallocate space used in passing parameters
         staa   result
```

```
;          bra      $              ; uncomment this instruction for CodeWarrior
           swi
; ***********************************************************************************
; The following subroutine uses the binary search algorithm to search an array of
; 8-bit elements to find if there is any element that matches the supplied key.  The array
; base address, array count, and key are passed in the stack.
; ***********************************************************************************
binSearch  pshx                    ; save X in the stack
           pshb                    ; save B in the stack
           leas     −locVar,SP     ; allocate space for locVar variables
           movb     #0,min,SP      ; initialize min to 0
           ldaa     arrCnt,SP      ; initialize max to arCnt − 1
           deca                    ; "
           staa     max,SP         ; "
           ldx      arrBas,SP      ; use X as the pointer to the array
loop       ldab     min,SP         ; is search over yet?
           cmpb     max,SP         ; "
           lbhi     notfound       ; if min > max, then not found (unsigned comparison)
           addb     max,SP         ; compute mean
           lsrb                    ; "
           stab     mean,SP        ; save mean
           ldaa     b,x            ; get a copy of the element arr[mean]
           cmpa     key,SP         ; compare key with array[mean]
           beq      found          ; found it?
           bhi      searchLO       ; continue to search In lower half
           ldaa     mean,SP        ; prepare to search in upper half
           inca                    ; "
           staa     min,SP         ; "
           bra      loop
searchLO   ldaa     mean,SP        ; set up indices range for searching in the
           deca                    ; lower half
           staa     max,SP         ; "
           bra      loop
found      ldaa     #1
           bra      exit
notfound   ldaa     #0
exit       leas     locVar,SP
           pulb
           pulx
           rts
arr        db       1,3,6,9,11,20,30,45,48,60
           db       61,63,64,65,67,69,72,74,76,79
           db       80,83,85,88,90,110,113,114,120,123
;          org      $FFFE          ; uncomment this line for CodeWarrior
;          dc.w     start          ; uncomment this line for CodeWarrior
           end
```

▲

4.7 Mathematical Subroutines

Mathematical computation is one of the most important application areas for the micro-controller. Several examples will be used to illustrate parameter passing, local variable allocation, and result returning.

4.7.1 Subroutine for Performing Multiple-Byte Division

The HCS12 provides instructions for 16-bit by 16-bit signed and unsigned divisions and also for 32-bit by 16-bit signed and unsigned divisions. However, the HCS12 does not have instructions for performing higher-precision divisions; say, for example, 32-bit by 32-bit division. We will need to write subroutines to synthesize such operations.

The most popular method for performing high-precision divide operations is the repeated subtraction method. The conceptual hardware for implementing this algorithm is shown in Figure 4.10.

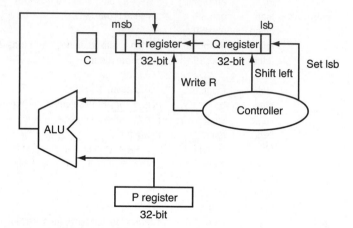

Figure 4.10 ■ Conceptual hardware for implementing the repeated subtraction method

Step 1
icnt ← n; R ← 0; Q ← dividend; P ← divisor.

Step 2
Shift the register pair (R, Q) 1 bit left.

Step 3
Subtract the register P from the register R; put the result back to R if the result is nonnegative.

Step 4
If the result of step 2 is negative, then set the least significant bit of Q to 0. Otherwise, set the least significant bit of Q to 1.

Step 5
icnt ← icnt − 1.

Step 6
If (icnt == 0), then stop;
else go to step 2.

Example 4.7

Write a subroutine that implements the division algorithm using the repeated subtraction method for a 32-bit unsigned dividend and divisor. The caller of this subroutine will pass the dividend and divisor in the stack and will allocate space in the stack for this subroutine to return the quotient and remainder. Also, write an instruction sequence to test this subroutine.

Solution: The stack frame of this subroutine is shown in Figure 4.11.

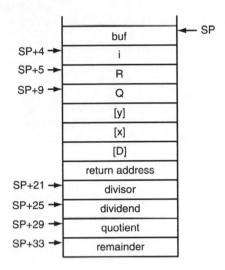

Figure 4.11 ■ Stack frame of Example 4.7

The divide subroutine and its test program are as follows:

```
buf        equ    0           ; distance of buf from the top of the stack
i          equ    4           ; distance of i from the top of the stack
R          equ    5           ; distance of R from the top of the stack
Q          equ    9           ; distance of Q from the top of the stack
divisor    equ    21          ; distance of divisor from the top of the stack
dividend   equ    25          ; distance of dividend from the top of the stack
quo        equ    29          ; distance of quo from the top of the stack
rem        equ    33          ; distance of rem from the top of the stack
locVar     equ    13          ; number of bytes for local variables
dvdendHI   equ    $42         ; dividend to be tested
dvdendLO   equ    $4c15       ; "
dvsorHI    equ    $0          ; divisor to be tested
dvsorLO    equ    $64         ; "

           org    $1000
quotient   ds.b   4           ; memory locations to hold the quo
remain     ds.b   4           ; memory locations to hold the remainder
           org    $1500       ; starting address of the program
start      lds    #$1500      ; initialize stack pointer
           leas   −8,SP       ; make a hole of 8 bytes to hold the result
```

```
                        ldd         #dvdendLO
                        pshd
                        ldd         #dvdendHI
                        pshd
                        ldd         #dvsorLO
                        pshd
                        ldd         #dvsorHI
                        pshd
                        jsr         div32           ; call the divide subroutine
; the following instruction deallocates the stack space used by the divisor and dividend
                        leas        8,SP
; the following four instructions get the quotient from the stack
                        puld
                        std         quotient
                        puld
                        std         quotient + 2
; the following four instructions get the remainder from the stack
                        puld
                        std         remain
                        puld
                        std         remain + 2
;                       bra         $               ; uncomment this instruction for CodeWarrior
                        swi

; ****************************************************************************
; The following subroutine divides an unsigned 32-bit integer by another unsigned
; 32-bit integer
; ****************************************************************************
;
div32           pshd
                pshx
                pshy
                leas        −locVar,sp          ; allocate space for local variables
                ldd         #0
                std         R,sp                ; initialize register R to 0
                std         R + 2,sp            ; "
                ldd         dividend,sp         ; place dividend in register Q
                std         Q,sp                ; "
                ldd         dividend + 2,sp     ; "
                std         Q+2,sp              ; "
                ldaa        #32                 ; initialize loop count
                staa        i,sp                ; "
loop            lsl         Q + 3,sp            ; shift register pair Q and R to the left
                rol         Q + 2,sp            ; by 1 bit
                rol         Q + 1,sp            ; "
                rol         Q,sp                ; "
                rol         R + 3,sp            ; "
                rol         R + 2,sp            ; "
                rol         R + 1,sp            ; "
                rol         R,sp                ; "
```

```
; the following eight instructions subtract the divisor from register R
            ldd         R + 2,sp
            subd        divisor + 2,sp
            std         buf + 2,sp
            ldaa        R + 1,sp
            sbca        divisor + 1,sp
            staa        buf + 1,sp
            ldaa        R,sp
            sbca        divisor,sp
            bcs         smaller
; the following six instructions store the difference back to R register
            staa        R,sp
            ldaa        buf + 1,sp
            staa        R + 1,sp
            ldd         buf + 2,sp
            std         R + 2,sp
            bset        Q + 3,sp,$01    ; set the least significant bit of Q register to 1
            bra         looptest
smaller     bclr        Q + 3,sp,$01    ; set the least significant bit of Q register to 0
looptest    dec         i,sp
            lbne        loop
; the following four instructions copy the remainder into the hole in the stack
            ldd         R,sp
            std         rem,sp
            ldd         R + 2,sp
            std         rem + 2,sp
; the following four instructions copy the quotient into the hole in the stack
            ldd         Q,sp
            std         quo,sp
            ldd         Q + 2,sp
            std         quo + 2,sp
            leas        locVar,sp       ; deallocate local variables
            puly
            pulx
            puld
            rts
;           org         $FFFE           ; uncomment this line for CodeWarrior
;           dc.w        start           ; uncomment this line for CodeWarrior
            end
```

4.7.2 Finding the Square Root

There are several methods available for finding the square root of a number q. One of the methods is based on successive approximation shown in Figure 4.12. The square root of a $2n$-bit number is n bits long. The successive approximation method computes the square root of a 32-bit integer in the following manner:

Step 1

sar ← 0; mask ← $8000; lpcnt ← 16; tcmp ← 0

Step 2
temp ← sar OR mask

Step 3
If ((temp ** 2) > num) sar ← temp;

Step 4
mask ← mask srl 1 (shift right logically one place);

Step 5
lpcnt ← lpcnt − 1

Step 6
If (lpcnt == 0) stop; else go to step 2.

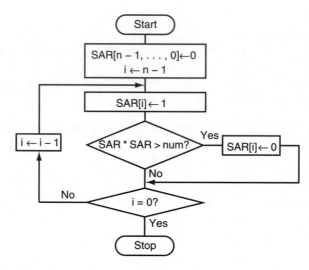

Figure 4.12 ■ Successive-approximation method for finding square root

Example 4.8

Write a subroutine to implement the square root algorithm. This subroutine must be able to find the square root of a 32-bit unsigned integer. The parameter q (for which we want to find the square root) is pushed into the stack and the square root is returned in double accumulator D.

Solution: Four local variables are needed for implementing the square root algorithm.

- *mask*. set a bit in the SAR register to be 1
- *sar*. successive approximation register
- *temp*. the value of (*mask* OR *sar*)
- *lpcnt*. number of loop iterations to be performed

The stack frame of this algorithm is shown in Figure 4.13.

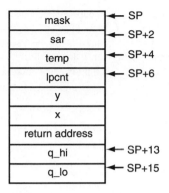

Figure 4.13 ■ Stack frame of Example 4.8

The subroutine for finding the square root of a 32-bit integer and its testing program is as follows:

```
                #include        "c:\miniide\hcs12.inc"
mask            equ             0               ; stack offset of the variable mask from SP
sar             equ             2               ; stack offset of the variable sar from SP
temp            equ             4               ; stack offset of the variable temp from SP
lpcnt           equ             6               ; stack offset of 1-byte loop count from SP
q_hi            equ             13              ; stack offset of the upper and lower halves of the
q_lo            equ             15              ; number Q we want to find the square root from SP
testHi          equ             $00             ; upper half of the test number (q)
testLo          equ             $7BC4           ; lower half of the test number (q)
locVar          equ             7
                org             $1000
sqroot          ds.w            1               ; square root

                org             $1500
start           lds             #$1500
                movw            #testLo,2,-SP    ; push testHi into stack
                movw            #testHi,2,-SP    ; push testLo into stack
                jsr             findsqr
                std             sqroot           ; save the returned square root
                leas            4,SP             ; deallocate the space used in passing parameters
;               bra             $                ; uncomment this line for CodeWarrior
                swi

; ********************************************************************************
; The following subroutine computes the closest integer square root for the incoming
; unsigned integer pushed in the stack (q_hi and q_lo).
; ********************************************************************************

findsqr         pshx
                pshy
                leas            -locVar,SP       ; allocate space for local variables
                movw            #0,sar,SP        ; initialize sar to 0
                movw            #$8000,mask,SP   ; initialize mask to $8000
```

```
                movb       #16,lpcnt,SP     ; initialize loop count to 16
      iloop     ldd        mask,SP          ; get the mask
                oraa       sar,SP           ; set a bit in sar to 1
                orab       sar + 1,SP       ; "
                std        temp,SP          ; save a copy of your guess
                tfr        D,Y              ; compute sar * sar
                emul                        ; "
                cpy        q_hi,SP          ; is our guess correct?
                bhi        nextb            ; "
                cpd        q_lo,SP          ; "
                bhi        nextb            ; "
                ldd        temp,SP          ; yes, our guess is correct
                std        sar,SP           ; so, transfer temp to sar
      nextb     lsr        mask,SP          ; shift mask to the right one place
                ror        mask + 1,SP      ; "
                dec        lpcnt,SP
                bne        iloop
                ldd        sar,SP           ; put sar in D as the approximated square root
                leas       locVar,SP        ; deallocate local variables
                puly
                pulx
                rts
      ;         org        $FFFE            ; uncomment this line for CodeWarrior
      ;         dc.w       start            ; uncomment this line for CodeWarrior
                end
```

This subroutine will find the exact square root if the given number has one. If the given number does not have an exact integer square root, then the number returned by the subroutine may not be the closest approximation. The algorithm in Figure 4.12 will compute a **sar** that satisfies the relationship **sar * sar < q**. This may have the undesirable effect that the last *sar* value may not be as close to the real square root as is the value of *sar* + 1. However, this can be fixed easily by comparing the following two expressions:

1. $(sar + 1)^2 - q$
2. $q - sar^2$

If the first expression is smaller, then $(sar + 1)$ is a better choice. Otherwise, we should choose *sar* as the approximation to the real square root. This will be left as an exercise problem. We will make the subroutine findsqr into a file (findsqr.asm) so that it can be reused in the next example.

4.7.3 Subroutine for Prime Testing

Testing whether a number is a prime may not be extremely useful for embedded applications. However, it can serve as a good example for software reuse.

The most efficient method for testing whether a number is a prime is to divide the number by all the prime numbers from 2 to its integral square root. Since the prime numbers from 2 to the square root of the given number are not available, we will be satisfied with dividing the given number by all the integers from 2 to its integral square root.

The algorithm for testing whether a number is a prime is as follows:

Step 1
Let *num*, *i*, and *isprime* represent the number to be tested, the loop index, and the flag to indicate if *num* is a prime.

Step 2
isprime ← 0; tlimit ← square root of num;

Step 3
for i = 2 to tlimit do
      if ((num % i) == 0) then return;
isprime ← 1;
return;

The following example implements this algorithm:

Example 4.9
▼

Write a subroutine that can test whether an unsigned integer is a prime number.

Solution: The subroutine that determines whether a number is a prime and its test program is as follows:

```
test_hi     equ         $0638        ; number to be tested for prime
test_lo     equ         $F227        ; "
            org         $1000
isprime     ds.b        1
            org         $1500
start       lds         #$1500       ; set up stack pointer
            movw        #test_lo,2,−SP  ; push the lower half of the test number
            movw        #test_hi,2,−SP  ; push the upper half of the test number
            jsr         PrimeTest
            staa        isprime
;           bra         $            ; uncomment this line for CodeWarrior
            swi
; ************************************************************************************
; The following subroutine tests whether an integer is a prime. The number to be tested
; is 32-bit and is pushed into the stack with the lower half first. The result is returned in A
; and will be 1 if the number is a prime.
;
; ************************************************************************************
ii          equ         0            ; stack offset from SP of loop index
tlimit      equ         2            ; stack offset from SP of test limit
pNumHi      equ         10           ; stack offset from SP of upper half of test number
pNumLo      equ         12           ; stack offset from SP of lower half of test number
pLocal      equ         4            ; number of bytes used by local variables
; ************************************************************************************
PrimeTest   pshx
            pshy
            leas        −pLocal,SP   ; allocate space for local variables
            ldaa        pNumLo + 1,SP  ; check if the number is even (if bit 0 is 0)
            anda        #$01         ; "
            beq         nonPRI       ; "
```

```
            ldd      pNumHi,SP
            cpd      #0
            bne      testPR         ; upper half nonzero, then enter normal test
            ldd      pNumLo,SP      ; if upper half is 0, then test lower half
            cpd      #0             ; is lower half equal to 0?
            beq      nonPri         ; 0 is not a prime
            cpd      #1             ; is lower half equal to 1
            beq      nonPri         ; 1 is not a prime
testPR      ldd      pNumLo,SP      ; find the square root of Num
            ldx      pNumHi,SP      ; "
            pshd                    ; "
            pshx                    ; "
            jsr      findsqr        ; "
            leas     4,SP           ; deallocate space for passing parameters
            std      tlimit,SP      ; save returned value as the prime test limit
            movw     #2,ii,SP       ; initialize test divide number to 3
divLoop     ldd      ii,SP
            cpd      tlimit,SP      ; has test divided all numbers up to tlimit?
            bhi      isPRI          ; the number is prime
            ldd      pNumLo,SP      ; divide Num by ii
            ldx      pNumHi,SP      ; "
            ldy      ii,SP          ; "
            leas     −8,SP          ; "
            pshd                    ; " (push pNumLo)
            pshx                    ; " (push pNumHi)
            pshy                    ; " (push ii)
            movw     #0,2,−SP       ; " (push 0 to the stack)
            jsr      div32          ; " (call the divide subroutine)
            leas     14,SP          ; deallocate the space used by outgoing parameters
            puld                    ; get the lower 2 bytes of the remainder
            cpd      #0             ; is remainder equal to 0?
            beq      nonPRI         ; If remainder equals 0, then Num is not a prime
            ldd      ii,SP          ; test divide the next higher integer
            addd     #1             ; "
            std      ii,SP          ; "
            bra      divLoop
isPRI       ldaa     #1
            bra      exitPT
nonPRI      ldaa     #0
exitPT      leas     pLocal,SP
            puly
            pulx
            rts
            #include "c:\miniide\findsqr.asm"
            #include "c:\miniide\div32.asm"
;           org      $FFFE          ; uncomment this line for CodeWarrior
;           dc.w     start          ; uncomment this line for CodeWarrior
            end
```

4.8 Using the D-Bug12 Functions to Perform I/O Operations

The D-Bug12 monitor provides a few subroutines to support I/O operations. One can utilize these I/O routines to facilitate program developments on a demo board that contains the D-Bug12 monitor. A summary of these routines is in Table 4.2.

These user-accessible routines are written in C. All except the first parameter are passed to the user-callable functions on the stack. Parameters must be pushed onto the stack in the reverse order (right to left) so that they are listed in the function declaration. The first parameter is passed to the function in accumulator D. If a function has only a single parameter, then the parameter is passed in accumulator D. Parameters of type *char* must be converted to an *integer* (16-bit). Parameters of type *char* will occupy the lower-order byte (higher address) of a word pushed onto the stack or accumulator B if the parameter is passed in D.

Parameters pushed onto the stack before the function is called remain on the stack when the function returns. It is the responsibility of the *caller* to remove passed parameters from the stack.

All 8- and 16-bit function values are returned in accumulator D. A value of type *char* returned in accumulator D is located in the 8-bit accumulator B. A *Boolean* function returns a zero value for false and a nonzero value for true.

None of the CPU register contents, except the stack pointer, are preserved by the called functions. If any of the register values need to be preserved, they should be pushed onto the stack before any of the parameters have been pushed and restored after deallocating the parameters.

Subroutine	Function	Pointer Address
far main ()	Start of D-Bug12	$EE80
getchar ()	Get a character from SCI0 or SCI1	$EE84
putchar ()	Send a character out to SCI0 or SCI1	$EE86
printf ()	Formatted string output-translates binary values to string	$EE88
far GetCmdLine ()	Get a line of input from the user	$EE8A
far sscanhex()	Convert ASCII hex string to a binary integer	$EE8E
isxdigit ()	Check if a character (in B) is a hex digit	$EE92
toupper()	Convert lowercase characters to uppercase	$EE94
isalpha ()	Check if a character is alphabetic	$EE96
strlen ()	Returns the length of a NULL-terminated string	$EE98
strcpy ()	Copy a NULL-terminated string	$EE9A
far out2hex ()	Output 8-bit number as two ASCII hex characters	$EE9C
far out4hex ()	Output a 16-bit number as four ASCII hex characters	$EEAO
SetUserVector ()	Set up a vector to a user's interrupt service routine	$EEA4
far WriteEEByte()	Write a byte to the on-chip EEPROM memory	$EEA6
far EraseEE ()	Bulk erase the on-chip EEPROM memory	$EEAA
far ReadMem ()	Read data from the HCS12 memory map	$EEAE
far WriteMem ()	Write data to the HCS12 memory map	$EEB2

Table 4.2 ■ D-Bug12 monitor (version 4.x.x) routines

4.8.1 Calling D-Bug12 Functions from Assembly Language

Calling the functions from assembly language is a simple matter of pushing the parameters onto the stack in the proper order and loading the first or only function parameter into accumulator D. The function can then be called with a jsr instruction. The code following the jsr instruction should remove any parameters pushed onto the stack. If a single parameter was pushed onto the stack, a simple pulx or puly instruction is one of the most efficient ways to remove the parameter from the stack. If two or more parameters are pushed onto the stack, the leas instruction is the most efficient way to remove the parameters. Any of the CPU registers that were saved on the stack before the function parameters should be restored with a corresponding pull instruction.

For example, the WriteEEByte() function has two parameters: The first parameter is the address of the memory location to which the data is to be written; the second is the data itself. An example of the instruction sequence to call this function to write the value #$55 into EEPROM is as follows:

```
WriteEEByte      equ        $EEA6
                 .
                 .
                 ldd        #$55                 ; write $55 to EEPROM
                 pshd
                 ldd        #EEAddress           ; EEPROM address to write data
                 jsr        [WriteEEByte,pcr]    ; call the routine
                 leas       2,sp                 ; remove the parameter from stack
                 beq        EEWError             ; 0 return value means error
                 .
                 .
```

The addressing mode used in the jsr instruction of this example is a form of indexed indirect addressing that uses the program counter as an index register. The PCR mnemonic used in place of an index register name stands for *program counter relative* addressing. In reality, the HCS12 does not support PCR. Instead, the PCR mnemonic is used to instruct the assembler to calculate an offset to the address specified by the label WriteEEByte. The offset is calculated by subtracting the value of the PC at the address of the first object code byte of the next instruction (in this example, leas 2,sp) from the address supplied in the indexed offset field (WriteEEByte). When the jsr instruction is executed, the opposite occurs. The HCS12 adds the value of the PC at the first object code byte of the next instruction to the offset embedded in the instruction object code. The indirect addressing, indicated by the square brackets, specifies that the address calculated as the sum of the index register (in this case the PC) and the 16-bit offset contains a pointer to the destination of the jsr.

The MiniIDE software supports this syntax. However, if you are using an assembler that does not support program-counter-relative-indexed addressing, the following two-instruction sequence can be used:

```
        ldx        WriteEEByte        ; load the address of WriteEEByte()
        jsr        0,x                ; call the subroutine
```

If the name of a library function is preceded by the keyword *far*, then it is located in the expanded memory and must be called by using the call instruction. Other library functions can be called by executing the jsr instruction.

4.8.2 Descriptions of Callable Functions

For each of the callable functions, the prototype declaration and the pointer address (where the starting address of the function is stored) are listed.

void far main (void);
Pointer address: $EE80

This function simply restarts the D-Bug12 monitor, which will reinitialize all of D-Bug12's internal tables and variables. Any previously set breakpoints will be lost. This function will not be useful to user application development.

int getchar (void);
Pointer address: $EE84

This function retrieves a single character from the control terminal SCI. If an unread character is not available in the receive data register when this function is called, it will wait until one is received. Because the character is returned as an integer, the 8-bit character is placed in accumulator B.

Adding the following instruction sequence in your program will read a character from the SCI0 port:

```
getchar      equ     $EE84
             ...
             jsr     [getchar,PCR]
             ...
```

int putchar(int);
Pointer address: $EE86

This function outputs a single character to the control terminal SCI. If the control SCI's transmit data register is full when the function is called, putchar() will wait until the transmit data register is empty before sending the character. No buffering of characters is provided. Putchar() returns the character that was sent. However, it does not detect any error conditions that may occur in the process and therefore will never return EOF (end of file). Adding the following instruction sequence in your program will output the character A to the control SCI (when the program is running on a demo board, the character A will be displayed on the monitor screen):

```
putchar      equ     $EE86
             ...
             ldd     #'A'
             jsr     [putchar,PCR]
             ...
```

int printf(char *format, . . .);
Pointer address: $EE88

This function is used to convert, format, and print its arguments on the standard output (the output device could be the monitor screen, printer, LCD, etc.) under the control of the format string pointed to by *format*. It returns the number of characters that were sent to standard output (sent through serial port SCI0). All except floating-point data types are supported.

The format string can contain two basic types of objects: ASCII characters that are copied directly from the format string to the display device, and conversion specifications that cause succeeding printf() arguments to be converted, formatted, and sent to the display device. Each conversion specification begins with a percent sign (%) and ends with a single conversion character. Optional formatting characters may appear between the percent sign and end with a single conversion character in the following order:

[−] [<FieldWidth>] [.] [<Precision>] [h | l]

These optional formatting characters are explained in Table 4.3.

The *FieldWidth*, or *Precision*, field may contain an asterisk (*) character instead of a number. The asterisk will cause the value of the argument in the argument list to be used instead.

Character	Description
− (minus sign)	Left justifies the converted argument.
FieldWidth	Integer number that specifies the minimum field width for the converted argument. The argument will be displayed in a field at least this wide. The displayed argument will be padded on the left or right if necessary.
. (period)	Separates the field width from the precision.
Precision	Integer number that specifies the maximum number of characters to display from a string or the minimum number of digits for an integer.
h	To have an integer displayed as a short.
l (letter ell)	To have an integer displayed as a long.

Table 4.3 ■ Optional formatting characters

The formatting characters supported by the printf() function are listed in Table 4.4. If the conversion character following the percent sign is not one of the formatting characters shown in this table or the characters shown in Table 4.3, the behavior of the printf() function is undefined.

The printf() function can be used to print a message. One example is as follows:

```
CR          equ       $0D
LF          equ       $0A
printf      equ       $EE88
            ...
            ldd       #prompt
            jsr       [printf,PCR]
            ...
prompt      db        "Flight simulation",CR,LF,0
```

This instruction sequence will cause the message *Flight simulation* to be displayed, and the cursor will be moved to the beginning of the next line.

Character	Argument Type; Displayed As
d, i	int; signed decimal number
o	int; unsigned octal number (without a leading zero)
x	int; unsigned hex number using abcdef for 10, . . . , 15
X	int; unsigned hex number using ABCDEF for 10, . . . , 15
u	int; unsigned decimal number
c	int; single character
s	char *; display from the string until a '\0' (NULL)
p	void *; pointer (implementation-dependent representation)
%	no argument is converted; print a %

Table 4.4 ■ Printf() conversion characters

Suppose labels *m*, *n*, and *gcd* represent three memory locations and the memory locations (2 bytes) starting with the label *gcd* hold the greatest common divisor of two numbers stored at memory locations starting with labels *m* and *n*. By adding the following instruction sequence, the relationship among *m*, *n*, and *gcd* can be displayed on the PC monitor screen when the program is executed on a demo board with D-Bug12 monitor:

```
CR          equ     $0D
LF          equ     $0A
printf      equ     $EE88
            ...
            ldd     gcd
            pshd
            ldd     n
            pshd
            ldd     m
            pshd
            ldd     #prompt
            jsr     [printf,PCR]
            leas    6,sp
            ...
promptdb            "The greatest common divisor of %d and %d is %d",CR,LF,0
            ...
```

This example is probably not easy to follow. In reality, this example is equivalent to the following C statement:

```
printf("The greatest common divisor of %d and %d is %d\n", m, n, gcd);
```

This function call has four parameters: the first parameter is a pointer to the formatting string and should be placed in D (done by the ldd #prompt instruction). The other parameters (*m*, *n*, and *gcd*) should be pushed into the stack in the order from right to left. There are three formatting characters corresponding to these three variables to be output.

int far GetCmdLine(char *CmdLineStr, int CmdLineLen);

Pointer address: $EE8A

This function is used to obtain a line of input from the user. GetCmdLine() accepts input from the user one character at a time by calling getchar(). As each character is received it is echoed back to the user terminal by calling putchar() and placed in the character array pointed to by **CmdLineStr**. A maximum of CmdLineLen − 1 characters may be entered. Only printable ASCII characters are accepted as input with the exception of the ASCII backspace character ($08) and the ASCII carriage return ($0D). All other nonprintable ASCII characters are ignored by the function.

The ASCII backspace character ($08) is used by the GetCmdLine() function to delete the previously received character from the command line buffer. When GetCmdLine() receives the backspace character, it will echo the backspace to the terminal, print the ASCII space character ($20), and then send a second backspace character to the terminal device. At the same time, the character is deleted from the command line buffer. If a backspace character is received when there are no characters in CmdLineStr, the backspace character is ignored.

The reception of an ASCII carriage return character ($0D) terminates the reception of characters from the user. The carriage return is not placed in the command line buffer. Instead, a NULL character ($00) is placed in the next available buffer location.

Before returning, all the entered characters are converted to uppercase. GetCmdLine()
always returns an error code of **noErr.**

We can use this function to request the user to enter a string from the keyboard when running programs on a demo board with the D-Bug12 monitor. Usually, the program would output
a message so that the user knows when to enter the string. The following instruction sequence
will ask the user to enter a string from the keyboard:

```
printf          equ         $EE88
GetCmdLine      equ         $EE8A
cmdlinelen      equ         100
CR              equ         $0D
LF              equ         $0A
                ...
prompt          db          "Please enter a string: ",CR,LF,0
                ...
inbuf           ds.b        100
                ...
                ldd         #prompt                 ; output a prompt to remind the user to
                jsr         [printf,PCR]            ; enter a string
                ldd         #cmdlinelen             ; push the CmdLineLen
                pshd                                ;          "
                ldd         #inbuf
                call        [GetCmdLine,PCR]        ; read a string from the keyboard
                puld                                ; clean up the stack
```

char * far sscanhex(char *HexStr, unsigned int *BinNum);
Pointer address: $EE8E

The sscanhex() function is used to convert a hex string to a binary number. The hex string
pointed to by **HexStr** may contain any number of ASCII hex characters. However, the converted
value must be no greater than $FFFF. The string must be terminated by either a space or a
NULL character.

Suppose one has entered an ASCII string that represents a number, then the following
instruction sequence will convert it into a hex number (or binary number):

```
sscanhex        equ         $EE8E
                ...
HexStr          ds.b        10                      ; input buffer to hold a hex string
BinNum          ds.b        2                       ; to hold the converted number
                ...
                ldd         #BinNum
                pshd
                ldd         #HexStr
                call        [sscanhex,PCR]
                leas        2,sp                    ; deallocates space used by outgoing parameters
                ...
```

int isxdigit(int c);
Pointer address: $EE92

The isxdigit() function tests the character passed in **c** for membership in the character set
[0..9, a..f, A..F]. If the character c is in this set, the function returns a nonzero (true) value.

Otherwise, a value of zero is returned. The following instruction sequence illustrates the use of this function:

```
isxdigit          equ         $EE92
                  ...
c_buf             ds.b        20              ; buffer that holds data to be validated
                  ...
                  clra                        ; clear accumulator A
                  ldab        c_buf           ; get one character
                  jsr         [isxdigit,PCR]

                  ...
```

int toupper(int c);
Pointer address: $EE94

If c is a lowercase character, toupper() will return the corresponding uppercase letter. If the character is in uppercase, it simply returns c. The following instruction utilizes this function to convert a character contained in B to uppercase:

```
toupper           equ         $EE94
                  ...
c_buf             ds.b        20              ; buffer that contains string to be converted to
                  ...                         ; uppercase
                  ldab        c_buf           ; get one character to convert
                  clra
                  jsr         [toupper,PCR]

                  ...
```

int isalpha(int c);
Pointer address: $EE96

This function tests the character passed in c for membership in the character set [a..z, A..Z]. If the character c is in this set, the function returns a nonzero value. Otherwise, it returns a zero. This function would also be useful for validating an input string. The following instruction sequence illustrates the use of this function:

```
isalpha           equ         $EE96
                  ...
c_buf             ds.b        20
                  ...
                  ldab        c_buf
                  clra
                  jsr         [isalpha,PCR]   ; check whether the character in B is alphabetic

                  ...
```

unsigned int strlen(const char *cs);
Pointer address: $EE98

The strlen() function returns the length of the string pointed to by cs. The following instruction sequence counts the number of characters contained in the string pointed to by cs:

```
strlen            equ         $ EE98
                  ...
cs                db          "....."
                  ...
```

```
                ldd        #cs
                jsr        [strlen,PCR]
                ...
```

char *strcpy(char *s1, char *s2);
Pointer address: $EE9A

This function copies the string pointed to by **s2** into the string pointed to by **s1** and returns a pointer to s1. The following instruction sequence copies the string pointed to by s1 to the memory location pointed to by s2:

```
strcpy          equ        $EE9A
                ...
s1              db         "......"
s2              ds.b       ...

                ...
                ldd        #s2
                pshd
                ldd        #s1
                jsr        [strcpy,PCR]
                leas       2,sp
                ...
```

void far out2hex(unsigned int num);
Pointer address: $EE9C

This function displays the lower byte of **num** on the terminal screen as two hex characters. The upper byte of num is ignored. The function out2hex() simply calls printf() with a format string of "%2.2X."

The following instruction sequence outputs the number in accumulator B as two hex digits to the terminal screen:

```
out2hex         equ        $EE9C
                ...
data            ds.b       20

                ...
                ldab       data
                clra
                call       [out2hex,PCR]
                ...
```

void far out4hex(unsigned int num);
Pointer address: $EEA0

This function displays num on the control terminal as four hex characters. The function out4hex() simply calls printf() with a format string of "%4.4X." The following instruction sequence outputs the 16-bit number stored at memory location num as four hex digits:

```
out4hex         equ        $EEA0
                ...
num             db         ...

                ...
                ldd        num
                call       [out4hex,PCR]
                ...
```

int SetUserVector(int VectNum, Address UserAddress);
Pointer address: $EEA4

This function will be discussed in Chapter 6.

Boolean far WriteEEByte (Address EEAddress, Byte EEData);
Pointer address: $EEA6

The WriteEEByte() function provides a mechanism to program individual bytes of the on-chip EEPROM. It does not perform any range checking on the passed **EEAddress.** A user's program can determine the start address and size of the on-chip EEPROM array by examining the data contained in the custom data area fields **CustData.EEBase** and **CustData. EESize.**

A byte-erase operation is performed before the programming operation and a verification operation is performed after the programming operation. If the EEPROM data does not match EEData, false (0 value) is returned by the function.

Int far EraseEE(void);
Pointer address: $EEAA

This function performs a bulk erase operation to the on-chip EEPROM without having to manipulate the EEPROM programming control registers. After the bulk erase operation is performed, the memory range described by CustData.EEBase and CustData.EESize is checked for erasure. If any of the bytes does not contain 0xFF, a nonzero error code is returned.

int far ReadMem(Address StartAddress, Byte *MemDataP, unsigned int NumBytes);
Pointer address: $EEAE

This function is used internally by D-Bug12 for all memory read access.

int WriteMem(Address StartAddress, Byte *MemDataP, unsigned int NumBytes);
Pointer address: $EEB2

The WriteMem() function is used internally by D-Bug12 for all memory write accesses. If a byte is written to the memory range described by CustData.EEBase and CustData.EESize, WriteMem() calls the WriteEEByte() function to program the data into the on-chip EEPROM memory. A nonzero error code is returned if a problem occurs while writing to target memory.

4.8.3 Using the D-Bug12 Functions

A useful program usually consists of several functions. An example is given in this section.

Example 4.10

Write a program that invokes appropriate functions to find the prime number between 1000 and 2000. Output eight prime numbers in one line. To do this, we need to

1. Invoke the **PrimeTest** subroutine to test if an integer is a prime. The PrimeTest subroutine will in turn call subroutines **FindSqr** and **div32.**

2. Invoke the printf() function to output the prime number.

3. Write a loop to test all the integers between 1000 and 2000.

Solution: The logic structure of the program is as follows:

Step 1

Output the message "The prime numbers between 1000 and 2000 are as follows:".

Step 2

For every number between 1000 and 2000, do the following:

1. Call the **PrimeTest()** subroutine to see if it is a prime.

2. Output the number (call printf()) if it is a prime.

3. If there are already eight prime numbers in the current line, then also output a carry return.

The assembly program is as follows:

```
CR       equ      $0D
LF       equ      $0A

upper    equ      2000        ; upper limit for testing prime
lower    equ      1000        ; lower limit for testing prime
printf   equ      $EE88       ; location where the address of printf() is stored

         org      $1000
out_buf  ds.b     10
PRIcnt   ds.b     1           ; prime number count
k        ds.b     2
tmp      ds.b     2
         org      $1500
start    ldx      #upper
         stx      tmp
         pshx
         ldx      #lower
         stx      k           ; initialize k to 1000 for prime testing
         pshx
         ldd      #form0
         jsr      [printf,PCR]
         leas     4,sp
         clr      PRIcnt
again    ldd      k
         cpd      #upper
         bhi      Pstop       ; stop when k is greater than upper
         pshd
         ldd      #0
         pshd
         jsr      primetest   ; test if k is prime
         leas     4,sp        ; deallocate space used by outgoing parameters
         tsta
         beq      next_k      ; test next integer if k is not prime
         inc      PRIcnt      ; increment the prime count
         ldd      k
         pshd
         ldd      #form1
         jsr      [printf,PCR]  ; output k
         leas     2,sp
```

```
              ldaa      PRIcnt
              cmpa      #8              ; are there eight prime numbers in the current line?
              blo       next_k
; output a CR, LF if there are already eight prime numbers in the current line
              ldd       #form2
              jsr       [printf,PCR]
              clr       PRIcnt
next_k        ldx       k
              inx
              stx       k
              lbra      again
; stop        bra       $               ; uncomment this line for CodeWarrior
Pstop         swi                       ; comment this line for CodeWarrior
              #include "c:\miniide\primetest.asm"
form0         db        CR,LF,"The prime numbers between %d and %d are as follows: "
              db        CR,LF,CR,LF,0
form1         db        " %d ",0
form2         db        " ",CR,LF,0
;             org       $FFFE           ; uncomment this line for CodeWarrior
;             dc.w      start           ; uncomment this line for CodeWarrior
              end
```

The program execution output should look like the following:

```
>load
**
>g 1500
```

The prime numbers between 1000 and 2000 are as follows:

1009	1013	1019	1021	1031	1033	1039	1049
1051	1061	1063	1069	1087	1091	1093	1097
1103	1109	1117	1123	1129	1151	1153	1163
1171	1181	1187	1193	1201	1213	1217	1223
1229	1231	1237	1249	1259	1277	1279	1283
1289	1291	1297	1301	1303	1307	1319	1321
1327	1361	1367	1373	1381	1399	1409	1423
1427	1429	1433	1439	1447	1451	1453	1459
1471	1481	1483	1487	1489	1493	1499	1511
1523	1531	1543	1549	1553	1559	1567	1571
1579	1583	1597	1601	1607	1609	1613	1619
1621	1627	1637	1657	1663	1667	1669	1693
1697	1699	1709	1721	1723	1733	1741	1747
1753	1759	1777	1783	1787	1789	1801	1811
1823	1831	1847	1861	1867	1871	1873	1877
1879	1889	1901	1907	1913	1931	1933	1949
1951	1973	1979	1987	1993	1997	1999	User Bkpt Encountered

```
PP  PC       SP        X         Y        D = A:B      CCR = SXHI NZVC
30 1560      1500      07D1      15DE     07:D1              1001 0000
xx:1560      34                  PSHX
>
```

4.9 Subroutines for Creating Time Delay

We have learned how to create a time delay using program loops. These program loops can be converted into subroutines so that they can be called from any place in a program.

Example 4.11

Write a subroutine that can create a time delay of 100 ms.

Solution: By adding an rts instruction to the end of the instruction sequence of Example 2.25, a subroutine that creates a delay of 100 ms can be created.

```
delay100ms
          pshx
          ldx       #60000      ; 2 E cycles
iloop     psha                  ; 2 E cycles
          pula                  ; 3 E cycles
          psha                  ; 2 E cycles
          pula                  ; 3 E cycles
          psha                  ; 2 E cycles
          pula                  ; 3 E cycles
          psha                  ; 2 E cycles
          pula                  ; 3 E cycles
          psha                  ; 2 E cycles
          pula                  ; 3 E cycles
          psha                  ; 2 E cycles
          pula                  ; 3 E cycles
          psha                  ; 2 E cycles
          pula                  ; 3 E cycles
          nop                   ; 1 E cycle
          nop                   ; 1 E cycle
          dbne      x,iloop     ; 3 E cycles
          pulx
          rts
```

This subroutine will be more useful if it is made parameterized; that is, the delay time can be set to be a multiple of 100 ms instead of a fixed value. The following subroutine can generate a time delay that is a multiple of 100 ms with the multiple passed in Y:

```
delayby100ms
          pshx
eloop3    ldx       #60000      ; 2 E cycles
iloop3    psha                  ; 2 E cycles
          pula                  ; 3 E cycles
          psha                  ; 2 E cycles
          pula                  ; 3 E cycles
          psha                  ; 2 E cycles
          pula                  ; 3 E cycles
          psha                  ; 2 E cycles
          pula                  ; 3 E cycles
          psha                  ; 2 E cycles
```

```
      pula              ; 3 E cycles
      psha              ; 2 E cycles
      pula              ; 3 E cycles
      psha              ; 2 E cycles
      pula              ; 3 E cycles
      nop               ; 1 E cycle
      nop               ; 1 E cycle
      dbne    x,iloop3  ; 3 E cycles
      dbne    y,eloop3  ; 3 E cycles
      pulx
      rts
```

▲

The subroutines that can generate a delay that is a multiple of an other value (for example, 1 ms or 10 ms) can be created in the same manner. The delay.asm file in the complementary CD provides the following delay subroutines that the user can call to generate delays that are a multiple of 50 μs, 1 ms, 10 ms, and 100 ms with the multiple also passed in Y:

- delayby50μs
- delayby1ms
- delayby10ms
- delayby100ms

4.10 Introduction to Parallel I/O Port and Simple I/O Devices

So far we have considered how to write general-purpose assembly programs which do not involve any microcontroller peripheral functions. Dealing with peripheral functions and I/O devices will add great fun to the learning of assembly language programming. The purpose of this section is to introduce the parallel I/O port, simple I/O devices, and their applications.

An HCS12 device may have from 48 to 144 signal pins arranged in 3 to 12 I/O ports. Most signal pins serve multiple functions. When a pin is not used by any peripheral function, it can be used for general-purpose input/output such as driving an LED or a switch. An I/O port consists of a set of I/O pins and the registers required to control its operations. The I/O pins associated with an I/O port may be configured for input or output and can be used to output values to **LEDs or LCDs** or input values from DIP switches or keypad directly.

The pin assignment of the HCS12 D subfamily is shown in Figure 4.14. Most I/O ports have eight pins although a few do not. A summary of the HCS12 I/O ports is listed in Table 4.5.

4.10.1 Addressing the I/O Port Register

To perform input and output, the CPU sends data to or from a register instead of dealing with I/O pins. Each register is assigned an address. The user accesses a register by specifying the address assigned to it. For example, the Port A data register is assigned to address 0. The following instruction will output the value $35 to Port A:

```
      movb    #$35,0    ; address 0 is Port A data register
```

However, it is quite difficult to read the program that performs I/O operations that refers to the address of a port data register. Instead, the user can use an equ directive to make the instruction more readable:

```
PTA   equ     0
      movb    #$35,PTA
```

The HCS12 assembler collects all the equate directives into an *include file* called *hcs12.inc* so that the user can access any peripheral register by referring to an appropriate symbolic name. In order to use symbolic names to access peripheral registers, the user should add the following statement to his or her program, preferably as the first line of the program:

#include "c:\...\hcs12.inc"   ; specify the directory that contains the hcs12.inc file

Although different assemblers may provide different include files, this author has prepared an include file (hcs12.inc) in the complementary CD that enables the user to access peripheral registers and their bits by referring to their names.

4.10.2 I/O Port Direction Configuration

Each I/O port has a data direction register (DDRx, x is the port name) that allows the user to configure data transfer direction. Setting a bit in the DDRx register to 1 configures the associated pin for output whereas setting a bit in DDRx to 0 configures the associated pin for input. For example, the instruction

    movb   #$FF,DDRA   ; configure Port A for output

configures Port A for output.

    movb   #0,DDRA   ; configure Port A for input

configures Port A for input.

The user can configure a few pins of an I/O port for input and the remaining pins for output. For example, the instruction

    movb   #$55,DDRB

configures the even-numbered pins of Port B for output and the odd-numbered pins for input.

Port Name	No. of Pins	Pin Name
A	8	PA7~PA0
B	8	PB7~PB0
E	8	PE7~PE0
H	8	PH7~PH0
J	4	PJ7~PJ0
K	7	PK4~PK0
M	8	PM7~PM0
P	8	PP7~PP0
S	8	PS3~PS0
T	8	PT7~PT0
PAD1, PAD0	16	PAD15~PAD0
L	8	PL7~PL0
U	8	PU7~PU0
V	8	PV7~PV0
W	8	PW7~PW0

Table 4.5 ■ Number of pins available in each parallel port

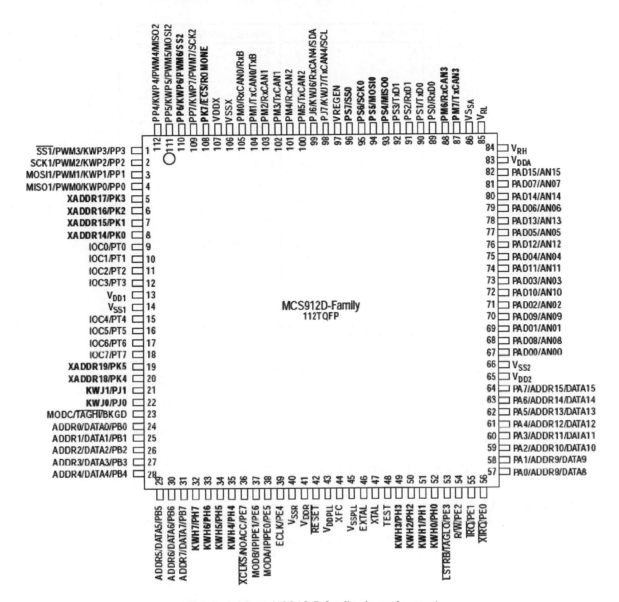

Figure 4.14 ■ HCS12 D family pin assignment

4.10.3 I/O Port Data Register

When outputting data, write the value to the port data register. When inputting data, read from the port data register. For example, the following instruction outputs $12 to Port B:

```
movb      #$12,PTB
```

Port Name	Data Register Name
A	PORTA[1]
B	PORTB[1]
E	PORTE[1]
K	PORTK[1]
H	PTH
J	PTJ
M	PTM
P	PTP
S	PTS
T	PTT
PAD1, PAD0	PORTAD0, PORTAD1
L	PTL[2]
U	PTU[2]
V	PTV[2]
W	PTW[2]

Note: 1. PORTA, PORTB, PORTE, and PORTK are also
referred to as PTA, PTB, PTE, and PTK in the hcs12.inc
and hcs12.h files
2. Port L, U, V, and W are available in H-family devices only

Table 4.6 ■ HCS12 parallel I/O port data register names

The following instruction reads a value from Port E and places the value in accumulator A:

ldaa   PTE

The HCS12 port data register names are listed in Table 4.6.

4.11 Simple I/O Devices

Many embedded systems only require simple output devices such as switches, light-emitting diodes, keypads, and seven-segment displays.

4.11.1 Interfacing with LEDs

The LED is often used to indicate the system operation mode, whether power is turned on, whether system operation is normal, and so forth. An LED can illuminate when it is forward biased and has sufficient current flowing through it. The current required to light an LED may range from a few to more than 10 mA. The forward voltage drop across the LED can range from about 1.6 V to more than 2.2 V.

Figure 4.15 suggests three methods for interfacing with LEDs. Methods (a) and (b) are recommended for use with LEDs that need only 1 or 2 mA to produce enough brightness. The circuit (c) is recommended for use with LEDs that need larger current to light. Resistors R_1, R_2, and R_3 are referred to as *current-limiting resistors* because they set the magnitude of the current flowing through the LED. The current-limiting resistor for circuits (a) and (b) should be larger (between 1.5 kΩ to 2 kΩ).

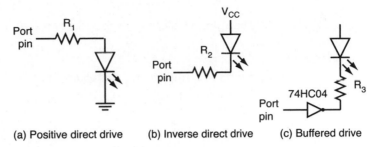

(a) Positive direct drive   (b) Inverse direct drive   (c) Buffered drive

Figure 4.15 ■ An LED connected to a CMOS inverter through a current-limiting resistor

Example 4.12

Use the HCS12 Port B to drive eight LEDs. Light each of them for half a second in turn and repeat, assuming that the HCS12 has a 24-MHz E-clock. When the Port J pin 1 is low, the LEDs are enabled to light. This circuit is used in the Dragon12 demo board.

Solution: The circuit for driving eight LEDs using Port B is shown in Figure 4.16.

To turn on the LED driven by pins PB7~PB0 one at a time, we output the values $80, $40, . . . , and $01 to Port B in turn. The procedure to achieve the desired LED display pattern is as follows:

Step 1
Place the values $80, $40, . . . , and $01 in a table. Use the index register X to point to the start of this table. Pull the PJ1 pin low to enable LEDs to light.

Step 2
Output the value pointed to by X to Port B. Increment the pointer X.

Step 3
Wait for half a second.

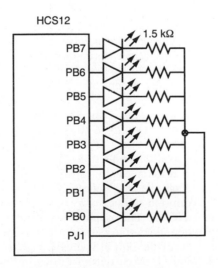

Figure 4.16 ■ Circuit connection for Example 4.12

Step 4

If X points to the end of the table, reset X to point to the start of the table.

Step 5

Go to step 2.

The assembly program that implements this algorithm is as follows:

```
                #include "C:\miniide\hcs12.inc"
                org       $1000
lpcnt           ds.b      1
                org       $1500
start           movb      #$FF,DDRB        ; configure port B for output
                bset      DDRJ,$02         ; configure PJ1 pin for output
                bclr      PTJ,$02          ; enable LEDs to light
forever         movb      #16,lpcnt        ; initialize LED pattern count
                ldx       #led_tab         ; Use X as the pointer to LED pattern table
led_lp          movb      1,x+,PORTB       ; turn on one LED
                ldy       #5               ; wait for half a second
                jsr       delayby100ms     ; "
                dec       lpcnt            ; reach the end of the table yet?
                bne       led_lp
                bra       forever          ; start from beginning
led_tab         dc.b      $80,$40,$20,$10,$08,$04,$02,$01
                dc.b      $01,$02,$04,$08,$10,$20,$40,$80
                #include  "C:\miniide\delay.asm"
;               org       $FFFE            ; uncomment this line for CodeWarrior
;               dc.w      start            ; uncomment this line for CodeWarrior
                end
```

4.11.2 Interfacing with Seven-Segment Displays

Seven-segment displays are often used when the embedded product needs to display only a few digits. Seven-segment displays are mainly used to display decimal digits and a small subset of letters.

Although an HCS12 device has enough current to drive a seven-segment display, it is not advisable to do so when an HCS12-based embedded product needs to drive many other I/O devices. In Figure 4.17, Port B drives a common-cathode seven-segment display through the buffer chip 74HC244. The V_{OH} (output high voltage) value of the 74HC244 is about 5 V. Adding a 470-Ω resistor will set the display segment current to about 6.4 mA, which should be sufficient to light an LED segment. The light patterns corresponding to 10 BCD digits are shown in Table 4.7. Depending on how segments a–g are connected to the I/O pins there are two different values for displaying the same decimal digit. In the Dragon12 demo board, segments a–g are connected to from the pin PB0 to pin PB6.

When an application needs to display multiple BCD digits, the time-multiplexing technique is often used. An example of the circuit that displays six BCD digits is shown in Figure 4.18. In Figure 4.18, the common cathode of a seven-segment display is connected to the Y output of the hex buffer chip 74HC367. The output low voltage (V_{OL}) of a 74HC367 output pin (Y_i, $i = 0, \ldots, 5$) is about 0.33 V when the output current (the current actually flowing into the pin) is larger than a few mA. When a port B pin output voltage is high, an LED segment current will be about 3.33 mA [$\approx (5 - 2 - 0.33)/800$]. Total current that flows into the Y_i pin is about 23.3 mA (= 7 × 3.33) and is within the rated value of the DC output current.

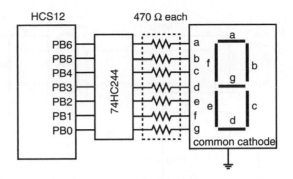

Figure 4.17 ■ Driving a single seven-segment display

Decimal Digit	Segments							Corresponding Hex Number	
	a	b	c	d	e	f	g	Figure 4.17 Circuit	Dragon12 Demo Board
0	1	1	1	1	1	1	0	$7E	$3F
1	0	1	1	0	0	0	0	$30	$06
2	1	1	0	1	1	0	1	$6D	$5B
3	1	1	1	1	0	0	1	$79	$4F
4	0	1	1	0	0	1	1	$33	$66
5	1	0	1	1	0	1	1	$5B	$6D
6	1	0	1	1	1	1	1	$5F	$7D
7	1	1	1	0	0	0	0	$70	$07
8	1	1	1	1	1	1	1	$7F	$7F
9	1	1	1	1	0	1	1	$7B	$6F

Table 4.7 ■ Decimal to seven-segment decoder

The circuit in Figure 4.18 can display up to six digits by utilizing the time-multiplexing technique, in which each seven-segment display is lighted in turn briefly and then turned off. When one display is lighted, all other displays are turned off. Within one second, each seven-segment display is lighted and then turned off many times. Because of the *persistence of vision*, all six displays will appear to be lighted simultaneously.

Example 4.13

Write an instruction sequence to display 7 on the seven-segment display #5 in Figure 4.18.

Solution: To display 7 on display #5, we need to
- Output the hex value $07 to Port B
- Set the PP5 pin to low
- Set pins PP4 through PP0 to high

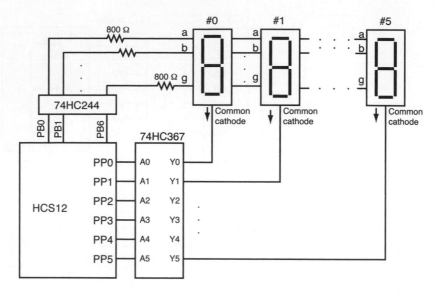

Figure 4.18 ■ Port B and port P together drive six seven-segment displays (MC9S12DG256)

The instruction sequence is as follows:

```
        #include    "c:\miniide\hcs12.inc"
seven   equ         $07
        ...
        movb        #$FF,DDRB
        movb        #$3F,DDRP
        movb        #$1F,PTP        ; enable display #5 to light
        movb        #seven,PTB      ; send out the segment pattern of 7
```

Example 4.14

Write a program to display 123,456 on the six seven-segment displays shown in Figure 4.18.

Solution: The digits 1, 2, 3, 4, 5, and 6 are displayed on display #5, #4, . . . , and #0, respectively. The values to be output to Port B and Port P to display one digit at a time are shown in Table 4.8. This table can be created by the following assembler directives:

```
display   dc.b      $06,$1F        ; value to display 1 on seven-segment display #5
          dc.b      $5B,$2F        ; value to display 2 on seven-segment display #4
          dc.b      $4F,$37
          dc.b      $66,$3B
          dc.b      $6D,$3D
          dc.b      $7D,$3E
```

The algorithm for displaying 123,456 on the six seven-segment displays in Figure 4.18 is as follows:

Step 1
Set ptr to point to the first byte of the display table.

Seven-Segment Display	Displayed BCD Digit	Port B	Port P
#5	1	$06	$1F
#4	2	$5B	$2F
#3	3	$4F	$37
#2	4	$66	$3B
#1	5	$6D	$3D
#0	6	$7D	$3E

Table 4.8 ■ Table of display patterns for Example 4.14

Step 2
Output the byte pointed to by ptr to Port B and then increment ptr by 1.

Step 3
Output the byte pointed to by ptr to Port P and then increment ptr by 1.

Step 4
Wait for 1 ms.

Step 5
If (ptr == (display + 12)), then go to step 1;
else, go to Step 2.

The assembly program that implements this algorithm is as follows:

```
            #include    "c:\miniide\hcs12.inc"
            org         $1500
start       lds         #$1500
            movb        #$FF,DDRB
            movb        #$3F,DDRP
forever     ldx         #DispTab        ; set X to point to the display table
loopi       movb        1,x+,PTB        ; output segment pattern
            movb        1,x+,PTP        ; output display select
            ldy         #1
            jsr         delayby1ms      ; wait for 1 ms
            cpx         #DispTab + 12   ; reach the end of the table?
            bne         loopi
            bra         forever
            #include    "c:\miniide\delay.asm"
DispTab     dc.b        $06,$1F
            dc.b        $5B,$2F
            dc.b        $4F,$37
            dc.b        $66,$3B
            dc.b        $6D,$3D
            dc.b        $7D,$3E
;           org         $FFFE           ; uncomment this line for CodeWarrior
;           dc.w        start           ; uncomment this line for CodeWarrior
            end
```

4.11.3 Generating a Digital Waveform Using I/O Pins

A periodic digital waveform can be easily generated by manipulating I/O pin voltage level and inserting an appropriate delay between the two voltage levels. For example, a 1-kHz periodic square wave can be generated from the PT5 pin using the following algorithm:

Step 1
Configure the PT5 pin for output.

Step 2
Pull the PT5 pin to high.

Step 3
Wait for 0.5 ms.

Step 4
Pull the PT5 pin to low.

Step 5
Wait for 0.5 ms.

Step 6
Go to step 2.

Example 4.15

Write an assembly program to generate a 1-kHz periodic square wave from the PT5 pin.

Solution: The following assembly program implements the algorithm described previously:

```
        #include     "c:\miniide\hcs12.inc"
        org          $1500
start   lds          #$1500
        bset         DDRT,BIT5        ; configure PT5 pin for output
forever bset         PTT,BIT5         ; pull PT5 pin to high
        ldy          #10              ; wait for 0.5 ms
        jsr          delayby50us      ; "
        bclr         PTT,BIT5         ; pull PT5 pin to low
        ldy          #10              ; wait for 0.5 ms
        jsr          delayby50us      ; "
        bra          forever
        #include     "c:\miniide\delay.asm"
;       org          $FFFE            ; uncomment for CodeWarrior
;       dc.w         start            ; uncomment for CodeWarrior
        end
```

By connecting the PT5 pin to a speaker (or a buzzer) and making the frequency in the audible range, a sound can be made. Since the frequency of the square wave generated in Example 4.15 is in the audible range, a sound can be heard if a speaker is connected to the PT5 pin.

By alternating the frequency of the generated waveform from the I/O pin between two different values, a two-tone siren can be generated. The duration of the siren tone is variable. The siren would sound more urgent if the tone duration were shorter. The following example generates a two-tone siren that alternates between 250 Hz and 500 Hz with each tone lasting half of a second.

Example 4.16

Write a program to generate a two-tone siren that alternates between 250 Hz and 500 Hz with each siren tone lasting half of a second.

Solution:

```
            #include     "c:\miniide\hcs12.inc"
            org          $1500
start       lds          #$1500
            bset         DDRT,BIT5        ; configure PT5 pin for output
forever     ldx          #250             ; repeat 500 Hz waveform 250 times
tone1       bset         PTT,BIT5         ; pull PT5 pin to high
            ldy          #1
            jsr          delayby1ms
            bclr         PTT,BIT5
            ldy          #1
            jsr          delayby1ms
            dbne         x,tone1
            ldx          #125             ; repeat 250 Hz waveform for 125 times
tone2       bset         PTT,BIT5
            ldy          #2
            jsr          delayby1ms
            bclr         PTT,BIT5
            ldy          #2
            jsr          delayby1ms
            dbne         x,tone2
            bra          forever
            #include     "c:\miniide\delay.asm"
;           org          $FFFE            ; uncomment this line for CodeWarrior
;           dc.w         start            ; uncomment this line for CodeWarrior
            end
```

4.11.4 Interfacing with DIP Switches

A switch is probably the simplest input device we can find. To make input more efficient, a set of eight switches organized as a *dual inline package* (DIP) is often used. A DIP package can be connected to any input port with eight pins, such as Port A, Port B, and Port H, as shown in Figure 4.19. When a switch is closed, the associated Port A pin input is 0. Otherwise, the associated Port A pin has a value of 1. Each Port A pin is pulled up to high via a 10-kΩ resistor when the associated switch is open.

Example 4.17

Write a sequence of instructions to read the value from an eight-switch DIP connected to Port A of the HCS12 into accumulator A.

Solution:

```
#include    "c:\miniide\hcs12.inc"
            movb         #$0,DDRA         ; configure Port A for input
            ldaa         PTA
            ...
```

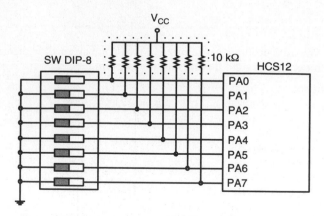

Figure 4.19 ■ Connecting a set of eight DIP switches to port A of the HCS12

4.12 Tips for Program Debugging Involving Subroutine Calls

Program debugging becomes much more difficult when a program calls subroutines. To make program debugging easier, we test each individual subroutine thoroughly to make sure each subroutine works correctly and returns to its caller.

4.12.1 What to Do When the Program Gets Stuck

It is a common problem that the program gets stuck in one of the subroutines and cannot return to the caller. The procedure for debugging in this situation is as follows:

Step 1
Write an instruction sequence to call each subroutine to find out if the program gets stuck in that subroutine. This can be done by setting a breakpoint at the instruction immediately after the jsr or bsr instruction. One can use the D-Bug12 asm command to find out the address of the breakpoint to be set. In CodeWarrior, this can be done much easier.

Step 2
Find out why the program gets stuck in the subroutine. There are at least the following four causes:

1. Forgetting to restore registers pushed onto the stack before returning to the caller
2. Forgetting to deallocate local variables before returning to the caller
3. Some infinite loops in the subroutine
4. Calling other subroutines that do not return

The first two causes can be identified by simply looking at the program. To determine if the program gets stuck because of forgetting to restore registers, check to see if the subroutine has several push instructions at its entrance. If it does, then it should have the same number of pull instructions in the reverse order before it returns (using the rts instruction). For the second cause, check to see if the subroutine has the instruction **leas –k,SP** to allocate space to local variables at the entry point of the subroutine. If it does, then the subroutine should have the instruction **leas k,SP** or its equivalent to deallocate the stack space used by local variables before returning to the caller. The first two causes result in the incorrect return address to be popped out from the stack. Sometimes a subroutine has several returning points. Make sure to restore registers saved in the stack and deallocate local variables before each rts instruction.

If the first two causes are not present, it is still possible that the subroutine gets stuck in some loop. To find out if the program gets stuck in a certain loop, insert a breakpoint after the last instruction of the suspicious loop. If the breakpoint is never reached, then you know that the loop has some problems. Once the infinite loop is identified, you should be able to figure out what's wrong.

To make sure that the subroutine does not get stuck after calling other subroutines, make sure that the subroutines called by the current routine do not have the first three problems described in this step.

4.12.2 Handling the Stack Variable Access with Care

The HCS12 does not have a frame pointer. When a subroutine with local variables needs to call another subroutine, the programmer must pay attention to this issue. For example, Example 4.9 calls the **findSqr** and **div32** subroutines. It uses the following instruction sequence to invoke findSqr:

```
testPR     ldd      pNumLo,SP     ; find the square root of Num
           ldx      pNumHi,SP     ; "
           pshd                   ; "
           pshx                   ; "
           jsr      findsqr       ; "
```

The subroutine will not generate a correct result if the previous instruction sequence is changed to

```
testPR     ldd      pNumLo,SP
           pshd
           ldx      pNumHi,SP
           pshx
           jsr      findsqr
```

In the modified instruction sequence, the pshd instruction changes the stack offset of the original **pNumHi** slot and hence the intended value is not passed to the callee.

4.12.3 General Debugging Strategy

Subroutines can be classified into two categories: *intermediate* and *leaf* subroutines. An intermediate subroutine may call other subroutines, whereas a leaf subroutine does not call any other subroutines. Making sure that a subroutine returns to its caller does not guarantee that it produces correct results. You need to use the methods described in Section 3.9.5 to debug each leaf subroutine to make sure it works correctly. After making sure that each leaf subroutine works correctly, start to debug the intermediate subroutines. Make sure that each intermediate subroutine does not get stuck and works correctly using the methods discussed in Sections 4.12.1 and 3.9.5. After each intermediate subroutine has been debugged, perform the top-level program debugging. Again, the method discussed in Section 3.9.5 can be used.

4.13 Summary

Experience shows that when designing a complicated embedded system, designers spend a smaller percentage (10 to 20%) of their time on hardware design but a much higher percentage (80 to 90%) of their time on software development and debugging. Since it takes much longer to get the software designed right, it is important for us to learn the right way for software development. The *top-down design with hierarchical refinement* approach is considered the most effective system development methodology. For example, using the top-down design with hierarchical refinement approach, the two-tone siren can be generated in the following manner:

In the first iteration, the algorithm for generating the two-tone siren can be outlined as follows:

Iteration 1

Step 1
Generate a 500-Hz periodic square wave for half of a second.

Step 2
Generate a 250-Hz periodic square wave for half of a second.

Step 3
Go to step 1.

In the second iteration, we need to work out the details for generating 500-Hz and 250-Hz periodic square waves for half of a second.

Iteration 2.1   (generation of 500-Hz square wave)
The period of a 500-Hz square wave is 2 ms, and therefore there are 250 periods of the waveforms in half of a second.

Step 1
lpcnt1 ← 250

Step 2
Pull the PT5 pin to high.

Step 3
Wait for 1 ms.

Step 4
Pull the PT5 pin to low.

Step 5
Wait for 1 ms.

Step 6
lpcnt1 ← lpcnt1 − 1;

Step 7
If (lpcnt1 ≠ 0), go to step 2; else, continue to the next step.

Iteration 2.2
The period of a 250-Hz square wave is 4 ms, and therefore there are 125 periods of the waveforms in half of a second.

Step 1
lpcnt1 ← 125

Step 2
Pull the PT5 pin to high.

Step 3
Wait for 2 ms.

Step 4
Pull the PT5 pin to low.

Step 5
Wait for 2 ms.

Step 6
lpcnt1 ← lpcnt1 − 1;

Step 7
If (lpcnt1 ≠ 0), go to step 2; else, continue to the next step.

After the second iteration, the only thing left is how to generate the 2-ms and 1-ms time delays. These time delays can be generated by calling the existing delay subroutines.

The subroutine is the mechanism that allows the same sequence of code to be reused (called) from many places of the program. This is the start of the concept of *software reuse*. In addition, the subroutines created for certain programs can also be called (reused) in other programs. To make subroutines more reusable, the following principles must be followed:

- Avoid using global memory to hold variables that are supposed to be local to the subroutine. Stack is the best place for holding local variables.
- Save CPU registers used in the subroutine unless they are used to pass incoming parameters or to return computation results. By doing this, the caller does not need to be concerned about what registers to save before calling the subroutine.
- Describe clearly how to pass incoming parameters and how to return results using appropriate comments.

The HCS12 provides instructions bsr, jsr, and call for making subroutine calls. The instructions bsr and jsr will save the return address in the stack before jumping to the subroutine. The call instruction will save the contents of the PPAGE register in the stack in addition to the return address. The call instruction is provided to call subroutines located in expanded memory. All subroutines should have rts (rtc for subroutines in expanded memory) as the last instruction. The rts instruction will pop the return address onto the PC register from the stack, and program control will be returned to the point that called the subroutine. The rtc instruction should be used by subroutines that are in expanded memory. The rtc instruction will restore the PPAGE value pushed onto the stack in addition to the return address.

Parameters can be passed in registers, program memory, the stack, or the global memory. The result computed by the subroutine can be returned in CPU registers, the stack, or the global memory. Local variables must be allocated in the stack so that they are not accessible to the caller and other program units. Local variables come into being only when the subroutine is being executed. The HCS12 provides instructions to facilitate the access to variables in the stack. The leas instruction is most effective for local variable allocation and deallocation.

The D-Bug12 monitor provides many functions to support I/O programming on demo boards that include the D-Bug12 monitor. The **printf** subroutine is a very useful subroutine because it provides complicated output data formatting.

The HCS12 has many signal pins. These signal pins are divided into I/O ports. An I/O port consists of a set of signal pins and the registers required for the I/O operation. Since an I/O pin can be used for input and output, the user must configure the direction of the I/O pin before using it in data transfer. This is achieved by writing an appropriate value into a data direction register. Most signal pins serve multiple functions. A signal pin can be used for general-purpose I/O when it is not being used by any peripheral module.

The HCS12 I/O ports can drive many different types of I/O devices directly. It is important to consider the electrical and timing compatibility issues when driving I/O devices. These two issues will be discussed in more detail in Chapter 7. The function of each signal pin will also be discussed in more detail in Chapter 7.

The LED is a simple output device. It is often used to indicate whether power is turned on, whether the device is functioning properly, and so on. Today, LEDs also replace incandescent light bulbs because LEDs are much more energy efficient.

Seven-segment displays are mainly used to display small amounts of information using decimal digits and a smaller set of characters. Due to the availability of a wide range of sizes, seven-segment displays have become one of the most versatile display devices.

DIP switches are often used to provide information to the startup program of many embedded systems. After being powered up, the MCU reads the DIP switch's setting and configures the embedded system accordingly. Users can change the setting of DIP switches to change the system configuration.

4.14 Exercises

E4.1 Assuming that we have the following instruction sequence to be executed by the HCS12, what will be the contents of the topmost 4 bytes of the stack after the execution of these instructions?

```
lds     #$1500
ldaa    #$56
staa    1, −SP
ldab    #22
staa    1, −SP
ldy     #0
sty     2, −SP
```

E4.2 Write instructions to perform the following operation:

1) Push the value $2301 (word value) onto the stack

2) Push the value $34 (byte value) onto the stack

3) Pop the top word of the stack and save it at memory location $1000~$1001.

4) Push the 16-bit value stored at $1000~$1001 into the stack

E4.3 Revise the subroutine in Example 4.2 so that it can convert a 16-bit signed integer to a BCD string.

E4.4 Write a subroutine to find the greatest common divisor of two 16-bit unsigned integers. These two 16-bit numbers are passed in the stack. The subroutine returns the gcd in double accumulator D.

E4.5 Write a subroutine to convert all the lowercase letters in a string to uppercase. The pointer to the string is passed to this subroutine in X.

E4.6 The label *array_x* is the starting address of an array of 100 8-bit elements. Trace the following code sequence and describe what the subroutine *sub_x* does:

```
ldx     #array_x
ldaa    #100
jsr     sub_x
...
```

```
sub_x          deca
               ldab      0,x
               inx
loop           cmpb      0,x
               ble       next
               ldab      0,x
next           inx
               deca
               bne       loop
               rts
```

E4.7 Write a subroutine that can multiply two 32-bit unsigned integers. Both the multiplicand and the multiplier are passed to this subroutine in the stack. The caller pushes the multiplicand into the stack first and then pushes the multiplier. The pointer to the buffer to hold the product is passed in index register X.

E4.8 Write a subroutine that can count the number of characters and words contained in a given string. The pointer to the string to be examined is passed in X. The character count and word count are returned in Y and B, respectively.

E4.9 Write a subroutine that can find whether a given word is contained in a string. The pointer to the word to be found and the pointer to the string to be searched are passed in X and Y, respectively. This subroutine would return a 1 in B if the word is found in the string. Otherwise, a 0 is returned in B.

E4.10 Draw the stack frame and enter the value of each stack slot (if it is known) at the end of the following instruction sequence:

```
               leas      −2,sp
               clrb
               ldaa      #20
               psha
               ldaa      #$E0
               psha
               ldx       #$7000
               pshx
               jsr       sub_abc
               ...
sub_abc        pshd
               leas      −12,sp
               ...
```

E4.11 Draw the stack frame and enter the value of each stack slot (if it is known) at the end of the following instruction sequence:

```
               leas      −8,sp
               ldd       #$1020
               psha
               ldx       #$800
               pshx
               bsr       xyz
               ...
xyz            pshd
               pshx
               leas      −10,sp
               ...
```

E4.12 Write a subroutine to convert all the uppercase letters in a string into lowercase. The starting address of the string is passed to this subroutine in index register X.

E4.13 Write a subroutine to generate a 16-bit random number. The result is returned to the caller in D. Pass any appropriate parameters to this subroutine in the stack.

E4.14 Write a subroutine to compute the *least common multiple* of two 16-bit integers. Incoming parameters are passed in the stack and the result should be returned in a (Y, D) pair with the upper 16-bit in Y and the lower 16-bit in D.

E4.15 Write a subroutine to convert an 8-bit signed integer into an ASCII string that represents a decimal number. The 8-bit integer and the pointer to the buffer to hold the ASCII string are passed to this subroutine in accumulator B and index register X, respectively.

E4.16 Give an instruction sequence to call the out4hex() function to output the 16-bit integer stored in memory location $1000~$1001.

E4.17 Give an instruction sequence that outputs the prompt "Please enter a string" and reads the string entered by the user from the keyboard and then echoes the string on the screen again.

E4.18 Write a subroutine that will convert a 32-bit signed integer into a BCD ASCII string so that it can be output to the console terminal (SCI) and appear as a BCD string on the screen. The 32-bit integer to be converted and the pointer to the buffer to hold the resultant string are passed to this subroutine via the stack.

E4.19 Write an instruction sequence to configure Port A and Port B for input and output, respectively; read the value of Port A and output the value to Port B.

E4.20 Give an instruction to configure the pins 7, 5, 1, and 0 of Port B for output and the remaining pins for input.

E4.21 For the circuit shown in Figure 4.18, write a program to display 135790 from the left to the right one digit at a time with each digit lasting for about half of a second.

E4.22 Write a program to display the following patterns on the six seven-segment displays in Figure 4.18 continuously with each pattern lasting for half of a second:

       123456

       234567

       345678

       456789

       567890

       678901

       789012

       890123

       901234

       012345

4.15 Lab Exercises and Assignments

L4.1 *Program entering, assembling, and downloading.* Enter, assemble, and download the following program for execution on the demo board using MiniIDE:

```
CR              equ      $0D
LF              equ      $0A
printf          equ      $EE88
getcmdline      equ      $EE8A
```

```
cmdlinelen      equ       40
                org       $1000
inbuf           ds.b      20
err_flag        ds.b      1
sign_flag       ds.b      1
                org       $1500
                lds       #$1500
                ldd       #prompt1
                jsr       [printf,PCR]        ; output a prompt to remind the user to enter
                ldd       #cmdlinelen         ; an integer
                pshd
                ldd       #inbuf
                call      [getcmdline,PCR]    ; read in a string that represents an integer
                leas      2,sp
                ldd       #prompt3            ; move cursor to the next line
                jsr       [printf,PCR]        ; "
                ldd       #inbuf
                pshd
                ldd       #prompt2
                jsr       [printf,PCR]        ; output the number that you entered
                leas      2,sp
                ldd       #prompt3
                jsr       [printf,PCR]        ; move cursor to the next line
                swi
prompt1         db        "Please enter a number: ",CR,LF,0
prompt2         db        "The entered number is: %s ",0
prompt3         db        " ",CR,LF,0
                end
```

When you see the message "Please enter a number:," enter an integer followed by a carriage return. The screen output should be similar to what appears here.

```
>g 1500
Please enter a number:

The entered number is: 1234
User Bkpt Encountered
```

PP PC	SP	X	Y	D = A:B	CCR = SXHI NZVC
30 1533	1500	1500	156C	00:03	1001 0000
xx:1533	50		NEGB		
>					

L4.2 *Temperature conversion.* Write a subroutine that will convert the temperature in Fahrenheit to Celsius accurate to one decimal digit. Write a main program that will

1. Prompt the user to enter a temperature in Fahrenheit by displaying the message **"Please enter a temperature in Fahrenheit:"**.

2. Call the GetCmdLine() function to read in the temperature.

3. Call a subroutine to convert the input string (representing a decimal number) into a binary number.

4. Call the temperature conversion subroutine to convert it to Celsius.

5. Output the current temperature to the screen in the following format:

   Current temperature:

   xxxx°F   yyyy.y°C

6. Output the next message:

   Want to continue? (y/n)

7. Call the getchar() function to read in one character. If the character entered by the user is y, then repeat the process. Otherwise, return to the D-Bug12 monitor by executing the swi instruction.

Note: The ASCII code of the degree character ° is 176 (or $B0).

L4.3 *Seven-segment display shifting.* Write a program to display the following patterns on the four seven-segment displays on the Dragon12 demo board (or any other demo board) with each pattern lasting for half of a second:

> 1234
>
> 2345
>
> 3456
>
> 4567
>
> 5678
>
> 6789
>
> 7890
>
> 8901
>
> 9012
>
> 0123

The patterns will be displayed continuously.

L4.4 *Three-tone siren generation.* Write a program to generate a three-tone siren using the PT5 pin with each tone lasting for half of a second. The frequencies of these three tones are 200 Hz, 500 Hz, and 1 kHz.

L4.5 *I/O Routine application and time-of-day display.* The complementary CD includes a file that contains a set of I/O subroutines that you can call. By adding the file stdio0.asm to your program (using the "#include c:\miniIDE\stdio0.asm" statement), you can output a string to the terminal window and read a string from the keyboard.

Study the subroutines contained in stdio0.asm and write a program to perform the following operations:

1. Output a message by calling printf() (or the supplied subroutine puts) to ask the user to enter the current time in the format of hh:mm:ss using the PC keyboard (include the colon character).

2. Read in the time of day by calling the gets subroutine and store the time of day in a buffer.

3. Display the time of day in the format of "Current time is hh:mm:ss" where hh, mm, and ss are the time components that you just entered.

4. Wait for one second and update the current time-of-day stored in the buffer.

5. Backspace eight places (write a loop and call the putch subroutine to output the backspace character).

6. Redisplay the new time of day.

7. Go to step 4.

5

C Language Programming

5.1 Objectives

After completing this chapter, you should be able to

- Explain the overall structure of a C language program

- Use the appropriate operators to perform desired operations in C language

- Understand the basic data types and expressions of C language

- Write program loops in C language

- Write functions and make function calls in C language

- Use arrays and pointers for data manipulation

- Perform basic I/O operations in C language

- Use the CodeWarrior IDE to enter, compile, and debug C programs

- Use ImageCraft ICC12 IDE to enter, compile, and debug C programs

- Write C programs to interface with simple I/O devices such as LEDs, seven-segment displays, and DIP switches

5.2 Introduction to C

This chapter is not intended to provide a complete coverage of C language. Instead, it provides a summary of those C language constructs that will be used in this book. You will be able to deal with the basic HCS12 interface programming if you fully understand the contents of this chapter. In addition to providing a tutorial to C language, this chapter will also provide tutorials on using CodeWarrior IDE and ImageCraft C compiler to enter, compile, and debug C programs.

C language is gradually replacing assembly language in many embedded applications because it has several advantages over assembly language. The most important one is that it allows the user to work on program logic at a level higher than assembly language, and thus programming productivity is greatly improved.

A C program, whatever its size, consists of functions and variables. A function contains statements that specify the operations to be performed. The types of statements in a function could be a *declaration*, *assignment*, *function call*, *control*, or *null*. A variable stores a value to be used during the computation. The *main( )* function is required in every C program and is the one to which control is passed when the program is executed. A simple C program is as follows:

(1)	`#include <stdio.h>`	—*include HCS12 header file*
(2)	`/* this is where program execution begins */`	
(3)	`void main (void)`	—*defines a function named* **main** *that receives*
		—*no argument values and returns no value*
(4)	`{`	—*statements of main are enclosed in braces*
(5)	`    int  a, b, c;`	—*defines three variables of type* **int**
(6)	`    a = 3;`	—*assigns 3 to variable* **a**
(7)	`    b = 5;`	—*assigns 5 to variable* **b**
(8)	`    c = a + b;`	—*adds* **a** *and* **b** *together and assigns it to* **c**
(9)	`    printf("a + b = %d \n", c);`	—*calls library function* printf *to print the result*
(10)	`    return 0;`	—*returns 0 to the caller of main*
(11)	`}`	—*the end of* **main** *function*

The first line of the program

```
#include <stdio.h>
```

causes the file stdio.h to be included in the program. This line appears at the beginning of many C programs. The header file stdio.h contains the prototype declarations of all I/O routines that can be called by the user program and the constant declarations that can be used by the user program. C language requires that a function prototype be declared before that function can be called if a function is not defined when it is called. The inclusion of the stdio.h file allows the function printf() be invoked in the program.

The second line is a comment. A comment explains what will be performed and will be ignored by the compiler. A comment in C language starts with /* and ends with */. Everything in between is ignored. Comments provide documentation to the program and enhance readability. Comments affect only the size of the text file and do not increase the size of the executable code. Many commercial C compilers also allow the use of two slashes (//) for commenting out a single line.

The third line main() is where program execution begins. The opening brace on the fourth line marks the start of main() function's code. Every C program must have one and only one main() function. Program execution is also ended with the main function. The fifth line declares three integer variables *a*, *b*, and *c*. In C, all variables must be declared before they can be used.

The sixth line assigns 3 to the variable *a*. The seventh line assigns 5 to the variable *b*. The eighth line computes the sum of variables *a* and *b* and assigns it to the variable *c*. You will see that assignment statements are major components in C programs.

The ninth line calls the library function printf() to print the string *a* + *b* = followed by the value of *c* and move the cursor to the beginning of the next line. The tenth line returns a 0 to the caller of main(). The closing brace in the eleventh line ends the main() function.

5.3 Types, Operators, and Expressions

Variables and constants are the basic objects manipulated in a program. Variables must be declared before they can be used. A variable declaration must include the name and type of the variable and may optionally provide its initial value. A variable name may start with a letter (**A** through **Z** or **a** through **z**) or an underscore character followed by zero or more letters, digits, or underscore characters. Variable names cannot contain arithmetic signs, dots, apostrophes, C keywords, or special symbols such as @, #, ?, and so on. Adding the underscore character (_) may sometimes improve the readability of long variables. Don't begin variable names with an underscore, however, since library routines often use such names. C language is case sensitive. Upper- and lowercase letters are distinct.

5.3.1 Data Types

There are only a few basic data types in C: void, char, int, float, and double. A variable of type void represents nothing. The type void is used most commonly with functions and can indicate that the function does not return any value or does not have incoming parameters. A variable of type char can hold a single byte of data. A variable of type *int* is an integer, which is normally the natural size (word length) for a particular machine. The type *float* refers to a 32-bit, single-precision, floating-point number. The type *double* represents a 64-bit, double-precision, floating-point number. In addition, there are a number of qualifiers that can be applied to these basic types. *Short* and *long* apply to integers. These two qualifiers will modify the lengths of integers. An integer variable is 16-bit by default for many C compilers including the CodeWarrior C and the GNU C compiler. The modifier *short* does not change the length of an integer. The modifier *long* doubles a 16-bit integer to 32 bits. The keyword *unsigned* should be used if the variables are never negative to improve the efficiency of the generated code.

5.3.2 Variable Declarations

All variables must be declared before their use. A declaration specifies a type and contains a list of one or more variables of that type, as in

```
int     i, j, k;
char    cx, cy;
```

A variable may also be initialized when it is declared, as in

```
int     i = 0;
char    echo = 'y';        /* the ASCII code of letter y is assigned to variable echo. */
```

5.3.3 Constants

There are four kinds of constants: *characters, integers, floating-point numbers*, and *strings*. A character constant is an integer, written as one character within single quotes, such as 'x'. A character constant is represented by the ASCII code of the character. A string constant is a sequence of zero or more characters surrounded by double quotes, as in

"HCS12DG256 is a microcontroller made by Freescale"

or

""   /* an empty string */

Each individual character in the string is represented by its ASCII code.

An integer constant such as 3241 is an int. A long constant is written with a terminal l (lowercase letter "el") or L, as in 44332211L. The following constant characters are predefined in C language (can be embedded in a string):

\a	alert (bell) character	\\	backslash
\b	backspace	\?	question mark
\f	form feed	\'	single quote
\n	new line	\"	double quote
\r	carriage return	\ooo	octal number
\t	horizontal tab	\xhh	hexadecimal number
\v	vertical tab		

As in assembly language, a number in C can be specified in different bases. The method to specify the base of a number is to add a prefix to the number. The prefixes for different bases are

Base	Prefix	Example	
decimal	none	1357	
octal	0	04723	; preceded by a zero
hexadecimal	0X	0X2A	

5.3.4 Arithmetic Operators

There are seven arithmetic operators.

+	add and unary plus
−	subtract and unary minus
*	multiply
/	divide
%	modulus (or remainder)
++	increment
−−	decrement

The expression

a % b

produces the remainder when a is divided by b. The % operator cannot be applied to float or double. The ++ operator adds 1 to the operand, and the −− operator subtracts 1 from the operand. The / operator performs a division and truncates the quotient to an integer when both operands are integers.

Example 5.1

What value will be assigned to *ck* for the following statement?

ck = 230/13;

Solution: The integral part of 230/13 is 17. Therefore, *ck* will receive the value of 17 after the previous statement is executed.

Example 5.2

What value will be assigned to *cx* for the following statement?

cx = 330 % 19;

Solution: The remainder of 330/19 is 7. Therefore, *cx* receives the value of 7 after the execution of the previous statement.

Example 5.3

Assume that *ax* is a six-digit (decimal) integer. Write a few C statements to separate *ax* into two parts and assign the upper three digits to the variable *bx* and the lower three digits to the variable *cx*.

Solution: A six-digit number can be written

$$d_5 d_4 d_3 d_2 d_1 d_0 = d_5 d_4 d_3 \times 1000 + d_2 d_1 d_0$$

Therefore we can divide a six-digit integer into two halves by dividing the given number by 1000. The following two statements will achieve the desired operation:

bx = ax/1000;
cx = ax % 1000;

5.3.5 Bitwise Operators

C provides six operators for bit manipulations; these may be applied only to integral operands, that is, char, short, int, and long, whether they are *signed* or *unsigned*.

&	AND
|	OR
^	XOR
~	NOT
>>	right shift
<<	left shift

The & operator is often used to clear one or more bits to 0. For example, the statement

```
PORTC    = PORTC & 0xAA;      /* PORTC is 8 bits */
```

clears the even bits of PORTC to 0.

The | operator is often used to set one or more bits to 1. For example, the statement

```
PORTB    = PORTB | 0xAA;      /* PORTB is 8 bits */
```

sets the odd bits of PORTB to 1.

The XOR operator can be used to toggle one or multiple bits. For example, the statement

```
abc    = abc ^ 0xF0;      /* abc is of type char */
```

toggles the upper 4 bits of the variable *abc*.

The >> operator shifts the involved operand to the right for the specified number of places. For example, the statement

```
xyz = abc >> 3;
```

shifts the variable *abc* to the right three places and assigns it to the variable *xyz*.

The << operator shifts the involved operand to the left for the specified number of places. For example, the statement

```
xyz = xyz << 4;
```

shifts the variable *xyz* to the left four places.

The assignment operator = is often combined with the operator. For example, the statement

```
PTP = PTP & 0xBD;
```

can be rewritten as

```
PTP &= 0xBD;
```

and the statement

```
PORTB = PORTB | 0x40;
```

can be rewritten as

```
PORTB |= 0x40;
```

5.3.6 Relational and Logical Operators

Relational operators are used in expressions to compare the values of two operands. If the result of the comparison is true, then the value of the expression is 1. Otherwise, the value of the expression is 0. Here are the relational and logical operators.

==	equal to (two "=" characters)
!=	not equal to
>	greater than
>=	greater than or equal to
<	less than
<=	less than or equal to
&&	and
||	or
!	not (one's complement)

Here are some examples of relational and logical operators.

```
if (!(ADCTL & 0x80))
    statement₁;          // if bit 7 is 0, then execute statement₁
```

```
if (i > 0 && i < 10)
        statement₂;          // if 0 < i < 10 then execute statement₂
if (a1 == a2)
        statement₃;          // if a1 equals a2 then execute statement₃
```

5.3.7 Precedence of Operators

Precedence refers to the order in which operators are processed. C language maintains a precedence for all operators, shown in Table 5.1. Operators at the same level are evaluated from left to right. A few examples that illustrate the precedence of operators are listed in Table 5.2.

Precedence	Operator	Associativity
Highest	( ) [ ] .	left to right
	! ~ ++ −− + − * & (type) sizeof	right to left
	* / %	left to right
	+ −	left to right
	<< >>	left to right
	< <= > >=	left to right
	== !=	left to right
	&	left to right
	^	left to right
			left to right
	&&	left to right
	||	left to right
	?:	right to left
	= += −= *= /= %= &= ^= |= <<= >>=	right to left
Lowest	'	left to right

Table 5.1 ■ Table of precedence of operators

Expression	Result	Note
15 − 2 * 7	1	* has higher precedence than +
(13 − 4) * 5	45	
(0x20 | 0x01) != 0x01	1	
0x20	0x01 != 0x01	0x20	!= has higher precedence than	
1 << 3 + 1	16	+ has higher precedence than <<
(1 << 3) + 1	9	

Table 5.2 ■ Examples of operator precedence

5.4 Control Flow

Control-flow statements specify the order in which computations are performed. In C language, the semicolon is a statement terminator. Braces { } are used to group declarations and statements together into a *compound statement,* or *block,* so that they are syntactically equivalent to a single statement.

5.4.1 If Statement

The *if statement* is a conditional statement. The statement associated with the *if statement* is executed on the basis of the outcome of a condition. If the condition evaluates to non-zero, the statement is executed. Otherwise, it is skipped. The syntax of the *if statement* is

```
if (expression)
        statement;
```

Here is an example of an *if statement*.

```
if (a > b)
        sum += 2;
```

The value of *sum* will be incremented by 2 if the variable *a* is greater than the variable *b*.

5.4.2 If-Else Statement

The *if-else* statement handles conditions where a program requires one statement to be executed if a condition is nonzero and a different statement if the condition is zero. The syntax of an *if-else* statement is

```
if (expression)
        statement₁
else
        statement₂
```

The *expression* is evaluated. If it is true (nonzero), $statement_1$ is executed. If it is false (zero), $statement_2$ is executed. Here is an example of the *if-else* statement.

```
        if (a != 0)
                r = b;
        else
                r = c;
```

The *if-else* statement can be replaced by the **?:** operator. The statement

```
        r = (a != 0)? b : c;
```

is equivalent to the previous *if-else* statement.

5.4.3 Multiway Conditional Statement

A multiway decision can be expressed as a cascaded series of *if-else* statements. Such a series looks like this.

```
if (expression₁)
        statement₁
else if (expression₂)
        statement₂
else if (expression₃)
        statement₃
...
else
        statementₙ
```

Here is an example of a three-way decision.

```
if (abc > 0) return 5;
else if (abc == 0) return 0;
else return −5;
```

5.4.4 Switch Statement

The *switch* statement is a multiway decision based on the value of a control expression. The syntax of the *switch* statement is

```
switch (expression) {
    case const_expr₁:
        statement₁;
        break;
    case const_expr₂:
        statement₂;
        break;
    . . .
    default:
        statementₙ;
}
```

As an example, consider the following program fragment:

```
switch (i) {
    case 1:
        pay = 100;
        break;
    case 2:
        pay = 200;
        break;
    case 3:
        pay = 300;
        break;
    case 4:
        pay = 400;
        break;
    case 5:
        pay = 500;
        break;
    default:
        pay = 0;
}
```

The variable *pay* receives a value that is equal to the value of $i \times 100$. The keyword *break* forces the program flow to drop out of the *switch* statement so that only the statements under the corresponding *case-label* are executed. If any *break* statement is missing, then all the statements from that case-label until the next *break* statement within the same *switch* statement will be executed.

5.4.5 For-Loop Statement

The syntax of a *for-loop* statement is

```
for (expr1; expr2; expr3)
    statement;
```

where *expr1* and *expr3* are assignments or function calls, and *expr2* is a relational expression. For example, the following *for loop* computes the sum of the squares of integers from 1 to 9:

```
sum = 0;
for (i = 1; i < 10; i++)
    sum = sum + i * i;
```

The following *for loop* prints out the numbers that are smaller than 100 and indivisible by 13:

```
for (i = 1; i < 100; i++)
        if (i % 13) printf("%d ", i);
```

5.4.6 While Statement

During the time an expression is nonzero, the *while* loop repeats a statement or block of code. The value of the expression is checked prior to each execution of the statement. The syntax of a *while* statement is

```
while (expression)
        statement;
```

The *expression* is evaluated. If it is nonzero (true), *statement* is executed and *expression* is reevaluated. This cycle continues until *expression* becomes zero (false), at which point execution resumes after *statement*. The *statement* may be a NULL statement. A NULL statement does nothing and is represented by a semicolon. Consider the following program fragment:

```
intCnt = 5;
while (intCnt);
```

The CPU will do nothing before the variable *intCnt* is decremented to 0. In microprocessor applications, the decrement of *intCnt* is often triggered by external events such as interrupts.

5.4.7 Do-While Statement

The *while* and *for* loops test the termination condition at the beginning of a statement. By contrast, the *do-while* statement tests the termination condition at the end of the statement; the body of the statement is executed at least once. The syntax of the *do-while* statement is

```
do
        statement
while (expression);
```

The following *do-while* statement displays the integers 9 down to 1:

```
int digit = 9;
do
        printf("%d", digit--);
while (digit >= 1);
```

5.4.8 Goto Statement

Execution of a *goto* statement causes control to be transferred directly to the labeled statement, which must be located in the same function as the *goto* statement. The use of the *goto* statement interrupts the normal sequential flow of a program and thus makes it harder to follow and decipher. For this reason, the use of *goto*'s is not considered good programming style, and it is recommended that you do not use them in your program.

The syntax of the *goto* statement is

```
goto    label
```

An example of the use of a *goto* statement is

```
if (x > 100)
        goto        severe_error;
...
severe_error:
    printf("Variable x is out of bound!\n");
```

5.5 Input and Output

Input and output facilities are not part of C language itself. However, input and output are fairly important in applications. The ANSI standard defines a set of library functions that must be included so that they can exist in a compatible form on any system where C exists. Some of the functions deal with file input and output. Others deal with text input and output. In this section we will look at the following four input and output functions:

1. **int *getchar* ( ).** This function returns a character when it is called. The following program fragment returns a character and assigns it to the variable *xch*:

   ```
   char xch;
   xch = getchar ();
   ```

2. **int *putchar* (int).** This function outputs a character on the standard output device. The following statement outputs the letter *a* from the standard output device:

   ```
   putchar ('a');
   ```

3. **int *puts* (const char *s).** This function outputs the string pointed to by **s** on the standard output device. The following statement outputs the string "Learning microcontroller is fun!" from the standard output device:

   ```
   puts ("Learning microcontroller is fun! \n");
   ```

4. **int *printf* (*formatting string*, arg$_1$, arg$_2$, . . . , arg$_n$).** This function converts, formats, and prints its arguments on the standard output under control of *formatting string*. arg$_1$, arg$_2$, . . . , arg$_n$ are arguments that represent the individual output data items. The arguments can be written as constants, single variable or array names, or more complex expressions. The formatting string is composed of individual groups of characters, with one character group associated with each output data item. The character group corresponding to a data item must start with **%**. In its simplest form, an individual character group will consist of the percent sign followed by a *conversion character* indicating the type of the corresponding data item.

Multiple character groups can be contiguous or separated by other characters, including white-space characters. These other characters are simply transferred directly to the output device where they are displayed. A subset of the more frequently used conversion characters is given in Table 5.3. Between the % character and the conversion character there may be, in order,

- A minus sign, which specifies *left adjustment* of the converted argument.
- A number, which specifies the minimum *field width*. The converted argument will be printed in a field at least this wide. If necessary, it will be padded on the left (or right, if left adjustment is called for) to make up the field width.
- A period, which separates the field width from the precision.
- A number that specifies the maximum number of characters to be printed from a string or the number of digits after the decimal point of a floating-point value or the minimum number of digits for an integer.
- An *h* if the integer is to be printed as a *short,* or an *l* (letter "el") if as a *long*.

Several valid printf calls are

```
printf ("this is a challenging course! \n");      /* outputs only a string */
printf ("%d %d %d", x1, x2, x3);                   /* outputs variables x1, x2, x3 using a minimal number of digits
                                                      with one space separating each value */
```

printf("Today's temperature is %4.1d \n", temp); /* display the string *Today's temperature is* followed by the value of temp. Display one fractional digit and use at least four digits for the value. */

Conversion Character	Meaning
c	Data item is displayed as a single character.
d	Data item is displayed as a signed decimal number.
e	Data item is displayed as a floating-point value with an exponent.
f	Data item is displayed as a floating-point value without an exponent.
g	Data item is displayed as a floating-point value using either e-type or f-type conversion, depending on value; trailing zeros, trailing decimal point will not be displayed.
i	Data item is displayed as a signed decimal integer.
o	Data item is displayed as an octal integer, without a leading zero.
s	Data item is displayed as a string.
u	Data item is displayed as an unsigned decimal integer.
x	Data item is displayed as a hexadecimal integer, without the leading 0x.

Table 5.3 ■ Commonly used conversion characters for data output

5.6 Functions and Program Structure

Every C program consists of one or more functions. If a program consists of multiple functions, their definitions cannot be embedded within another. The same function can be called from several different places within a program. Generally, a function will process information passed to it from the calling portion of the program and return a single value. Information is passed to the function via special identifiers called *arguments* (also called *parameters*) and returned via the *return* statement. Some functions, however, accept information but do not return anything (for example, the library function *printf*).

The syntax of a function definition is

```
return_type function_name (declarations of arguments)
{
        declarations and statements
}
```

The declaration of an argument in the function definition consists of two parts: the *type* and the *name* of the variable. The return type of a function is *void* if it does not return any value to the caller. An example of a function that converts a lowercase letter to an uppercase letter is

```
char lower2upper (char cx)
{
    if (cx >= 'a' && cx <= 'z') return (cx − ('a' − 'A'));
    else return cx;
}
```

A character is represented by its ASCII code. A letter is in lowercase if its ASCII code is between 97 (0x61) and 122 (0x7A). To convert a letter from lowercase to uppercase, subtract its ASCII code by the difference of the ASCII codes of letters *a* and *A*.

To call a function, simply put down the name of the function and replace the argument declarations by actual arguments or values and terminate it with a semicolon.

Example 5.4

▼

Write a function to find the square root of a 32-bit integer using the successive approximation method described in Section 4.7.2.

Solution: The C function that computes the square root of a 32-bit unsigned integer is as follows:

```
// ************************************************************************
// This function computes the square root of a 32-bit unsigned integer
// using the successive approximation method.
// Incoming parameter: a 32-bit number of which the square root is to be found.
// ************************************************************************
unsigned long int FindSqr (unsigned long int xval)
{
        unsigned    long    int      mask,temp,sar;
        unsigned    int      ix;
        mask = 0x8000;         // Initialize mask for making bit value guessing
        sar  = 0;
        for (ix = 0; ix < 16; ix++){
        temp = sar | mask;
        if((temp * temp) <= xval)
        sar   = temp;
        mask >>= 1;
        }
        if ((xval − (sar * sar)) > ((sar+1)*(sar+1) − xval))
        sar   += 1;
        return sar;
}
```

▲

Example 5.5

▼

Write a function to test whether a 32-bit nonnegative integer is a prime number.

Solution: The integer 1 is not a prime number. A number is a prime if it cannot be divided by any integer between 2 and its square root. The prime test function needs to call the **FindSqr** function to find the test divide limit. The prime test function is

```
unsigned int PrimeTest(unsigned long int xval)
{
    unsigned long int TLimit;
    unsigned int ptest;
    if ((xval == 1) || (xval == 2))
    return 0;
```

```
TLimit = FindSqr(xval);    // call FindSqr() to find the limit for test divide
for (ptest = 2; ptest <= TLimit; ptest ++){
if ((xval % ptest) == 0)  // is remainder zero?
        return 0;
}
return 1;
}
```

Example 5.6

Write a program to find out the number of prime numbers between 100 and 1000.

Solution: One can find the number of prime numbers between 100 and 1000 by calling the function written in Example 5.5.

```
#include "c:\cwHCS12\include\hcs12.h"
unsigned long int FindSqr (unsigned long int xval);
unsigned int PrimeTest(unsigned long int xval);
unsigned int prime_count;
void main(void) {
    unsigned long int  i;
    prime_count = 0;
    for (i = 100; i <= 1000; i++) {
      if (PrimeTest(i))
          prime_count++;
    }
    while(1);
}
unsigned int PrimeTest(unsigned long int xval)
{
unsigned long int TLimit;
unsigned int ptest;
if (xval = 1)
return 0;
TLimit = FindSqr(xval);
for (ptest = 2; ptest <= TLimit; ptest++){
if ((xval % ptest) == 0)
return 0;
}
return 1;
}
// include FindSqr() function here
```

This program consists of three functions. It is very common for a C program to consist of many functions. Functions of the same nature can be placed in the same file and be reused.

5.6.1 Function Prototype

A function cannot be called before it has been defined. This dilemma is solved by using the function prototype statement. The syntax for a function prototype statement is

```
return_type function_name (declarations of arguments);
```

The statement

    unsigned int PrimeTest(unsigned long int xval);

the third line in the program of Example 5.6, is a function prototype statement.

To call a function, simply write the name of the function and replace the argument declarations by actual arguments or values and terminate it with a semicolon.

5.6.2 Creating Header Files

When placing multiple functions of the same nature in one file and reusing them in other programs, the user needs also to include the prototype declaration of these functions in the program. A common approach is to create a header file to contain the prototype declarations. For example, the complementary CD provides a file called *delay.c* which contains several delay functions. The prototype declarations for those delay functions are placed in the *delay.h* file. The user will need to include the delay.h file in the program in order to invoke those delay functions. The delay.c file contains the following functions:

- *delayby10μs()*. This function creates a delay that is a multiple of 10 μs. The multiple is passed to this function.

- *delayby50μs()*. This function creates a delay that is a multiple of 50 μs. The multiple is passed to this function.

- *delayby1ms()*. This function creates a delay that is a multiple of 1 ms. The multiple is passed to this function.

- *delayby10ms()*. This function creates a delay that is a multiple of 10 ms. The multiple is passed to this function.

- *delayby100ms()*. This function creates a delay that is a multiple of 100 ms. The multiple is passed to this function.

For example, the following statement creates a delay of 200 ms:

    delayby100ms(2);

5.7 Pointers, Arrays, Structures, and Unions

5.7.1 Pointers and Addresses

A *pointer* is a variable that holds the address of a variable. Pointers are used frequently in C, as they have a number of useful applications. For example, pointers can be used to pass information back and forth between a function and its reference (calling) point. In particular, pointers provide a way to return multiple data items from a function via function arguments. Pointers also permit references to other functions to be specified as arguments to a given function. This has the effect of passing functions as arguments to the given function.

Pointers are also closely associated with arrays and therefore provide an alternative way to access individual array elements. The syntax for declaring a pointer type is

    type_name *pointer_name;

For example,

    int *ax;

declares that the variable *ax* is a pointer to an integer.

    char *cp;

declares that the variable *cp* is a pointer to a character.

To access the value pointed to by a pointer, use the *dereferencing* operator *. For example,

```
int a, *b;        // b is a pointer to int
...
a = *b;
```

assigns the value pointed to by *b* to variable *a*.

We can assign the address of a variable to a pointer by using the unary operator &. The following example shows how to declare a pointer and how to use & and *:

```
int x, y;
int *ip;        // ip is a pointer to an integer

ip = &x;        // assigns the address of the variable x to ip
y = *ip;        // y gets the value of x
```

5.7.2 Arrays

Many applications require the processing of multiple data items that have common characteristics (e.g., a set of numerical data, represented by $x_1, x_2, \ldots, x_n$). In such situations it is more convenient to place data items into an *array*, where they will all share the same name. The individual data items can be characters, integers, floating-point numbers, and so on. They must all be of the same type and the same storage class.

Each array element is referred to by specifying the array name followed by one or more *subscripts*, with each subscript enclosed in brackets. Each subscript must be expressed as a nonnegative integer. Thus, the elements of an *n*-element array x are $x[0], x[1], \ldots, x[n-1]$. The number of subscripts determines the dimensionality of the array. For example, *x[i]* refers to an element of a one-dimensional array. Similarly, *y[i][j]* refers to an element of a two-dimensional array. Higher-dimensional arrays can be formed by adding additional subscripts in the same manner. However, higher-dimensional arrays are not used very often in 8- and 16-bit microcontroller applications. In general, a one-dimensional array can be expressed as

```
data-type array_name[expression];
```

A two-dimensional array is defined as

```
data-type array_name[expr1][expr2];
```

An array can be initialized when it is defined. This is a technique used in table lookup, which can speed up the computation process.

Example 5.7

▼

Write the bubble sort function to sort an array of integers.

Solution: The algorithm for bubble sort is already described in Chapter 4. Here is the C language version.

```
void     swap (int *px, int *py);
void     bubble (int a[], int n)   /* n is the array count */
{
        int i, j;
        for (i = 0; i < n − 1; i++)
            for (j = 0; j > n − i − 2; j++)
```

```
                                        if (a[j] > a[j+1])
                                                swap (&a[j], &a[j+1]);
        }
        void swap (int *px, int *py)
        {
                int temp;
                temp = *px;
                *px = *py;
                *py = temp;
        }
```

▲

5.7.3 Pointers and Arrays

In C, there is a strong relationship between pointers and arrays. Any operation that can be achieved by array subscripting can also be done with pointers. The pointer version will in general be faster but somewhat harder to understand. For example,

```
    int  ax[20];
```

defines an array *ax* of 20 integral numbers. The notation *ax*[i] refers to the *i*th element of the array. If *ip* is a pointer to an integer, declared as

```
    int *ip;
```

then the assignment

```
    ip = &ax[0];
```

makes *ip* contain the address of *ax*[0]. Now the statement

```
    x = *ip;
```

will copy the contents of *ax*[0] into *x*. If *ip* points to *ax*[0], then *ip* + 1 points to *ax*[1], and *ip* + *i* points to *ax*[i], and so on.

5.7.4 Passing Arrays to a Function

An array name can be used as an argument to a function, thus permitting the entire array to be passed to the function. To pass an array to a function, the array name must appear by itself, without brackets or subscripts, as an actual argument within the function call. When declaring a one-dimensional array as a formal argument, the array name is written with a pair of empty square brackets. The size of the array is not specified within the formal argument declaration. If the array is two-dimensional, then there should be two pairs of empty brackets following the array name.

The following program outline illustrates the passing of an array from the main portion of the program to a function:

```
    int average (int n, int arr[]);
    void main ( )
    {
            int n, avg;                     /* variable declaration */
            int arr[50];                    /* array definition */
            . . .
            avg = average(n, arr);          /* function call */
            . . .
    }
```

```
int average (int k, int brr[])              /* function definition */
{
        . . .
}
```

Within *main* we see a call to the function *average*. This function call contains two arguments—the integer variable *n* and the one-dimensional integer array *arr*. Note that *arr* appears as an ordinary variable within the function call. In the first line of the function definition, we see two formal arguments, *k* and *brr*. The formal argument declarations establish *k* as an integer variable and *brr* as a one-dimensional integer array. Note that the size of *brr* is not defined in the function definition. As formal parameters in a function definition,

```
int  brr[];         and        int *brr;
```

are equivalent.

5.7.5 Initializing Arrays

C allows the initialization of arrays. Standard data-type arrays may be initialized in a straightforward manner. The syntax for initializing an array is

```
array_declarator = { value-list }
```

The following statement shows a five-element integer array initialization:

```
int arr[5] = {10, 20, 30, 40, 50};
```

The element arr[0] has the value of 10 and the element arr[4] has the value of 50.

A string (character array) can be initialized in two ways. One method is to make a list of each individual character:

```
char strgx[5] = {'w', 'x', 'y', 'z', 0};
```

The second method is to use a string constant.

```
char myname [6] = "Edison";
```

A null character is automatically appended at the end of "Edison." When initializing an entire array, the array size (which is one more than the actual length) must be included.

```
char prompt [24] = "Please enter an integer:";
```

5.7.6 Structures

A structure is a group of related variables that can be accessed through a common name. Each item within a structure has its own data type, which can be different from those of the other data items. The syntax of a structure declaration is

```
struct struct_name {                    /* struct_name is optional */
        type1       member1;
        type2       member2;
        . . .
};
```

The *struct_name* is optional and, if it exists, defines a *structure tag*. A *struct* declaration defines a type. The right brace that terminates the list of members may be followed by a list of variables, just as for any basic type. The following example is for a card catalog in a library:

```
struct catalog_tag {
        char    author [40];
        char    title [40];
        char    pub [40];
```

```
        unsigned    int date;
        unsigned    char rev;
    } card;
```

where the variable *card* is of type *catalog_tag*.

A structure definition that is not followed by a list of variables reserves no storage; it merely describes a template or the shape of a structure. If the declaration is tagged (i.e., has a name), however, the tag can be used later in definitions of instances of the structure. For example, suppose we have the following structure declaration:

```
struct point {
        int x;
        int y;
};
```

We can then define a variable *pt* of type *point* as follows:

```
    struct point pt;
```

A member of a particular structure is referred to in an expression by a construction of the form

structure-name.member   or   structure-pointer → member

The structure member operator . connects the structure name and the member name. As an example, the square of the distance of a point to the origin can be computed as follows:

```
long integer sq_distance;
. . .
sq_distance = pt.x * pt.x | pt.y * pt.y;
```

Structures can be nested. One representation of a circle consists of the center and radius, as shown in Figure 5.1.

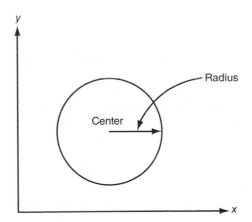

Figure 5.1 ■ A circle

This circle can be defined by

```
struct circle {
        struct     point    center;
        unsigned   int      radius;
};
```

5.7.7 Unions

A *union* is a variable that may hold (at different times) objects of different types and sizes, with the compiler keeping track of size and alignment requirements. Unions provide a way to manipulate different kinds of data in a single area of storage, without embedding any machine-dependent information in the program. The syntax of the union is

```
union  union_name {
        type-name1      element1;
        type-name2      element2;
        . . .
        type-namen      elementn;
};
```

The field *union_name* is optional. When it exists, it is also called *union-tag*. We can declare a union variable at the same time we declare a union type. The union variable name should be placed after the right brace }. In order to represent the current temperature using both the integer and string, we can use the following declaration:

```
union u_tag {
        int i;
        char c[4];
} temp;
```

Four characters must be allocated to accommodate the larger of the two types. Integer type is good for internal computation, whereas string type is suitable for output. Of course, some conversion may be needed before making a certain kind of interpretation. Using this method, the variable *temp* can be interpreted as an integer or a string, depending on the purpose. Syntactically, members of a union are accessed as

```
union-name.member        or        union-pointer → member
```

just as for structures.

5.8 Writing C Programs to Perform Simple I/O

The parallel I/O ports have been briefly discussed in Section 4.10. Before performing an I/O operation, the user needs to configure the I/O port for either input or output. By writing a 1 to a bit in the data direction register (DDRx, x is the port name), the associated I/O pin is configured for output. By writing a 0 to a bit in the data direction register, the associated I/O pin is configured for input.

For example, the following statement configures Port B for output:

```
DDRB  = 0xFF;
```

The following statement configures Port B for input:

```
DDRB  = 0;
```

The user may also configure a few pins of a port for output and other pins for input. For example, the following statement configures the upper four pins of Port B for output and the lower four pins of Port B for input:

```
DDRB  = 0xF0;
```

To output a value to an output port, simply assign the value to its associated port data register. For example, the following statement outputs the value 0x35 to Port B:

```
PTB    = 0x35;  // the user can also use PORTB instead of PTB
```

To read a value from an input port, simply assign the associated port data register to the destination variable. The following statement reads the value from Port A and assigns the value to the variable *xyz*:

```
xyz    = PTA;    // the user can also use PORTA instead of PTA
```

Example 5.8

Write a program to drive the LED circuit in Figure 4.16 and display one LED at a time from the one driven by pin 7 toward the one driven by pin 0 and then reverse. Repeat this operation forever. Each LED is lighted for 200 ms assuming that the HCS12 uses an 8-MHz crystal oscillator to generate a system clock.

Solution: The values to drive Port B to turn on one LED at a time should be placed in a lookup table. The program reads one value at a time from the table and outputs it to Port B and then waits for 200 ms.

The creation of time delays requires the user to set the E-clock frequency properly. The clock setting issue will be discussed in Chapter 6. However, we will invoke the *SetClk8* function in Chapter 6 to set the E-clock to 24 MHz. The 200-ms time delay can be created by calling the *delayby100ms()* function and passing 2 as the parameter. The C program that performs the desired operation is as follows:

```
#include "c:\cwHCS12\include\hcs12.h"
void SetClk8(void);
void delayby100ms(int k);
void main (void)
{
        unsigned char led_tab[16]=  {0x80,0x40,0x20,0x10,0x08,0x04,0x02,0x01,
                                     0x01,0x02,0x04,0x08,0x10,0x20,0x40,0x80};

        char i;
        DDRB       = 0xFF;    // configure Port B for output
        DDRJ     |= 0x02;     // configure PJ1 pin for output
        PTJ      &= 0xFD;     // enable LEDs to light
        SetClk8();                    // enable PLL and set E-clock to 24-MHz using a 4-MHz crystal oscillator
        while (1) {
                for (i = 0; i < 16; i++) {
                        PTB = led_tab[i];    // output a new LED pattern
                        delayby100ms(2);    // wait for 200 ms
                }
        }
}
// Include the SetClk8() function here (to be discussed in Chapter 6)
// Include the delayby100ms() function here (to be discussed in Chapter 8)
```

Example 5.9

Write a program to display the following patterns from the seven-segment display circuit shown in Figure 4.18 from display #0 to #5 and repeat with each pattern lasting for 600 ms:

123456
234567
345678
456789
567890
678901
789012
890123
901234
012345

Solution: Since there are 10 different segment pattern sequences and two adjacent sequences are offset by 1, we can overlap these 10 sequences as shown in Figure 5.2.

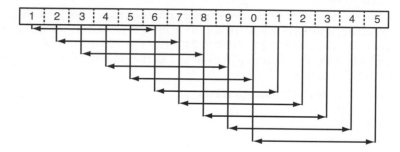

Figure 5.2 ■ Seven-segment display patterns overlapped

Figure 5.1 gives an idea how to produce such a display pattern.

Let *SegPat*, *i*, *j*, and *k* represent the segment pattern array, the start index of the pattern array for the sequence in effect, the number of times remaining for the current display sequence to be repeated, and the display to be lighted, respectively.

The algorithm of the program is as follows:

Step 1
i ← 0;

Step 2
j ← 0

Step 3
k ← 0

Step 4
Output SegPat[i+k] to Port B, and turn on seven-segment display #k.

Step 5
Wait for 1 ms; k ← k + 1;

Step 6
If (k < 6) go to step 4.

Step 7
j ← j + 1;

Step 8
If (j < 100) go to step 3.

Step 9
i ← i + 1;

Step 10
If (i < 10)
    go to step 2;
else go to step 1.

The program that can display the specified seven-segment pattern sequence is as follows:

```
#include "c:\cwHCS12\include\hcs12.h"
#include "c:\cwHCS12\include\delay.h"
void SetClk8(void);
unsigned char SegPat[16]    = {0x06, 0x5B, 0x4F, 0x66, 0x6D, 0x7D, 0x07, 0x7F, 0x67, 0x3F,
                               0x06, 0x5B, 0x4F, 0x66, 0x6D};
unsigned char digit[6]      = {0xFE, 0xFD, 0xFB, 0xF7, 0xEF, 0xDF};

void main(void) {
        int i, j, k;
        SetClk8(); // set E-clock frequency to 24 MHz
        DDRB = 0xFF; //configure Port B for output
        DDRP = 0xFF; //configure Port P for output
        while(1) {
            for (i = 0; i < 10; i++) { // pattern array start index
                for (j = 0; j < 100; j++) { // repeat loop for each pattern sequence
                    for (k = 0; k < 6; k++) { // select the display # to be lighted
                    PTB             = SegPat[i+k]; // output segment pattern
                    PTP             = digit[k];      // output digit select value
                    delayby1ms(1);                   // display one digit for 1 ms
                    }
                }
            }
        }
}
// include SetClk8(void) here
```

5.9 Miscellaneous Items

5.9.1 Automatic, External, Static, and Volatile

A variable defined inside a function is an *internal variable* of that function. These variables are called *automatic* because they come into existence when the function is entered and disappear when it is left. Internal variables are equivalent to local variables in assembly language. *External variables* are defined outside of any function and are thus potentially available to many functions. Because external variables are globally accessible, they provide an alternative to functional arguments and return values for communicating data between functions. Any function may access an external variable by referring to it by name, if the name has been declared somehow. External variables are also useful when two functions must share some data, yet neither calls the other.

The use of *static* with a local variable declaration inside a block or a function causes the variable to maintain its value between entrances to the block or function. Internal *static* variables are local to a particular function just as automatic variables are, but unlike automatic variables, they remain in existence rather than coming and going each time the function is activated. When a variable is declared *static* outside of all functions, its scope is limited to the file that contains the definition. A function can also be declared as static. When a function is declared as static, it becomes invisible outside of the file that defines the function.

A *volatile* variable has a value that can be changed by something other than the user code. A typical example is an input port or a timer register. These variables must be declared as *volatile* so the compiler makes no assumptions on their values while performing optimizations. The keyword *volatile* prevents the compiler from removing apparently redundant references through the pointer.

5.9.2 Scope Rules

The functions and external variables that make up a C program need not all be compiled at the same time; the source text of the program may be kept in several files, and previously compiled routines may be loaded from libraries.

The scope of a name is the part of the program within which the name can be used. For a variable declared at the beginning of a function, the scope is the function in which the name is declared. Local (internal) variables of the same name in different functions are unrelated.

The scope of an external variable or a function lasts from the point at which it is declared to the end of the file being compiled. Consider the following program segment:

```
...
void f1 (...)
{

        ...

}
int a, b, c;
void f2 (...)
{

        ...

}
```

Variables a, b, and c are accessible to function f_2 but not to f_1.

When a C program is split into several files, it is convenient to put all global variables into one file so that they can be accessed by functions in different files. Functions residing in different files that need to access global variables must declare them as external variables. In addition, the prototypes of certain functions can be placed in one file so that they can be called by functions in other files. The following example is a skeletal outline of a two-file C program that makes use of external variables:

In *file1*:

```
extern int  xy;
extern  long arr[];
main ( )
```

```
        {
                . . .
        }
        void foo (int abc) { . . . }
        long soo (void) { . . . }
```

In *file2:*

```
    int  xy;
    long  arr[100];
```

5.9.3 Type Casting

Type casting causes the program to treat a variable of one type as though it contains data of another type. The format for type casting is

```
    (type) variable
```

For example, the following expression converts the variable *kk* to a long integer:

```
    int kk;
    . . .
    (long) kk
```

Type casting can avoid many errors caused by size mismatch among operands. For example, in the following program segment:

```
    long  result;
    int      x1, x2;
    . . .
    result = x1 * x2;
```

if the product of $x1$ and $x2$ is larger than $2^{16} - 1$, it will be truncated to 16 bits. Then the variable *result* will receive an incorrect value. To fix the error, use type casting to force $x1$ and $x2$ to long integers, as follows:

```
    result = ((long) x1) * ((long) x2);
```

This technique is used in several examples in this text.

Another example of the use of type casting is in pointer type. Sometimes one needs to treat the contents of a structure type variable as a string. The most convenient way to do it is to recast the pointer to a structure-type variable into a pointer to a string (character type). For the declarations

```
    struct personal {
        char name      [10];
        char addr [20];
        char sub1[5];
        char sub2[5];
        char sub3[5];
        char sub4[5];
    } ptr1;
    char *cp;
```

we can use the following statement to treat the variable *ptr1* as a string:

```
        cp = (char *) &ptr1;
```

5.10 Using the C Compiler

There are many C compilers that support the HCS12 microcontroller. The freeware GNU C compiler has a port for HCS12 and Eric Engler wrote an IDE (referred to as EGNU) that provides a simple development environment for the GNU C compiler. The combination of the EGNU IDE and the GNU C compiler for the HCS12 is free to the user. The EGNU IDE contains a text editor, a project manager, and a terminal program and works with D-Bug12 monitor. Freescale provides a special edition of CodeWarrior IDE that supports the entering, compilation, and source-level debugging of C programs for the HCS12 up to the 32-kB size limit. CodeWarrior supports the serial monitor and several BDM debuggers. The ImageCraft C compiler is also popular in academic institutions. The tutorials for using CodeWarrior and ImageCraft C compilers will be given in this chapter whereas the tutorial for using the GNU C compiler and EGNU IDE is given in Appendix E.

5.10.1 Issue in Accessing Peripheral Registers

An important part of microcontroller programming is to access peripheral registers.

An address is assigned to each peripheral register. A C program may assign a value to a peripheral register or read the contents of a peripheral register. As we have done in Section 5.8, we assign a value to the name of a register in order to write to a register. Referring to the C language syntax, the peripheral name actually refers to the contents of a memory location. On the other hand, a peripheral register may be 8-bit (unsigned character type) or 16-bit (unsigned integer type). Therefore, a peripheral register should be declared using one of the following methods:

- #define reg_name *(volatile unsigned char *) reg_addr
- #define reg_name *(volatile unsigned int *) reg_addr

The phrase "volatile unsigned char *" typecasts *reg_addr* to a pointer to unsigned character whereas "volatile unsigned int *" typecasts *reg_addr* to a pointer to unsigned integer. After applying the dereferencing operator *, we can assign a value to the register name to change its value. The keyword *volatile* is used because the register contents may be changed by external events instead of program assignment.

The HCS12 allows all peripheral registers as a block to be remapped to other memory locations. To support this, the *reg_addr* is written as the sum of a *base address* and an *offset*. By default, the register block base address is set to 0x0000. The previous declaration can be rewritten as

```
#define      reg_base      0x0000
#define      reg_name1     *(volatile unsigned char *) (reg_base + offset1)
#define      reg_name2     *(volatile unsigned int *) (reg_base + offset2)
```

This method is used in ImageCraft C compiler MCU header file. For example, the Port A data register PORTA is defined as

```
#define      REG_BASE      0x0000
#define      PORTA      (*(volatile unsigned char *)(REG_BASE + 0x00000000))
```

Of course, there are other methods for defining the peripheral register name. The header file *hcs12.h* provided in the complementary CD uses macro substitution as follows:

```
#define   IOREGS_BASE   0x0000
#define   _IO8(off)        *(unsigned char  volatile *)(IOREGS_BASE + off)
#define   _IO16(off)       *(unsigned short volatile *)(IOREGS_BASE + off)
#define   PORTA          _IO8(0x00)          // port A data register
```

```
#define    PTA        _I08(0x00)              // alternate name for PORTA
#define    ATD0DR0    _I016(0x90)             // ADC result 0 register (a 16-bit register)
```

The header file is usually stored under the include directory. For example, the hcs12.h file is stored in the directory c:\iccv712\include (c:\cwHCS12\include and c:\egnu\include). The directory *cwHCS12* is the installation directory for CodeWarrior used by this author whereas the include directory is created under the *cwHCS12* directory. One should add one of the following statements to the C program if one is developing application programs that involve peripheral registers:

```
#include "c:\iccv712\include\hcs12.h"
#include "c:\cwHCS12\include\hcs12.h"
#include "c:\egnu\include\hcs12.h"
```

5.10.2 Peripheral Register Bit Definitions

In addition to the association of register names and addresses, the header file hcs12.h also contains the bit values for many registers. Each bit is associated with a value that is equal to its position weight. For example, the seventh bit of the ATD0CTL2 register is defined as follows:

```
#define       ADPU     0x80
```

With this definition, you can use the following statement to set the ADPU bit of the ATD0CTL2 register:

```
ATD0CTL2  |= ADPU;        // set the ADPU bit (bit 7)
```

To clear the ADPU bit, use the following statement:

```
ATD0CTL2  &= ~ADPU;       // clear the ADPU bit
```

All register names and all bit names are in uppercase. The user can change the case to satisfy his or her preference. The header file provided by ImageCraft C compiler and the hcs12.h file provided in the complementary CD use the same bit names as defined by Freescale. Header files for other HCS12 members can be derived by editing the hcs12.h file.

5.10.3 Inline Assembly Instructions

Most C compilers allow the user to add inline assembly instructions in the C program. The syntax for inline assembly is

```
asm("<string>");
```

For example, adding the following statement will cause the program control to return to the D-Bug12 monitor when the program is running on a demo board with the D-Bug12 monitor:

```
asm("swi");
```

Adding the following statement will enable interrupt globally:

```
asm("cli");
```

5.11 Using the CodeWarrior IDE to Develop C Programs

CodeWarrior supports program debugging for demo boards programmed with the serial monitor or through the BDM debugger. CodeWarrior can be started by clicking on its icon.

5.11.1 Entering C Programs in CodeWarrior

The startup screen of CodeWarrior is shown in Figure 5.3. The user has four options to choose from. If the user wants to enter one or more C functions before creating a new project, she or he

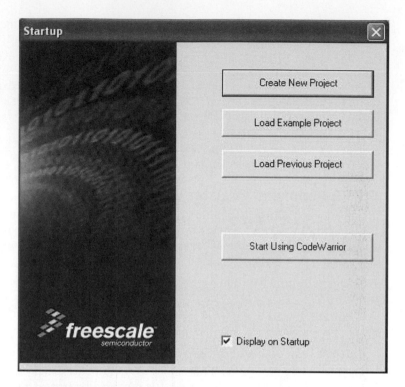

Figure 5.3 ■ CodeWarrior startup screen

should click on the "Start Using CodeWarrior" button. After clicking on the "Start Using Code-Warrior" button, the "Tips of the Day" will appear. Click on **Close** to get rid it. The user can also choose to read a few tips of the day. The next screen will look like that in Figure 5.4.

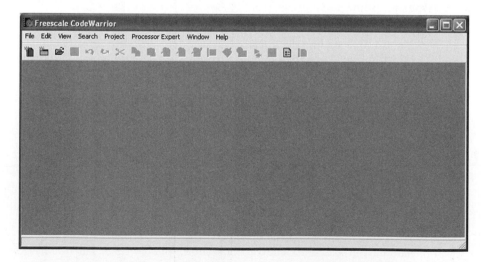

Figure 5.4 ■ CodeWarrior screen after getting rid of "Tips of the Day"

In the following, we will enter a program that finds the maximum and minimum elements of an array of integers.

To enter a new program file, press the **File** menu and select **New Text File** from the Code-Warrior window as shown in Figure 5.5. An empty screen with the keyword **untitled** on the top bar will appear and the user can start to type in the program. The entered program that finds the maximum and minimum elements of the given array is shown in Figure 5.6.

Save the entered program by pressing the **File** menu and select **Save As** as shown in Figure 5.7. The user should enter the filename and press **Save,** and the file will now be saved.

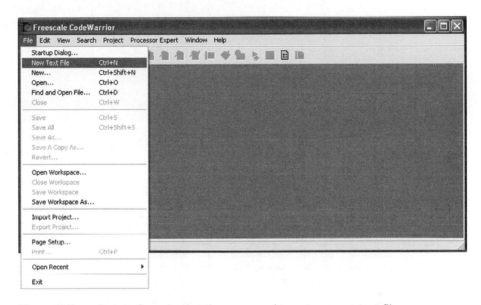

Figure 5.5 ■ Screen for selecting the command to enter a new text file

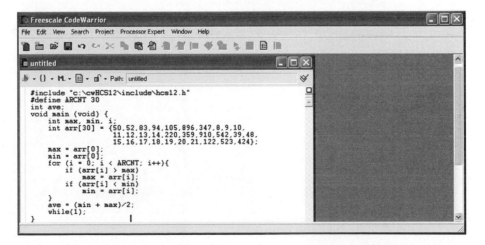

Figure 5.6 ■ Program for finding the maximum and minimum elements of an array

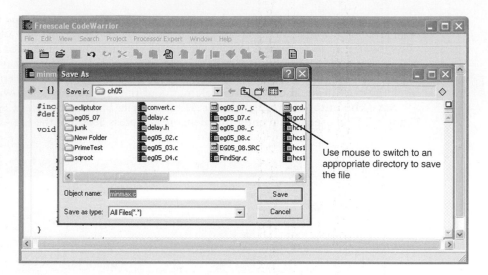

Figure 5.7 ■ Save the entered program in a file called "minimax.c"

5.11.2 Create a New Project

Like most IDEs, CodeWarrior also uses the project to manage programming effort. A new project can be created by pressing the **File** menu and selecting **New** . . . as shown in Figure 5.8. After selecting New . . . , a popup dialog prompts the user to enter the project name as shown in Figure 5.9. The user also needs to select the directory to hold the project information. This is done by clicking on the button **Set** in Figure 5.9. After that another dialog box allows the user to browse and select the directory to hold the project.

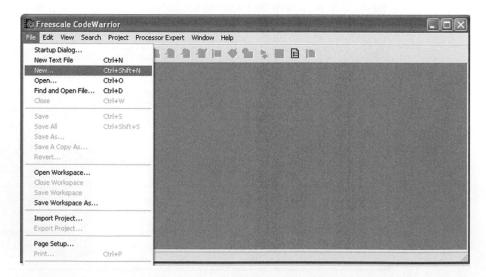

Figure 5.8 ■ Select "New . . ." under the File menu to create a new project

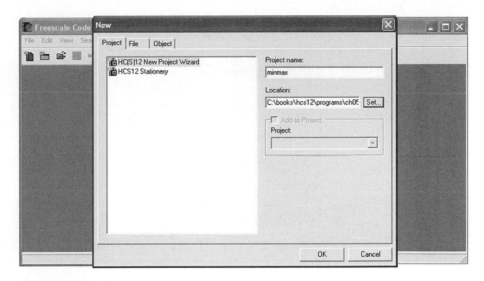

Figure 5.9 ■ Popup dialog that allows the user to enter the project name and project directory

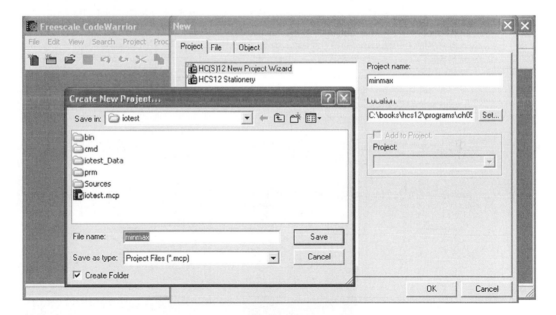

Figure 5.10 ■ Popup dialog box for setting project directory

After selecting the *project* directory, the user should click on the **Save** button to save the project information. As shown in Figure 5.10, the user enters *minmax* as the name of the project to save. After this, the screen will change back to that in Figure 5.9. The user should click on OK and the screen will change to that in Figure 5.11. Click on Next and the screen for selecting the HCS12 device will appear as shown in Figure 5.12.

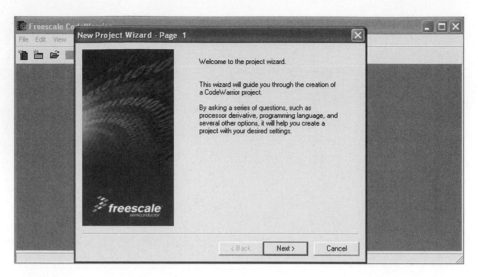

Figure 5.11 ■ CodeWarrior screen after saving a newly created project

The user should select an appropriate HCS12 device and click on the **Next** button. The screen will then change to that in Figure 5.13. The user should select the programming language using the dialog screen in Figure 5.13 and click on **Next.**

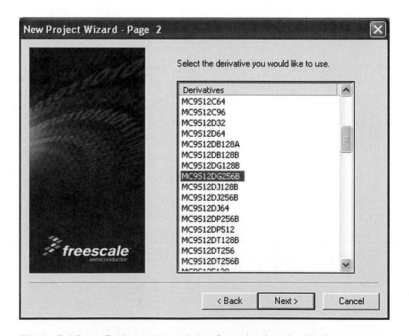

Figure 5.12 ■ Project wizard dialog for selecting the device

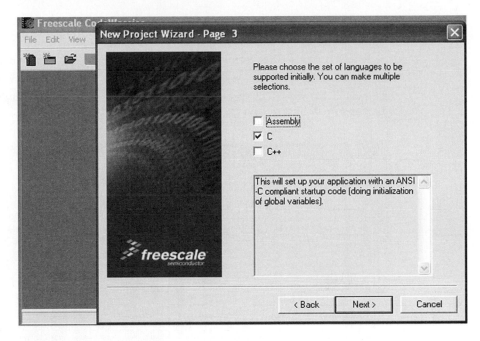

Figure 5.13 ■ Project wizard dialog for selecting programming language

After selecting the programming language, there are six more dialogs before reaching the dialog for selecting the connection method shown in Figure 5.14.

1. Use processor expert dialog: select no.
2. Want to use the OSEKturbo operating system in the project dialog: select no.

Figure 5.14 ■ Project wizard dialog for selecting connection method

3. Want to create a project set up for PC-lint dialog: select no.

4. Level of startup code dialog: select ANSI startup code.

5. Select floating-point format supported dialog: select none to reduce overhead.

6. Memory model selection dialog: select banked.

In Figure 5.14, we select three connection methods so that we can switch among them. Click on the **Finish** button, and the screen is changed to that in Figure 5.15.

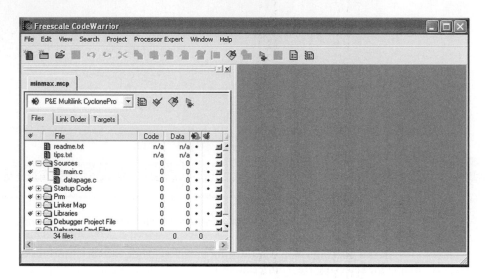

Figure 5.15 ■ CodeWarrior screen after a new project is created

In Figure 5.15, the P&E Multilink CyclonePro connection method is selected. However, the user can switch to two other connection methods depending on the debugging method that is used.

5.11.3 Adding Source Code to the Project

In the left pane of Figure 5.15, there are two files under the *Sources* directory: *main.c* and *datapage.c*. Both are generated by CodeWarrior. The datapage.c file contains paged data access runtime routines that handle the data access across different pages in expanded memory. Users do not need to be concerned about this file.

The main.c file contains the template of the required main function. The user can modify this function to perform the desired operation or delete it from the project and add the desired main function into the project. We will adopt the second approach.

To delete a file, press the right mouse button on the filename to bring up the popup dialog for deleting the file as shown in Figure 5.16. After removing main.c from the project, we press the right mouse button on the **Sources** directory and select **Add Files**. A new popup dialog as shown in Figure 5.17 will be brought up to allow the user to select a file to be added into the Sources directory. We can now use the mouse to browse the file directory tree to select the file to be added. In this tutorial, we add *minmax.c* that was created earlier to the Sources directory. The screen will change to that in Figure 5.18 after selecting the file (minmax.c) to be added to the project. After clicking on **OK,** the minmax.c file will be added under the Sources directory.

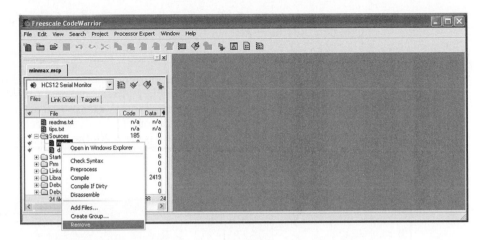

Figure 5.16 ■ Popup dialog for removing a file

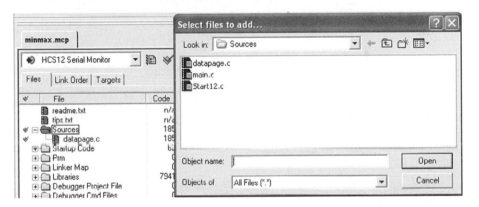

Figure 5.17 ■ Popup dialog for adding a file into the project

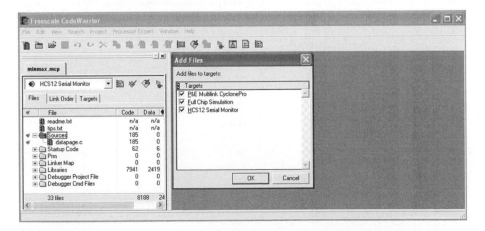

Figure 5.18 ■ CodeWarrior screen after selecting a file to be added into the project

The contents of a file can be displayed in the right pane of the CodeWarrior window by double clicking on the filename.

5.11.4 Building the Project

A project can be built by pressing the **Project** menu and selecting **Make** (shown in Figure 5.19) or simply pressing the **F7** function key. After a project build, CodeWarrior will display error messages and warnings if there is any syntax error or potential problem, as shown in Figure 5.20. The user must fix program errors in order to get the project built. CodeWarrior will not display anything if there are no errors. In Figure 5.20, the warning is caused by the statement of while(1); in Figure 5.19; this is fine.

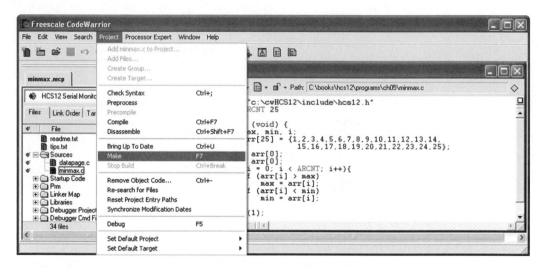

Figure 5.19 ■ Select "Make" to build (compile the programs contained in) the project

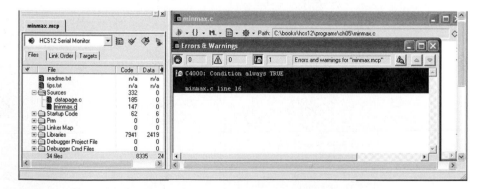

Figure 5.20 ■ Error and warnings after building a project

5.11.5 Executing and Debugging the Program with CodeWarrior

Program debugging is carried out by the debugger. The CodeWarrior debugger is started by pressing the function key **F5** or pressing the **Project** menu and selecting **Debug** from the CodeWarrior IDE window. The initial response of the CodeWarrior debugger varies depending on the chosen connection method.

INITIAL DEBUGGER RESPONSE FOR THE SERIAL MONITOR

The debugger startup screen for the serial monitor is shown in Figure 3.34. The user should select the appropriate communication port and then click on **OK.** After the debugger resets the MCU successfully, the screen will change to that in Figure 5.21. If the screen is not the same as that in Figure 5.21, the user should press the **Reset** button on the demo board.

INITIAL RESPONSE UNDER THE P&E MULTILINK

If the user uses a P&E Multilink or a CyclonePro debugger to debug the program running on the Dragon12-Plus demo board, the initial debugger screen after pressing the F5 function will look like that in Figure 3.36. This screen asks the user whether it is OK to erase the flash memory and download the program onto the memory. Click on **OK** and the screen will change to that shown in Figure 5.21. The debugger screen layout is explained in Section 3.9.5.

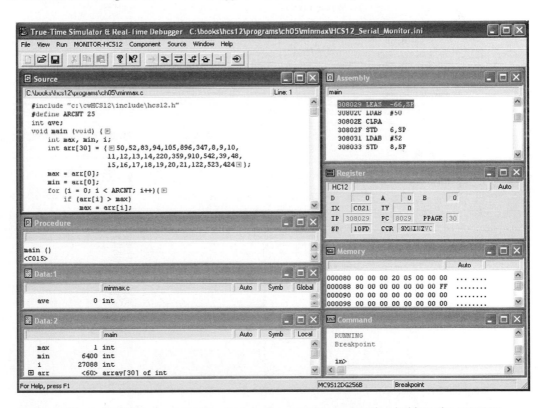

Figure 5.21 ■ CodeWarrior debugger screen after program is downloaded into the memory

CODEWARRIOR DEBUGGER SCREEN

The debugger screen is divided into eight small windows. Not all the windows are useful in the debug session. The user can get rid of those windows that are of no interest by clicking on the × symbol of that window's upper right corner. In this tutorial, we get rid of the Procedure window and resize the Source so that more statements of the main function can be displayed at the same time.

The windows within the debugger screen can be moved around and resized. The **Data:1** window displays the global variable (*ave*) and its value. The **Data:2** window displays the local variables (*max*, *min*, *i*, and *arr*) and their values.

The Assembly window shows how the main function is translated into assembly language. The main function has four local variables (one is an array with 30 elements) and needs 66 bytes to hold them. The first instruction, LEAS −66,SP allocates space in the stack to these local variables. The following instructions pushed the initial values of the array into the stack (pushing one value requires three instructions):

```
ldab    #50
clra
std     6,SP
...
```

DEBUGGER COMMANDS

CodeWarrior debugger provides commands to support debug activities. By pressing the right mouse button on any statement of the main function or other function, one can see all the debug commands supported by the debugger as shown in Figure 5.22.

If we single-step (by clicking on the corresponding icon shown in Figure 3.35) the program a few times, we can see that the debugger steps over three instructions to push one array element into the stack.

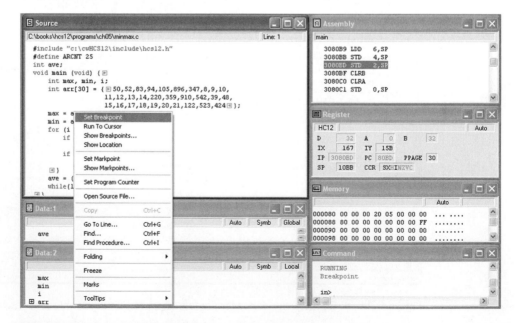

Figure 5.22 ■ CodeWarrior debugger commands

Now place the mouse on the statement min = arr[0];, press the right mouse button, and select **Run To Cursor**. We can see that the value of max in the **Data:2** window changes to 50, and the instruction **STD 2,SP** is highlighted in the Assembly window (shown in Figure 5.23). The value of *max* is correct.

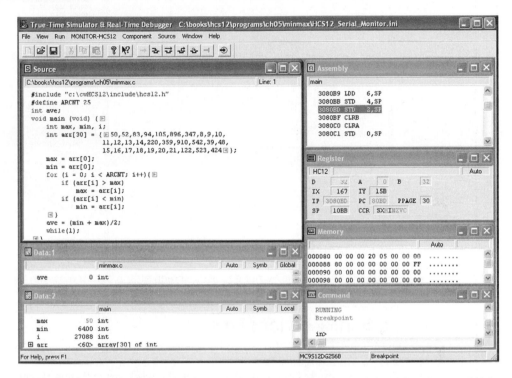

Figure 5.23 ■ CodeWarrior debugger screen after program executes until "min = arr[0];" statement

Now press the mouse on the statement **while(1)**; and execute the **Run To Cursor** command again; the debugger window will change to that in Figure 5.24. The values of variables *min*, *max*, *i*, and *ave* are changed to 8, 910, 30, and 459, respectively, and are correct. The variables *min* and *max* receive the minimum and maximum values of the given array as expected whereas the value of *ave* equals the average of *min* and *max*.

This completes the CodeWarrior debugger tutorial. It is highly recommended that the reader experiment with other debug commands to become more familiar with the use of the debugger.

5.12 Using the ImageCraft C Compiler

The ImageCraft ICC12 is a simple IDE that combines a text editor, a project manager, a C compiler, and a terminal program. The terminal program allows the ICC12 to work with the Dragon12-Plus programmed with the D-Bug12 monitor.

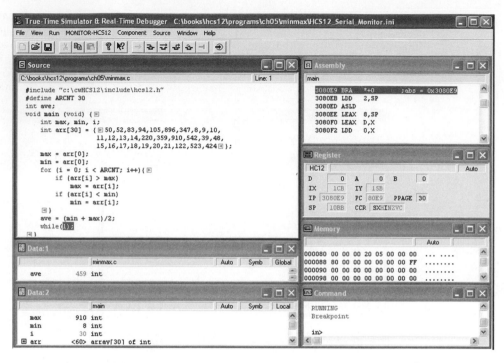

Figure 5.24 ■ CodeWarrior debugger screen after program runs into the "while(1);" statement

5.12.1 Starting the ICC12 IDE

The ImageCraft ICC12 IDE can be started by clicking on its icon. The startup window of ICC12 is shown in Figure 5.25.

The user enters his or her program using the work space. When the program is successfully compiled into a **.s19** file, the user also uses the work space as the terminal window to download

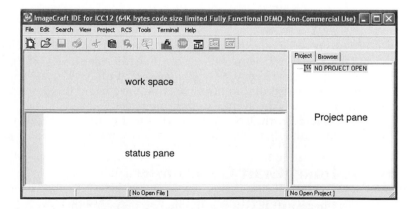

Figure 5.25 ■ ImageCraft ICC12 IDE startup window

the file onto the demo board for execution. Error messages and warnings generated by the ICC12 IDE are displayed in the status pane. The project file structure is displayed in the Project pane.

When the user uses the ICC12 IDE for the first time, he or she needs to set up the compiler options. The compiler options dialog can be brought up by pressing the **Project** menu in Figure 5.25 and selecting Options . . . . A dialog box as shown in Figure 5.26 will appear to allow the user to set all the compiler options to appropriate values. The setting in Figure 5.26 will work for the Dragon12-Plus demo board. When the user is satisfied with the compiler options, he or she should click on **OK** to get rid of the dialog.

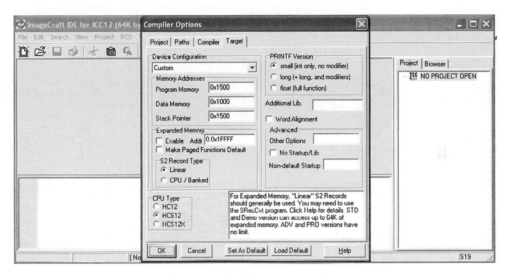

Figure 5.26 ■ Dialog for setting ICC12 compiler options

Compiler options need only be set once. The ICC12 C compiler will remember the options until the user changes them.

5.12.2 Creating a New Project

A new project can be created by pressing the **Project** menu and selecting **New.** A popup dialog box as shown in Figure 5.27 will appear to allow the user to enter the project name and select the project directory (specified in the space to the right of **Save in:**). If we set the project name to **ICCTUTOR** and save it in the directory of *ch05* and then click on **Save,** the screen will change to that in Figure 5.28.

5.12.3 Adding Files to the Project

A new file can be created by pressing on the **File** menu and selecting **New** to convert the work-space pane in Figure 5.25 into an editor work space. We will use the program that finds the greatest common devisor (gcd) as an example to illustrate the project build process for the ICC12 IDE. The program is shown in Figure 5.29. The printf function provided by ImageCraft requires the putchar function to work. However, putchar was not provided by the ICC12 IDE. A version must be created and added to your project.

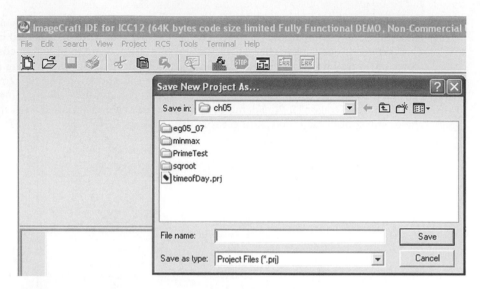

Figure 5.27 ■ Dialog for creating a new project

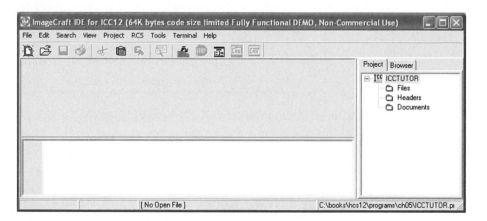

Figure 5.28 ■ ICC12 IDE screen after a new project is created

We can add an existing file into the project or enter a new program and then add it into the project. There are three directories in the Project pane. To add an existing file into the project, press the right mouse button on **Files** in the Project pane to bring up the dialog for adding files into the project (shown in Figure 5.30). After adding all the files needed in the project, the ICC12 IDE screen changes to that in Figure 5.31. We can examine the contents of a file by double clicking on the file name.

5.12.4 Building the Project

A project can be built (compiled and linked) by pressing the **Project** menu and selecting **Make Project** or clicking on the function key **F9.** Error messages and warnings will be displayed

Figure 5.29 ■ The GCD function and its test program

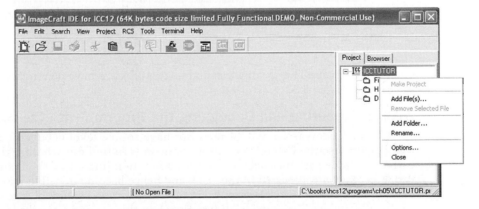

Figure 5.30 ■ Add a file into the project "ICCTUTOR"

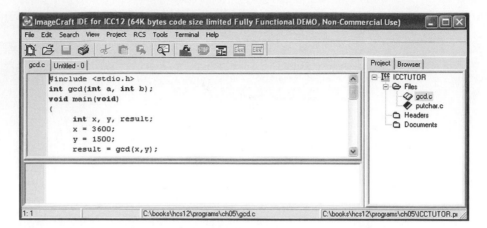

Figure 5.31 ■ ICC12 window after adding files to the project "ICCTUTOR"

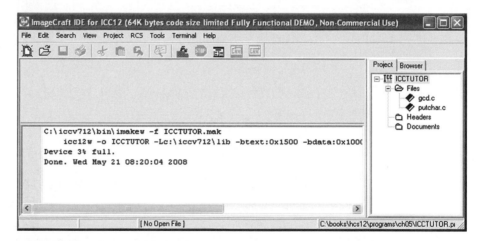

Figure 5.32 ■ ICC12 project build message without errors

on the status pane. The gcd function and its test program have no errors, and the build message is shown in Figure 5.32.

5.12.5 Executing and Debugging the Program with the ICC12 IDE

Unlike CodeWarrior, ICC12 does not have source-level debugging capability. The user will need to use the D-Bug12 monitor commands to find out whether the program executes correctly. Since variables declared within a function (including the main function) are local variables, they are allocated in the stack and hence become invisible when the function is exited. In order to view the program execution result, the variables of interest must be declared as global variables (i.e., outside all functions). In this example, the declaration of the variable *result* should be moved out of the main function.

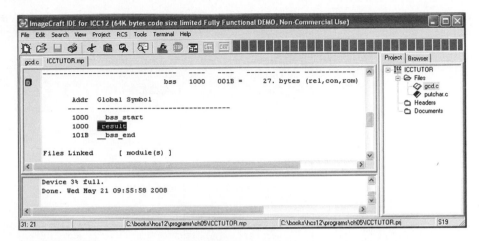

Figure 5.33 ■ Find out the address of program variables by examining the map file

After making *result* a global variable, remake the project. The project build process generates a map file. For example, the map file of the *icctutor* project is *icctutor.mp*. The user can scroll around the file *icctutor.mp* and find the address of *result* (changed to *_result*). As shown in Figure 5.33, the variable *result* is assigned to the memory location $1000.

ACTIVATE THE TERMINAL WINDOW

Before the user can download the program onto the demo board, she or he needs to switch the work space to be used as the terminal window because the work space can be used as the editor space or terminal window. Activate the terminal window by pressing the **Terminal** menu and select **Show Terminal Window.** The ICC12 screen with the terminal window activated is shown in Figure 5.34.

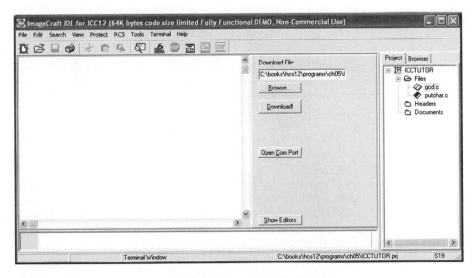

Figure 5.34 ■ ICC12 screen with terminal window activated

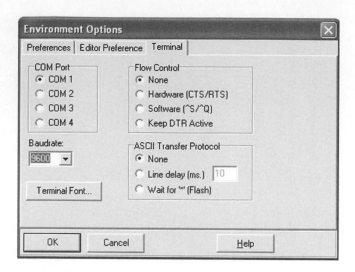

Figure 5.35 ■ Dialog for configuring Terminal setting

When using the terminal window for the first time, the user needs to perform the appropriate configuration. The user should press the **Tools** menu and select **Environment Options** to bring out the dialog as shown in Figure 5.35. The baud rate should be set to the same as the demo board (9600 for D-Bug12 monitor). The choices for the flow control and ASCII Transfer Protocol should be set to **None.**

DOWNLOADING THE PROGRAM FOR EXECUTION

The terminal environment needs to only be configured once. The ICC12 IDE will save the setting for future use. Before the PC can communicate (via the ICC12 terminal window) with the D-Bug12 monitor on the demo board, the user should click on the **Open Com Port** button. Clicking on the **Open Com Port** button requests the PC to configure the selected COM port (COM 1 in Figure 5.35) to operate with the same setting as that of the demo board.

The user can then press the **Enter** key to bring up the D-Bug12 monitor prompt as shown in Figure 5.36. To download a program file onto the demo board for execution, type the command **Load** and then press the **Enter** key. After this, the user can use the mouse to browse the directory to select the appropriate *.s19* (*icctutor.s19* in this example) to be downloaded.

Clicking on the **Open** button in Figure 5.36 will cause Windows to transfer the selected file into the main memory from the hard disk. Clicking on the **Download!** button in Figure 5.36 will start the download operation. After the icctutor.s19 file is downloaded, the ICC12 IDE will make a sound and the screen will change to that in Figure 5.37. Due to a bug in ICC12, the completion of the file download will not bring up the D-Bug12 monitor prompt.

To bring back the D-Bug12 monitor prompt, press the **Reset** button on the demo board. To execute the downloaded program, type **g 1500** and then press the **Enter** key. To display the value of the result variable, type the **md 1000** command. The screen will change to that shown in Figure 5.38 after the previous commands are entered. The contents of the memory location at $1000 and $1001 are $012C (= 300_{10}) and are correct.

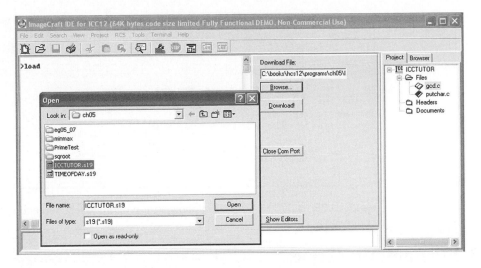

Figure 5.36 ■ Select a file (ICCTUTOR.s19) to be downloaded

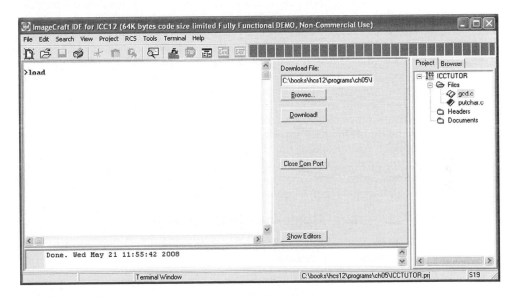

Figure 5.37 ■ ICC12 screen after a successful program download

Because the ICC12 IDE cannot update the variable values at the breakpoint like the Code-Warrior IDE, the user is advised to use the print function call or output data on the LCD or other output devices to verify the program execution result.

If the user wants to edit the *gcd.c* file, he or she can double-click on gcd.c under the *Files* directory in the Project pane. This completes the tutorial of the ICC12 IDE.

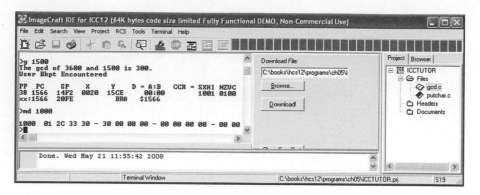

Figure 5.38 ■ ICC12 terminal window after program execution and the "md 1000" command

5.13 Programming Style

Programming style refers to a set of rules or guidelines used when writing a program. It is believed that by following a particular programming style the resultant program will be more readable and easier to debug. The programming style used in a particular program may be derived from the **coding standards** or **code conventions** of a company or other computing organization, as well as the preferences of the author of the code. Programming styles are often designed for a specific programming language (or language family), but some rules are commonly applied to many languages. (Style considered good in C source code may not be appropriate for BASIC source code, and so on.)

The essence of good programming style is **communication**. Any program that will be used must be maintained by somebody and that somebody must be able to understand the code by reading it. Any program that needs debugging will be easier to debug if the creator carefully explains what's going on. Within the program text, programmers have three primary tools for communicating their intentions: **comments**; **naming** of variables, constants, and subroutines; and **program appearance** (spacing, alignment, and indentation).

5.13.1 General Guidelines to Comments

The most useful things to know about program documentation are

- What to comment
- When to comment

In general, the programmer should include comments explaining what every function does, what every variable does, and an explanation for every tricky expression or section of code. "When" is easy—*comment it before you write it*. Whenever you declare a variable, include a comment. Whenever you start to write a subroutine, first write a comment explaining it. That will help to make clear in your mind what you are about to do. If you find the explanation for a subroutine difficult to write, it's a sure sign that you have done a poor job of structuring the program. *Avoid obscure programming language constructs and reliance on language-specific precedence rules.* It is often better to force precedence by use of parentheses since this leaves no doubt as to meaning. In general, if you had to look up some rule or definition, your readers most likely will too. Whenever you need to write a difficult expression or some other tricky business and find it difficult to do, you should expect that it will be difficult to understand. So, add a comment.

5.13.2 Program Documentation

All programs should include, at or near the beginning of the program (the **main()** function in C), a comment block. The comment block should include at least the following information:

- The programmer's name
- The date of creation
- The operating system and IDE for which the program was written
- Hardware environment (circuit connection) to run the program
- Program structure (organization)
- Algorithm and data structures of the program
- How to run the program

Below is an example of a program comment header block.

```
// ********************************************************************************
// Program: RtiMultiplex7segs
// Author: Han-Way Huang
// Build Environment: CodeWarrior IDE under Windows XP
// Date: 07/09/2008
// Hardware connection: shown in Figure 4.18 of Huang's HCS12 text
// Operation: This program shifts seven-segment patterns of 4 consecutive BCD digits
//       by using time-multiplexing technique with each pattern lasting for 0.5 s. The
//       time-multiplexing operation is controlled by the real-time interrupt. The patterns are
//       1 2 3 4
//       2 3 4 5
//       3 4 5 6
//       4 5 6 7
//       5 6 7 8
//       6 7 8 9
//       7 8 9 0
//       8 9 0 1
//       9 0 1 2
//       0 9 1 2
// ********************************************************************************
```

5.13.3 Function (Subroutine) Documentation

Each C function should have a clearly defined task. Writing the comment describing that task before writing the function can help to keep you straight. Functions should be short enough so that their meanings can be grasped as a unit by the reader. Normally, one page of code represents the limit of intellectual complexity that a reader can grasp.

Similar to the main() function, every function should be preceded by a comment describing the purpose and use of that subroutine. Below is an example of a subroutine comment header block.

```
// ********************************************************************************
// Function: FindSquareRoot
// Purpose: Computes and returns the square root of a 32-bit unsigned integer
// Parameter: A 32-bit unsigned integer of which the square root is to be found
// Called by: PrimeTest()
// Returned value: the square root of a 32-bit unsigned integer
// Side effects: none
// ********************************************************************************
```

One of the greatest sources of bugs is in the interface between functions. It is a common mistake to have type mismatches between the formal parameters and the actual parameters. It is also common to mis-order parameters, or to leave some out. The programmer should always be scrupulous in checking for such problems when writing a program and do it early on. One way to minimize the problem is to keep the number of parameters to any one function at a minimum. Seven or more parameters for a function is probably too many—having this many parameters is an indication that the programmer has structured his or her program in the wrong way.

5.13.4 Code Appearance

In addition to comments, program readability can be improved by

- Proper spacing
- Proper indentation
- Vertical alignment

PROPER SPACING

Adding spaces makes your program statements easier to read. For example, it is quite obvious that the second one of the following two statements is easier to read:

```
for (i = 0; i < 15; i++)        &        for (i = 0; i < 15; i++)
```

PROPER INDENTATION

Indentation is especially helpful in improving the readability of nested **if-elseif** statements and nested program loops. With proper indentation, the logical hierarchies of nested if-elseif statements and nested program loops become much clearer. For example, a programmer may write the main loop of the seven-segment patterns shifting program as follows:

```
while(TRUE) {
for (i = 0; i < 10; i++) {      // pattern array start index
for (j = 0; j < 100; j++) {      // repeat loop for each pattern sequence
for (k = 0; k < 6; k++) {      // select the display # to be lighted
PTB   = SegPat[i+k];      // output segment pattern
PTP   = digit[k];      // output digit select value
delayby1ms(1);      // display one digit for 1 ms
}
}
}
    }
```

Obviously, the readability of these statements is very poor. It is quite difficult to identify the hierarchical relationship of these for statements. The readability would be much improved if these for statements are indented as follows:

```
while(TRUE) {
    for (i = 0; i < 10; i++) { // pattern array start index
        for (j = 0; j < 100; j++) { // repeat loop for each pattern sequence
            for (k = 0; k < 6; k++) {      // select the display # to be lighted
                PTB   = SegPat[i+k];      // output segment pattern
                PTP   = digit[k];      // output digit select value
```

                    delayby1ms(1);   // display one digit for 1 ms
                  }
              }
              }
          }

VERTICAL ALIGNMENT

It is often helpful to align similar elements vertically, to make typo-generated bugs more obvious. Compare the array definitions given in Figure 5.39; it is obvious that the reader will be able to locate any missing elements more easily with version b than with version a.

```
(a) non-vertically aligned array

// values to generate 1 cycle of sine wave
unsigned rom char upper[60] = {0x38, 0x38, 0x39, 0x3A, 0x3B, 0x3C,
    0x3C, 0x3D, 0x3D, 0x3E, 0x3E, 0x3F, 0x3F, 0x3F, 0x3F,
0x3F, 0x3F, 0x3F, 0x3F, 0x3F, 0x3E, 0x3E, 0x3D, 0x3D,
0x3C, 0x3C, 0x3B, 0x3A, 0x39, 0x38, 0x38, 0x37, 0x36, 0x35, 0x34, 0x34,
    0x33, 0x32, 0x32, 0x31, 0x38, 0x37, 0x36, 0x35, 0x34, 0x34,
0x30, 0x30, 0x30, 0x30, 0x30, 0x31, 0x31, 0x32, 0x32, 0x33,
    0x34, 0x34, 0x35, 0x36, 0x37};

(b) vectically aligned array

// values to generate 1 cycle of sine wave
unsigned rom char upper[60] = {
            0x38, 0x38, 0x39, 0x3A, 0x3B, 0x3C, 0x3C, 0x3D, 0x3D, 0x3E,
            0x3E, 0x3F, 0x3F, 0x3F, 0x3F, 0x3F, 0x3F, 0x3F, 0x3F, 0x3F,
            0x3E, 0x3E, 0x3D, 0x3D, 0x3C, 0x3C, 0x3B, 0x3A, 0x39, 0x38,
            0x38, 0x37,0x36,0x35,0x34,0x34,0x33,0x32,0x32,0x31,
            0x31, 0x30, 0x30, 0x30, 0x30, 0x30, 0x30, 0x30, 0x30, 0x30,
            0x31, 0x31, 0x32, 0x32, 0x33, 0x34, 0x34, 0x35, 0x36, 0x37};
```

Figure 5.39 ■ Influence of vertical alignment of array on readability

5.13.5 Naming of Variables, Constants, and Functions

The names of variables, constants, and functions should spell out their meaning or purpose.

A variable name may have one word or multiple words. Use lowercase when the name contains only one word. For example, *sum*, *limit*, and *average* are examples of single-word variable names. A multiple-word variable name should be in mixed case starting with lowercase. For example,

    inBuf, outBuf, squareRoot, arrayMax

Function names should follow the same principle. A few examples of function names are

sevenSegShift(), putcSPI(), putsSPI(), openLCD(), and putsLCD()

Named constants should be all uppercase using underscores to separate words. A few examples of constant names are

SONG_TOTAL_NOTES, HI_DELAY_COUNT, LO_DELAY_COUNT

5.14 Tips for C Program Debugging

As in assembly language, C program errors can be divided into two categories:

- Syntax/semantics errors
- Logical errors

Many syntax/semantics errors can be discovered if the programmer follows the rules of C language.

- *Undeclared variables.* A variable must be declared before it can be used. Most errors of this type can be discovered easily.
- *Variable and symbol names are case sensitive.* Sometimes, the programmer has declared a variable but still has an undeclared variable error with the same variable. This error could occur because one or more characters of the variable name are in the wrong case. For example, in the program segment

```
int        A1, A2, xyClk;
. . .
A1 = A1 * xycLk;
```

the compiler would output the error message "undefined variable xycLk."

- *Missing parenthesis (or brace or bracket).* A missing parenthesis can cause many other errors. This type of error can be avoided by entering the matching parenthesis whenever the left parenthesis is entered before entering statements between the parentheses.
- *Mismatch of function name in function prototype declaration and function definition.* The EGNU will generate the error message "unknown error—see Make Log." When the user clicks on **Make log** on the status pane of EGNU, the message "undefined reference to . . ." will appear. ICC12 and CodeWarrior C compilers would output the error message "reference functions without prototype."

Logical errors are harder to debug than syntax and semantic errors. Whenever the program behavior is not what we expect, the first step is to read the program carefully to find out where the most likely problem spot is. After identifying the potential problem spot, we can set a breakpoint and examine the program execution result.

If the user has a source-level debugger (CodeWarrior) to debug the program, the following actions can be taken to solve the problem:

- *Set breakpoints.* Breakpoints allow the user to examine program execution results at the suspicious point.
- *Set up a watch list (in Data:1 and Data:2 windows in CodeWarrior IDE).* A watch list is used together with the breakpoints. A watch list consists of pairs of program

variables of interest and their values at a certain time. It allows the user to find out program execution results quickly at any breakpoint.

■ *Trace program execution.* Users can find out the execution result of a few instructions by tracing instruction execution. Without a source-level debugger, it is very difficult to trace the program execution unless the D-bug12 monitor commands are used.

The freeware EGNU and ICC12 IDE provide very little support for debugging. A commercial source-level debugger for C language can cost thousands of dollars. Without a source-level debugger, many debugging activities cannot be performed easily. CodeWarrior IDE provides very useful and helpful features to support programming debugging.

There are several purposes for setting breakpoints.

■ *To determine whether a segment of code has ever been entered by the CPU.* The programmer can use the embedded assembly instruction **asm(swi)** to find out with a demo board programmed with the D-Bug12 monitor. Using the swi instruction allows the programmer to find out where the program execution gets stuck and identify the error.

■ *To determine whether the execution result is correct up to the breakpoint.* Without a source-level debugger, the programmer can output the program execution result to the LCD or terminal monitor to find out if the program executes correctly up to the breakpoint. This also serves as a watch list for the program execution.

The following guidelines can help reduce and identify logic errors:

■ Make sure the precedence of operators has been observed.
■ Match the size of the source operands and that of the destination variables. Use type casting when necessary. Type casting has been used in several examples in this text.
■ Walk through the program algorithm carefully before converting it into the program code. An incorrect algorithm is often the cause of program bugs.

Use enough data to test the program. The program must be tested with normal inputs, maximum and minimum inputs, and also the illegal inputs to make sure that it operates correctly under all circumstances. When the problem to be solved gets complicated, a structured programming approach should be used to organize the program. The guidelines described in Section 2.4 should be followed to develop the program algorithm and convert the algorithm into program code. Each individual function should be tested thoroughly before the whole program is tested. A comprehensive discussion of structured programming and testing is beyond the scope of this text but can be found in many textbooks on software engineering.

5.15 Summary

A C program consists of one or more functions and variables. The **main ( )** function is required in every C program. It is the entry point of a C program. A function contains statements that specify the operations to be performed. The types of statements in a function could be *declaration*, *assignment*, *function call*, *control*, and *null*.

A *variable* stores a value to be used during the computation. A variable must be declared before it can be used. The declaration of a variable consists of the name and the type of the variable. There are four basic data types in C: *char*, *int*, *float*, and *double*. Several qualifiers can be added to the variable declarations. They are *short*, *long*, *signed*, and *unsigned*.

Constants are often needed in forming a statement. There are four types of constants: *integers, characters, floating-point numbers,* and *strings.*

There are seven *arithmetic operators*: +, −, *, /, %, ++, and −−. There are six *bitwise operators*: &, |, ^, ~, >>, and <<. Bitwise operators can be applied only to integers. *Relational operators* are used in control statements. They are ==, !=, >, >=, <, <=, &&, ||, and !.

The *control-flow statements* specify the order in which computations are performed. Control-flow statements include if-else statements, multiway conditional statements, switch statements, for-loop statements, while statements, and do-while statements.

Every C program consists of one or more functions. If a program consists of multiple functions, their definitions cannot be embedded within another. The same function can be called from several different places within a program. Generally, a function will process information passed to it from the calling portion of the program and return a single value. Information is passed to a function via special identifiers called *arguments* (also called *parameters*) and returned via the *return* statement. Some functions, however, accept information but do not return anything (for example, the library function **printf**).

A *pointer* holds the address of a variable. Pointers can be used to pass information back and forth between a function and its reference (calling) point. In particular, pointers provide a way to return multiple data items from a function via function arguments. Pointers also permit references to other functions to be specified as arguments to a given function. Two operators are related with pointers: * and &. The * operator returns the value of the variable pointed to by the pointer. The & operator returns the address of a variable.

Data items that have common characteristics are placed in an *array*. An array may be one-dimensional or multidimensional. The dimension of an array is specified by the number of square bracket pairs [ ] following the array name. An array name can be used as an argument to a function, thus permitting the entire array to be passed to the function. To pass an array to a function, the array name must appear by itself, without brackets or subscripts. An alternative way to pass arrays to a function is to use pointers.

A variable defined inside a function is an *internal variable* of that function. *External variables* are defined outside of any function and are thus potentially available to many functions. The *scope* of a name is the part of the program within which the name can be used. The scope of an external variable or a function lasts from the point at which it is declared to the end of the file being compiled.

Tutorials for using the CodeWarrior IDE and ImageCraft ICC12 are provided at the end of this chapter. The tutorial for using the EGNU IDE is in Appendix E. All three IDEs are window driven, but only CodeWarrior provides source-level debugging capability.

Neither ImageCraft nor EGNU IDEs provides many library functions to the user. Three sets of library functions are provided in the complementary CD and are listed in Table E.1 of Appendix E. These three sets of library functions are stored in files *stdio0.c, delay.c,* and *convert.c.*

5.16 Exercises

E5.1 Assume that $ax = 83$ and $bx = 11$. What is the value of ax/bx?

E5.2 Assume that $ax = 97$ and $bx = ax \% 23$. What is the value of bx?

E5.3 Assume that $ax = 0x39$ and $bx = ax \wedge 0x79$. What is the value of bx?

E5.4 Assume that $ax = 0x9D$ and $bx = ax << 2$. What is the value of bx?

E5.5 Assume that $ax = 0x6B$ and $bx = ax \& 0xDE$. What is the value of bx?

E5.6 Write a C function to test whether an integer is a multiple of 8. If the number to be tested is a multiple of 8, the function returns a 1 to the caller. Otherwise, it returns a 0 to the caller.

E5.7 Write a C program to find the median and mode of an array of integers. When the array has an even number of elements, the median is defined as the average of the middle two elements. Otherwise, it is defined as the middle element of the array. The mode is the element that occurs most frequently. You need to sort the array in order to find the median.

E5.8 Write a function that tests if a given number is a multiple of 10. A 1 is returned if the given number is a multiple of 10. Otherwise, a 0 is returned.

E5.9 Write a function that computes the least common multiple (lcm) of two integers m and n.

E5.10 What is a function prototype? What is the difference between a function prototype and function declaration?

E5.11 Write a *switch* statement that will examine the value of an integer variable xx and store one of the following messages in an array of seven characters, depending on the value assigned to xx (terminate the message with a NULL character):

a) Cold if $xx == 1$

b) Chilly if $xx == 2$

c) Warm if $xx == 3$

d) Hot if $xx == 4$

E5.12 Write a C program to clear the screen and then move the cursor to the middle of a line and output the message "Microcontroller is fun to use!" The screen can be cleared by outputting the character \f (form-feed character).

E5.13 Write a function that will convert an uppercase letter to lowercase.

E5.14 Write a C program that swaps the first column of a matrix with the last column, swaps the second column of the matrix with the second-to-last column, and so on.

E5.15 Write a loop to compute the sum of the squares of the first 100 odd integers.

E5.16 An *Armstrong number* is a number of n digits that is equal to the sum of each digit raised to the nth power. For example, 153 (which has three digits) equals $1^3 + 5^3 + 3^3$. Write a function to store all three-digit Armstrong numbers in an array.

E5.17 Write a C function to perform a binary search on a sorted array. The binary search algorithm is given in Example 4.6. The starting address, the key, and the array count are parameters to this function. Both the key and array count are integers.

E5.18 Write a program to find the first five numbers that when divided by 2, 3, 4, 5, and 6, leave a remainder of 1 and, when divided by 7, have no remainder.

E5.19 Take a four-digit number. Add the first two digits to the last two digits. Now, square the sum. Surprise, you've got the original number again. Of course, not all four-digit numbers have this property. Write a C program to find three numbers that have this special property.

E5.20 Write a program to find six prime numbers that are closest to 10,000, with three of them being less than 10,000 and the other three being larger than 10,000, and print them out.

E5.21 For the seven-segment display circuit shown in Figure 4.18, write a C program to display 1, 2, 3, 4, 5, and 6 one digit at a time from left to right. Each digit is displayed for half a second.

5.17 Lab Exercises and Assignments

L5.1 Enter, compile, and download the following C program onto a demo board for execution using the procedure described in Section 5.10 (you need to add *delay.c*, *convert.c*, and *stdio.c* into the project):

```
#include "c:\cwHCS12\include\hcs12.h"
#include "c:\cwHCS12\include\delay.h"
```

```
#include "c:\cwHCS12\include\convert.h"
#include "c:\cwHCS12\include\stdio.h"
int main (void)
{
    const char *ptr = " seconds passed!";
    char buf[20];
    int  cnt;
    DDRB = 0xFF; /* configure port B for output */
    cnt  = 0;       /* initialize count to 0 */
    while(1) {
        delayby100ms(5);
        int2alpha(cnt,buf);
        putsr(&buf[0]);
        puts(ptr);
        cnt++;
        PORTB = cnt;
        newline();
    }
    return 0;
}
```

L5.2 Write a C program that will generate every third integer, beginning with $i = 2$ and continuing for all integers that are less than 300. Calculate the sum of those integers that are divisible by 5. Store those integers and their sum in an array and as an integer variable, respectively.

L5.3 Write a C function that calculates the *least common multiple* (lcm) of two integers m and n. Integers m and n are parameters to this function. Also write a main program to test this function with several pairs of integers. Use the library functions to convert and output the results.

L5.4 Use the seven-segment display circuit shown in Figure 4.18 and write a program to display the following sequence of digits with each sequence lasting for half a second:

       1
      2 1
     3 2 1
    4 3 2 1
   5 4 3 2 1
  6 5 4 3 2 1   (6)
  7 6 5 4 3 2
  8 7 6 5 4 3
  9 8 7 6 5 4
  0 9 8 7 6 5
  1 0 9 8 7 6
  2 1 0 9 8 7
  3 2 1 0 9 8
  4 3 2 1 0 9
  5 4 3 2 1 0   (15)

After displaying these 15 sequences once, repeat sequence 6 to 15 forever.

L5.5 Write a program to generate a four-tone siren using the PP5 pin with each tone lasting for half a second. The frequencies of these four tones are 100 Hz, 250 Hz, 500 Hz, and 1000 Hz. The PP5 pin is connected to a buzzer via a jumper on the Dragon12-Plus demo board.

6

Interrupts, Clock Generation, Resets, and Operation Modes

6.1 Objectives

After completing this chapter, you should be able to

- Explain the difference between interrupts and resets
- Describe the handling procedures for interrupts and resets
- Raise one of the HCS12 maskable interrupts to the highest priority
- Enable and disable maskable interrupts
- Use one of the low-power modes to reduce the power consumption
- Use COP watchdog timer reset to detect software failure
- Set up the interrupt vector jump table for demo boards that have the D-Bug12 monitor
- Use RTI to generate periodic interrupts
- Write interrupt-driven application programs
- Distinguish the HCS12 operation modes

6.2 Fundamental Concepts of Interrupts

Interrupts and resets are among the most useful mechanisms that a computer system provides. With interrupts and resets, I/O operations are performed more efficiently, errors are handled more smoothly, and CPU utilization is improved. This chapter begins with a general discussion of interrupts and resets and then focuses on the specific features of the HCS12 interrupts and resets. Examples are given to demonstrate how interrupts can be used to effectively trigger I/O operations.

6.2.1 What Is an Interrupt?

An interrupt is an event that requires the CPU to stop normal program execution and perform some service related to the event. An interrupt can be generated internally (inside the chip) or externally (outside the chip). An external interrupt is generated when the external circuitry asserts an interrupt signal to the CPU. An internal interrupt can be generated by the hardware circuitry inside the chip or caused by software errors. Most microcontrollers have timers, I/O interface functions, and the CPU incorporated on the same chip. These subsystems can generate interrupts to the CPU. Abnormal situations that occur during program execution, such as illegal opcodes, overflow, divide by zero, and underflow, are called *software interrupts*. The terms *traps* and *exceptions* are also used to refer to software interrupts.

A good analogy for an interrupt is how you act when you are sitting in front of a desk to read this book and the phone rings. You probably act like this.

1. Remember the page number or place a bookmark on the page that you are reading, close the book, and put it aside.

2. Pick up the phone and say, "Hello, this is so and so."

3. Listen to the voice over the phone to find out who is calling or ask who is calling if the voice is not familiar.

4. Talk to that person.

5. Hang up the phone when you finish talking.

6. Open the book and turn to the page where you placed the bookmark and resume reading this book.

The phone call example spells out a few things that are similar to how the microprocessor handles the interrupt.

1. As a student, you spend most of your time studying. Answering the phone call happens only occasionally. Similarly, the microprocessor is executing application programs most of the time. Interrupts will only force the microprocessor to stop executing the application program briefly and take some necessary actions.

2. Before picking up the phone, you finish reading the sentence and then place a bookmark to remind yourself of the page number that you are reading so that you can resume reading after finishing the conversation over the phone. Most microprocessors will finish the instruction they are executing and save the address of the next instruction in memory (usually in the stack) so that they can resume the program execution later.

3. You find out who the person is by listening to the voice over the phone, or you ask questions so that you can decide what to say. Similarly, the microprocessor needs to identify the cause of the interrupt before it can take appropriate actions. This is built into the microprocessor hardware.

4. After identifying the person who called you, you start the phone conversation with that person on some appropriate subjects. Similarly, the microprocessor will take some actions appropriate to the interrupt source.

5. When finishing the phone conversation, you hang up the phone, open the book to the page where you placed the bookmark, and resume reading. Similarly, after taking some actions appropriate to the interrupt, the microprocessor will jump back to the next instruction, after the interrupt occurred, and resume program execution. This can be achieved easily because the address of the instruction to be resumed was saved in memory (the stack). Most microprocessors do this by executing a *return-from-interrupt* instruction.

6.2.2 Why Are Interrupts Used?

Interrupts are useful in many applications, such as the following:

- *Coordinating I/O activities and preventing the CPU from being tied up during the data transfer process.* The CPU needs to know if the I/O device is ready before it can proceed. Without the interrupt capability, the CPU will need to check the status of the I/O device continuously or periodically. The interrupt mechanism is often used by the I/O device to inform the CPU that it is ready for data transfer. CPU time can thus be utilized more efficiently because of the interrupt mechanism. Interrupt-driven I/O operations will be explained in more detail in later chapters.

- *Performing time-critical applications.* Many emergent events, such as power failure and process control, require the CPU to take action immediately. The interrupt mechanism provides a way to force the CPU to divert from normal program execution and take immediate actions.

- *Providing a graceful way to exit from an application when a software error occurs.* The service routine for a software interrupt may also output useful information about the error so that it can be corrected.

- *Reminding the CPU to perform routine tasks.* There are many microprocessor applications that require the CPU to perform routine work, such as the following:

  1. *Keeping track of time of day.* Without the timer interrupt, the CPU will need to use program loops in order to update the current time. The CPU cannot do anything else without a timer interrupt in this application. The periodic timer interrupts prevent the CPU from being tied up.

  2. *Periodic data acquisition.* Some applications are designed to acquire data periodically.

  3. *Task switching in a multitasking operating system.* In a modern computer system, multiple application programs are resident in the main memory, and the CPU time is divided into many short slots (one slot may be from 10 to 20 ms). A multitasking operating system assigns a program to be executed for one time slot. At the end of a time slot or when a program is waiting for the completion of an I/O operation, the operating system takes over and assigns another program for execution. This technique is called *multitasking*. Multitasking can dramatically improve the CPU utilization and is implemented by using periodic timer interrupts.

6.2.3 Interrupt Maskability

Depending on the situation and application, some interrupts may not be desired or needed and should be prevented from affecting the CPU. Most microprocessors and microcontrollers have the option of ignoring these interrupts. These types of interrupts are called *maskable interrupts*. There are other types of interrupts that the CPU cannot ignore and must take immediate actions for; these are *nonmaskable interrupts*. A program can request the CPU to service or ignore a maskable interrupt by setting or clearing an *enable bit*. When an interrupt is enabled, the CPU will respond to it. When an interrupt is disabled, the CPU will ignore it. An interrupt is said to be *pending* when it is active but not yet serviced by the CPU. A pending interrupt may or may not be serviced by the CPU, depending on whether or not it is enabled.

To make the interrupt system more flexible, a computer system normally provides a global and local interrupt masking capability. When none of the interrupts are desirable, the processor can disable all the interrupts by clearing the global interrupt enable bit (or setting the global interrupt mask bit for some other processor). In other situations, the processor can selectively enable certain interrupts while at the same time disabling other undesirable interrupts. This is achieved by providing each interrupt source an enable bit in addition to the global interrupt mask. Whenever any interrupt is undesirable, it can be disabled while at the same time allowing other interrupt sources to be serviced (attended) by the processor. Today, almost all commercial processors are designed to provide this two-level (or even three-level) interrupt-enabling capability.

6.2.4 Interrupt Priority

If a computer is supporting multiple interrupt sources, then it is possible that several interrupts would be pending at the same time. The CPU has to decide which interrupt should receive service first in this situation. The solution is to prioritize all interrupt sources. An interrupt with higher priority always receives service before interrupts at lower priorities. Many microcontrollers, including the HCS12, prioritize interrupts in hardware. For those microcontrollers that do not prioritize interrupts in hardware, the software can be written to handle certain interrupts before others. By doing this, interrupts are essentially prioritized. For most microprocessors and microcontrollers, interrupt priorities are not programmable.

6.2.5 Interrupt Service

The CPU provides service to an interrupt by executing a program called an *interrupt service routine*. After providing service to an interrupt, the CPU must resume normal program execution. How can the CPU stop the execution of a program and resume it later? It achieves this by saving the program counter and the CPU status information before executing the interrupt service routine and then restoring the saved program counter and CPU status before exiting the interrupt service routine. The complete interrupt service cycle involves

1. Saving the program counter value
2. Saving the CPU status (including the CPU status register and some other registers) in the stack (This step is optional for some microcontrollers and microprocessors.)
3. Identifying the source of the interrupt
4. Resolving the starting address of the corresponding interrupt service routine
5. Executing the interrupt service routine
6. Restoring the CPU status from the stack
7. Restoring the program counter from the stack
8. Resuming the interrupted program

For all maskable hardware interrupts, the microprocessor starts to provide service when it completes the execution of the current instruction (the instruction being executed when the interrupt occurred). For some nonmaskable interrupts, the CPU may start the service without completing the current instruction. Many software interrupts are caused by an error in instruction execution that prevents the instruction from being completed. The service to this type of interrupt is simply to output an error message and abort the program.

6.2.6 Interrupt Vector

The term *interrupt vector* refers to the starting address of the interrupt service routine. In general, interrupt vectors are stored in a table called an *interrupt-vector table*. The starting address of each entry (holds one interrupt vector) in the interrupt vector table is called the *vector address*. The interrupt-vector table is fixed for some microprocessors and may be relocated for other microprocessors.

The CPU needs to determine the interrupt vector before it can provide service. One of the following methods can be used by a microprocessor or microcontroller to determine the interrupt vector:

1. *Predefined*. In this method, the starting address of the service routine is predefined when the microcontroller is designed. The processor uses a table to store all the interrupt service routines. The Intel 8051 microcontrollers use this approach. Each interrupt is allocated the same number of bytes to hold its service routine. The Intel 8051 allocates eight words to each interrupt service routine. When the service routine requires more than eight words, the solution is to place a jump instruction in the predefined location to jump to the actual service routine.

2. *Fetch the vector from a predefined memory location*. In this approach, the interrupt vector of each interrupt source is stored at a predefined location in the interrupt-vector table, where the microprocessor can get it directly. The Freescale HCS12 and most other Freescale microcontrollers use this approach.

3. *Execute an interrupt acknowledge cycle to fetch a vector number in order to locate the interrupt vector*. During the interrupt acknowledge cycle, the microprocessor performs a read bus cycle, and the external I/O device that requested the interrupt places a number on the data bus to identify itself. This number is called the *interrupt-vector number*. The CPU can figure out the starting address of the interrupt service routine by using this number. The CPU needs to perform a read cycle in order to obtain it. The Freescale 68000 and Intel x86 family microprocessors support this method. The Freescale 68000 family of microprocessors also uses the second method. This method is not used by microcontrollers because of the incurred latency.

6.2.7 Interrupt Programming

Interrupt programming deals with how to provide service to the interrupt. There are three steps in interrupt programming.

Step 1
Initialize the interrupt-vector table. (This step is not needed for microprocessors that have predefined interrupt vectors.) This can be done by using the assembler directive org (or its equivalent) as follows:

```
org      $xxxx          ; xxxx is the vector table address
dc.w     service_1      ; store the starting address of interrupt source 1
dc.w     service_2      ;
   .
   .
   .
dc.w     service_n
```

where *service_i* is the starting address of the service routine for interrupt source *i*. The HCS12 uses this method to set up interrupt vectors. The assembler syntax and the number of bytes needed to store an interrupt vector on your particular microcontroller may be different.

Step 2

Write the interrupt service routine. An interrupt service routine should be as short as possible. For some interrupts, the service routine may only output a message to indicate that something unusual has occurred. A service routine is similar to a subroutine—the only difference is the last instruction. An interrupt service routine uses the *return-from-interrupt* (or *return-from-exception*) instruction instead of *return-from-subroutine* instruction to return to the interrupted program. The following instruction sequence is an example of an interrupt service routine (in HCS12 instructions):

```
irq_isr    ldx     #msg
           jsr     puts                      ; call puts to output a string pointed by X
           rti                               ; return from interrupt
msg        fcc     "This is an error"
```

The service routine may or may not return to the interrupted program, depending on the cause of the interrupt. It makes no sense to return to the interrupted program if the interrupt is caused by a software error such as divide by zero or overflow, because the program is unlikely to generate correct results under these circumstances. In such situations, the service routine would return to the monitor program or the operating system instead. Returning to a program other than the interrupted program can be achieved by changing the saved program counter (in the stack) to the desired value. Execution of the *return-from-interrupt* instruction will then return CPU control to the new address.

Step 3

Enable the interrupts to be serviced. An interrupt can be enabled by clearing the global interrupt mask and setting the local interrupt enable bit in the I/O control register. It is a common mistake to forget enabling interrupts when writing interrupt-driven application programs.

6.2.8 Overhead of Interrupts

Although the interrupt mechanism provides many advantages, it also involves some overhead. The overhead of the HCS12 interrupt includes

1. Saving the CPU registers, including accumulators (A:B), index registers X and Y, and the condition code register (CCR), and fetching the interrupt vector. This takes at least 9 E-clock cycles.

2. The execution time of the RTI instruction. This instruction restores all the CPU registers that have been stored in the stack by the CPU during the interrupt and takes from 8 to 11 E-clock cycles to complete for the HCS12.

3. Execution time of instructions of the interrupt service routine. This depends on the type and the number of instructions in the service routine.

The total overhead is thus at least 17 to 20 E-clock cycles, which amounts to almost 1 μs for a 24-MHz E-clock. We need to be aware of the overhead involved in interrupt processing when deciding whether to use the interrupt mechanism.

6.3 Resets

The initial values of some CPU registers, flip-flops, and the control registers in I/O interface chips must be established before the computer can operate properly. Computers provide a reset mechanism to establish initial conditions.

There are at least two types of resets in each microprocessor: the *power-on reset* and the *manual reset*. A power-on reset allows the microprocessor to establish the initial values of registers and flip-flops and to initialize all I/O interface chips when power to the microprocessor is turned on. A manual reset without power-down allows the computer to get out of most error conditions (if hardware hasn't failed) and reestablish the initial conditions. The computer will *reboot* itself after a reset.

The starting address of the reset service routine either is a fixed value or is stored at a fixed location (for HCS12). The reset service routine is stored in the read-only memory of all microprocessors so that it is always ready for execution. At the end of the service routine, control should be transferred to either the monitor program or the operating system.

Like nonmaskable interrupts, resets are also nonmaskable. However, resets are different from the nonmaskable interrupts in that no registers are saved by resets because resets establish the values of registers.

6.4 HCS12 Exceptions

The HCS12 exceptions can be classified into the following categories:

- *Maskable interrupts.* These include the $\overline{\text{IRQ}}$ pin interrupt and all peripheral function interrupts. Since different HCS12 members implement a different number of peripheral functions, they have different numbers of maskable interrupts.

- *Nonmaskable interrupts.* These include the $\overline{\text{XIRQ}}$ pin interrupt, the swi instruction interrupt, and the unimplemented opcode trap.

- *Resets.* These include the power-on reset, the RESET pin manual reset, the COP (computer operating properly) reset, and the clock monitor reset. For other microcontrollers, the COP reset is also called the *watchdog* reset.

6.4.1 Maskable Interrupts

Since different HCS12 members implement a different number of peripheral functions, they have a different number of maskable interrupts. The I flag in the CCR register is the global mask of all maskable interrupts. Whenever the I flag is 1, all maskable interrupts are disabled. All maskable interrupts have a local enable bit that allows them to be selectively enabled. They are disabled (I flag is set to 1) when the HCS12 gets out of the reset state.

As with any other microcontroller, all HCS12 exceptions are prioritized. The priorities of resets and nonmaskable interrupts are not programmable. However, we can raise one of the maskable interrupts to the highest level within the group of maskable interrupts so that it can get quicker attention from the CPU. The relative priorities of the other sources remain the same. The bits 7 to 1 of the HPRIO register select the maskable interrupt at the highest priority within the group of maskable interrupts. The contents of the HPRIO register are shown in Figure 6.1.

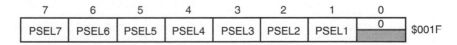

Figure 6.1 ■ Highest-priority I interrupt register

The priorities and vector addresses of all HCS12 exceptions are listed in Table 6.1. The vector number column specifies the order of a specific interrupt source in the table. It is provided

Vector Address	Vector Number	Interrupt Source	CCR Mask	Local Enable	HPRIO Value to Elevate to Highest I Bit
$FFFE	0	Reset	none	none	–
$FFFC	1	Clock monitor failure reset	none	PLLCTL(CME, SCME)	–
$FFFA	2	COP failure reset	none	COP rate select	–
$FFF8	3	Unimplemented instruction trap	none	none	–
$FFF6	4	swi	none	none	–
$FFF4	5	XIRQ	X bit	none	–
$FFF2	6	IRQ	I bit	IRQCR(IRQEN)	$F2
$FFF0	7	Real-time interrupt	I bit	CRGINT(RTIE)	$F0
$FFEE	8	Enhanced capture timer channel 0	I bit	TIE(C0I)	$EE
$FFEC	9	Enhanced capture timer channel 1	I bit	TIE(C1I)	$EC
$FFEA	10	Enhanced capture timer channel 2	I bit	TIE(C2I)	$EA
$FFE8	11	Enhanced capture timer channel 3	I bit	TIE(C3I)	$E8
$FFE6	12	Enhanced capture timer channel 4	I bit	TIE(C4I)	$E6
$FFE4	13	Enhanced capture timer channel 5	I bit	TIE(C5I)	$E4
$FFE2	14	Enhanced capture timer channel 6	I bit	TIE(C6I)	$E2
$FFE0	15	Enhanced capture timer channel 7	I bit	TIE(C7I)	$E0
$FFDE	16	Enhanced capture timer overflow	I bit	TSCR2(TOI)	$DE
$FFDC	17	Pulse accumulator A overflow	I bit	PACTL(PAOVI)	$DC
$FFDA	18	Pulse accumulator A input edge	I bit	PACTL(PAI)	$DA
$FFD8	19	SPI0	I bit	SP0CR1(SPIE, SPTIE)	$D8
$FFD6	20	SCI0	I bit	SC0CR2(TIE,TCIE,RIE,ILIE)	$D6
$FFD4	21	SCI1	I bit	SC1CR2(TIE,TCIE,RIE,ILIE)	$D4
$FFD2	22	ATD0	I bit	ATD0CTL2(ASCIE)	$D2
$FFD0	23	ATD1	I bit	ATD1CTL2(ASCIE)	$D0
$FFCE	24	Port J	I bit	PTJIF(PTJIE)	$CE
$FFCC	25	Port H	I bit	PTHIF(PTHIE)	$CC
$FFCA	26	Modulus down counter underflow	I bit	MCCTL(MCZI)	$CA
$FFC8	27	Pulse accumulator B overflow	I bit	PBCTL(PBOVI)	$C8
$FFC6	28	CRG PLL lock	I bit	CRGINT(LOCKIE)	$C6
$FFC4	29	CRG self-clock mode	I bit	CRGINT(SCMIE)	$C4
$FFC2	30	BDLC	I bit	DLCBCR1(IE)	$C2
$FFC0	31	IIC Bus	I bit	IBCR(IBIE)	$C0
$FFBE	32	SPI1	I bit	SP1CR1(SPIE, SPTIE)	$BE
$FFBC	33	SPI2	I bit	SP2CR2(SPIE, SPTIE)	$BC
$FFBA	34	EEPROM	I bit	EECTL(CCIE, CBEIE)	$BA
$FFB8	35	Flash	I bit	FCTL(CCIE, CBEIE)	$B8
$FFB6	36	CAN0 wake-up	I bit	CAN0RIER(WUPIE)	$B6
$FFB4	37	CAN0 errors	I bit	CAN0RIER(CSCIE, OVRIE)	$B4
$FFB2	38	CAN0 receive	I bit	CAN0RIER(RXFIE)	$B2
$FFB0	39	CAN0 transmit	I bit	CAN0TIER(TXEIE2-TXEIE0)	$B0
$FFAE	40	CAN1 wake-up	I bit	CAN1RIER(WUPIE)	$AE
$FFAC	41	CAN1 errors	I bit	CAN1RIER(CSCIE, OVRIE)	$AC
$FFAA	42	CAN1 receive	I bit	CAN1RIER(RXFIE)	$AA
$FFA8	43	CAN1 transmit	I bit	CAN1TIER(TXEIE2-TXEIE0)	$A8
$FFA6	44	CAN2 wake-up	I bit	CAN2RIER(WUPIE)	$A6
$FFA4	45	CAN2 errors	I bit	CAN2RIER(CSCIE, OVRIE)	$A4
$FFA2	46	CAN2 receive	I bit	CAN2RIER(RXFIE)	$A2
$FFA0	47	CAN2 transmit	I bit	CAN2TIER(TXEIE2-TXEIE0)	$A0
$FF9E	48	CAN3 wake-up	I bit	CAN3RIER(WUPIE)	$9E
$FF9C	49	CAN3 errors	I bit	CAN3RIER(CSCIE, OVRIE)	$9C
$FF9A	50	CAN3 receive	I bit	CAN3RIER(RXFIE)	$9A
$FF98	51	CAN3 transmit	I bit	CAN3TIER(TXEIE2-TXEIE0)	$98
$FF96	52	CAN4 wake-up	I bit	CAN4RIER(WUPIE)	$96
$FF94	53	CAN4 errors	I bit	CAN4RIER(CSCIE, OVRIE)	$94
$FF92	54	CAN4 receive	I bit	CAN4RIER(RXFIE)	$92
$FF90	55	CAN4 transmit	I bit	CAN4TIER(TXEIE2-TXEIE0)	$90
$FF8E	56	Port P interrupt	I bit	PTPIF(PTPIE)	$8E
$FF8C	57	PWM emergency shutdown	I bit	PWMSDN(PWMIE)	$8C

Table 6.1 ■ Interrupt vector map

for the convenience of setting up the whole interrupt-vector table. To raise a maskable interrupt source to the highest priority, simply write the low byte of the vector address of this interrupt to the HPRIO register. For example, to raise the capture timer channel 7 interrupt to the highest priority, write the value of $E0 to the HPRIO register.

All HCS12 reset and interrupt vectors are stored in a table (shown in Table 6.1) located at $FF8C to $FFFF. In Table 6.1, exceptions that have higher vector addresses are at higher priorities. Not all the exceptions are available in all HCS12 members.

IRQ̄ PIN INTERRUPT

The $\overline{\text{IRQ}}$ pin (multiplexed with the PE1 pin) is the only external maskable interrupt signal. The $\overline{\text{IRQ}}$ pin interrupt can be edge triggered or level triggered. The triggering method is selected by programming the IRQE bit of the Interrupt Control Register (IRQCR). The $\overline{\text{IRQ}}$ pin interrupt has a local enable bit IRQEN bit, which is bit 6 of the IRQCR. The contents of the IRQCR register are shown in Figure 6.2.

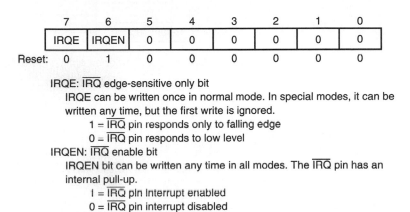

IRQE: $\overline{\text{IRQ}}$ edge-sensitive only bit
    IRQE can be written once in normal mode. In special modes, it can be
    written any time, but the first write is ignored.
        1 = $\overline{\text{IRQ}}$ pin responds only to falling edge
        0 = $\overline{\text{IRQ}}$ pin responds to low level
IRQEN: $\overline{\text{IRQ}}$ enable bit
    IRQEN bit can be written any time in all modes. The $\overline{\text{IRQ}}$ pin has an
    internal pull-up.
        1 = $\overline{\text{IRQ}}$ pin interrupt enabled
        0 = $\overline{\text{IRQ}}$ pin interrupt disabled

Figure 6.2 ■ Interrupt control register (IRQCR)

The advantage of making the $\overline{\text{IRQ}}$ interrupt *level sensitive* (active low) is that it allows multiple external interrupt sources to be tied to this pin. Whenever one of the interrupt sources (that are tied to the $\overline{\text{IRQ}}$ pin) is low, an interrupt request will be detected by the HCS12. The user of this method must make sure that the $\overline{\text{IRQ}}$ signal is de-asserted (goes high) before the HCS12 exits the interrupt service routine if there are no other pending interrupts connected to the $\overline{\text{IRQ}}$ pin.

The major advantage of making the $\overline{\text{IRQ}}$ interrupt *edge sensitive* (falling edge) is that the user does not need to be concerned about the duration of the assertion time of the $\overline{\text{IRQ}}$ signal. However, this approach is not appropriate for a noisy environment. In a noisy environment, any noise spike could generate an undesirable interrupt request on the $\overline{\text{IRQ}}$ pin.

INTERRUPT RECOGNITION

Once enabled, an interrupt request can be recognized at any time after the I mask bit is cleared. When an interrupt service request is recognized, the CPU responds at the completion of the instruction being executed. Interrupt latency varies according to the number of cycles required to complete the current instruction. The HCS12 has implemented a few instructions to support fuzzy-logic rule evaluation. These instructions, which take a much longer time to complete, include fuzzy-logic rule evaluation (REV), fuzzy-logic rule evaluation weighted (REVW), and weighted-average (WAV) instructions. The HCS12 does not wait until the completion of these instructions to service the interrupt request. These instructions will resume at the point when they were interrupted.

Before the HCS12 starts to service an interrupt, it will set the I mask to disable other maskable interrupts. When the CPU begins to service an interrupt, the instruction queue is refilled, a return address is calculated, and then the return address and the contents of all CPU registers (except SP) are saved in the stack in the order shown in Figure 6.3.

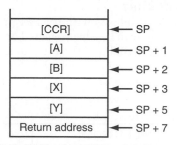

Figure 6.3 ■ Stack order on entry to interrupts

THE RTI INSTRUCTION

RTI is used to terminate interrupt service routines. RTI is an 8-cycle instruction when no other interrupt is pending and an 11-cycle instruction when another interrupt is pending. In either case, the first 5 cycles are used to restore the CCR, B:A, X, Y, and return address from the stack. The HCS12 then clears the I mask to enable further maskable interrupts.

If no other interrupt is pending at this point, three program words are fetched to refill the instruction queue from the area of the return address, and processing proceeds from there. If another interrupt is pending after registers are restored, a new vector is fetched, and the stack pointer is adjusted to point at the CCR value that was just recovered (SP = SP – 9). This makes it appear that the registers have been stacked again. After the SP is adjusted, three program words are fetched to refill the instruction queue, starting at the address the vector points to. Processing then continues with execution of the instruction that is now at the head of the queue.

6.4.2 Nonmaskable Interrupts

There are three interrupts in this category: $\overline{\text{XIRQ}}$ pin (multiplexed with the PE0 pin), swi instruction, and unimplemented opcode trap.

$\overline{\text{XIRQ}}$ PIN INTERRUPT

The $\overline{\text{XIRQ}}$ pin interrupt is disabled during a system reset and on entering the interrupt service routine for an $\overline{\text{XIRQ}}$ interrupt.

During reset, both the I and X bits in the CCR register are set. This disables maskable interrupts and interrupt requests made by asserting the $\overline{\text{XIRQ}}$ pin (pulled to low). After minimum system initialization, software can clear the X bit using an instruction such as andcc #$BF. Software cannot reset the X bit from 0 to 1 once it has been cleared, and hence the interrupt requests made via the $\overline{\text{XIRQ}}$ pin become nonmaskable.

When a nonmaskable interrupt is recognized, both the X and I bits are set after CPU registers are saved. The X bit is not affected by maskable interrupts. The execution of an RTI instruction at the end of the $\overline{\text{XIRQ}}$ service routine will restore the X and I bits to the preinterrupt request state.

UNIMPLEMENTED OPCODE TRAP

The HCS12 uses up to 16 bits (2 pages) to encode the opcode. All 256 combinations in the page 1 opcode map have been used. However, only 54 of the 256 positions on page 2 of the opcode map are used. If HCS12 attempts to execute one of the 202 unused opcodes on page 2, an unimplemented opcode trap occurs. The 202 unimplemented opcodes are essentially interrupts that share a common interrupt-vector address, $FFF8:$FFF9. The HCS12 uses the next address after an unimplemented page 2 opcode as a return address.

SOFTWARE INTERRUPT INSTRUCTION (SWI)

Execution of the swi instruction causes an interrupt without an interrupt request signal. swi is not inhibited by the global mask bits in the CCR. So far, we have been using this instruction to jump back to the D-Bug12 monitor.

The swi instruction is commonly used in the debug monitor to implement *breakpoints* and to transfer control from a user program to the debug monitor. A breakpoint in a user program is a memory location where we want program execution to be stopped and information about instruction execution (in the form of register contents) to be displayed. To implement breakpoints, the debug monitor sets up a breakpoint table. Each entry of the table holds the address of the breakpoint and the opcode byte at the breakpoint. The monitor also replaces the opcode byte at the breakpoint with the opcode of the swi instruction. When the instruction at the breakpoint is executed, it causes an swi interrupt. The service routine of the swi interrupt will look up the breakpoint table and take different actions depending on whether the saved PC value is in the breakpoint table:

Case 1
The saved PC value is not in the breakpoint table. In this case, the service routine will simply replace the saved PC value (in the stack) with the address of the monitor program and return from the interrupt.

Case 2
The saved PC is in the breakpoint table. In this case, the service routine will

1. Replace the swi opcode with the opcode in the breakpoint table

2. Display the contents of the CPU registers

3. Replace the saved PC value (in the stack) with the address of the monitor program

4. Return from the interrupt (return to the debug monitor because of step 3)

6.4.3 Interrupts in D-Bug12 EVB Mode

Most users would use the EVB mode of the D-Bug12 monitor to develop applications on a demo board having the D-Bug12 monitor. However, the D-bug12 monitor's EVB mode does not allow the use of the on-chip flash memory, which prevents using the default interrupt-vector table to hold interrupt vectors. To allow the user to develop interrupt-driven applications, the D-Bug12 monitor provides an SRAM-based interrupt-vector table. This table starts at $3E00 and has 64 entries of 2 bytes each. The contents of this table are shown in Table 6.2. Initially, all entries in the table contain an address of $0000. Storing a value other than $0000 in any of the RAM interrupt-vector table entries causes execution of the interrupt service routine pointed to by the address when an associated interrupt occurs.

Interrupt Source	RAM Vector Address	Interrupt Source	RAM Vector Address
Reserved $FF80	$3E00	IIC bus	$3E40
Reserved $FF82	$3E02	BDLC	$3E42
Reserved $FF84	$3E04	SCME	$3E44
Reserved $FF86	$3E06	CRG clock	$3E46
Reserved $FF88	$3E08	Pulse accumulator B overflow	$3E48
Reserved $FF8A	$3E0A	Modulus down counter underflow	$3E4A
PWM emergency shutdown	$3E0C	Port H interrupt	$3E4C
Port P interrupt	$3E0E	Port J interrupt	$3E4E
MSCAN 4 transmit	$3E10	ATD1	$3E50
MSCAN 4 receive	$3E12	ATD0	$3E52
MSCAN 4 errors	$3E14	SCI1	$3E54
MSCAN 4 wake-up	$3E16	SCI0	$3E56
MSCAN 3 transmit	$3E18	SPI0	$3E58
MSCAN 3 receive	$3E1A	Pulse accumulator A input edge	$3E5A
MSCAN 3 errors	$3E1C	Pulse accumulator A overflow	$3E5C
MSCAN 3 wake-up	$3E1E	Timer overflow	$3E5E
MSCAN 2 transmit	$3E20	Timer channel 7	$3E60
MSCAN 2 receive	$3E22	Timer channel 6	$3E62
MSCAN 2 errors	$3E24	Timer channel 5	$3E64
MSCAN 2 wake-up	$3E26	Timer channel 4	$3E66
MSCAN 1 transmit	$3E28	Timer channel 3	$3E68
MSCAN 1 receive	$3E2A	Timer channel 2	$3E6A
MSCAN 1 errors	$3E2C	Timer channel 1	$3E6C
MSCAN 1 wake-up	$3E2E	Timer channel 0	$3E6E
MSCAN 0 transmit	$3E30	Real-time interrupt	$3E70
MSCAN 0 receive	$3E32	$\overline{IRQ}$	$3E72
MSCAN 0 errors	$3E34	$\overline{XIRQ}$	$3E74
MSCAN 0 wake-up	$3E36	swi	$3E76
Flash	$3E38	Unimplemented instruction trap	$3E78
EEPROM	$3E3A	N/A	$3E7A
SPI2	$3E3C	N/A	$3E7C
SPI1	$3E3E	N/A	$3E7E

Table 6.2 ■ D-Bug12 RAM interrupt vector address

If an unmasked interrupt occurs and a table entry contains the default address of $0000, program execution is returned to D-Bug12. The D-Bug12 would display a message indicating the source of the interrupt and also display the CPU registers at the point where the program was interrupted. The only exception to this is the SCI0 interrupt. Even though there is an entry for SCI0 in Table 6.2, one cannot use the SCI0 interrupt because it has been used by D-Bug12 for all of its communications. The hcs12.inc (and hcs12.h) file also provides the mnemonic name for each interrupt source so that the user can use the mnemonic name to specify the SRAM vector address. The complete list of these mnemonic names is shown in Table 6.3.

Interrupt Source	RAM Vector Address	Vector Number	Interrupt Source	RAM Vector Address	Vector Number
UserRsrv0x80	$3E00	0	UserIIC	$3E40	32
UserRsrv0x82	$3E02	1	UserDLC	$3E42	33
UserRsrv0x84	$3E04	2	UserSCME	$3E44	34
UserRsrv0x86	$3E06	3	UserCRG	$3E46	35
UserRsrv0x88	$3E08	4	UserPAccBOv	$3E48	36
UserRsrv0x8a	$3E0A	5	UserModDwnCtr	$3E4A	37
UserPWMShDn	$3E0C	6	UserPortH	$3E4C	38
UserPortP	$3E0E	7	UserPortJ	$3E4E	39
UserMSCAN4Tx	$3E10	8	UserAtoD1	$3E50	40
UserMSCAN4Rx	$3E12	9	UserAtoD0	$3E52	41
UserMSCAN4Errs	$3E14	10	UserSCI1	$3E54	42
UserMSCAN4Wake	$3E16	11	UserSCI0	$3E56	43
UserMSCAN3Tx	$3E18	12	UserSPI0	$3E58	44
UserMSCAN3Rx	$3E1A	13	UserPAccEdge	$3E5A	45
UserMSCAN3Errs	$3E1C	14	UserPAccOvf	$3E5C	46
UserMSCAN3Wake	$3E1E	15	UserTimerOvf	$3E5E	47
UserMSCAN2Tx	$3E20	16	UserTimerCh7	$3E60	48
UserMSCAN2Rx	$3E22	17	UserTimerCh6	$3E62	49
UserMSCAN2Errs	$3E24	18	UserTimerCh5	$3E64	50
UserMSCAN2Wake	$3E26	19	UserTimerCh4	$3E66	51
UserMSCAN1Tx	$3E28	20	UserTimerCh3	$3E68	52
UserMSCAN1Rx	$3E2A	21	UserTimerCh2	$3E6A	53
UserMSCAN1Errs	$3E2C	22	UserTimerCh1	$3E6C	54
UserMSCAN1Wake	$3E2E	23	UserTimerCh0	$3E6E	55
UserMSCAN0Tx	$3E30	24	UserRTI	$3E70	56
UserMSCAN0Rx	$3E32	25	UserIRQ	$3E72	57
UserMSCAN0Errs	$3E34	26	UserXIRQ	$3E74	58
UserMSCAN0Wake	$3E36	27	UserSWI	$3E76	59
UserFlash	$3E38	28	UserTrap	$3E78	60
UserEEPROM	$3E3A	29	N/A	$3E7A	−1
UserSPI2	$3E3C	30	N/A	$3E7C	−1
UserSPI1	$3E3E	31	N/A	$3E7E	−1

Note: Vector number is used by the SetUserVector function to set up the interrupt vector.

Table 6.3 ■ Mnemonic names for D-Bug12 RAM interrupt-vector addresses

It is very simple to set up the SRAM interrupt vector. Assuming that the service routine for the $\overline{\text{IRQ}}$ interrupt starts with the label **irqISR**, then one of the following two methods can be used to establish the SRAM table entry for the $\overline{\text{IRQ}}$ interrupt:

Method 1

```
org     $3E72           ; one can also use "org UserIRQ"
dc.w    irqISR
```

Method 2

```
ldd     #irqISR
std     $3E72           ; one can also use "std UserIRQ"
```

Example 6.1

▼

Assume that the $\overline{\text{IRQ}}$ pin of the HCS12DP256 is connected to a 1-Hz digital waveform and Port B is connected to eight LEDs. Write a program to configure Port B for output and enable the $\overline{\text{IRQ}}$ interrupt, and also write the service routine for the $\overline{\text{IRQ}}$ interrupt. The service routine for the $\overline{\text{IRQ}}$ interrupt simply increments a counter and outputs it to Port B. This program is to be executed on an HCS12 demo board programmed with the D-Bug12 monitor.

Solution: The program is as follows:

```
        #include     "c:\miniide\hcs12.inc"
                org         $1000
count           ds.b        1                    ; reserve 1 byte for count
                org         $1500
                lds         #$1500               ; set up the stack pointer
                movw        #IRQISR,UserIRQ      ; set up interrupt vector in SRAM
                clr         count
                movb        #$FF,DDRB            ; configure Port B for output
                bset        DDRJ,$02            ; configure PJ1 pin for output (required in Dragon12)
                bclr        PTJ,$02            ; enable LEDs to light (required in Dragon12)
                movb        count,PTB           ; display the count value on LEDs
                movb        #$C0,IRQCR          ; enable IRQ pin interrupt and select edge triggering
                cli                             ;        "
forever         nop
                bra         forever             ; wait for IRQ pin interrupt
; ********************************************************************************
; This is the IRQ service routine.
; ********************************************************************************
IRQISR          inc         count               ; increment count
                movb        count,PTB           ; and display count on LEDs
                rti
                end
```

▲

6.5 Interrupt Programming in C Language

Interrupt programming in C language varies with the C compiler. In this section, we examine how interrupt programming is done in CodeWarrior, ICC12 IDE, and EGNU IDE.

6.5.1 Interrupt Programming in CodeWarrior

CodeWarrior uses the keyword interrupt to inform its C compiler that a function is an interrupt service routine. Whenever the keyword interrupt appears before a function name, the CodeWarrior C compiler generates RTI as the last instruction of the function. The template for an interrupt service routine in CodeWarrior is as follows:

```
Interrupt void ISR_name (void)
{
        ...         // statements to service the interrupt
}
```

where **ISR_name** is the name of the interrupt service routine.

After writing the interrupt service routine, the user also needs to set up the interrupt vector. The template for setting up the interrupt vector table is given in Appendix G. There are two parts in setting up the interrupt vector.

Part 1. Declare the interrupt service routine to be external.

Part 2. Insert the name of the interrupt service routine into the appropriate place of the interrupt-vector table.

For example, the RTI interrupt vector can be set up as follows assuming that the name of its service routine is rtiISR:

```
extern void near rtiISR (void); // rtiISR is defined outside this file
#pragma CODE_SEG __NEAR_SEG NON_BANKED // interrupt section for this module
_interrupt    void        UnimplementedISR(void)
{
              for( ; ; );         // do nothing, but return from interrupt
}
#pragma CODE_SEG DEFAULT
typedef void (*near  tlsrFunc)(void);

const          tlsrFunc _vect[ ] @0xFF80 = { // interrupt-vector table starts from this line
                                          // interrupt-vector table
              UnimplementedISR,
              . . .
              . . .
              rtiISR,                        // RTI interrupt vector (at the address 0xFFF0)
              . . .
              . . .
};
```

The entry *UnimplementedISR* is provided as a catchall handler for all unintended interrupts. If the user is not concerned about unintended interrupts, then the interrupt vector table can be reduced to

```
const tlsrFunc _vect [ ] @0xFFF0 = { // 0xFFF0 is the address to store RTI vector
              rtiISR
};
```

6.5.2 Interrupt Handling in ImageCraft ICC12

The ICC12 C compiler uses the *#pragma* statement to indicate that a function is an interrupt service routine. For example, the following statements indicate that the function *rtiISR( )* is an interrupt service routine:

```
#pragma interrupt_handler rtiISR
void riISR (void)
{
              . . .
}
```

A *pragma* statement can declare several interrupt service routines separated by spaces.

When working with a demo board programmed with the D-Bug12 monitor, the user can use the **SetUserVector** function provided by the D-Bug12 monitor to set up the interrupt vector. The prototype declaration of the SetUserVector function is as follows:

```
int SetUserVector (int VectNum, Address UserAddress);       – at $EEA4
```

The function argument VectNum refers to the vector number associated with the interrupt source which is given in Table 6.3. The argument UserAddress is the interrupt vector of the interrupt source and is represented by the name of the interrupt service routine. One way to set up the interrupt vector in ICC12 when working with a demo board programmed with the D-Bug12 monitor is to use the inline assembly instructions.

```
asm("ldd #_rtiISR");    // push address of RTI interrupt service routine
asm("pshd");            //   "
asm("ldd #56");         // place vector number of RTI in D
asm("ldx $EEA4");       // place the SetUserVector function address in X
asm("jsr 0,X");         // jump to SetUserVector function to set up RTI
                        // interrupt vector (in SRAM)
```

6.5.3 Interrupt Programming in EGNU IDE

In the EGNU IDE, we must take the following actions in order to use interrupts:

1. *Use the interrupt attribute of the GCC compiler.* We must include the following statement as one of the first few statements in the program (there are two '_' characters before and after "attribute"):

   ```
   #define INTERRUPT __attribute__((interrupt))
   ```

2. *Apply the interrupt attribute to the interrupt service routine.* Assuming that rtiISR() and oc5ISR() are two interrupt service routines, then we add the following two prototype declaration statements (for rtiISR() and oc5ISR()) in the program:

   ```
   void INTERRUPT rtiISR(void);
   void INTERRUPT oc5ISR(void);
   ```

3. *Include the vectors12.h header file* so that one can use mnemonic names to store the starting addresses of interrupt service routines in the vector table (in SRAM). Include the following statements as one of the first few statements in the program:

   ```
   #include "c:\egnu\include\vectors12.h"
   ```

4. *Store interrupt vectors in the SRAM vector table.* This can be done by assignment statements.

   ```
   UserRTI = (unsigned short)&rtiISR;
   UserTimerCh5 = (unsigned short)&oc5ISR;
   ```

5. *Write the actual interrupt service routine.*

The C language version of the $\overline{\text{IRQ}}$ interrupt program given in Example 6.1 is as follows:

```
#include        "c:\egnu\include\hcs12.h"
#include        "c:\egnu\include\vectors12.h"
#define         INTERRUPT __attribute__((interrupt))
void            INTERRUPT IRQISR(void);
unsigned char cnt;

void main(void)
{
        UserIRQ = (unsigned short)&IRQISR;
        DDRB    = 0xFF;             // configure Port B for output
        cnt     = 0;
        DDRJ    |= BIT1;            // configure PJ1 pin for output
        PTJ     &= ~BIT1;           // enable LEDs to light (required for Dragon12 demo board)
        IRQCR   = 0xC0;             // enable IRQ interrupt on falling edge
```

```
        asm("cli");              // enable interrupt globally
        while(1);                // wait for interrupt forever
    }
    void INTERRUPT IRQISR(void)
    {
        cnt++;                   // increment the count value
        PTB        = cnt;        // display the count value on LEDs
    }.
```

6.6 Clock and Reset Generation Block (CRG)

This block is responsible for generating the clock signals required by the HCS12 instruction execution and all peripheral operations and providing default values for all on-chip registers. The block diagram of a CRG is shown in Figure 6.4.

A microcontroller needs a clock signal to operate. The clock signal has the waveform of a square wave. Most people use a crystal oscillator to generate the clock signal. However, the output of a crystal oscillator is a sinusoidal waveform that cannot be used to drive digital circuitry directly. Most microcontrollers and microprocessors have an on-chip oscillator to square up the incoming sinusoidal waveform so that it can be used as the clock signal. Many crystal oscillators also include the circuitry to square up the sinusoidal waveform. The on-chip oscillator circuitry can be bypassed for this type of crystal oscillator (also called external oscillators in Freescale literature).

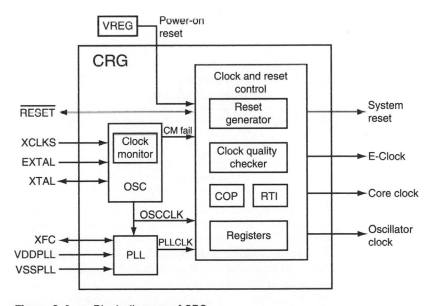

Figure 6.4 ■ Block diagram of CRG

The phase-locked-loop (PLL) is a circuit technique that can accept a low-frequency clock signal and produce a high-frequency clock output. In addition, the PLL has a feedback circuitry to stabilize the frequency of its output signal. The PLL has two operation modes: *acquisition mode* and *tracking mode*. When the PLL is first powered up, the frequency of its output is far from its *target frequency*. During this period, the PLL is in acquisition mode in which the PLL can make large adjustments to quickly reach the target frequency. After the PLL has attained its target output frequency, it enters tracking mode, in which it makes only small adjustments so as not to deviate from the target frequency. This mode is useful when the incoming clock signals (square waveform) deviate slightly because the frequency of the crystal oscillator output may change slightly due to a change in temperature, humidity, or barometric pressure.

6.6.1 Modes of CRG Operation

The CRG block can operate in one of the following four modes:

- *Run mode*. All functional parts of the CRG block are running in this mode.
- *Wait mode*. This mode allows the user to disable the system and core clocks by programming the individual bits in the CLKSEL register.
- *Stop mode*. Depending on the setting of the PSTP bit of the CLKSEL register, *stop* mode can be differentiated between *full-stop* (PSTP = 0) and *pseudo-stop* mode (PSTP = 1). In *full-stop* mode, the oscillator is disabled, and thus all system and core clocks are stopped. In *pseudo-stop* mode, the oscillator continues to run and most of the system and core clocks are stopped. If the respective enable bits are set, the COP and Real-Time Interrupt (RTI) modules will continue to run.
- *Self-clock mode*. This mode is entered if both the clock monitor enable bit (the CME bit of the PLLCTL register) and the *self-clock* mode enable bit (the SCME bit of PLLCTL) are set and the clock monitor detects a loss of clock (external oscillator or crystal). As soon as the self-clock mode is entered, the CRG starts to perform a clock quality check. The self-clock mode remains active until the clock check indicates that the required quality (frequency and amplitude) of the incoming clock signal is met. The self-clock mode should be used for safety purposes only. It provides reduced functionality to the microcontroller (MCU) in case a loss of clock is causing severe system conditions.

6.6.2 CRG Signals

- *VDDPLL and VSSPLL*. The PLL is a critical component for deriving a clock signal with stable frequency. These two pins provide the operating voltage (VDDPLL) and ground (VSSPLL) for the PLL circuitry and allow the supply voltage to the PLL to be independent of the power supply to the rest of the circuit.
- *XFC*. A passive external loop filter must be placed on the XFC pin. The filter is a second-order, low-pass filter to eliminate the VCO input ripple. The value of the external filter network and the reference frequency determine the speed of the corrections and the stability of the PLL. If the PLL usage is not required, the XFC pin must be tied to VDDPLL. The PLL loop filter connection recommended by Freescale is shown in Figure 6.5.
- *EXTAL and XTAL*. These two pins allow the user to connect an external crystal oscillator or a CMOS compatible clock to control the internal clock generator circuitry. The circuit connection for an external crystal oscillator is shown in

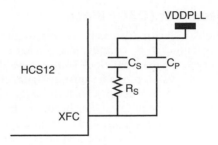

Figure 6.5 ■ PLL loop filter connections

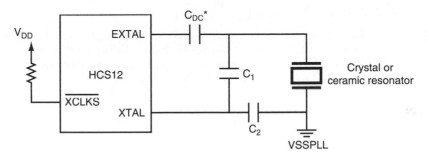

Figure 6.6 ■ Common crystal connections

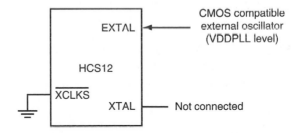

Figure 6.7 ■ External oscillator connections

Figure 6.6; the circuit connection that uses a CMOS compatible external clock is shown in Figure 6.7.

- $\overline{RESET}$. This pin is an active low, bidirectional reset pin. As an input, this signal initializes the microcontroller to a known state. As an output, it indicates that a system reset (internal to MCU) has been triggered.

- $\overline{XCLKS}$. This signal is an input that controls whether a crystal in combination with the internal oscillator or an external clock source on the EXTAL pin (oscillator circuitry is bypassed) is used to provide the clock signal required for the MCU operation. The $\overline{XCLKS}$ signal is sampled on the rising edge of $\overline{RESET}$. When this signal is high, the crystal and the internal oscillator provide the OSC CLK signal. When this signal is low, the external clock provides the OSC_CLK signal.

6.6.3 The Oscillator (OSC) Block

The OSC block in Figure 6.4 has two external pins, EXTAL and XTAL. The oscillator input pin, EXTAL, is intended to be connected to either a crystal or an external clock source. The selection of the crystal or external clock source depends on the $\overline{\text{XCLKS}}$ signal, which is sampled during the reset. The XTAL pin is an output signal that provides crystal circuit feedback and can be buffered to drive other devices with the same voltage amplitude.

A buffered EXTAL signal, OSCCLK, becomes the internal reference clock. The oscillator is enabled on the basis of the PSTP bit and the stop condition. The oscillator is disabled when the MCU is in stop mode except when the pseudo-stop mode is enabled.

To improve noise immunity, the oscillator is powered by the VDDPLL and VSSPLL power supply pins.

6.6.4 Phase-Locked-Loop (PLL)

The PLL is used to run the microcontroller with a clock frequency different from the incoming OSCCLK signal and provides high stability to the system clock. In Figure 6.4, the frequency of the PLLCLK signal is determined by the synthesizer (SYNR) and the reference divide (REFDV) registers using the following equation:

$$\text{PLLCLK} = 2 \times \text{OSCCLK} \times \frac{(\text{SYNR} + 1)}{(\text{REFDV} + 1)} \tag{6.1}$$

The contents of the SYNR register and the REFDV register are shown in Figures 6.8 and 6.9, respectively.

	7	6	5	4	3	2	1	0
	0	0	SYN5	SYN4	SYN3	SYN2	SYN1	SYN0
Reset:	0	0	0	0	0	0	0	0

Figure 6.8 ■ The CRG synthesizer register (SYNR)

	7	6	5	4	3	2	1	0
	0	0	0	0	REFDV3	REFDV2	REFDV1	REFDV0
Reset:	0	0	0	0	0	0	0	0

Figure 6.9 ■ The CRG reference divider register (REFDV)

The PLL operates in either the acquisition mode or tracking mode, depending on the difference between the output frequency and the target frequency. The PLL can change between acquisition and tracking modes either automatically or manually. The functional diagram of the PLL is shown in Figure 6.10.

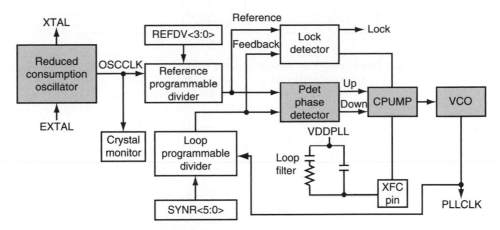

Figure 6.10 ■ PLL functional diagram

PLL OPERATION

In Figure 6.10, the OSCCLK clock is fed through the reference programmable divider and is divided into a range from 1 to 16 (REFDV + 1) to output the reference clock. The VCO output clock (PLLCLK) is fed back through the programmable loop divider and is in a range of 2 to 128 in increments of [2 × (SYNR + 1)] to output the feedback clock.

The phase detector then compares the feedback clock with the reference clock. Correction pulses are generated on the basis of the phase difference between these two signals. The loop filter then slightly alters the DC voltage on the external filter capacitor connected to the XFC pin on the basis of the width and direction of the correction pulse. The filter can make fast or slow corrections depending on its mode. The values of the external filter network and the reference frequency determine the speed of the corrections and the stability of the PLL.

ACQUISITION AND TRACKING MODES

The lock detector compares the frequencies of the feedback clock and the reference clock. Therefore, the speed of the lock detector is directly proportional to the final reference frequency. The circuit determines the mode of the PLL and the lock condition on the basis of this comparison.

The PLL filter can be manually or automatically configured into one of two possible operating modes: *acquisition* mode and *tracking* mode. In acquisition mode, the filter can make large frequency corrections to the Voltage Controlled Output (VCO) circuit. This mode is used at PLL startup or when the PLL has suffered a severe noise hit and the VCO frequency is far off the desired frequency. In tracking mode, the filter makes only small corrections to the frequency of the VCO. PLL jitter is much lower in tracking mode, but the response to noise is also slower. The PLL enters tracking mode when the VCO frequency is nearly correct.

The operation of the PLL is controlled by four registers: CRGINT, CRGFLG, CLKSEL, and PLLCTL. The contents of these four registers are shown in Figures 6.11, 6.12, 6.13, and 6.14.

The CRG interrupt register (CRGINT) enables or disables the interrupts associated with the CRG module. The CRG Flag Register (CRGFLG) holds the status flags of the CRG module. The CLKSEL register selects the clock source for the PLL. The PLLCTL register provides the overall control to the PLL module.

7	6	5	4	3	2	1	0
RTIE	0	0	LOCKIE	0	0	SCMIE	0

Reset: 0 0 0 0 0 0 0 0

RTIE: Real-time interrupt enable bit
  0 = Interrupt requests from RTI are disabled.
  1 = Interrupt requests from RTI are enabled.
LOCKIE: Lock interrupt enable bit
  0 = LOCK interrupt requests are disabled.
  1 = LOCK interrupt requests are enabled.
SCMIE: Self-clock mode interrupt enable bit
  0 = SCM interrupt requests are disabled.
  1 = Interrupt will be requested whenever the SCMIF bit is set.

Figure 6.11 ■ The CRG interrupt enable register (CRGINT)

7	6	5	4	3	2	1	0
RTIF	PORF	0	LOCKIF	LOCK	TRACK	SCMIF	SCM

Reset: 0 0 0 0 0 0 0 0

RTIF: real-time interrupt flag
  The RTIF flag is set to 1 at the end of the RTI period. This flag can only be cleared by writing a 1 to it. When the RTIE bit is 1, the setting of this bit will cause an interrupt.
  0 = RTI time-out has not occurred.
  1 = RTI time-out has occurred.
PORF: power-on reset flag
  This flag is set to 1 when a power-on reset occurs. It can only be cleared by writing a 1 to it.
  0 = Power-on reset has not occurred.
  1 = Power-on reset has occurred.
LOCKIF: PLL lock interrupt flag
  This flag is set to 1 when the LOCK status bit changes. This flag can only be cleared by writing a 1 to it.
  0 = No change in the LOCK bit.
  1 = The LOCK bit has changed.
LOCK: lock status bit
  This bit reflects the current state of PLL lock condition. This bit is cleared in self-clock mode.
  0 = PLL VCO is not within the desired tolerance of the target frequency.
  1 = PLL VCO is within the desired tolerance of the target frequency.
TRACK: track status bit
  This bit reflects the current state of PLL lock condition. This bit is cleared in self-clock mode.
  0 = Acquisition mode status.
  1 = Tracking mode status.
SCMIF: self-clock mode interrupt flag
  This bit is set to 1 when the SCM status bit changes. This flag can only be cleared by writing a 1 to it.
  0 = No change in SCM bit.
  1 = SCM bit has changed.
SCM: self-clock mode status bit
  SCM reflects the current clocking mode.
  0 = MCU is operating normally with OSCCLK available.
  1 = MCU is operating in self-clock mode with OSCCLK in an unknown state. All clocks are derived from PLLCLK running at its minimum frequency, f_{SCM}.

Figure 6.12 ■ The CRG flag register (CRGFLG)

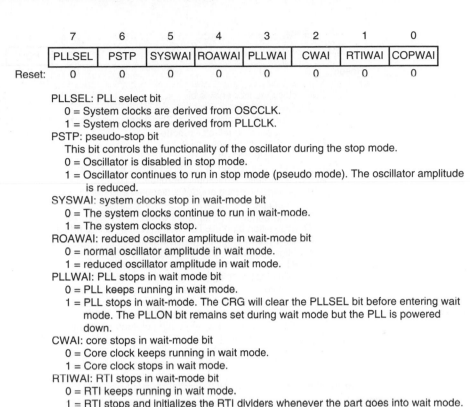

Figure 6.13 ■ The CRG clock select register (CLKSEL)

	7	6	5	4	3	2	1	0
	PLLSEL	PSTP	SYSWAI	ROAWAI	PLLWAI	CWAI	RTIWAI	COPWAI
Reset:	0	0	0	0	0	0	0	0

PLLSEL: PLL select bit
  0 = System clocks are derived from OSCCLK.
  1 = System clocks are derived from PLLCLK.
PSTP: pseudo-stop bit
  This bit controls the functionality of the oscillator during the stop mode.
  0 = Oscillator is disabled in stop mode.
  1 = Oscillator continues to run in stop mode (pseudo mode). The oscillator amplitude
      is reduced.
SYSWAI: system clocks stop in wait-mode bit
  0 = The system clocks continue to run in wait-mode.
  1 = The system clocks stop.
ROAWAI: reduced oscillator amplitude in wait-mode bit
  0 = normal oscillator amplitude in wait mode.
  1 = reduced oscillator amplitude in wait mode.
PLLWAI: PLL stops in wait mode bit
  0 = PLL keeps running in wait mode.
  1 = PLL stops in wait-mode. The CRG will clear the PLLSEL bit before entering wait
      mode. The PLLON bit remains set during wait mode but the PLL is powered
      down.
CWAI: core stops in wait-mode bit
  0 = Core clock keeps running in wait mode.
  1 = Core clock stops in wait mode.
RTIWAI: RTI stops in wait-mode bit
  0 = RTI keeps running in wait mode.
  1 = RTI stops and initializes the RTI dividers whenever the part goes into wait mode.
COPWAI: COP stops in wait-mode bit
  0 = COP keeps running in wait mode.
  1 = COP stops and initializes the COP dividers whenever the part goes into wait
      mode.

The PLL can change the bandwidth or operational mode of the loop filter manually or automatically. In *automatic bandwidth control* mode (the AUTO bit of the PLLCTL register is set to 1), the lock detector automatically switches between acquisition and tracking modes. The automatic bandwidth control mode is also used to determine when the PLL clock (PLLCLK) is safe to use as the source for the system and core clocks. When the LOCK bit of the CRGFLG register is set to 1, the PLLCLK signal can be used as the system clock safely. The setting of the LOCK bit can be detected by using the interrupt or polling method. Any change of the LOCK bit will cause the LOCKIF flag bit to be set to 1 and may optionally request an interrupt to the CPU.

The PLL circuit can also operate in *manual* mode (AUTO bit = 0). The manual mode is used by systems that do not require an indicator of the lock condition for proper operation. Such systems typically operate well below the maximum system frequency (f_{sys}) and require fast startup. To operate in manual mode, we must observe the following procedure:

1. Assert the ACQ bit (in PLLCTL register) before turning on the PLL manual mode. This configures the filter in acquisition mode.

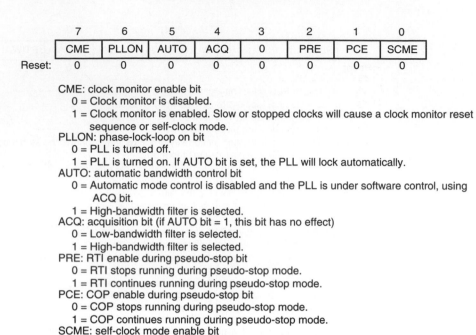

	7	6	5	4	3	2	1	0
	CME	PLLON	AUTO	ACQ	0	PRE	PCE	SCME
Reset:	0	0	0	0	0	0	0	0

CME: clock monitor enable bit
  0 = Clock monitor is disabled.
  1 = Clock monitor is enabled. Slow or stopped clocks will cause a clock monitor reset
     sequence or self-clock mode.
PLLON: phase-lock-loop on bit
  0 = PLL is turned off.
  1 = PLL is turned on. If AUTO bit is set, the PLL will lock automatically.
AUTO: automatic bandwidth control bit
  0 = Automatic mode control is disabled and the PLL is under software control, using
     ACQ bit.
  1 = High-bandwidth filter is selected.
ACQ: acquisition bit (if AUTO bit = 1, this bit has no effect)
  0 = Low-bandwidth filter is selected.
  1 = High-bandwidth filter is selected.
PRE: RTI enable during pseudo-stop bit
  0 = RTI stops running during pseudo-stop mode.
  1 = RTI continues running during pseudo-stop mode.
PCE: COP enable during pseudo-stop bit
  0 = COP stops running during pseudo-stop mode.
  1 = COP continues running during pseudo-stop mode.
SCME: self-clock mode enable bit
  0 = Detection of crystal clock failure causes clock monitor reset.
  1 = Detection of crystal clock failure forces the MCU in select-clock mode.

Figure 6.14 ■ The CRG PLL control register (PLLCTL)

2. After turning on the PLL, wait for a given time (t_{acq}) before entering tracking mode (clear ACQ to 0). The parameter t_{acq} can be found in the HCS12DP256 block user guide.

3. After entering the tracking mode, wait for t_{al} ns before selecting the PLLCLK signal as the source for system and core clocks (by setting the PLLSEL bit of the PLLCTL register to 1).

6.6.5 System Clock Generation

The clock generator is illustrated in Figure 6.15. This circuit generates the system and core clocks used in the microcontroller. When dealing with external memory or peripheral modules, the bus clock, referred to as E-clock in this book, is used. The E-clock is derived by dividing the SYSCLK clock by 2. If the PLLCLK is chosen to be the SYSCLK, then the frequency of the E-clock is half that of PLLCLK.

When the MCU enters the self-clock mode, the oscillator clock source is switched to the PLLCLK running at its minimum frequency, f_{SCM}.

Either the oscillator output OSCCLK (when PLLSEL = 0) or the PLL output PLLCLK (PLLSEL = 1) can be selected as SYSCLK. The oscillator can be completely bypassed and turned off by selecting an external clock source instead. The clock monitor, PLL, RTI, COP, and all clock signals based on OSCCLK are driven by this external clock instead of the output of the oscillator. As Figure 6.15 shows, SYSCLK can be a buffered version of the external clock input and the E-clock can be derived by dividing the external clock by 2.

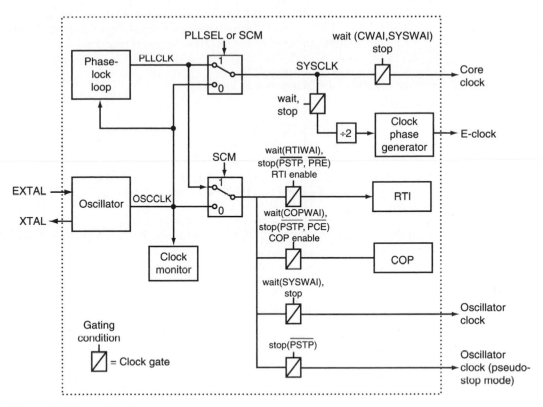

Figure 6.15 ■ HCS12 clock generation circuit

Example 6.2

There is a system that derives its E-clock from the PLL circuit, and an external clock of 8 MHz is selected. The desired E-clock is 24 MHz. Write an instruction sequence to perform the desired configuration.

Solution: Since the E-clock frequency is higher than the external clock's, we need to use the PLL circuit. According to Figure 6.15, the frequency of SYSCLK would be 48 MHz. The frequencies of OSCCLK and PLLCLK are 8 MHz and 48 MHz, respectively. According to Equation 6.1,

$$48 \text{ MHz} = 2 \times 8 \text{ MHz} \times [\text{SYNR} + 1]/[\text{REFDV} + 1]$$

One of the alternatives could be to set SYNR and REFDV to 2 and 0, respectively. The following subroutine achieves the desired configuration:

```
SetClk8    movb    #$02,SYNR      ; set SYNR to 2
           movb    #$0,REFDV      ; set REFDV to 0
           movb    #$80,CLKSEL    ; enable PLL, keep SYSCLK running in wait mode,
                                  ; keep RTI, COP, PLL, and core running in wait mode
```

```
            movb       #$60,PLLCTL           ; disable clock monitor, enable PLL, set automatic
                                             ; bandwidth control, disable RTI and COP in pseudo-stop
            brclr      CRGFLG,LOCK,*          ; wait until PLL locks into the target frequency
            rts
```

The C language version of the function is as follows:

```
    void SetClk8 (void)
    {
            SYNR    = 0x02;              // use PLL and 4-MHz crystal to generate 24-MHz system clock
            REFDV   = 0;                // "
            CLKSEL  = 0x80;             // enable PLL, keep SYSCLK running in wait mode
            PLLCTL  = 0x60;             // enable PLL, set automatic bandwidth control
            while (!(CRGFLG & 0x08));   // wait until PLL locks into the target frequency
    }
```

In addition to this instruction sequence, we also need to tie the $\overline{\text{XCLKS}}$ pin to ground to bypass the crystal oscillator.

▲

Example 6.3
▼

There is a system that uses a 4-MHz crystal oscillator to derive a 24-MHz E-clock. Write a subroutine to perform the required configuration.

Solution: The frequency of SYSCLK would be 48 MHz. The frequencies of OSCCLK and PLLCLK are 4 MHz and 48 MHz, respectively. According to Equation 6.1,

$$48 \text{ MHz} = 2 \times 4 \text{ MHz} \times [\text{SYNR} + 1]/[\text{REFDV} + 1]$$

One of the alternatives could be to set SYNR and REFDV to 5 and 0, respectively. The following subroutine achieves the desired configuration:

```
    SetClk4    movb    #$05,SYNR            ; set SYNR to 5
               movb    #$0,REFDV            ; set REFDV to 0
               movb    #$80,CLKGSEL         ; enable PLL, keep SYSCLK running in wait mode,
                                            ; keep RTI, COP, PLL, and core running in wait mode
               movb    #$60,PLLCTL          ; disable clock monitor, enable PLL, set automatic
                                            ; bandwidth control, disable RTI and COP in pseudo-stop
               brclr   CRGFLG,LOCK,*        ; wait until PLL locks into the target frequency
               rts
```

We also need to pull the $\overline{\text{XCLKS}}$ pin to high to select the crystal oscillator to generate the OSCCLK signal.

▲

6.6.6 Clock Monitor

The clock monitor circuit is based on an internal resistor-capacitor (RC) time delay so that it can operate without any MCU clocks. If no OSCCLK edges are detected within this RC time delay, the clock monitor indicates failure; this asserts self-clock mode or generates a system reset depending on the state of the SCME bit. If the clock monitor is disabled or the presence of clocks is detected, no failure is indicated. The clock monitor function is enabled or disabled by the CME control bit.

6.7 Real-Time Interrupt

The main function of the RTI circuit is to generate hardware interrupts periodically. If enabled, this interrupt will occur at the rate selected by the RTICTL register. The contents of this register are shown in Figure 6.16. The possible interrupt periods (in number of OSCCLK cycles) are listed in Table 6.4.

The time-multiplexing technique has been used to display multiple seven-segment displays at the same time in Chapter 4. However, the method used in Chapter 4 requires the CPU to call a delay function to generate the desired time delay to switch the digit to be displayed; this would prevent the CPU from performing other operations. One solution to this problem is to remind the CPU to switch digits to be displayed periodically. The RTI function serves this purpose perfectly well. The next three examples use the RTI program to perform seven-segment pattern shifting.

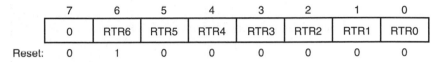

7	6	5	4	3	2	1	0
0	RTR6	RTR5	RTR4	RTR3	RTR2	RTR1	RTR0

Reset: 0   1   0   0   0   0   0   0

Figure 6.16 ■ CRG RTI control register (RTICTL)

RTR[3:0]	RTR[6:4]							
	000 (off)	001 (2^{10})	010 (2^{11})	011 (2^{12})	100 (2^{13})	101 (2^{14})	110 (2^{15})	111 (2^{16})
0000 ($\div 1$)	off*	2^{10}	2^{11}	2^{12}	2^{13}	2^{14}	2^{15}	2^{16}
0001 ($\div 2$)	off*	2×2^{10}	2×2^{11}	2×2^{12}	2×2^{13}	2×2^{14}	2×2^{15}	2×2^{16}
0010 ($\div 3$)	off*	3×2^{10}	3×2^{11}	3×2^{12}	3×2^{13}	3×2^{14}	3×2^{15}	3×2^{16}
0011 ($\div 4$)	off*	4×2^{10}	4×2^{11}	4×2^{12}	4×2^{13}	4×2^{14}	4×2^{15}	4×2^{16}
0100 ($\div 5$)	off*	5×2^{10}	5×2^{11}	5×2^{12}	5×2^{13}	5×2^{14}	5×2^{15}	5×2^{16}
0101 ($\div 6$)	off*	6×2^{10}	6×2^{11}	6×2^{12}	6×2^{13}	6×2^{14}	6×2^{15}	6×2^{16}
0110 ($\div 7$)	off*	7×2^{10}	7×2^{11}	7×2^{12}	7×2^{13}	7×2^{14}	7×2^{15}	7×2^{16}
0111 ($\div 8$)	off*	8×2^{10}	8×2^{11}	8×2^{12}	8×2^{13}	8×2^{14}	8×2^{15}	8×2^{16}
1000 ($\div 9$)	off*	9×2^{10}	9×2^{11}	9×2^{12}	9×2^{13}	9×2^{14}	9×2^{15}	9×2^{16}
1001 ($\div 10$)	off*	10×2^{10}	10×2^{11}	10×2^{12}	10×2^{13}	10×2^{14}	10×2^{15}	10×2^{16}
1010 ($\div 11$)	off*	11×2^{10}	11×2^{11}	11×2^{12}	11×2^{13}	11×2^{14}	11×2^{15}	11×2^{16}
1011 ($\div 12$)	off*	12×2^{10}	12×2^{11}	12×2^{12}	12×2^{13}	12×2^{14}	12×2^{15}	12×2^{16}
1100 ($\div 13$)	off*	13×2^{10}	13×2^{11}	13×2^{12}	13×2^{13}	13×2^{14}	13×2^{15}	13×2^{16}
1101 ($\div 14$)	off*	14×2^{10}	14×2^{11}	14×2^{12}	14×2^{13}	14×2^{14}	14×2^{15}	14×2^{16}
1110 ($\div 15$)	off*	15×2^{10}	15×2^{11}	15×2^{12}	15×2^{13}	15×2^{14}	15×2^{15}	15×2^{16}
1111 ($\div 16$)	off*	16×2^{10}	16×2^{11}	16×2^{12}	16×2^{13}	16×2^{14}	16×2^{15}	16×2^{16}

* Denotes the default value out of reset. This value disables RTI.

Table 6.4 ■ RTI period (in units of OSCCLK cycle)

Example 6.4

▼

Write an assembly program to use the RTI to time-multiplex four seven-segment displays using the circuit shown in Figure 4.18. Turn on one display at a time and light each display for about 1 ms, then switch to the next display. Use display #0 to display #3. Use CodeWarrior and a demo board programmed with a serial monitor to implement the circuit.

Solution: The algorithm in time-multiplexing four seven-segment displays is similar to that described in Example 5.8 but uses a RTI to trigger the switch of display. There are 10 display sequences.

1234

2345

. . .

0123

For each sequence, digits are indexed from 0 to 3. By repeating each sequence 100 times and turning on one digit for 1 ms, a sequence will last for 0.4 s. Within 0.4 s, there will be 400 RTIs. Let *base*, *dindex*, and *timeCnt* represent the starting address of a sequence, index to digits within a sequence, and the repetition count of a sequence, respectively. The following assembly program performs the seven-segment display shifting:

```
            include "c:\miniide\hcs12.inc"
            org       $1000
base        ds.w      1                    ; base address of the current four-digit sequence
timeCnt     ds.w      1                    ; repetition count of the current sequence of four digits
dindex      ds.b      1                    ; digit index in a sequence (0, 1, 2, or 3)
            org       $1500
start       lds       #$1500               ; set up stack pointer
            jsr       SetClk8              ; set E-clock to 24 MHz
            movw      #Seg7Pat,base        ; initialize the base address of a new sequence of digits
            clr       dindex               ; the offset digit to be displayed from the base
            movb      #$FF,DDRB            ; configure Port B for output
            movb      #$0F,DDRP            ; configure Port P lower 4 pins for output
            movw      #400,timeCnt         ; allow four digits to multiplex for 0.4 s
            movb      #$40,RTICTL          ; set RTI to about 1 ms
            bset      CRGINT,RTIE          ; enable RTI locally
            cli                            ; enable interrupt globally
            bra       $                    ; wait for interrupt to occur
; ********************************************************************************
SetClk8     movb      #$02,SYNR            ; set SYNR to 2
            movb      #$0,REFDV            ; set REFDV to 0
            movb      #$80,CLKSEL          ; enable PLL, keep SYSCLK running in wait mode,
                                           ; keep RTI, COP, PLL, and core running in wait mode
            movb      #$60,PLLCTL          ; disable clock monitor, enable PLL, set automatic
                                           ; bandwidth control, disable RTI and COP in pseudo-
                                           ; stop
```

```
                    brclr       CRGFLG,LOCK,*        ; wait until PLL locks into the target frequency
                    rts
; **********************************************************************************
rtiISR              movb        #RTIF,CRGFLG         ; clear the RTIF flag
                    ldx         base                 ; load the base address stored in base
                    ldab        dindex               ; place the digit offset
                    movb        b,X,PTB              ; output the segment pattern
                    cmpb        #0
                    beq         digit0               ; turn on display #0
                    cmpb        #1
                    beq         digit1               ; turn on display #1
                    cmpb        #2
                    beq         digit2               ; turn on display #2
                    movb        #$07,PTP             ; turn on display #3
                    bra         update
digit2              movb        #$0B,PTP             ; turn on digit 2
                    bra         update
digit1              movb        #$0D,PTP             ; turn on digit 1
                    bra         update
digit0              movb        #$0E,PTP             ; turn on digit 0
update              inc         dindex               ; switch to the next digit
                    ldaa        dindex               ; check if we need to reset digit index to 0
                    cmpa        #4                   ;  "
                    bne         nextc
                    movb        #0,dindex
nextc               ldy         timeCnt              ; decrement repletion count
                    dey                              ;  "
                    sty         timeCnt              ;  "
                    cpy         #0                   ; do we need to change to a new sequence?
                    bne         next                 ; the current four digits need not be changed yet
                    movw        #400,timeCnt         ; restore the time count to 400
                    ldx         base                 ; shift to the next four digits (shift by 1)
                    inx                              ;  "
                    stx         base                 ;  "
                    cpx         #Seg7Pat+10          ; is this the last sequence? (0,1,2,3)
                    bne         next                 ;  "
                    movw        #Seg7Pat,base        ; go back to the first sequence again
next                rti
; **********************************************************************************
Seg7Pat             dc.b        $06, $5B, $4F, $66, $6D, $7D, $07, $7F, $6F, $3F, $06, $5B, $4F
; **********************************************************************************
                    org    $FFF0
                    dc.w   rtiISR                    ; RTI vector stored here
                    org    $FFFE
                    dc.w   start                     ; reset vector
                    end
```

▲

Example 6.5

▼

Write a C program to use the RTI to time-multiplex four seven-segment displays using the circuit shown in Figure 4.18 and shift the seven-segment display pattern as described in Example 6.4. Turn on one display at a time and light each display for about 1 ms, then switch to the next display. Use display 0 to display 3. Use CodeWarrior and a demo board programmed with serial monitor to implement the circuit.

Solution: To implement the digit sequence shifting program in C language, we make the following arrangement:

- Place the segment patterns in one array (segPat[]). This array is the overlapping arrangement of 10 display sequences as shown in Figure 5.2.
- Place digit select values in one array (digit[]).
- Use a variable (seq) as an index to the segment pattern array that identifies the first digit of the current sequence.
- Use a variable (ix) as an index to the digits within one sequence. The range of this variable is from 0 to 3.
- Use a variable (count) to specify the repetition count of a sequence.

The main function that performs the initialization and the RTI service routine that performs time multiplexing is

```
#include        "c:\cwHCS12\include\hcs12.h"
#include        "c:\cwHCS12\include\SetClk.h"
int     seq;                            // start index to segPat[] of a sequence of digits (0 to 9)
int     ix;                             // index of digits of a sequence (0 to 3)
int     count;                          // repetition count of a sequence
char segPat[13] = {0x06, 0x5B, 0x4F, 0x66, 0x6D, 0x7D, 0x07, 0x7F, 0x67, 0x3F, 0x06, 0x5B, 0x4F};
char digit[4]  = {0xFE, 0xFD, 0xFB, 0xF7};

void main (void) {
        seq    = 0;                     // initialize the start index to segPat[] for the display sequence
        ix     = 0;                     // initialize the index of a new sequence
        count  = 400;                   // initialize the RTI count of a sequence
        SetClk8();                      // set E-clock to 24 MHz from an 8-MHz crystal oscillator
        RTICTL   = 0x40;                // RTI interval set to 2**10 OSCCLK cycles
        DDRB     = 0xFF;                // configure Port B for output
        DDRP     = 0xFF;                // configure Port P for output
        CRGINT| = RTIE;                 // enable RTI
        asm("CLI");                     // enable interrupt globally
        while(1);
}
// RTI service routine
interrupt void rtiISR(void) {
        CRGFLG = 0x80;                  // clear RTIF bit
        PTB = segPat[seq+ix];           // send out digit segment pattern
        PTP = digit[ix];                // turn on the display
        ix++;                           // increment the index to digits of a sequence
        if (ix == 4)                    // make sure the index to digits of a sequence is from 0 to 3
        ix = 0;                         //    "
```

```
          count−−;
          if(count == 0){          // is time for the current sequence expired?
          seq++;                   // change to a new sequence of digits
          count = 400;             // reset repetition count
          }
          if(seq == 10)            // is this the last sequence?
          seq = 0;                 // reset start index of a sequence
     }
}
```

This programming project also contains the vectors.c file.

```
extern void near rtiISR(void);
#pragma CODE_SEG __NEAR_SEG NON_BANKED

#pragma CODE_SEG DEFAULT          // Change code section to DEFAULT.

typedef void (*near tlsrFunc)(void);
const tlsrFunc _vect[] @0xFFF0 = {
          rtiISR
};
```

▲

Example 6.6

▼

Modify the C program in Example 6.5 so that it can be compiled using the ICC12 C compiler and run in a demo board programmed with the D-Bug12 monitor.

Solution: Since the ICC12 IDE does not support flash memory programming, it can only support program execution in SRAM. We will use inline assembly instructions to set up an interrupt vector for RTI. The C program for ICC2 to implement seven-segment display shifting described in Example 6.5 is

```
#include        "c:\cwHCS12\include\hcs12.h"
#include        "c:\cwHCS12\include\SetClk.h"
void rtiISR(void);
int seq,ix,count;
char segPat[13] = {0x06, 0x5B, 0x4F, 0x66, 0x6D, 0x7D, 0x07, 0x7F, 0x67, 0x3F, 0x06, 0x5B, 0x4F};
char digit[4]  = {0xFE, 0xFD, 0xFB, 0xF7};
void main (void) {
          asm("ldd #_rtiISR");          // set up RTI vector by calling SetUserVector
          asm("pshd");                  // function using inline assembly instructions
          asm("ldd #56");               //    "
          asm("ldx $EEA4");             //    "
          asm("jsr 0,x");               //    "
          seq    = 0;
          ix     = 0;
          count  = 400;
          SetClk8();
          RTICTL = 0x40;                // RTI interval set to 2**10 OSCCLK cycles
          DDRB   = 0xFF;                // configure Port B for output
```

```
            DDRP     = 0xFF;              // configure Port P for output
            CRGINT   |= RTIE;             // enable RTI interrupt
            asm("CLI");                   // enable interrupt globally
            while(1);
    }

    #pragma interrupt_handler rtiISR
    void rtiISR(void) {
            CRGFLG = 0x80;                // clear RTIF bit
            PTB = segPat[seq+ix];         // output the segment pattern
            PTP = digit[ix];              // output digit select value
            ix++;                         // increment digit index within a sequence
            if (ix == 4)
            ix = 0;
            count--;                      // decrement repetition count
            if(count == 0){               // if repetition count is 0, then change to the next sequence
            seq++;
            count = 400;
            }
            if(seq == 10)                 // Reach the last sequence?
            seq = 0;
    }
```

The required modification to this program for it to compile with the EGNU IDE is given in Appendix E.

▲

6.8 Computer Operating Properly

The COP (a free-running watchdog timer) enables the user to find out if a user program is running and sequencing properly. When the COP times out, it resets the CPU. The CPU reset by the COP is an indication that the software is no longer being executed in the intended sequence. The software that utilizes the COP function must include an instruction sequence to prevent the COP from timing out.

To prevent the COP module from resetting the MCU, one must write the value $55 followed by $AA into the ARMCOP register. Other instructions can be inserted between the values $55 and $AA, but the sequence ($55, $AA) must be completed prior to the COP timeout period to avoid a COP reset. Writing any value other than $55 or $AA into this register will reset the MCU.

The functioning of the COP module and its timeout period are controlled by the COPCTL register. The contents of this register are shown in Figure 6.17. The COP module is disabled if the CR2:CR0 bits of this register are set to 000. The COP has a windowed option for its operation. The windowed COP operation is enabled by setting the WCOP bit of the COPCTL register. In this mode, writes to the ARMCOP register to clear the COP timer must occur in the last 25 percent of the selected timeout period. A premature write will immediately reset the MCU.

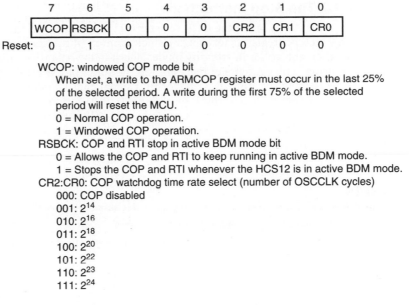

WCOP: windowed COP mode bit
When set, a write to the ARMCOP register must occur in the last 25% of the selected period. A write during the first 75% of the selected period will reset the MCU.
0 = Normal COP operation.
1 = Windowed COP operation.
RSBCK: COP and RTI stop in active BDM mode bit
0 = Allows the COP and RTI to keep running in active BDM mode.
1 = Stops the COP and RTI whenever the HCS12 is in active BDM mode.
CR2:CR0: COP watchdog time rate select (number of OSCCLK cycles)
000: COP disabled
001: 2^{14}
010: 2^{16}
011: 2^{18}
100: 2^{20}
101: 2^{22}
110: 2^{23}
111: 2^{24}

Figure 6.17 ■ CRG COP control register (COPCTL)

6.9 Low-Power Modes

When a microcontroller is performing normal operations, power consumption is unavoidable. However, the microcontroller in an embedded system may not always be performing useful operations. Under this situation, it would be ideal for the power consumption to be reduced to the minimum. This issue is especially important for those embedded products powered by batteries. The HCS12 has two low-power modes that can reduce power consumption dramatically: *wait* and *stop* modes.

6.9.1 The Wai Instruction

The wait mode is entered by executing the **wai** instruction. The wai instruction pushes all CPU registers (except the stack pointer) and the return address onto the stack and enters a wait state. During the wait state, CPU clocks are stopped (clock signals that drive the ALU and register file), but other clocks in the microcontroller (clock signals that drive peripheral functions) continue to run.

The CPU leaves the wait state when it senses one or more of the following events:

- Maskable interrupts that are not masked
- Nonmaskable interrupts
- Resets

On leaving the wait state, the HCS12 CPU sets the appropriate interrupt mask bit(s), fetches the vector corresponding to the exception sensed, and continues instruction execution at the location the vector points to.

6.9.2 The Stop Instruction

When the S bit in the CCR register is cleared and a stop instruction is executed, the HCS12 saves all CPU registers (except the stack pointer) in the stack, stops all system clocks, and puts the microcontroller in *standby* mode. The standby operation minimizes the system power consumption. The contents of registers and the states of I/O pins remain unchanged.

Asserting the $\overline{\text{RESET}}$, $\overline{\text{XIRQ}}$, or $\overline{\text{IRQ}}$ signal ends standby mode. If it is the $\overline{\text{XIRQ}}$ signal that ends the stop mode and the X mask bit is 0, instruction execution resumes with a vector fetch for the $\overline{\text{XIRQ}}$ interrupt. If the X mask bit is 1 ($\overline{\text{XIRQ}}$ disabled), a two-cycle recovery sequence is used to adjust the instruction queue, and execution continues with the next instruction after the **stop** instruction.

6.10 Resets

There are four possible sources of resets.

- Power-on (POR) and low-voltage detector (LVD) reset
- $\overline{\text{RESET}}$ pin
- COP reset
- Clock monitor reset

The COP and clock monitor resets have been discussed earlier. Power-on, low-voltage detector, and $\overline{\text{RESET}}$ pin resets share the same reset vector. The COP reset and the clock monitor reset each have a separate vector.

6.10.1 Power-On Reset

The HCS12 has a circuitry to detect when the V_{DD} supply to the MCU has reached a certain level and asserts reset to the internal circuits. The detector circuit is triggered by the slew rate. As soon as a power-on reset is triggered, the CRG module performs a quality check on the incoming clock signal. Start of the reset sequence is delayed until the clock check indicates a valid clock signal or the clock check was unsuccessful, and the CRG module enters self-clock mode.

6.10.2 External Reset

The HCS12 distinguishes between internal and external resets by sensing how quickly the signal on the $\overline{\text{RESET}}$ pin rises to logic high after it has been asserted. When the HCS12 senses any of the four reset conditions, internal circuitry drives the $\overline{\text{RESET}}$ pin low for 128 SYSCLK cycles (this number might be increased by 3 to 6 SYSCLK cycles), then releases. Sixty-four SYSCLK cycles later, the CPU samples the state of the signal applied to the $\overline{\text{RESET}}$ pin. If the signal is still low, an external reset has occurred. If the signal is high, the reset has been initiated internally by either the COP system or the clock monitor.

The power supply to an embedded system may drop below the required level. If the microcontroller keeps working under this situation, the contents of the EEPROM might be corrupted. The common solution is to pull the $\overline{\text{RESET}}$ signal to low so that the microcontroller cannot execute instructions. A low-voltage-inhibit (LVI) circuit, such as the Freescale MC34064, can be used to protect against the EEPROM corruption. Figure 6.18 shows an example of the reset circuit with a manual reset and LVI circuit.

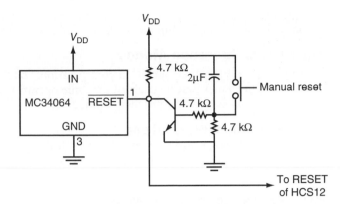

Figure 6.18 ■ A typical external reset circuit

6.11 HCS12 Operation Modes

As shown in Table 6.5, the HCS12 can operate in eight different modes. Each mode has a different default memory map and external bus configuration. After reset, most system resources can be mapped to other addresses by writing to the appropriate control registers.

The states of the BKGD, MODB, and MODA pins when the RESET signal is low determine the operation mode after the CPU leaves the reset state. The SMODN, MODB, and MODA bits in the MODE register show the current operation mode and provide limited mode switching during the operation. The states of the BKGD, MODB, and MODA pins are latched into these bits on the rising edge of the RESET signal. During reset, an active pull-up (on-chip transistor) is connected to the BKGD pin (as input) and active pull-downs (on-chip transistors) are connected to the MODB and MODA pins. If an open circuit occurs on any of these pins, the device will operate in normal single-chip mode.

The two basic types of operation modes are

1. *Normal modes.* Some registers and bits are protected against accidental changes.

2. *Special modes.* Greater access for special purposes such as testing and emulation to protected control registers and bits is allowed.

BKGD	MODB	MODA	Mode	Port A	Port B
0	0	0	Special single chip	General-purpose I/O	General-purpose I/O
0	0	1	Special expanded narrow	ADDR[15:8]DATA[7:0]	ADDR[7:0]
0	1	0	Special peripheral	ADDR/DATA	ADDR/DATA
0	1	1	Special expanded wide	ADDR/DATA	ADDR/DATA
1	0	0	Normal single chip	General-purpose I/O	General-purpose I/O
1	0	1	Normal expanded narrow	ADDR[15:8]DATA[7:0]	ADDR[7:0]
1	1	0	Reserved (forced to peripheral)	—	—
1	1	1	Normal expanded wide	ADDR/DATA	ADDR/DATA

Table 6.5 ■ HCS12 mode selection

The background debug mode (BDM) is a system development and debug feature and is available in all modes. In special single-chip mode, BDM is active immediately after reset.

6.11.1 Normal Operation Modes

These modes provide three operation configurations. Background debugging is available in all three modes, but must first be enabled for some operations by means of a BDM command. BDM can then be made active by another command.

NORMAL EXPANDED WIDE MODE

In this mode, Ports A and B are used as the multiplexed 16-bit address and data buses. ADDR[15..8] and DATA[15..8] are multiplexed on Port A. ADDR[7..0] and DATA[7..0] are multiplexed on Port B.

NORMAL EXPANDED NARROW MODE

The 16-bit external address bus uses Port A for the high byte and Port B for the low byte. The 8-bit external data bus uses Port A. ADDR[15..8] and DATA[7..0] are multiplexed on Port A.

NORMAL SINGLE-CHIP MODE

Normal single-chip mode has no external buses. Ports A, B, and E are configured for general-purpose I/O. Port E bits 1 and 0 are input only with internal pull-ups and the other 22 pins are bidirectional I/O pins that are initially configured as high-impedance inputs. Port E pull-ups are enabled on reset. Port A and B pull-ups are disabled on reset.

6.11.2 Special Operation Modes

Special operation modes are commonly used in factory testing and system development.

SPECIAL EXPANDED WIDE MODE

This mode is for emulation of normal expanded wide mode and emulation of normal single-chip mode with a 16-bit bus. The bus-control pins of Port E are all configured for their bus-control output functions rather than general-purpose I/O.

SPECIAL EXPANDED NARROW MODE

This mode is for emulation of normal expanded narrow mode. External 16-bit data is handled as two back-to-back bus cycles, one for the high byte followed by one for the low byte. Internal operations continue to use full 16-bit data paths.

SPECIAL SINGLE-CHIP MODE

This mode can be used to force the microcontroller to active BDM to allow a system debug through the BKGD pin. The HCS12 CPU does not fetch the reset vector or execute application code as it would in other modes. Instead, the active background mode is in control of CPU execution, and BDM firmware waits for additional serial commands through the BKGD pin. There are no external address and data buses in this mode. The microcontroller operates as a stand-alone device, and all program and data space are on-chip. External port pins can be used for general-purpose I/O.

SPECIAL PERIPHERAL MODE

The HCS12 CPU is not active in this mode. An external master can control on-chip peripherals for testing purposes. It is not possible to change to or from this mode without going through reset. Background debugging should not be used while the microcontroller is in special peripheral mode, as internal bus conflicts between the BDM and the external master can cause improper operation of both modes.

6.12 Summary

Interrupt is a special event that requires the CPU to stop normal program execution and provide a certain service to the event. The interrupt mechanism has many applications, including coordinating I/O activities, exiting from software errors, and reminding the CPU to perform routine work, and so on.

Some interrupts are *maskable* and can be ignored by the CPU. Other interrupts are *nonmaskable* and cannot be ignored by the CPU. Nonmaskable interrupts are often used to handle critical and emergent events such as process control and power failure.

Multiple interrupts may be pending at the same time. The CPU needs to decide which one to service first. The solution to this issue is to *prioritize* all of the interrupt sources. The pending interrupt with the highest priority will receive service before other pending interrupts.

The CPU provides service to an interrupt request by executing an *interrupt service routine*. The current program counter value is saved in the stack before the CPU executes the service routine so that CPU control can be returned to the interrupted program when the interrupt service routine is completed.

In order to provide service to the interrupt, the CPU must have some way to find out the starting address of the interrupt service routine. There are three methods to determine the starting address (called *interrupt vector*) of the interrupt service routine.

1. Each interrupt vector is predefined when the microcontroller is designed. In this method, the CPU simply jumps to the predefined location to execute the service routine.

2. Each interrupt vector is stored in a predefined memory location. When an interrupt occurs, the CPU fetches the interrupt vector from that predefined memory location.

The HCS12 uses this approach.

3. The interrupt source provides an interrupt vector number to the CPU so that the CPU can figure out the memory location where the interrupt vector is stored. The CPU needs to perform a read bus cycle to obtain the interrupt vector number.

There are three steps in the interrupt programming.

Step 1
Initialize the interrupt vector table that holds all the interrupt vectors. This step is not needed for those microcontrollers that use the first method to resolve the interrupt vector.

Step 2
Write the interrupt service routine.

Step 3
Enable the interrupt to be serviced.

Users who own a demo board with a resident D-Bug12 monitor would use the EVB mode for application development. The D-Bug12 monitor has occupied the default memory space where the interrupt vectors are to be stored. To enable the user to use interrupts in their applications, D-Bug12 allows vectors to be stored in a table in the SRAM.

Clock signals are critical to the proper operation of the MCU. The HCS12 has a phase-locked-loop (PLL) circuit that can be used to generate the system clock and E-clock with a frequency higher than that of the external crystal oscillator. The PLL can generate a high-frequency clock signal using a low-frequency crystal. In addition, the PLL provides stability to the clock signals.

Reset is a mechanism for

1. Setting up operation mode for the microcontroller
2. Setting up initial values for control registers
3. Exiting from software errors and some hardware errors

All HCS12 microcontrollers have the same number of resets and nonmaskable interrupt sources despite the fact that they may not have the same number of maskable interrupts. The HCS12 has two low-power modes that are triggered by the execution of wai and stop instructions. Power consumption will be reduced dramatically in either low-power mode. The HCS12 has a COP timer reset mechanism to detect the software error. A software program that behaves properly will reset the COP timer before it times out and prevent it from resetting the CPU.

The HCS12 has a clock monitor reset mechanism that can detect the slowing down or loss of clock signals. Whenever the clock frequency gets too low, the clock monitor will detect it and reset the CPU.

The real-time interrupt (RTI) mechanism, when enabled, generates periodic interrupts to remind the CPU to perform routine work such as time-multiplexing seven-segment displays, environment monitoring, task switching in a multitasking operating system, and so on.

The HCS12 has seven different operation modes divided into two basic categories: normal modes and special modes. Normal modes are used for embedded applications, whereas special modes are used in fabrication testing and development debugging activities.

6.13 Exercises

E6.1 What is the name given to a routine that is executed in response to an interrupt?

E6.2 What are the advantages of using interrupts to handle data inputs and outputs?

E6.3 What are the requirements for interrupt processing?

E6.4 How do you enable other interrupts when the HCS12 is executing an interrupt service routine?

E6.5 Why would there be a need to promote one of the maskable interrupts to the highest priority among all maskable interrupts?

E6.6 Write the assembler directives to initialize the $\overline{\text{IRQ}}$ interrupt vector located at $3000 for the EVB mode of the D-Bug12 monitor.

E6.7 What is the last instruction in most interrupt service routines? What does this instruction do?

E6.8 Suppose that the HCS12 is executing the following instruction segment and the $\overline{\text{IRQ}}$ interrupt occurs when the TSY instruction is being executed. What will be the contents of the top 10 bytes in the stack?

```
org     $2000
lds     #$2000
clra
ldx     #$0
bset    10,X $48
ldab    #$40
inca
tap
pshb
tsy
adda    #10
```

E6.9 Suppose that the OSCCLK clock frequency is 5 MHz. Compute the COP watchdog timer timeout period for all the possible combinations of the CR2, CR1, and CR0 bits in the COPCTL register.

E6.10 Assume that the interrupt vector for the timer overflow is $3000. Write the assembler directives to initialize its vector table entry on a demo board with the D-Bug12 monitor.

E6.11 Write an instruction sequence to clear the X and I bits in the CCR. Write an instruction sequence to set the S, X, and I bits in the CCR.

E6.12 Why does the HCS12 need to be reset when the power supply is too low?

E6.13 Write an instruction sequence to prevent the COP timer from timing out and resetting the microcomputer.

E6.14 Suppose you want to generate a 25-MHz E-clock; propose a set of values for the SYNR and REFDV registers to achieve this goal using a 5-MHz crystal oscillator and the PLL circuit.

E6.15 Suppose you want to generate a 24-MHz E-clock; propose a set of values for the SYNR and REFDV registers to achieve this goal using a 3-MHz crystal oscillator and the PLL circuit.

E6.16 Suppose you want to generate a 20-MHz E-clock; propose a set of values for the SYNR and REFDV registers to achieve this goal using a 4-MHz crystal oscillator and the PLL circuit.

E6.17 Suppose you want to generate a 24-MHz E-clock; propose a set of values for the SYNR and REFDV registers to achieve this goal using a 6-MHz crystal oscillator and the PLL circuit.

E6.18 Set up the interrupt vectors for the enhanced capture timer Ch1, enhanced capture timer Ch0, RTI, and IRQ to work with the CodeWarrior IDE by modifying the *vectors.c* file given in Example 6.5.

E6.19 Write a program to drive the LED circuit in Figure 4.16 and display one LED at a time from the one driven by pin 7 toward the one driven by pin 0 and then reverse. Repeat this operation forever. Each LED is lighted for about 400 ms assuming that the HCS12 uses an 8-MHz crystal oscillator to generate a system clock. Use the RTI to trigger the change of the LED light patterns. It may take several RTIs to trigger one change of the LED pattern for this problem.

E6.20 Write a program to generate a periodic square waveform that is about 500 Hz (roughly) using the PT0 pin and the RTI.

6.14 Lab Exercises and Assignments

L6.1 $\overline{IRQ}$ *input interrupt experiment.* Use the 555 timer (or take the signal from a function generator) to generate a digital waveform with frequency equal to approximately 1 Hz. The circuit connection of the 555 timer is illustrated in Figure L6.1. Connect the 555 timer output (pin 3) to the $\overline{IRQ}$ pin. The $\overline{IRQ}$ interrupt service routine would output the message "Interrupt k." Enter the program (both assembly and C) to a file, assemble or compile, and download the S-record file onto the demo board for execution.

L6.2 *Simple interrupts.* Connect the $\overline{IRQ}$ pin of the demo board to a debounced switch that can generate a negative-going pulse. Write a main program and an $\overline{IRQ}$ interrupt service routine. The main program initializes the variable *irq_cnt* to 10, stays in a loop, and keeps checking the value of *irq_cnt*. When *irq_cnt* is decremented to 0, the main program jumps back to monitor. The $\overline{IRQ}$ service routine simply decrements *irq_cnt* by 1 and returns.

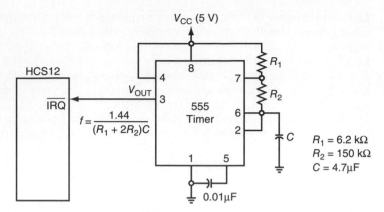

Figure L6.1 ■ HCS12 $\overline{\text{IRQ}}$ pin interrupt circuit

The lab procedure is as follows:

Step 1
Connect the $\overline{\text{IRQ}}$ pin of the demo board to a debounced switch that can generate a clean negative-going pulse. A debounced switch is available on the SSE256 demo board.

Step 2
Enter the main program and $\overline{\text{IRQ}}$ service routine, assemble them, and then download them to the single-board computer. Remember to enable the $\overline{\text{IRQ}}$ interrupt in your program.

Step 3
Pulse the switch 10 times. If everything works properly, you should see the D-Bug12 monitor prompt after 10 pulses applied to the $\overline{\text{IRQ}}$ pin.

L6.3 *RTI Interrupts and LED flashing.* Write a C (or assembly) program to flash the LEDs on your demo board with the following patterns:

1. Turn on all LEDs for about 0.4 s and turn them off also for about 0.4 s.

2. Repeat step 1 three more times.

3. Turn on one LED at a time from left to right with each LED turned on for about 0.4 s.

4. Repeat step 3 three more times.

5. Turn on one LED at a time from right to left with each LED turned on for about 0.4 s.

6. Repeat step 5 three more times.

7. Turn LEDs driven by pins RB7 and RB0 on and off four times. The on-time and off-time are each about 0.4 s.

8. Turn LEDs driven by pins RB6 and RB1 on and off four times. The on-time and off-time are each about 0.4 s.

9. Turn LEDs driven by pins RB5 and RB2 on and off four times. The on-time and off-time are each about 0.4 s.

10. Turn LEDs driven by pins RB4 and RB3 on and off four times. The on-time and off-time are each about 0.4 s.

11. Turn LEDs driven by pins RB3 and RB4 on and off four times. The on-time and off-time are each about 0.4 s.

12. Turn LEDs driven by pins RB2 and RB5 on and off four times. The on-time and off-time are each about 0.4 s.

13. Turn LEDs driven by pins RB1 and RB6 on and off four times. The on-time and off-time are each about 0.4 s.

14. Turn LEDs driven by pins RB0 and RB7 on and off four times. The on-time and off-time are each about 0.4 s.

15. Go to step 1 and repeat.

You are required to use the RTI to generate the required delay and trigger the change of LED patterns.

7

Advanced Parallel I/O

7.1 Objectives

- Explain I/O addressing issue

- Explain I/O synchronization methods

- Verify voltage compatibility when interfacing a microcontroller with peripheral devices

- Verify current compatibility when driving peripheral devices

- Configure the LCD

- Write program to display messages on the LCD

- Write program to enter data using keypads or keyboards

- Explain keypad- or keyboard-debouncing method

- Explain the principles of operation of stepper motors

- Write programs to drive stepper motors

- Write programs to generate waveforms using the D/A converter

- Explain the operation, application, and programming of key-wake-up ports

7.2 I/O Related Issues

A brief introduction to the I/O operation and I/O devices was given in Chapter 4. In this chapter, the issues related to I/O addressing, I/O synchronization, electrical characteristics compatibility, and HCS12 I/O ports configuration are explored. More complicated I/O devices such as liquid crystal displays (LCDs), keypads and keyboards, stepper motors, and D/A converters will be explained.

7.3 I/O Addressing Issue

I/O devices are also called *peripheral devices* in the sense that they are outside the core of a computer system. To perform an I/O operation, the CPU needs to specify the I/O device that it intends to deal with. This involves two issues.

- *Address space.* This issue is related to the question of whether I/O devices should be treated the same as memory devices, that is, whether the I/O devices should occupy the same *address space* (some people like to call it *memory space*) as the memory devices (SRAM, DRAM, EEPROM, or flash memory). Both approaches have been used by microprocessor and microcontroller vendors. The current trend is for I/O devices and memory components to share the same memory space.

- *Addressing modes and instructions.* I/O devices may have their own addressing modes and instruction set or share the same addressing modes and instruction set with the memory devices. In the first approach, the microprocessor may use the following instructions for input and output:

  out 3: sends data in accumulator to I/O device at address 3.

  in 5: inputs a byte from input device 5 to the accumulator.

  In the second approach, the microprocessor uses the same instructions and addressing modes to perform input and output operations. Again, the current trend is the second approach.

Traditionally, Motorola microprocessors and microcontrollers have used the same addressing modes and instruction sets to access I/O and memory devices. Memory and I/O devices share a single memory space.

7.4 I/O Synchronization

As described in Section 1.3.2, microprocessors either cannot provide the current required by the peripheral devices or operate at a voltage level different from those of peripheral devices. Therefore they usually communicate with peripheral devices via interface chips. For the microcontroller, the functions of most of these interface chips are built into the same chip as the microcontroller. When transferring data in the parallel format (multiple bits at a time), the microcontroller reads and writes data through the parallel port. When transferring data in the serial format (1 bit in one clock cycle), the microcontroller reads and writes data through the serial interface such as a serial communication interface (SCI), serial peripheral interface (SPI), inter-integrated circuit (I²C), or controller area network (CAN).

Because data transfers go through the interface chip (or logic for the MCU), the synchronization issue occurs in two places. One is between the processor and the interface chip (or logic for the MCU). The other is between the interface chip (or logic for the MCU) and the peripheral devices.

7.4.1 Synchronization Issue for Parallel Ports

The design of parallel ports of today's microcontrollers (including the HCS12) is to allow the data written into the data register to appear on the output pins directly and allow the read operation from the data register to obtain the instantaneous voltage levels on the input port pins. There is no concern about the synchronization issue. This is quite different from the era of microprocessors. To perform an I/O operation, the microprocessor needs to make sure that interface chip for the input device has new data before it reads it or that the interface chip for the output device can handle new data before it sends new data to it. This is achieved by either polling or interrupt. The interface chip is designed to have status flags to indicate whether it has new data or can accept new data for output. The interface chip may use a strobe signal or handshake signals to achieve synchronization between the interface chip and the peripheral device. This type of synchronization is no longer needed in today's microcontrollers.

7.4.2 Synchronization Issue for Serial Interface

For the serial interface, the data transfer rate between the processor and the interface logic is much faster than that between the interface logic and the peripheral device. The processor needs to make sure that there is new data in the interface logic (usually held in a data register) before reading it. It also needs to make sure that the interface logic can handle more data before sending new data to the interface logic. This is achieved by using either the polling or the interrupt method to make sure that the new I/O operation can be started.

The synchronization between the interface logic and the peripheral device is achieved by following certain data transfer protocols. For example, the interface logic and the peripheral device will use the same clock signal to synchronize data transfer in the synchronous protocol. The SPI and I²C module use this approach. For data transfers that use asynchronous protocols, both the transmitter and receiver will agree on a common data transfer rate and the receiver will use a sampling clock signal with a frequency that is a multiple (16, 32, or 64) of the data rate to detect the incoming data. This is used in the SCI module.

7.5 The HCS12 Parallel Ports

As mentioned in Section 4.10, the user configures an I/O port for input or output by programming the associated data direction register. To output, the user writes data to the port data register. To input, the user reads data from the port data register. Most I/O ports have additional registers to control their operations, and most I/O pins service multiple purposes. This section discusses the pins of each parallel I/O port and their registers in more detail.

7.5.1 Port A and Port B

In expanded mode, both Port A and Port B are used as time-multiplexed address or data pins. When configured in single-chip mode, these two ports are used as general-purpose I/O ports. Each Port A or Port B pin can be configured as an input or output pin. When the HCS12 is configured in expanded mode, Port A carries the time-multiplexed upper address and data signals (A15/D15~A8/D8), whereas Port B carries the time-multiplexed lower address and data signals (A7/D7~A0/D0).

7.5.2 Port E

As shown in Figure 7.1, Port E pins are used for bus control and interrupt service request signals. When a pin is not used for one of these specific functions, it can be used as a general-purpose I/O. However, two of the Port E pins (PE[1:0]) can only be used for input, and the states of these pins can be read in the port data register even when they are used for $\overline{IRQ}$ and $\overline{XIRQ}$.

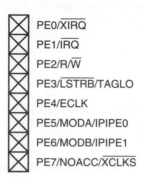

Figure 7.1 ■ Port E pins and their alternate functions

The PE7 pin serves three functions.

1. A general-purpose I/O pin
2. A signal that indicates that the MCU is not accessing external memory
3. A signal that selects between the external clock or crystal oscillator signal as the clock input to the HCS12

The PE6 and PE5 pins serve three functions.

1. General-purpose I/O pins (in single-chip mode)
2. Signals that set the operation mode of the HCS12 after reset
3. Instruction queue tracking signals (in expanded mode)

The PE4 pin has two functions.

1. A general-purpose I/O pin (in single-chip mode)
2. The E-clock output (in expanded mode)

The PE3 pin has three functions.

1. A general-purpose I/O pin (in single-chip mode)
2. The LCD front-plane segment driver output pin (in H subfamily)
3. The low-byte strobe signal to indicate the type of access (in expanded mode)

The PE2 pin has two functions.

1. A general-purpose I/O pin (in single-chip mode)
2. A signal to indicate whether the current bus cycle is a read cycle or a write cycle (in expanded mode)

The PE1 pin has two functions.

1. A general-purpose I/O pin
2. The $\overline{IRQ}$ input

The PE0 pin has two functions.

1. A general-purpose I/O pin
2. The $\overline{XIRQ}$ input

In addition to the DDRE and PORTE registers, Port E also has the following registers:

- Port E assignment register (PEAR)
- Mode register (MODE)
- Pull-up control register (PUCR)
- Reduced drive register (RDRIV)
- External bus interface control register (EBICTL)

PORT E ASSIGNMENT REGISTER (PEAR)

In expanded mode, this register assigns the function of each Port E pin. The contents of the register PEAR are shown in Figure 7.2. The PEAR register is not accessible for reads or writes in peripheral mode.

7	6	5	4	3	2	1	0	
NOACCE	0	PIPOE	NECLK	LSTRE	0	RDWE	0	Reset value = $000A

NOACCE: *No Access output enable*. Can be read/written any time.
    0 = PE7 is used as general-purpose I/O pin.
    1 = PE7 is output and indicates whether the cycle is a CPU free cycle.
PIPOE: *Pipe signal output enable*.
    In normal modes, write once. Special modes, write anytime except the first
    time. This bit has no effect in single-chip modes.
    0 = PE[6:5] are general-purpose I/O.
    1 = PE[6:5] are outputs and indicate the state of the instruction queue.
NECLK: *No external E-clock*. Can be read anytime.
    In expanded modes, writes to this bit have no effect. E-clock is required for
    demultiplexing the external address. NECLK can be written once in
    normal single-chip mode and can be written anytime in special single-chip mode.
    0 = PE4 is the external E-clock.
    1 = PE4 is a general-purpose I/O pin.
LSTRE: *Low strobe ($\overline{LSTRB}$) enable*. Can be read anytime.
    In normal modes, write once. Special modes, write anytime. This bit has no
    effect in single-chip modes or normal expanded narrow mode.
    0 = PE3 is a general-purpose I/O pin.
    1 = PE3 is configured as the $\overline{LSTRB}$ bus-control output, provided the
        HCS12 is not in single-chip or normal expanded narrow modes.
RDWE: *Read/write enable*. Can be read anytime.
    In normal modes, write once. Special modes, write anytime except the
    first time. This bit has no effect in single-chip modes.
    0 = PE2 is a general-purpose I/O pin.
    1 = PE2 is configured as the R/$\overline{W}$ pin. In single-chip modes, RDWE has
        no effect and PE2 is a general-purpose I/O pin.
        R/W is used for external writes. After reset in normal expanded mode, it
        is disabled. If needed, it should be enabled before any external writes.

Figure 7.2 ■ PEAR register

MODE REGISTER

This register establishes the operation mode and other miscellaneous functions (i.e., internal visibility and emulation of Ports E and K). The contents of this register are shown in Figure 7.3.

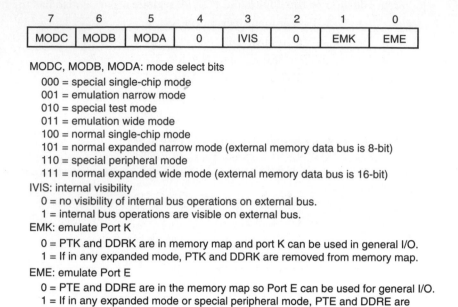

7	6	5	4	3	2	1	0
MODC	MODB	MODA	0	IVIS	0	EMK	EME

MODC, MODB, MODA: mode select bits
    000 = special single-chip mode
    001 = emulation narrow mode
    010 = special test mode
    011 = emulation wide mode
    100 = normal single-chip mode
    101 = normal expanded narrow mode (external memory data bus is 8-bit)
    110 = special peripheral mode
    111 = normal expanded wide mode (external memory data bus is 16-bit)
IVIS: internal visibility
    0 = no visibility of internal bus operations on external bus.
    1 = internal bus operations are visible on external bus.
EMK: emulate Port K
    0 = PTK and DDRK are in memory map and port K can be used in general I/O.
    1 = If in any expanded mode, PTK and DDRK are removed from memory map.
EME: emulate Port E
    0 = PTE and DDRE are in the memory map so Port E can be used for general I/O.
    1 = If in any expanded mode or special peripheral mode, PTE and DDRE are
        removed from memory map, which allows the user to emulate the function
        of these registers externally.

Figure 7.3 ■ The MODE register

PULL-UP CONTROL REGISTER (PUCR)

This register is used to select the pull-up resistors for the pins associated with the core part. The MC9S12DG256 has Ports A, B, E, and K in its core part. This register can be written any time; its contents are shown in Figure 7.4.

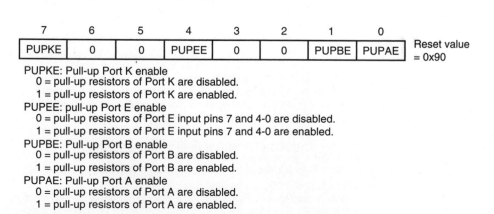

7	6	5	4	3	2	1	0	
PUPKE	0	0	PUPEE	0	0	PUPBE	PUPAE	Reset value = 0x90

PUPKE: Pull-up Port K enable
    0 = pull-up resistors of Port K are disabled.
    1 = pull-up resistors of Port K are enabled.
PUPEE: pull-up Port E enable
    0 = pull-up resistors of Port E input pins 7 and 4-0 are disabled.
    1 = pull-up resistors of Port E input pins 7 and 4-0 are enabled.
PUPBE: Pull-up Port B enable
    0 = pull-up resistors of Port B are disabled.
    1 = pull-up resistors of Port B are enabled.
PUPAE: Pull-up Port A enable
    0 = pull-up resistors of Port A are disabled.
    1 = pull-up resistors of Port A are enabled.

Figure 7.4 ■ Pull-Up Control register (PUCR)

REDUCED DRIVE REGISTER (RDRIV)

This register is used to select reduced drive for the pins associated with the core ports, which gives reduced power consumption and reduced RFI with a slight increase in transition time, a feature used on ports that have a light loading. This register is not in the memory map in expanded mode. The contents of the register are shown in Figure 7.5.

7	6	5	4	3	2	1	0	
RDPK	0	0	RDPE	0	0	RDPB	RDPA	Reset value = 0x00

RDPK: reduced drive of Port K
  0 = all Port K pins have full drive enabled.
  1 = all Port K pins have reduced drive enabled.
RDPE: reduced drive of Port E
  0 = all Port E pins have full drive enabled.
  1 = all Port E pins have reduced drive enabled.
RDPB: reduced drive of Port B
  0 = all Port B pins have full drive enabled.
  1 = all Port B pins have reduced drive enabled.
RDPA: reduced drive of Port A
  0 = all Port A pins have full drive enabled.
  1 = all Port A pins have reduced drive enabled.

Figure 7.5 ■ Reduced Drive register (RDRIV)

EXTERNAL BUS INTERFACE CONTROL REGISTER (EBICTL)

Only the bit 0 (ESTR) of this register is implemented. It controls the stretching of the external E-clock. When the ESTR bit is set to 0, the E-clock is free running and does not stretch (lengthen). When the HCS12 is interfacing with a slower memory device, then the E-clock can be lengthened (have its high interval stretched) by setting this bit.

7.5.3 Port K

Port K has a Port K data register (PORTK or PTK) and a data direction register (DDRK). In the expanded mode, Port K carries the expanded address XADDR14~XADDR19, emulated chip-select ($\overline{ECS}$), and external chip-select ($\overline{XCS}$) signals. The PK6 pin is available only in the H subfamily. The functions of Port K pins are shown in Figure 7.6. At the rising edge of the

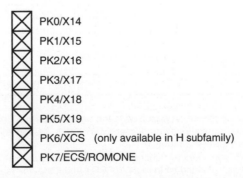

PK0/X14
PK1/X15
PK2/X16
PK3/X17
PK4/X18
PK5/X19
PK6/$\overline{XCS}$ (only available in H subfamily)
PK7/$\overline{ECS}$/ROMONE

Figure 7.6 ■ Port K pins and their alternate functions

RESET signal, the state of the PK7 pin is latched to the ROMON bit of the MISC register. If this bit is 1, the on-chip flash memory is enabled in the expanded mode. This bit is forced to 1 in single-chip modes.

7.5.4 Port T

Port T has a port I/O register (PTT), port data direction register (DDRT), port input register (PTIT), reduced drive register (RDRT), pull device enable register (PERT), and port polarity select register (PPST).

The PTIT register allows us to read back the status of Port T pins. This register can also be used to detect overload or short circuit conditions on output pins. Reading the port data register achieves the same purpose.

The RDRT register configures the drive strength of each port T output pin as either full or reduced current. If a port T pin is used as input, this bit is ignored. The reduced drive feature for each pin can be enabled individually. The contents of RDRT are shown in Figure 7.7.

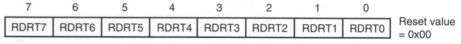

RDRT[7:0]: reduced drive Port T
   0 = full drive strength at output.
   1 = associated pin drives at about 1/3 of the full drive strength.

Figure 7.7 ■ Port T Reduced Drive register (RDRT)

The PERT register configures whether a pull-up or pull-down device is enabled if the port is used as input. Each pin's pull-up or pull-down device can be enabled individually. No pull-up or pull-down device is enabled out of reset. The contents of PERT are shown in Figure 7.8.

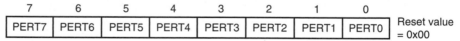

PERT[7:0]: pull device enable Port T
   0 = pull-up or pull-down is disabled.
   1 = either pull-up or pull-down is enabled.

Figure 7.8 ■ Port T Pull Device Enable register (PERT)

The PPST register selects whether a pull-down or a pull-up device is connected to the pin. This register has an effect on input pins only. The contents of PPST are shown in Figure 7.9.

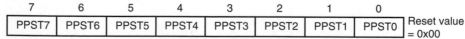

PPST[7:0]: pull device enable Port T
   0 = a pull-up device is connected to the associated Port T pin, if enabled
      by the associated bit in register PERT and if the port is used as input
      or as wired-OR output.
   1 = a pull-down device is connected to the associated Port T pin, if enabled
      by the associated bit in register PERT and if the port is used as input.

Figure 7.9 ■ Port T Polarity Select register (PPST)

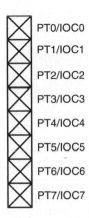

Figure 7.10 ■ Port T pins and their alternate functions

The pin functions of Port T are shown in Figure 7.10. In addition to being used as general I/O pins, Port T pins can also be used as input capture or output compare action pins.

7.5.5 Port S

Port S has a Port S Wired-OR Mode register (WOMS) in addition to all the registers associated with Port T (PTS, DDRS, PTIS, RDRS, PERS, and PPSS). Each bit of this register configures the associated output pin as wired-OR. The contents of this register are shown in Figure 7.11. The contents of PTIS, RDRS, PERS, and PPSS are identical to those of PTIT, RDRT, PERT, and PPST, respectively. As shown in Figure 7.12, Port S pins can be used as general I/O pins or serial interface signals.

7	6	5	4	3	2	1	0	
WOMS7	WOMS6	WOMS5	WOMS4	WOMS3	WOMS2	WOMS1	WOMS0	Reset value = 0x00

WOMS[7:0]: wired-OR mode Port S
    0 = output buffers operate as push-pull outputs.
    1 = output buffers operate as open-drain outputs.

Figure 7.11 ■ Port S wired-OR mode register (WOMS)

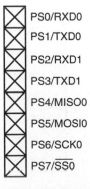

Figure 7.12 ■ Port S pins and their alternate functions

7.5.6 Port M

In addition to having all the equivalent registers (PTM, DDRM, PTIM, RDRM, PERM, PPSM, and WOMM) contained in Port S, Port M also has a Module Routing register (MODRR). This register configures the rerouting of CAN0, CAN4, SPI0, SPI1, and SPI2 on defined port pins. The contents of the MODRR register are shown in Figure 7.13. The functions of Port M pins are illustrated in Figure 7.14.

7	6	5	4	3	2	1	0	
0	MODRR6	MODRR5	MODRR4	MODRR3	MODRR2	MODRR1	MODRR0	Reset value = 0x00

CAN0 routing

MODRR1	MODRR0	RXCAN0	TXCAN0
0	0	PM0	PM1
0	1	PM2[1]	PM3[1]
1	0	PM4[2]	PM5[2]
1	1	reserved	

SPI0 routing

MODRR4	MISO0	MOSI0	SCK0	$\overline{SS0}$
0	PS4	PS5	PS6	PS7
1	PM2[5]	PM4[6]	PM5[6]	PM3[5]

CAN4 routing

MODRR3	MODRR2	RXCAN4	TXCAN4
0	0	PJ6	PJ7
0	1	PM4[3]	PM5[3]
1	0	PM6[4]	PM7[4]
1	1	reserved	

SPI1 routing

MODRR5	MISO1	MOSI1	SCK1	$\overline{SS1}$
0	PP0	PP1	PP2	PP3
1	PH0	PH1	PH2	PH3

SPI2 routing

MODRR6	MISO2	MOSI2	SCK2	$\overline{SS2}$
0	PP4	PP5	PP7	PP6
1	PH4	PH5	PH6	PH7

Notes: 1. Routing to this pin takes effect only if CAN1 is disabled.
2. Routing to this pin takes effect only if CAN2 is disabled.
3. Routing to this pin takes effect only if CAN2 is disabled and CAN0 is not routed here.
4. Routing to this pin takes effect only if CAN3 is disabled.
5. Routing to this pin takes effect only if CAN1 is disabled, and CAN0 is disabled if routed here.
6. Routing to this pin takes effect only if CAN2 is disabled; CAN0 is disabled if routed here and CAN4 is disabled if routed here.

Figure 7.13 ■ Module Routing register (MODRR)

PM0/RXCAN0/RXB
PM1/TXCAN0/TXB
PM2/RXCAN1/RXCAN0/MISO0
PM3/TXCAN1/TXCAN0/$\overline{SS0}$
PM4/RXCAN2/RXCAN0/RXCAN4/MOSI0
PM5/TXCAN2/TXCAN0/TXCAN4/SCK0
PM6/RXCAN3/RXCAN4
PM7/TXCAN3/TXCAN4

Figure 7.14 ■ Port M pins and their alternate functions

Example 7.1

Give an instruction to configure the MODRR register to achieve the following port routing:

1. CAN0: use pins PM1 and PM0
2. CAN1: use pins PM3 and PM2
3. CAN2: use pins PM5 and PM4
4. CAN3: use pins PM7 and PM6
5. I²C: use PJ7 and PJ6
6. SPI0: use pins PS7~PS4
7. SPI1: use pins PH3~PH0
8. SPI2: use pins PH7~PH4

Solution: For this routing requirement, all we need to do is to prevent CAN4 from using any port pins and keep the default routing after reset. The following instruction will satisfy the requirement:

```
movb    #$60,MODRR    ; CAN4 must be disabled
```

Example 7.2

Give an instruction to configure the MODRR register to achieve the following port routing:

1. CAN0: use pins PM1 and PM0
2. CAN1: use pins PM3 and PM2
3. CAN2: disabled
4. CAN3: disabled
5. I²C: use PJ7 and PJ6
6. SPI0: use pins PS7~PS4
7. SPI1: use pins PP3~PP0
8. SPI2: use pins PH7~PH4

Solution: This routing requirement can be satisfied by the following instruction:

```
movb    #$40,MODRR    ; CAN2~CAN4 must be disabled
```

7.5.7 Ports H, J, and P

These three I/O ports have the same set of registers associated with them. All of the pins associated with these three ports have edge-triggered interrupt capability in the wired-OR fashion. The SPI function pins can be routed to Ports H and P. The rerouting of SPI functions is done by programming the MODRR register. Each of these three ports has eight associated registers.

1. Port I/O register (PTH, PTJ, PTP)
2. Port Input register (PTIH, PTIJ, PTIP)

3. Port Data Direction register (DDRH, DDRJ, DDRP)
4. Port Reduced Drive register (RDRH, RDRJ, RDRP)
5. Port Pull Device Enable register (PERH, PERJ, PERP)
6. Port Polarity Select register (PPSH, PPSJ, PPSP)
7. Port Interrupt Enable register (PIEH, PIEJ, PIEP)
8. Port Interrupt Flag register (PIFH, PIFJ, PIFP)

All except the last two registers have their equivalents in Port T. The contents of the Port H Interrupt Enable register and Port H Interrupt Flag register are shown in Figures 7.15 and 7.16, respectively.

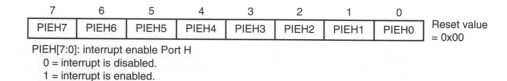

7	6	5	4	3	2	1	0	
PIEH7	PIEH6	PIEH5	PIEH4	PIEH3	PIEH2	PIEH1	PIEH0	Reset value = 0x00

PIEH[7:0]: interrupt enable Port H
  0 = interrupt is disabled.
  1 = interrupt is enabled.

Figure 7.15 ■ Port H Interrupt Enable register (PIEH)

7	6	5	4	3	2	1	0	
PIFH7	PIFH6	PIFH5	PIFH4	PIFH3	PIFH2	PIFH1	PIFH0	Reset value = 0x00

PIFH[7:0]: interrupt flag Port H
  0 = no active edge pending.
  1 = active edge has occurred (writing a 1 clears the associated flag).

Figure 7.16 ■ Port H Interrupt Flag register (PIFH)

Port H is associated with two SPI modules. Port J is associated with the fifth CAN and the I²C module. Port P is associated with the PWM and two SPI modules. In all modes, Port P pins PP[7:0] can be used either for general-purpose I/O or with the PWM and SPI subsystems. The pins are shared between the PWM channels and the SPI1 and SPI2 modules. If the PWM function is enabled, these pins become PWM output channels, with the exception of pin 7, which can be PWM input or output. If SPI1 or SPI2 are enabled and PWM is disabled, the respective pin configuration is determined by several status bits in the SPI module. Both Port H and Port P have eight pins, whereas Port J has only four pins.

The interrupt enable as well as the sensitivity to rising or falling edges can be individually configured on a per-pin basis. If a pin's pull-down device is enabled, then the interrupt is rising edge triggered. Otherwise, it is falling edge triggered. All 8 bits or pins of the port share the same interrupt vector. Interrupts can be used with the pins configured as inputs or outputs.

An interrupt is generated when a bit in the Port Interrupt Flag register and its corresponding port interrupt enable bit are both set. This feature can be used to wake up the CPU when it is in the *stop* or *wait* mode. Each Port P pin can also be used as an edge-sensitive interrupt source. A digital filter on each pin prevents pulses shorter than a specified time from generating an interrupt. The minimum time varies over process conditions, temperature, and voltage.

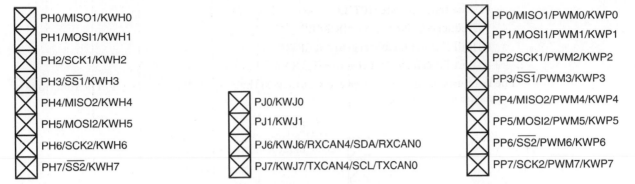

PH0/MISO1/KWH0
PH1/MOSI1/KWH1
PH2/SCK1/KWH2
PH3/$\overline{SS1}$/KWH3
PH4/MISO2/KWH4
PH5/MOSI2/KWH5
PH6/SCK2/KWH6
PH7/$\overline{SS2}$/KWH7

PJ0/KWJ0
PJ1/KWJ1
PJ6/KWJ6/RXCAN4/SDA/RXCAN0
PJ7/KWJ7/TXCAN4/SCL/TXCAN0

PP0/MISO1/PWM0/KWP0
PP1/MOSI1/PWM1/KWP1
PP2/SCK1/PWM2/KWP2
PP3/$\overline{SS1}$/PWM3/KWP3
PP4/MISO2/PWM4/KWP4
PP5/MOSI2/PWM5/KWP5
PP6/$\overline{SS2}$/PWM6/KWP6
PP7/SCK2/PWM7/KWP7

Figure 7.17 ■ Port H pins and their alternate functions

Figure 7.18 ■ Port J pins and their alternate functions

Figure 7.19 ■ Port P pins and their alternate functions

The functions of Port H, J, and P pins are shown in Figures 7.17, 7.18, and 7.19, respectively.

7.5.8 Ports AD0 and AD1

Many HCS12 devices have implemented two 8-channel A/D converters (AD1 and AD0). For those devices with a single 8-channel A/D converter, the module is referred to as AD instead of AD0. These two ports are analog input interfaces to the analog-to-digital subsystem. When analog-to-digital functions are not enabled, these two ports are available for general-purpose I/O. Since these two ports cannot be used as output, there are no data direction registers associated with them.

The ports have no resistive input loads and no reduced drive controls. PTAD0 and PTAD1 are data registers of Port AD0 and AD1, respectively. Each A/D module has an ATD Digital Input Enable register. In order to use an A/D pin as a digital input, its associated bit in this register needs to be set to 1. The contents of this register are shown in Figure 7.20.

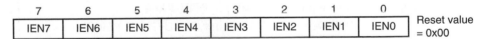

7	6	5	4	3	2	1	0	
IEN7	IEN6	IEN5	IEN4	IEN3	IEN2	IEN1	IEN0	Reset value = 0x00

IENx: ATD digital input enable on channel x
  0 = disable digital input buffer to PTADx pin.
  1 = enable digital input buffer to PTADx pin.

Figure 7.20 ■ ATD Input Enable register (ATD0DIEN and ATD1DIEN)

7.5.9 Port L

This port is available in the HCS12's H subfamily. In addition to being a general I/O port, Port L is connected to the LCD driver. The HCS12 H subfamily is designed to support LCD interfacing and motor control. Port L has the following registers:

- Port I/O register (PTL)
- Port Data Direction register (DDRL)

- Port Input register (PTIL)
- Reduced Drive register (RDRL)
- Pull Device Enable register (PERL)
- Port Polarity Select register (PPSL)

The contents of these registers are identical to their equivalents in Port T. The functions of Port L pins are shown in Figure 7.21.

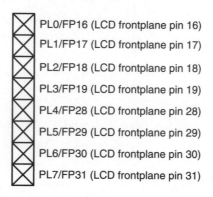

PL0/FP16 (LCD frontplane pin 16)

PL1/FP17 (LCD frontplane pin 17)

PL2/FP18 (LCD frontplane pin 18)

PL3/FP19 (LCD frontplane pin 19)

PL4/FP28 (LCD frontplane pin 28)

PL5/FP29 (LCD frontplane pin 29)

PL6/FP30 (LCD frontplane pin 30)

PL7/FP31 (LCD frontplane pin 31)

Figure 7.21 ■ Port L pins and their alternate functions

7.5.10 Ports U, V, and W

These three ports are available in HCS12's H subfamily. In addition to being general I/O ports, these three ports are provided for motor control. The following registers support the operations of these three ports:

- Port I/O register (PTU, PTV, PTW)
- Port Data Direction register (DDRU, DDRV, DDRW)
- Port Input register (PTIU, PTIV, PTIW)
- Pull Device Enable register (PERU, PERV, PERW)
- Port Polarity Select register (PPSU, PPSV, PPSW)
- Port Slew Rate register (SRRU, SRRV, SRRW)

The functions of the first five of these registers are identical to their equivalents in Port L. The contents of the Port Slew Rate register are illustrated in Figure 7.22.

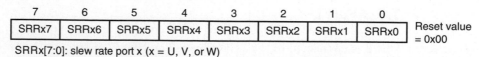

7	6	5	4	3	2	1	0	
SRRx7	SRRx6	SRRx5	SRRx4	SRRx3	SRRx2	SRRx1	SRRx0	Reset value = 0x00

SRRx[7:0]: slew rate port x (x = U, V, or W)
   0 = disable slew rate control.
   1 = enable slew rate control.

Figure 7.22 ■ Port Slew Rate register (SRRU, SRRV, and SRRW)

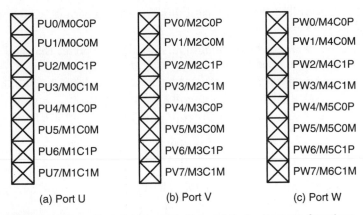

(a) Port U   (b) Port V   (c) Port W

Figure 7.23 ■ Port U, V, and W pins and their alternate functions

The functions of these port pins are shown in Figure 7.23a–c, respectively.

7.6 Electrical Characteristic Considerations for I/O Interfacing

Most embedded systems require the use of logic chips, peripheral devices in addition to the microcontroller, to perform their function. Because these chips may use different types of integrated circuit (IC) technologies, there is a concern that the resultant embedded system may not function properly.

The major concern in interfacing IC chips that are made with different technologies is whether they are electrically compatible. There are two issues involved in electrical compatibility.

- *Voltage-level compatibility*. Is the high output level of an IC chip high enough to be considered as a high for the input of another IC chip? Is the low output level of an IC chip low enough to be considered as a low for the input of another IC chip?
- *Current drive capability*. Does the output of an IC chip have enough current to drive its load? Can the output circuit of an IC chip sink the currents of its load?

Signal timing is also an important factor for making sure that the digital circuit functions correctly. The main concern about timing is whether the signal from one chip becomes valid early enough to be used by other chips. This is a *timing compatibility* issue. When the operating frequency becomes very high, the *transmission line effect* and *ground bounce effect* will need to be considered too. However, we are not dealing with high-frequency systems in this text. These two issues will not be discussed.

7.6.1 Voltage-Level Compatibility

There are many IC technologies in use today. Some of them are bipolar; others are unipolar. A bipolar IC technology has both the electron and hole currents in any moment. However, in a unipolar IC technology, the current in any part of the circuit is either the electron current or the hole current. The unipolar CMOS technology is the dominant IC technology in use today.

The voltage-level compatibility issue arises because IC technologies differ in the following four voltages:

- *Input high voltage* (V_{IH}). This is the voltage that will be treated as a logic 1 when applied to the input of a digital circuit.
- *Input low voltage* (V_{IL}). This is the voltage that will be treated as a logic 0 when applied to the input of a digital circuit.
- *Output high voltage* (V_{OH}). This is the voltage level when a digital circuit outputs a logic 1.
- *Output low voltage* (V_{OL}). This is the voltage level when a digital circuit outputs a logic 0.

In order for the digital circuit X to be able to drive circuit Y correctly, the following conditions must be satisfied:

- $V_{OHX} \geq V_{IHY}$ (the output high voltage of circuit X must be higher than the input high voltage of circuit Y). The difference between V_{OH} and V_{IH} of the same technology is referred to as the *noise margin high* (NMH).
- $V_{OLX} \leq V_{ILY}$ (the output low voltage of circuit X must be lower than the input low voltage of circuit Y). The difference between V_{IL} and V_{OL} of the same technology is referred to as the *noise margin low* (NML).

The input and output voltage levels of a few popular logic families are listed in Table 7.1, from which one can draw the following conclusions:

- There is no problem using CMOS logic chips to drive bipolar logic chips at the same power supply level.
- The HCS12 has no problem driving the CMOS logic chips and being driven by the CMOS logic chips at the same power supply level.

Logic Family	V_{DD}	V_{IH}	V_{OH}	V_{IL}	V_{OL}
HCS12[3]	5 V	3.25 V	4.2 V	1.75 V	0.8 V
S[4]	5 V	2 V	3.0~3.4 V[1]	0.8 V	0.4~0.5 V[2]
LS[4]	5 V	2 V	3.0~3.4 V[1]	0.8 V	0.4~0.5 V[2]
AS[4]	5 V	2 V	3.0~3.4 V[1]	0.8 V	0.35 V
F[4]	5 V	2 V	3.4 V	0.8 V	0.3 V
HC[3]	5 V	3.5 V	4.9 V	1.5 V	0.1 V
HCT[3]	5 V	3.5 V	4.9 V	1.5 V	0.1 V
ACT[3]	5 V	2 V	4.9 V	0.8 V	0.1 V
ABT[5]	5 V	2 V	3 V	0.8 V	0.55 V
BCT[5]	5 V	2 V	3.3 V	0.8 V	0.42 V
FCT[5]	5 V	2 V	2.4 V	0.8 V	0.55 V

[1] V_{OH} value will get lower when output current is larger.
[2] V_{OL} value will get higher when output current is larger. The V_{OL} values of different logic gates are slightly different.
[3] HCS12, HC, HCT, and ACT are based on the CMOS technology.
[4] S, LS, AS, and F logic families are based on the bipolar technology.
[5] ABT, BCT, and FCT use the BiCMOS technology.

Table 7.1 ■ Input and output voltage levels of common logic families

- The BiCMOS logic is not suitable for driving the HCS12 and other CMOS logic chips at the same power supply level.
- The bipolar logic ICs are not suitable for driving the CMOS logic ICs or the HCS12 microcontroller at the same power supply level.

7.6.2 Current Drive Capability

A microcontroller needs to drive other peripheral I/O devices in an embedded system. The second electrical compatibility issue is whether the microcontroller can supply (when the output voltage is high, also called source) or sink (when the output voltage is low) the current needed by the I/O devices that it interfaces with. Depending on the voltage level of an output pin, the current may flow out from (supply the current) or into (sink the current) the pin. The designer must make sure that the following two requirements are satisfied:

- Each I/O pin can supply (flowing out from the pin) and sink (flowing into the pin) the current needed by the I/O devices that it interfaces with.
- The total current required to drive I/O devices does not exceed the maximum current rating of the microcontroller.

Each logic chip has the following four currents that are involved in the current drive calculation:

- *Input high current* (I_{IH}). This is the input current (flowing into the input pin) when the input voltage is high.
- *Input low current* (I_{IL}). This is the input current (flowing out of the input pin) when the input voltage is low.
- *Output high current* (I_{OH}). This is the output current (flowing out of the output pin) when the output voltage is high.
- *Output low current* (I_{OL}). This is the output current (flowing into the output pin) when the output voltage is low.

The current capabilities of several common logic families and the HCS12 are listed in Table 7.2. In the CMOS technology, the gate terminal (one of the three terminals in an N or P transistor) draws a significant current only when they are charged up toward V_{CC} or pulled

Logic Family	V_{CC}	I_{IH}	I_{IL}	I_{OH}	I_{OL}
HCS12[2,3]	5 V	2.5 µA	2.5 µA	25 mA	25 mA
S	5 V	50 µA	1.0 mA	1 mA	20 mA
LS	5 V	20 µA	0.2 mA	15 mA	24 mA
AS	5 V	20 µA	0.5 mA	15 mA	64 mA
F	5 V	20 µA	0.5 mA	1 mA	20 mA
HC[3]	5 V	1 µA	1 µA	25 mA	25 mA
HCT[3]	5 V	1 µA	1 µA	25 mA	25 mA
ACT[3]	5 V	1 µA	1 µA	24 mA	24 mA
ABT[3]	5 V	1 µA	1 µA	32 mA	64 mA
BCT	5 V	20 µA	1 mA	15 mA	64 mA
FCT[3]	5 V	1 µA	1 µA	15 mA	64 mA

[1]Values are based on the 74xx244 of Texas Instrument (xx is the technology name).
[2]The total HCS12 supply current is 65 mA.
[3]The values for I_{IH} and I_{IL} are input leakage currents.

Table 7.2 ■ Current capabilities of common logic families[1]

down toward GND level. After that, the gate terminal draws only leakage currents. Bipolar technology is different from the CMOS technology in that a DC current always flows into or out of the base terminal of the transistor of a bipolar logic chip.

To determine whether a pin can supply and sink currents to all the peripheral pins that it drives directly, the designer needs to check the following two requirements:

1. The I_{OH} of an I/O pin of the microcontroller is equal to or greater than the sum of currents flowing into all peripheral pins that are connected directly to the microcontroller I/O pins.

2. The I_{OL} of an I/O pin of the microcontroller is equal to or greater than the sum of currents flowing out of all peripheral pins that are connected directly to the microcontroller I/O pins.

In addition, the designer must also make sure that the total current needed to drive the peripheral signal pins does not exceed the total current that the microcontroller can supply.

One question that arises here is what should be done if an I/O pin cannot supply (or sink) the current needed to drive the peripheral pins? A simple solution is to add buffer chips (for example, 74ABT244) that can supply enough current between the microcontroller and the peripheral chips. This technique is widely used in microcontroller applications and is illustrated in Example 4.13.

Example 7.3

You are given the three seven-segment display circuits shown in Figure 7.24a to 7.24c. Perform an appropriate analysis to find out if any circuit has a current incompatibility problem.

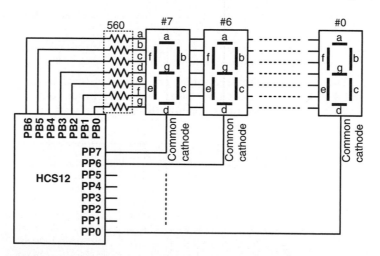

Figure 7.24a ■ HCS12 seven-segment display circuit 1

Solution: In Figure 7.24a, the V_{OH} of the 74HC244 is 5 V with 5-V power supply. The V_{OL} of HCS12 is 0.8 V. Assume the voltage drop of the LED is 1.8 V, then the current flowing into each segment is $(4.2 - 1.8 - 0.8) \div 560 = 2.86$ mA. This value is within the current source capability of each HCS12 I/O pin. The total maximum current that the HCS12 must supply is $7 \times (4.2 - 1.8 - 0.8) \div 560 = 20$ mA. This value is within the current-source capability of the HCS12 (25 mA). An HCS12 I/O pin can also sink this amount of current. Therefore, there is no current drive incompatibility problem.

In Figure 7.24b, the V_{OH} of the 74HC244 is 5 V with 5-V power supply. The V_{OL} of 74HC244 is 0.1 V. Assume the voltage drop of the LED is 1.8 V, then the current flowing through each

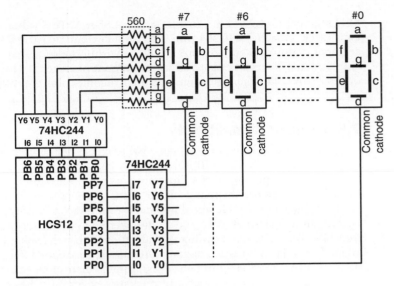

Figure 7.24b ■ HCS12 seven-segment display circuit 2

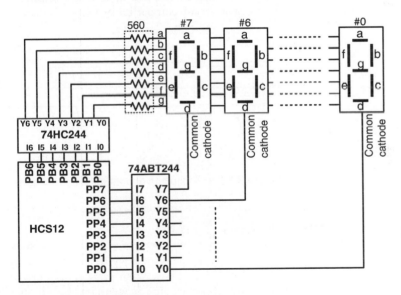

Figure 7.24c ■ HCS12 seven-segment display circuit 3

LED segment is $(4.9 - 1.8 - 0.1) \div 560 = 5.36$ mA. This value is well within the current drive capability of each 74HC244 pin. The total current that the 74HC244 needs to drive for the segment pattern is about 37.5 mA, which is within the total drive capability (70 mA) of the 74HC244 from TI. However, each digit select pin (connected to a single 74HC244 pin) needs to sink about 37.5 mA current, which is above the pin current sink capability of the SN74HC244 (35 mA) from TI. Therefore, the current drive compatibility problem exists. However, this problem can be solved by using a slightly larger current-limiting resistor (for example, 680 Ω).

In Figure 7.24c, the V_{OH} of the 74HC244 is 5 V with 5-V power supply. The V_{OL} of 74ABT244 is 0.55 V. Assume the voltage drop of the LED is 1.8 V, then the current flowing through each

LED segment is $(4.9 - 1.8 - 0.55) \div 560 = 4.55$ mA. This value is well within the current drive capability of each 74HC244 pin. The total current that the 74HC244 needs to drive for the segment pattern is about 32 mA which, is within the total drive capability of the 74HC244. A single 74ABT244 pin needs to sink this amount of current, which is also within the capability of the 74ABT244. There is no current drive incompatibility problem.

7.6.3 Timing Compatibility

If an I/O pin is driving a peripheral pin that does not contain latches or flip-flops, then timing is not an issue. A latch or flip-flop usually has a control signal or clock signal to control the latching of an input signal. As illustrated in Figure 7.25, the D input to the D flip-flop must be valid for t_{su} ns before the rising edge of the CLK signal and remain valid for at least t_{hd} ns after the rising edge of the CLK signal in order for its value to be correctly copied to the output signal Q. The timing parameters t_{su} and t_{hd} are referred to as the *setup* and *hold* time requirements of the D flip-flop. The main timing consideration is that the setup and hold time requirements for all latches and flip-flops in a digital system must be satisfied in order for the system to work correctly. A signal may pass through several intermediate chips before it is used by the final latch or flip-flop. The time delays of all intermediate devices must be added when considering the timing analysis. Timing requirement analysis can be very complicated and is best illustrated using examples. Two examples of timing compatibility analysis are given in Section 14.8.

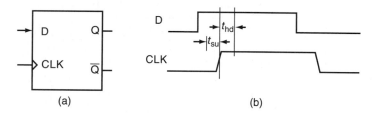

Figure 7.25 ■ D flip-flop and its latching timing requirement

7.7 Liquid Crystal Displays (LCDs)

Although seven-segment displays are easy to use, they are bulky and quite limited in the set of characters that they can display. When more than a few letters and digits are to be displayed, seven-segment displays become inadequate. Liquid crystal displays (LCDs) come in handy when the application requires the display of many characters.

A liquid crystal display has the following advantages:

- High contrast
- Low power consumption
- Small footprint
- Ability to display both characters and graphics

The basic construction of an LCD is shown in Figure 7.26. The most common type of LCD allows light to pass through when activated. A segment is activated when a low-frequency bipolar

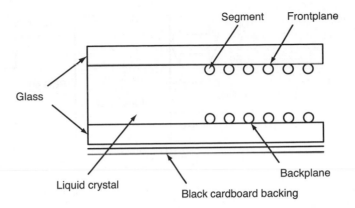

Figure 7.26 ■ A liquid crystal display (LCD)

signal in the range of 30 to 1000 Hz is applied to it. The polarity of the voltage must alternate, or the LCD will not be able to change very quickly.

When a voltage is applied across the segment, an electrostatic field is set up that aligns the crystals in the liquid. This alignment allows light to pass through the segment. If no voltage is applied across a segment, the crystals appear to be opaque because they are randomly aligned. Random alignment is assured by the AC excitation voltage applied to each segment. In a digital watch, the segments appear darker when they are activated because light passes through the segment to a black cardboard backing that absorbs all light. The area surrounding the activated segment appears to grow brighter in color because the randomly aligned crystals reflect much of the light. In a backlit computer display, the segment appears to grow brighter because of a light placed behind the display; the light is allowed to pass through the segment when it is activated.

In recent years, the price of LCD displays has dropped to such an acceptable level that all PC vendors bundle LCD displays instead of CRT displays with their PC systems. Notebook computers used LCD as displays right from the beginning. Because of the price reduction of LCDs, the prices of notebook computers also have dropped sharply, and more and more computer users have switched from desktop to notebook computers.

Although LCDs can display graphics and characters, only character-based LCDs are discussed in this text. LCDs are often sold in a module that consists of the LCD and its controller. The Hitachi HD44780 (with two slightly different versions: HD44780U and HD44780S) is one of the most popular LCD display controllers in use today. The following section examines the operation and programming of this controller.

7.8 The HD44780U LCD Controller

The block diagram of an LCD kit that incorporates the HD44780U controller is shown in Figure 7.27. The pin assignment shown in Table 7.3 is the industry standard for character-based LCD modules with a maximum of 80 characters. The pin assignment shown in Table 7.4 is the industry standard for character-based LCD modules with more than 80 characters.

The DB7~DB0 pins are used to exchange data with the microcontroller. The E pin is an enable signal to the kit. The R/W signal determines the direction of data transfer. The RS signal selects the register to be accessed. When the RS signal is high, the *data register* is selected.

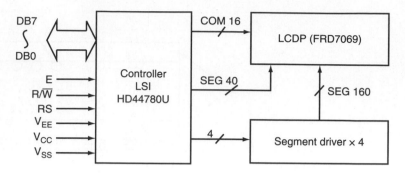

Figure 7.27 ■ Block diagram of an HD44780U-based LCD kit

Pin No.	Symbol	I/O	Function
1	V_{SS}	—	Power supply (GND)
2	V_{CC}	—	Power supply (+5 V)
3	V_{EE}	—	Contrast adjust
4	RS	I	0 = instruction input, 1 = data input
5	R/$\overline{W}$	I	0 = write to LCD, 1 = read from LCD
6	E	I	Enable signal
7	DB0	I/O	Data bus line 0
8	DB1	I/O	Data bus line 1
9	DB2	I/O	Data bus line 2
10	DB3	I/O	Data bus line 3
11	DB4	I/O	Data bus line 4
12	DB5	I/O	Data bus line 5
13	DB6	I/O	Data bus line 6
14	DB7	I/O	Data bus line 7

Table 7.3 ■ Pin assignment for displays with less than 80 characters

Otherwise, the *instruction register* is selected. The V_{EE} pin is used to control the brightness of the display and is often connected to a potentiometer. The V_{EE} input should not be set to the maximum value (= V_{CC}) for an extended period of time to avoid burning the LCD.

An LCD module can be used as a memory-mapped device and be enabled by an address decoder. The E signal is normally connected to the address decoder output qualified by other control signals to meet the timing requirement. The R/$\overline{W}$ pin can be connected to the same pin of the microcontroller. The RS pin can be connected to the least significant bit of the address pin (A0) from the microcontroller. This approach is applicable only for those microcontrollers that support external memory. The LCD programming in this approach is generally easier and more straightforward.

An LCD module can also be interfaced directly with an I/O port. In this configuration, the designer will need to use I/O pins to control the signals E, R/$\overline{W}$, and RS. Programming will be

Pin No.	Symbol	I/O	Function
1	DB7	I/O	Data bus line 7
2	DB6	I/O	Data bus line 6
3	DB5	I/O	Data bus line 5
4	DB4	I/O	Data bus line 4
5	DB3	I/O	Data bus line 3
6	DB2	I/O	Data bus line 2
7	DB1	I/O	Data bus line 1
8	DB0	I/O	Data bus line 0
9	E1	I	Enable signal row 0 and 1
10	R/$\overline{\text{W}}$	I	0 = write to LCD, 1 = read from LCD
11	RS	I	0 = instruction input, 1 = data input
12	V_{EE}	—	Contrast adjust
13	V_{ss}	—	Power supply (GND)
14	V_{cc}	—	Power supply (+5 V)
15	E2	I	Enable signal row 2 and 3
16	N.C	—	

Table 7.4 ■ Pin assignment for displays with more than 80 characters

slightly more cumbersome than the memory-mapped approach due to the need to manipulate these three signals. The HD44780U provides a set of instructions for the user to set up the LCD parameters. The operations performed by these instructions are summarized in Table 7.5. The meanings of certain bits in these instructions are explained in Table 7.6.

The HD44780U can be configured to control one-line, two-line, and four-line LCDs. The mappings of the character positions on the LCD screen and the DDRAM addresses are not sequential and are shown in Table 7.7.

7.8.1 Display Data RAM

Display data RAM (DDRAM) stores display data represented in 8-bit character codes. Its extended capacity is 80 × 8 bits, or 80 characters. The area in DDRAM that is not used for display can be used as general data RAM. The relationships between DDRAM addresses and positions on the LCD are shown in Table 7.7.

7.8.2 Character Generator ROM (CGROM)

The character generator ROM generates 5 × 8 or 5 × 10 dot character patterns from 8-bit character codes. It can generate 208 5 × 8 dot character patterns and 32 5 × 10 dot character patterns.

7.8.3 Character Generator RAM (CGRAM)

The user can rewrite character patterns into the CGRAM by programming. For 5 × 8 fonts, eight character patterns can be written, and for 5 × 10 fonts, four character patterns can be written.

Instruction	Code										Description	Execution Time
	RS	R/$\overline{W}$	B7	B6	B5	B4	B3	B2	B1	B0		
Clear display	0	0	0	0	0	0	0	0	0	1	Clears display and returns cursor to the home position (address 0).	1.64 ms
Cursor home	0	0	0	0	0	0	0	0	1	*	Returns cursor to home position without changing DDRAM contents. Also returns display being shifted to the original position.	1.64 ms
Entry mode set	0	0	0	0	0	0	0	1	I/D	S	Sets cursor move direction (I/D); specifies to shift the display (S). These operations are performed during data read/write.	40 μs
Display on/off control	0	0	0	0	0	0	1	D	C	B	Sets on/off of all display (D), cursor on/off (c), and blink of cursor position character (B).	40 μs
Cursor/ display shift	0	0	0	0	0	1	S/C	R/L	*	*	Sets cursor-move or display-shift (S/C), shift direction (R/L). DDRAM contents remain unchanged.	40 μs
Function set	0	0	0	0	1	DL	N	F	*	*	Sets interface data length (DL), number of display line (N), and character font (F).	40 μs
Set CGRAM address	0	0	0	1	CGRAM address						Sets the CGRAM address. CGRAM data are sent and received after this setting.	40 μs
Set DDRAM address	0	0	1	DDRAM address							Sets the DDRAM address. DDRAM data are sent and received after this setting.	40 μs
Read busy flag and address counter	0	1	BF	CGRAM/DDRAM address							Reads busy flag (BF) indicating internal operation being performed and reads CGRAM or DDRAM address counter contents (depending on previous operation).	0 μs
Write CGRAM or DDRAM	1	0	writes data								Writes data to CGRAM or DDRAM.	40 μs
Read from CGRAM or DDRAM	1	1	read data								Reads data from CGRAM or DDRAM.	40 μs

Table 7.5 ■ HD44780U instruction set

Bit Name	Settings	
I/D	0 = decrement cursor position	1 = increment cursor position
S	0 = no display shift	1 = display shift
D	0 = display off	1 = display on
C	0 = cursor off	1 = cursor on
B	0 = cursor blink off	1 = cursor blink on
S/C	0 = move cursor	1 = shift display
R/L	0 = shift left	1 = shift right
DL	0 = 4-bit interface	1 = 8-bit interface
N	0 = 1/8 or 1/11 duty (1 line)	1 = 1/16 duty (2 lines)
F	0 = 5 × 8 dots	1 = 5 × 10 dots
BF	0 = can accept instruction	1 = internal operation in progress

Table 7.6 ■ LCD instruction bit names

Display Size	Visible	
	Character Positions	DDRAM Addresses
1 × 8	00,...,07	0x00,...,0x07
1 × 16	00,...,15	0x00,...,0x0F
1 × 20	00,...,19	0x00,...,0x13
1 × 24	00,...,23	0x00,...,0x17
1 × 32	00,...,31	0x00,...,0x1F
1 × 40	00,...,39	0x00,...,0x27

Table 7.7a ■ DDRAM address usage for a one-line LCD

Display Size	Visible	
	Character Positions	DDRAM Addresses
2 × 16	00,...,15	0x00,...,0x0F + 0x40..0x4F
2 × 20	00,...,19	0x00,...,0x13 + 0x40..0x53
2 × 24	00,...,23	0x00,...,0x17 + 0x40..0x57
2 × 32	00,...,31	0x00,...,0x1F + 0x40..0x5F
2 × 40	00,...,39	0x00,...,0x27 + 0x40..0x67

Table 7.7b ■ DDRAM address usage for a two-line LCD

Display Size	Visible	
	Character Positions	DDRAM Addresses
4 × 16	00..15	0x00..0x0F + 0x40..0x4F + 0x14..0x23 + 0x54..0x63
4 × 20	00..19	0x00..0x13 + 0x40..0x53 + 0x14..0x27 + 0x54..0x67
4 × 40	00..39 on 1st controller and 00..39 on 2nd controller	0x00..0x27 + 0x40..0x67 on 1st controller and 0x00..0x27 + 0x40..0x67 on 2nd controller

Note: Two LCD controllers are needed to control LCD displays with 4 × 40 characters.

Table 7.7c ■ DDRAM address usage for a four-line LCD

7.8.4 Registers

The HD44780U has two 8-bit registers, an *instruction register* (IR) and a *data register* (DR). The IR stores *instruction codes*, such as **display clear** and **cursor move,** and *address information* for display data RAM (DDRAM) and character generator RAM (CGRAM). The microcontroller writes commands into this register to set up the LCD operation parameters. To write data into the DDRAM or CGRAM, the microcontroller writes data into the DR. Data written into the DR will be automatically written into DDRAM or CGRAM by an internal operation. The DR is also used for data storage when reading data from DDRAM or CGRAM. When address information is written into the IR, data is read and then stored in the DR from DDRAM or CGRAM by an internal operation. The microcontroller can then read the data from the DR. After a read operation, data in DDRAM or CGRAM at the next address is sent to the DR and the microcontroller does not need to send another address. The IR and DR are distinguished by the RS signal. The IR is selected when the RS input is low. The DR is selected when the RS input is high. Register selection is illustrated in Table 7.8.

RS	R/$\overline{\text{W}}$	Operation
0	0	IR write as an internal operation (display clear, etc.).
0	1	Read busy flag (DB7) and address counter (DB0 to DB6).
1	0	DR write as an internal operation (DR to DDRAM or CGRAM).
1	1	DR read as an internal operation (DDRAM or CGRAM to DR).

Table 7.8 ■ Register selection

Busy Flag (BF)

The HD44780U has a busy flag (BF) to indicate whether the current internal operation is complete. When BF is 1, the HD44780U is still busy with an internal operation. When RS = 0 and R/$\overline{\text{W}}$ is 1, the busy flag is output to the DB7 pin. The microprocessor can read this pin to find out if the HD44780U is still busy.

Address Counter (AC)

The HD44780U uses a 7-bit address counter to keep track of the address of the next DDRAM or CGRAM location to be accessed. When an instruction is written into the IR, the address information contained in the instruction is transferred to the AC register. The selection of DDRAM or CGRAM is determined by the instruction. After writing into (reading from) DDRAM or CGRAM, the content of the AC register is automatically incremented (decremented) by 1. The contents of the AC register is output to the DB6..DB0 pins when the RS signal is low and the R/$\overline{\text{W}}$ signal is high.

7.8.5 Instruction Description

The functions of LCD instructions are discussed in this section.

CLEAR DISPLAY

This instruction writes the space code 0x20 into all DDRAM locations. It then sets 0 into the address counter and returns the display to its original status if it was shifted. In other words, the display disappears and the cursor or blinking goes to the upper left corner of the display. It also sets the I/D bit to 1 (increment mode) in entry mode.

RETURN HOME

This instruction sets DDRAM address 0 into the address counter and returns to its original status if it was shifted. The DDRAM contents are not changed. The cursor or blinking goes to the upper left corner of the display.

ENTRY MODE SET

The **I/D** bit of this instruction controls the incrementing (I/D = 1) or decrementing (I/D = 0) of the DDRAM address. The cursor or blinking will be moved to the right or left depending on whether this bit is set to 1 or 0. The same applies to writing and reading of CGRAM.

The **S** bit of this instruction controls the shifting of the LCD display. The display shifts if S = 1. Otherwise, the display does not shift. If S is 1, it will seem as if the cursor does not move but the display does. The display does not shift when reading from DDRAM. Also, writing into or reading from CGRAM does not shift the display.

DISPLAY ON/OFF CONTROL

This instruction has three bit parameters: **D**, **C**, and **B**. When the D bit is set to 1, the display is turned on; otherwise it is turned off. The cursor is turned on when the C bit is set to 1. The character indicated by the cursor will blink when the B bit is set to 1.

CURSOR OR DISPLAY SHIFT

This instruction shifts the cursor position to the right or left without writing or reading display data. The shifting is controlled by 2 bits, as shown in Table 7.9. This function is used to correct or search the display. When the cursor gets to the end of a line, it will be moved to the beginning of the next line.

S/C	R/L	Operation
0	0	Shifts the cursor position to the left. (AC is decremented by 1)
0	1	Shifts the cursor position to the right. (AC is incremented by 1)
1	0	Shifts the entire display to the left. The cursor follows the display shift.
1	1	Shifts the entire display to the right. The cursor follows the display shift.

Table 7.9 ■ LCD shift function

When the displayed data is shifted repeatedly, each line moves only horizontally. The second line of the display does not shift into the first row. The contents of the address counter will not change if the only action performed is a display shift.

FUNCTION SET

This instruction allows the user to set the interface data length, select the number of display lines, and select the character fonts. There are three bit variables in this instruction:

> **DL:** Data is sent or received in 8-bit length (DB7 to DB0) when DL is set to 1 and in 4-bit length (DB7 to DB4) when DL is 0. When the pin count is at a premium for the application, the 4-bit data length should be chosen even though it is cumbersome to perform the programming.

> **N:** This bit sets the number of display lines. When set to 0, one-line display is selected. When set to 1, two-line display is selected.

> **F:** When set to 0, the 5×8 font is selected. When set to 1, the 5×10 font is selected.

SET CGRAM ADDRESS

This instruction contains the CGRAM address to be set into the address counter.

SET DDRAM ADDRESS

This instruction allows the user to set the address of the DDRAM in the address counter. This instruction is used whenever the user wants to set the cursor to a certain position on the LCD screen.

READ BUSY FLAG AND ADDRESS

This instruction reads the busy flag (BF) and the address counter. The BF flag indicates whether the LCD controller is still executing the previously received instruction.

7.8.6 Interfacing the HD44780U to the HCS12 Microcontroller

The data transfer between the HD44780U and the MCU can be done in 4 or 8 bits at a time. When in 4-bit mode, data are carried on the upper four data pins (DB7~DB4). The upper 4 bits are sent over DB7~DB4 first and followed immediately by the lower 4 bits.

For those HCS12 members that do not support external memory, the designer must use I/O ports to interface with the LCD module. For those HCS12 members that support external memory, the designer has the choice of using I/O ports to interface with the LCD module or treating the LCD as a memory device. This chapter will treat the LCD only as an I/O device.

The LCD circuit connection for the Dragon12-Plus (4-bit data bus) is illustrated in Figure 7.28. The R/W̄ signal to the LCD kit in the Dragon12 demo board is grounded; this prevents the user from polling the BF flag to determine whether the LCD internal operation has been completed.

Certain timing parameters must be satisfied in order to access the LCD successfully. The read and write timing diagrams are shown in Figures 7.29 and 7.30, respectively. The values of timing parameters depend on the frequency of the operation. HD44780U-based LCDs can operate at either 1 MHz (cycle time of E signal) or 2 MHz. The values of timing parameters at these two frequencies are shown in Tables 7.10 and 7.11, respectively.

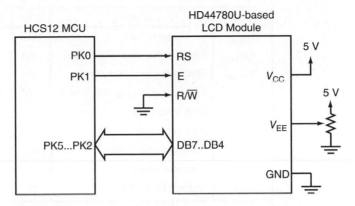

Figure 7.28 ■ LCD interface example (4-bit bus, used in Dragon12)

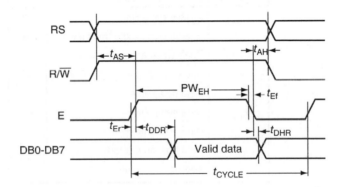

Figure 7.29 ■ HD44780U LCD controller read timing diagram

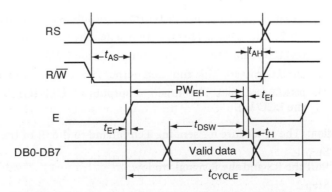

Figure 7.30 ■ HD44780U LCD controller write timing diagram

Symbol	Meaning	Min.	Typ.	Max.	Unit
t_{CYCLE}	Enable cycle time	1000	–	–	ns
PW_{EH}	Enable pulse width (high level)	450	–	–	ns
t_{Er}, t_{Ef}	Enable rise and decay time	–	–	25	ns
t_{AS}	Address setup time, RS, R/$\overline{W}$, E	60	–	–	ns
t_{DDR}	Data delay time	–	–	360	ns
t_{DSW}	Data setup time	195	–	–	ns
t_H	Data hold time (write)	10	–	–	ns
t_{DHR}	Data hold time (read)	5	–	–	ns
t_{AH}	Address hold time	20	–	–	ns

Table 7.10 ■ HD44780U bus timing parameters (1-MHz operation)

Symbol	Meaning	Min.	Typ.	Max.	Unit
t_{CYCLE}	Enable cycle time	500	–	–	ns
PW_{EH}	Enable pulse width (high level)	230	–	–	ns
t_{Er}, t_{Ef}	Enable rise and decay time	–	–	20	ns
t_{AS}	Address setup time, RS, R/$\overline{W}$, E	40	–	–	ns
t_{DDR}	Data delay time	–	–	160	ns
t_{DSW}	Data setup time	80	–	–	ns
t_H	Data hold time (write)	10	–	–	ns
t_{DHR}	Data hold time (read)	5	–	–	ns
t_{AH}	Address hold time	10	–	–	ns

Table 7.11 ■ HD44780U bus timing parameters (2-MHz operation)

Example 7.4

▼

Write a set of LCD routines that perform the following functions for the Dragon12 demo board:

1. cmd2LCD (char cmd). This function sends the command **cmd** to the LCD kit.
2. openLCD (void). This function initializes the LCD.
3. putcLCD (char cx). This function outputs the character **cx** to the LCD kit.
4. putsLCD (char *ptr). This function outputs a NULL-terminated string pointed to by **ptr** to the LCD kit.

Solution: The procedure for sending a command to the IR of the LCD is as follows:

Step 1
Pull the RS and the E signals to low.

Step 2
Pull the R/$\overline{W}$ signal to low.

Step 3
Pull the E signal to high.

Step 4
Output data to the output port attached to the LCD data bus. We need to configure the I/O port for output before writing data to the LCD kit.

Step 5
Pull the E signal to low and make sure that the internal operation is complete.

The procedure for writing a byte to the LCD data register is as follows:

Step 1
Pull the RS signal to high.

Step 2
Pull the R/$\overline{W}$ signal to low.

Step 3
Pull the E signal to high.

Step 4
Output data to the I/O port attached to the LCD data bus.

Step 5
Pull the E signal to low and make sure that the internal operation is complete.

These procedures need to be repeated once for an LCD kit that has a 4-bit interface.

The following constant definitions will be used in the specified functions:

```
lcd_dat   equ   PTK          ; LCD data pins (PK5~PK2)
lcd_dir   equ   DDRK         ; LCD data direction port
lcd_E     equ   $02          ; E signal pin
lcd_RS    equ   $01          ; RS signal pin
```

All of the LCD commands take a much longer time to complete than an instruction does. There is always a possibility that the LCD is still busy with its internal operation when we want to send a new command to the LCD. There are two methods for solving this problem.

1. The **cmd2LCD** function calls a subroutine to make sure that the LCD is idle before proceeding with the new command. After performing the desired operation, the cmd2lcd function simply returns without waiting for the internal LCD operation to complete. This approach cannot work with the Dragon12 demo board.

2. The **cmd2LCD** function performs the desired operation, waits for 40 μs (or slightly longer), and then returns to the caller. By waiting for 40 μs, all except two instructions (clear display and cursor home) will be completed. For these two instructions, the caller needs to call a delay subroutine to wait until the internal LCD operation is completed before proceeding with other operations.

The function that sends a command to the LCD kit using the second approach is as follows:

```
cmd2LCD    psha                          ; save the command in stack
           bclr   lcd_dat,lcd_RS         ; select the instruction register
           bset   lcd_dat,lcd_E          ; pull the E signal high
           anda   #$F0                   ; clear the lower 4 bits
           lsra                          ; match the upper 4 bits with the LCD
```

```
            lsra                        ; data pins
            oraa    #lcd_E              ; maintain the E signal value
            staa    lcd_dat             ; send the command, along with the RS and E signals
            nop                         ; extend the duration of the E pulse
            nop                         ;    "
            nop                         ;    "
            bclr    lcd_dat,lcd_E       ; pull the E signal low
            pula                        ; retrieve the LCD command
            anda    #$0F                ; clear the upper 4 bits
            lsla                        ; match the lower 4 bits with the LCD
            lsla                        ; data pins
            bset    lcd_dat,lcd_E       ; pull the E signal high
            oraa    #lcd_E              ; maintain the E signal value
            staa    lcd_dat             ; send the lower 4 bits of command with E and RS
            nop                         ; extend the duration of the E pulse
            nop                         ;    "
            nop                         ;    "
            bclr    lcd_dat,lcd_E       ; clear the E signal to complete the write operation
            ldy     #1                  ; adding this delay will complete the internal
            jsr     delayby50us         ; operation for most instructions
            rts
```

Before using the LCD, the user must configure it properly. The configuration of the LCD involves at least the following four LCD instructions:

1. Entry mode set. The common setting for this instruction is to move the cursor to the right after reading or writing a character from or to the LCD.

2. Display on/off. The common setting for this instruction is to turn on the display, cursor, and cursor blinking.

3. Function set. This instruction sets the number of rows for display, the font size, and the width of the interface data (4 or 8 bits).

4. Clear display. Before outputting any data, it is always a good idea to clear the LCD screen and move the cursor to the home position (upper left corner).

The following function performs the LCD configuration:

```
openLCD     movb    #$FF,lcd_dir        ; configure Port K for output
            ldy     #10                 ; wait for LCD to be ready
            jsr     delayby10ms         ;    "
            ldaa    #$28                ; set 4-bit data, two-line display, 5 × 8 font
            jsr     cmd2lcd             ;    "
            ldaa    #$0F                ; turn on display, cursor, and blinking
            jsr     cmd2lcd             ;    "
            ldaa    #$06                ; move cursor right (entry mode set instruction)
            jsr     cmd2lcd             ;    "
            ldaa    #$01                ; clear display screen and return to home position
            jsr     cmd2lcd             ;    "
            ldy     #2                  ; wait until clear display command is complete
            jsr     delayby1ms          ;    "
            rts
```

The function that outputs a character to the LCD and makes sure that the write operation is complete is as follows:

```
putcLCD     psha                        ; save a copy of the data
            bset    lcd_dat,lcd_RS      ; select lcd Data register
            bset    lcd_dat,lcd_E       ; pull E to high
            anda    #$F0                ; mask out the lower 4 bits
            lsra                        ; match the upper 4 bits with the LCD
            lsra                        ; data pins
            oraa    #$03                ; keep signal E and RS unchanged
            staa    lcd_dat             ; send the upper 4 bits and E, RS signals
            nop                         ; provide enough duration to the E signal
            nop                         ;    "
            nop                         ;    "
            bclr    lcd_dat,lcd_E       ; pull the E signal low
            pula                        ; retrieve the character from the stack
            anda    #$0F                ; clear the upper 4 bits
            lsla                        ; match the lower 4 bits with the LCD
            lsla                        ; data pins
            bset    lcd_dat,lcd_E       ; pull the E signal high
            oraa    #$03                ; keep E and RS unchanged
            staa    lcd_dat
            nop
            nop
            nop
            bclr    lcd_dat,lcd_E       ; pull E low to complete the write cycle
            ldy     #1                  ; wait until the write operation is
            jsr     delayby50us         ; complete
            rts
```

The function that outputs a NULL-terminated string pointed to by the index register X is as follows:

```
putsLCD     ldaa    1,x+                ; get one character from the string
            beq     donePS              ; reach NULL character?
            jsr     putcLCD
            bra     putsLCD
donePS      rts
```

Example 7.5

▼

Write an assembly program to test the previous four subroutines by displaying the following messages on two lines:

       hello world!

       LCD is working!

Solution: The program is as follows:

```
#include     "c:\miniide\hcs12.inc"
lcd_dat      equ     PTK              ; LCD data pins (PK5~PK2)
lcd_dir      equ     DDRK             ; LCD data direction port
```

```
lcd_E           equ     $02                     ; E signal pin
lcd_RS          equ     $01                     ; RS signal pin
                org     $1500
                lds     #$1500                  ; set up stack pointer
                jsr     openLCD                 ; initialize the LCD
                ldx     #msg1
                jsr     putsLCD
                ldaa    #$C0                    ; move to the second row
                jsr     cmd2LCD                 ;   "
                ldx     #msg2
                jsr     putsLCD
                swi
msg1            dc.b    "hello world!",0
msg2            dc.b    "LCD is working!",0
                #include  "c:\miniIDE\delay.asm"           ; include delay routines here
;
; include the previous four LCD functions
;
;               org     $FFFE                   ; uncomment this line for CodeWarrior
                dc.w    start                   ; uncomment this line for CodeWarrior
                end
```

Example 7.6

Write the C language versions of the previous four LCD functions and a program to test them.

Solution: The C functions for the LCD kit on the Dragon12 demo board and their test program are as follows:

```
#include        "c:\cwHCS12\include\hcs12.h"
#include        "c:\cwHCS12\include\delay.h"
#include        "c:\cwHCS12\include\lcd_util.h"
#define         LCD_DAT   PORTK    // Port K drives LCD data pins, E, and RS
#define         LCD_DIR   DDRK     // Direction of LCD port
#define         LCD_E     0x02     // E signal
#define         LCD_RS    0x01     // RS signal
#define         LCD_E_RS  0x03     // assert both E and RS signals
void main (void)
{
        char *msg1 = "hello world!";
        char *msg2 = "LCD is working!";
        openLCD();
        putsLCD(msg1);
        cmd2LCD(0xC0);        // move cursor to 2nd row, 1st column
        putsLCD(msg2);
        asm("swi");
        while(1);
}
void cmd2LCD    (char cmd)      [unsigned]
{
        char    temp;
```

```
        temp = cmd;                    // save a copy of the command
        cmd &=0xF0;                    // clear out the lower 4 bits
        LCD_DAT  &= (~LCD_RS);         // select LCD instruction register
        LCD_DAT  |= LCD_E;             // pull E signal to high
        cmd >>= 2;                     // shift to match LCD data pins
        LCD_DAT  = cmd | LCD_E;        // output upper 4 bits, E, and RS
        asm ("nop");                   // dummy statements to lengthen E
        asm ("nop");                   //       "
        asm ("nop");
        LCD_DAT  &= (~LCD_E);          // pull E signal to low
        cmd = temp & 0x0F;             // extract the lower 4 bits
        LCD_DAT  |= LCD_E;             // pull E to high
        cmd <<= 2;                     // shift to match LCD data pins
        LCD_DAT  = cmd | LCD_E;        // output upper 4 bits, E, and RS
        asm("nop");                    // dummy statements to lengthen E
        asm("nop");                    //       "
        asm("nop");
        LCD_DAT  &= (~LCD_E);          // pull E-clock to low
        delayby50us(1);                // wait until the command is complete
}
void openLCD(void)
{
        LCD_DIR  = 0xFF;               // configure LCD_DAT port for output
        delayby10ms(10);
        cmd2LCD(0x28);                 // set 4-bit data, 2-line display, 5×7 font
        cmd2LCD(0x0F);                 // turn on display, cursor, blinking
        cmd2LCD(0x06);                 // move cursor right
        cmd2LCD(0x01);                 // clear screen, move cursor to home
        delayby1ms(2);                 // wait until clear display command is complete
}
void putcLCD(char cx)
{
        char temp;
        temp = cx;
        LCD_DAT  |= LCD_RS;            // select LCD data register
        LCD_DAT  |= LCD_E;             // pull E signal to high
        cx  &= 0xF0;                   // clear the lower 4 bits
        cx >>= 2;                      // shift to match the LCD data pins
        LCD_DAT  = cx|LCD_E_RS;        // output upper 4 bits, E, and RS
        asm("nop");                    // dummy statements to lengthen E
        asm("nop");                    //       "
        asm("nop");
        LCD_DAT  &= (~LCD_E);          // pull E to low
        cx  = temp & 0x0F;             // get the lower 4 bits
        LCD_DAT  |= LCD_E;             // pull E to high
        cx  <<= 2;                     // shift to match the LCD data pins
        LCD_DAT  = cx|LCD_E_RS;        // output lower 4 bits, E, and RS
        asm("nop");                    // dummy statements to lengthen E
        asm("nop");                    //       "
        asm("nop");
        LCD_DAT  &= (~LCD_E);          // pull E to low
```

```
            delayby50us(1);
    }
    void putsLCD (char *ptr)
    {
            while (*ptr) {
                putcLCD(*ptr);
                ptr++;
            }
    }
```

The assembly and C language versions of the LCD functions have been grouped into files to be included in the user's program and are provided on the complementary CD.

7.9 Interfacing Parallel Ports to a Keypad

A keypad is another commonly used input device. Like a keyboard, a keypad is arranged as an array of switches, which can be mechanical, membrane, capacitive, or Hall effect in construction. In mechanical switches, two metal contacts are brought together to complete an electric circuit. In membrane switches, a plastic or rubber membrane presses one conductor onto another; this type of switch can be made very thin. Capacitive switches comprise two plates of a parallel plate capacitor; pressing the key cap effectively increases the capacitance between the two plates. Special circuitry is needed to detect this change in capacitance. In Hall-effect key switches, the motion of the magnetic flux lines of a permanent magnet perpendicular to a crystal is detected as voltage appearing between the two faces of the crystal; it is this voltage that *registers* a switch closure.

Mechanical keypads and keyboards are most popular due to their low cost and strength of construction. However, mechanical switches have a common problem called *contact bounce*. Instead of producing a single, clean pulse output, pressing a mechanical switch generates a series of pulses because the switch contacts do not come to rest immediately. This phenomenon is illustrated in Figure 7.31.

When the key is not pressed, the voltage output to the computer is 5 V. In order to detect which key has been pressed, the microcontroller needs to scan every key switch of the keypad. A human being cannot press and release a key switch in less than 20 ms. During this interval, the microprocessor can scan the same key switch closure tens or even hundreds of thousands of times, interpreting each low signal as a new input when in fact only one input should be sent.

Because of the contact bounce and the disparity in speed between the microprocessor and human key pressing, a *debouncing* process is needed. A keypad input program can be divided into three stages.

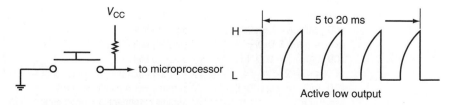

Figure 7.31 ■ Key switch contact bounce

1. *Keypad scanning* to find out which key was pressed
2. *Key switch debouncing* to make sure a key is indeed pressed
3. *Table lookup* to find the ASCII code of the key that was pressed

7.9.1 Keypad Scanning

Keypad scanning is usually performed row by row and column by column. A 16-key keypad can easily be interfaced with any available I/O port. Figure 7.32 shows a 16-key keypad organized into four rows with each row consisting of four switches.

For the keypad input application, the upper four pins (PA7~PA4) of Port A should be configured for output, whereas the lower four pins (PA3~PA0) of Port A should be configured for input.

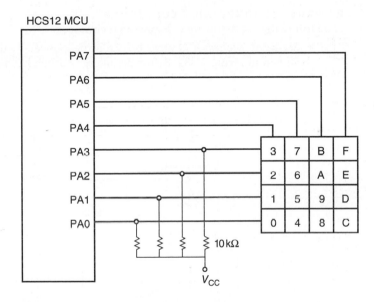

Figure 7.32 ■ Sixteen-key keypad connected to the HCS12

The rows and columns of a keypad are simply conductors. In Figure 7.32, Port A pins PA3~PA0 are pulled up to high by pull-up resistors. Whenever a key switch is pressed, the corresponding row and column are shorted together. In order to distinguish the row being scanned from those not being scanned, the row being scanned is driven low, whereas the other rows are driven high. The row selection of the 16-key keypad is shown in Table 7.12.

PA7	PA6	PA5	PA4	Selected Keys
1	1	1	0	0, 1, 2, and 3
1	1	0	1	4, 5, 6, and 7
1	0	1	1	8, 9, A, and B
0	1	1	1	C, D, E, and F

Table 7.12 ■ Sixteen-key keypad row selections

7.9.2 Keypad Debouncing

Contact bounce is due to the dynamics of a closing contact. The signal falls and rises a few times within a period of about 5 ms as a contact bounces. Since a human being cannot press and release a switch in less than 20 ms, a debouncer will recognize that the switch is closed after the voltage is low for about 10 ms and will recognize that the switch is open after the voltage is high for about 10 ms.

Both hardware and software solutions to the key bounce problem are available. Hardware solutions to contact bounce include an analog circuit that uses a resistor and a capacitor to smooth the voltage and two digital solutions that use set-reset latches or CMOS buffers and double-throw switches. Dedicated scanner chips that perform keypad scanning and debouncing are also available. National Semiconductor 74C922 and 74C923 are two examples.

HARDWARE DEBOUNCING TECHNIQUES

The following are hardware debouncing techniques:

- *Set-reset latches.* A key switch can be debounced by using the set-reset latch shown in Figure 7.33a. Before being pressed, the key is touching the set input and

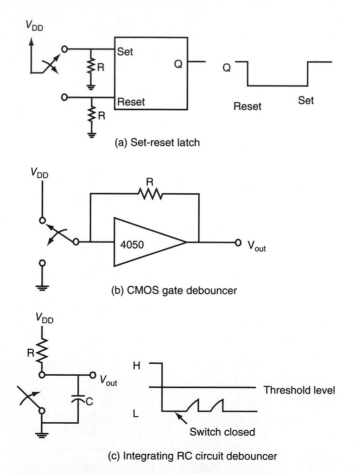

(a) Set-reset latch

(b) CMOS gate debouncer

(c) Integrating RC circuit debouncer

Figure 7.33 ■ Hardware debouncing techniques

the Q voltage is high. When pressed, the key moves toward the reset position. When the key touches the reset position, the Q voltage will go low. When the key is bouncing and touching neither the set nor the reset input, both set and reset inputs arc pulled low by the pull-down resistors. Since both set and reset are low, the Q voltage will remain low and the key will be recognized as pressed.

- *Noninverting CMOS buffer with high input impedance.* The CMOS buffer output is identical to its input. When the switch is pressed, the input of the buffer chip 4050 is grounded and hence V_{OUT} is forced to low. When the key switch is bouncing (not touching the input), the resistor R keeps the output voltage low. This is due to the high input impedance of 4050, which causes a negligible voltage drop on the feedback resistor. Thus the output is debounced. This solution is shown in Figure 7.33b.

- *Integrated debouncers.* The RC constant of the integrator determines the rate at which the capacitor charges up toward the supply voltage once the ground connection via the switch has been removed. As long as the capacitor voltage does not exceed the logic 0 threshold value, the V_{OUT} signal will continue to be recognized as a logic 0. This solution is shown in Figure 7.33c.

SOFTWARE DEBOUNCING TECHNIQUES

A simple software debouncing method is *wait and see;* that is, wait for 10 ms and reexamine the same key to see if it is still pressed. A key is considered pressed if its output voltage is still low 10 ms after it is detected low. If the output voltage is high, the program will consider the signal as noise or the key as still bouncing. In either case, the program will continue to perform the scanning. For most applications, this simple wait-and-see method is adequate.

7.9.3 ASCII Code Lookup

For an application that needs a keyboard, the easiest way to find out the ASCII code of the pressed key is to perform a table lookup. However, table lookup is not necessary for the keypad because ASCII code lookup can be embedded in the program that performs the scanning and debouncing.

Example 7.7

Write a program to perform keypad scanning and debouncing and return the ASCII code of the pressed key to the caller.

Solution: To implement keypad scanning, debouncing, and ASCII code lookup requires the use of four local variables.

- *maskr.* the mask to select a row to be scanned
- *maskc.* the mask to select a column to be scanned
- *column.* the column (0 to 3) being scanned
- *row.* the row (0 to 3) being scanned

The following assembly routine will perform scanning and debouncing and return the ASCII code of the pressed key in accumulator A to the caller:

```
keypad    equ    PTA
maskr     equ    0              ; the mask to scan a row
maskc     equ    1              ; the mask to scan a column
```

```
column    equ     2                   ; the column being scanned
row       equ     3                   ; the row being scanned
; ******************************************************************************
; The following subroutine reads a character entered from the keypad. It performs
; scanning and then debouncing when it detects a pressed key. It returns the ASCII
; code in A after making sure the key is indeed pressed. It uses stack to allocate four
; 1-byte local variables.
; ******************************************************************************
getkey    leas    −4,SP               ; allocate 4 bytes for local variables
          movb    #$F0,DDRA           ; configure keypad Port A pin directions
gkloope   clr     row,SP              ; start from row 0
          movb    #$EF,maskr,SP       ; mask for selecting row 0
nextR     clr     column,SP           ; start from column 0 in a row
          movb    #$01,maskc,SP       ; mask for scanning column 0
          movb    #$FF,keypad         ; prepare to scan
          ldaa    maskr,SP            ; select a row to scan
          anda    keypad              ;              "
          staa    keypad
gkloopi   ldaa    maskc,SP            ; select a column to check
          anda    keypad              ;              "
          beq     debnce              ; if low, then a key press has been detected
next      ldaa    #3
          cmpa    column,SP
          blo     inc_col             ; not reach the last column in a row yet, branch
          ldaa    #3
          cmpa    row,SP              ; not reach row 3 yet, branch and try next row
          blo     inc_row             ;              "
          bra     gkloope             ; restart from row 0, column 0
inc_col   inc     column,SP           ; move to next column
          lsl     maskc,SP            ; update the column mask to check the
          bra     gkloopi             ; goto scan next column
inc_row   inc     row,SP              ; move to next row
          lsl     maskr,SP            ; update the mask for scanning
          bset    maskr,SP,$01        ; always force bit 0 of maskr to 1
          bra     nextR
debnce    ldy     #1
          jsr     delayby10ms
          ldaa    maskc,SP            ; reexamine the same key
          anda    keypad              ;              "
          beq     getcode             ; if still low, get the ASCII code
          bra     next                ; not low, continue to scan
getcode   ldaa    row,SP              ; find out the key number that
          lsla                        ; has been pressed
          lsla                        ;              "
          adda    column,SP           ;              "
          cmpa    #10                 ; is [A] >= 10?
          blo     isdeci              ; jump if A < 10
          adda    #$37                ; the pressed key is greater than 9
          leas    4,SP
```

```
           rts
isdeci     leas      4,SP
           adda      #$30              ; compute the ASCII code
           rts
#include "c:\miniIDE\delay.asm"
```

The C language version of the function is as follows:

```
#define        keypad        PTA                              // keypad port
#define        keypad_dir    DDRA                             // keypad port direction register
// ****************************************************************************
// rmask is row mask, cmask is column mask, row is the row being scanned, col is the
// column being scanned
// ****************************************************************************
char getkey (void)
{
       char rmask, cmask, row, col;
       char temp, keycode;
       keypad_dir = 0xF0;   // configure lower four pins for input
       while (1) {
              rmask      = 0xEF;
              for (row = 0; row < 4; row++){
                     cmask      = 0x01;
                     keypad     &= rmask; // select the current row
                     for (col = 0; col < 4; col++){
                            if (!(keypad & cmask)){          // key switch detected pressed
                                   delayby10ms(1);
                                   if(!(keypad & cmask)){   // check the same key again
                                          keycode = row * 4 + col;
                                          if (keycode < 10)
                                                 return (0x30 + keycode);
                                          else
                                                 return (0x37 + keycode);
                                   }
                            }
                            cmask = cmask << 1;
                     }
                     rmask = (rmask << 1) | 0x0F;
              }
       }
}
```

7.10 Using the D/A Converter

A digital-to-analog converter (DAC) converts fixed-point binary numbers into an electric voltage (more often) or current. Normally the output is a linear function of the input number. These numbers are often obtained from a sampling process (for example, digital audio) at uniform intervals. To restore to the original physical quantity, these numbers are fed to

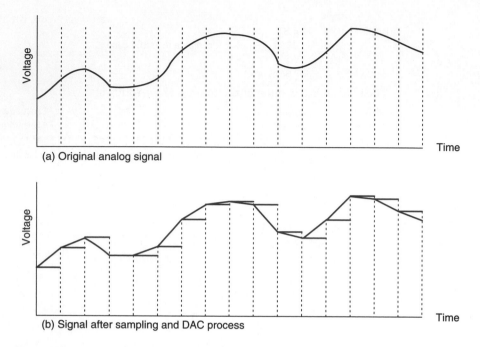

Figure 7.34 ■ Analog signal and the resulting signal after sampling, ADC, and DAC process

the DAC also at the uniform intervals. As a result, all practical DACs output a sequence of piecewise constant values or rectangular pulses. An example of this process is illustrated in Figure 7.34.

A DAC has many applications. Examples are digital gain and offset adjustment, programmable voltage and current sources, programmable attenuators, digital audio, closed-loop positioning, robotics, and so on. In the last few years, digital video is also getting more and more popular. As more and more flat panel displays come with DVI and HDMI interfaces, digital video will become the norm in a few years. Although there are a few microcontrollers incorporating the D/A converter on the chip, most microcontrollers still need to use an off-chip D/A converter to perform the D/A conversion function. The HCS12 is no exception. A D/A converter may use a serial or parallel interface to obtain digital code from the microprocessor or microcontroller.

There are several factors to consider when choosing a DAC.

- *Resolution.* This is the number of possible output levels the DAC is designed to reproduce. The resolution is usually stated as the number of bits it uses and is the base 2 logarithm of the number of levels. For example, an 8-bit DAC can represent 256 levels.

- *Dynamic range.* This is a measurement of the difference between the largest and smallest signals the DAC can reproduce represented in decibels. This characteristic is related to the DAC resolution and noise floor (the sum of all the noise sources).

- *Number of channels.* A DAC may have more than one output channel to satisfy the needs of the applications that require more than one channel.

- *Type of output.* The output of a DAC may be in the form of voltage or current to meet the requirement of the target application.
- *Monotonicity.* This refers to the ability of the DAC's analog output to increase with digital code.

In this section, we use the 8-bit DAC AD7302 from Analog Devices to illustrate the use of the DAC.

7.10.1 The 8-bit AD7302 DAC

The AD7302 is a dual-channel (two DACs on the same chip), 8-bit, DAC chip from Analog Devices that has a parallel interface with the microcontroller. The AD7302 converts an 8-bit digital value into an analog voltage. The block diagram of the AD7302 is shown in Figure 7.35. The AD7302 is designed to be a memory-mapped device. In order to send data to the AD7302, the $\overline{CS}$ signal must be pulled to low. On the rising edge of the $\overline{WR}$ signal, the values on pins D7–D0 will be latched into the Input register. When the signal $\overline{LDAC}$ is low, the data in the Input register is transferred to the DAC register and a new D/A conversion is started. The AD7302 needs a reference voltage to perform the D/A conversion. The reference voltage can come from either the external REFIN input or the internal V_{DD}. The $\overline{A}/B$ signal selects the channel (A or B) to perform the D/A conversion. The PD pin puts the AD7302 in power-down mode and reduces the power consumption to 1 μW.

The AD7302 operates from a single +2.7- to 5.5-V supply and typically consumes 15 mW at 5 V, making it suitable for battery-powered applications. Each digital sample takes about 2 μs to convert. The output voltage ($V_{OUT}A$ or $V_{OUT}B$) from either DAC is given by

$$V_{OUT}A/B = 2 \times V_{REF} \times (N/256)$$

where V_{REF} is derived internally from the voltage applied at the REFIN pin or V_{DD}. If the voltage applied to the REFIN pin is within 1 V of the V_{DD}, $V_{DD}/2$ is used as the reference voltage automatically. Otherwise, the voltage applied at the REFIN pin is used as the reference voltage. The range of V_{REF} is from 1 V to $V_{DD}/2$. N is the decimal equivalent of the code loaded to the DAC register, ranging from 0 to 255.

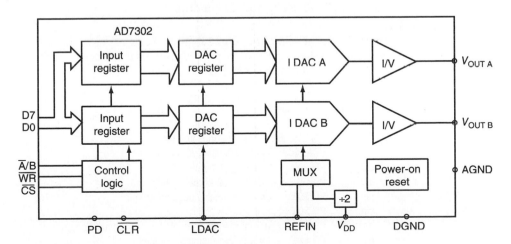

Figure 7.35 ■ Functional block diagram of the AD7302

7.10.2 Interfacing the AD7302 with the HCS12

Interfacing the AD7302 with the HCS12 can be very simple. Both the $\overline{CS}$ and $\overline{LDAC}$ signals can be tied to ground permanently. The value to be converted must be sent to the AD7302 via a parallel port (connect to pins D7~D0). An output pin can be used (as the $\overline{WR}$ signal) to control the transferring of data to the Input register. A typical connection between the HCS12 and the AD7302 is shown in Figure 7.36.

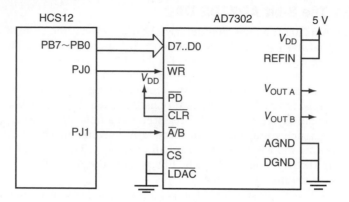

Figure 7.36 ■ Circuit connection between the AD7302 and the HCS12

Example 7.8

Write a program to generate a sawtooth waveform from the $V_{OUT}A$ pin.

Solution: The procedure for generating a sawtooth waveform is as follows:

Step 1
Configure PB7~PB0, PJ0, and PJ1 for output.

Step 2
Output the digital value from 0 to 255 and repeat. For each value, pull the PJ0 to low and then to high so that the value on pins PB7~PB0 can be transferred to the AD7302. Pull the signal PJ1 to low during the process.

The assembly program is as follows:

```
        #include  "c:\miniide\hcs12.inc"
        org       $1500
start   movb      #$FF,DDRB ; configure Port B for output
        bset      DDRJ,$03  ; configure PJ1~PJ0 for output
        bclr      PTJ,$02   ; select V_OUT A output
loop    inc       PORTB     ; increase the output by one step
        bclr      PTJ,$01   ; generate a rising edge on PJ0 pin
        bset      PTJ,$01   ;         "
        bset      PTJ,$01   ; add nine more bset instructions to provide 2 µs
        bset      PTJ,$01   ; for D/A conversion to complete
        bset      PTJ,$01   ;         "
        bset      PTJ,$01   ;         "
```

```
        bset    PTJ,$01   ; "
        bset    PTJ,$01   ; "
        bset    PTJ,$01   ; "
        bset    PTJ,$01   ; "
        bset    PTJ,$01   ; "
        bra     loop      ; to complete the D/A conversion
;       org     $FFFE     ; uncomment this line for CodeWarrior
;       dc.w    start     ; uncomment this line for CodeWarrior
        end
```

The C language version of the program is as follows:

```c
#include "c:\cwHCS12\include\hcs12.h"
void main(void)
{
        DDRB    = 0xFF;          // configure PORTB for output
        DDRJ    |= 0x03;         // configure pins PJ1~PJ0 for output
        PTJ     &= 0xFD;         // pull the signal A/B to low to select channel A
        while (1) {
                PORTB += 1;
                PTJ &= 0xFE;     // generate a rising edge
                PTJ |= 0x01;     //           "
                PTJ |= 0x01;     // use dummy statements to provide 2 µs
                PTJ |= 0x01;     // time for D/A conversion to complete
                PTJ |= 0x01;
                PTJ |= 0x01;
                PTJ |= 0x01;
                PTJ |= 0x01;
                PTJ |= 0x01;
                PTJ |= 0x01;
                PTJ |= 0x01;
        }
}
```

The AD7302 can be used to generate many interesting waveforms. Several exercise problems are given on the applications of this DAC at the end of the chapter.

7.11 Stepper Motor Control

Stepper motors are digital motors. They are convenient for applications where a high degree of positional control is required. Printers, tape drives, disk drives, and robot joints, for example, are typical applications of stepper motors.

7.11.1 Principles of Rotation for the Stepper Motor

In its simplest form, a stepper motor has a permanent magnet rotor and a stator consisting of two coils. The rotor aligns with the stator coil that is energized. By changing which coil is energized, as illustrated in the following figures, the rotor is turned.

In Figure 7.37a–d, the permanent magnet rotor lines up with the coil pair that is energized. The direction of the current determines the polarity of the magnetic field, and thus the angular position of the rotor. Energizing coil pair C3–C4 causes the rotor to rotate 90 degrees. Again, the direction of the current determines the magnetic polarity and thus the angular position of the rotor. In this example, the direction of the current causes the rotor to rotate in a clockwise direction, as shown in Figure 7.37b.

Next, coils C1–C2 are energized again, but with a current opposite to that in step 1. The rotor moves 90 degrees in a clockwise direction, as shown in Figure 7.37c. The last full step moves the rotor another 90 degrees in a clockwise direction. Note that again the coil pair C3–C4 is energized, but with a current opposite to that in step 2.

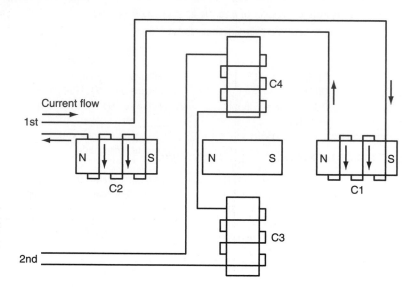

Figure 7.37a ■ Stepper motor full step 1

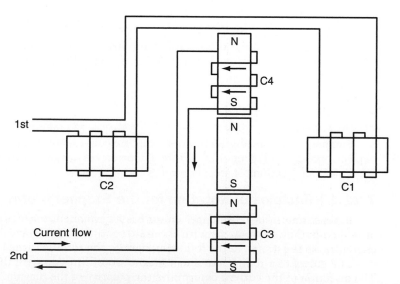

Figure 7.37b ■ Stepper motor full step 2

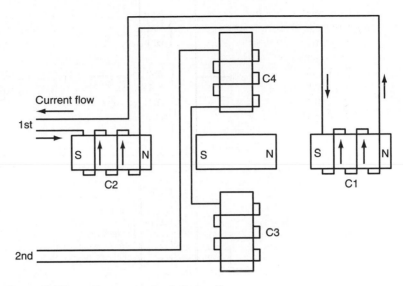

Figure 7.37c ■ Stepper motor full step 3

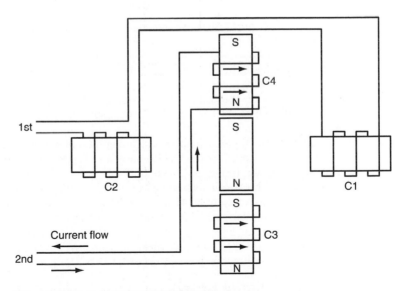

Figure 7.37d ■ Stepper motor full step 4

We can also rotate the stepper motor in the counterclockwise direction. This can be done by reversing the polarities of coils C3 and C4 in Figure 7.37b and d. Figure 7.38 shows the counterclockwise sequence.

The stepper motor may also be operated with half steps. A *half step* occurs when the rotor (in a four-pole step) is moved to eight discrete positions (45 degrees). To operate the stepper motor in half steps, sometimes both coils may have to be on at the same time. When two coils in close

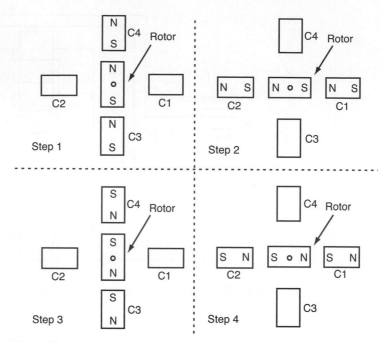

Figure 7.38 ■ Full-step counterclockwise operation of step motor

proximity are energized, there is a resultant magnetic field whose center will depend on the relative strengths of the two magnetic fields. Figure 7.39 illustrates the half-stepping sequence.

The step sizes of stepper motors vary from approximately 0.72 to 90 degrees. However, the most common step sizes are 1.8, 7.5, and 15 degrees. The steps of 90 or 45 degrees are too crude for many applications.

The actual stator (the stationary electromagnets) of a real motor has more segments on it than previously indicated. One example is shown in Figure 7.40. The rotor is also a little bit different and is also shown in Figure 7.40.

In Figure 7.40, the stator has eight individual sections (coils) on it and hence the angle between two adjacent sections is 45 degrees. The rotor has six sections on it and hence there are 60 degrees between two adjacent sections. Using the principle of a Vernier mechanism, the actual movement of the rotor for each step would be 60–45 degrees, or 15 degrees. Interested readers should try to figure out how these sections are energized to rotate the motor in the clockwise and counterclockwise directions.

7.11.2 Stepper Motor Drivers

Driving a stepper motor involves applying a series of voltages to the coils of the motor. A subset of coils is energized at the same time to cause the motor to rotate one step. The pattern of coils energized must be followed exactly for the motor to work correctly. The pattern will vary depending on the mode used on the motor. A microcontroller can easily time the duration that the coil is energized, and hence control the speed of the stepper motor in a precise manner.

The circuit in Figure 7.41 shows how the transistors are used to switch the current to each of the four coils of the stepper motor. The diodes in Figure 7.41 are called *fly back*

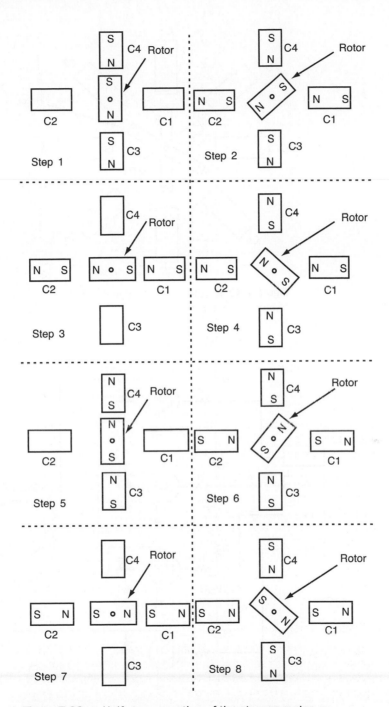

Figure 7.39 ■ Half-step operation of the stepper motor

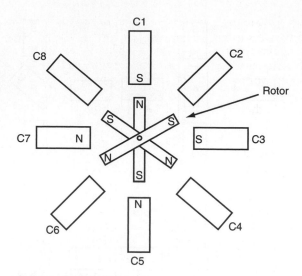

Figure 7.40 ■ Actual internal construction of stepper motor

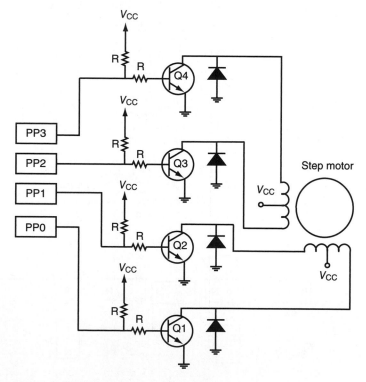

Figure 7.41 ■ Driving a stepper motor

diodes and are used to protect the transistors from reverse bias. The transistor loads are the windings in the stepper motor. The windings are inductors, storing energy as a magnetic field. When the current is cut off, the inductor dispenses its stored energy in the form of an electric current. This current attempts to flow through the transistor, reverse biasing its collector-emitter pair. The diodes are placed to prevent this current from going through the transistors.

For higher-torque applications, the normal full-step sequence is used as shown in Table 7.13. For lower-torque applications, the half-step mode is used and its sequence is shown in Table 7.14. To control the motor, the microcontroller must output the values in the table in the sequence shown. Tables 7.13 and 7.14 are circular in that after the last step, the next output must be the first step. The values may be output in the order shown to rotate the motor in one direction, or in reverse order to rotate in the reverse direction. It is essential that the order be preserved even if the motor is stopped for a while. The next step to restart the motor must be the next sequential step following the last step used. The mechanical inertia of the motor will require a short delay (usually 5 to 20 ms) between two steps to prevent the motor from missing steps.

Step	Q4 PP3	Q3 PP2	Q2 PP1	Q1 PP0	Value
1	on	on	off	off	1100
2	off	on	on	off	0110
3	off	off	on	on	0011
4	on	off	off	on	1001

Table 7.13 ■ Full-step sequence for clockwise rotation

Step	Q4 PP3	Q3 PP2	Q2 PP1	Q1 PP0	Value
1	on	off	off	off	1000
2	on	on	off	off	1100
3	off	on	off	off	0100
4	off	on	on	off	0110
5	off	off	on	off	0010
6	off	off	on	on	0011
7	off	off	off	on	0001
8	on	off	off	on	1001

Table 7.14 ■ Half-step sequence for clockwise rotation

Example 7.9

Suppose that the pins PP3..PP0 are used to drive the four transistors as shown in Figure 7.41. Write a subroutine to rotate the stepper motor clockwise one cycle using the half-step sequence.

Solution: The assembly language subroutine is as follows:

```
#include    "c:\miniIDE\hcs12.inc"
step1       equ     $08
step2       equ     $0C
step3       equ     $04
step4       equ     $06
step5       equ     $02
step6       equ     $03
step7       equ     $01
step8       equ     $09
HStep       movb    #$FF,DDRP        ; configure PTP for output
            movb    #step1,PTP
            bsr     delay10ms
            movb    #step2,PTP
            bsr     delay10ms
            movb    #step3,PTP
            bsr     delay10ms
            movb    #step4,PTP
            bsr     delay10ms
            movb    #step5,PTP
            bsr     delay10ms
            movb    #step6,PTP
            bsr     delay10ms
            movb    #step7,PTP
            bsr     delay10ms
            movb    #step8,PTP
            bsr     delay10ms
            rts
; the following subroutine waits for 10 ms
delay10ms   movb    #$90,TSCR1       ; enable TCNT and fast flags clear
            movb    #$06,TSCR2       ; configure the prescale factor to 64
            bset    TIOS,IOS0        ; enable OC0
            ldd     TCNT
            addd    #3750            ; start an output compare operation
            std     TC0              ; with 10-ms time delay
            brclr   TFLG1,$01,*
            rts
            end
```

7.12 Key Wake-Ups

Most embedded products are powered by batteries. In order for batteries to last longer, most microcontrollers have incorporated power-saving modes such as the *wait* mode or the *stop* mode in the HCS12. Whenever there is no activity from the end user over a period of time, the application software would put the embedded product in one of the power-saving modes and reduce the power consumption. Whenever the end user wants to use the product, a keystroke would wake up the microcontroller and put the embedded product back to normal operation mode. As you learned in Chapter 6, both the reset and unmasked interrupts can put the microcontroller back to normal operation mode. However, reset is not recommended because it will restart the microcontroller, which would delay the response to the user request. An unmasked interrupt does not have this drawback. All of the HCS12 members have incorporated the key wake-up feature that will issue an interrupt to wake up the MCU when it is in the *stop* or *wait* mode. After the appropriate configuration, an active edge on these pins will generate an interrupt to the MCU and force the MCU to exit from the low-power mode.

7.12.1 Key-Wake-Up Registers

As described in Section 7.5.7, Ports H, J, and P have the key-wake-up capability. Each port has eight associated registers. The Port Pull Device Enable register and the Port Polarity Select register together allow the user to choose to use either the rising edge or falling edge to wake up the MCU. The Port Interrupt Enable register and the Port Interrupt Flag register together allow the user to use key pressing to wake up the MCU.

7.12.2 Key-Wake-Up Initialization

In order to use the HCS12 key-wake-up feature, the user needs to initialize the wake-up port properly. The procedure for using the key-wake-up feature is as follows:

Step 1
Set the direction of the key-wake-up bits to input by writing zeros to the *related* Data Direction register.

Step 2
Select the rising edge or the falling edge of the wake-up pin to interrupt the MCU by programming the related registers.

Step 3
Write the service routine for the key-wake-up interrupt and initialize the key-wake-up interrupt vector.

Step 4
Clear any flags that have been set in the Key-Wake-Up Flag register.

Step 5
Enable the key-wake-up bit by setting the appropriate bits in the Wake-Up Interrupt Enable register.

Step 6
Clear the global interrupt mask (the I bit of the CCR register).

7.12.3 Considerations for the Key-Wake-Up Application

The main application of the key-wake-up feature is to support the power-saving modes of the HCS12. Application software puts the microcontroller in low-power mode by executing a stop or a wai instruction when the inactivity of the end user has exceeded the preset time.

Many applications are designed to be a wait loop that waits for the user to enter a request for service. When a request is entered, the application calls an appropriate routine to provide the service. After the service is done, the routine returns to the wait loop. After providing service to a user request, the application software starts a timer. If the user enters another command before the timer times out, the application software resets the timer and responds to the user request. If the timer times out before the user makes another service request, the application software puts the microcontroller in low-power mode to save power. As long as there is no user request for service, the microcontroller will stay in the low-power mode. The timer timeout interval could be a few minutes or longer depending on how much power the user wants to save. Either the timer **output compare** or **modulus down counter** function can be used to implement the timeout interval. When the user presses the key, the microcontroller will exit the low-power mode and continue to execute the instruction following the stop (or wai) instruction and another cycle of the normal application loop is started.

In order to be used in a key-wake-up application, port pins must be configured for input. Ports H, J, and P can be configured to use the rising or falling edge to wake up the microcontroller. The choice of signal edge will dictate the choice of the pull-up or pull-down resistive device. As shown in Figure 7.42, a rising edge results when a high voltage is applied to a pull-down resistor; a falling edge results when a low voltage is connected to a pull-up resistor. Therefore, the designer should enable the pull-down resistor when a rising edge is selected and enable a pull-up resistor when a falling edge is chosen to wake up the microcontroller.

Since the purpose of the key-wake-up feature is to enable the microcontroller to resume normal operation mode, the interrupt service routine need only perform minimal operation. The minimal operation to be performed would be to clear the interrupt flag set by the wake-up interrupt.

The logic flow of an embedded application that incorporates key-wake-up interrupt is illustrated in Figure 7.43.

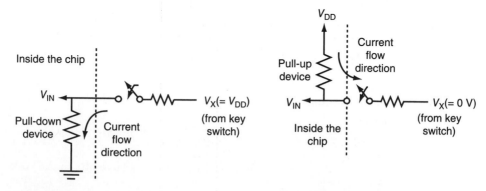

Figure 7.42 ■ (a) Pull-down resistor creates rising edge; (b) Pull-up resistor creates falling edge

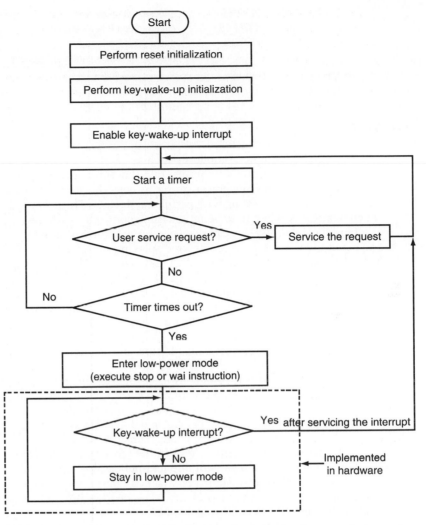

Figure 7.43 ■ Logic flow of embedded applications that incorporate key-wake-up feature

---

Example 7.10

▼

Write an instruction sequence to configure the Port P upper four pins for the key-wake-up feature. Program Port P so that pins PP7~PP4 generate an interrupt whenever there is a falling edge applied to any one of these four pins.

Solution:

```
#include     "c:\miniide\hcs12.inc"
             ...
             bclr    DDRP,$F0        ; configure PP<7:4> pins for input
             bset    PERP,$F0        ; enable PP7~PP4 pins' pull device
```

```
bclr    PPSP,$F0          ; choose pull-up device
movb    #$FF,PIFP         ; clear the Port P key-wake-up flags
bset    PIEP,$F0          ; enable Port P interrupt
cli                       ; enable key-wake-up interrupt globally
```

▲

7.13 Summary

The on-chip peripheral modules such as parallel ports, SCI, SPI, I²C, and CAN are implemented mainly to perform synchronization functions during I/O operations. Since these peripheral modules use the same power source as does the processor, an external interface chip may still be needed when dealing with peripheral devices. The user needs to pay attention to electrical compatibility when interfacing the microcontroller with other peripheral chips.

Liquid crystal displays can display a lot of information while consuming very little power and hence are the device of choice for displaying a large amount of information. For simple embedded systems, a character-based LCD is adequate. The Hitachi HD44780 is the most popular character-based LCD controller. The LCD is often sold as a kit that consists of the LCD panel, LCD controller, and other required electronic circuits.

Keypads and keyboards are the most important input devices for computer systems. The input process for keypads and keyboards consists of three steps: key scanning, key debouncing, and ASCII code lookup. The purpose of key debouncing is to make sure that one keystroke is only recognized as one character. In addition, the debouncing technique can also be used to generate a single pulse with one key press.

A digital-to-analog converter converts a digital value into an electric quantity in the form of current or voltage. Voltage output is more popular for the DAC. The DAC has many applications. Examples include digital gain and offset adjustment, programmable voltage and current sources, programmable attenuators, digital audio, and digital video. When choosing a DAC, the user needs to consider the resolution, dynamic range, monotonicity, number of channels, and type of output.

A stepper motor is a digital motor in the sense that each step of the rotation rotates a fixed number of degrees. It is most suitable for applications that require a high degree of positional control, such as plotters, disk drives, magnetic tape drives, robot joints, and so on. The resolution of one step of a stepper motor can be as small as 0.72 and as large as 90 degrees. The simplest stepper motor has two pairs of coils.

Driving a stepper motor involves applying a series of voltages to the coils of the motor. A subset of coils is energized at the same time to cause the motor to rotate one step. The pattern of coils energized must be followed exactly for the motor to work correctly. The pattern will vary depending on the mode used on the motor. A microcontroller can easily time the duration during which the coil is energized and hence control the speed of the stepper motor in a precise manner.

Saving power is a major concern in most battery-powered embedded products. When the user is not using an embedded system, the microcontroller should be switched to a low-power mode. The HCS12 has two low-power modes: *wait* mode and *stop* mode. The HCS12 consumes the least power in the stop mode. The stop (wait) mode can be entered by executing the stop (wai) instruction. To facilitate the exit of the stop or wait mode, the HCS12 provides the key-wake-up feature. Whenever the HCS12 is in one of the low-power modes and a selected signal edge arrives at one of the key-wake-up port pins, an interrupt request will be generated and the HCS12 will be waked up. The service for the key-wake-up interrupt is simply to clear the interrupt flag and resume the execution of the instruction following the wai or stop instruction.

The signal edge to wake up the microcontroller could be rising or falling. When the rising edge is selected, the pull-down device should be enabled. When the falling edge is chosen, the pull-up device should be enabled. Ports H, J, and P provide the key-wake-up capability.

7.14 Exercises

E7.1 Can the AS logic family drive the HCS12 input pins? Can the HCS12 output drive the AS logic input?

E7.2 Can the ACT logic family drive the HCS12 input? Can the HCS12 output drive the ACT logic input?

E7.3 Can the ABT logic family drive the HCS12 input? Can the HCS12 output drive the ABT logic input?

E7.4 Can the BCT logic family drive the ABT logic input? Can the ABT logic drive the BCT logic input?

E7.5 For the circuit shown in Figure 4.15a, what resistor value should be used to allow 5 mA current flow through the LED assuming the voltage drop across the LED is 2.1 V when it is forward biased and the port pin output is 4.9 V?

E7.6 For the circuit shown in Figure 4.15a, what resistor value should be used to allow 5 mA current flow through the LED assuming the voltage drop across the LED is 1.7 V when it is forward biased and the port pin output is 4.8 V?

E7.7 Write an assembly and a C program to display the following information in two rows in the LCD connected to the demo board:

> Date: 10 10 1952

> Time: 10:20:10

E7.8 Calculate the period of the sawtooth waveform generated in Example 7.7. What are the voltages that correspond to the digital values 20, 30, 50, 127, and 192?

E7.9 Write a program to generate a sine waveform from the $V_{OUT}A$ pin in Figure 7.36.

E7.10 Write a program to generate a 1-kHz periodic square wave from the $V_{OUT}A$ pin and a 2-kHz periodic square wave from the $V_{OUT}B$ pin in Figure 7.36.

E7.11 Write a program to generate a periodic square wave from the $V_{OUT}A$ pin and let the frequency switch between 1 and 4 kHz every 5 s in Figure 7.36.

E7.12 Write a program to generate the waveform shown in Figure 7E.13 from the pin $V_{OUT}A$ in Figure 7.36.

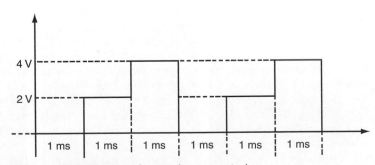

Figure 7E.13 ■ A waveform to be generated

E7.13 Write an assembly and a C program to generate a triangular waveform from the $V_{OUT}A$ pin in Figure 7.36. What is the period of the waveform generated by your program?

E7.14 Write an instruction sequence to rotate the stepper motor shown in Figure 7.41 clockwise one cycle using the full-step sequence.

E7.15 Write a C program to rotate the stepper motor one cycle in the counterclockwise direction using the half-step sequence by using 5 ms as the delay between two half steps. The circuit connection of the stepper motor is shown in Figure 7.41.

E7.16 Write a C program to rotate the stepper motor shown in Figure 7.41 clockwise one cycle using the full-step sequence with 5 ms as the delay between two steps.

E7.17 Write a C function to initialize the Port H key-wake-up function for the HCS12 that interrupts the microcontroller on the falling edge (automatic) of any Port H pin. Also write an interrupt service routine for this interrupt that simply clears the key-wake-up interrupt flags and returns.

7.15 Lab Exercises and Assignments

L7.1 Write an assembly or C program to generate a periodic square wave with a frequency that alternates between 1 and 2 kHz every 4 s using the DAC circuit shown in Figure 7.36.

L7.2 Connect a stepper motor circuit as shown in Figure 7.41. The stepper motor mini board from Futurlec (www.futurlec.com) as shown in Figure L7.2 may work well with the circuit given in Figure 7.40. Write a program to perform the following operations:

Step 1
Rotate the stepper motor using full-step sequence clockwise for 5 s.

Step 2
Rotate the stepper motor using full-step sequence counterclockwise for 5 s.

Step 3
Rotate the stepper motor using half-step sequence clockwise for 5 s.

Step 4
Rotate the stepper motor using half-step sequence counterclockwise for 5 s.

Experiment with intrastep delays of 2, 5, and 10 ms. The IND, INC, INB, and INA pins of the stepper motor should be connected to PP3, PP2, PP1, and PP0 pins, respectively.

Figure L7.2 ■ A hobby stepper motor kit from Futurlec

L7.3 Keypad input. Use the circuit shown in Figure 7.7 to perform keypad input practice. The procedure is as follows:

Step 1
Configure the upper four pins of Port P3 for output and the lower four pins of P3 for input.

Step 2
Initialize the LCD properly.

Step 3
Output the message "Enter an integer:" on the first row of the LCD. After seeing this message, you enter a number on the keypad. Use the F key (or other key) to terminate the number. Your program reads in the number, converts it into a binary, and saves it in a buffer.

Step 4
Output the message "Enter another integer:" on the first row of the LCD. After seeing this message, you enter a number on the keypad. Use the F key to terminate the number. Your program reads in the number, converts it into a binary, and saves it in a buffer.

Step 5
Compute the gcd of these two numbers and display it on the LCD screen as follows:

> The gcd of xxxx and
> yyyy is zzzz.

where xxxx and yyyy are the numbers that you entered from the keypad and zzzz is the gcd of these two numbers.

8

Timer Functions

8.1 Objectives

After completing this chapter, you should be able to

- Explain the overall structure of the HCS12 timer system

- Use the **input-capture** function to measure the duration of a pulse or the period of a square wave

- Use the input-capture function to measure the duty cycle of a waveform or the phase difference of two waveforms having the same frequency

- Use the input-capture function to measure the frequency of a signal

- Use the **output-compare** function to create a time delay or generate a pulse or periodic waveform

- Use the forced output-compare function

- Use the **pulse accumulator** function to measure the frequency of an unknown signal and count the number of events occurring in an interval

- Configure PWM channels to generate waveforms of certain frequencies and duty cycles

8.2 Why Are Timer Functions Important?

There are many applications that require a dedicated timer system, including

- Time delay creation and measurement
- Period and pulse-width measurement
- Frequency measurement
- Event counting
- Arrival time comparison
- Time-of-day tracking
- Waveform generation
- Periodic interrupt generation

These applications will be very difficult to implement without a dedicated timer system. The HCS12 implements a very complicated timer system to support the implementation of these applications.

At the heart of the HCS12 timer system is the 16-bit timer counter (TCNT). This counter can be started or stopped, as you like. One of the timer functions is called input-capture. The input-capture function copies the contents of the 16-bit timer to a latch when the specified *event* arrives. An event is represented by a signal edge, which could be a rising or a falling edge. By capturing the timer value, many measurements can be made. Some of them include

- Pulse-width measurement
- Period measurement
- Duty cycle measurement
- Event arrival time recording
- Time reference

Another timer function is called output-compare. The output-compare circuit compares the 16-bit timer value with that of the output-compare register in each clock cycle and performs the following operations when they are equal:

- (optionally) Triggers an action on a pin (set to high, set to low, or toggle its signal level)
- Sets a flag in a register
- (optionally) Generates an interrupt request

The output-compare function is often used to generate a time delay, trigger an action at some future time, and generate a digital waveform. The key to using the output-compare function is to make a copy of the 16-bit timer, add a delay to it, and store the sum in an *output-compare register*. The HCS12 has eight output-compare channels, which share the signal pins and registers with input-capture channels.

The third timer function is the pulse accumulator. This circuit is often used to count the events arriving in a certain interval or measure the frequency of an unknown signal.

Certain HCS12 members (for example, A family, C family, and H family) implement a *Standard Timer Module* (TIM) that consists of eight channels of input-capture or output-compare functions and a 16-bit pulse accumulator, PACA. Other HCS members add additional features to the TIM module, and the resultant module is referred to as *Enhanced Captured Timer Module* (ECT). The design goal of the ECT is to support the requirements of automotive, process control, and other applications that need additional timer features.

8.3 Standard Timer Module

The HCS12 standard timer module is illustrated in Figure 8.1. The block diagram in Figure 8.1 shows that the timer module has eight channels of input-capture and output-compare modules. Each channel can be configured to perform input-capture or output-compare, but not both at the same time. The 16-bit counter serves as the base timer for these eight channels. The 16-bit counter needs a clock signal to operate. The clock signal is derived by dividing the E-clock by a prescaler.

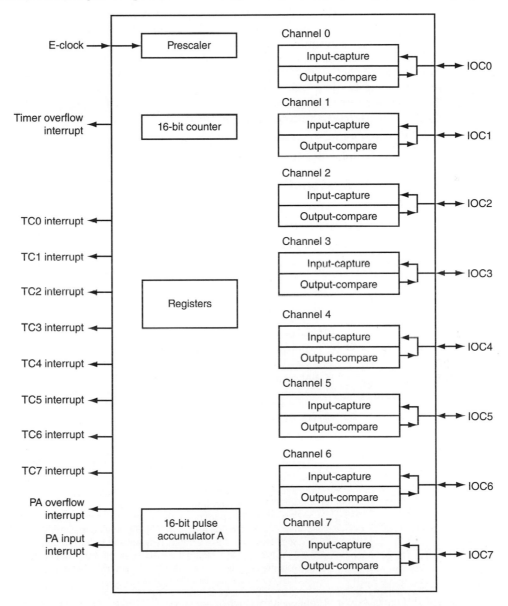

Figure 8.1 ■ HCS12 standard timer (TIM) block diagram

The timer module has a 16-bit pulse accumulator, which can be used to count the number of events that have occurred or to measure the frequency of an unknown signal. The timer module shares the use of Port T pins PT7~PT0. The signal pins IOC7~IOC0 correspond to PT7~PT0. The 16-bit pulse accumulator shares the use of the PT7 pin. The PT7 pin is referred to as the PAI pin when it is used as the pulse-accumulator input.

8.4 Timer Counter Register

The timer counter register (TCNT), a 16-bit register, is required for the functioning of all input-capture and output-compare functions. The user must access this register in one access rather than two separate accesses, to its high byte and low byte. Because the TCNT does not stop during the access operation, the value accessed in a 16-bit read won't be the same as two separate accesses to its high byte and low byte.

There are three registers related to the operation of the TCNT. They are

1. Timer System Control register 1 (TSCR1)
2. Timer System Control register 2 (TSCR2)
3. Timer Interrupt Flag 2 register (TFLG2)

8.4.1 Timer System Control Register 1

The contents of the TSCR1 register are shown in Figure 8.2. The timer counter must be enabled before it can count. Setting bit 7 of the TSCR1 register enables the TCNT to count (up).

	7	6	5	4	3	2	1	0
	TEN	TSWAI	TSFRZ	TFFCA	0	0	0	0
Value after reset	0	0	0	0	0	0	0	0

TEN: timer enable bit

    0 = disables timer; this can be used to save power consumption.
    1 = allows timer to function normally.

TSWAI: timer stop while in wait mode bit

    0 = allows timer to continue running during wait mode.
    1 = disables timer when MCU is in wait mode.

TSFRZ: timer and modulus counter stop while in freeze mode

    0 = allows timer and modulus counter to continue running while in freeze mode.
    1 = disables timer and modulus counter when MCU is in freeze mode.

TFFCA: timer fast flag clear all bits

    0 = allows timer flag clearing to function normally.
    1 = for TFLG1, a read from an input-capture or a write to the output-compare channel causes the corresponding channel flag, CnF, to be cleared. For TFLG2, any access to the TCNT register clears the TOF flag. Any access to the PACN3 and PACN2 registers clears the PAOVF and PAIF flags in the PAFLG register. Any access to the PACN1 and PACN0 registers clears the PBOVF flag in the PBFLG register.

Figure 8.2 ■ Timer System Control register 1 (TSCR1)

The timer counter can be stopped (by setting bit 6 of TSCR1) during the wait mode to save more power. If the timer function is not needed, it can also be stopped during the freeze mode.

All timer interrupt flags can be cleared by writing a 1 to them. However, there is a faster way to clear timer flags. When bit 4 of the TSCR1 register is set to 1, a read from or a write to the appropriate TC register will clear the corresponding flag in the TFLG1 register; any access to the TCNT register will clear the TOF flag, and any access to the PACNT register will clear the PAOVF and PAIF flags in the PAFLG register. This feature reduces the software overhead in a separate clear sequence.

8.4.2 Timer System Control Register 2 (TSCR2)

Timer system control register 2 is another register that controls the operation of the timer counter. Its contents are shown in Figure 8.3. An interrupt will be requested when the TCNT overflows (when TCNT rolls over from $FFFF to $0000) and bit 7 of TSCR2 is set to 1.

The timer counter needs a clock signal to operate. The clock input to the timer counter could be the E-clock prescaled by a factor or the PAI pin input prescaled by a factor. The user has the option to choose the prescale factor. When bit 3 and bit 2 of the PACTL register are 00, the clock to the TCNT is the E-clock prescaled by a factor. The lowest 3 bits of the TSCR2 register specify the prescale factor for the E-clock, as shown in Table 8.1.

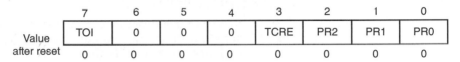

	7	6	5	4	3	2	1	0
	TOI	0	0	0	TCRE	PR2	PR1	PR0
Value after reset	0	0	0	0	0	0	0	0

TOI: timer overflow interrupt enable bit
  0 = interrupt inhibited.
  1 = interrupt requested when TOF flag is set.
TCRE: timer counter reset enable bit
  0 = counter reset inhibited and counter free runs.
  1 = counter reset by a successful output-compare 7.
  If TC7 = $0000 and TCRE = 1, TCNT stays at $0000
  continuously. If TC7 = $FFFF and TCRE = 1, TOF will never
  be set when TCNT rolls over from $FFFF to $0000.

Figure 8.3 ■ Timer System Control register 2

PR2	PR1	PR0	Prescale Factor
0	0	0	1
0	0	1	2
0	1	0	4
0	1	1	8
1	0	0	16
1	0	1	32
1	1	0	64
1	1	1	128

Table 8.1 ■ Timer counter prescale factor

8.4.3 Timer Interrupt Flag 2 Register (TFLG2)

Only bit 7 (TOF) of this register is implemented. When TCNT rolls over from $FFFF to $0000, bit 7 of this register is set to 1. This flag can be cleared by writing a 1 to it.

8.5 Input-Capture Function

Some applications need to know the arrival time of events. In a computer, *physical time* is represented by the count value in a counter, and the occurrence of an event is represented by a signal edge (either the rising or falling edge). The time when an event occurs can be recorded by latching the count value when a signal edge arrives, as illustrated in Figure 8.4.

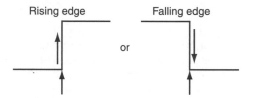

Figure 8.4 ■ Events presented by signal edges

The HCS12 timer system has eight input-capture channels that implement this operation. Each input-capture channel includes a 16-bit input-capture register, an input pin, input edge-detection logic, and an interrupt generation circuit. In the HCS12, physical time is represented by the count in the TCNT.

8.5.1 Input-Capture/Output-Compare Selection

Since input-capture and output-compare functions share signal pins and registers, they cannot be enabled simultaneously. When one is enabled, the other is disabled. The selection is done by the Timer Input-Capture/Output-Compare Select register (TIOS), as shown in Figure 8.5.

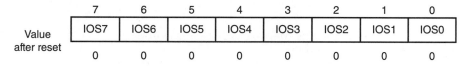

7	6	5	4	3	2	1	0
IOS7	IOS6	IOS5	IOS4	IOS3	IOS2	IOS1	IOS0

Value after reset

| 0 | 0 | 0 | 0 | 0 | 0 | 0 | 0 |

IOS[7:0]: input-capture or output-compare channel configuration bits
  0 = the corresponding channel acts as an input-capture.
  1 = the corresponding channel acts as an output-compare.

Figure 8.5 ■ Timer Input-Capture/Output-Compare Select register (TIOS)

Example 8.1

Write an instruction sequence to enable the output-compare channels 7~4 and input-capture channels 3~0.

Solution: The following instruction sequence will achieve the desired configuration:

```
#include        "c:\miniide\hcs12.inc"
        . . .
        movb    #$F0,TIOS
```

8.5.2 Pins for Input-Capture

Port T has eight signal pins (PT7~PT0) that can be used as input-capture/output-compare or general I/O pins. PT3~PT0 can also be used as the *pulse accumulator input* (PA3~PA0). The user must make sure that the PT3~PT0 pins are enabled for one and only one of these three functions (**OCn, ICn,** and **PAn**). When these pins are not used for timer functions, they can also be used as general-purpose I/O pins. When being used as general I/O pins, the user must use the DDRT register to configure their direction (input or output). When a Port T pin is used as a general-purpose I/O pin, its value is reflected in the corresponding bit in the PTT register.

8.5.3 Registers Associated with Input-Capture

The user needs to specify what signal edge to capture. The edge selection is done via the Timer Control registers 3 and 4, as shown in Figure 8.6.

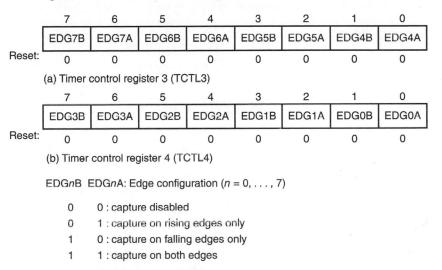

EDG*n*B EDG*n*A: Edge configuration (*n* = 0, . . . , 7)

0	0 : capture disabled
0	1 : capture on rising edges only
1	0 : capture on falling edges only
1	1 : capture on both edges

Figure 8.6 ■ Timer Control registers 3 and 4

When an input-capture channel is selected but capture is disabled, the associated pin can be used as a general-purpose I/O pin.

An input-capture channel can optionally generate an interrupt request on the arrival of a selected edge if it is enabled. The enabling of an interrupt is controlled by the Timer Interrupt Enable register (TIE). The enabling of input-capture 7 through input-capture 0 interrupt is controlled by bit 7 through bit 0 of TIE. When a selected edge arrives at the input-capture pin, the corresponding flag in the Timer Flag register 1 (TFLG1) is set. The contents of TIE and TFLG1 are shown in Figures 8.7 and 8.8, respectively.

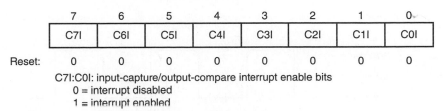

Figure 8.7 ■ Timer Interrupt Enable register (TIE)

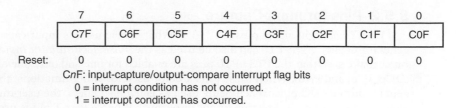

7	6	5	4	3	2	1	0
C7F	C6F	C5F	C4F	C3F	C2F	C1F	C0F

Reset: 0   0   0   0   0   0   0   0

C*n*F: input-capture/output-compare interrupt flag bits
0 = interrupt condition has not occurred.
1 = interrupt condition has occurred.

Figure 8.8 ■ Timer Interrupt Flag register 1 (TFLG1)

To clear a flag in the TFLG1 register, write a 1 to it. However, there is a better way to clear the flag that incurs less overhead. Setting bit 4 (TFFCA) of the TSCR1 register allows clearing a flag by reading the corresponding input-capture register or writing a new value into the output-compare register. This operation is needed for the normal operation of the input-capture or output-compare function.

Each input-capture channel has a 16-bit register (TCx, x = 0 to 7) to hold the count value when the selected signal edge arrives at the pin. This register is also used as the output-compare register when the output-compare function is selected instead.

8.5.4 Input-Capture Applications

There are many applications for the input-capture function. Examples include the following:

- *Event arrival-time recording*. Some applications, for example, swimming competitions, need to compare the arrival times of several different swimmers. The input-capture function is very suitable for this application. The number of events that can be compared is limited by the number of input-capture channels.

- *Period measurement*. To measure the period of an unknown signal, the input-capture function should be configured to capture the timer values corresponding to two consecutive rising or falling edges, as illustrated in Figure 8.9.

- *Pulse-width measurement*. To measure the width of a pulse, two adjacent rising and falling edges are captured, as shown in Figure 8.10.

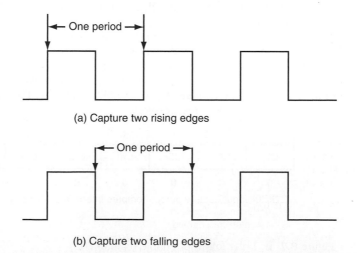

(a) Capture two rising edges

(b) Capture two falling edges

Figure 8.9 ■ Period measurement by capturing two consecutive edges

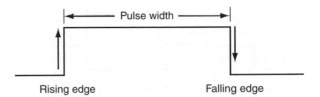

Figure 8.10 ■ Pulse-width measurement using input-capture

- *Interrupt generation.* All input-capture pins can serve as edge-sensitive interrupt sources. Once enabled, interrupts will be generated on the selected edge(s).
- *Event counting.* An event can be represented by a signal edge. An input-capture channel can be used in conjunction with an output-compare function to count the number of events that occur during an interval. An event counter can be set up and incremented by the input-capture interrupt service routine. This application is illustrated in Figure 8.11.

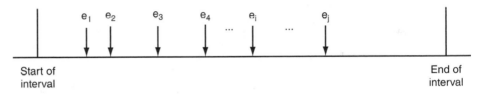

Figure 8.11 ■ Using an input-capture function for event counting

- *Time reference.* In this application, an input-capture function is used in conjunction with an output-compare function. For example, if the designer wants to activate an output signal a certain number of clock cycles after detecting an input event, the input-capture function would be used to record the time at which the edge is detected. A number corresponding to the desired delay would be added to this captured value and stored to an output-compare register. This application is illustrated in Figure 8.12.

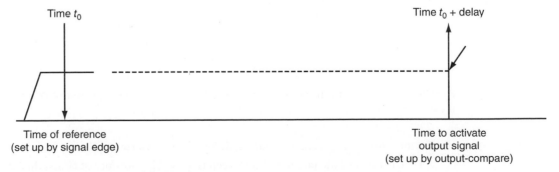

Figure 8.12 ■ A time-reference application

- *Duty cycle measurement.* The *duty cycle* is the percent of time that the signal is high within a period in a periodic digital signal. The measurement of the duty cycle is illustrated in Figure 8.13.

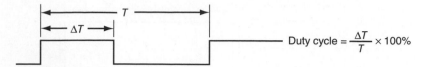

Figure 8.13 ■ Definition of duty cycle

■ *Phase difference measurement.* The *phase difference* is defined as the difference of arrival times (in percentage of a period) of two signals that have the same frequency but do not coincide in their rising and falling edges. The definition of the phase difference is illustrated in Figure 8.14.

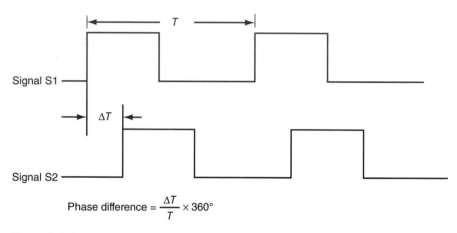

Figure 8.14 ■ Phase difference definition for two signals

The unit used in most of the measurements is the number of clock cycles. When it is desirable, the unit should be converted into an appropriate unit, such as seconds.

Example 8.2

Period measurement. Use the input-capture channel 0 to measure the period of an unknown signal. The period is known to be shorter than 128 ms. Assume that the E-clock frequency is 24 MHz. Use the number of clock cycles as the unit of the period.

Solution: Since the input-capture register is 16-bit, the longest period of the signal that can be measured with the prescaler to TCNT set to 1 is

$2^{16} \div 24\ \text{MHz} = 2.73\ \text{ms}$

To measure a period that is equal to 128 ms, we have two options.

1. Set the prescale factor to 1 and keep track of the number of times that the timer counter overflows.
2. Set the prescale factor to 64 and do not keep track of the number of times that the timer counter overflows.

In this example, we adopt the second approach to make the programming easier. The result of this measurement is in number of clock cycles, and the period of each clock cycle is 2.67 μs.

The circuit connection for the period measurement is shown in Figure 8.15, and the logic flow for the period measurement is shown in Figure 8.16.

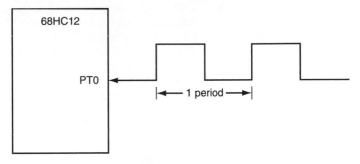

Figure 8.15 ■ Period measurement signal connection

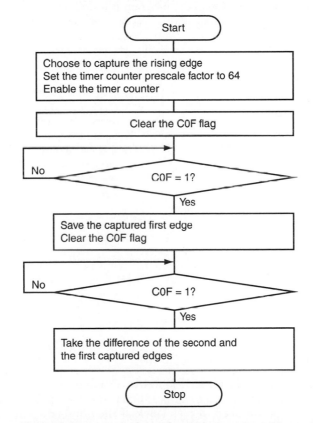

Figure 8.16 ■ Logic flow of period measurement

The assembly program that measures the period using this method is as follows:

```
#include    "c:\miniide\hcs12.inc"
            org    $1000
edge_1st    rmb    2              ; memory to hold the first edge
period      rmb    2              ; memory to store the period
```

```
org       $1500
movb      #$90,TSCR1        ; enable timer counter and enable fast timer flag clear
bclr      TIOS,IOS0         ; enable input-capture 0
movb      #$06,TSCR2        ; disable TCNT overflow interrupt, set prescaler to 64
movb      #$01,TCTL4        ; capture the rising edge of PT0 signal
movb      #COF,TFLG1        ; clear the COF flag
brclr     TFLG1,COF,*       ; wait for the arrival of the first rising edge
ldd       TC0               ; save the  first edge and clear the COF flag
std       edge_1st
brclr     TFLG1,COF,*       ; wait for the arrival of the second edge
ldd       TC0
subd      edge_1st          ; compute the period
std       period
swi
end
```

The C language version of the program is as follows:

```c
#include "c:\cwHCS12\include\hcs12.h"

void main(void)
{
      unsigned int edge1, period;
      TSCR1    = 0x90;           // enable timer counter, enable fast flag clear
      TIOS     &= ~IOS0;         // enable input-capture 0
      TSCR2    = 0x06;           // disable TCNT overflow interrupt, set prescaler to 64
      TCTL4    = 0x01;           // capture the rising edge of the PT0 pin
      TFLG1    = 0x01;           // clear the COF flag */
      while (!(TFLG1 & COF));    // wait for the arrival of the first rising edge
      edge1    = TC0;            // save the first captured edge and clear the COF flag
      while (!(TFLG1 & COF));    // wait for the arrival of the second rising edge
      period   = TC0 - edge1;
      while(1);
}
```

Example 8.3

Write a program to measure the pulse width of a signal connected to the PT0 pin. Assume that the E-clock frequency is 24 MHz.

Solution: We set the prescale factor to 32 and use the clock cycle as the unit of measurement. The period of one clock cycle is 1.33 µs. Since the pulse width could be much longer than 2^{16} clock cycles, we need to keep track of the number of times that the TCNT overflows. Each TCNT overflow adds 2^{16} clock cycles to the pulse width.

Let

$ovcnt$　　= TCNT counter overflow count

$diff$　　 = the difference of two consecutive edges

$edge1$　 = the captured time of the first edge

$edge2$　 = the captured time of the second edge

The pulse width can be calculated by the following equations:

Case 1

$edge2 \geq edge1$
  pulse width $= ovcnt \times 2^{16} + diff$

Case 2

$edge2 < edge 1$
  pulse width $= (ovcnt - 1) \times 2^{16} + diff$

In case 2, the timer overflows at least once even if the pulse width is shorter than $2^{16} - 1$ clock cycles. Therefore, we need to subtract 1 from the timer overflow count in order to get the correct result. The pulse width is obtained by appending the difference of the two captured edges to the TCNT overflow count. The logic flow of the program is shown in Figure 8.17.

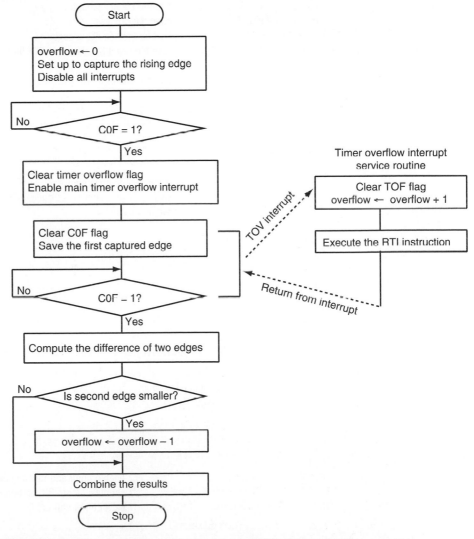

Figure 8.17 ■ Logic flow for measuring pulse width of slow signals

The assembly program that implements this algorithm is as follows:

```
#include "c:\miniide\hcs12.inc"
            org     $1000
edge1       ds.w    1
ovcnt       ds.w    1
PW          ds.w    1
            org     $1500
            movw    #tov_isr,UserTimerOvf; set up TCNT overflow interrupt vector
            lds     #$1500              ; set up stack pointer
            movw    #0,ovcnt
            movb    #$90,TSCR1          ; enable TCNT and fast timer flag clear
            movb    #$05,TSCR2          ; disable TCNT interrupt, set prescaler to 32
            bclr    TIOS,IOS0           ; select IC0
            movb    #$01,TCTL4          ; capture the rising edge
            movb    #$01,TFLG1          ; clear the C0F flag
            brclr   TFLG1,C0F,*         ; wait for the first rising edge
            movw    TC0,edge1           ; save the first edge, clear the C0F flag
            movb    #$80,TFLG2          ; clear the TOF flag
            bset    TSCR2,$80           ; enable TCNT overflow interrupt
            cli                         ;         "
            movb    #$02,TCTL4          ; capture the falling edge on PT0 pin
            brclr   TFLG1,C0F,*         ; wait for the arrival of the falling edge
            ldd     TC0
            subd    edge1
            std     PW
            bcc     next                ; is the second edge smaller?
            ldx     ovcnt               ; second edge is smaller, so decrement
            dex                         ; overflow count by 1
            stx     ovcnt               ;         "
next        swi
tov_isr     movb    #$80,TFLG2          ; clear the TOF flag
            ldx     ovcnt
            inx
            stx     ovcnt
            rti
            end
```

The C language version of this program to be compiled by the CodeWarrior C compiler is as follows:

```
#include         "c:\cwHCS12\include\hcs12.h"
unsigned         diff, edge1, overflow;
unsigned         long pulse_width;
void main(void)
{
        overflow    = 0;
        TSCR1       = 0x90;             // enable timer and fast flag clear
        TSCR2       = 0x05;             // set prescaler to 32, no timer overflow interrupt
        TIOS        &= ~IOS0;           // select input-capture 0
```

```
TCTL4     = 0x01;                    // prepare to capture the rising edge
TFLG1     = COF;                     // clear COF flag
while(!(TFLG1 & COF));               // wait for the arrival of the rising edge
TFLG2     = 0x80;                    // clear TOF flag
TSCR2     |= 0x80;                   // enable TCNT overflow interrupt
asm("cli");
edge1     = TC0;                     // save the first edge
TCTL4     = 0x02;                    // prepare to capture the falling edge
while (!(TFLG1 & COF));              // wait for the arrival of the falling edge
diff      = TC0 − edge1;
if (TC0 < edge1)
overflow   −= 1;
pulse_width = (long)overflow * 65536u + (long)diff;
while (1);
}
interrupt void tovisr(void)
{
      TFLG2     = 0x80;              /* clear the TOF flag */
      overflow++;
}
```

The timer overflow vector is set up by using the following function:

```
extern void near tovisr(void);
#pragma CODE_SEG __NEAR_SEG NON_BANKED
#pragma CODE_SEG DEFAULT                 // Change code section to DEFAULT.
typedef void (*near tlsrFunc)(void);
const tlsrFunc _vect[] @0xFFDE = {
      tovisr
};
```

The required modification for setting up an interrupt vector for the ICC12 C compiler is similar to Example 6.6.

▲

8.6 Output-Compare Function

The HCS12 has eight output-compare channels. Each channel consists of

- A 16-bit comparator
- A 16-bit compare register TCx, $x = 0..7$ (also used as input-capture register)
- An output action pin (PTx—can be pulled up to high, pulled down to low, or toggled)
- An interrupt request circuit
- A **forced-compared** function (CFORCx)
- Control logic

8.6.1 Operation of the Output-Compare Function

One of the major applications of an output-compare function is performing an action at a specific time in the future (when the 16-bit timer counter reaches a specific value). The action might be to toggle a signal, turn on a switch, turn off a valve, and so on. To use an output-compare function, the user

1. Makes a copy of the current contents of the TCNT register
2. Adds to this copy a value that can generate the desired delay
3. Stores the sum into an output-compare register (TCx)

The delay value to be added is dependent on the prescaler to TCNT. The prescaler needs to be set before the delay value is chosen. The user has the option of specifying the action to be activated on the selected output-compare pin by programming the TCTL1 and TCTL2 registers. The comparator compares the value of the TCNT and that of the specified output-compare register (TCx) in every clock cycle (the clock input to TCNT). If they are equal, the specified action on the output-compare pin is activated and the associated status bit in TFLG1 is set to 1. An interrupt request is generated if it is enabled. The 16-bit output-compare register can be read and written any time.

8.6.2 Registers Related to the Output-Compare Function

The actions that can be activated on an output-compare pin are

- Pull up to high
- Pull down to low
- Toggle

The action of an OC pin can be selected by programming the TCTL1 and TCTL2 registers as shown in Figure 8.18. When either OMn or OLn is 1, the pin associated with OCn becomes an output tied to OCn regardless of the state of the associated DDRT bit.

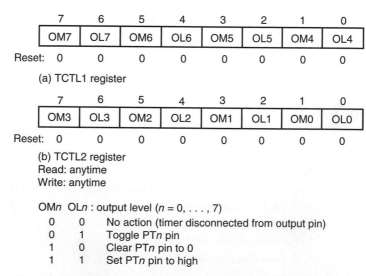

Figure 8.18 ■ Timer Control registers 1 and 2 (TCTL1 and TCTL2)

A successful compare sets the corresponding flag bit in the 8-bit TFLG1 register. An interrupt is generated if it is enabled. An output-compare interrupt is enabled by setting the corresponding bit in the TIE register. The same bit enables either the input-capture or output-compare interrupt, depending on which one is selected.

The OC7 channel has the option to reset TCNT when the TC7 register matches the TCNT. This option is enabled by setting the TCRE bit of the TSCR2 register. This feature makes the TCNT a programmable modulus counter.

8.6.3 Applications of the Output-Compare Function

An output-compare function can be programmed to perform a variety of functions. Generation of a single pulse, a square wave, and a specific delay are among the most popular applications.

To generate a periodic square waveform, the user may use the output-compare function to continuously toggle the selected Port T pin with appropriate delay added in each output-compare operation. The first output-compare operation must be started by the main function whereas the remaining output-compare operations may be performed by the main function or the output-compare interrupt service routine depending on the approach. The following example uses the interrupt-driven approach.

Example 8.4

Generate an active high 1-kHz digital waveform with a 30 percent duty cycle from the PT5 pin. Use the interrupt-driven method to check the success of the output-compare operation. The frequency of the E-clock is 24 MHz.

Solution: An active high 1-kHz waveform with a 30 percent duty cycle is shown in Figure 8.19.

Figure 8.19 ■ A 1-kHz 30 percent duty cycle waveform

The logic flow of this problem is illustrated in Figure 8.20. Suppose we set the prescale factor to 8 so that the period of the clock input to TCNT is set to 1/3 μs. Then the intervals of the PT0 signal to be high and low in one period would be 900 and 2100 clock cycles, respectively.

The algorithm first pulls the PT5 pin to high in a very short period of time using the output-compare operation and then changes the pin action to toggle and enable the OC5 interrupt. The OC5 interrupt simply starts the next output-compare operation.

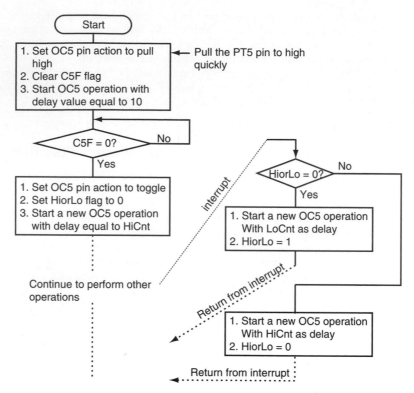

Figure 8.20 ■ The program logic flow for digital waveform generation

The following assembly program implements this algorithm using the **OC5** function:

```
#include        "c:\miniide\hcs12. inc"
HiCnt  equ      1200                          ; delay count for high interval of the waveform
LoCnt  equ      1800                          ; delay count for low interval of the waveform
       org      $1000
HIorLO ds.b     1                             ; flag to select HiCnt (1) or LoCnt (0)
       org      $1500
       lds      #$1500                        ; establish stack pointer
       movw     #OC5ISR,UserTimerCh5          ; set up OC0 interrupt vector
       movb     #$90,TSCR1                    ; enable TCNT and fast timer flag clear
       movb     #$03,TSCR2                    ; set TCNT clock prescaler to 8
       bset     TIOS,BIT5                     ; enable OC5
       movb     #$0C,TCTL1                    ; configure the OC5 pin action to pull high
       ldd      TCNT                          ; start OC5 with delay count equal to 10 and
       addd     #10                           ; pin action to pull high
       std      TC5
       brclr    TFLG1,C5F,*                   ; wait until C5F is set to 1 and PT5 pin pulled high
       ldd      TC5                           ; start another OC5 operation with
       addd     #HiCnt                        ; delay count set to HiCnt
       std      TC5                           ; "
```

```
                movb        #$04,TCTL1          ; change pin action to toggle
                clr         HIorLO              ; LoCnt will be the delay count next time
                bra         $                   ; prepare to perform other operations
;****************************************************************************************
; OC5 interrupt service routine starts a new OC5 operation with delay count equal to
; HiCnt or LoCnt depending on whether HIorLO flag is 1 or 0. After starting a new OC5
; operation, toggle HIorLO flag.
;****************************************************************************************
OC5ISR
                tst         HIorLO              ; which delay count should be added?
                beq         addLow              ; if 0 then select LoCnt
                ldd         TC5                 ; select HiCnt as the delay count for
                addd        #HiCnt              ; the new OC5 operation
                std         TC5                 ;          "
                clr         HIorLO              ; toggle HIorLo flag
                rti
addLow          ldd         TC5                 ; select LoCnt as the delay count for
                addd        #LoCnt              ; the new OC5 operation
                std         TC5                 ;          "
                movb        #1,HIorLO           ; toggle HIorLO flag
                rti
                end
```

The C language version of the program is as follows:

```
#include "c:\cwHCS12\include\hcs12.h"
#define  HiCnt     1200
#define  LoCnt     1800
char HiorLo;
void main (void)
{
            TSCR1       = 0x90;        // enable TCNT and fast timer flag clear
            TSCR2       = 0x03;        // disable TCNT interrupt, set prescaler to 8
            TIOS        | = OC5;       // enable OC5 function
            TCTL1       = 0x0C;        // set OC5 action to pull high
            TFLG1       = 0xFF;        // clear all CxF flags
            TC5         = TCNT + 10;
            while(TFLG1 & C5F);        // wait until C5F is set
            TCTL1       = 0x04;        // set OC5 pin action to toggle
            TC5         += HiCnt;      // start an new OC5 operation
            HiorLo      = 0;           // add LoCnt for the next OC5 operation
            TIE         = 0x20;        // enable OC5 interrupt locally
            asm("cli");                // enable interrupt globally
            while(1);
}
interrupt void oc5ISR (void)
{
        if(HiorLo){
            TC5         += HiCnt;
            HiorLo      = 0;
        }
```

```
        else{
                TC5 += LoCnt;
                HiorLo = 1;
        }
}
```

The OC5 interrupt vector is set up using the following function (in vectors.c):

```
extern void near oc5ISR(void);
#pragma CODE_SEG __NEAR_SEG NON_BANKED
#pragma CODE_SEG DEFAULT            // Change code section to DEFAULT.
typedef void (*near tlsrFunc)(void);
const tlsrFunc _vect[] @0xFFE4 = {
        oc5ISR
};
```

Using the OC5 interrupt to restart the subsequent OC5 operations allows the HCS12 to continue to perform other operations.

▲

Example 8.5

▼

Write a function to generate a time delay that is a multiple of 1 ms. Assume that the E-clock frequency is 24 MHz. The number of milliseconds is passed in Y. Also write an instruction sequence to test this function.

Solution: There are many ways to create a 1-ms time delay using the output-compare function. One method is

- Set the prescaler to TCNT to 8.
- Perform the number of output-compare operations given in Y with each operation creating a 1-ms time delay. The number to be added to the TC0 register is 3000 $(3000 \times 8 \div 24,000,000 = 1 \text{ ms})$.

The corresponding assembly function is as follows:

```
delayby1ms      pshd
                movb        #$90,TSCR1              ; enable TCNT and fast flag clear
                movb        #$03,TSCR2              ; configure prescaler to 8
                bset        TIOS,OC0               ; enable OC0
                ldd         TCNT
again0          addd        #3000                  ; start an output-compare operation
                std         TC0                    ; with 1-ms time delay
wait_lp0        brclr       TFLG1,OC0,wait_lp0
                ldd         TC0
                dbne        y,again0
                puld
                rts
```

The C language version of the function and its test program are as follows:

```
#include "c:\cwHSC12\include\hcs12.h"
void delayby1ms(int k);
void main (void)
```

383

```
{
        unsigned char count;
        DDRB = 0xFF;        /* configure Port b for output */
        count = 0;
        PTB = count;
        while(1) {
                delayby1ms(200);
                count++;
                PTB = count;
        }
}
void delayby1ms(int k)
{
        int ix;
        TSCR1 = 0x90;        /* enable TCNT and fast timer flag clear */
        TSCR2 = 0x03;        /* disable timer interrupt, set prescaler to 8*/
        TIOS |= OC0;         /* enable OC0 */
        TC0 = TCNT + 3000;
        for(ix = 0; ix < k; ix++) {
                while(!(TFLG1 & C0F));
                TC0 += 3000;
        }
        TIOS &= ~OC0;        /* disable OC0 */
}
```

Example 8.6

Generate a sequence of pulses using an OC function. Write a program to generate a number of pulses with the specified high interval duration (12 ms) and low interval duration (8 ms). Use the interrupt-driven approach so that the CPU can perform other operations.

Solution: Let the number of pulses to be generated, the high interval duration, low interval duration, the flag to select *DelayHi* or *DelayLo*, and number of **OC0** operations to be performed be *NN*, *DelayHi*, *DelayLo*, *HiorLo*, and *pcnt*, respectively.

The algorithm for generating a sequence of pulses with the specified duration of high interval and low interval is as follows:

Step 1:
Pull the PT0 pin high quickly using the OC0 operation.

Step 2
Change the OC0 pin action to toggle. Start the next OC0 operation with delay equal to *DelayHi*.

Step 3
$pcnt \leftarrow 2 * NN - 1. HiorLo \leftarrow 0.$

Step 4
Enable OC0 interrupt.

Step 5

The main program continues to perform other operations.

The interrupt service routine of OC0 performs the following operations:

If (*HiorLo* == 1)

        Start an OC0 operation using *DelayHi* as the delay count

        HiorLo ← 0

        Return from interrupt

Else

        Start an OC0 operation using *DelayLo* as the delay count

        HiorLo ← 1

        Return from interrupt.

The assembly program that implements this algorithm is as follows:

```
#include         "c:\miniide\hcs12.inc"
DelayHi   equ    18000              ; pulse high interval duration
DelayLo   equ    12000              ; pulse low interval duration
NN        equ    10                 ; number of pulses to be created
          org    $1000
pcount    ds.b   1                  ; number of OC0 operations remaining to be performed
HiorLo    ds.b   1                  ; flag to choose DelayHi(1) or DelayLo(0)
          org    $1500
          lds    #$1500
          movw   #oc0ISR,UserTimerCh0   ; set up OC0 interrupt vector
          movb   #$90,TSCR1
          movb   #$04,TSCR2
          bset   TIOS,OC0               ; enable OC0
          movb   #COF,TFLG1             ; clear COF flag
          movb   #$03,TCTL2             ; set OC0 pin action to pull high
          ldd    TCNT                   ; use OC0 operation to pull TC0 pin high
          addd   #12                    ; quickly
          std    TC0                    ;    "
          brclr  TFLG1,COF,*            ; wait until COF flag is set
          movb   #$01,TCTL2             ; set OC0 pin action to toggle
          ldd    TC0                    ; start next OC0 operation with
          addd   #DelayHi               ; delay set to DelayHi
          std    TC0                    ;    "
          movb   2*NN-1,pcount          ; prepare to perform pcount OC0 operations
          clr    HiorLo                 ; next OC0 operation use DelayLo as delay
          bra    $
oc0ISR    ldaa   HiorLo                 ; check the flag to choose delay count
          beq    pulseLo                ; if flag is 0, then go and use DelayLo
          ldd    TC0                    ; start an OC0 operation and use
          addd   #DelayHi               ; DelayHi as delay count
          std    TC0                    ;    "
          movb   #0,HiorLo              ; toggle the flag
          bra    decCnt
pulseLo   ldd    TC0                    ; start an OC0 operation and use
          addd   #DelayLo               ; DelayLo as delay count
          std    TC0                    ;    "
          movb   #1,HiorLo              ; toggle the flag
```

```
decCnt      dec     pcount
            bne     quit
            movb    #0,TIE          ; disable OC0 and
            bclr    TIOS,$01        ; its interrupt
quit        rti
            end
```

The C language version of the program to be compiled by CodeWarrior is as follows:

```
#include        "c:\cwHCS12\include\hcs12.h"
#include        "c:\cwHCS12\include\SetClk.h"
#define     DelayHi     18000       // high time of the pulses to be created
#define     DelayLo     12000       // low time of the pulses to be created
#define     NN          10          // number of pulses to be created
int         pcnt;                   // pulse count
char   HiorLo;                      // flag to choose
void main(void)
{
        SetClk8();
        TSCR1   = 0x90;             // enable TCNT and faster timer flag clear
        TSCR2   = 0x04;             // set TCNT clock input prescaler to 16
        TFLG1   = C0F;              // clear C0F flag
        TIOS    |= OC0;             // enable OC0
        TCTL2   = 0x03;             // set OC0 pin action to be pull high
        TC0     = TCNT + 16;        // pull PT0 pin high quickly
        while(!(TFLG1 & C0F));      //    "
        pcnt    = 2 * NN − 1;       // prepare to create NN pulses
                                    //    (need to toggle 2*NN − 1 times)

        TCTL2   = 0x01;             // set OC0 pin action to be toggle
        TC0     += DelayHi;         // start the second OC0 operation
        HiorLo  = 0;                // next time use DelayLo as delay count of OC0 operation
        TIE     |= C0I;             // enable TC0 interrupt
        asm("cli");                 //    "
        while (1);                  // do nothing or do something else
}
interrupt void tc0ISR(void) {
        if(HiorLo){
                TC0     += DelayHi;
                HiorLo  = 0;
        } else {
                TC0     += DelayLo;
                HiorLo  = 1;
        }
        pcnt−−;
        if(pcnt  == 0){
                TIE     = 0;        // disable OC0 interrupt
                TIOS    &= 0xFE;    // disable OC0
        }
}
```

8.6.4 Making Sound Using the Output-Compare Function

Using the output-compare function to make sound is easy. A sound can be made by creating a digital waveform of appropriate frequency and using it to drive a speaker or a buzzer. A small speaker of 8-Ω resistance that consumes between 10 and 20 mW can produce clear sound. The next example illustrates how to use an output-compare channel to generate a siren.

Example 8.7

Describe the circuit for making a sound and write a program that uses an output-compare channel to generate a siren that oscillates between 300 and 1200 Hz.

Solution: A simple 8-mW speaker (or a buzzer on a demo board) has two terminals: one terminal is for signal input, whereas the other terminal is for ground connection. The circuit connection for siren generation is shown in Figure 8.21.

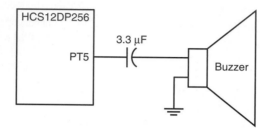

Figure 8.21 ■ Circuit connection for a buzzer

The algorithm for generating the siren is as follows:

Step 1
Enable an appropriate OC channel (OC5 in this example) to drive the speaker circuit.

Step 2
Start an output-compare operation and enable its interrupt with a delay count equal to half the period of the siren.

Step 3
Wait a certain amount of time (say half a second). During the waiting period, interrupts will be requested by the output-compare match many times. The interrupt service routine simply starts the next output-compare operation and then returns.

Step 4
At the end of the delay, choose a different delay count for the output-compare operation so the siren sound with a different frequency can be generated.

Step 5
Wait for the same amount of time as in step 3. Again, the interrupt caused by the output-compare match will be requested many times. At the end of the delay, switch back to the delay count used in step 2.

Step 6
Go to step 2.

The assembly program that implements this algorithm for an HCS12DP256 running with a 24-MHz E-clock is as follows:

```
#include "c:\miniide\hcs12.inc"
hi_freq     equ     1250            ; delay count for 1200 Hz (with 1:8 prescaler)
lo_freq     equ     5000            ; delay count for 300 Hz (with 1:8 prescaler)
toggle      equ     $04             ; value to toggle the TC5 pin
            org     $1000
delay       ds.w    1               ; store the delay for output-compare operation
            org     $1500
            lds     #$1500
            movw    #oc5_isr,UserTimerCh5 ; initialize the interrupt vector entry
            movb    #$90,TSCR1      ; enable TCNT, fast timer flag clear
            movb    #$03,TSCR2      ; set main timer prescaler to 8
            bset    TIOS,OC5        ; enable OC5
            movb    #toggle,TCTL1   ; select toggle for OC5 pin action
            ldd     #hi_freq
            std     delay           ; use high-frequency delay count first
            ldd     TCNT            ; start the high-frequency sound
            addd    delay           ;       "
            std     TC5             ;       "
            bset    TIE,OC5         ; enable OC5 interrupt
            cli                     ;       "
forever     ldy     #5              ; wait for half a second
            jsr     delayby100ms    ;       "
            movw    #lo_freq,delay  ; switch to low-frequency delay count
            ldy     #5
            jsr     delayby100ms
            movw    #hi_freq,delay  ; switch to high-frequency delay count
            bra     forever
oc5_isr     ldd     TC5
            addd    delay
            std     TC5
            rti
#include  c:\miniide\delay.asm"
            end
```

In this program, both the OC0 and OC5 channels rely on the TCNT for timing control. One should use the same prescaler throughout all the timing operations. The C language version of the program to generate the siren to be compiled by CodeWarrior is as follows:

```
#include "c:\cwHCS12\include\hcs12.h"
#include "c:\cwHCS12\include\delay.h"
#define HiFreq    1250
#define LoFreq    5000
int         dlycnt;                 // delay count for OC5 operation
void        oc5ISR(void);
void        main(void)
{
            TSCR1   = 0x90;         // enable TCNT and fast timer flag clear
            TSCR2   = 0x03;         // set prescaler to TCNT to 1:8
```

```
        TIOS      | = BIT5;              // enable OC5
        TCTL1     = 0x04;               // select toggle for OC5 pin action
        dlycnt    = HiFreq;            // use high-frequency delay count first
        TC5       = TCNT + dlycnt;     // start an OC5 operation
        TIE       | = BIT5;            // enable TC5 interrupt
        asm("cli");
        while(1) {
          delayby100ms(5);             // wait for half a second
          dlycnt   = LoFreq;           // switch to low-frequency tone
          delayby100ms(5);             // wait for half a second
          dlycnt   = HiFreq;           // switch to high-frequency tone
        }
}
interrupt void oc5ISR(void)
{
        TC5 += dlycnt;
}
```

The required interrupt-vector setup function (**vectors_0807.c**) is as follows:

```
extern void near oc5ISR(void);
#pragma CODE_SEG __NEAR_SEG NON_BANKED
#pragma CODE_SEG DEFAULT            // Change code section to DEFAULT.
typedef void (*near tIsrFunc)(void);
const tIsrFunc _vect[] @0xFFE4 = {
        oc5ISR
};
```

A siren can be considered as a song with only two notes. A song can be created by modifying the siren program. To make the switching of frequency easier, the whole score should be stored in a table. A note has two components: one is the frequency and the other is the duration. The duration of a quarter note is about 0.4 s. The durations of other notes can be derived proportionally. The complete list of the music notes and their frequencies is in Appendix F.

To play a song from the speaker, one places the frequencies and durations of all the notes in a music score in a table. For every note, the user program uses the output-compare function to generate the digital waveform with the specified frequency and duration. The following example illustrates this idea.

It is common for a song to have several contiguous identical notes in different places. When playing multiple contiguous identical notes, the buzzer will generate one long note instead of multiple notes. This problem can be avoided by inserting a short note with an inaudible frequency. The following example uses this method:

Example 8.8

Use the circuit shown in Figure 8.21 and write a program to play "The Star-Spangled Banner." Assume that the demo board is running with a 24-MHz E-clock.

Solution: The following assembly program plays the national anthem:

```
#include "c:\miniide\hcs12.inc"
G3    equ    7653    ; delay count to generate G3 note (with 1:8 prescaler)
B3    equ    6074    ; delay count to generate B3 note (with 1:8 prescaler)
```

```
C4       equ     5733                    ; delay count to generate C4 note (with 1:8 prescaler)
C4S      equ     5412                    ; delay count to generate C4S note
D4       equ     5108                    ; delay count to generate D4 note (with 1:8 prescaler)
E4       equ     4551                    ; delay count to generate E4 note (with 1:8 prescaler)
F4       equ     4295                    ; delay count to generate F4 note (with 1:8 prescaler)
F4S      equ     4054                    ; delay count to generate F4S note (with 1:8 prescaler)
G4       equ     3827                    ; delay count to generate G4 note (with 1:8 prescaler)
A4       equ     3409                    ; delay count to generate A4 note (with 1:8 prescaler)
B4F      equ     3218                    ; delay count to generate B4F note (with 1:8 prescaler)
B4       equ     3037                    ; delay count to generate B4 note (with 1:8 prescaler)
C5       equ     2867                    ; delay count to generate C5 note (with 1:8 prescaler)
D5       equ     2554                    ; delay count to generate D5 note (with 1:8 prescaler)
E5       equ     2275                    ; delay count to generate E5 note (with 1:8 prescaler)
F5       equ     2148                    ; delay count to generate F5 note (with 1:8 prescaler)
ZZ       equ     20                      ; delay count to generate an inaudible note
notes    equ     118
toggle   equ     $04                     ; value to toggle the TC5 pin
         org     $1000
delay    ds.w    1                       ; store the delay for output-compare operation
rep_cnt  ds.b    1                       ; repeat the song this many times
ip       ds.b    1                       ; remaining notes to be played
         org     $1500
; establish the SRAM vector address for OC5
         movw    #oc5_isr,UserTimerCh5
         lds     #$1500
         movb    #$FF,DDRB
         movb    #$90,TSCR1              ; enable TCNT, fast timer flag clear
         movb    #$03,TSCR2              ; set main timer prescaler to 8
         bset    TIOS,OC5                ; enable OC5
         movb    #toggle,TCTL1           ; select toggle for OC5 pin action
         ldx     #score                  ; use as a pointer to score table
         ldy     #duration               ; points to duration table
         movb    #1,rep_cnt              ; play the song twice
         movb    #notes,ip
         movw    2,x+,delay              ; start with 0th note
         ldd     TCNT                    ; play the first note
         addd    delay                   ;    "
         std     TC5                     ;    "
         bset    TIE,C5I                 ; enable OC5 interrupt
         cli                             ;    "
forever  pshy                            ; save duration table pointer in stack
         ldy     0,y                     ; get the duration of the current note
         jsr     delayby10ms             ;    "
         puly                            ; get the duration pointer from stack
         iny                             ; move the duration pointer
         iny                             ;    "
         ldd     2,x+                    ; get the next note, move pointer
         std     delay                   ;    "
         dec     ip
         bne     forever
```

```
              dec      rep_cnt
              beq      done           ; if not finished playing, reestablish
              ldx      #score         ; pointers and loop count
              ldy      #duration      ;           "
              movb     #notes,ip      ;           "
              movw     0,x,delay      ; get the first note delay count
              ldd      TCNT           ; play the first note
              addd     delay          ;           "
              std      TC5
              bra      forever
done          swi
```
; ***
; The OC5 interrupt service routine simply starts a new OC5 operation.
; ***
```
oc5_isr       ldd      TC5
              addd     delay
              std      TC5
              rti
```
; ***
; The following subroutine creates a time delay which is equal to [Y] times
; 10 ms. The timer prescaler is 1:8.
; ***
```
delayby10ms
              bset     TIOS,OC0       ; enable OC0
              ldd      TCNT
again1        addd     #30000         ; start an output-compare operation
              std      TC0            ; with 10-ms time delay
              brclr    TFLG1,C0F,*
              ldd      TC0
              dbne     y,again1
              bclr     TIOS,OC0       ; disable OC0
              rts
```
; ***
; This table determines the frequency of each note.
; ***
```
score         dc.w     D4,B3,G3,B3,D4,G4,B4,A4,G4,B3,C4S
              dc.w     D4,ZZ,D4,ZZ,D4,B4,A4,G4,F4S,E4,F4S,G4,ZZ,G4,D4,B3,G3
              dc.w     D4,B3,G3,B3,D4,G4,B4,A4,G4,B3,C4S,D4,ZZ,D4,ZZ,D4
              dc.w     B4,A4,G4,F4S,E4,F4S,G4,ZZ,G4,D4,B3,G3,B4,ZZ,B4
              dc.w     B4,C5,D5,ZZ,D5,C5,B4,A4,B4,C5,ZZ,C5,ZZ,C5,B4,A4,G4
              dc.w     F4S,E4,F4S,G4,B3,C4S,D4,ZZ,D4,G4,ZZ,G4,ZZ,G4,F4S
              dc.w     E4,ZZ,E4,ZZ,E4,A4,C5,B4,A4,G4,ZZ,G4,F4S,D4,ZZ,D4
              dc.w     G4,A4,B4,C5,D5,G4,A4,B4,C5,A4,G4
```
; ***
; Each of the following entries multiplied by 10 ms gives the duration of a note.
; ***
```
duration      dc.w     30,10,40,40,40,80,30,10,40,40,40
              dc.w     80, 3,20,3,20,60,20,40,80,20,20,40,3,40,40,40,40
              dc.w     30,10,40,40,40,80,30,10,40,40,40,80,3,20,3,20
              dc.w     60,20,40,80,20,20,40,3,40,40,40,40,20,3,20
```

```
        dc.w      40,40,40,3,80,20,20,40,40,40,3,80,3,40,60,20,40
        dc.w      80,20,20,40,40,40,80,3,40,40,3,40,3,20,20
        dc.w      40, 3,40,3,40,40,20,20,20,20,3,40,40,20,3,20
        dc.w      60,20,20,20,80,20,20,60,20,40,80
        end
```

The C language version of the program to be compiled by CodeWarrior and also support a serial monitor is as follows:

```
#include "c:\cwHCS12\include\hcs12.h"
#include "c:\cwHCS12\include\SetClk.h"
#define      G3      7653
#define      B3      6074
#define      C4      5733
#define      C4S     5412
#define      D4      5108
#define      E4      4551
#define      F4      4295
#define      F4S     4054
#define      G4      3827
#define      A4      3409
#define      B4F     3218
#define      B4      3037
#define      C5      2867
#define      D5      2554
#define      E5      2275
#define      F5      2148
#define      ZZ      20          // delay count to create an inaudible sound
#define      toggle  0x04        // value to toggle OC5 pin
int delay;
unsigned int score[] = {
                D4,B3,G3,B3,D4,G4,B4,A4,G4,B3,C4S,
                D4,ZZ,D4,ZZ,D4,B4,A4,G4,F4S,E4,F4S,G4,ZZ,G4,D4,B3,G3,
                D4,B3,G3,B3,D4,G4,B4,A4,G4,B3,C4S,D4,ZZ,D4,ZZ,D4,
                B4,A4,G4,F4S,E4,F4S,G4,ZZ,G4,D4,B3,G3,B4,ZZ,B4,
                B4,C5,D5,ZZ,D5,C5,B4,A4,B4,C5,ZZ,C5,77,C5,B4,A4,G4,
                F4S,E4,F4S,G4,B3,C4S,D4,ZZ,D4,G4,ZZ,G4,ZZ,G4,F4S,
                E4,ZZ,E4,ZZ,E4,A4,C5,B4,A4,G4,ZZ,G4,F4S,D4,ZZ,D4,
                G4,A4,B4,C5,D5,G4,A4,B4,C5,A4,G4,0};
unsigned int dur[] = {
                30,10,40,40,40,80,30,10,40,40,40,
                80,3,20,3,20,60,20,40,80,20,20,40,3,40,40,40,40,
                30,10,40,40,40,80,30,10,40,40,40,80,3,20,3,20,
                60,20,40,80,20,20,40,3,40,40,40,60,20,3,20,
                40,40,40,3,80,20,20,40,40,40,3,80,3,40,60,20,40,
                80,20,20,40,40,40,80,3,40,40,3,40,3,20,20,
                40,3,40,3,40,40,20,20,20,20,3,40,40,20,3,20,
                60,20,20,20,80,20,20,60,20,40,80};
void delayby10ms(int kk);
void main(void) {
        int       j;
        SetClk8();                         // set E clock to 24 MHz
```

```
        TSCR1    = 0x90;              // enable TCNT to fast timer flag clear
        TSCR2    = 0x03;              // set TCNT clock prescaler to 8
        TIOS     |= OC5;             // enable OC5
        TCTL1    = toggle;            // set OC5 pin action to toggle
        delay    = score[0];
        j        = 0;
        TC0      = TCNT + delay;      //start an OC0 operation
        TIE      |= C5I;             //enable TC5 interrupt
        asm("cli");
        while(score[j]){
            delay = score[j];
            delayby10ms(dur[j]);
            j++;
        }
        TIOS &= 0xDF; //disable OC5
        while(1);
    }
    void delayby10ms(int kk) {
        char ix;
        TIOS  |= 0x40;               //enable OC6
        TC6   = TCNT + 30000;        //start OC6 operation with 20-ms delay
        for(ix = 0; ix < kk; ix++){
            while(!(TFLG1 & 0x40));
            TC6 += 30000;
        }
        TIOS  &= 0xBF; //disable OC6
    }
    interrupt void tc5ISR(void) {
        TC5 += delay;
    }
```

The function for setting up the interrupt vector for OC5 is as follows:

```
    extern void near tc5ISR(void);
    #pragma        CODE_SEG __NEAR_SEG NON_BANKED
    #pragma        CODE_SEG DEFAULT         // Change code section to DEFAULT.
    typedef        void (*near tIsrFunc)(void);
    const tIsrFunc _vect[] @0xFFE4 = {
        tc5ISR
    };
```

There are at least two drawbacks in using this method to play a song.

1. A periodic square waveform is only an approximation to an actual note. Those who have learned communication theory know that a periodic square waveform also contains harmonics with higher frequencies. The solution to this problem is beyond the scope of this text.

2. There is no loudness control. Using the PWM function rather than the output-compare function and adjusting the duty cycle of each note is one of the possible solutions to this problem.

8.6.5 Using OC7 to Control Multiple Output-Compare Functions

The output-compare function OC7 is special because it can control up to eight output-compare functions at the same time. The register OC7M specifies the output-compare channels to be controlled by OC7. The value that any PTx ($x = 0, \ldots, 7$) pin can assume when the value of TC7 equals that of TCNT is specified by the OC7D register. To control an output-compare pin using OC7, the user sets the corresponding bit in the OC7M register. When a successful OC7 compare is made, each affected pin assumes the value of the corresponding bit of the OC7D register. The contents of the OC7M and OC7D registers are shown in Figures 8.22 and 8.23.

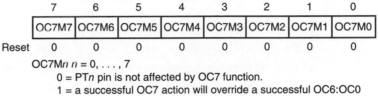

	7	6	5	4	3	2	1	0
	OC7M7	OC7M6	OC7M5	OC7M4	OC7M3	OC7M2	OC7M1	OC7M0
Reset	0	0	0	0	0	0	0	0

OC7Mn $n = 0, \ldots, 7$
   0 = PTn pin is not affected by OC7 function.
   1 = a successful OC7 action will override a successful OC6:OC0
       compare action during the same cycle and the OCn action taken
       will depend on the corresponding OC7D bit.

Figure 8.22 ■ Output-Compare 7 Mask register (OC7M)

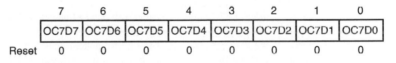

	7	6	5	4	3	2	1	0
	OC7D7	OC7D6	OC7D5	OC7D4	OC7D3	OC7D2	OC7D1	OC7D0
Reset	0	0	0	0	0	0	0	0

Figure 8.23 ■ Output-Compare 7 Data register (OC7D)

For the OC7 pin, the HCS12 document does not specify what will happen when both the TCTL1 register and the OC7M:OC7D register pair specify the OC7 pin action on a successful OC7 compare. Avoid using this combination.

For OC0:OC6, when the OC7Mn ($n = 0, \ldots, 6$) bit is set, a successful OC7 action will override a successful OC6:OC0 compare action during the same cycle; therefore, the OCn action taken will depend on the corresponding OC7D bit. This feature allows a PTn pin to be controlled by two output-compare functions simultaneously [OC7 and OCn ($n = 0, \ldots, 6$)].

Example 8.9
▼

What value should be written into OC7M and OC7D if one wants pins PT2, PT3, and PT4 to assume the values of 1, 0, and 1, respectively, when OC7 compare succeeds?

Solution: Bits 4, 3, and 2 of OC7M must be set to 1, and bits 4, 3, 2 of OC7D should be set to 1, 0, and 1, respectively. The following instruction sequence set up these values:

```
movb    #$1C,OC7M
movb    #$14,OC7D
```

The following C statements achieve the same goal:

```
OC7M = 0x1C;
OC7D = 0x14;
```

▲

The ability to control multiple output-compare pins is very useful in applications that require multiple actions to be triggered simultaneously in the near future.

Example 8.10

▼

Suppose that there is an application that requires the following operations to be triggered 50 ms later:

- Turn off the light controlled by the TC5 pin
- Turn on the temperature sensor controlled by the TC4 pin
- Turn off the heater controlled by the TC3 pin
- Turn on the music controlled by the TC2 pin

Write an instruction sequence to perform the desired operation.

Solution: The values to be written into OC7M and OC7D are $3C and $14, respectively. Set the prescaler to TCNT to 64. Then the value to be added to create a 50-ms delay can be computed as follows:

Delay $= (24,000,000 \div 64) \times 0.05 = 18,750$

The following instruction sequence can perform the desired operations 50 ms later:

```
movb    #$90,TSCR1      ; enable TCNT and fast timer flag clear
movb    #$06,TSCR2      ; set prescaler to 64
movb    #$3C,OC7M       ; allow OC7 to control TC5, TC4, TC3, and TC2 pins
movb    #$14,OC7D       ; pin actions for TC5, TC4, TC3, and TC2
ldd     TCNT            ; start an OC7 operation with 50 ms as delay
addd    #18750          ;    "
std     TC7             ; start the OC7 operation
```

▲

Two output-compare functions can control the same pin simultaneously. Thus, OC7 can be used in conjunction with one or more other output-compare functions to achieve even more time flexibility. We can generate a digital waveform with a given duty cycle by using the OC7 and any other output-compare function.

Example 8.11

▼

Use OC7 and OC0 together to generate a 2-kHz digital waveform with a 40 percent duty cycle on the PT0 pin. Assume that the E-clock frequency is 24 MHz.

Solution: We set the prescale factor to 8. The period of a 2-kHz waveform is 0.0005 s. The period of this signal corresponds to $0.0005 \times (24,000,000 \div 8) = 1500$ cycles. The high interval in one period of the waveform is 600 clock cycles, whereas the low interval is 900 clock cycles.

The idea in using two OC functions in generating the waveform is as follows:

- Use OC0 (or OC7) to pull the PT0 pin to high every 1500 clock cycles.
- Use OC7 (or OC0) to pull the PT0 to low after it is high for 600 clock cycles.

We use the interrupt-driven approach to implement this algorithm. After starting the first action on OC7 and OC0, the program will stay in a wait loop to wait for interrupts to be requested from OC7 and OC0. The interrupt service routine will simply start the next output-compare operation with 15,000 as the delay count. The assembly program that implements this idea is as follows:

```
          #include "c:\miniide\hcs12.inc"
high_cnt  equ     600                       ; interval during which pulse is high
delay     equ     1500                      ; a delay corresponds to the period of 2 kHz
          org     $1500
          lds     #$1500
          movw    #oc7_isr,UserTimerCh7     ; set up OC7 interrupt vector
          movw    #oc0_isr,UserTimerCh0     ; set up OC0 interrupt vector
          movb    #$90,TSCR1                ; enable TCNT and fast timer flag clear
          movb    #$03,TSCR2                ; set prescaler to 8
          movb    #$81,TIOS                 ; enable OC7 and OC0
          movb    #$01,OC7M                 ; allow OC7 to control OC0 pin
          movb    #$01,OC7D                 ; OC7 action on PT0 pin to pull high
          movb    #$02,TCTL2                ; select pull low as the OC0 action
          movb    #$81,TIE                  ; enable OC7 and OC0 to interrupt
          ldd     TCNT
          addd    #delay
          std     TC7                       ; start an OC7 operation
          addd    #high_cnt
          std     TC0                       ; start an OC0 operation
          cli                               ; enable interrupt
loop      bra     loop                      ; infinite loop to wait for interrupt
          swi
oc7_isr   ldd     TC7                       ; start the next OC7 action with 15,000
          addd    #delay                    ; clock cycles delay, also clear the C7F
          std     TC7                       ; flag
          rti
oc0_isr   ldd     TC0                       ; start the next OC0 action with 15,000
          addd    #delay                    ; clock cycles delay, also clear the C0F
          std     TC0                       ; flag
          rti
          end
```

The C language version of this program is very straightforward and hence is left as an exercise.

8.6.6 Forced Output-Compare

There may be applications in which the user requires an output-compare pin action to occur immediately instead of waiting for a match between TCNT and the proper output-compare register. This situation arises in the spark-timing control in some automotive engine control applications. To use the forced output-compare mechanism, the user would write to the CFORC register with 1s in the bit positions corresponding to the output-compare channels to be forced. At the next timer count after the write to CFORC, the forced channels will trigger their programmed pin actions to occur.

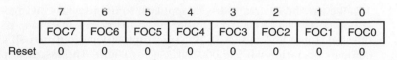

Figure 8.24 ■ Contents of the CFORC register

The forced actions are synchronized to the timer counter clock input. The forced output-compare signal causes pin action but does not affect the timer flag or generate an interrupt. Normally, the force mechanism would not be used in conjunction with the automatic pin action that toggles the corresponding output-compare pin. The contents of CFORC are shown in Figure 8.24. CFORC always reads as all zeroes.

Example 8.12

Suppose that the contents of the TCTL1 and TCTL2 registers are $D6 and $6E, respectively. The content of the TFLG1 register is $00. What would occur on pins PT7 to PT0 on the next clock cycle if the value $7F is written into the CFORC register?

Solution: The TCTL1 and TCTL2 registers configure the output-compare actions as shown in Table 8.2. Since the content of the TFLG1 register is 0, none of the started output-compare operations have succeeded yet.

Because the CFORC register specifies that the output-compare channels 6 to 0 are to be forced immediately, the actions specified in the fourth column in Table 8.2 occur immediately.

Register	Bit Positions	Value	Action to Be Triggered
TCTL1	7 6	1 1	Set the PT7 pin to high
	5 4	0 1	Toggle the PT6 pin
	3 2	0 1	Toggle the PT5 pin
	1 0	1 0	Pull the PT4 pin to low
TCTL2	7 6	0 1	Toggle the PT3 pin
	5 4	1 0	Pull the PT2 pin to low
	3 2	1 1	Set the PT1 pin to high
	1 0	1 0	Pull the PT0 pin to low

Table 8.2 ■ Pin actions on PT7–PT0 pins

8.7 Pulse Accumulator

The HCS12 standard timer system has a 16-bit pulse accumulator, PACA, whereas the Enhanced Captured Timer system has four 8-bit pulse accumulators (PAC3, . . . , PAC0). Two adjacent

8-bit pulse accumulators can be concatenated into a single 16-bit pulse accumulator. PAC3 and PAC2 can be concatenated into the 16-bit pulse accumulator A (PACA), whereas PAC1 and PAC0 can be concatenated into the 16-bit pulse accumulator B (PACB). The block diagrams of four 8-bit pulse accumulators and two 16-bit pulse accumulators are shown in Figures 8.25 and 8.26, respectively.

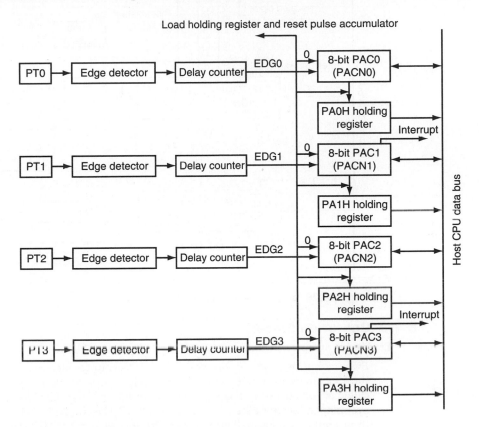

Figure 8.25 ■ Block diagram of four 8-bit pulse accumulators

There are four possible configurations for these pulse accumulators.

- Two 16-bit pulse accumulators PACA and PACB
- One 16-bit pulse accumulator PACA and two 8-bit pulse accumulators PAC1 and PAC0
- One 16-bit pulse accumulator PACB and two 8-bit pulse accumulators PAC3 and PAC2
- Four 8-bit accumulators PAC3~PAC0

8.7.1 Signal Pins

Four 8-bit pulse accumulators PAC3~PAC0 are sharing the signal pins PT3~PT0 with the lower four input-capture modules IC3~IC0, one pin per pulse accumulator. However, when concatenated into a 16-bit pulse accumulator, PACA and PACB use signal pins PT7 and PT0, respectively.

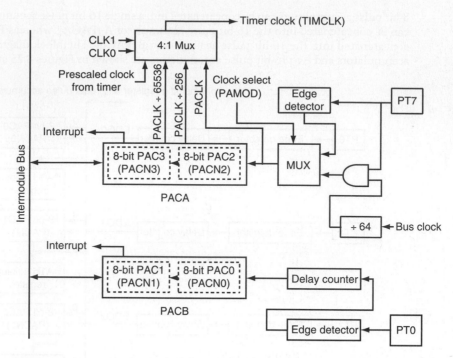

Figure 8.26 ■ 16-bit accumulator block diagram

8.7.2 Operation Modes

The 16-bit pulse accumulator PACA has two operation modes.

1. *Event-counting mode.* In this mode, the PACA counter increments on the active edge of the PT7 pin. Many microcontroller applications involve counting things. These things are called *events*, but in real applications they might be anything: pieces on an assembly line, cycles of an incoming signal, or units of time. To be counted by the accumulator, these things must be translated into rising or falling edges on the PT7 pin. A trivial example of an event might be to count pieces on the assembly line: a light emitter and detector pair could be placed across a conveyor so that as each piece passes the sensor, the light beam is interrupted and a logic-level signal that can be connected to the PT7 pin is produced.

2. *Gated-time-accumulation mode.* The 16-bit PACA counter is clocked by a free-running E ÷ 64 signal, subject to the PT7 signal being active. One common use of this mode is to measure the duration of a single pulse. The counter is set to 0 before the pulse starts, and the resultant pulse time is read directly when the pulse is finished.

The 16-bit pulse accumulator B and all four 8-bit pulse accumulators have only one mode: event counting. That is, their counters are incremented on the active edge of their associated pins. The active edges of PACB and PAC3, . . . , PACO0 are identical to those of timer capture functions IC0 (PT0 pin) and IC3, . . . , ICO (PT3, . . . , PTO pins), respectively. The active edges of the PT3~PT0 pins are defined by the TCTL4 register.

8.7.3 Interrupt Sources

The 16-bit PACA has two interrupt sources.

1. PT7-edge interrupt
2. PACA counter overflow

Only two (PAC3 and PAC1) of the four 8-bit pulse accumulators may request interrupt to the MCU. Interrupts are requested whenever PAC3 or PAC1 rolls over from $FF to $00. The 16-bit PACB may request an interrupt to the MCU whenever its upper 8-bit counter (PAC1) rolls over from $FF to $00.

8.7.4 Registers Related to Pulse Accumulator

The operation of the 16-bit PACA is controlled by the PACTL register. The status of the PACA is recorded in the PAFLG register. The PACA has a 16-bit counter. Since PACA is formed by concatenating PAC3 and PAC2, we can use the instruction ldd PAC3 to copy the contents of the 16-bit counter. To be backward compatible with the 68HC12, the register name PACNT is added to the header file so that one can also use the instruction ldd PACNT to copy the contents of the 16-bit counter. The contents of the PACTL and PAFLG registers are shown in Figures 8.27 and 8.28, respectively.

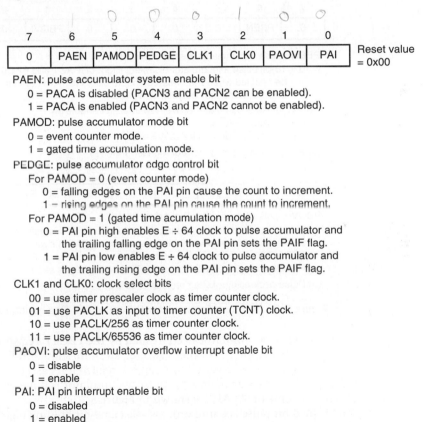

PAEN: pulse accumulator system enable bit
 0 = PACA is disabled (PACN3 and PACN2 can be enabled).
 1 = PACA is enabled (PACN3 and PACN2 cannot be enabled).
PAMOD: pulse accumulator mode bit
 0 = event counter mode.
 1 = gated time accumulation mode.
PEDGE: pulse accumulator edge control bit
 For PAMOD = 0 (event counter mode)
  0 = falling edges on the PAI pin cause the count to increment.
  1 = rising edges on the PAI pin cause the count to increment.
 For PAMOD = 1 (gated time acumulation mode)
  0 = PAI pin high enables E ÷ 64 clock to pulse accumulator and
   the trailing falling edge on the PAI pin sets the PAIF flag.
  1 = PAI pin low enables E ÷ 64 clock to pulse accumulator and
   the trailing rising edge on the PAI pin sets the PAIF flag.
CLK1 and CLK0: clock select bits
 00 = use timer prescaler clock as timer counter clock.
 01 = use PACLK as input to timer counter (TCNT) clock.
 10 = use PACLK/256 as timer counter clock.
 11 = use PACLK/65536 as timer counter clock.
PAOVI: pulse accumulator overflow interrupt enable bit
 0 = disable
 1 = enable
PAI: PAI pin interrupt enable bit
 0 = disabled
 1 = enabled

Figure 8.27 ■ Pulse Accumulator Control register (PACTL)

The PACTL register also controls the clock source for the timer counter (TCNT). When the CLK1 and CLK0 bits are not 00, the PACLK signal (from the PT7 pin) is prescaled by 1, 256, or 65,536 and used as the clock input to the timer counter.

Bits 1 and 0 of the PAFLG register keep track of the status of the operation of the PACA, as shown in Figure 8.28. Any access to the PAC3 or PAC2 register will clear all the flag bits in the PAFLG register if the TFFCA bit in the TSCR1 register is set to 1.

7	6	5	4	3	2	1	0	
0	0	0	0	0	0	PAOVF	PAIF	Reset value = 0x00

PAOVF: pulse accumulator overflow flag

    This flag is set when PACNT overflows from $FFFF to $0000 and can be cleared by writing a 1 to it.

PAIF: PT7 pin edge flag

    When in event-counting mode, this bit is set when the selected edge on the PT7 pin is detected.

    When in gated-accumulation mode, the selected trailing edge sets this flag.

Figure 8.28 ■ Pulse Accumulator Flag register (PAFLG)

7	6	5	4	3	2	1	0	
0	PBEN	0	0	0	0	PBOVI	0	Reset value = 0x00

PBEN: pulse accumulator B system enable bit

    0 = 16-bit pulse accumulator disabled. Eight-bit PAC1 and PAC0 can be enabled when their related enable bits in ICPACR are set.

    1 = pulse accumulator B system enabled.

PBOVI: pulse accumulator B overflow interrupt enable bit

    0 = interrupt inhibited.

    1 = interrupt requested if PBOVF is set.

(a) Pulse accumulator B control register (PBCTL)

7	6	5	4	3	2	1	0	
0	0	0	0	0	0	PBOVF	0	Reset value = 0x00

PBOVF: pulse accumulator B overflow flag

    This bit is set when the 16-bit pulse accumulator B overflows from $FFFF to $0000 or when 8-bit accumulator 1 (PAC1) overflows from $FF to $00. It is cleared by writing 1 to it or by accessing PACN1 and PACN0 when the TFFCA bit in the TSCR1 register is set.

(b) Pulse accumulator B flag register (PBFLG)

Figure 8.29 ■ Pulse Accumulator B Control and Flag registers

The PACB pulse accumulator is controlled by the PBCTL register, and the PBFLG register records its status. The contents of these two registers are shown in Figure 8.29.

Each of the 8-bit pulse accumulators can be enabled if its associated 16-bit pulse accumulator is disabled. The enabling of an 8-bit pulse accumulator is done by programming the ICPAR register. The contents of ICPAR are shown in Figure 8.30.

Each of the 8-bit pulse accumulator also has an 8-bit holding register (PA3H~PA0H).

7	6	5	4	3	2	1	0	
0	0	0	0	PA3EN	PA2EN	PA1EN	PA0EN	Reset value = 0x00

PAxEN: 8-bit pulse accumulator x enable bit

    0 = pulse accumulator x disabled

    1 = pulse accumulator x enabled

Figure 8.30 ■ Input Control Pulse Accumulator Control register (ICPACR)

8.7.5 Operations of the Enhanced Pulse Accumulators

The 16-bit PACA has two operation modes: event counting and gated time accumulation. When in the event-counting mode, the PACA counter increases on the selected edge of the PT7 signal. The PAIF flag of the PAFLG register is set to 1 whenever the selected signal edge is detected on the PT7 pin. The setting of this flag may request an interrupt to the MCU if the PAI bit of the PACTL register is set to 1. When in gated-time-accumulation mode, the 16-bit counter is enabled to increment by the E/64 clock signal if the selected signal level is applied on the PT7 pin.

All other pulse accumulators (PACB, PAC3~PAC0) count the number of active edges at their associated pins. Whenever the PAC3 or PAC1 rolls over from $FF to $00, its associated flag (PAOVF or PBOVF) will be set to 1 and may optionally request an interrupt to the MCU. Pulse accumulators PAC2 and PAC0 do not have the interrupt capability. The user can prevent 8-bit pulse accumulators counting further than $FF by setting the PACMX bit in the ICSYS register. In this case, a value of $FF means that 255 counts or more have occurred. The contents of the ICSYS register are illustrated in Figure 8.31.

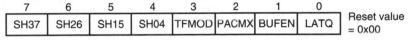

SHxy: share input action of input-capture channel *x* and *y* bits
  0 = normal operation.
  1 = the channel input *x* causes the same action on the channel *y*.
TFMOD: timer-flag-setting mode bit
  0 = the timer flags C3F:C0F in TFLG1 are set when a valid input-
      capture transition on the corresponding port pin occurs.
  1 = if in the queue mode (BUFEN = 1 and LATQ = 0), the timer flags
      C3F:C0F in TFLG1 are set only when a latch on the corresponding
      holding register occurs. If the queue mode is not engaged, the timer
      flags C3F:C0F are set the same way as for TFMOD = 0.
PACMX: 8-bit pulse accumulator maximum count bit
  0 = normal operation. When the 8-bit pulse accumulator has reached
      $FF, with the next active edge, it will be incremented to $00.
  1 = when the 8-bit pulse accumulator has reached the value $FF, it
      will not be incremented further. The value $FF indicates a count
      of 255 or more.
BUFEN: IC buffer enable bit
  0 = input-capture and pulse accumulator holding registers are
      disabled.
  1 = input-capture and pulse accumulator holding registers are
      enabled.
LATQ: input-capture latch or queue mode select bit
  The BUFEN bit should be set to enable IC and the pulse accumulator's
  holding registers. Otherwise, the LATQ latching mode is disabled.
  0 = the queue mode of input-capture is enabled.
  1 = the latch mode is enabled. Latching function occurs when modulus
      down counter reaches 0 or a 0 is written into the count register
      MCCNT. With a latching event, the contents of IC registers and
      8-bit pulse accumulators are transferred to their holding registers.
      The 8-bit pulse accumulators are cleared.

Figure 8.31 ■ Input Control System Control register (ICSYS)

8.7.6 Pulse Accumulator Applications

The pulse accumulator has a few interesting applications, such as interrupting after N events, frequency measurement, and pulse duration measurement.

Example 8.13

Suppose certain events are converted into pulses and connected to the PT7 pin. Write a program so that the pulse accumulator generates an interrupt to the HCS12 when N events have occurred. Assume that N is less than 65,536.

Solution: By writing the two's complement of N into PACNT, it will overflow after N events and generate an interrupt. The assembly program is as follows:

```
#include    "c:\miniide\hcs12.inc"
N           equ     1350
            org     $1500
            lds     #$1500              ; set up stack pointer
            movw    #paov_isr,UserPAccOvf   ; set up PAOV interrupt vector
            ldd     #N                  ; place the two's complement in PACNT
            coma                        ;  "
            comb                        ;  "
            addd    #1                  ;  "
            std     PACNT               ;  "
            movb    #$52,PACTL          ; enable PACA, event counting mode, active
                                        ; edge is rising
            cli                         ; enable PAOV interrupt
;           . . .
            swi
paov_isr    movb    #PAOVF,PAFLG        ; clear the PAOVF flag
            end
```

The C language version of the program for CodeWarrior C compiler is as follows:

```
#include    "c:\cwHCS12\include\hcs12.h"
#define     NN  1350
void main(void) {
        PACNT = ~N + 1;
        PACTL= 0x52;                    //enable PACTL, event count mode,
                                        //  increment on rising edge
        asm("cli");                     //enable PACNT overflow interrupt
        //  . . .
}
interrupt void paovISR(void) {
        PAFLG = PAOVF;                  //clear the flag and return
}
```

The pulse accumulator system can be used to measure the signal frequency. The procedure is as follows:

Step 1
Connect the unknown signal to the PT7 pin.

Step 2
Set up the PACA to operate in event-counting mode.

Step 3
Use one of the output-compare functions to create a 1-s time interval.

Step 4
Use a memory location to keep track of the number of pulse accumulator counter overflow interrupts.

Step 5
Enable the PAOV interrupt.

Step 6
Disable the PAOV interrupt at the end of 1 s.

Example 8.14

Write a program to measure the frequency of a signal connected to the PT7 pin using the algorithm described previously.

Solution: We use the OC0 function to create a 1-s delay. Fifty OC0 operations will be performed, with each OC0 operation creating a 20-ms delay. The service routine for the PACNT overflow interrupt will increase the overflow count by 1. Let *paov_cnt* represent the PACNT overflow count. At the end of 1 s, the frequency is equal to the following expression:

$$\text{Frequency} = paov\_cnt \times 2^{16} + PACNT$$

The assembly program that implements this procedure is as follows:

```
#include    "c:\miniide\hcs12.inc"
            org     $1000
oc_cnt      ds.b    1
paov_cnt    ds.b    2                       ; PACNT overflow count
frequency   ds.b    4                       ; signal frequency
            org     $1500
            lds     #$1500
            movw    #paov_isr,UserPAccOvf   ; set up PAOV interrupt vector
            movb    #50,oc_cnt              ; prepare to perform 50 OC0 actions
            ldd     #0
            std     PACNT                   ; let PACNT count up from 0
            std     paov_cnt                ; initialize PACNT overflow count to 0
            std     frequency               ; initialize frequency to 0
            std     frequency+2             ;    "
```

```
            movb    #$90,TSCR1              ; enable TCNT and fast timer flag clear
            bset    TIOS,OC0               ; select OC0 function
            movb    #$03,TSCR2             ; set prescaler to TCNT to 8
            bclr    DDRT,$80              ; configure PT7 for input
; configure PA function: enable PA, select event-counting mode, rising edge
; of PAI signal increments the PACNT counter, enable PAOV interrupt
            movb    #$52,PACTL
            cli                             ; enable PAOV interrupt
            ldd     TCNT
sec_loop    addd    #60000
            std     TC0
            brclr   TFLG1,COF,*            ; wait for 20 ms here
            ldd     TC0
            dec     oc_cnt
            bne     sec_loop
            movb    #0,PACTL               ; disable PA function
            sei                            ; disable interrupt
            ldd     PACNT
            std     frequency+2
            ldd     paov_cnt
            std     frequency
            swi
paov_isr    movb    #PAOVF,PAFLG           ; clear the PAOVF flag
            ldx     paov_cnt              ; increment PACNT overflow
            inx                            ; count by 1
            stx     paov_cnt              ;       "
            end
```

The C language version of the program is as follows:

```
#include     "c:\cwHCS12\include\hcs12.h"
unsigned     long int frequency;
unsigned     int paov_cnt;
void main (void)
{
      int oc_cnt;
      PACNT      = 0;
      frequency  = 0;
      paov_cnt   = 0;
      TSCR1      = 0x90;          // enable TCNT and fast flag clear
      TIOS       = OC0;           // select OC0 function
      TSCR2      = 0x03;          // set prescale factor to 8
      PACTL      = 0x52;          // enable PA function, enable PAOV interrupt
      DDRT      &= 0x7F;          // configure the PT7 pin for input
      asm("cli");                 // enable interrupt globally
      oc_cnt     = 50;
      TC0        = TCNT + 60000u;
      while (oc_cnt) {
            while(!(TFLG1 & COF));
            TC0 = TC0 + 60000u;
```

```
                oc_cnt— —;
        }
        PACTL     = 0x00;              // disable PA function
        asm("sei");
        frequency = (long)paov_cnt * 65536l + (long)PACNT;
        asm("swi");
}
interrupt void paovISR (void)
{
        PAFLG     = PAOVF;             // clear PAOVF flag
        paov_cnt  = paov_cnt + 1;
}
```

The function that sets up the interrupt vector for pulse accumulator A is as follows:

```
extern void near paovISR(void);
#pragma CODE_SEG __NEAR_SEG NON_BANKED
#pragma CODE_SEG DEFAULT             // Change code section to DEFAULT.
typedef void (*near tlsrFunc)(void);
const tlsrFunc _vect[] @0xFFDC = {
        paovISR
};
```

The pulse accumulator module can be set up to measure the duration of a pulse using the gated time accumulation mode. When the active level is applied to the PT7 pin, the PACNT can count and will stop counting on the trailing edge of the PT7 signal. The clock input to the pulse accumulator is E ÷ 64. The procedure for measuring the duration of a pulse is as follows:

Step 1
Set up the pulse accumulator system to operate in the gated time accumulation mode, and initialize PACNT to 0.

Step 2
Select the falling edge as the active edge (for measuring positive pulse). In this setting, the pulse accumulator counter will increment when the signal connected to the PAI pin is high and generate an interrupt to the HCS12 on the falling edge.

Step 3
Enable the PAI active edge interrupt and wait for the arrival of the active edge of PAI.

Step 4
Stop the pulse accumulator counter when the interrupt arrives.

Without keeping track of the PACNT overflows, the longest pulse width (E = 24 MHz) that can be measured is

$$pulse\_width = 2^{16} \times 64T_E = 2^{16} \times 64 \times 1/24 \ \mu s = 174.763 \ ms$$

To measure a longer pulse width, we need to keep track of the number of times that the PACNT counter overflows in the duration of the pulse. Let *paov_cnt* be the overflow count of the PACNT counter, then

$$pulse\_width = [(2^{16} \times paov\_cnt) + PACNT] \times 64T_E$$

Example 8.15

Write a program to measure the duration of an unknown signal connected to the PAI pin.

Solution: The assembly program that implements the previous algorithm is as follows:

```
#include  "c:\miniide\hcs12.inc"
          org       $1000
paovCnt   ds.b      1                       ; use to keep track of the PACNT overflow count
pw        ds.b      3                       ; hold the signal pulse width
          org       $1500
          movw      #paovISR,UserPAccOvf    ; set up PAOV interrupt vector
          ldd       #0
          std       PACNT                   ; let PACNT count up from 0
          clr       paovCnt                 ; initialize PACNT overflow count to 0
          movb      #$0,TSCR2               ; set TCNT timer prescaler to 1
; configure PA function: enable PA, select gated-time—accumulator mode, high level
; of the PAI signal enables PACNT counter, enable PAOV interrupt
          movb      #$62,PACTL
          bclr      DDRT,$80                ; configure PAI pin for input
          cli                               ; enable PAOV interrupt
          brclr     PAFLG,PAIF,*            ; wait for the arrival of the falling edge of PAI
          movb      #0,PACTL                ; disable PA function
          sei                               ; disable interrupt
          ldd       PACNT
          std       pw
          ldaa      paovCnt
          staa      pw
          swi
paovISR   movb      #PAOVF,PAFLG            ; clear PAOVF flag
          inc       paovCnt                 ; increment PACNT overflow count by 1
          end
```

The C language version of the program for the CodeWarrior C compiler is as follows:

```
#include "c:\cwHCS12\include\hcs12.h"
unsigned int paovCnt;
long unsigned int pw;
void main(void)
{
     PACNT = 0;          // let PACNT count up from 0
     TSCR1 = 0x90;       // enable TCNT and fast timer flag clear
     paovCnt = 0;
     pw = 0;
     TSCR2 = 0x00;       // set TCNT prescaler to 1
     DDRT = 0x00;        // configure all timer port pins for input
// configure PA function: enable PA, select gated-time-accumulator mode,  high level
// of PAI enables PACNT to count, enable PAOV interrupt
     PACTL = 0x62;
     asm("cli");
```

```
                while(!(PAFLG & PAIF)); // wait for the arrival of the PAI falling edge
                PACTL = 0x00;        // disable PA system
                asm("sei");
                pw = (long)paovCnt * 65536l + (long)PACNT;
                asm ("swi");
        }
        interrupt void paovISR (void)
        {
                PAFLG     = PAOVF; // clear PAOVF flag
                paovCnt   = paovCnt + 1;

        }
```

The function for setting up an interrupt vector for the pulse accumulator A is identical to that of Example 8.14.

▲

Example 8.16

▼

Write an instruction sequence to enable the 8-bit pulse accumulators PAC1 and PAC3 and let them increase on the rising edge of their associated pins. Disable their overflow interrupts.

Solution: To enable PAC1 and PAC3, the PAEN and PBEN bits must be cleared and the PA3EN and PA1EN bits set. To select the rising edge as their active edge, the value $44 needs to be written into the TCTL4 register and also the PT3 and PT1 pins must be configured for input-capture. To disable overflow interrupt, clear the PAOVI and PBOVI bits.

The following instruction sequence will perform the required configuration:

```
    bclr      PACTL,$42       ; disable 16-bit PACA, disable overflow interrupt
    bclr      PBCTL,$42       ; disable 16-bit PACB, disable overflow interrupt
    movb      #$44,TCTL4      ; select the rising edges as active edge
    bset      ICPAR,$0A       ; enable PAC3 and PAC1
    bclr      DDRT,$0A        ; configure PT3 and PT1 for input
```

▲

8.8 Modulus Down Counter

The HCS12 timer system contains a modulus down counter that can be used as a time base to generate periodic interrupts. It can also be used to latch the value of the IC registers and the pulse accumulators to their holding registers. The action of latching can be periodic or only once.

The modulus down counter has a prescaler, which divides the E-clock and uses its output as the clock input to the down counter. The prescaler can be 1, 4, 8, and 16. The operation of the modulus down counter is controlled by the MCCTL register. The MCFLG register records the status of the modulus down counter. The contents of the MCCTL and MCFLG registers are shown in Figures 8.32 and 8.33, respectively.

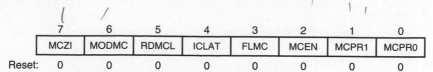

7	6	5	4	3	2	1	0
MCZI	MODMC	RDMCL	ICLAT	FLMC	MCEN	MCPR1	MCPR0

Reset: 0   0   0   0   0   0   0   0

MCZI: modulus counter underflow interrupt enable bit
    0 = modulus counter underflow interrupt is disabled.
    1 = modulus counter underflow interrupt is enabled.
MODMC: modulus mode enable bit
    0 = the counter counts once from the value written to it and will stop at $0000.
    1 = modulus mode is enabled. When the counter reaches $0000, the counter is
       loaded with the latest value written into to the modulus count register.
RDMCL: read modulus down counter load bit
    0 = reads of the modulus count register will return the present value of the count
       register.
    1 = reads of the modulus count register will return the contents of the load
       register (i.e., the reload value is returned).
ICLAT: input-capture force latch action bit
    This bit has effect only when both the LATQ and BUFEN bits in ICSYS are set.
    0 = no effect.
    1 = forces the contents of the input-capture registers TC0 to TC3 and their corres-
       ponding 8-bit pulse accumulators to be latched into the associated holding registers.
       The pulse accumulators will be cleared when the latch action occurs.
FLMC: force load register into the modulus counter count register bit
    This bit has effect only when MCEN = 1.
    0 = no effect.
    1 = loads the load register into the modulus counter count register. This also resets
       the modulus counter prescaler.
MCEN: modulus down counter enable bit
    0 = modulus counter is disabled and preset to $FFFF.
    1 = modulus counter is enabled.
MCPR1 and MCPR0: modulus counter prescaler select bits
    0 0 = prescale rate is 1.
    0 1 = prescale rate is 4.
    1 0 = prescale rate is 8.
    1 1 = prescale rate is 16.

Figure 8.32 ■ Modulus Down Counter register (MCCTL)

7	6	5	4	3	2	1	0
MCZF	0	0	0	POLF3	POLF2	POLF1	POLF0

Reset: 0   0   0   0   0   0   0   0

MCZF: modulus counter underflow interrupt flag
    This flag is set when the modulus down counter reaches 0. Writing
    1 to this bit clears the flag.
POLF3:POLF0: first input-capture polarity status bits
    These are read-only bits. Writing to these bits has no effect. Each
    status bit gives the polarity of the first edge that has caused an input-
    capture to occur after capture latch has been read.
    0 = the first input-capture has been caused by a falling edge.
    1 = the first input-capture has been caused by a rising edge.

Figure 8.33 ■ Modulus Down Counter Flag register (MCFLG)

8.8.1 Using the Modulus Down Counter to Generate Periodic Interrupts

Periodic interrupts can be generated by selecting the modulus mode and enabling its interrupt. The interrupt period is equal to the value written into the load register multiplied by the prescaler. When writing a value into MCCNT, the same value is also written into the load register. In the modulus mode, when the MCCNT is decremented to zero, the value in the load register will be reloaded into the MCCNT register and another cycle of down counting is started.

Example 8.17

▼

Write an instruction sequence to generate periodic interrupt every 10 ms.

Solution: One possible value to be written into the MCCTL register is $C0, which will

- Enable MCCNT
- Enable MCCNT interrupt
- Enable modulus mode
- Set prescaler to 16

The instruction sequence to achieve the desired setting is as follows:

```
movb    #$C7,MCCTL
movw    #15000,MCCNT    ; place the value that will be decremented
                        ; to 0 in 10 ms
cli                     ; enable interrupt
```

▲

8.8.2 Using the Modulus Down Counter to Generate Time Delays

The method for generating a time delay by using the modulus down counter is similar to that for generating periodic interrupts. However, one can choose to use either the *modulus* mode or *nonmodulus* mode. The following example use the nonmodulus mode to generate a time delay that is equal to 10 ms multiplied by the value in register Y.

Example 8.18

▼

Write a subroutine to generate a time delay that is equal to 10 ms multiplied by the value in register Y.

Solution: The assembly program that can generate a delay that is equal to 10 ms multiplied by the value in register Y is as follows:

```
delay10ms   bset    TSCR1,TFFCA     ; enable timer fast flag clear
            movb    #$07,MCCTL      ; enable modulus down counter with 1:16 as prescaler
            movw    #15000,MCCNT    ; load the value to be down-counted
            brclr   MCFLG,MCZF,*
            bclr    MCCTL,$04       ; disable modulus down counter
            dbne    y,delay10ms
            rts
```

▲

The C language version of the function to be compiled by the CodeWarrior is as follows:

```
void delayby10ms(int k)
{
    int i;
    TSCR1 |= TFFCA;                      // enable fast timer flag clear
    for (i = 0; i < k; i++) {
        MCCTL = 0x07;                    // enable modulus down counter with 1:16 as prescaler
        MCCNT = 15000;                   // let modulus down counter count down from 15,000
        while(!(MCFLG & MCZF));
        MCCTL &= ~0x04;                  // disable modulus down counter
    }
}
```

By storing the previous delay function in the appropriate directory, one can include it in the program. The following program calls the previous subroutine to create a delay of 200 ms, increments a count value, and then outputs this count value to Port B:

```
#include   "c:\miniIDE\hcs12.inc"
           org      $1000
count      ds.b     1
           org      $1500
           movb     #$FF,DDRB
           clr      count
           movb     count,PORTB
loops      ldy      #20                  ; create 200-ms delay
           jsr      delayby10ms          ;    "
           inc      count                ; increment count by 1
           movb     count,PORTB          ; update LED displays
           bra      loops
#include   "c:\miniIDE\delay.asm"
           end
```

Delay functions that create time delays by a multiple of other delay times can be created by modifying the previous delay function. The file *delay.asm* (and *delay.c*) contains the following delay functions and is contained in the complementary CD:

1. **delayby10μs**
2. **delayby50μs**
3. **delayby1ms**
4. **delayby10ms**
5. **delayby100ms**

8.9 Enhanced Capture Timer (ECT) Module

Some of the HCS12 members (e.g., MC9S12DG256) implement an enhanced capture timer (ECT) module that has the features of the standard timer (TIM) module enhanced by additional features in order to enlarge the field of applications. These additional features are

- One 16-bit buffer register for each of the four input-capture (IC) channels.
- Four 8-bit pulse accumulators. Each of these 8-bit pulse accumulators has an associated 8-bit buffer. Two of these 8-bit pulse accumulators can be concatenated into a single 16-bit pulse accumulator.

- A 16-bit modulus down counter with 4-bit prescaler.
- Four user-selectable delay counters for increasing input noise immunity.

8.9.1 Enhanced Capture Timer Modes of Operation

The enhanced capture timer has eight input-capture/output-compare (IC/OC) channels, the same as on the HCS12 standard timer module. Four IC channels (IC7, ..., IC4) are the same as the standard timer with one capture register that memorizes the timer value captured by an action on the associated input pin. Four other IC channels (IC3, ..., IC0), in addition to the capture register, also have one buffer, called the *holding register*. This permits the register to memorize two different timer values without generating any interrupt. This feature can reduce software overhead in applications that require capturing two edges in order to perform further computation. In addition, the ECT module provides the option of preventing a captured value from being overwritten before it was read or transferred to the holding register. This option is controlled by the Input Control Overwrite (ICOVW) register. The contents of this register are shown in Figure 8.34. This capability will be useful when external events occur at a rate that the CPU cannot read quickly enough.

	7	6	5	4	3	2	1	0	
	NOVW7	NOVW6	NOVW5	NOVW4	NOVW3	NOVW2	NOVW1	NOVW0	Reset value = 0x00

NOVW*n*: no input-capture overwrite
0 = the contents of the related capture register or holding register can be overwritten when a new input-capture or latch occurs.
1 = the related capture register or holding register cannot be written by an event unless it is empty. This will prevent the captured value to be overwritten until it is read or latched in the holding register.

Figure 8.34 ■ Input Control Overwrite register (ICOVW)

Four 8-bit pulse accumulators are associated with the four IC buffered channels. Each pulse accumulator has a holding register to memorize its value by an action on its external input. Each pair of pulse accumulators can be used as a 16-bit pulse accumulator.

8.9.2 Why the Enhanced Capture Timer Module?

There are applications that require the capture of two consecutive edges (could be both rising or both falling, or one rising and the other falling) at very high frequencies. In Example 8.2, the following instructions (faster than movw TCx,edge_ist by one E cycle) are executed before we have time to wait for the arrival of the second edge (after we detect the first edge):

```
ldd    TCx
std    ...
```

It takes five E-clock cycles to execute these two instructions, which set the upper limit on the signal frequency that can be dealt with. By providing the capability of setting the interrupt flag (or interrupting the CPU) after two signal edges have been captured, the upper limit of the signal frequency that can be handled can be significantly improved.

The input-capture function of the original standard timer module allows the newly captured value to overwrite the old one even if the CPU has not read the old value yet. This can cause a problem when the event frequency is very high. The enhanced input-capture function allows the user to prevent the overwriting of captured values by enabling the nonoverwrite feature.

8.9.3 The Operation of the Enhanced Input-Capture Function

The enhanced input-capture function has all the registers in the input-capture function of the standard timer module. New registers are added to implement the additional features. An input-capture register is *empty* if its value has been read or latched into its associated holding register. A holding register is *empty* if its value has been read. An enhanced input-capture channel can be configured to operate in either *latch* mode or *queue* mode.

Figures 8.35 and 8.36 illustrate the registers related to the operation of the *latch* mode and *queue* mode, respectively. In both diagrams, channels IC0 to IC3 are identical and IC4 to IC7 are identical. Only one channel in each group is shown in the figure. The latch mode and queue mode are selected by setting and clearing the LATQ bit of the ICSYS register (shown in Figure 8.31).

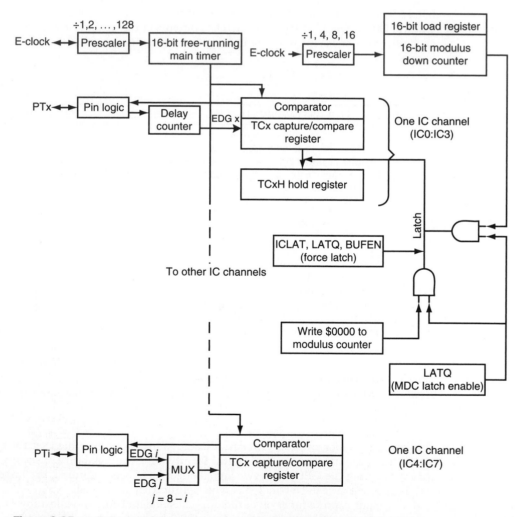

Figure 8.35 ■ Enhanced input-capture function block diagram in latch mode

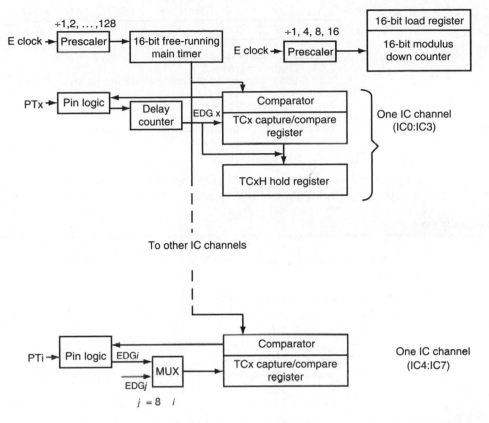

Figure 8.36 ■ Enhanced input-capture function block diagram in queue mode (channels IC0...IC3 block diagram)

In the latch mode, the latching function occurs when the modulus down counter reaches 0 or a zero is written into the count register MCCNT. With a latching event, the contents of the IC registers and 8-bit pulse accumulators are transferred to their holding registers. After this, the 8-bit pulse accumulators are cleared. In the queue mode, the main timer value is copied into the IC register (TCx register) by a valid input pin transition. With a new occurrence of a capture, the value of the IC register is transferred to its holding register and the IC register holds the new timer value. The queue mode and latch mode can be entered only when the holding registers are enabled; this can be done by setting the BUFEN bit of the ICSYS register.

The HCS12 is often used in noisy environments such as automotive applications. Due to the noise, false signal edges often occur in input-capture pins. To distinguish between true edge and false edge (usually very short), a delay counter is added to the enhanced input-capture function. Any detected edge with a duration shorter than the preprogrammed value is ignored. The duration of the delay is controlled by the DLYCT register. The contents of the DLYCT register are shown in Figure 8.37. If enabled, after detection of a valid edge on the input-capture pin, the delay counter counts the preselected number of E-clock cycles; then it generates a pulse to latch the TCNT value into the input-capture register (TCx). The pulse is generated only if the level of

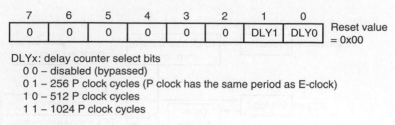

DLYx: delay counter select bits
  0 0 – disabled (bypassed)
  0 1 – 256 P clock cycles (P clock has the same period as E-clock)
  1 0 – 512 P clock cycles
  1 1 – 1024 P clock cycles

Figure 8.37 ■ Delay Counter Control register (DLYCT)

input signal, after the preset delay, is the opposite of the level before the transition. This avoids reaction to narrow pulses caused by noise. After counting, the counter is cleared automatically.

Example 8.19

Modify the program in Example 8.2 to take advantage of the queue mode of the enhanced input-capture function.

Solution: The modified program is as follows:

```
#include "c:\miniide\hcs12.inc"
              org      $1000
period        ds.w     1                    ; memory to store the period
              org      $1500
start         movb     #$90,TSCR            ; enable timer counter and fast timer flag clear
              bclr     TIOS,IOS0            ; select input-capture 0
              movb     #$04,TSCR2           ; disable TCNT overflow interrupt, set prescale
                                            ; factor to 16
              movb     #$01,TCTL4           ; choose to capture the rising edge of PT0 pin
              movb     #$0A,ICSYS           ; enable timer flag-setting mode, IC buffer, and queue
                                            ; mode
              clr      DLYCT               ; disable delay counter
              bset     ICOVW,NOVW0         ; no input-capture overwrite for IC0
              ldd      TC0                 ; empty the input-capture register TC0
              ldd      TC0H                ; empty the holding register TC0H
              brclr    TFLG1,$FE,*         ; wait for the arrival of the second rising edge
              ldd      TC0
              subd     TC0H                ; subtract the first edge from the second edge
              std      period
              swi
              end
```

The C language version of the program is as follows:

```
#include "c:\cwHCS12\include\hcs12.h"
void main(void)
{
       unsigned    int      period;
       TSCR1     = 0x90;              /* enable timer counter, enable fast timer flag clear*/
       TIOS      &= ~IOS0;            /* enable input-capture 0 */
```

```
        TSCR2    = 0x04;                /* set prescale factor to 16 */
        TCTL4    = 0x01;                /* capture the rising edge of PT0 pin */
/* enable timer flag-setting mode, IC buffer, and queue mode */
        ICSYS    = 0x0A;
        DLYCT    = 0x00;                /* disable delay counter */
        ICOVW    |= NOVW0;              /* disable input-capture overwrite */
        period   = TC0;                /* empty TC0 and clear the COF flag */
        period   = TC0H;               /* empty the TC0H register */
/* wait for the arrival of the second rising edge */
        while (!(TFLG1 & COF));
        period   = TC0 - TC0H;
        while(1);
}
```

Example 8.20

Suppose we want to measure the pulse width of a signal connected to the PT0 pin in a noisy environment. Write a program to perform the operation. Ignore any noise pulse shorter than 256 E-clock cycles.

Solution: Since the range of the pulse width is unknown, we need to consider the timer overflow. The program in Example 8.3 is modified to perform the measurement as follows:

```
#include        "c:\miniide\hcs12.inc"
                org     $1000
edge1           rmb     2
overflow        rmb     2
pulse_width     rmb     2
                org     $1500
                movw    #tov_isr,UserTimerOvf ; set up timer overflow interrupt vector
                ldd     #0
                std     overflow
                movb    #$90,TSCR1      ; enable TCNT and fast timer flag clear
                movb    #$04,TSCR2      ; set prescaler to TCNT to 16
                bclr    TIOS,IOS0       ; enable input-capture 0
                movb    #$01,DLYCT      ; set delay count to 256 E cycles
                movb    #$01,ICOVW      ; prohibit overwrite to TC0 register
                movb    #$0,ICSYS       ; disable queue mode
                movb    #$01,TCTL4      ; capture the rising edge on PT0 pin
                movb    #COF,TFLG1      ; clear the COF flag
                brclr   TFLG1,COF,*     ; wait for the arrival of the first rising edge
                movb    #$80,TFLG2      ; clear the TOF flag
                bset    TSCR2,TOI       ; enable TCNT overflow interrupt
                cli                     ;           "
                movw    TC0,edge1       ; clear the COF flag and save the captured first edge
                movb    #$02,TCTL4      ; capture the falling edge on PT0 pin        "
                brclr   TFLG1,COF,*     ; wait for the arrival of the falling edge
                ldd     TC0
```

```
                    subd       edge1
                    std        pulse_width
                    bcc        next
; second edge is smaller, so decrement overflow count by 1
                    ldx        overflow
                    dex
                    stx        overflow
          next      swi
          tov_isr   movb       #TOF,TFLG2        ; clear the TOF flag
                    ldx        overflow          ; increment TCNT overflow count
                    inx                          ;        "
                    stx        overflow          ;        "
                    rti
                    end
```

▲

The modification to the C program in Example 8.3 is also minor and hence is left as an exercise problem.

8.10 Pulse-Width Modulation (PWM) Function

There are many applications that require the generation of digital waveforms. The output-compare function has been used to generate digital waveforms with any duty cycle in Section 8.6. However, the generation of waveforms using the output-compare function requires frequent attention from the MCU. Most microcontrollers designed in the last few years have incorporated the **Pulse-Width Modulation (PWM)** function to simplify the task of waveform generation.

The MC9S12DG256 and many other HCS12 members implement an 8-channel, 8-bit PWM function. As shown in Figure 8.38, each channel has a period register, a duty cycle register, a control register, and a dedicated counter to support the waveform generation. The clock signal is critical to the setting of the frequency of the generated waveform. The clock source of the counter is programmable through a two-stage circuitry.

The two most important characteristics of a PWM waveform are the period (or frequency) and the duty cycle of the waveform. The clock source and the period register together determine the period of the generated waveform whereas the clock select chain sets the frequency of the clock source to the PWM counter. The *period* of the PWM waveform is set by placing an appropriate value into the period register and setting the clock select block properly. The *duty cycle* is determined by the ratio of the duty register and the period register.

8.10.1 PWM Clock Select

There are four possible clock sources for the PWM function: clock A, clock B, clock SA (scaled A), and clock SB (scaled B). These four clocks are derived from the E-clock. Clocks A and B are derived by dividing the E-clock by a factor of 1, 2, 4, 8, 16, 32, 64, or 128. Clock SA (SB) is derived by dividing clock A (B) by an 8-bit reloadable counter.

Each PWM channel has the option of selecting one of two clocks, either the prescaled clock (clock A or B) or the scaled clock (clock SA or SB). Figure 8.39 illustrates the block diagram of the four different clocks and how the scaled clocks are created.

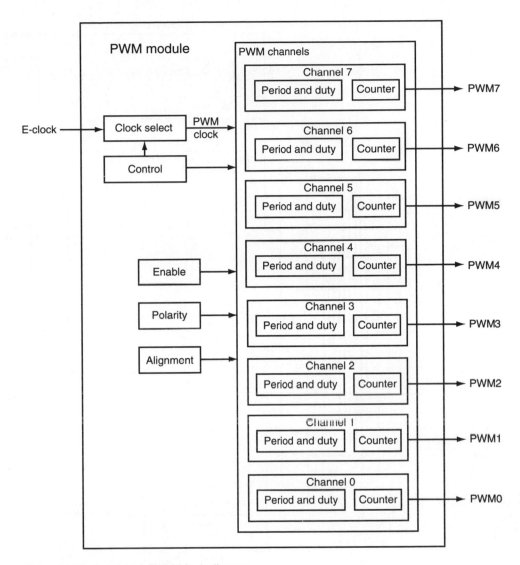

Figure 8.38 ■ HCS12 PWM block diagram

PRESCALE

The input clock to the PWM prescaler is the E-clock. It can be disabled whenever the MCU is in freeze mode by setting the PFRZ bit in the PWMCTL register. If this bit is set, whenever the MCU is in freeze mode the active clock to the prescaler is disabled. This is useful for emulation in order to freeze the PWM. The input clock can also be disabled when all eight PWM channels are disabled (PWME7–0 = 0). This is useful for reducing power consumption by disabling the prescale counter.

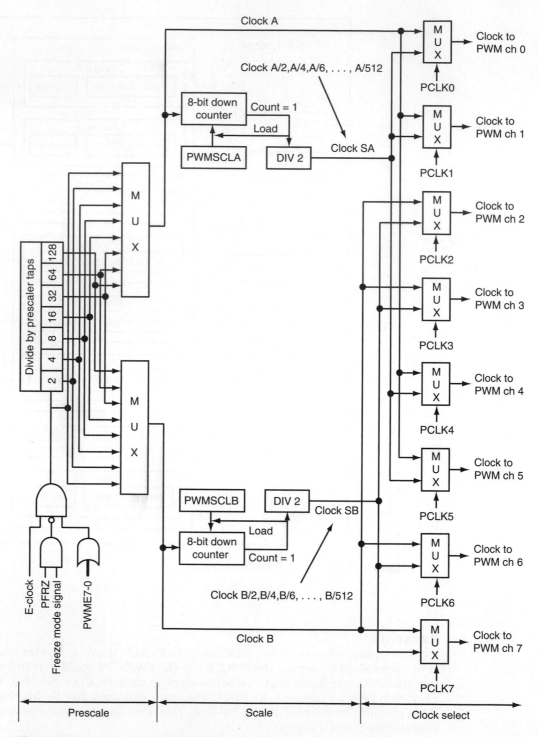

Figure 8.39 ■ PWM clock select block diagram

Clock A and clock B are scaled values of the input clock. The possible values for clock A and clock B are derived by dividing the E-clock by 1, 2, 4, 8, 16, 32, 64, and 128. The value selected for clock A and clock B is determined by the PCKA2~PCKA0 and PCKB2~PCKB0 bits in the PWMPRCLK register, respectively. The contents of the PWMCTL and PWMPRCLK registers are shown in Figure 8.40 and Figure 8.41, respectively. Tables 8.3 and 8.4 show the available prescale values for clock A and clock B.

	7	6	5	4	3	2	1	0
	CON67	CON45	CON23	CON01	PSWAI	PFRZ	0	0
Reset:	0	0	0	0	0	0	0	0

CONjk: concatenate channels j and k (j = 0, 2, 4, or 6; $k = j + 1$)
    0 = channels j and k are separate 8-bit PWMs.
    1 = channels j and k are concatenated to create one 16-bit PWM channel. Channel j
        becomes the high-order byte and channel k becomes the low-order byte. Channel
        k output pin is used as the output for this 16-bit PWM. Channel k clock select bit
        determines the clock source, channel k polarity bit determines the polarity,
        channel k enable bit enables the output,and channel k center-aligned enable bit
        determines the output mode.
PSWAI: PWM stops in wait mode
    0 = allow the clock to the prescaler to continue while in wait mode.
    1 = stop the input clock to the prescaler whenever the MCU is in wait mode.
PFRZ: PWM counters stop in freeze mode
    0 = allow PWM to continue while in freeze mode.
    1 = disable PWM input clock to the prescaler whenever the part is in freeze mode.

Figure 8.40 ■ PWM Control register (PWMCTL)

	7	6	5	4	3	2	1	0
	0	PCKB2	PCKB1	PCKB0	0	PCKA2	PCKA1	PCKA0
Reset:	0	0	0	0	0	0	0	0

Figure 8.41 ■ PWM Prescale Clock Select register (PWMPRCLK)

PCKB2	PCKB1	PCKB0	Value of Clock B
0	0	0	E-clock
0	0	1	E-clock/2
0	1	0	E-clock/4
0	1	1	E-clock/8
1	0	0	E-clock/16
1	0	1	E-clock/32
1	1	0	E-clock/64
1	1	1	E-clock/128

Table 8.3 ■ Clock B prescaler selects

PCKA2	PCKA1	PCKA0	Value of Clock A
0	0	0	E-clock
0	0	1	E-clock/2
0	1	0	E-clock/4
0	1	1	E-clock/8
1	0	0	E-clock/16
1	0	1	E-clock/32
1	1	0	E-clock/64
1	1	1	E-clock/128

Table 8.4 ■ Clock A prescaler selects

CLOCK SCALE

The SA clock takes clock A as one of its inputs and divides it further with a user-programmable value (from 1 to 256) and then divides it by 2. The SB clock is derived similarly, but with clock B as its input.

In Figure 8.39, clock A is an input to an 8-bit down counter. This down counter loads a user-programmable scale value from the scale register (PWMSCLA). When the down counter is decremented to 1, two things happen: A pulse is output and the 8-bit counter is reloaded. The output signal from this circuit is further divided by 2. In other words, the clock SA is derived by the following equation:

Clock SA = clock A/(2 * PWMSCLA)

When PWMSCLA equals $00, the PWMSCLA value is considered a full-scale value of 256. Similarly,

Clock SB = clock B/(2 * PWMSCLB)

CLOCK SELECT

Each PWM channel has a choice of two clock signals to use as the clock source for that channel. The clock source selection is done by the PWMCLK register. The contents of this register are shown in Figure 8.42.

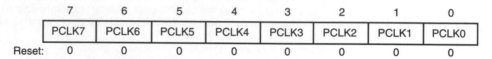

	7	6	5	4	3	2	1	0
	PCLK7	PCLK6	PCLK5	PCLK4	PCLK3	PCLK2	PCLK1	PCLK0
Reset:	0	0	0	0	0	0	0	0

PCLKx: PWM channel *x* clock select (*x* = 7, 6, 3, 2)
   0 = clock B as the clock source
   1 = clock SB as the clock source
PCLKy: PWM channel *y* clock select (*y* = 5, 4, 1, 0)
   0 = clock A as the clock source
   1 = clock SA as the clock source

Figure 8.42 ■ PWM Clock Select register (PWMCLK)

8.10.2 PWM Channel Timers

The main part of the PWM module consists of the timers. Each of the PWM channels has an 8-bit counter, an 8-bit period register, and an 8-bit duty cycle register.

The waveform output period is controlled by a match between the period register and the value in the counter. The duty cycle is controlled by a match between the duty cycle register and the counter value that causes the state of the output to change during the period. The starting polarity of the output is selectable on a per-channel basis and is selected by programming the PWMPOL register. The contents of the PWMPOL register are shown in Figure 8.43.

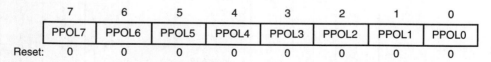

	7	6	5	4	3	2	1	0
	PPOL7	PPOL6	PPOL5	PPOL4	PPOL3	PPOL2	PPOL1	PPOL0
Reset:	0	0	0	0	0	0	0	0

PPOLx: PWM channel x polarity
   0 = PWM channel x output is low at the start of a period, then goes high when the
      duty count is reached.
   1 = PWM channel x output is high at the start of a period, then goes low when the
      duty count is reached.

Figure 8.43 ■ PWM Polarity register (PWMPOL)

The block diagram of a PWM channel is shown in Figure 8.44. A PWM channel must be enabled to work. It is enabled by setting a bit in the PWME register. The contents of the PWME register are shown in Figure 8.45. There is an edge-synchronizing circuit (labeled as GATE in Figure 8.44) to guarantee that the clock will only be enabled or disabled at an edge.

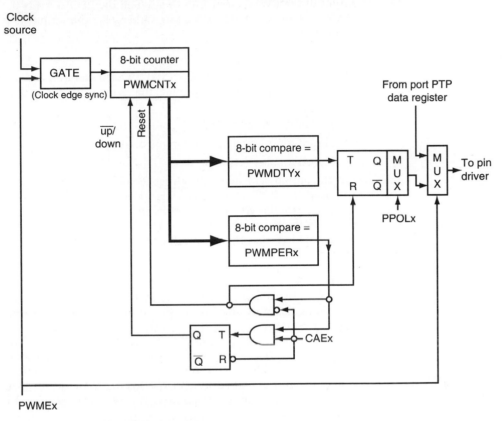

Figure 8.44 ■ PWM channel block diagram

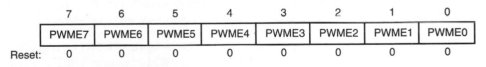

	7	6	5	4	3	2	1	0
	PWME7	PWME6	PWME5	PWME4	PWME3	PWME2	PWME1	PWME0
Reset:	0	0	0	0	0	0	0	0

PWMEx: PWM channel x enable
0 = PWM channel x disabled.
1 = PWM channel x enabled.

Figure 8.45 ■ PWM Enable register (PWME)

8.10.3 PWM Waveform Properties

There are four major properties in a PWM waveform: polarity, alignment, period, and duty cycle.

PWM POLARITY

The *polarity* of a PWM waveform refers to the voltage level (high or low) that a PWM waveform starts with in each period. Each PWM channel has a polarity bit to allow starting a waveform cycle with a high or low level. This is shown in Figure 8.44 as a MUX select of either the Q or $\overline{Q}$ output of the PWM output flip-flop. When the PPOLx bit is 0, the Q output is selected and the PWM output will start with a low. Otherwise, the output $\overline{Q}$ is selected and the PWM output starts with a high.

PWM PERIOD AND DUTY

Dedicated period and duty registers exist for each channel and are double buffered so that if they change while the channel is enabled, the change will not take effect until one of the following occurs:

- The effective period ends.
- The counter is written (counter is reset to $00).
- The channel is disabled.

In this way, the output will always be either the old waveform or the new waveform, not some variation in between. If the channel is not enabled, then writes to the period and duty registers will go directly to the latches and the buffer. A change in duty or period can be forced into effect immediately by writing the new values to the duty and/or period registers and then writing to the counter. This forces the counter to reset and the new duty and/or period values to be latched.

PWM COUNTERS

Each channel has a dedicated 8-bit up-and-down counter that runs at the rate of the selected clock source. The counter is compared to the duty and period registers in each clock cycle. When the counter matches the duty register, the output flip-flop changes state, causing the PWM waveform to also change state. A match between the PWM counter and the period register behaves differently depending on what output mode is selected.

Any value written to the counter causes the counter to reset to $00 and start to count up, both the duty and period registers to be loaded with values from their buffers, and the output to change according to the polarity bit. When the channel is disabled, the counter stops.

PWM WAVEFORM ALIGNMENT

The PWM timer provides the choice of two types of outputs: *left aligned* and *center aligned*. A left-aligned waveform has two line segments in each period whereas a center-aligned waveform has three line segments in one period. They are selected with the CAEx bits in the PWMCAE register. The contents of the PWMCAE register are shown in Figure 8.46. If the CAEx bit is 0, the corresponding PWM output will be left aligned.

	7	6	5	4	3	2	1	0
	CAE7	CAE6	CAE5	CAE4	CAE3	CAE2	CAE1	CAE0
Reset:	0	0	0	0	0	0	0	0

CAEx: center-aligned enable bit for channel x
  0 = PWM channel x output is left aligned.
  1 = PWM channel x output is center aligned.

Figure 8.46 ■ PWM Center Align Enable register (PWMCAE)

LEFT-ALIGNED OUTPUT

In the *left-aligned output* mode, the 8-bit counter is configured as an up counter only. When the PWM counter equals the duty register, the output flip-flop changes state; this causes the PWM waveform to also change state. A match between the PWM counter and the period register resets the counter and the output flip-flop. The counter counts from 0 to the value in the period register – 1.

The waveform of the left-aligned mode is shown in Figure 8.47. The frequency of the PWM output is given by the following equation:

PWMx frequency = clock (A, B, SA, or SB) ÷ *PWMPERx*

The duty cycle of the waveform depends on the selected polarity.

Polarity = 0,
PWMx duty cycle = [(*PWMPERx* − *PWMDTYx*) ÷ *PWMPERx*] × 100%

Polarity = 1,
PWMx duty cycle = [*PWMDTYx* ÷ *PWMPERx*] × 100%

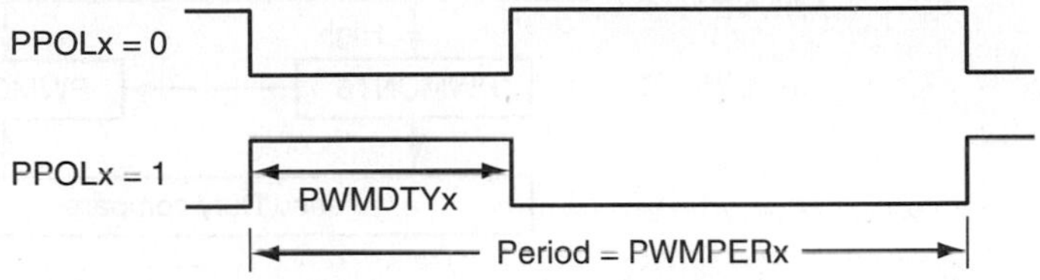

Figure 8.47 ■ PWM left-aligned output waveform

CENTER-ALIGNED MODE

In this mode, the 8-bit PWM counter operates as an up-and-down counter and is set to count up whenever the counter is equal to $00. The counter compares with two registers, a duty register and a period register, in each clock cycle. When the counter matches the duty register, the output flip-flop changes state, causing the PWM waveform to also change state. A match between the PWM counter and the period register changes the counter direction from an up count to a down count. When the PWM counter decrements and matches the duty register again, the output flip-flop changes state, causing the PWM output to also change state. When the PWM counter decrements and reaches zero, the counter direction changes from a down count back to an up count and the period and duty registers are reloaded from their buffers. Since the PWM counter counts from 0 up to the value in the period register and then back down to 0, the effective period is PWMPERx × 2.

The output waveform of the center-aligned mode is shown in Figure 8.48. The frequency of the center-aligned PWM output can be calculated using the following expression:

PWMx frequency = clock(A, B, SA, or SB) ÷ (2 × *PWMPERx*)

The duty cycle of the waveform depends on the selected polarity.

Polarity = 0,
PWMx duty cycle = [(*PWMPERx* − *PWMDTYx*) ÷ *PWMPERx*] × 100%

Polarity = 1,
PWMx duty cycle = [*PWMDTYx* ÷ *PWMPERx*] × 100%

PWM 16-BIT FUNCTIONS

Two 8-bit PWM modules can be concatenated into one 16-bit PWM module. The concatenation of the PWM channels is controlled by the CON bits of the PWMCTL register. The 16-bit mode PWM system is illustrated in Figure 8.49.

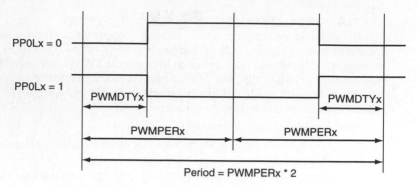

Figure 8.48 ■ PWM center-aligned output waveform

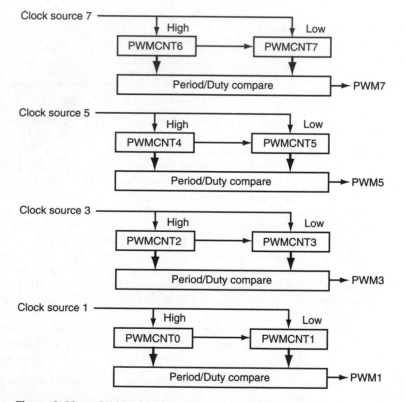

Figure 8.49 ■ PWM 16-bit mode

When PWM channels k and $k + 1$ are concatenated, channel k is the high-order channel, whereas channel $k + 1$ is the low-order channel. When using the 16-bit concatenated mode, the clock source is determined by the low-order 8-bit channel clock select control bits. That is, it is determined by channel $k + 1$ ($k = 0, 2, 4,$ or 6) when channel k and $k + 1$ are concatenated. The resulting PWM is output to the pin of the corresponding low-order 8-bit channel. The polarity of the resulting PWM output is controlled by the PPOLx bit of the corresponding low-order 8-bit channel as well.

Once concatenated mode is enabled, then enabling or disabling the corresponding 16-bit PWM channel is controlled by the low-order PWMEx bit. In this case, the PWMEx bits of the high-order bytes have no effect and their corresponding PWM outputs are disabled. In concatenated mode, writes to the 16-bit counter by using a 16-bit access or writes to either the low- or high-order byte of the counter will reset the 16-bit counter. Reads of the 16-bit counter must be made by a 16-bit access to maintain data coherency. Either left-aligned or center-aligned output mode can be used in concatenated mode and is controlled by the low-order CAEx bit. The high-order CAEx bit has no effect.

PWM BOUNDARY CASES

Table 8.5 summarizes the boundary conditions for the PWM regardless of the output mode (applicable to both the 8-bit and 16-bit PWM modes).

PWMDTYx	PWMPERx	PPOLx	PWMx Output
$00 (indicates no duty)	>$00	1	Always low
$00 (indicates no duty)	>$00	0	Always high
xx	$00[1] (indicates no period)	1	Always high
xx	$00[1] (indicates no period)	0	Always low
>= PWMPERx	xx	1	Always high
>= PWMPERx	xx	0	Always low

Note: 1. Counter = $00 and does not count.

Table 8.5 ■ PWM boundary cases

EMERGENCY PWM SHUTDOWN

The PWM system can be shut down under emergency conditions. The emergency shutdown is controlled by the PWMSDN register. The contents of this register are shown in Figure 8.50.

Example 8.21

Write an instruction sequence to program the PWM channel 0 to output a waveform with 50 percent duty cycle and 100-kHz frequency. Assume that the E-clock is 24 MHz.

Solution: To achieve the 50 percent duty cycle, 100-kHz PWM output, we use the following parameters:
- Clock source prescale factor set to 2
- Clock A selected as the clock input to PWM channel 0

- Left-aligned mode
- The value 120 written into the PWMPER0 register (frequency is 24 MHz ÷ 120 ÷ 2 = 100 kHz)
- The value 60 written into the PWMDTY0 register

The following instruction sequence will perform the configuration:

```
#include "c:\miniide\hcs12.inc"
    ...
        movb    #0,PWMCLK       ; select clock A as the clock source for PWM0
        movb    #1,PWMPRCLK     ; set clock A prescaler to 2
        movb    #1,PWMPOL       ; channel 0 output high at the start of the period
        movb    #0,PWMCAE       ; select left-aligned mode
        movb    #$0C,PWMCTL     ; 8-bit mode, stop PWM in wait and freeze mode
        movb    #120,PWMPER0    ; set period value
        movb    #60,PWMDTY0     ; set duty value
        movb    #0,PWMCNT0      ; reset the PWM0 counter
        bset    PWMEN,PWME0     ; enable PWM channel 0
```

▲

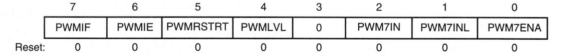

7	6	5	4	3	2	1	0
PWMIF	PWMIE	PWMRSTRT	PWMLVL	0	PWM7IN	PWM7INL	PWM7ENA

Reset: 0   0   0   0   0   0   0   0

PWMIF: PWM interrupt flag
    0 = no change on PWM7IN input.
    1 = change on PWM7IN input.
PWMIE: PWM interrupt enable
    0 = PWM interrupt is disabled.
    1 = PWM interrupt is enabled.
PWMRSTRT: PWM restart
    The PWM can only be restarted if the PWM channel input 7 is de-asserted. After writing a logic 1 to this bit, the PWM channels start running after the corresponding counter passes the next "counter == 0" phase.
PWMLVL: PWM shutdown output level
    0 = PWM outputs are forced to 0.
    1 = PWM outputs are forced to 1.
PWM7IN: PWM channel 7 input status
    This bit reflects the current status of the PWM7 pin.
PWM7INL: PWM shutdown active input level for channel 7
    0 = active level is low.
    1 = active level is high.
PWM7ENA: PWM emergency shutdown enable
    If this bit is logic 1, the pin associated with channel 7 is forced to input and the emergency shutdown feature is enabled. All the other bits in this register are meaningful only if PWM7ENA = 1.
    0 = PWM emergency feature is disabled.
    1 = PWM emergency feature is enabled.

Figure 8.50 ■ PWM Shutdown register (PWMSDN)

Example 8.22

Assume that the E-clock frequency is 24 MHz. Write an instruction sequence to generate a square wave with a period of 20 μs and 60 percent duty cycle using PWM channel 0. Use center-aligned mode.

Solution: Select clock A with prescale factor 2 as the clock source to PWM0, then

$$20 \text{ μs} = 2/24 \text{ μs} \times \text{PWMPER0} \times 2 \Rightarrow \text{PWMPER0} = 120$$

Select the PWM0 waveform to start with high level. Then the value to be loaded into the PWMDTY0 is computed as follows:

$$60\% = (\text{PWMDTY0} \div \text{PWMPER0}) \times 100\% \Rightarrow \text{PWMDTY0} = 72$$

We need to choose the following parameters for this waveform:

- Clock source prescale factor set to 2
- Clock A selected as the clock input to PWM channel 0
- Center-aligned mode with PPOL0 equal to 1
- The value 120 written into the PWPER0 register
- The value 72 written into the PWDTY0 register

The following instruction sequence will configure the PWM0 properly:

```
movb    #0,PWMCLK        ; select clock A as the clock source
movb    #1,PWMPOL        ; set PWM0 output to start with high level
movb    #1,PWMPRCLK      ; set the PWM0 prescaler to clock A to 2
movb    #1,PWMCAE        ; select PWM0 center-aligned mode
movb    #$0C,PWMCTL      ; select 8-bit mode, stop PWM in wait mode
movb    #120,PWMPER0     ; set period value
movb    #72,PWMDTY0      ; set duty value
bset    PWME,PWME0       ; enable PWM channel 0
```

The following C language statements will configure PWM0 properly:

```
PWMCLK      = 0;            // select clock A as the clock source to PWM0
PWMPOL      = 1;            // PWM0 output start with high level
PWMCTL      = 0x0C;         // select 8-bit PWM0
PWMCAF      = 1;            // PWM0 center aligned
PWMPRCLK    = 1;            // set clock A prescaler to 2
PWPER0      = 120;
PWDTY0      = 72;
PWEN        |= PWME0;       // enable PWM0
```

Example 8.23

Assume that the E-clock frequency is 24 MHz. Write an instruction sequence to generate a 50-Hz digital waveform with 80 percent duty cycle using the 16-bit mode from the PWM1 pin output.

Solution: The ratio of 24 MHz and 50 Hz is 480,000. One of the possible breakdowns of this number is 16 times 30,000. Therefore, one of the possible settings is as follows:

- Select clock A as the clock source.
- Set the prescaler to 16.
- Select left-aligned mode.
- Select polarity of the waveform to be 1.
- Load the value of 30,000 into the PWMPER0:PWMPER1.
- Load the value of 24,000 into the PWMDTY0:PWMDTY1.

The following instruction sequence will achieve the desired setting:

```
movb  #0,PWMCLK       ; select clock A as the clock source
movb  #2,PWMPOL       ; set PWM0:PWM1 output to start with high level
movb  #4,PWMPRCLK     ; set prescaler to 16
movb  #$1C,PWMCTL     ; concatenate PWM0:PWM1, stop PWM in wait mode
movb  #0,PWMCAE       ; select left align mode
movw  #30000,PWMPER0  ; set period to 30000
movw  #24000,PWMDTY0  ; set duty to 24000
bset  PWME,PWME1      ; enable PWM0:PWM1
```

The PWM function can be used in many applications that require the average value of output voltages. The lamp dimmers and DC motor speed control are two examples of the PWM applications.

Example 8.24

Using PWM in dimming the light. Suppose we are using the PWM0 of the HCS12 to control the brightness of a light bulb. The circuit connection is shown in Figure 8.51. Write a program so that the light is turned down to 10 percent brightness gradually in 5 s. The E-clock frequency is 24 MHz.

Solution: We dim the light in the following manner: Use the PWM0 output to control the brightness of the light bulb. Set the duty cycle to 100 percent at the beginning and then dim the brightness by 10 percent in the first second, and then 20 percent per second in the following 4 s. Use 100 as the initial duty and period values.

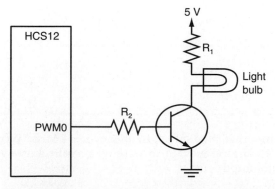

Figure 8.51 ■ Using PWM0 function to dim the light

Since the brightness of the light is proportional to the average current that flows through the bulb, we can dim the light bulb by reducing the duty cycle of the PWM output from 100 percent down to 10 percent in 5 s. We will reduce the duty cycle in steps. In 1 s, we reduce the duty value 10 times. Therefore, we reduce the duty value by 1 every 100 ms in the first second and reduce the duty value by 2 every 100 ms in the following 4 s.

The assembly program that implements this idea is as follows:

```
            #include   "c:\miniide\hcs12.inc"
            org        $1000
dim_cnt     rmb        1
            org        $1500
            movb       #0,PWMCLK          ; select clock A as the clock source
            movb       #1,PWMPOL          ; make waveform to start with high level
            movb       #$0C,PWMCTL        ; select 8-bit mode
            movb       #2,PWMPRCLK        ; set clock prescaler to 4
            movb       #0,PWMCAE          ; select left-aligned mode
            movb       #100,PWMPER0       ; set period value (PWM frequency doesn't matter)
            movb       #100,PWMDTY0       ; set duty value
            bset       PWME,PWME0         ; enable PWM channel 0
; The following instruction segment reduces duty count by 1 per 100 ms
            movb       #10,dim_cnt
loop1       ldy        #1
            jsr        delayby100ms
            dec        PWMDTY0            ; decrement duty by 1
            dec        dim_cnt
            bne        loop1
; The following instruction segment reduces duty count by 2 per 100 ms in 4 s
            movb       #40,dim_cnt
loop2       ldy        #1                 ; wait for 100 ms
            jsr        delayby100ms       ;    "
            dec        PWMDTY0            ; decrement duty cycle by 2%
            dec        PWMDTY0            ; per 100 ms
            dec        dim_cnt
            bne        loop2
            swi
            #include   "c:\miniide\delay.asm"; include delayby100ms here
            end
```

The C language version of this program is as follows:

```
#include       "c:\cwHCS12\include\hcs12.h"
#include       "c:\cwHCS12\include\delay.h"
void main ()
{
    int   dim_cnt;
    PWMCLK   = 0;                 // select clock A as the clock source
    PWMPOL   = 1;                 // make waveform to start with high level
    PWMCTL   = 0x0C;              // select 8-bit mode
    PWMPRCLK = 2;                 // set clock prescaler to 4
    PWMCAE   = 0;                 // select left aligned mode
    PWMPER0  = 100;               // set period of PWM0 to 0.1 ms
    PWMDTY0  = 100;               // set duty cycle to 100%
```

```
PWME        |= 0x01;                    // enable PWM0 channel
                                        // reduce duty cycle 1% per 100 ms in the first second
for (dim_cnt = 0; dim_cnt < 10 ; dim_cnt++) {
        delayby100ms(1);
        PWMDTY0 − −;
}
                                        // reduce duty cycle 2% per 100 ms in the next 4 s
for (dim_cnt = 0; dim_cnt < 40; dim_cnt++) {
        delayby100ms(1);
        PWMDTY0 − = 2;
}
while(1);
}
```

8.11 DC Motor Control

DC motors are used extensively in control systems as positional devices because their speeds and torques can be precisely controlled over a wide range. The DC motor has a permanent magnetic field and its armature is a coil. When a voltage and a subsequent current flow are applied to the armature, the motor begins to spin. The voltage level applied across the armature determines the speed of rotation.

The microcontroller can digitally control the angular velocity of a DC motor by monitoring the feedback lines and driving the output lines. Almost every application that uses a DC motor requires it to reverse its direction of rotation or vary its speed. Reversing the direction is done by changing the polarity of the voltage applied to the motor. Changing the speed requires varying the voltage level of the input to the motor, and that means changing the input level to the motor driver. In a digitally controlled system, the analog signal to the driver must come from some form of D/A converter. However, adding a D/A converter to the circuit increases the chip count, which means increasing the system cost and power consumption. The other alternative is to vary the pulse width of a digital signal input to the motor. By varying the pulse width, the average voltage delivered to the motor changes and so does the speed of the motor. The HCS12 PWM subsystem can be used to control the DC motor.

The HCS12 can interface with a DC motor through a driver, as shown in Figure 8.52. This circuit takes up only three I/O pins. The pin that controls the direction can be an ordinary I/O pin,

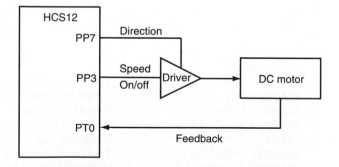

Figure 8.52 ■ Simplified circuit for DC motor control

but the pin that controls the speed must be a PWM pin. The pin that receives the feedback must be an input-capture pin.

Although some DC motors can operate at 5 V or less, the HCS12 cannot supply the necessary current to drive a motor directly. The minimum current required by any practical DC motor is much higher than any microcontroller can supply. Depending on the size and rating of the motor, a suitable driver must be selected to take control signals from the HCS12 and deliver the necessary voltage and current to the motor.

8.11.1 Drivers

Standard motor drivers are available in many current and voltage ratings. Examples are the L293 from ST microelectronics and SN754410 from TI. These two chips are pin compatible. The SN754410 has four channels and can output up to 1 A of current per channel with a supply of 36 V. It has a separate logic supply and takes a logic input (0 or 1) to enable or disable each channel. The SN754410 also includes clamping diodes needed to protect the driver from the back electromagnetic frequency (EMF) generated during the motor reversal. The pin assignment and block diagram of the SN754410 are shown in Figure 8.53. There are two supply voltages: V_{CC1} and V_{CC2}. V_{CC1} is the logic supply voltage, which can be from 4.5 to 36 V (normally 5.0 V). V_{CC2} is the analog supply voltage and can be as high as 36 V.

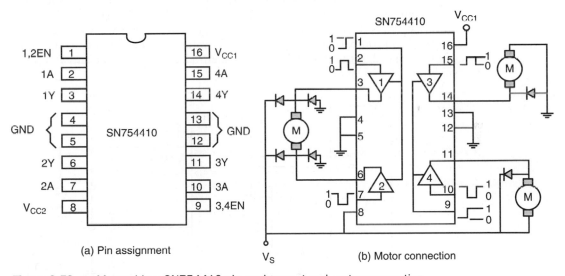

(a) Pin assignment   (b) Motor connection

Figure 8.53 ■ Motor driver SN754410 pin assignment and motor connection

8.11.2 Feedback

The DC motor controller needs information to adjust the voltage output to the motor driver circuit. The most important information is the speed of the motor, which must be fed back from the motor by a sensing device. The sensing device may be an optical encoder, infrared detector, Hall-effect sensor, and so on. Whatever the means of sensing is, the result is a signal, which is fed back to the microcontroller. The microcontroller can use the feedback to determine the speed and position of the motor. Then it can make adjustments to increase or decrease the speed, reverse the direction, or stop the motor.

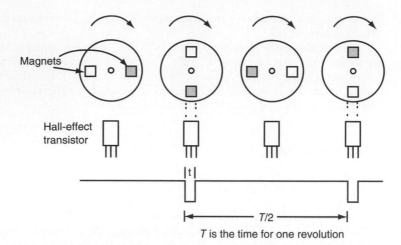

Figure 8.54 ■ The output waveform of the Hall-effect transistor

Assume that a Hall-effect transistor is mounted on the armature (stator) of a DC motor and two magnets are mounted on the shaft (rotor). As shown in Figure 8.54, every time the magnet passes by the Hall-effect transistor, a pulse is generated. The input-capture function of the HCS12 can capture the passing time of the pulse. The time between two captures is half of a revolution. Thus the motor speed can be calculated. By storing the value of the capture registers each time and comparing it with its previous value, the controller can constantly measure and adjust the speed of the motor. Using this method, a motor can be run at a precise speed or be synchronized with another event.

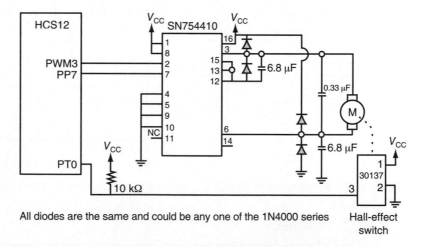

Figure 8.55 ■ Schematic of an HCS12-based motor-control system

The schematic of a motor-control system is illustrated in Figure 8.55. The PWM output from the PWM3 pin is connected to one end of the motor, and the PP7 pin is connected to the other end of the motor. The circuit is connected so that the motor will rotate clockwise when the voltage of the PP7 pin is 0 while the PWM output is nonzero (positive). The direction of

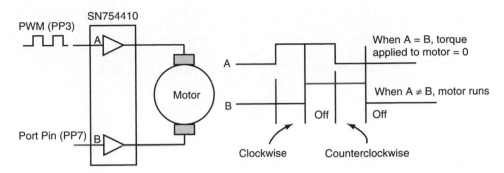

Figure 8.56 ■ The TI SN754410 motor drive

motor rotation is illustrated in Figure 8.56. By applying appropriate voltages on PP7 and PP3 (PWM3), the motor can rotate clockwise, counterclockwise, or even stop. Input-capture channel 0 is used to capture the feedback from the Hall-effect transistor.

When a motor is first turned on, it cannot reach a steady speed immediately. A certain amount of startup time should be allowed for the motor to get to speed. A smaller motor can reach steady speed faster than a larger one. It is desirable for the motor speed to be constant for many applications. However, when a load is applied to the motor, it will be slowed down. To keep the speed constant, the duty cycle of the voltage applied to the motor should be increased. When the load gets lighter, the motor will accelerate and run faster than desired. To slow down the motor, the duty cycle of the applied voltage should be reduced.

The response time will be slow if the change to the duty cycle is small. However, a large variation in duty cycle tends to cause the motor to overreact and oscillate. There are control algorithms for motors that you can find in textbooks on control. A DC motor cannot respond to the change of duty cycle instantaneously because of its inertia. A certain amount of time should be allowed for the motor to speed up or slow down before the effect of the change of duty cycle is measured.

Example 8.25

Write a C function to measure the motor speed (in rpm) assuming that the HCS12 is running with a 16-MHz E-clock.

Solution: To measure the motor speed, one needs to capture two consecutive rising edges. Let the difference of two consecutive edges be *diff* and the period of the timer be set to 1 μs; then the motor speed (rpm) is

Speed = $60 \times 10^6 \div (2 \times diff)$

The C function that measures the motor speed is as follows:

```
#include "c:\cwHCS12\include\hcs12.h"
unsigned int motor_speed (void)
{
        unsigned int edge1, diff, rpm;
        long int temp;
        TSCR1    = 0x90;          /* enable TCNT and fast flag clear */
        TIOS    &= IOS0;          /* select IC0 function */
        TSCR2    = 4;             /* set TCNT prescale factor to 16 */
```

```
TCTL4      = 0x01;                    /* select to capture the rising edge of PT0 */
TFLG1      = COF;                     /* cleared COF flag */
while (!(TFLG1 & COF));               /* wait for the first edge */
edge1      = TC0;
while (!(TFLG1 & COF));               /* wait for the second edge */
diff       = TC0 − edge1;
temp       = 10000001/(long)(2 * diff);
rpm        = temp * 60;
return rpm;
}
```

8.11.3 Electrical Braking

Once a DC motor is running, it picks up speed. Turing off the voltage to the motor does not make it stop immediately because the momentum will keep it rotating. After the voltage is turned off, the momentum will gradually wear out because of friction. If the application does not require an abrupt stop, then the motor can be brought to a gradual stop by removing the driving voltage.

An abrupt stop may be required by certain applications in which the motor must run a few turns and stop quickly at a predetermined point. This could be achieved by electrical braking. Electrical braking is done by reversing the voltage applied to the motor. The length of time that the reversing voltage is applied must be precisely calculated to ensure a quick stop while not starting the motor in the reverse direction. A discussion of good motor braking algorithms is outside the scope of this textbook. In a closed-loop system, the feedback can be used to determine where or when to start and stop braking and when to discontinue. In Figure 8.56, the motor can be braked by (1) reducing the PWM duty count to 0 or (2) setting port pin PP7 output to high for an appropriate amount of time.

8.12 Summary

Many applications require a dedicated timer. Without a timer the following applications will become very difficult or even impossible to implement:

- The measurement of pulse width, frequency, period, duty cycle, and phase difference
- The detection of certain events
- The creation of time delays
- The generation of waveforms
- The generation of a siren and playing of songs

Some HCS12 members implement an Enhanced Capture Timer (ECT) module, which has the features of the Standard Timer Module (TIM) enhanced by additional functions. The heart of the timer system is the 16-bit main timer, TCNT. This timer must be enabled in order to run. Its clock signal is derived by dividing the E-clock by a prescaler. The prescale factor can be from 1 to 128 for the standard timer module and the enhanced capture timer module.

The input-capture function can be programmed to latch the main timer value (TCNT) into the input-capture register on the arrival of an active edge and optionally generate an interrupt. The input-capture function is often used to measure the period, pulse width, duty cycle, and phase shift. It can also be used as a time reference and count the events that occur within an interval. There can be up to eight channels (IC0 to IC7) of input-capture function. The input-capture function of TIM has several limitations that make it unsuitable for high-frequency applications and a noisy environment. The ECT module adds a holding register to each of the four input-capture channels (IC0 to IC3). This enhancement adds the following capabilities:

- *Interrupt the CPU after two edges have been captured instead of each edge*. This capability reduces the software overhead.
- *Ignore the short pulse*. This capability enables the user to capture event arrival time and measure pulse width or period in a noisy environment.
- *Selective no-overwrite*. This capability allows the user to do measurements in very high frequency and not miss the true events because the CPU is busy with other chores.

There can be up to eight output-compare channels (OC0 to OC7). Each output-compare channel has a 16-bit register, a 16-bit comparator, and an output-compare action pin. The output-compare function is often used to create a time delay, to generate a waveform, to activate an operation at a predetermined future time, and so on. To use the output-compare function, one makes a copy of the main timer, adds a delay to this copy, and stores the sum into an output-compare register. The 16-bit comparator compares the contents of the main timer with that of the output-compare register. When they are equal, the corresponding timer flag will be set and an optional action on the associated signal pin will be triggered: pull to high, pull to low, or toggle. An output-compare operation can be forced to take effect immediately by writing a 1 into the corresponding bit in the FORC register. This action will not set the timer flag and will not generate an interrupt either.

The output-compare channel 7 can control up to eight output-compare channels at the same time. This capability allows the user to use two output-compare channels to control the same signal pin.

The TIM module has a 16-bit pulse-accumulator module (PACA). The PACA function has two operation modes: *event counting* and *gated time accumulation*. This function has been used to generate an interrupt after *N* events have occurred, measure the frequency of a signal, count events, measure the pulse width, and so on.

The ECT module adds four buffered 8-bit pulse accumulators. Each pulse accumulator has an 8-bit counter that will increment when an active edge arrives. Each pair of the pulse accumulators can be concatenated into a 16-bit pulse accumulator.

A set of delay functions are created and stored in the files *delay.asm* and *delay.c* and can be called by the end user. These functions utilize the modulus down counter to generate time delays.

Using the output-compare function to generate digital waveforms requires frequent attention from the CPU. This reduces the CPU time available for other applications. The pulse-width modulation (PWM) is designed to reduce the CPU load from waveform generation. The HCS12 provides eight channels of 8-bit PWM modules. The clock source, prescale factor, the polarity, the waveform alignment, the duty cycle, and the period of the waveform are all programmable. Two adjacent 8-bit PWM channels can be concatenated into one 16-bit PWM channel. The PWM function is used in motor control, light dimming, and any application that requires the control of the average voltage level.

8.13 Exercises

Assume that the E-clock frequency of the HCS12 is 24 MHz for the following questions unless it is specified otherwise:

E8.1 Write a program to configure all Port T pins to be used in capturing event arrival times. Use the interrupt-driven approach. Stay in a wait loop after completing the configuration. Exit the wait loop when all eight channels have arrived. Store the arrival times in memory locations starting from $1000.

E8.2 Use the input-capture channel 1 to measure the duty cycle of a signal. Write an assembly and a C program (in a subroutine format) to do the measurement.

E8.3 Assume that two signals having the same frequency are connected to the pins PT1 and PT0. Write an assembly and a C program to measure their phase difference.

E8.4 Write an assembly and a C program to generate a 2-kHz, 70 percent duty cycle waveform from the PT6 pin.

E8.5 Write an assembly and a C program to generate a 4-kHz, 80 percent duty cycle waveform from the PT5 pin.

E8.6 Write a subroutine that can generate a time delay from 1 to 100 s using the modulus down counter. The number of seconds is passed to the subroutine in accumulator A.

E8.7 What would be the output frequency of the PT0 signal generated by the following program segment?

```
clr     PTT
movb    #$01,DDRT
bset    TIOS,$81
movb    #$01,OC7M
movb    #$01,OC7D
movb    #$90,TSCR1
movb    #$02,TCTL2
movb    #$0,TIE
movb    #$08,TSCR2
ldd     #$0
std     TC0
ldd     #$1
std     TC7
```

E8.8 Write a program to generate a 25-Hz digital waveform with a 50 percent duty cycle on the PT0 pin as long as the voltage level on the PP7 pin is high. Your program should consist of two parts:

1. *Entry test.* As long as the PP7 pin is low, it stays in this loop.

2. *Waveform generation body.* This part generates a pulse with 20 ms high time and 20 ms low time, and at the end of a period tests the PP7 signal. If PP7 is still high, it generates the next pulse. Otherwise, it jumps to *entry test.*

E8.9 Write a program to wait for an event (rising edge) to arrive at the PT0 pin. After that, the program will wait for 100 ms and trigger a pulse 20 ms wide on the PT6 pin.

E8.10 In Example 8.2, we used the polling method to check for the arrival of edges. Write a program that uses the interrupt-driven approach to measure the period of an unknown signal. There will be two interrupts related to the PT0 active edges and zero or more TCNT overflow interrupts to be dealt with.

E8.11 Write a program to generate 10 pulses from the PT6 pin. Each pulse has 60 μs high time and 40 μs low time.

E8.12 Write a program to generate an interrupt to the HCS12 20 ms after the rising edge on the pin PT2 has been detected.

E8.13 Suppose that the contents of the TCTL1 and TCTL2 registers are $79 and $9B, respectively. The content of the TFLG1 register is $00. What would occur on pins PT7 to PT0 on the next clock cycle if the value $7F is written into the CFORC register?

E8.14 Write an instruction sequence to configure the modulus down counter so that it generates periodic interrupts to the microcontroller every 40 ms.

E8.15 Modify the programs in Example 8.8 to avoid the drawback in the generated song.

E8.16 Find the score of the song "Home, Sweet Home" and modify the program in Example 8.8 to play it.

E8.17 Write an instruction sequence to generate a 160-kHz digital waveform with 80 percent duty cycle from the PWM2 pin output. Use left-aligned mode.

E8.18 Write an instruction sequence to generate a 120-kHz digital waveform with 40 percent duty cycle from the PWM2 pin output. Use center-aligned mode.

E8.19 Write an instruction sequence to generate a 10-kHz digital waveform with 60 percent duty cycle from the PWM2 pin output. Use left-aligned mode.

E8.20 Write an instruction sequence to generate a 5-kHz digital waveform with 70 percent duty cycle from the PWM2 pin output. Use center-aligned mode.

E8.21 Write an instruction sequence to generate a 20-Hz digital waveform with 50 percent duty cycle using the 16-bit mode from the PWM1 pin output. Use left-aligned mode.

E8.22 Write an instruction sequence to generate a 10-Hz digital waveform with 60 percent duty cycle using the 16-bit mode from the PWM3 pin output. Use center-aligned mode.

E8.23 What is the slowest clock signal that can be generated from the PWM output?

8.14 Lab Exercises and Assignments

L8.1 *Frequency measurement.* Use the pulse-accumulator function to measure the frequency of an unknown signal. The procedure is as follows:

Step 1
Set the function generator output to square wave and adjust the output to between 0 and 5 V. Connect the signal to the PAI (PT7) pin.

Step 2
Also connect the signal to an oscilloscope or a frequency counter. This is for verification purposes.

Step 3
Output the message "Do you want to continue to measure the frequency? (y/n)".

Step 4
Set up the frequency of the signal to be measured and enter *y* or *n* to inform the microcontroller if you want to continue the measurement.

Step 5
Your program would read in the answer from the user. If the character read in is *n*, then stop. If the answer is *y*, then repeat the measurement. If the character is something else, then repeat the same question.

Step 6
Perform the measurement and display the frequency in Hz in decimal format on the screen and go back to step 3. Use as many digits as necessary. The output format should look like

The signal frequency is xxxxxx Hz.

Crank up the frequency until the measurement becomes inaccurate. What is the highest frequency that you can measure?

L8.2 *Pulse-width measurement.* Use the input-capture function to measure the pulse width. The procedure is as follows:

Step 1

Set the function generator output to be square wave and adjust the output to between 0 and 5 V. Connect the signal to the PAI (PT7) pin.

Step 2

Connect the signal to an oscilloscope or a frequency counter to verify your measurement.

Step 3

Output the message "Do you want to continue to measure the pulse width? (y/n)".

Step 4

Set up the appropriate period (frequency) of the signal to be measured and enter *y* or *n* to inform the microcontroller if you want to continue the measurement.

Step 5

Your program would read in the answer from the user. If the character read in is *n*, then stop. If the answer is *y*, then repeat the measurement. If the character is something else, then repeat the same question.

Step 6

Perform the measurement and display the period in μs in decimal format on the screen and go back to step 3. Use as many digits as necessary. The output format should look like

The signal period is xxxxxx microseconds.

Crank up the frequency until the measurement becomes inaccurate. What is the shortest period that you can measure?

L8.3 *Driving the DC motor and servomotor using the PWM module*

Motor kits used

- A cooling fan with DC motor (D24-B10A-04W4-000 from Globe motors) (shown in Figure L8.3a)
- A hobby servo-motor HS-311 made by Hitec (shown in Figure L8.3b)

A *servo* is a small device that incorporates a three-wire DC motor, a gear train, a potentiometer, an integrated circuit, and an output shaft bearing. The shaft of the servomotor can be positioned to specific angular positions by sending coded signals. As long as the coded signal exists on the input line, the servomotor will maintain the angular position of the shaft. If the coded signal changes, then the angular position of the shaft changes.

A common use of servomotors is in radio-controlled models like cars, airplanes, robots, and puppets. They are also used in powerful heavy-duty sailboats. Servos come in different sizes but use similar control schemes and are extremely useful in robotics. The motors are small and extremely powerful for their size. They also draw power proportional to the mechanical load. A lightly loaded servo, therefore, doesn't consume much energy.

A typical servo looks like a rectangular box with a motor shaft coming out of one end and a connector with three wires out of the other end. The three wires are the power, control, and ground. Servos work with voltages between 4 and 6 V. The control line is used to position the servo. Inexpensive servos have plastic gears, and more expensive servos have metal gears which are much more rugged but wear faster.

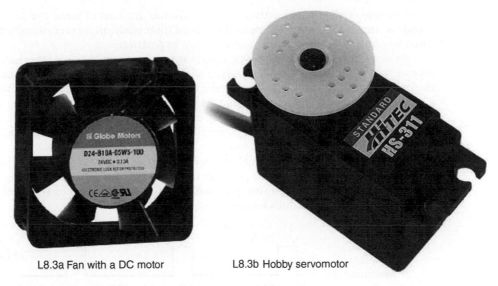

L8.3a Fan with a DC motor   L8.3b Hobby servomotor

Figure L8.3a, b ■ Photos of DC and servomotors

Servos are constructed from three basic pieces: a motor, a potentiometer that is connected to the output shaft, and a control board. The potentiometer allows the control circuitry to monitor the current angle of the servomotor. The motor, through a series of gears, turns the output shaft and the potentiometer simultaneously. The potentiometer is fed into the servo control circuit, and when the control circuit detects that the angle is not correct, it turns the motor the correct direction until the angle is correct. Normally a servo is used to control an angular motion between 0 and 180 degrees. It is not mechanically capable (unless modified) of turning any farther due to the mechanical stop build on the main output gear.

Servos are controlled by sending them a pulse of variable width. The control wire is used to send this pulse. As shown in Figure L8.3c, the pulse has a *minimum pulse*, a *maximum pulse*, and a *repetition rate*. Given the rotation constraints of the servo, neutral is defined to be the position where the servo has exactly the same amount of potential rotation in the clockwise direction as it does in the counterclockwise direction. It is important to note that different servos will have different constraints on their rotation but they all have a neutral position, and that position is always 1.5 ms.

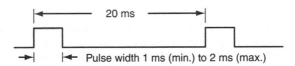

Figure L8.3c ■ Pulse pattern of a servomotor

The angle (indicated by the pointer on the white plate of the servomotor) is determined by the duration of a pulse that is applied to the control line. The servo expects to see a pulse every 20 ms. The length of the pulse will determine how far the motor turns. For example, a 1.5-ms pulse will make the motor turn to the 90-degree position (neutral position).

When a servo is commanded to move, it will move to the position and hold that position. If an external force pushes against the servo while the servo is holding a position, the servo will resist moving out of that position. The maximum amount of force the servo can exert is the torque rating of the servo. Servos will not hold their position forever though; the position pulse must be repeated to instruct the servo to stay in position.

As shown in Figure L8.3d, when a pulse is sent to a servo that is less than 1.5 ms, the servo rotates to a position and holds its output shaft some number of degrees counterclockwise from the neutral point. When the pulse is wider than 1.5 ms, the opposite occurs. The minimal width and the maximum width of pulse that will command the servo to turn to a valid position are functions of each servo. Different brands, and even different servos of the same brand, have different maximum and minimums. Generally the minimum pulse will be about 1 ms wide and the maximum pulse will be 2 ms wide.

Another parameter that varies from servo to servo is the turn rate. This is the time it takes for the servo to change from one position to another. The worst-case turning time is when the servo is holding at the minimum rotation and it is commanded to go to maximum rotation. This can take several seconds for very high torque servos.

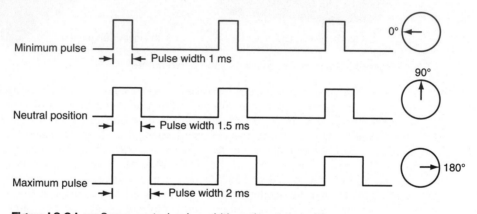

Figure L8.3d ■ Servo control pulse width and motor position

The diagram of the HS-311 servomotor to be used in this lab is shown in Figure L8.3b. There are three terminals connected to the control circuit (see its left).

- Black wire: ground
- Red wire: connected to V_{CC}
- Yellow wire: connected to signal (PWM output)

The HCS12 PWM cannot supply enough current to drive the DC motor or the servomotor directly. A motor driver chip such as the SN754410 from TI is needed to supply the current needed by the motor. The pin assignment of SN754410 is shown in Figure 8.53. The lab procedure is as follows:

Step 1
Connect the circuit properly. Connect the PWM0 and PWM1 pins to the 1A and 2A pins of the SN754410, respectively. Connect the 1Y and 2Y pins to the DC motor and servomotor control input, respectively. The servomotor needs the driving signal to be 50 Hz.

Step 2
Write a program that enables and configures the PWM function properly. The frequency of the PWM output for this lab is fixed (at what frequency?), but the duty cycle will be changed in response to the user input.

Step 3

Enter, assemble, and download the program onto the demo board for execution.

Step 4

Run the program. When the program runs, it outputs the message "duty cycle for DC:" to the UART port (displayed on the HyperTerminal window). You enter from 00 to 99 (using the PC keyboard) to set the duty cycle for the DC motor and the program reads. After that, the MCU will update the duty cycle for the DC motor immediately. The HCS12 next outputs the message "duty cycle for servo:" to the HyperTerminal window. The user will enter the new duty cycle for the digital waveform that drives the servomotor. Remember that the duty cycle for the servomotor is between 5 and 10 percent only. (See Figure L8.3d.) You can figure out a set of convenient values for specifying duty cycles for the servomotor. The program reads in the duty cycle and then performs the required computation to derive the appropriate value that should be written into the duty cycle registers. Your program will echo the new duty cycle on the LCD screen for the DC motor and servomotor, respectively. If you are using the serial monitor to communicate with the PC, then you will need two COM ports from the PC.

After that, the speed of the DC motor and the position of the servomotor would be changed. Your program will wait for 5 s and then ask you to enter the new duty cycles and repeat the same operation.

9

Serial Communication
Interface (SCI)

9.1 Objectives

After completing this chapter, you should be able to

- Explain the four aspects of the TIA-232 standard

- Explain the errors that occur in data transmission

- Establish null-modem connection

- Explain the operation of the SCI subsystem

- Wire the SCI pins to the TIA-232 connector

- Program the SCI subsystem to perform data transmission and reception

9.2 Fundamental Concept of Serial Communications

The need to exchange data between the MCU and peripheral devices can be satisfied by using parallel data transfer (multiple bits in one transfer operation). However, there are a few drawbacks.

- Parallel data transfer requires many I/O pins. This requirement prevents the microcontroller from interfacing with as many devices as desired in the application.
- Many I/O devices do not have a high enough data rate to justify the use of parallel data transfer.
- Data synchronization for parallel transfer is difficult to achieve over a long distance. This requirement is one of the reasons that data communications always use serial transfer.
- Higher cost.

The *serial communication interface (SCI)* was designed to transfer data in an asynchronous mode that utilizes the industrial standard TIA-232 protocol. The TIA-232 was originally called RS-232 because it was a *recommended standard*. You have been using this interface to communicate with and download programs onto the demo board for execution. Only two wires are used by the SCI function.

9.3 The RS-232 Standard

The RS-232 standard was established in 1960 by the Electronic Industry Association (EIA) for interfacing between a computer and a modem. It has experienced several revisions since then. The latest revision, TIA-232F, was published in July 1997. TIA stands for Telecommunication Industry Association. In this revision, the prefix has been changed to TIA. This change represents no change in the standard, but was made to allow users to identify the source of the standard. In data communication terms, both computers and terminals are called *data terminal equipment (DTE)*, whereas modems, bridges, and routers are referred to as *data communication equipment (DCE)*.

There are four aspects to the TIA-232 standard.

1. *Electrical specifications.* Specify the voltage level, rise time and fall time of each signal, achievable data rate, and the distance of communication.
2. *Functional specifications.* Specify the function of each signal.
3. *Mechanical specifications.* Specify the number of pins and the shape and dimensions of the connectors.
4. *Procedural specifications.* Specify the sequence of events for transmitting data, based on the functional specifications of the interface.

9.3.1 TIA-232E Electrical Specification

The following electrical specifications of the TIA-232E are of interest to us:

1. *Data rates.* The TIA-232 standard is applicable to data rates of up to 20,000 bits per second (the usual upper limit is 19,200 baud). Fixed baud rates are not set by the TIA-232E standard. However, the commonly used values are 300, 1200, 2400, 9600, and 19,200 baud. Other accepted values that are not often used are 110 (mechanical teletype machines), 600, and 4800 baud.
2. *Signal state voltage assignments.* Voltages of −3 to −25 V with respect to signal ground are considered logic 1 (the *mark* condition), whereas voltages of +3 to +25 V

are considered logic 0 (the *space* condition). The range of voltages between -3 and $+3$ V is considered a transition region for which a signal state is not assigned.

3. *Signal transfer distance.* The signal should be able to transfer correctly within 15 m. Greater distance can be achieved with good design.

9.3.2 TIA-232E Functional Specification

The TIA-232E standard specifies 22 signals. A summary of these signals is given in Table 9.1. These signals can be divided into six categories.

1. *Signal ground and shield.*

2. *Primary communications channel.* This is used for data interchange and includes flow control signals.

3. *Secondary communications channel.* When implemented, this is used for control of the remote modem, requests for retransmission when errors occur, and governance over the setup of the primary channel.

Pin No.	Circuit	Description
1	–	Shield
2	BA	Transmitted data
3	BB	Received data
4	CA/CJ	Request to send/ready for receiving[1]
5	CB	Clear to send
6	CC	DCE ready
7	AB	Signal common
8	CF	Received line signal detector
9	–	Reserved for testing
10	–	Reserved for testing
11	–	Unassigned[2]
12	SCF/CI	Secondary received line signal detection/data rate selector (DCE source)[3]
13	SCB	Secondary clear to send
14	SBA	Secondary transmitted data
15	DB	Transmitter signal element timing (DCE source)
16	SBB	Secondary received data
17	DD	Receiver signal element timing
18	LL	Local loopback
19	SCA	Secondary request to send
20	CD	DTE ready
21	RL/CG	Remote loopback/signal quality detector
22	CE	Ring indicator
23	CH/CI	Data signal rate selector (DTE/DCE source)[3]
24	DA	Transmitter signal element timing (DTE source)
25	TM	Test mode

1. When hardware flow control is required, circuit CA may take on the functionality of circuit CJ. This is one change from the former TIA-232.
2. Pin 11 is unassigned. It will not be assigned in future versions of TIA-232. However, in international standard ISO 2110, this pin is assigned to select transmit frequency.
3. For designs using interchange circuit SCF, interchange circuits CH and CI are assigned to pin 23. If SCF is not used, CI is assigned to pin 12.

Table 9.1 ■ Functions of EIA-232E signals

4. *Modem status and control signals.* These signals indicate modem status and provide intermediate checkpoints as the telephone voice channel is established.

5. *Transmitter and receiver timing signals.* If a synchronous protocol is used, these signals provide timing information for the transmitter and receiver, which may operate at different baud rates.

6. *Channel test signals.* Before data is exchanged, the channel may be tested for its integrity and the baud rate automatically adjusted to the maximum rate that the channel could support.

SIGNAL GROUND

Pins 7 and 1 and the shell are included in this category. Cables provide separate paths for each, but internal wiring often connects pin 1 and the cable shell or shield to the signal ground on pin 7. All signals are referenced to a common ground as defined by the voltage on pin 7. This conductor may or may not be connected to protective ground inside the DCE device.

PRIMARY COMMUNICATION CHANNEL

Pin 2 carries the *transmit data (TxD)* signal, which is active when data is transmitted from the DTE device to the DCE device. When no data is transmitted, the signal is held in the mark condition (logic 1, negative voltage).

Pin 3 carries the *received data (RxD)*, which is active when the DTE device receives data from the DCE device. When no data is received, the signal is held in the mark condition.

Pin 4 carries the *request to send (RTS)* signal, which is asserted (logic 0, positive voltage) to prepare the DCE device for accepting transmitted data from the DTE device. Such preparation might include enabling the receive circuits or setting up the channel direction in half-duplex applications. When the DCE is ready, it acknowledges by asserting the CTS signal.

Pin 5 carries the *clear to send (CTS)* signal, which is asserted (logic 0) by the DCE device to inform the DTE device that transmission may begin. RTS and CTS are commonly used as handshaking signals to moderate the flow of data into the DCE device.

SECONDARY COMMUNICATION CHANNEL

Pin 14 is the *secondary transmitted data (STxD)*. Pin 16 is the *secondary received data (SRxD)*. Pin 19 carries the *secondary request to send (SRTS)* signal. Pin 13 carries the *secondary clear to send (SCTS)* signal. These signals are equivalent to the corresponding signals in the primary communications channel. The baud rate, however, is typically much slower in the secondary channel, for increased reliability.

MODEM STATUS AND CONTROL SIGNALS

This group includes the following signals:

Pin 6—DCE ready (DSR). When originating from a modem, this signal is asserted (logic 0) when all the following three conditions are satisfied:

1. The modem is connected to an active telephone line that is *off-hook*.
2. The modem is in data mode, not voice or dialing mode.
3. The modem has completed dialing or call setup functions and is generating an answer tone.

If the line goes off-hook, a fault condition is detected, or a voice connection is established, the DCE ready signal is de-asserted (logic 1).

Pin 20—DTE ready (DTR). This signal is asserted (logic 0) by the DTE device when it wishes to open a communications channel. If the DCE device is a modem, the assertion of DTR prepares the modem to be connected to the telephone circuit and, once connected, maintains the connection. When DTR is de-asserted, the modem is switched to *on-hook* to terminate the connection (same as placing the phone back on the telephone socket).

Pin 8—Received line signal detector, also called *carrier detect (CD)*. This signal is relevant when the DCE device is a modem. It is asserted (logic 0) by the modem when the telephone line is off-hook, a connection has been established, and an answer tone is being received from the remote modem. The signal is de-asserted when no answer tone is being received or when the answer tone is of inadequate quality to meet the local modem's requirements.

Pin 12—Secondary received line signal detector (SCD). This signal is equivalent to the CD (pin 8) signal but refers to the secondary channel.

Pin 22—Ring indicator (RI). This signal is relevant when the DCE device is a modem and is asserted (logic 0) when a ringing signal is being received from the telephone line. The assertion time of this signal is approximately equal to the duration of the ring signal, and it is de-asserted between rings or when no ringing is present.

Pin 23—Data signal rate selector. This signal may originate in either the DTE or the DCE devices (but not both) and is used to select one of two prearranged baud rates. The assertion condition (logic 0) selects the higher baud rate.

TRANSMITTER AND RECEIVER TIMING SIGNALS

This group consists of the following signals:

Pin 15—Transmitter signal element timing, also called *transmitter clock (TC)*. This signal is relevant only when the DCE device is a modem and is operating with a synchronous protocol. The modem generates this clock signal to control exactly the rate at which data is sent on TxD (pin 2) from the DTE device to the DCE device. The logic 1–to–logic 0 (negative to positive transition) transition on this line causes a corresponding transition to the next data element on the TxD line. The modem generates this signal continuously, except when it is performing internal diagnostic functions.

Pin 17—Receiver signal element timing, also called *receiver clock (RC)*. This signal is similar to TC, except that it provides timing information for the DTE receiver.

Pin 24—Transmitter signal element timing, also called *external transmitter clock (ETC)*, with timing signals provided by the DTE device for use by a modem. This signal is used only when TC and RC (pins 15 and 17) are not in use. The logic 1–to–logic 0 transition indicates the time center of the data element. Timing signals will be provided whenever the DTE is turned on regardless of other signal conditions.

CHANNEL TEST SIGNALS

This group consists of the following signals:

Pin 18—Local loopback (LL). This signal is generated by the DTE device and is used to place the modem into a test state. When LL is asserted (logic 0, positive voltage), the modem redirects its modulated output signal, which is normally fed into the telephone line, back into its receive circuitry. This enables data generated by the DTE to be echoed back through the local modem to check the condition of the modem circuitry. The modem asserts its test mode signal on pin 25 to acknowledge that it has been placed in LL condition.

Pin 21—Remote loopback (RL). This signal is generated by the DTE device and is used to place the remote modem into a test state. When RL is asserted (logic 0), the remote modem redirects its received data back to its transmitted data input, thereby remodulating the received data and returning it to its source. When the DTE initiates such a test, transmitted data is passed through the local modem, the telephone line, the remote modem, and back, to exercise the channel and confirm its integrity. The remote modem signals the local modem to assert test mode on pin 25 when the remote loopback test is underway.

Pin 25—Test mode (TM). This signal is relevant only when the DCE device is a modem. When asserted (logic 0), it indicates that the modem is in an LL or RL condition. Other internal self-test conditions may also cause the TM signal to be asserted, depending on the modem and the network to which it is attached.

9.3.3 TIA-232E Mechanical Specification

The TIA-232E uses a 25-pin D-type connector, as shown in Figure 9.1a. Since only a small subset of the 25 signals is actually used, a 9-pin connector (DB9) is used in most PCs. The signal assignment of DB9 is shown in Figure 9.1b. The DB9 is not part of the TIA-232E standard.

Signal Direction	Signal Name			Signal Name	Signal Direction
to DCE	Secondary transmitted data	14	1	Protective ground	Both
to DTE	Transmit clock	15	2	Transmitted data	to DCE
to DTE	Secondary received data	16	3	Received data	to DTE
to DTE	Receiver clock	17	4	Request to send	to DCE
	Unassigned	18	5	Clear to send	to DTE
to DCE	Secondary request to send	19	6	Data set ready	to DTE
to DCE	Data terminal ready	20	7	Signal ground	Both
to DTE	Signal quality detect	21	8	Carrier detect	to DTE
to DTE	Ring indicator	22	9	Reserved	
Both	Data rate select	23	10	Reserved	
to DCE	Transmit clock	24	11	Unassigned	
	Unassigned	25	12	Secondary carrier detect	to DTE
			13	Secondary clear to send	to DTE

Figure 9.1a ■ EIA-232E DB25 connector and pin assignment

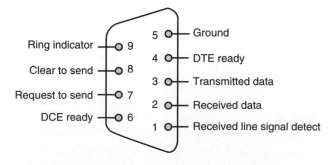

Ring indicator — 9	5 — Ground
Clear to send — 8	4 — DTE ready
Request to send — 7	3 — Transmitted data
DCE ready — 6	2 — Received data
	1 — Received line signal detect

Figure 9.1b ■ EIA-232 DB9 connector and signal assignment

9.3.4 TIA-232E Procedural Specification

The sequence of events that occurs during data transmission using the TIA-232E is easier to understand by studying examples. Two examples are used to explain the procedure.

In the first example, two DTEs are connected with a point-to-point link using a modem. The modem requires only the following circuits to operate:

- Signal ground (AB)
- Transmitted data (BA)
- Received data (BB)
- Request to send (CA)
- Clear to send (CB)
- Data set ready (CC)
- Carrier detect (CF)

Before the DTE can transmit data, the DSR circuit must be asserted to indicate that the modem is ready to operate. This signal should be asserted before the DTE attempts to make a request to send data. The DSR pin can simply be connected to the power supply of the DCE to indicate that it is switched on and ready to operate. When a DTE is ready to send data, it asserts the RTS signal. The modem responds, when ready, with the CTS signal asserted, indicating that data may be transmitted over circuit BA. If the arrangement is half-duplex, then the assertion of the RTS signal also inhibits the receive mode. The DTE sends data to the local modem bit serially. The local modem modulates the data into the carrier signal and transmits the resultant signal over the dedicated communication lines. Before sending out modulated data, the local modem sends out a carrier signal to the remote modem so that it is ready to receive the data. The remote modem detects the carrier and asserts the CD signal. The assertion of the CD signal tells the remote DTE that the local modem is transmitting. The remote modem receives the modulated signal, demodulates it to recover the data, and sends it to the remote DTE over the RxD pin. The circuit connections are illustrated in Figure 9.2.

The next example involves two computers exchanging data through a public telephone line. One of the computers (initiator) must dial the phone (automatically or manually) to establish

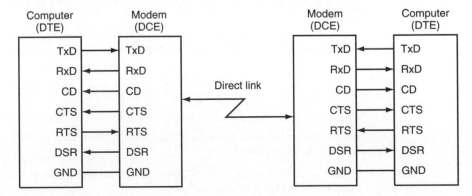

Figure 9.2 ■ Point-to-point asynchronous connection

the connection, just like people talking over the phone. Two additional leads are required for this application.

- Data terminal ready (DTR)
- Ring indicator (RI)

The data transmission in this setting can be divided into three phases.

PHASE 1

Establishing the connection. The following events occur in this phase:

1. The transmitting computer asserts the DTR signal to indicate to the local modem that it is ready to make a call.
2. The local modem opens the phone line and dials the destination number. The number can be stored in the modem or transmitted to the modem by the computer via the TxD pin.
3. The remote modem detects a ring on the phone line and asserts the RI signal to inform the remote computer that a call has arrived.
4. The remote computer asserts the DTR signal to accept the call.
5. The remote modem answers the call by sending a carrier signal to the local modem via the phone line. It also asserts the DSR signal to inform the remote computer that it is ready for data transmission.
6. The local modem asserts both the DSR and CD signals to indicate that the connection is established and it is ready for data communication.
7. For full-duplex data communication, the local modem also sends a carrier signal to the remote modem. The remote modem then asserts the CD signal.

PHASE 2

Data transmission. The following events occur during this phase:

1. The local computer asserts the RTS signal when it is ready to send data.
2. The local modem responds by asserting the CTS signal.
3. The local computer sends data bits serially to the local modem over the TxD pin. The local modem then modulates its carrier signal to transmit the data to the remote modem.
4. The remote modem receives the modulated signal from the local modem, demodulates it to recover the data, and sends it to the remote computer over the RxD pin.

PHASE 3

Disconnection. Disconnection requires only two steps.

1. When the local computer has finished the data transmission, it drops the RTS signal.
2. The local modem then de-asserts the CTS signal and drops the carrier (equivalent to hanging up the phone).

The circuit connection for this example is shown in Figure 9.3. A timing signal is not required in an asynchronous transmission.

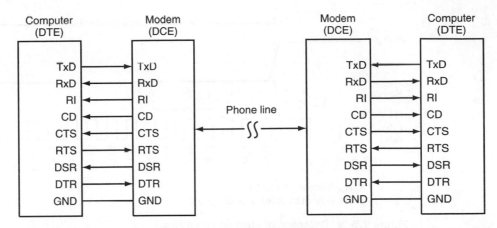

Figure 9.3 ■ Asynchronous connection over public phone line

9.3.5 Data Format

In asynchronous data transfer, data is transferred character by character. Each character is preceded by a start bit (a low), followed by 8 or 9 data bits, and terminated by a stop bit. The data format of a character is shown in Figure 9.4.

Start bit	0	1	2	3	4	5	6	7	Stop bit

Figure 9.4 ■ The format of a character

As shown in Figure 9.4, the least significant bit is transmitted first, and the most significant bit is transmitted last. The stop bit is high. The start bit and stop bit identify the start and end of a character.

Since there is no clock information in the asynchronous format, the receiver uses a clock signal with a frequency that is a multiple (usually 16) of the data rate to sample the incoming data in order to detect the arrival of the start bit and determine the logical value of each data bit. A clock, with a frequency that is 16 times the data rate, can tolerate a frequency difference in the clocks slightly over 3 percent at the transmitter and receiver.

To detect the arrival of a start bit, the SCI waits for the falling edge after the RxD pin has been *idle* (*high*) for at least three sampling times. It will then look at the third, fifth, and seventh samples after the first low sample (these are called verification samples) to determine if a valid start bit has arrived. This process is illustrated in Figure 9.5. If the majority of these three samples are low, then a valid start bit is detected. Otherwise, the SCI will restart the process. After detecting a valid start bit, the SCI will start to shift in the data bits.

To determine the data bit value, the SCI uses a clock with a frequency about 16 times (most often) that of the data rate to sample the RxD signal. If the majority of the eighth, ninth, and tenth samples are 1s, then the data bit is determined to be 1. Otherwise, the data bit is determined to be 0.

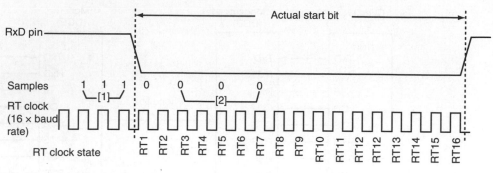

1. A 0 following three 1s.
2. Majority of samples 3, 5, and 7 are 0s.

Figure 9.5 ■ Detection of start bit (ideal case)

The stop bit is high. Using this format, it is possible to transfer data character by character without any gap.

The term *baud rate* is defined as the number of signal changes per second. Since the RS-232 standard uses a *non-return-to-zero* (NRZ) encoding method, baud rate is identical to bit rate. In an NRZ code, a logic 1 bit is sent as a high value and a logic 0 bit is sent as a low value. When a logic 1 follows another logic 1 during a data transfer, the voltage does not drop to zero before it goes high.

Example 9.1

Sketch the output of the letter *g* when it is transmitted using the format of 1 start bit, 8 data bits, and 1 stop bit.

Solution: Letters are represented in ASCII code. The ASCII code of letter *g* is $67 (= 01100111). Since the least significant bit goes out first in the TIA-232 protocol, the format of the output of letter *g* is as shown in Figure 9.6.

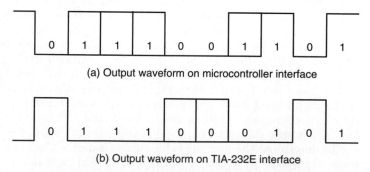

(a) Output waveform on microcontroller interface

(b) Output waveform on TIA-232E interface

Figure 9.6 ■ Data format for letter *g*

9.3.6 Data Transmission Errors

The following errors may occur during the data transfer process using asynchronous serial transmission:

- *Framing error.* A framing error occurs when a received character is improperly framed by the start and stop bits; it is detected by the absence of the stop bit. This error indicates a synchronization problem, faulty transmission, or a break condition. A *break* is defined as the transmission or reception of a logic 0 for one frame or more.

- *Receiver overrun.* One or more characters in the data stream were received but were not read from the buffer before subsequent characters were received.

- *Parity error.* A parity error occurs when an odd number of bits change value. It can be detected by a parity error detecting circuit.

9.3.7 Null Modem Connection

When two DTE devices are located side by side and use the TIA-232E interface to exchange data, there is really no reason to use two modems to connect them. However, the TIA-232E standard does not allow the direct connection of two DTEs. In order to make this scheme work, a *null modem* is needed. The null modem interconnects leads in such a way as to fool both DTEs into thinking that they are connected to modems. The null modem connection is shown in Table 9.2.

Signal Name	DTE 1		DTE 2		Signal Name
	DB25 Pin	DB9 Pin	DB9 Pin	DB25 Pin	
FG (frame ground)	1	–	–	1	FG
TD (transmit data)	2	3	2	3	RD
RD (receive data)	3	2	3	2	TD
RTS (request to send)	4	7	8	5	CTS
CTS (clear to send)	5	8	7	4	RTS
SG (signal ground)	7	5	5	7	SG
DSR (data set ready)	6	6	4	20	DTR
CD (carrier detect)	8	1	4	20	DTR
DTR (data terminal ready)	20	4	1	8	CD
DTR (data terminal ready)	20	4	6	6	DSR

Table 9.2 ■ Null modem connection

In Table 9.2, the signals of DTE1 and DTE2 that are to be wired together are listed in the same row. The transmitter timing and receiver timing signals are not needed in asynchronous data transmission. A ring indicator is not needed either because the transmission is not through a public phone line.

9.4 The HCS12 Serial Communication Interface

An HCS12 device may have one or two serial communication interfaces. These two SCI modules are referred to as SCI0 and SCI1, respectively. The SCI0 module shares the use of the Port S pins PS1 (TxD0) and PS0 (RxD0); the SCI1 shares the use of the Port S pins PS3 (TxD1) and PS2 (RxD1). The block diagram of the HCS12 SCI module is shown in Figure 9.7.

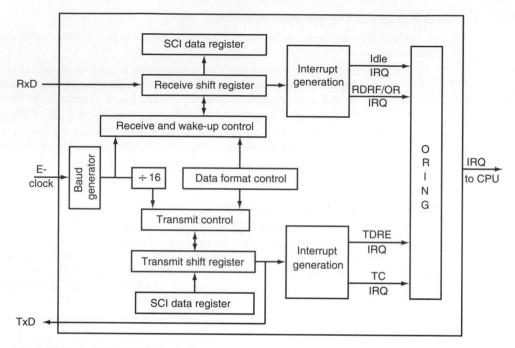

Figure 9.7 ■ HCS12SCI block diagram

The operation of an SCI module involves the following registers:

- *Two baud rate registers.* SCIxBDH and SCIxBDL (x = 0 or 1)
- *Two control registers.* SCIxCR1 and SCIxCR2 (x = 0 or 1)
- *Two status registers.* SCIxSR1 and SCIxSR2 (x = 0 or 1)
- *Two data registers.* SCIxDRH and SCIxDRL (x = 0 or 1)

The HCS12 SCI interface uses a data format of 1 start bit, 8 or 9 data bits, and 1 stop bit. When the SCI is configured to use 9 data bits, one of the bits can be used as the parity bit. The collection of the start bit, data bits, and the stop bit is called a *frame.* The SCI function has the capability to send a break to attract the attention of the other party of the data communication. The SCI function supports hardware parity for transmission and reception. When enabled, a parity bit is generated in hardware for transmitted data and received data. Received parity errors are flagged in hardware. The SCI module supports two wake-up methods, *idle line* wake-up and *address mark* wake-up; this allows the HCS12 to operate in a *multiple-node* environment.

9.5 SCI Baud Rate Generation

The HCS12 SCI modules use a clock signal that is 16 times the data rate to detect the arrival of the start bit and determine the logic value of data bits. The HCS12 SCI modules use a 13-bit counter to generate this clock signal. This circuit is called a *baud rate generator.* To set the baud rate to a certain value, one needs to write an appropriate value to the SCIxBDH:SCIxBDL register

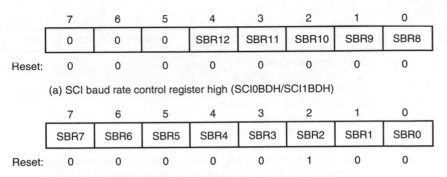

7	6	5	4	3	2	1	0
0	0	0	SBR12	SBR11	SBR10	SBR9	SBR8

Reset: 0 0 0 0 0 0 0 0

(a) SCI baud rate control register high (SCI0BDH/SCI1BDH)

7	6	5	4	3	2	1	0
SBR7	SBR6	SBR5	SBR4	SBR3	SBR2	SBR1	SBR0

Reset: 0 0 0 0 0 1 0 0

(b) SCI baud rate control register low (SCI0BDL/SCI1BDL)

Figure 9.8 ■ SCI baud rate control register

pair. The upper 3 bits of the SCIxBDH register are tied to zeros. The contents of these two registers are shown in Figure 9.8.

This baud rate generator divides down the E-clock to derive the clock signal for reception and transmission. The value (referred to as *SBR*) to be written into the SCIxBDH:SCIxBDL register pair can be derived by rounding the following expression to an integer value:

$$SBR = f_E \div 16 \div \text{baud rate}$$

The divide factors for the baud rate generator for the 16-MHz and 24-MHz E clocks are listed in Table 9.3.

Desired SCI Baud Rate	Baud Rate Divisor for $f_E = 16$ MHz	Baud Rate Divisor for $f_E = 24$ MHz
300	3333	5000
600	1667	2500
1200	833	1250
2400	417	625
4800	208	313
9600	104	156
14,400	69	104
19,200	52	78
38,400	26	39

Table 9.3 ■ Baud rate generation

9.6 The SCI Operation

The operation of the SCI module is controlled by two control registers: SCI0CR1 (SCI1CR1) and SCI0CR2 (SCI1CR2). Their contents are shown in Figures 9.9 and 9.10, respectively. The SCI module allows full duplex, asynchronous, non-return-to-zero (NRZ) serial communication between the CPU and remote devices, including other CPUs. The SCI transmitter and receiver

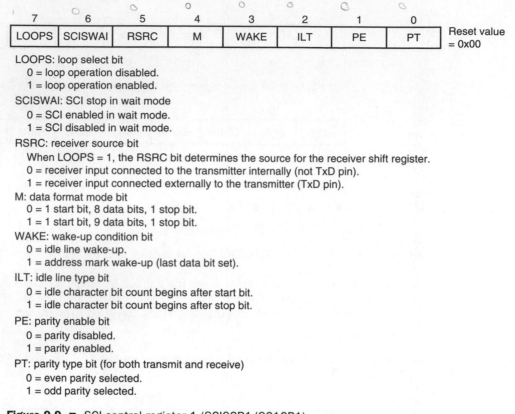

7	6	5	4	3	2	1	0	
LOOPS	SCISWAI	RSRC	M	WAKE	ILT	PE	PT	Reset value = 0x00

LOOPS: loop select bit
  0 = loop operation disabled.
  1 = loop operation enabled.

SCISWAI: SCI stop in wait mode
  0 = SCI enabled in wait mode.
  1 = SCI disabled in wait mode.

RSRC: receiver source bit
  When LOOPS = 1, the RSRC bit determines the source for the receiver shift register.
  0 = receiver input connected to the transmitter internally (not TxD pin).
  1 = receiver input connected externally to the transmitter (TxD pin).

M: data format mode bit
  0 = 1 start bit, 8 data bits, 1 stop bit.
  1 = 1 start bit, 9 data bits, 1 stop bit.

WAKE: wake-up condition bit
  0 = idle line wake-up.
  1 = address mark wake-up (last data bit set).

ILT: idle line type bit
  0 = idle character bit count begins after start bit.
  1 = idle character bit count begins after stop bit.

PE: parity enable bit
  0 = parity disabled.
  1 = parity enabled.

PT: parity type bit (for both transmit and receive)
  0 = even parity selected.
  1 = odd parity selected.

Figure 9.9 ■ SCI control register 1 (SCI0CR1/SC1CR1)

operate independently, although they use the same baud rate generator. The CPU monitors the status of the SCI, writes the data to be transmitted, and processes received data.

9.6.1 Character Transmission

The block diagram of the SCI transmitter is shown in Figure 9.11. To transmit data, the MCU writes the data bits to the SCI data registers (SCIxDRH/SCIxDRL), which in turn are transferred to the transmit shift register. The transmit shift register then shifts a frame out through the TxD pin, after it has prefaced the bits with a start bit and appended a stop bit. The SCI data registers are the write-only buffers between the internal data bus and the transmit shift register.

The SCI also sets a flag, the transmit data register empty flag (TDRE), every time it transfers data from the buffer (SCIxDRH/L) to the transmit shift register. The transmit driver routine may respond to this flag by writing another byte to the transmit buffer, while the shift register is still shifting out the first byte.

There are three major steps in the SCI transmission.

1. *Configure the SCI transmission.*
    a. Select a baud rate. The user must write an appropriate value to the SCI baud registers to set the baud rate. Writing to the SCIxBDH register has no effect without also writing to the SCIxBDL register.

7	6	5	4	3	2	1	0	
TIE	TCIE	RIE	ILIE	TE	RE	RWU	SBK	Reset value = 0x00

TIE: transmit interrupt enable bit
    0 = TDRE interrupt disabled.
    1 = TDRE interrupt enabled.
TCIE: transmit complete interrupt enable bit
    0 = TC interrupt disabled.
    1 = TC interrupt enabled.
RIE: receiver full interrupt enable bit
    0 = RDRF and OR interrupts disabled.
    1 = RDRF and OR interrupt enabled.
ILIE: idle line interrupt enable bit
    0 = IDLE interrupt disabled.
    1 = IDLE interrupt enabled.
TE: transmitter enable bit
    0 = transmitter disabled.
    1 = transmitter enabled.
RE: receiver enable
    0 = receiver disabled.
    1 = receiver enabled.
RWU: receiver wake-up bit
    0 = normal SCI receiver.
    1 = enables the wake-up function and inhibits further receiver
        interrupts. Normally, hardware wakes up the receiver by
        automatically clearing this bit.
SBK: send break bit
    0 = no break characters.
    1 = generate a break code, at least 10 or 11 contiguous 0s. As long
        as SBK remains set, the transmitter sends 0s.

Figure 9.10 ■ SCI control register 2 (SCI0CR2/SCI1CR2)

    b. Write to the SCIxCR1 register to configure the word length, parity, and other configuration bits (LOOPS, RSRC, M, WAKE, ILT, PR, PT).

    c. Enable the transmitter, interrupt, receive, and wakeup as required by writing to the SCIxCR2 register bits (TIE, TCIE, RIE, ILIE, TE, RE, RWU, and SK). A preamble will now be shifted out of the transmitter shift register.

2. Set a *transmit procedure for each character.*

    a. Poll the TDRE flag by reading the SCIxSR1 register or responding to the TDRE interrupt.

    b. If the TDRE flag is set, write the data to be transmitted to SCIxDRH/L, where the ninth bit is written to the T8 bit in the SCIxDRH register if the SCI is in 9-bit data format. A new transmission will not result until the TDRE flag has been cleared.

3. *Repeat step 2 for each subsequent transmission.*

    The contents of the SCIxSR1 and SCIxSR2 registers are shown in Figures 9.12 and 9.13, respectively.

    Setting the TE bit from 0 to 1 automatically loads the transmit shift register with a preamble of 10 logic 1s (if $M = 0$) or 11 logic 1s (if $M = 1$). After the preamble shifts out, control logic transfers the data from the SCI data register into the transmit shift register. A logic 0

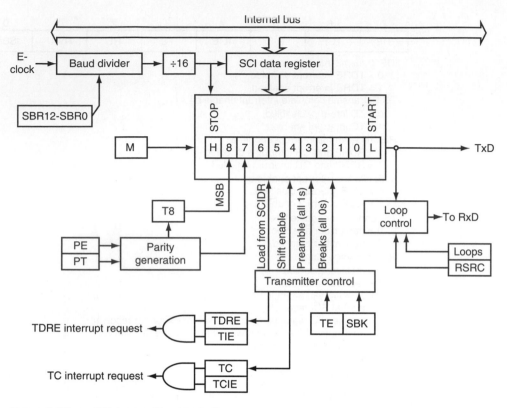

Figure 9.11 ■ SCI transmitter block diagram

start bit automatically goes into the most significant bit position. Hardware supports odd or even parity. When parity is enabled, the most significant bit of the data character is the parity bit. The transmit data register empty flag, TDRE, in the SCIxSR1 register becomes set when the SCI data register transfers a byte to the transmit shift register. The TDRE flag indicates that the SCI data register can accept new data from the internal data bus. If the transmit interrupt enable bit, TIE, in the SCIxCR2 register is also set, the TDRE flag generates a transmit interrupt request.

When the transmit shift register is not transmitting a frame, the TxD output signal goes to the idle state, logic 1. If at any time software clears the TE bit in the SCIxCR2 register, the TxD signal goes idle. If software clears the TE bit while a transmission is in progress, the frame in the transmit shift register continues to shift out. To avoid accidentally cutting off the last frame in a message, always wait for TDRE to go high after the last frame before clearing the TE bit.

Use the following procedure if it is desirable to separate messages with preambles with minimum idle line time:

1. Write the last byte of the first message to the transmit data register.

2. Wait for the TDRE flag to go high, indicating the transfer of the last frame to the transmit shift register.

3. Queue a preamble by clearing and then setting the TE bit.

4. Write the first byte of the second message to the transmit data register.

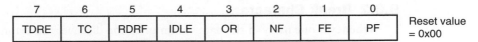

7	6	5	4	3	2	1	0
TDRE	TC	RDRF	IDLE	OR	NF	FE	PF

Reset value
= 0x00

TDRE: transmit data register empty flag
    0 = no byte was transferred to the transmit shift register.
    1 = transmit data register is empty.
TC: transmit complete flag
    0 = transmission in progress.
    1 = no transmission in progress.
RDRF: receiver data register full flag
    0 = SCIxDR empty.
    1 = SCIxDR full.
IDLE: idle line detected flag
    0 = RxD line active.
    1 = RxD line becomes idle.
OR: overrun error flag
    0 = no overrun.
    1 = overrun detected.
NF: noise error flag
    Set during the same cycle as the RDRF bit but not set in the case.
    of an overrun (OR).
    0 = no noise.
    1 = noise.
FE: framing error flag
    Set when a 0 is detected where a stop bit was expected.
    0 – no framing error.
    1 = framing error.
PF: parity error flag
    0 = parity correct.
    1 = incorrect parity detected.

Figure 9.12 ■ SCI status register 1 (SCI0SR1/SCI1SR1)

7	6	5	4	3	2	1	0
0	0	0	0	0	BK13	TXDIR	RAF

Reset value
= 0x00

BK13: break transmit character length
    0 = break character is 10- or 11-bit long.
    1 = break character is 13- or 14-bit long.
TXDIR: transmit pin data direction in single-wire mode
    0 = TxD pin to be used as an input in single-wire mode.
    1 = TxD pin to be used as an output in single-wire mode.
RAF: receiver active flag
    RAF is set when the receiver detects a logic 0 during the RT1 time
    period of the start bit search. RAF is cleared when the receiver detects
    an idle character.
    0 = no reception in progress.
    1 = reception in progress.

Figure 9.13 ■ SCI status register 2 (SCI0SR2/SCISR2)

The TDRE bit in the SCIxSR1 register is cleared by reading the SCIxSR1 register and followed by writing a byte into the SCIxDRL register. All status flags related to reception are cleared by reading the SCI status register followed by reading the SCIxDRL register.

9.6.2 Break Characters

Whenever one party in the data communication discovers an error, it can send break characters to discontinue the communication and start over again. To send a break character, the user sets the SBK bit in the SCIxCR1 register to 1. As long as the SBK bit is 1, transmitter logic continuously loads break characters into the transmit shift register. After software clears the SBK bit, the shift register finishes transmitting the last break character and then transmits at least one logic 1. The automatic logic 1 at the end of a break character guarantees the recognition of the start bit of the next frame.

The SCI module recognizes a break character when a start bit is followed by 8 or 9 logic 0 data bits and a logic 0 where the stop bit should be. Receiving a break character has these effects on SCI registers.

- Sets the framing error flag FE
- Sets the receive data register full flag RDRF
- Clears the SCI data registers (SCIxDRH/L)
- May set the overrun flag OR, noise flag NF, parity error flag PE, or receiver active flag RAF

9.6.3 Idle Characters

An idle character contains all 1s and has no start, stop, or parity bit. The length of the idle character depends on the M bit in the SCIxCR1 register. The preamble is a synchronizing idle character that begins the first transmission initiated after setting the TE bit from 0 to 1.

If the TE bit is cleared during a transmission, the TxD signal becomes idle after the completion of the transmission in progress. Clearing and then setting the TE bit during a transmission queues an idle character to be sent after the frame currently being transmitted.

9.6.4 Character Reception

The block diagram of the SCI receiver is shown in Figure 9.14. The SCI receiver can accommodate either 8-bit or 9-bit data characters. The state of the M bit in the SCI control register 1 determines the length of data characters. When receiving 9-bit data, the R8 bit of the SCIxDRH register holds the ninth bit.

During an SCI reception, the receive shift register shifts in a frame from the RxD pin. The SCI data register is the read-only buffer between the internal bus and the receive shift register. After a complete frame is shifted into the receive shift register, the data portion of the frame is transferred to the SCI data register. The receive data register full flag, RDRF, in the SCIxSR1 register becomes set, indicating that the receive byte can be read. If the receive interrupt enable bit RIE in the SCIxCR2 register is also set, then an interrupt is requested to the MCU.

The receiver uses the method illustrated in Figure 9.5 to detect the arrival of the start bit and uses the majority function of the samples RT8, RT9, and RT10 to determine the logic value of a bit.

9.6.5 Receiver Wake-Up

The SCI module supports the HCS12 to operate in a multiple-receiver system. When a message is not intended for this MCU, the SCI module will put itself in a standby state to ignore the rest of the message. This is done by setting the RWU bit of the SCIxCR2 register. In the standby state, the SCI module will still load the receive data into the SCIxDRH/L registers, but it will not set the RDRF flag.

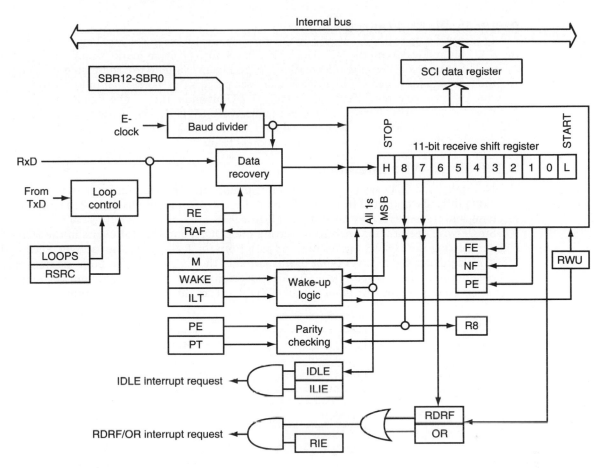

Figure 9.14 ■ SCI receiver block diagram

The transmitting device can address messages to selected receivers by including addressing information in the initial frame or frames of each message. One can choose to use idle line or address mark to wake up the receiver to compare the address information of the incoming frame.

IDLE LINE WAKE-UP

When the idle line wake-up method is chosen, an idle condition on the RxD pin clears the RWU bit in the SCIxCR2 register and wakes up the SCI. The initial frame or frames of every message contain addressing information. All receivers evaluate the addressing information, and receivers for which the message is addressed process the frames that follow. Any receiver for which a message is not addressed can set its RWU bit and return to the standby state. The RWU bit remains set and the receiver remains on standby until another idle character appears on the RxD signal.

Idle line wake-up requires that messages be separated by at least one idle character and that no message contain idle characters. The idle character that wakes a receiver does not set the receiver idle bit, IDLE, or the receive data register full flag, RDRF.

The idle-line-type bit, ILT, determines whether the receiver begins counting logic 1s as idle character bits after the start bit or after the stop bit. The ILT bit is in the SCIxCR1 register.

ADDRESS MARK WAKE-UP

In this wake-up method, a logic 1 in the most significant bit (msb) position of a frame clears the RWU bit and wakes up the SCI. The logic 1 in the msb position marks a frame as an address frame that contains addressing information. All receivers evaluate the addressing information, and the receivers for which the message is addressed process the frames that follow. Any receiver for which a message is not addressed can set its RWU bit and return to the standby state.

The logic 1 of an address frame clears the receiver's RWU bit before the stop bit is received and sets the RDRF flag. Address mark wake-up allows messages to contain idle characters but requires that the msb be reserved for use in address frames.

9.6.6 Single-Wire Operation

Normally, the SCI uses two pins for transmitting and receiving. In single-wire operation, the RxD pin is disconnected from the SCI module. The SCI module uses the TxD pin for both receiving and transmitting, as illustrated in Figure 9.15.

Single-wire operation is enabled by setting the LOOPS and the RSRC bits in the SCIxCR1 register. Setting the LOOPS bit disables the path from the RxD pin to the receiver. Setting the RSRC bit connects the receiver input to the output of the TxD pin driver. Both the transmitter and receiver must be enabled. The TXDIR bit determines whether the TxD pin is going to be used as an input (TXDIR = 0) or an output (TXDIR = 1) in this mode of operation.

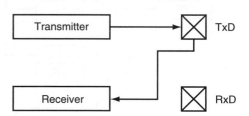

Figure 9.15 ■ Single-wire operation

9.6.7 Loop Operation

In this operation, the transmitter output goes to the receiver input. As shown in Figure 9.16, the RxD signal is disconnected from the SCI. The loop operation is enabled by setting the LOOPS bit and clearing the RSRC bit in the SCIxCR1 register. Setting the LOOPS bit disables the path from the RxD pin to the receiver. Clearing the RSRC bit connects the transmitter output to the receiver input. Both the transmitter and receiver must be enabled.

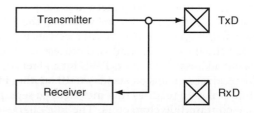

Figure 9.16 ■ Loop operation

9.7 Operation of the SCI in Different MCU Modes

When the MCU is in normal run mode, the SCI module performs normal operation.

When the MCU enters the *wait* mode, the SCI's operation depends on the state of the SCISWAI bit in the SCIxCR1 register.

1. If the SCISWAI bit is 0, the SCI operates normally when the CPU is in *wait* mode.
2. If the SCISWAI bit is 1, the SCI clock generation ceases and the SCI module enters a power-conservation state when the CPU is in *wait* mode. Setting the SCISWAI bit does not affect the state of the receiver enable bit (RE) or the transmitter enable bit (TE).

If the SCISWAI bit is set, any transmission or reception in progress stops at *wait* mode entry. The transmission or reception resumes when either an internal or external interrupt brings the CPU out of *wait* mode. Exiting *wait* mode by reset aborts any transmission or reception in progress and resets the SCI module.

When the MCU enters the *stop* mode, the SCI becomes inactive. The **stop** instruction does not affect the SCI register states, but the SCI module clock will be disabled. The SCI operation resumes from where it left off after an external interrupt brings the CPU out of *stop* mode. Exiting *stop* mode by reset aborts any transmission or reception in progress and resets the SCI.

9.8 Flow Control of USART in Asynchronous Mode

The SCI module will transmit data as fast as the baud rate allows. In some circumstances, the software that is responsible for reading the data from the SCIxDRH/L register may not be able to do so as fast as the data is being received. In this case, there is a need for the HCS12 MCU to tell the transmitting device to suspend transmission of data temporarily. Similarly, the HCS12 MCU may need to be told to suspend transmission temporarily. This is done by means of flow control. There are two common methods of flow control, XON/XOFF and hardware. Here, *X* stands for *transmission*.

The XON/XOFF flow control can be implemented completely in software with no external hardware, but full-duplex communication is required. When incoming data need to be suspended, an XOFF byte is transmitted back to the other device that is transmitting the data being received. To start the other device transmitting again, an XON byte is transmitted. XON and XOFF are standard ASCII control characters. This means that when sending raw data instead of ASCII text, care must be taken to ensure that XON and XOFF characters are not accidentally sent with the data. The ASCII codes of XON and XOFF are 0x11 and 0x13, respectively. The XON character is called the Device Control 1 (DC1) character; the XOFF character is called the Device Control 3 (DC3) character.

Hardware flow control uses extra signals to control the flow of data. To implement hardware flow control on an HCS12 device, extra I/O pins must be used. Generally, an output pin is controlled by the receiving device to indicate that the transmitting device should suspend or resume transmissions. The transmitting device tests an input pin before a transmission to determine whether data can be sent.

Example 9.2

▼

Write an instruction sequence to configure the SCI0 to operate with the following parameters:

- 9600 baud (E-clock is 24 MHz)
- 1 start bit, 8 data bits, and 1 stop bit format
- No interrupt
- Address mark wake-up
- Disable wake-up initially
- Long idle line mode
- Enable receive and transmit
- No loop back
- Disable parity
- SCI stops in *wait* mode

Solution: The following instruction sequence will achieve the desired configuration:

```
movb    #$00,SCI0BDH    ; set up baud rate to 9600
movb    #156,SCI0BDL    ;          "
movb    #$4C,SCI0CR1    ; select 8 data bits, address mark wake-up
movb    #$0C,SCI0CR2    ; enable transmitter and receiver
```

The equivalent C statements are as follows:

```
SCI0BDH = 0x00;
SCI0BDL = 0x9C;
SCI0CR1 = 0x4C;
SCI0CR2 = 0x0C;
```

▲

9.9 Interfacing SCI with TIA-232

Because the SCI circuit uses 0 and 5 V (or other lower voltages) to represent logic 0 and 1, respectively, it cannot be connected to the TIA-232 interface circuit directly. A voltage translation circuit, called the *TIA-232 transceiver*, is needed to translate the voltage levels of the SCI signals (RxD and TxD) to and from those of the corresponding TIA-232 signals.

TIA-232 transceiver chips are available from many vendors. The LT1080/1081 from Linear Technology, ST232 from ST Microelectronics, ICL232 from Intersil, MAX232 from MAXIM, and DS14C232 from National Semiconductor are TIA-232 transceiver chips that can operate with a single 5-V power supply and generate TIA-232 compatible outputs. These chips are also pin compatible. In this section, we discuss the use of DS14C232 from National Semiconductor to perform the voltage translation. The pin assignment and the use of each pin are shown in Figure 9.17.

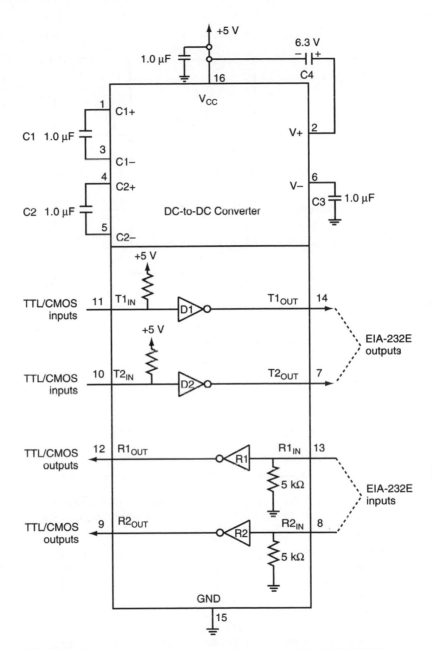

Figure 9.17 ■ Pin assignments and connections of the DS14C232

Adding a TIA-232 transceiver chip allows the HCS12 to use the SCI interface to communicate with the TIA-232 interface circuit. An example of such a circuit is shown in Figure 9.18. A null-modem wiring is followed in this circuit.

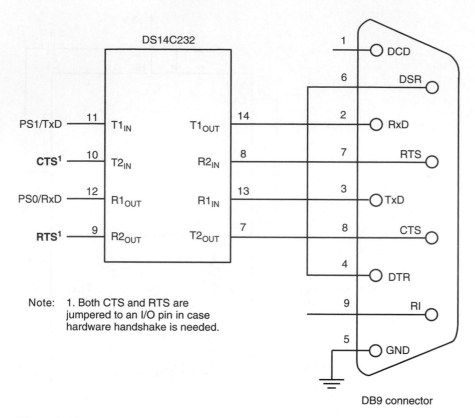

Figure 9.18 ■ Diagram of SCI and EIA-233DB9 connector wiring in SSE256 demo board

Example 9.3

▼

Write a subroutine to send a break to the communication port controlled by the SCI0 interface. The duration of the break is approximately 24,000 E-clock cycles, or 1 ms at 24 MHz.

Solution: A break character is represented by 10 or 11 consecutive 0s and can be sent out by setting the bit 0 of the SCI0CR2 register. As long as the bit 0 of the SCI0CR2 register remains set, the SCI will keep sending out the break character. The following assembly subroutine will perform the desired operation:

```
#include    "c:\miniide\hcs12.inc"
sendbrk     bset    SCI0CR2,SBK        ; turn on send break
            ldy     #1
            jsr     delayby1ms
            bclr    SCI0CR2,SBK        ; turn off send break
            rts
#include    "c:\miniide\delay.asm"
```

The C language version of the function is as follows:

```
#include      "c:\cwHCS12\include\hcs12.h"
#include      "c:\cwHCS12\include\delay.h"
void send_break (void)
{
    SCI0CR2 |= SBK;     // start to send break
    delayby1ms(1);
    SCI0CR2 &= ~SBK;  // stop sending break
}
```

▲

Example 9.4

▼

Write a subroutine to output the character in accumulator A to the SCI0 module using the polling method.

Solution: A new character should be sent out only when the transmit data register is empty. When the polling method is used, the subroutine will wait until bit 7 of the SCI0SR1 register is set before sending out the character in accumulator A. The assembly language function is as follows:

```
#include     "c:\miniide\hcs12.inc"
putcSCI0     brclr    SCI0SR1,TDRE,*      ; wait for TDRE to be set
             staa     SCI0DRL            ; output the character
             rts
```

The C language version of the function is as follows:

```
void putcSCI0 (char cx)
{
    while (!(SCI0SR1 & TDRE));
    SCI0DRL = cx;
}
```

▲

Example 9.5

▼

Write a subroutine to read a character from the SCI0 using the polling method. The character will be returned in accumulator A.

Solution: Using the polling method, the subroutine will wait until the RDRF bit (bit 5) of the SCI0SR1 register becomes set and then read the character held in the SCI0DRL register. The subroutine that reads a character from SCI0 is as follows:

```
#include     "c:\miniide\hcs12.inc"
getcSCI0     brclr    SCI0SR1,RDRF,*      ; wait until RDRF bit is set
             ldaa     SCI0DRL            ; read the character
             rts
```

The C language version of the function is as follows:

```
char  getcSCI0 (void)
{
        while(!(SCI0SR1 & RDRF));
        return (SCI0DRL);
}
```

Example 9.6

Write a subroutine that outputs a NULL-terminated string pointed to by index register X to the SCI0 using the polling method.

Solution: The subroutine will call putcSCI0() repeatedly until all characters of the string have been outputted.

```
putsSCI0    ldaa    1,x+        ; get a character and move the pointer
            beq     done        ; is this the end of the string?
            jsr     putcSCI0
            bra     putsSCI0
done        rts
```

The C language version of the program is as follows:

```
void    putsSCI0 (char *cx)
{
        while (!(*cx)) {
            putcSCI0(*cx);
            cx++;
        }
}
```

Example 9.7

Write a subroutine that inputs a string from the SCI0 module. The string is terminated by the carriage return character and must be stored in a buffer pointed to by index register X. This subroutine must allow the user to backspace to erase errors.

Solution: The subroutine will call getsSCI0() repeatedly until the carriage return character is inputted.

```
getsSCI0    jsr     getch
            cmpa    #CR         ; is it a carriage return?
            beq     qi
            staa    0,x         ; save the character and increment the pointer
            jsr     putch       ; echo it back to SCI0
            cmpa    #BS         ; is it a backspace character?
```

```
            bne     nc              ; no, continue
            dex                     ; decrement the input buffer pointer
            ldaa    #WS             ; wipe out the previous character
            jsr     putch
            ldaa    #BS
            jsr     putch
            bra     gets
nc          inx
            bra     gets
qi          clr     0,x             ; terminate the string by a NULL character
            rts
```

The C language version of the function is as follows:

```
int gets(char *ptr)
{
    char    cx;
    while ((cx = getch()) != 0x0D) {
        *ptr = cx;
        putch(cx);
        if (cx == 0x08) {               // is it a backspace character?
                ptr--;                  // move back the string pointer
                putch(0x20);            // output a space
                putch(0x08);            // output another backspace character
        }
        ptr++;                          // move buffer pointer
    }
    *ptr = 0;                           // terminate the string with a NULL character
    return 0;
}
```

9.10 Summary

When high-speed data transfer is not necessary, using serial data transfer enables the chip designer to make the most use of the limited number of I/O pins. Serial data transfer can be performed asynchronously or synchronously. The SCI interface is an asynchronous interface that is designed to be compatible with the TIA-232 standard. The TIA-232 standard has four aspects: electrical, mechanical, procedural, and functional. For a short distance, modems are not needed for two computers to communicate. A connection called *null modem* is used instead. This is achieved by connecting signals in such a way as to fool two computers into thinking that they are connected through modems.

Since there is no common clock signal for data transfer synchronization, devices participating in communication must agree on a common data rate before data exchange is started. Users of the TIA-232 standard follow a common data format: Each character is framed with a start bit and a stop bit. A character can have 8 or 9 data bits. The ninth bit is often used as a parity bit for error checking or an address mark to alert the receiver to perform address comparison. The receiver uses a clock signal with a frequency that is 16 times the data rate to sample the RxD pin to detect the start bit and determine the logic values of data bits.

Errors could happen during the data transmission process. The most common errors include framing, receiver overrun, and parity errors. A framing error occurs when the start and stop bits improperly frame a received character. A framing error is detected by a missing stop bit. A receiver-overrun error occurs when one or multiple characters are received but not read by the processor. A parity error occurs when an odd number of bits change value.

Some HCS12 members have two identical serial communication interface (SCI) subsystems. Since the TIA-232 standard uses a voltage level different from those to represent logic 1 and logic 0, a transceiver is required to do the voltage translation so that the SCI subsystem can interface with the TIA-232 circuit. Due to the widespread use of the TIA-232 standard, transceiver chips are available from many vendors. The DS14C232 from National Semiconductor is used as an example to illustrate the SCI hardware interfacing.

9.11 Exercises

E9.1 Sketch the output of the letters k and p at the TxD pin when they are transmitted using the format of 1 start bit, 8 data bits, and 1 stop bit.

E9.2 Write an instruction sequence to configure the SCI1 to operate with the following parameters:

- 19200 baud (E-clock is 24 MHz)
- One start bit, 8 data bits, and 1 stop bit format
- Enable both transmit and receive interrupts
- Idle line wake-up
- Disable wake-up initially
- Long idle line mode
- Enable receive and transmit
- No loop back
- Enable parity

E9.3 Write a subroutine to send a break to the communication port controlled by SCI1. The duration of the break must be approximately 480,000 E-clock cycles.

E9.4 Write a subroutine to output the contents of accumulator A as two hex digits to channel SCI0.

E9.5 Modify Example 9.4 so that the **putcSCI0** routine will expand the CR character into the CR/LF pair and expand the LF character into the LF/CR pair. You can add a flag to indicate whether the expansion should be performed.

E9.6 Add an echo flag (1 byte) and modify Example 9.5 so that the received character will be echoed back to the SCI0 when the flag is 1. Otherwise, no echo will be performed.

E9.7 Write a subroutine to input two hex digits from the SCI0 module and echo them back to SCI0. When this routine is run on a demo board, it will allow you to input two hex digits from the keyboard and echo them on the terminal screen.

E9.8 Modify the **getcSCI0** routine in Example 9.5 so that it will handle the backspace character like this:

- Check if the input character is the backspace character (ASCII code $08).
- If the entered character is the backspace character, then echo it, output a space character, and output another backspace character. Why would you do this?

9.12 Lab Exercises and Assignments

L9.1 Write a program to be run on your demo board. This program and the user interact as follows:

1. The program outputs the message "Please enter your age:" and waits for the user to enter his or her age.

2. The user enters his or her age (followed by a carriage return) and the program reads it.

3. The program outputs the message "Please enter your height in inches:" and waits for the user to enter his or her height.

4. The user enters his or her height in inches (followed by a carriage return) and the program reads it.

5. The program outputs the message "Please enter your weight in lbs:" and waits for the user to enter his or her weight.

6. The user enters his or her weight in lbs (followed by a carriage return) and the program reads it.

7. The program outputs the following messages on the terminal screen and exits:

   You are *xxx* years old.
   You are *kk* ft *mm* inches tall.
   You weigh *zzz* lbs.

L9.2 If your demo board has two TIA-232 connectors, then perform this experiment.

1. Connect one of the TIA-232 connectors to the PC so that you can download your program onto the demo board.

2. Connect the second TIA-232 connector to another demo board.

3. Write a program to be run on two demo boards. The program outputs one character per second to the SCI port that is connected to another demo board and then reads in characters from the same SCI port.

4. The program displays the SCI activities on the LCD screen as follows:

The first row displays what has been sent out: *snd:*xxxxx . . .

The second row displays what has been received: *rvd:*yyyyy . . .

Only the last 16 characters sent and received will be displayed on the LCD screen if a 16 × 2 LCD screen is used.

10

The SPI Function

10.1 Objectives

After completing this chapter, you should be able to

- Describe the HCS12 SPI module
- Configure the SPI operation parameters
- Interface with peripheral devices with SPI interface
- Use the SPI function to interface with the shift register 74HC595
- Use SPI to interface with the digital temperature sensor TC72
- Use SPI to interface with the 12-bit D/A converter MCP4922
- Use SPI to interface with the matrix LED display driver MAX6952

10.2 Introduction to the SPI Function

The serial peripheral interface (SPI) allows the HCS12 to communicate synchronously with peripheral devices and other microcontrollers. The SPI system in the HCS12 can operate as a master or as a slave. When the SPI module is configured as a master, it is responsible for generating the clock signal (SCK) during an SPI data transfer. The SPI subsystem is mainly used in interfacing with peripherals such as TTL shift registers, LED/LCD display drivers, phase-locked-loop (PLL) chips, memory components with serial interface, or A/D and D/A converter chips that do not need a very high data rate.

The SPI must be enabled to operate. When the SPI module is enabled, the four associated SPI port pins are dedicated to the SPI function as

- Slave select ($\overline{SS}$)
- Serial clock (SCK)
- Master out/slave in (MOSI)
- Master in/slave out (MISO)

The main element of the SPI system is the SPI data register. The 8-bit data register in the master and the 8-bit data register in the slave are linked by the MOSI and the MISO pins to form a distributed 16-bit register. When a data transfer is performed, this 16-bit register is serially shifted 8 bit positions by the SCK clock from the master; data is exchanged between the master and the slave. Data written to the master SPI data register becomes the output data for the slave, and data read from the master SPI data register after a transfer operation is the input data from the slave.

A write to the SPI data register puts data into the transmit buffer if the previous transmission was complete. When a transfer is complete, received data is moved into a receiver data register. Data may be read from this double-buffered system any time before the next transfer is completed. This 8-bit register acts as the SPI receive data register for reads and as the SPI transmit data register for writes. A single SPI register address is used for reading data from the read data buffer and for writing data to the shifter.

There are four possible clock formats to choose from. The user selects one of these four clock formats for data transfer by programming the CPOL and CPHA bits of the SPI control register 1. The CPOL bit simply selects a noninverted (idle low) or inverted (idle high) clock. The CPHA bit is used to accommodate two fundamentally different protocols by shifting the clock a half-cycle or by not shifting the clock.

The SPI function on the HCS12 has been modified slightly from that in the 68HC11 with the expectation of improving its applicability. An HCS12 device may have from one to three identical SPI modules (SPI0, SPI1, and SPI2).

10.2.1 SPI Signal Pins

The SPI0 function shares the use of the upper four Port S pins: MISO0 (PS4), MOSI0 (PS5), SCK0 (PS6), and $\overline{SS0}$ (PS7). Out of reset, the SPI1 and SPI2 share the use of the lower four Port P pins and upper four Port P pins, respectively. However, the SPI1 and SPI2 pins can also be rerouted to the lower four Port H pins and upper four Port H pins by programming the MODRR register. The assignment of SPI1 and SPI2 signal pins is described in Section 7.5.6. There is no need to configure the pin directions when the SPI function is enabled.

- MISOx (x = 0, 1, or 2): *master in slave out* (serial data input). This pin is used to transmit data out of the SPI module when it is configured as a slave and receive data when it is configured as a master.

- MOSIx (x = 0, 1, or 2): *master out slave in* (serial data output). This pin is used to transmit data out of the SPI module when it is configured as a master and receive data when it is configured as a slave.
- SCKx (x = 0, 1, or 2): *serial clock*. This pin is used to carry the clock signal that synchronizes SPI data transfer. It is an output if the SPI is configured as a master but an input if the SPI is configured as a slave.
- $\overline{SS}$x (x = 0, 1, or 2): *slave select*. When configured as a slave, this pin must be pulled low for the SPI module to operate.

10.3 Registers Related to the SPI Subsystem

Most of the SPI operational parameters are set by two SPI control registers: SPIxCR1 (x = 0, 1, or 2) and SPIxCR2 (x = 0, 1, or 2). Their contents are shown in Figures 10.1 and 10.2, respectively. The SPI must be enabled before it can start data transfer. Setting bit 6 of the SPIxCR1 register will enable the SPI subsystem. Bit 1 (SSOE) of the SPIxCR1 register allows the user to use the $\overline{SS}$ pin to select the slave device for data transfer automatically. However, this feature is not useful if the user wants to use the SPI to interface with multiple slave devices at the same time. The selection of $\overline{SS}$ for input and output is shown in Table 10.1. Some slave devices may

7	6	5	4	3	2	1	0	
SPIE	SPE	SPTIE	MSTR	CPOL	CPHA	SSOE	LSBFE	Reset value = 0x04

SPIE: SPI interrupt enable bit
  0 = SPI interrupts are disabled.
  1 = SPI interrupts are enabled.
SPE: SPI system enable bit
  0 = SPI disabled.
  1 = SPI enabled and pins PS4–PS7 are dedicated to SPI function.
SPTIE: SPI transmit interrupt enable
  0 = SPTEF interrupt disabled.
  1 = SPTEF interrupt enabled.
MSTR: SPI master/slave mode select bit
  0 = slave mode.
  1 = master mode.
CPOL: SPI clock polarity bit
  0 = active high clocks selected; SCK idle low.
  1 = active low clocks selected; SCK idle high.
CPHA: SPI clock phase bit
  0 = The first SCK edge is issued one-half cycle into the 8-cycle transfer operation.
  1 = The SCK edge is issued at the beginning of the 8-cycle transfer operation.
SSOE: slave select output enable bit
  The $\overline{SS}$ output feature is enabled only in master mode by asserting the
  SSOE bit and the MODFEN bit of the SPIxCR2 register.
LSBF: SPI least significant bit first enable bit
  0 = data is transferred most significant bit first.
  1 = data is transferred least significant bit first.

Figure 10.1 ■ SPI control register 1 (SPIxCR1, x = 0, 1, or 2)

MODFEN	SSOE	Master Mode	Slave Mode
0	0	$\overline{SS}$ not used by SPI	$\overline{SS}$ input
0	1	$\overline{SS}$ not used by SPI	$\overline{SS}$ input
1	0	SS input with MODF feature	$\overline{SS}$ input
1	1	$\overline{SS}$ output	$\overline{SS}$ input

Table 10.1 ■ $\overline{SS}$ input/output selection

7	6	5	4	3	2	1	0	
0	0	0	MODFEN	BIDIROE	0	SPSWAI	SPC0	Reset value = 0x08

MODFEN: mode fault enable bit
   0 = disable the MODF error.
   1 = enable setting the MODF error.
BIDIROE: output enable in the bidirectional mode of operation
   0 = output buffer disabled.
   1 = output buffer enabled.
SPSWAI: SPI stop in wait mode
   0 = SPI clock operates normally in stop mode.
   1 = stop SPI clock generation in wait mode.
SPC0: serial pin control bit 0
   With the MSTR bit in the SPIxCR1 register, this bit enables bidirectional pin
   configuration, as shown in Table 10.2.

Figure 10.2 ■ SPI control register 2 (SPIxCR2, x = 0, 1, or 2)

Pin Mode		SPC0	MSTR	MISO[1]	MOSI[2]	SCK[3]	$\overline{SS}$[4]
A	Normal	0	0	Slave out	Slave in	SCK in	SS in
B	Normal	0	1	Master in	Master out	SCK out	SS I/O
C	Bidirectional	1	0	Slave I/O	—	SCK in	SS in
D	Bidirectional	1	1	—	Master I/O	SCK out	SS I/O

[1] Slave output is enabled if BIDIROE bit = 1, $\overline{SS}$ = 0, and MSTR = 0 (C).
[2] Master output is enabled if BIDIROE bit = 1 and MSTR = 1 (D).
[3] SCK output is enabled if MSTR = 1 (B,D).
[4] $\overline{SS}$ output is enabled if MODFEN = 1, SSOE = 1, and MSTR = 1 (B,D).

Table 10.2 ■ Bidirectional pin configurations

transfer data with the least significant bit first. Bit 0 (LSBF) of the SPIxCR1 register allows the user to have this flexibility.

The SPI data shift rate (also called *baud rate*) is programmable. The baud rate is set by programming the SPIxBR register to appropriate values. The contents of the SPIxBR register are shown in Figure 10.3. The method for computing the baud rate is also given in Figure 10.3.

7	6	5	4	3	2	1	0	
0	SPPR2	SPPR1	SPPR0	0	SPR2	SPR1	SPR0	Reset value = 0x00

SPPR2~SPPR0: SPI baud rate preselection bits
SPR2~SPR0: SPI baud rate selection bits

BaudRateDivisor = $(SPPR + 1) \times 2^{(SPR + 1)}$
Baud Rate = E-Clock ÷ BaudRateDivisor

Figure 10.3 ■ SPI baud rate register (SPIxBR, x = 0, 1, or 2)

Example 10.1

Give a value to be loaded into the SPIxBR register to set the baud rate to 2 MHz for a 24-MHz E-clock.

Solution: 24 MHz ÷ 2 MHz = 12. By setting SPPR2~SPPR0 and SPR2~SPR0 to 010 and 001, respectively, we can set the baud rate to 2 MHz. The value to be loaded to the SPIxBR register is $21.

Example 10.2

What is the highest possible baud rate for the SPI with 24-MHz E-clock?

Solution: The highest SPI baud rate occurs when both SPPR2~SPPR0 and SPR2~SPR0 are set to 000. Under this condition, the BaudRateDivisor is 2, and hence the baud rate is 24 MHz/2 = 12 MHz.

The SPI has a status register that records the progress of data transfer and errors, as shown in Figure 10.4. The application program can check bit 7 of the SPIxSR register or wait for the SPI interrupt to find out if the SPI transfer has completed. When transferring data in higher frequency using the SPI format, using the polling method is more efficient due to the overhead involved in interrupt handling.

The setting of the SPIF flag may request an interrupt to the CPU if the SPIE bit of the SPIxCR1 register is also set to 1.

The SPTEF bit is set when there is room in the transmit data buffer. It is cleared by reading the SPIxSR register with SPTEF set, followed by writing a data value into the SPI data (SPIxDR) register. The SPTEF bit is set whenever the byte in the SPIxDR register is transferred to the transmit shift register. The setting of the SPTEF flag may request an interrupt to the CPU if the SPTIE bit of the SPIxCR1 register is also set to 1.

The MODF bit is set if the $\overline{SS}$ input becomes low while the SPI is configured as a master. The flag is cleared automatically by a read of the SPIxSR register followed by a write to the SPIxCR1 register. The MODF bit is set only if the MODFEN bit of the SPIxCR2 register is set.

The 8-bit SPIxDR register is both the input and output register for SPI data. A write to this register allows a byte to be queued and transmitted. For an SPI configured as a master, a queued

7	6	5	4	3	2	1	0	
SPIF	0	SPTEF	MODF	0	0	0	0	Reset value = 0x20

SPIF: SPI interrupt request bit
    SPIF is set after the eight SCK cycles in a data transfer, and it is
    cleared by reading the SP0SR register (with SPIF set) followed by
    a read access to the SPI data register.
    0 = transfer not yet complete.
    1 = new data copied to SPIxDR.
SPTEF: SPI data register empty interrupt flag
    0 = SPI data register not empty.
    1 = SPI data register empty.
MODF: mode error interrupt status flag
    0 = mode fault has not occurred.
    1 = mode fault has occurred.

Figure 10.4 ■ SPI status register (SPIxSR)

data byte is transmitted immediately after the previous transmission has completed. *Do not write to the SPIxDR register unless the SPTEF bit is 1.*

10.4 SPI Operation

Only a master SPI module can initiate transmission. A transmission begins by writing to the master SPI data register. Data is transmitted and received simultaneously. The serial clock (SCK) synchronizes shifting and sampling of the information on the two serial data lines. The $\overline{SS}$ line allows selection of an individual slave SPI device; slave devices that are not selected do not interfere with SPI bus activities. Optionally, on a master SPI device, the $\overline{SS}$ signal can be used to indicate multiple-master bus contention.

10.4.1 Transmission Formats

The CPHA and CPOL bits in the SPIxCR1 register allow the user to select one of the four combinations of serial clock phase and polarity. The clock phase control bit (CPHA) selects one of two fundamentally different transmission formats. Clock phase and polarity should be identical for the master SPI device and the communicating slave device.

When the CPHA bit is set to 0, the first edge on the SCK line is used to clock the first data bit of the slave into the master and the first data bit of the master into the slave. In some peripheral devices, the first bit of the slave's data is available at the slave data out pin as soon as the slave is selected. In this format, the first SCK edge is not issued until a half-cycle into the 8-cycle transfer operation. The first edge of SCK is delayed a half cycle by clearing the CPHA bit.

The SCK output from the master remains in the inactive state for a half SCK period before the first edge appears. A half SCK cycle later, the second edge appears on the SCK pin. When this second edge appears, the value previously latched from the serial data input is shifted into the least significant bit of the shifter. After this second edge, the next bit of the SPI transfer data is transmitted out of the MOSI pin of the master to the serial data input pin of the slave device. This process continues for a total of 16 edges on the SCK pin, with data being latched on odd-numbered edges and shifted (to the shift register) on even-numbered edges. Data reception is double-buffered. Data is shifted serially into the SPI shift register during the transfer and is transferred to the parallel SPI data register after the last bit is shifted in.

After the 16th (last) SCK edge, the SPIF flag in the SPIxSR register is set, indicating that the transfer is complete. The timing diagram for this transfer format (CPHA bit = 0) is shown in Figure 10.5.

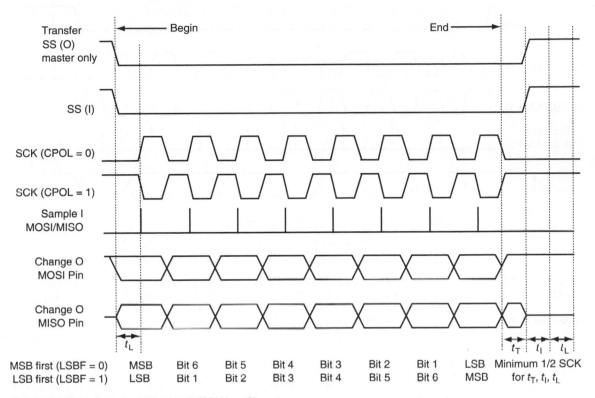

Figure 10.5 ■ SPI clock format 0 (CPHA = 0)

Some peripheral devices require the first SCK edge to appear before the first data bit becomes available at the data out pin; the second edge clocks data into the system. In this format, the first SCK edge is issued by setting the CPHA bit at the beginning of the 8-cycle transfer operation. The first edge of SCK occurs immediately after the half SCK clock cycle synchronization delay. This first edge commands the slave to transfer its most significant data bit to the serial data input pin of the master. A half SCK cycle later, the second edge appears on the SCK pin. This is the latching edge for both the master and slave. When the third edge occurs, the value previously latched from the serial data input pin is shifted into the least significant bit of the SPI shifter. After this edge, the next bit of the master data is shifted out of the serial data output pin of the master to the serial input pins on the slave.

This process continues for a total of 16 edges on the SCK line with data being latched on even-numbered edges and shifting taking place on odd-numbered edges. Again, the SPIF flag is set after the 16th SCK edge. The timing diagram of this transmission format is shown in Figure 10.6.

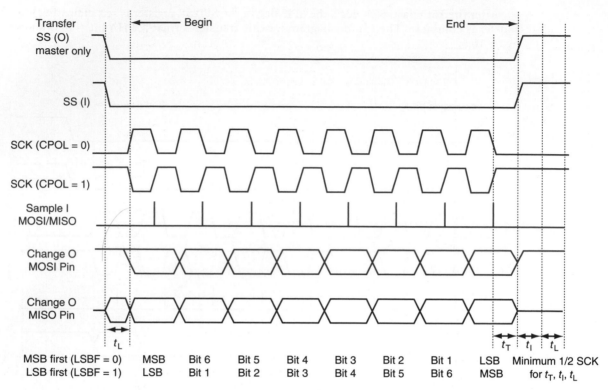

Figure 10.6 ■ SPI clock format 1 (CPHA = 1)

10.4.2 Bidirectional Mode (MOMI or SISO)

The SPI subsystem can be used in bidirectional mode. In this mode, the SPI uses only one serial data pin for the interface with external device(s). The MSTR bit of the SPIxCR1 register decides which pin is to be used. The MOSI pin becomes a serial data I/O (MOMI) pin for the master mode, and the MISO pin becomes a serial data I/O (SISO) pin for slave mode. The MISO pin in the master mode and the MOSI pin in the slave mode are not used by the SPI in bidirectional mode. The possible combinations are shown in Figure 10.7.

The direction of each serial I/O pin depends on the BIDIROE bit. If the pin is configured as an output, serial data from the shift register is driven out on the pin. The same pin is also the serial input to the shift register. If we want to read data from a peripheral device, then the BIDIROE bit should be cleared to 0.

The use of bidirectional mode is illustrated in Exercise 10.8.

10.4.3 Mode Fault Error

If the $\overline{SSx}$ input becomes low while the SPIx is configured as a master, it indicates a system error in which more than one master may be trying to drive the MOSIx and SCKx lines simultaneously. This condition is not permitted in normal operation. The MODF bit in the SPIxSR register is set automatically, provided that the MODFEN bit in the SPIxCR2 is set.

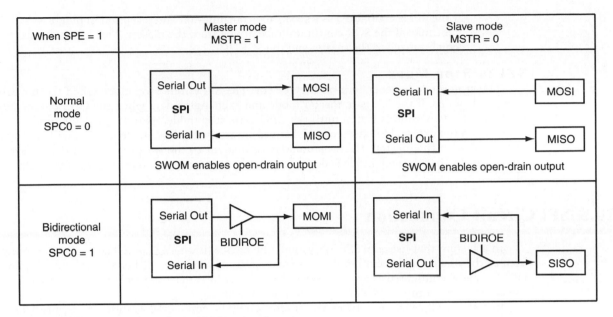

Figure 10.7 ■ Normal mode and bidirectional mode

In the special case where the MODFEN bit is cleared, the $\overline{SSx}$ pin is a general-purpose I/O pin for the SPI system configured in master mode. In this special case, the mode fault function is prohibited and the MODF flag remains cleared. In case the SPI system is configured as a slave, the $\overline{SSx}$ pin is a dedicated input pin. Mode fault error does not occur in slave mode.

When a mode fault error occurs, the MSTR bit in the SPIxCR1 is cleared, the MODF bit in the SPIxSR register is set, and the output enable for the SCKx, MISOx, and MOSIx pins are deasserted. If the mode fault error occurs in the bidirectional mode for an SPI system configured in master mode, output enable of the MOMI pin is cleared if it was set but SISO is not affected. No mode fault error occurs in the bidirectional mode for the SPI system configured in slave mode.

10.4.4 Low-Power Mode Options

The HCS12 has two low-power modes and the SPI module behaves differently in these two modes.

SPI IN WAIT MODE

SPI operation in wait mode depends on the state of the SPISWAI bit in the SPIxCR2 register.

- If the SPISWAI bit is cleared, the SPI operates normally when the CPU is in wait mode.
- If the SPISWAI bit is set, the SPI clock generation ceases and the SPI module enters a power conservation state when the CPU is in the wait mode. If the SPI is configured as a master, any transmission and reception in progress stops at wait mode entry. The transmission and reception resumes when the SPI exits wait mode.

If the SPI is configured as a slave, any transmission and reception in progress continues if the SCK continues to be driven from the master. This keeps the slave synchronized to the master and the SCK.

SPI IN STOP MODE

The stop mode is dependent on the system. The SPI enters stop mode when the module clock is disabled. If the SPI is in master mode and exchanging data when the CPU enters stop mode, the transmission is frozen until the CPU exits stop mode. After exiting stop mode, data to and from the external SPI device is exchanged correctly. In slave mode, the SPI will stay synchronized with the master. The stop mode is equivalent to the wait mode with the SPISWAI bit set except that the stop mode is dependent on the system and cannot be controlled with the SPISWAI bit.

10.5 SPI Circuit Connection

In a system that uses the SPI subsystem, one device (normally a microcontroller) is configured as the *master* and the other devices are configured as *slaves*. Either a peripheral chip or a HCS12 can be configured as a slave device. The master SPI device controls the data transfer and can control one or more SPI slave devices simultaneously.

In a single-slave configuration, the circuit would be connected as shown in Figure 10.8. If the $\overline{SSx}$ output enable feature is set, then the $\overline{SSx}$ pin will automatically go low to enable the slave device before the data transfer is started. This feature is not used in Figure 10.8.

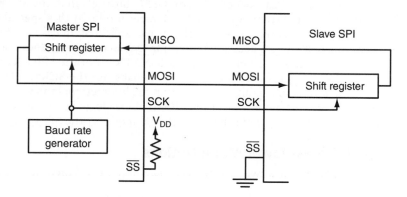

Figure 10.8 ■ Master/slave transfer block diagram

There could be several connection methods in a multislave SPI environment. One possibility is shown in Figure 10.9. In this connection method, the HCS12 can choose any peripheral device for data transfer. In Figure 10.9, Port P pins are used to drive the $\overline{SS}$ inputs of peripheral devices. Any other unused general-purpose I/O pins can be used for this purpose.

If we don't need the capability of selecting an individual peripheral device for data transfer, then the connection shown in Figure 10.10 can be used, which will save quite a few I/O pins. Figure 10.10 differs from Figure 10.9 in the following ways:

1. The MISO pin of each slave is wired to the MOSI pin of the slave device to its right. The MOSI pins of the master and slave 0 are still wired together.

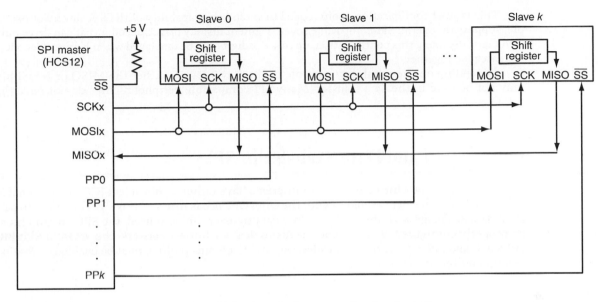

Figure 10.9 ■ Single-master and multiple-slave device connection (method 1)

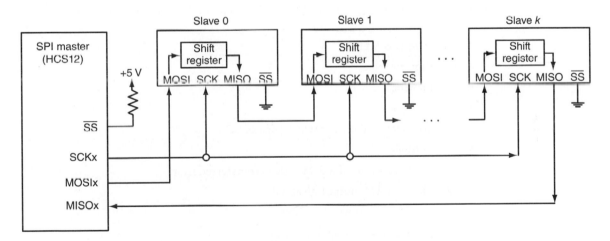

Figure 10.10 ■ Single-master and multiple-slave device connection (method 2)

2. The MISO pin of the master is wired to the same pin of the last slave device.

3. The $\overline{SS}$ inputs of all slaves are tied to ground to enable all slaves.

Thus the shift registers of the SPI master and slaves become a ring. The data of slave k is shifted to the master SPI, the data of the master is shifted to slave 0, the data of slave 0 is shifted to slave 1, and so on. In this configuration, a minimal number of pins control a large number of peripheral devices. However, the master does not have the freedom to select an arbitrary slave device for data transfer without going through other slave devices.

This type of configuration is often used to extend the capability of the SPI slave. For example, suppose there is an SPI-compatible seven-segment display driver/decoder that can drive only four digits. By using this configuration, up to $4 \times k$ digits can be displayed when k driver/decoders are cascaded together.

Depending on the capability and role of the slave device, either the MISO or MOSI pin may not be used in the data transfer. Many SPI-compatible peripheral chips do not have the MISO pin.

10.6 Configuration of and Data Transfer in SPI

Most peripheral chips with the SPI interface have either a chip-select (CS) or chip-enable (CE) signal to enable data exchange with the chip. One needs to assert this signal in order to perform data exchange with the device. Before data transfer can be started, the SPI function must be properly configured. The designer needs to deal with four registers: the WOMS, SPIxBR, SPIxCR1, and SPIxCR2 registers. In addition, the Port S pins pull-up must be enabled so that no external pull-up device is needed.

Example 10.3
▼

Configure the SPI0 module to operate with the following settings assuming that the E-clock is 24 MHz:

- Baud rate at 6 MHz
- Interrupt disabled
- SPI enabled
- SCK idle low
- Data to be latched on the rising edge of SCK
- Master mode
- Most significant bit of a data byte is transferred first
- Mode fault and $\overline{SS}$ output disabled
- Normal SPI operation
- SPI stopped in wait mode

Solution: Since 24 MHz/6 MHz = 4, we need to set SPPR2~SPPR0 and SPR2~SPR0 to 001 and 000, respectively. Therefore, the value to be written into the SPI0BR register would be $10. For other parameters, we need to write $50 and $02 into SPI0CR1 and SPI0CR, respectively. The WOMS register must also be cleared to enable the Port S pull-up.

The following instruction sequence will configure the SPI as desired:

```
movb   #$10,SPI0BR      ; set baud rate to 6 MHz (the value of 0x01 will also work)
movb   #$50,SPI0CR1     ; disable interrupt, enable SPI, SCK idle low,
                        ; data latched on rising edge, data transferred msb first
movb   #$02,SPI0CR2     ; disable bidirectional mode, stop SPI in wait mode
movb   #0,WOMS          ; enable Port S pull-up
```

The same setting can be achieved by the following C statements:

```
SPI0BR   = 0x10;    /* set baud rate to 6 MHz */
SPI0CR1  = 0x50;    /* enable SPI, data shift on SCK's rising edge, master mode */
SPI0CR2  = 0x02;    /* disable bidirectional mode, SPI stops in wait mode */
WOMS     = 0;       /* enable Port S pull-up */
```

Sending and reading data using the SPI interface is very straightforward. The following four functions are used most often:

1. *putcSPIx* (x = 0, 1, or 2). This function sends out a character through the SPI interface. In assembly language, the character to be output is passed in accumulator A.

2. *putsSPIx* (x = 0, 1, or 2). This function outputs a NULL-terminated string through the SPI interface. In assembly language, the string is pointed to by index register X.

3. *getcSPIx* (x = 0, 1, or 2). This function reads in a character from the SPI interface. In assembly language, the character is returned in accumulator A.

4. *getsSPIx* (x = 0, 1, or 2). This function reads in a string from the SPI interface. The buffer to hold the string is pointed to by index register X, whereas the number of characters to read is specified in accumulator B.

Example 10.4

Write the four common SPI data transfer functions in the assembly and C languages.

Solution: Before writing a byte to the SPI data register, the user must make sure that the SPTEF flag is 1. To make sure that the SPI transfer is completed, the program must wait until the SPIF flag is set to 1.

The following assembly function outputs the character in accumulator A to SPI0:

```
putcspi0  brclr  SPI0SR,SPTEF,*   ; wait until write operation is permissible
          staa   SPI0DR           ; output the character to SPI0
          brclr  SPI0SR,SPIF,*    ; wait until the byte is shifted out
          ldaa   SPI0DR           ; clear the SPIF flag
          rts
```

The following assembly function outputs a NULL-terminated string pointed to by index register X:

```
putsspi0  ldaa   1,x+             ; get 1 byte to be output to SPI port
          beq    doneps0          ; reach the end of the string?
          jsr    putcspi0         ; call subroutine to output the byte
          bra    putsspi0         ; continue to output
doneps0   rts
```

To read a byte from the SPI interface, a byte must be written into the SPIxDR register to trigger eight clock pulses to be sent out from the SCK pin. The following assembly function reads a character from the SPI0 interface and returns the character in accumulator A:

```
getcspi0  brclr  SPI0SR,SPTEF,*   ; wait until write operation is permissible
          staa   SPI0DR           ; trigger eight clock pulses for SPI transfer
          brclr  SPI0SR,SPIF,*    ; wait until a byte has been shifted in
          ldaa   SPI0DR           ; return the byte in A and clear the SPIF flag
          rts
```

The following assembly function reads a string from the SPI0 interface:

```
getsSPI0   tstb                    ; check the byte count
           beq   donegs0           ; return when byte count is zero
           jsr   getcspi0          ; call subroutine to read a byte
           staa  1,x+              ; save the returned byte in the buffer
           decb                    ; decrement the byte count
           bra   getsspi0
donegs0    clr   0,x               ; terminate the string with a NULL character
           rts
```

These four assembly functions are stored in the file *spi0util.asm* and are included on the complementary CD.

The C language versions of the previous four functions are as follows:

```
void putcspi0 (char cx)
{
        char      temp;
        while(!(SPI0SR & SPTEF));        // wait until write is permissible
        SPI0DR = cx;                     // output the byte to the SPI
        while(!(SPI0SR & SPIF));         // wait until write operation is complete
        temp = SPI0DR;                   // clear the SPIF flag
}
void putsspi0(char *ptr)
{
        while(*ptr) {            // continue until all characters have been outputted
                putcspi0(*ptr);
                ptr++;
        }
}
char getcspi0(void)
{
        while(!(SPI0SR & SPTEF));        // wait until write is permissible
        SPI0DR = 0x00;                   // trigger eight SCK pulses to shift in data
        while(!(SPI0SR & SPIF));         // wait until a byte has been shifted in
        return SPI0DR;                   // return the character
}
void getsspi0(char *ptr, char count)
{
        while(count) {          /* continue while byte count is nonzero */
                *ptr++ = getcspi0(); /* get a byte and save it in buffer */
                count--;
        }
        *ptr = 0;              /* terminate the string with a NULL */
}
```

These four C functions are stored in the file *spi0util.c* and are included in the complementary CD.

10.7 SPI-Compatible Chips

The SPI is a protocol proposed by Freescale to interface peripheral devices to a microcontroller. As long as a peripheral device supports the SPI interface protocol, it can be used with any microcontroller that implements the SPI subsystem. Many semiconductor manufacturers are producing SPI-compatible peripheral chips. The Freescale SPI protocol is compatible with the National Semiconductor Microwire protocol. Therefore, any peripheral device that is compatible with the SPI can also be interfaced with the Microwire protocol.

10.8 The 74HC595 Shift Register

As shown in Figure 10.11, the 74HC595 consists of an 8-bit shift register and an 8-bit D-type latch with three-state parallel outputs. The shift register accepts serial data and provides a serial output. The shift register also provides parallel data to the 8-bit latch. The shift register and the latch have different clock sources. This device also has an asynchronous reset input. The frequency of the shift clock can be as high as 100 MHz.

The functions of the pins in Figure 10.11 are as follows:

- DS: *Serial data input.* The data on this pin is shifted into the 8-bit shift register.
- SC: *Shift clock.* A low-to-high transition on this signal causes the data at the serial input pin to be shifted into the 8-bit shift register.
- $\overline{\text{Reset}}$. A low on this pin resets the shift register portion of this device only. The 8-bit latch is not affected.

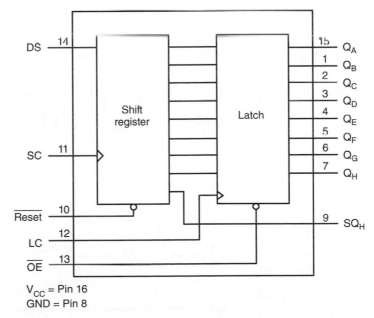

Figure 10.11 ■ The 74HC595 block diagram and pin assignment

- LC: *Latch clock.* The rising edge of this signal loads the contents of the shift register into the output latch.
- $\overline{OE}$: *Output enable.* A low on this pin allows the data from the latch to be presented at the output pins $Q_A, \ldots, Q_H$.
- Q_A to Q_H: Noninverted, tri-state latch outputs.
- SQ_H: *Serial data output.* This is the output of the eighth stage of the 8-bit shift register. This output does not have tri-state capability.

The 74HC595 is designed to shift in 8-bit data serially and then transfer it to the latch to be used as parallel data. The 74HC595 can be used to add parallel output ports to the HCS12. Both the connection methods shown in Figures 10.9 and 10.10 can be used in appropriate applications.

Example 10.5

Describe how to use two 74HC595s to drive eight common-cathode seven-segment displays, assuming that the E-clock frequency of the HCS12 is 24 MHz.

Solution: Two 74HC595s can be cascaded using the method shown in Figure 10.10. One 74HC595 is used to hold the seven-segment pattern, whereas the other 74HC595 is used to carry digit-select signals. The circuit connection is shown in Figure 10.12.

Since there are only seven segments, the Q_H bit of the segment-control 74HC595 is not needed. The PK7 pin is used to control the LC input of the 74HC595. The time-multiplexing technique illustrated in Example 4.14 will be used to display multiple digits in Figure 10.12.

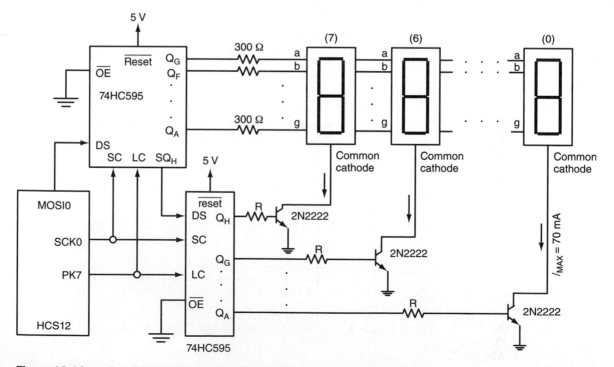

Figure 10.12 ■ Two 74HC595s together drive eight seven-segment displays

To light the digit on display 7, the voltage at Q_H of the digit-select 74HC595 must be driven to high. To light the digit on display 6, the voltage at Q_G of the digit-select 74HC595 must be driven to high, and so on.

There are four parts in the program to be written.

Part 1. Display table setup
The segment patterns and digit-select values are stored in a table so that the table lookup method can be used to display the desired pattern.

Part 2. SPI module initialization
Configure the SPI module to shift data at 12 MHz on the rising clock edge and most significant data bit first.

Part 3. Timer delay function
Include the library function *delayby1ms* to multiplex the digit per millisecond.

Part 4. SPI data transfer
Include the library function *putcspi0.asm* (or *putcspi0.c*) to output the display patterns and digit-select data.

The following program will display 87654321 on display 7 through display 0:

```
#include     "c:\miniide\hcs12.inc"
             org      $1000
icnt         ds.b     1                      ; loop count
             org      $1500
             lds      #$1500                 ; set up stack pointer
             bset     DDRK,$80               ; configure the PK7 pin for output
             jsr      openspi0               ; configure SPI0
forever      ldx      #disp_tab              ; use X as a pointer to the table
             movb     #8,icnt                ; set loop count to 8
loop         ldaa     1,x+                   ; send the digit-select byte to the 74HC595
             jsr      putcspi0               ;            "
             ldaa     1,x+                   ; send segment pattern to 74HC595
             jsr      putcspi0               ;            "
             bclr     PTK,BIT7               ; transfer data from shift register to output
             bsct     PTK,BIT7               ; latch
             ldy      #1                     ; display the digit for 1 ms
             jsr      delayby1ms             ;            "
             dec      icnt                   ;
             bne      loop                   ; if not reach digit 1, then next
             bra      forever                ; start from the start of the table
;******************************************************************************
; The following function configures the SPI module properly:
;******************************************************************************
openspi0     movb     #0,SPI0BR              ; set baud rate to 12 MHz
             movb     #$50,SPI0CR1           ; disable interrupt, enable SPI, SCK idle low,
                                             ; latch data on rising edge, transfer data msb first
             movb     #$02,SPI0CR2           ; disable bidirectional mode, stop SPI in wait mode
             movb     #0,WOMS                ; enable Port S pull-up
             rts
#include     "c:\miniide\delay.asm"
#include     "c:\miniide\spi0util.asm"
```

```
;************************************************************************************
; Each digit consists of 2 bytes of data. The first byte is
; digit select; the second byte is the digit pattern.
;************************************************************************************
disp_tab    dc.b     $80,$7F,$40,$70,$20,$5F,$10,$5B
            dc.b     $08,$33,$04,$79,$02,$6D,$01,$30
            end
```

The C language version of the program is as follows:

```c
#include  "c:\cwHCS12\include\hcs12.h"
#include  "c:\cwHCS12\include\spi0util.h"      // include spi0util.c into the project
#include  "c:\cwHCS12\include\delay.h"         // include delay.c into the project
void openspi0(void);
void main (void)
{
    unsigned char disp_tab[8][2] = {{0x80,0x7F},{0x40,0x70},{0x20,0x5F},{0x10,0x5B},
                                    {0x08,0x33},{0x04,0x79},{0x02,0x6D},{0x01,0x30}};

    char i;
    openspi0();            /* configure the SPI0 module */
    DDRK | = BIT7;         /* configure pin PK7 as output */
    while(1) {
        for (i = 0; i < 8; i++) {
            putcspi0(disp_tab[i][0]);          /* send out digit select value */
            putcspi0(disp_tab[i][1]);          /* send out segment pattern */
            PTK &= ~~BIT7;                     /* transfer values to latches of 74HC595s */
            PTK | = BIT7;                      /* " */
            delayby1ms(1);                     /* display a digit for 1 ms */
        }
    }
}
void openspi0(void)
{
    SPI0BR   = 0;       /* set baud rate to 12 MHz */
    SPI0CR1  = 0x50;    /* disable SPI interrupt, enable SPI, SCK idle low, shift
                           data on rising edge, shift data msb first */
    SPI0CR2  = 0x02;    /* disable bidirectional mode, disable SPI in wait mode */
    WOMS     = 0;       /* enable Port S pull up */
}
```

▲

This example illustrates one of the applications of the shift register 74HC595. The main drawback of the example program is that the CPU spends all its time on the task of multiplexing the displays. The program can be modified to be interrupt-driven to free the CPU from idle waiting. Several manufacturers produce LED display drivers to eliminate the display multiplexing task altogether. The MC14489 from Freescale and the MAX7221 from Maxim are two examples.

10.9 The TC72 Digital Thermometer

The TC72 from Microchip is a digital temperature sensor with the SPI interface. The TC72 has a 10-bit resolution; that is, it uses 10 bits to represent the ambient temperature. The pin assignment and functional block diagram are shown in Figure 10.13.

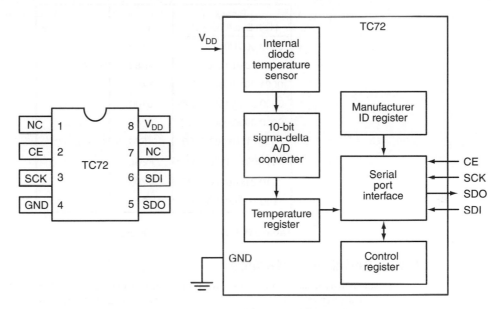

Figure 10.13 ■ TC72 pin assignment and functional block diagram

10.9.1 Functioning of TC72

The TC72 is capable of reading temperatures from $-55°C$ to $+125°C$. The TC72 can be used either in the *continuous temperature conversion mode* or the *one-shot conversion mode*. The continuous conversion mode measures temperature approximately every 150 ms and stores the data in temperature registers. The TC72 has an internal clock generator that controls the automatic temperature conversion sequence. In contrast, the one-shot mode performs a single temperature measurement and returns to the power-saving shutdown mode.

10.9.2 Temperature Data Format

Temperature data is represented by a 10-bit two's complement word with a resolution of 0.25°C per least significant bit. The analog-to-digital converter is scaled from $-128°C$ to $+127°C$ with 0°C represented as 0x0000, but the operating range of the TC72 is specified from $-55°C$ to $+125°C$. The 10-bit temperature value is stored in two 8-bit registers. Because the TC72 A/D converter is scaled from $-128°C$ to $+127°C$ and the conversion is represented in 10 bits, the temperature value is equal to the A/C conversion result divided by 4. The lowest 2 bits of the temperature are stored in the highest 2 bits of the lower byte of the temperature register. The temperature value is represented in the two's complement format. Whenever the most significant bit of the upper byte of the temperature register is 1, the temperature is negative. The magnitude of a negative temperature can be found by taking the two's complement of the temperature reading. After this is done, the upper 8 bits become the integer part of the temperature, whereas the least significant 2 bits become the fractional part of the temperature. A sample of the temperature readings and their corresponding temperature values are shown in Table 10.3.

Binary High Byte/Low Byte	Hex	Temperature
0010 0001/0100 0000	2140	33.25°C
0100 1010/1000 0000	4A80	74.5°C
0001 1010/1100 0000	1AC0	26.75°C
0000 0001/1000 0000	0180	1.5°C
0000 0000/0000 0000	0000	0°C
1111 1111/1000 0000	FF80	−0.5°C
1111 0010/1100 0000	F2C0	−13.25°C
1110 0111/0000 0000	E700	−24°C
1100 1001/0100 0000	C900	−55°C

Table 10.3 ■ TC72 Temperature output data

10.9.3 Serial Bus Interface

The serial interface consists of chip enable (CE), serial clock (SCK), serial data input (SDI), and serial data output (SDO). The CE input is used to select TC72 when there are multiple SPI slaves connected to the microcontroller. TC72 can operate as a SPI slave only.

The SDI input writes data into the TC72's control register, while the SDO output pin reads the temperature data from the temperature register and the status of the *shutdown bit* of the control register. The TC72 can shift data in or out using either the rising or the falling edge of the SCK input. The CE signal is active high. The SCK idle state is detected when the CE signal goes high. As shown in Figure 10.14, the clock polarity (CP) of SCK determines whether data is shifted on the rising or the falling edge. The highest SCK frequency (f_{SCK}) is 7.5 MHz.

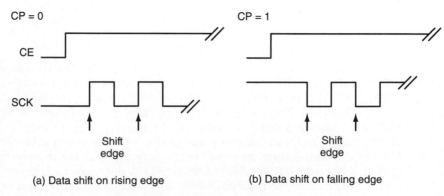

(a) Data shift on rising edge   (b) Data shift on falling edge

Figure 10.14 ■ Serial clock polarity

Data transfer to and from the TC72 consists of one address byte followed by one or multiple data (2 to 4) bytes. The most significant bit (A7) of the address byte determines whether a read (A7 = 0) or a write (A7 = 1) operation will occur. A multiple-byte read operation starts from a high address toward lower addresses. The user needs send in only the temperature result high byte address and read the temperature result high byte, low byte, and the control register. The timing diagrams for the single-data-byte write, single-data-byte read, and three-data-byte read are shown in Figure 10.15a, b, and c, respectively.

The procedure for reading the temperature result is as follows:

Step 1
Pull the CE pin to high to enable SPI transfer.

Step 2
Send the temperature result high-byte-read address (0x02) to the TC72. Wait until the SPI transfer is complete.

Step 3
Read the temperature result high byte. Write a dummy byte into the SSPBUF register to trigger eight pulses to be sent out from the SCK pin so that the temperature result high byte can be shifted in.

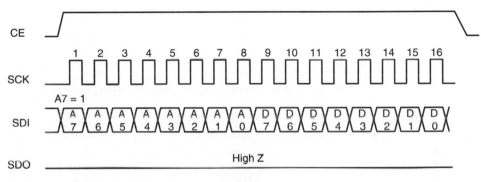

Figure 10.15a ■ Single-data-byte write operation

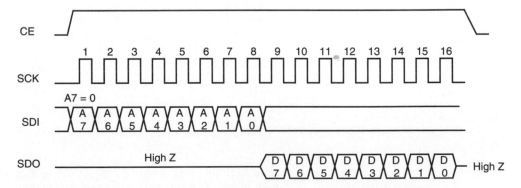

Figure 10.15b ■ Single-data-byte read operation

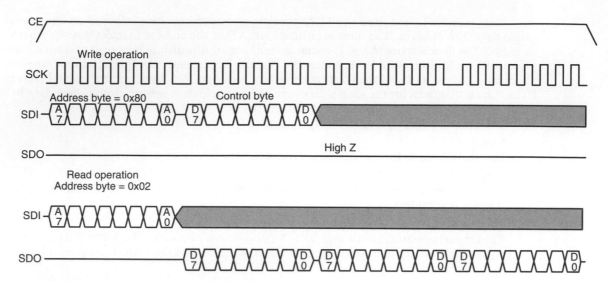

Figure 10.15c ■ SPI multiple-data-byte transfer

Step 4
Read the temperature result low byte. Likewise, write a dummy byte into the SSPBUF register to shift in the temperature low byte.

Step 5
Pull the CE pin to low so that a new transfer can be started.

10.9.4 Internal Register Structure

The TC72 has four internal registers: control register, LSB temperature, MSB temperature, and manufacturer ID. Their contents are shown in Table 10.4.

Register	Read Address	Write Address	Bit 7	Bit 6	Bit 5	Bit 4	Bit 3	Bit 2	Bit 1	Bit 0	Value on POR/BOR
Control	0x00	0x80	0	0	0	OS[1]	0	0	0	SHDN[2]	0x05
LSB temperature	0x01	N/A	T1	T0	0	0	0	0	0	0	0x00
MSB temperature	0x02	N/A	T9	T8	T7	T6	T5	T4	T3	T2	0x00
Manufacturer ID	0x03	N/A	0	1	0	1	0	1	0	0	0x54

[1] OS is one-shot
[2] SHDN is shutdown

Table 10.4 ■ Register for TC72

Control Register
The control register is read-/writable and is used to select the shutdown, continuous, or one-shot conversion operating mode. The temperature conversion mode selection logic is shown in Table 10.5.

At power-up, the SHDN bit is 1. Thus, the TC72 is in the shutdown mode at startup. The shutdown mode disables the temperature conversion circuitry; however, the serial I/O communication

Operation Mode	One-Shot Bit	Shutdown Bit
Continuous temperature conversion	0	0
Shutdown	0	1
Continuous temperature conversion	1	0
One-shot	1	1

Table 10.5 ■ Control register temperature conversion mode selection

port remains active. If the SHDN bit is 0, the TC72 will perform a temperature conversion approximately every 150 ms. In normal operation, a temperature conversion will be initialized by a write operation to the control register to select either the continuous temperature conversion or the one-shot operation mode. The temperature data will be available in the upper byte and lower byte of the temperature register approximately 150 ms after the control register is written into.

The one-shot mode performs a single temperature measurement and returns to the power-saving mode. After completion of the temperature conversion, the one-shot bit is reset to 0. The user must set the one-shot bit to 1 to initiate another temperature conversion.

TEMPERATURE REGISTER

The temperature register is a read-only register and contains a 10-bit two's complement representation of the temperature measurement. Bit 0 through bit 5 are always read as 0. After reset, the temperature register is reset to 0.

MANUFACTURING ID REGISTER

This register is read-only and is used to identify the temperature sensor as a Microchip component.

Example 10.6

▼

Describe the circuit connection between the HCS12 MCU and the TC72 for digital temperature reading, and write a C function to read the temperature every 200 ms. Convert the temperature value into a string so that it can be displayed on an appropriate output device. A pointer to the buffer to hold the string will be passed to this function. The E-clock frequency of the demo board is assumed to be 24 MHz.

Solution: A possible circuit connection is shown in Figure 10.16. A 0.1- to 1.0-μF capacitor should be added between the V_{DD} and the GND pins to filter out power noise.

The C function (**read_temp()**) that starts a temperature measurement and converts the temperature reading into a string is as follows:

```
#include "c:\cwHCS12\include\hcs12.h"
#include "c:\cwHCS12\include\spi0util.h"
#include "c:\cwHCS12\include\delay.h"
void openspi0(void);
void read_temp (char *ptr);
char buf[10];
void main (void)
{
      DDRM |= BIT1;        // configure the PM1 pin for output
      openspi0();          // configure SPI0 module
      read_temp(&buf[0]);
}
```

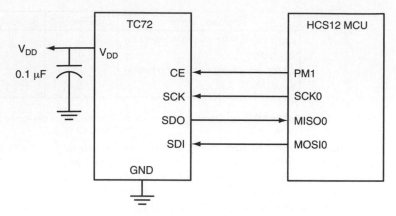

Figure 10.16 ■ Circuit connection between the TC72 and the HCS12

```
void openspi0(void)
{
    SPI0BR   = 0x10;                        // set baud rate to 6 MHz
    SPI0CR1  = 0x50;
    SPI0CR2  = 0x02;
    WOMS     = 0x00;                        // enable Port S pull-up
}
void int2alpha(unsigned int xx, char *ptr) {
        int quo;
        *(ptr+2) = xx % 10 + 0x30;
        quo = xx/10;
        if(quo != 0){
        *(ptr+1) = quo % 10 + 0x30;
        quo = quo/10;                       // tens digit
        }
        if(quo != 0)
        *ptr = quo + 0x30;
}
void read_temp (char *ptr)
{
        char hi_byte, lo_byte, temp, *bptr;
        unsigned int result;
        *ptr       = 0x20;                  // initialize the temperature buffer to 0.0
        *(ptr+1)  = 0x20;                   // and place two space characters before
        *(ptr+2)  = 0x30;                   // 0.0
        *(ptr+3)  = '.';                    //            "
        *(ptr+4)  = 0x30;                   //            "
        bptr       = ptr;
        PTM       |= BIT1;                  // enable TC72 data transfer
        putcspi0 (0x80);                    // send out TC72 control register write address
        putcspi0 (0x11);                    // perform one-shot conversion
        PTM       &= ~BIT1;                 // disable TC72 data transfer
        delayby100ms(2);                    // wait until temperature conversion is complete
```

```c
PTM |= BIT1;                          // enable TC72 data transfer
putcspi0(0x02);                       // send MSB temperature read address
hi_byte = getcspi0();                 // read the temperature high byte
lo_byte = getcspi0();                 // save the temperature low byte and clear SPIF
PTM &= ~BIT1;                         // disable TC72 data transfer
lo_byte &= 0xC0;                      // make sure the lower 6 bits are 0s
result = (int)hi_byte * 256 + (int)lo_byte;
if (hi_byte & 0x80) {                 // temperature is negative
        result = -result;            // take the two's complement of result
        result >>= 6;
        temp = result & 0x0003;      // place the lowest 2 bits in temp
        result >>= 2;                // get rid of fractional part
        *ptr++ = 0x2D;               // store the minus sign
        int2alpha(result, ptr);
}
else {                               // temperature is positive
        result >>= 6;
        temp = result & 0x0003;      // save fractional part
        result >>= 2;                // get rid of fractional part
        int2alpha(result, ptr);      // convert to ASCII string
}
while(*bptr){ // search the end of the string
        bptr++;
};
switch (temp){                       // add fractional digits to the temperature
        case 0:
                break;
        case 1:                      // fractional part is 25
                *bptr++ = 0x2E;      // add decimal point
                *bptr++ = 0x32;
                *bptr++ = 0x35;
                *bptr = '\0';
                break;
        case 2:                      // fractional part is 5
                *bptr++ = 0x2E;      // add decimal point
                *bptr++ = 0x35;
                *bptr = '\0';
                break;
        case 3:                      // fractional part is 75
                *bptr++ = 0x2E;      // add decimal point
                *bptr++ = 0x37;
                *bptr++ = 0x35;
                *bptr = '\0';
                break;
        default:
                break;
}
}
```

10.10 The D/A Converter MCP4922

The MCP4922 from Microchip is a 12-bit voltage output digital-to-analog converter (DAC) with a flexible four-wire serial interface. The four-wire serial interface allows glueless interface to SPI, QSPI, and Microwire serial ports.

The MCP4922 has an output settling time of 4.5 µs. A D/A conversion operation is started by writing a 16-bit serial string that contains 4 control and 12 data bits to the MCP4922. Designed for a wide range of supply voltages, the MCP4922 can operate from 2.7 to 5.5 V and is available in several types of packages including PDIP, SOIC, MSOP, and TSSOP.

10.10.1 Signal Pins

The pin assignment and functional block of the MCP4922 are shown in Figure 10.17. The MCP4922 device utilizes a resistive string architecture, which has the inherent advantages of low differential nonlinearity (DNL) error, low ratio metric temperature coefficient, and fast settling time. The MCP4922 includes double-buffered inputs, allowing simultaneous updates using the LDAC pin.

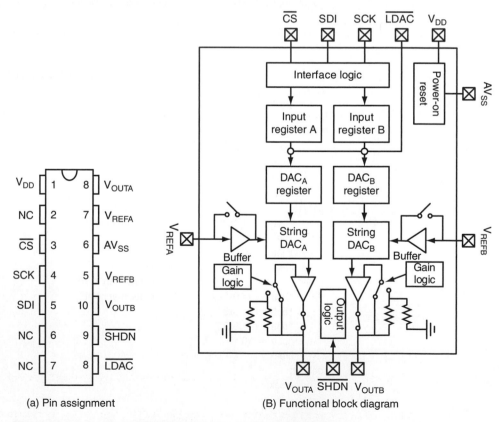

(a) Pin assignment   (B) Functional block diagram

Figure 10.17 ■ The MCP4922 DAC pins and block diagram

The functions of the MCP4922 signals are as follows:

AV_{SS}: analog ground.

$\overline{CS}$: chip select. This signal enables/disables inputs.

SDI: serial data input.

$\overline{LDAC}$: latch DAC input. When this signal is low, the input latch registers' contents are transferred to the DAC registers. This signal can be tied low if data transfer on the rising edge of the $\overline{CS}$ signal is desired.

$\overline{SHDN}$: Shut down input. When this signal is low, the MCP4922 is in the low-power standby mode.

DACx outputs (V_{OUTA}, V_{OUTB}): V_{OUTA} and V_{OUTB} are DAC outputs. The DAC output amplifiers drive these pins with a range of AV_{SS} to V_{DD}.

SCK: serial clock input. The MCP4922 can shift data with the SCK frequency up to 20 MHz.

V_{DD}: positive power supply.

10.10.2 Data Format

The microprocessor or microcontroller sends in a 16-bit frame that consists of 4 bits of command and 12 bits of data to the MCP4922 to start a D/A conversion. The contents of the 16-bit data frame are shown in Figure 10.18. The control bit SHDN allows the user to shut down the MCP4922 to save power when it is not needed.

15	14	13	12	11	10	9	8	7	6	5	4	3	2	1	0
A/B	BUF	$\overline{GA}$	SHDN	D11	D10	D9	D8	D7	D6	D5	D4	D3	D2	D1	D0

A/B: DAC_A or DAC_B select
  0 = write to DAC_A.
  1 = write to DAC_B.
BUF: V_{REF} input buffer control
  0 = unbuffered.
  1 = buffered.
$\overline{GA}$: output gain select bit
  0 = gain is 2 ($V_{OUT} = 2 \times V_{REF} \times D/4096$).
  1 = gain is 1 ($V_{OUT} = V_{REF} \times D/4096$).
$\overline{SHDN}$: output power down control bit
  0 = output buffer is disabled. Output is in high impedance.
  1 = output buffer is enabled.
D11:D0: DAC data bits

Figure 10.18 ■ Input data format for the MCP4922 DAC

10.10.3 MCP4922 Output Voltage

The output voltage of the MCP4922 is given by the following expression:

$$V_{OUT} = GA \times V_{REF} \times code \div 2^{12}$$

Where, GA is the selected gain (can be 1 or 2). Code is the value (represented by D11, . . . , D0) to be converted to voltage.

10.10.4 Format Data to Be Sent to the MCP4922

According to Figure 10.18, the user needs to represent the value to be converted in 12 bits and set the most significant 4 bits properly to select the channel, shutdown mode, gain, and reference voltage buffer mode.

Assume we want to generate a 3-V output from the V_{OUTA} pin, the value to be sent to the MCP4922 can be calculated as follows:

- Select channel A (set bit 15 to 0)
- Select unbuffered mode (set bit 14 to 0)
- Set gain to 1 (set bit 13 to 1)
- Enable output buffer (set bit 12 to 1)
- 3 V corresponds to the digital value (lowest 12 bits) of $2^{12} * 3/5 = 2458 = 0x99A$
- The value to be sent to MCP4922 is 0011 1001 1001 1010 = 0x399A

10.10.5 Interfacing the MCP4922 with the HCS12

A typical circuit connection for interfacing the HCS12 with the MCP4922 is shown in Figure 10.19. Two unused I/O port pins must be used to drive the $\overline{CS}$ and FS pins.

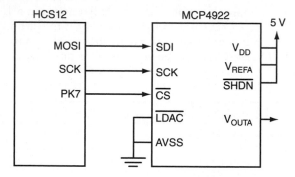

Figure 10.19 ■ Circuit connection between the HCS12 and MCP4922

Example 10.7

Write a program to generate the waveform from the V_{OUTA} pin shown in Figure 10.20 using the circuit shown in Figure 10.19 assuming that the E-clock frequency for the HCS12 is 24 MHz.

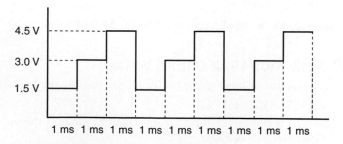

Figure 10.20 ■ Waveform to be generated

Solution: The values (bit 11 to bit 0 in Figure 10.18) corresponding to 1.5 V, 3.0 V, and 4.5 V are 1229 (0x4CD), 2458 (0x99A), and 3686 (0xE66), respectively.

The assembly program that can generate the specified waveform is as follows:

```
        #include  "c:\miniide\hcs12.lnc"
        org       $1500
        lds       #$1500
        bset      DDRK,BIT7          ; configure PK0 pin for output
        jsr       openSPI
forever ldd       #$34CD
        jsr       sendMCP4922        ; output 1.5 V from OUTA pin
        ldy       #1                 ; wait for 1 ms
        jsr       delayby1ms         ;          "
        ldd       #$39AA
        jsr       sendMCP4922        ; output 3.0 V from OUTA pin
        ldy       #1
        jsr       delayby1ms
        ldd       #$3E66
        jsr       sendMCP4922        ; output 4.5 V from OUTA pin
        ldy       #1
        jsr       delayby1ms
        bra       forever
;***********************************************************************************
; The following subroutine sends out the value to be converted by MCP4922.
; The value to be sent is passed in double accumulator D.
;***********************************************************************************
sendMCP4922
        bclr      PTK,BIT7
        jsr       putcSPI0
        tfr       B,A
        jsr       putcSPI0
        bset      PTK,BIT7
        rts
;***********************************************************************************
; The following subroutine enables SPI to shift data on the rising edge of SCK and
; force SCK to be idle low, and shift data at 6 MHz for the 24-MHz E-clock.
;***********************************************************************************
openSPI movb      #$10,SPI0BR        ; set SPI 0 baud rate to 6 MHz
        movb      #$50,SPI0CR1       ; disable interrupt, enable SPI, SCK idle low,
                                     ; data latched on the rising edge, msb first
        movb      #$02,SPI0CR2       ; disable bidirectional mode, stop SPI in wait mode
        movb      #0,WOMS            ; enable Port S pull-up (push-pull)
        rts
        #include  "c:\miniide\delay.asm"
        #include  "c:\miniide\spi0util.asm"
        end
```

The C language version of the program to be compiled by CodeWarrior for generating the specified waveform is as follows:

```
#include "c:\cwHCS12\Include\hcs12.h"
#include "c:\cwHCS12\include\delay.h"              // include delay.c in the project
```

```
#include "c:\cwHCS12\include\spiOutil.h"        // include spiOutil.c in the project
void sendMCP4922(char x1, char x2);
void openSPI0(void);
void main(void) {
        DDRK  |= BIT7;                           // configure PK7 pin for output
        openSPI0();
        while(1) {
                sendMCP4922(0x34,0xCD);          // generate 1.5 V from V_OUTA pin
                delayby1ms(1);                   // wait for 1 ms
                sendMCP4922(0x39,0xAA);          // generate 3.0 V from V_OUTA
                delayby1ms(1);
                sendMCP4922(0x3E,0x66);          // generate 4.5 V from V_OUTA
                delayby1ms(1);
        }
}
void openSPI0(void) {
        SPI0BR   = 0x10;                         // Set SPI0 baud rate to 6 MHz
        SPI0CR1  = 0x50;                         // enable SPI0, SCK idle low & shift on rising edge
        SPI0CR2  = 0x02;                         // disable bidirectional mode, stop SPI in wait mode
        WOMS     = 0;                            // enable port S pull-up
}
void sendMCP4922(char x1, char x2) {
        PORTK    &= 0x7F;                        // enable SPI transfer to MCP4922
        putcspi0(x1);                            // send out upper byte
        putcspi0(x2);                            // send out lower byte
        PTK      |= BIT7;                        // start DAC operation
}
```

Example 10.8

Write a program to generate the waveform shown in Figure 10.21 using the DAC circuit shown in Figure 10.19.

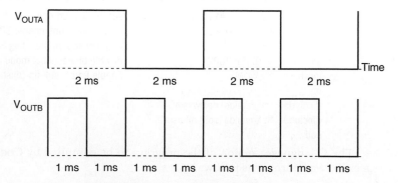

Figure 10.21 ■ Waveform to be generated

Solution: The procedure for generating the specified waveform is as follows:

Step 1
Configure SPI0 module properly.

Step 2
Output a high (5 V) from V_{OUTA} and also output a high from V_{OUTB}.

Step 3
Wait for 1 ms.

Step 4
Output a low from V_{OUTB}.

Step 5
Wait for 1 ms.

Step 6
Output a low from V_{OUTA} and output a high from V_{OUTB}.

Step 7
Wait for 1 ms.

Step 8
Output a low from V_{OUTB}.

Step 9
Wait for 1 ms.

Step 10
Go to step 2.

The value to be sent to MCP4922 to output a high from V_{OUTA} is 0x3FFF whereas the value to be sent to generate a low from the V_{OUTA} pin is 0x3000.

The value to be sent to MCP4922 to output a high from V_{OUTB} is 0xBFFF whereas the value to be sent to generate a low from the V_{OUTB} pin is 0xB000.

The assembly program for generating the specified waveform is as follows:

```
#include    "c:\miniide\hcs12.inc"
            org      $1500
            lds      #$1500        ; set up the stack pointer
            jsr      openSPI       ; configure the SPI0
            bset     DDRK,BIT7     ; configure PK7 pin for output
forever     ldd      #$3FFF        ; output a high from V_OUTA
            jsr      sendMCP4922   ;          "
            ldd      #$BFFF        ; output a high from V_OUTB
            jsr      sendMCP4922   ;          "
            ldy      #1            ; wait for 1 ms
            jsr      delayby1ms    ;          "
            ldd      #$B000
            jsr      sendMCP4922
            ldy      #1
            jsr      delayby1ms
            ldd      #$3000
            jsr      sendMCP4922
            ldd      #$BFFF
            jsr      sendMCP4922
            ldy      #1
```

```
          jsr      delayby1ms
          ldd      #$B000
          jsr      sendMCP4922
          ldy      #1
          jsr      delayby1ms
          lbra     forever
;   ****************************************************************************
; The following subroutine sends out the value to be converted by MCP4922.
; The value to be sent is passed in double accumulator D.
;   ****************************************************************************
sendMCP4922
          bclr     PTK,BIT7
          jsr      putcSPI0
          tfr      B,A
          jsr      putcSPI0
          bset     PTK,BIT7
          rts
;   ****************************************************************************
; The following subroutine enables SPI to shift data on the rising edge of SCK and
; force SCK to be idle low, and shift data at 6 MHz for the 24-MHz E-clock.
;   ****************************************************************************
openSPI   movb     #$10,SPI0BR        ; set SPI 0 baud rate to 6 MHz
          movb     #$50,SPI0CR1       ; disable interrupt, enable SPI, SCK idle low,
                                      ; data latched on the rising edge, msb first
          movb     #$02,SPI0CR2       ; disable bidirectional mode, stop SPI in wait mode
          movb     #0,WOMS            ; enable Port S pull-up (push-pull)
          rts
          #include "c:\miniide\delay.asm"
          #include "c:\miniide\spi0util.asm"
          end
```

▲

This program can be modified to be interrupt-driven so that the MCU can still perform other operations. The C language version of the program is straightforward and hence is left as an exercise problem.

10.11 Matrix LED Displays

Many organizations have the need to display important information at the entrance or some corners of their buildings. The information need not be displayed all at once but can be rotated. Temperature, date, humidity, and other data in turn are displayed at many crossroads. Schools display their upcoming games, the result of games yesterday, or other important events using large display panels. Many of these display panels use the matrix LED displays because of their brightness and versatility.

10.11.1 The Organization of Matrix LED Displays

Matrix LED displays are denoted by the number of their columns and rows. The most popular matrix LED display has five columns and seven rows (5×7). Other configurations, such as 5×8 and 8×8, are also available. It is obvious that the more rows and columns of LEDs are used, the better the resolution will be.

10.11.2 Colors of Matrix LED Displays

Like seven-segment displays, there are green, yellow, red, and bicolor (red/green) matrix LED displays.

10.11.3 Connection Method

Matrix LED displays can be organized as **cathode row** (anode column) or **anode row** (cathode column). In a cathode-row organization, the LEDs in a row have a common cathode. In an anode-row organization, the LEDs in a row have a common anode. The cathode-row organization and anode-row organizations are shown in Figures 10.22 and 10.23, respectively.

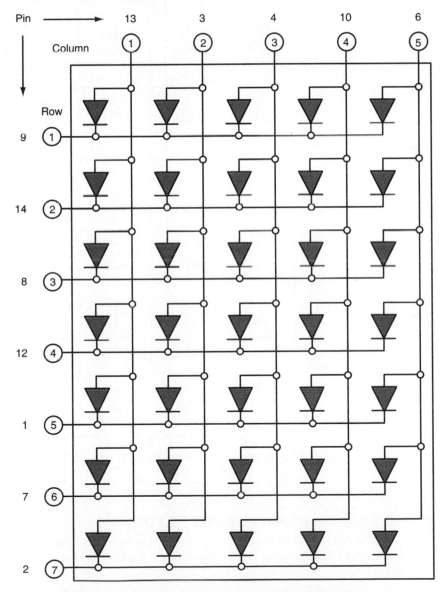

Figure 10.22 ■ Cathode-row matrix LEDs (Fairchild GMC8X75C)

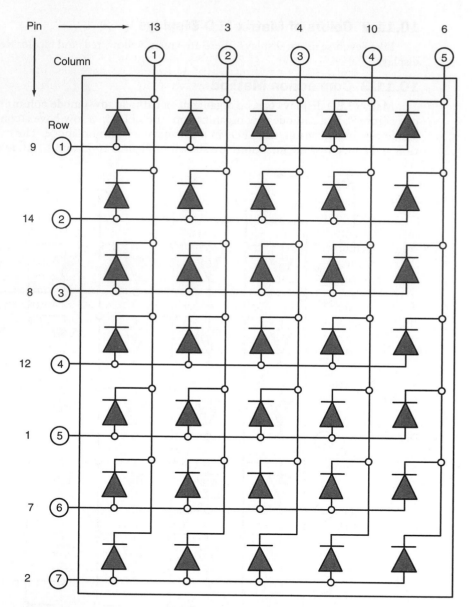

Figure 10.23 ■ Anode-row matrix LEDs (Fairchild GMA8X75C)

10.11.4 Dimension of Matrix LED Displays

The dimension of a matrix LED display is often indicated by its height. Common heights for matrix LED displays include 0.7, 1.2, 1.7, and 2.3 in. Larger matrix LED displays are also available.

10.11.5 Method of Driving Matrix LED Displays

Because of the internal connection of a matrix LED display, two parallel ports are needed to drive it. The matrix LED displays need to be scanned one row at a time, from top to bottom (or bottom to top). Usually, multiple matrix LED displays are needed in the application, so a time-multiplexing technique needs to be used, which is quite demanding on CPU time.

Because of the popularity of matrix LED displays, a few companies have produced driver chips for them. Among them, Maxim provides driver chips with an SPI (MAX6952) or I²C (MAX6953) interface for cathode-row matrix LEDs.

10.12 The MAX6952 Matrix LED Display Driver

The MAX6952 is designed to drive cathode-row matrix LED displays with a 5 × 7 organization. This chip can operate with a power supply from 2.7 to 5.5 V and can drive four monocolor or two bicolor cathode-row matrix LED displays. It has a built-in ASCII 104-character font and 24 user-definable characters. The built-in characters are in the Arial font, with the addition of the following common symbols: |, €, ¥, °, μ, ±, ↑, and ↓. The 24 user-definable characters are uploaded by the user into on-chip RAM through the serial interface and are lost when the device is powered down. It allows automatic blinking control for each segment and provides 16-step digital brightness control. Both the 36-pin SSOP and 40-pin packages are available. The device includes a low-power shutdown mode, segment blinking (synchronized across multiple drivers, if desired), and a test mode that forces all LEDs on.

10.12.1 Pin Functions

The pin functions are described in Table 10.6. The connection for four monocolor digits and two bicolor digits are shown in Tables 10.7 and 10.8, respectively. The typical circuit connection that drives four monocolor digits using the MAX6952 is shown in Figure 10.24.

The MAX6952 uses four pins to interface with the MCU: CLK, $\overline{CS}$, DIN, and DOUT. The $\overline{CS}$ signal must be low to clock data in and out of the device, and the DIN signal is shifted in on the rising edge of CLK. The MAX6952 can shift data at 26 MHz. When the MAX6952 is not being accessed, DOUT is not in high impedance, contrary to the SPI standard.

Multiple MAX6952 can be daisy-chained by connecting the DOUT pin of one device to the DIN pin of the next and driving the CLK and $\overline{CS}$ lines in parallel (Figure 10.25). Data at DIN propagates through the internal shift registers and appears at DOUT 15.5 clock cycles later, clocked out on the falling edge of CLK. When sending commands to daisy-chained MAX6952s, all devices are accessed at the same time. An access requires $16 \times n$ clock cycles, where n is the number of MAX6952s connected together. To update just one device in a daisy chain, the user can send the no-op command (0x00) to the others.

10.12.2 Internal Registers

The block diagram of the MAX6952 is shown in Figure 10.26. The MAX6952 contains a 16-bit shift register (in the serial interface block in Figure 10.26) into which DIN data are shifted on the rising edge of the CLK signal, when the $\overline{CS}$ signal is low. The 16 bits in the shift register are loaded into a latch on the rising edge of the $\overline{CS}$ signal. The 16 bits in the latch are then decoded and executed. The CLK input must be idle low. This signal must be taken low before data transfer is started.

Name	Pin		Function
	SSOP	PDIP	
O0 to O13	1, 2, 3, 6–14, 23, 24	1, 2, 3, 7–15, 26, 27	LED cathode drivers. O0 to O13 output sink current from the displays's cathode rows.
GND	4, 5, 6	4, 5, 6, 18	Ground
ISET	15	17	Segment current setting. Connect ISET to GND through series resistor R_{SET} to set the peak current.
BLINK	17	19	Blink clock output. Output is open-drain.
DIN	18	20	Serial data input. Data is loaded into the internal 16-bit shift register on the rising edge of the CLK.
CLK	19	21	Serial-clock input. On the rising edge of CLK, data is shifted into the internal shift register. On the falling edge of CLK, data is clocked out of DOUT. CLK input is active only when $\overline{CS}$ is low.
DOUT	20	22	Serial data output. Data clocked into DIN is output to DOUT 15.5 clock cycles later. Data is clocked out on the falling edge of CLK. Output is push-pull.
DOUT	21	23	Chip-select input. Serial data is loaded into the shift register while $\overline{CS}$ is low. The last 16 bits of serial data are latched on $\overline{CS}$'s rising edge.
OSC	22	24	Multiplex clock input. To use the internal oscillator, connect capacitor CSET from OSC to GND. To use the external clock, drive OSC with a 1-MHz to 8-MHz CMOS clock.
O14 to O23	25–31, 34, 35, 36	28–34, 38, 39, 40	LED anode drivers. O14 to O23 output source current to the display's anode columns.
V+	32, 33	35, 36, 37	Positive supply voltage. Bypass V+ to GND with a 47-µF bulk capacitor and a 0.1-µF ceramic capacitor.

Table 10.6 ■ MAX6952 4-digit matrix LED display driver pin functions

Digit	O0~O6	O7~O13	O14~O18	O19~O23
1	Digit 0 rows (cathodes) R1 to R7 Digit 1 rows (cathodes) R1 to R7	—	Digit 0 columns (anodes) C1 to C5	Digit 1 columns (anodes) C6 to C10
2	—	Digit 2 rows (cathodes) R1 to R7 Digit 3 rows (cathodes) R1 to R7	Digit 2 columns (anodes) C1 to C5	Digit 3 columns (anodes) C6 to C10

Table 10.7 ■ Connection scheme for four monocolor digits

Digit	O0~O6	O7~O13	O14~O18	O19~O23
1	Digit 0 rows (cathodes) R1 to R14	—	Digit 0 columns (anodes) C1 to C10	
			The 5 green anodes	The 5 red anodes
2	—	Digit 1 rows (cathodes) R1 to R14	Digit 1 columns (anodes) C1 to C10	
			The 5 green anodes	The 5 red anodes

Table 10.8 ■ Connection scheme for two bicolor digits

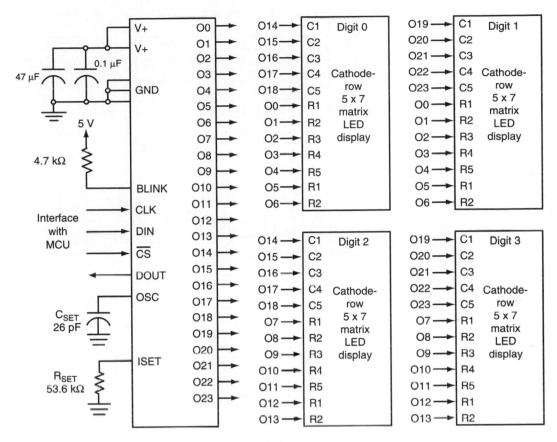

Figure 10.24 ■ MAX6952 driving four matrix LED displays

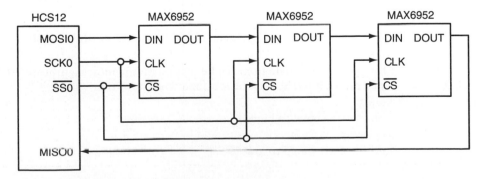

Figure 10.25 ■ MAX6952 daisy-chain connection

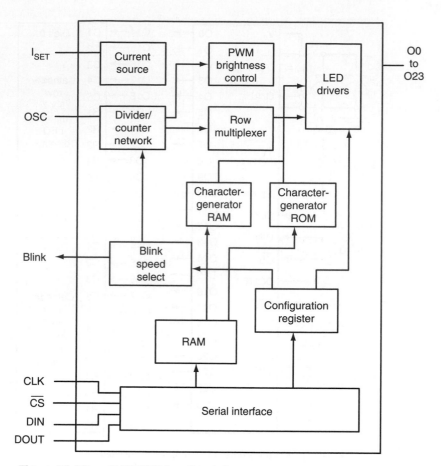

Figure 10.26 ■ MAX6952 functional diagram

The upper 8 bits of the shift register select the destination register to which the lower 8 bits of the shift register are to be transferred. The address map of the MAX6952 is shown in Table 10.9.

The procedure for writing the MAX6952 is as follows:

Step 1
Pull the CLK signal to low.

Step 2
Pull the $\overline{CS}$ signal to low to enable the internal 16-bit shift register.

Step 3
Shift in 16 bits of data from the DIN pin with the most significant bit first. The most significant bit (D15) must be low for a write operation.

Step 4
Pull the $\overline{CS}$ signal to high.

Step 5
Pull the CLK signal to low.

Register	Address (command byte)								Hex Code
	D15	D14	D13	D12	D11	D10	D9	D8	
No op	R/W̄	0	0	0	0	0	0	0	0x00
Intensity10	R/W̄	0	0	0	0	0	0	1	0x01
Intensity32	R/W̄	0	0	0	0	0	1	0	0x02
Scan limit	R/W̄	0	0	0	0	0	1	1	0x03
Configuration	R/W̄	0	0	0	0	1	0	0	0x04
User-defined fonts	R/W̄	0	0	0	0	1	0	1	0x05
Factory reserved (do not write into)	R/W̄	0	0	0	0	1	1	0	0x06
Display test	R/W̄	0	0	0	0	1	1	1	0x07
Digit 0 plane P0	R/W̄	0	1	0	0	0	0	0	0x20
Digit 1 plane P0	R/W̄	0	1	0	0	0	0	1	0x21
Digit 2 plane P0	R/W̄	0	1	0	0	0	1	0	0x22
Digit 3 plane P0	R/W̄	0	1	0	0	0	1	1	0x23
Digit 0 plane P1	R/W̄	1	0	0	0	0	0	0	0x40
Digit 1 plane P1	R/W̄	1	0	0	0	0	0	1	0x41
Digit 2 plane P1	R/W̄	1	0	0	0	0	1	0	0x42
Digit 3 plane P1	R/W̄	1	0	0	0	0	1	1	0x43
Write digit 0 plane P0 and plane P1 with same data (reads as 0x00)	R/W̄	1	0	0	0	0	0	0	0x60
Write digit 1 plane P0 and plane P1 with same data (reads as 0x00)	R/W̄	1	1	0	0	0	0	1	0x61
Write digit 2 plane P0 and plane P1 with same data (reads as 0x00)	R/W̄	1	1	0	0	0	1	0	0x62
Write digit 3 plane P0 and plane P1 with same data (reads as 0x00)	R/W̄	1	1	0	0	0	1	1	0x63

Table 10.9 ■ MAX6952 register address map

Any register data within the MAX6952 may be read by setting the D15 bit to 1. The procedure to read a register is as follows:

Step 1
Pull CLK to low.

Step 2
Pull the $\overline{CS}$ signal to low to enable the internal shift register.

Step 3
Clock 16 bits of data into the DIN pin with bit 15 first. Bit 15 must be 1. Bit 14 to bit 8 contain the address of the register to be read. Bits 7 through 0 contain dummy data.

Step 4
Pull the $\overline{CS}$ signal to high. Bits 7 to 0 of the serial shift register will be loaded with the data in the register addressed by bits 15 through 8.

Step 5
Pull CLK to low.

Step 6
Issue another read command (which can be a no-op), and examine the bit stream at the DOUT pin. The second 8 bits are the contents of the register addressed by bits 14 to 8 in step 3.

DIGIT REGISTERS

The MAX6952 uses eight digit registers to store the characters that the user wishes to display on the four 5 × 7 LED digits. These digit registers are implemented with two planes of 4 bytes, called P0 and P1. Each LED digit is represented by 2 bytes of memory, 1 byte in plane P0 and the other in plane P1. The digit registers are mapped so that a digit's data can be updated in plane P0 or plane P1 or both at the same time, as shown in Table 10.10.

Segment's Bit Setting in Plane P1	Segment's Bit Setting in Plane P10	Segment Behavior
0	0	Segment off
0	1	Segment on only during the first half of each blink period
1	0	Segment on only during the second half of each blink period
1	1	Segment on

Table 10.10 ■ Digit register mapping with blink globally enabled

If the blink function is disabled through the blink enable bit E in the configuration register, then the digit register data in plane P0 is used to multiplex the display. The digit register data in P1 is not used. If the blink function is enabled, then the digit register data in both plane P0 and P1 are alternately used to multiplex the display. Blinking is achieved by multiplexing the LED display using data planes P0 and P1 on alternate phases of the blink clock (shown in Table 10.10).

The data in the digit registers does not control the digit segments directly. Instead, the register data is used to address a character generator, which stores the data of a 128-character font. The lower 7 bits of the digit data (D6 to D0) select the character font. The most significant bit of the register data (D7) selects whether the font data is used directly (D7 = 0) or whether the font is inverted (D7 = 1). The inversion feature can be used to enhance the appearance of bicolor displays by displaying, for example, a red character on a green background.

CONFIGURATION REGISTER

The configuration register is used to enter and exit shutdown, select the blink rate, globally enable and disable the blink function, globally clear the digit data, and reset the blink timing. The contents of the configuration register are shown in Figure 10.27.

INTENSITY REGISTERS

Display brightness is controlled by four pulse-width modulators, one for each display digit. Each digit is controlled by a nibble of one of the two intensity registers, *Intensity10* and *Intensity32*. The upper nibble of the Intensity10 register controls the intensity of the matrix display 1, whereas the lower nibble of the same register controls the intensity of the matrix display 0. Matrix displays 3 and 2 are controlled by the upper and lower nibbles of the Intensity32 register, respectively. The modulator scales the average segment current in 16 steps from a maximum of 15/16 down to 1/16 of the peak current. The minimum interdigit blinking time is, therefore, 1/16 of a cycle. The maximum duty cycle is 15/16.

7	6	5	4	3	2	1	0
P	x	R	T	E	B	x	S

P: blink phase read back select
  0 = P1 blink phase.
  1 = P0 blink phase.
R: global clear digit data
  0 = digit data on both planes P0 and P1 are not affected.
  1 = clear digit data on both planes P0 and P1.
T: global blink timing synchronization
  0 = blink timing counters are unaffected.
  1 = blink timing counters are reset on the rising edge of $\overline{CS}$.
E: global blink enable/disable
  0 = blink function is disabled.
  1 = blink function is enabled.
B: blink rate selection
  0 = select slow blinking (refreshed for 1 s by plane P0, then 1 s
      by P1 at 4 MHz).
  1 = select fast blinking.
S: shutdown mode
  0 = shutdown mode.
  1 = normal operation.

Figure 10.27 ■ The MAX6952 configuration register

SCAN-LIMIT REGISTER

The scan-limit register sets how many monocolor digits are displayed, either two or four. A bicolor digit is connected as two monocolor digits. The multiplexing scheme drives digits 0 and 1 at the same time, then digits 2 and 3 at the same time. To increase the effective brightness of the displays, the MAX6952 drives only two digits instead of four. By doing this, the average segment current doubles, but this also doubles the number of MAX6952s required for driving a given number of digits.

The contents of the scan-limit register are shown in Figure 10.28. This register has only 1 bit implemented. When this bit is 0, only digits 0 and 1 are displayed. Otherwise, all four digits are displayed.

7	6	5	4	3	2	1	0
X	X	X	X	X	X	X	2 or 4

2 or 4: scan two digits (0 and 1) or all four digits.
  0 = display digits 0 and 1 only.
  1 = display digits 0, 1, 2, and 3.

Figure 10.28 ■ The MAX6952 scan-limit register

DISPLAY TEST REGISTER

The display test register switches the drivers between one of two modes: normal and display test. Display-test mode turns on all LEDs by overriding, but not altering, all control and digit registers (including the shutdown register). In display-test mode, eight digits are scanned

7	6	5	4	3	2	1	0
X	X	X	X	X	X	X	Test

Test: test bit
   0 = normal operation.
   1 = display test.

Figure 10.29 ■ The MAX6952 display test register

and the duty cycle is 7/16 (half power). The contents of the display test register are shown in Figure 10.29. Only bit zero of this register is implemented.

CHARACTER-GENERATOR FONT MAPPING

The character font is a 5 × 7 matrix. The character generator comprises 104 characters in ROM and 24 user-definable characters. The selection from a total of 128 characters is represented by the lower 7 bits of the 8-bit digit registers. The character map is in the Arial font for 96 characters in the range from 0x28 to 0x7F. The first 32 characters map the 24 user-defined positions (RAM00 to RAM23), plus 8 extra common characters in ROM. When the most significant bit is 0, the device will display the font normally. Otherwise, the chip will display the font inversely.

USER-DEFINED FONT REGISTER

The 24 user-definable characters are represented by 120 entries of 7-bit data, five entries per character, and are stored in the MAX6952's internal RAM. The 120 user-definable font data are written and read through a single register at the address 0x05. An auto-incrementing font address register pointer in the MAX6952 indirectly accesses the font data. The font address pointer can be written, setting one of 120 addresses between 0x00 and 0xF7, but cannot be read back. The font data is written to and read from the MAX6952 indirectly, using this font address pointer. Unused font locations can be used as general-purpose scratch RAM. Font registers are only 7 bits wide.

To define new fonts, the user first needs to set the font address pointer. This is done by placing the address in the font address pointer register and setting bit 7 to 1. After this, the user can write the font data in the lower 7 bits (to the font address pointer position) and clear bit 7.

The font address pointer autoincrements after a valid access to the user-definable font data. Auto-incrementing allows the 120 font data entries to be written and read back very quickly because the font pointer address needs to be set only once. When the last data location, 0xF7, is written into, the font address pointer increments to 0x80 automatically. If the font address pointer is set to an out-of-range address by writing data in the range from 0xF8 to 0xFF, then the address is set to 0x80 instead.

The memory mapping of user-defined font register 0x05 is detailed in Table 10.11. The behavior of the font pointer address is shown in Table 10.12. To display the font defined by the user, one must send in the RAM address from 0x00 through 0x17, corresponding to the font address pointer value that is 5 × RAM address (one character needs 5 bytes).

Address Code (hex)	Register Data	SPI Read or Write	Function
0x85	0x00–0x7F	Read	Read 7-bit user-definable font data entry from current font address. MSB of the register data is clear. Font address pointer is incremented after the read.
0x05	0x00–0x7F	Write	Write 7-bit user-definable font data entry to current font address. Font address pointer is incremented after the write.
0x05	0x80–0xFF	Write	Write font address pointer with the register data.

Table 10.11 ■ Memory mapping of user-defined font register 0x05

Font Pointer Address	Action
0x80–0xF6	Valid range to set the font address pointer. Pointer autoincrements after a font data read or write, while pointer address remains in this range.
0xF7	Font address resets to 0x80 after a font data read or write to this pointer address.
0xF8 to 0xFF	Invalind range to set the font address pointer. Pointer is set to 0x80.

Table 10.12 ■ Font pointer address behavior

10.12.3 Blinking Operation

The display blinking facility, when enabled, makes the LED drivers flip automatically between displaying the digit register data in planes P0 and P1. If the digit register data for any digit is different in two planes, then that digit appears to flip between two characters. To make a character appear to blink on or off, write the character to one plane and use the blank character (0x20) for the other plane. Once blinking has been configured, it continues automatically without further intervention. Blinking is enabled by setting the E bit of the configuration register.

The blink speed can be programmed to be fast or slow and is determined by the frequency of the multiplex clock, OSC, and by setting the B bit of the configuration register. The blink rate selection bit B of the configuration register sets either fast or slow blink speed for the whole display.

BLINK SYNCHRONIZATION

When multiple digits are displayed, one can choose to synchronize the blinking operation of these digits. Internally, blink synchronization is achieved by resetting the display multiplexing sequence. As long as all MAX6952s are daisy-chained with one device's DOUT connected to the DIN of the next device, global synchronization is achieved by toggling the $\overline{CS}$ pin for each device, either together or in quick succession.

BLINK OUTPUT

The blink output (the BLINK pin) indicates the blink phase and is high during the P0 period and low during the P1 period. Blink phase status can be read back as the P bit in the configuration register. Typical uses for this output are

- To provide an interrupt to the processor so that segment data can be changed synchronously to the blinking. For example, a clock application may have colon segments blinking every second between hour and minute digits, and the minute display is best changed in step with the colon segments. Also, if the rising edge of blink is detected, there is half a blink period to change the P1 data (P0 data drives the displays during this interval). Similarly, if the falling edge of blink is detected, there is half a blink period to change the P0 digit data.

- If OSC is driven with an accurate frequency, blink can be used as a seconds counter.

10.12.4 Choosing Values for R_{SET} and C_{SET}

The RC oscillator uses an external resistor, R_{SET}, and an external capacitor, C_{SET}, to set the oscillator frequency, f_{OSC}. The allowed range of f_{OSC} is 1 to 8 MHz. R_{SET} also sets the peak segment current. The recommended values for R_{SET} and C_{SET} set the oscillator to 4 MHz, which sets the slow and fast blink frequencies to 0.5 Hz and 1 Hz. The recommended value of R_{SET} also sets the peak current to 40 mA, which makes the segment current adjustable from 2.5 to 37.5 mA in 2.5-mA steps:

$$I_{SEG} = K_I/R_{SET} \text{ mA}$$
$$f_{OSC} = K_F/(R_{SET} \times C_{SET} + C_{STRAY}) \text{ MHz}$$

where

$K_I = 2144$
$K_F = 6000$
R_{SET} = external resistor in kΩ
C_{SET} = external capacitor in pF
C_{STRAY} = stray capacitance from OSC pin to GND in pF, typically 2 pF

The recommended value for R_{SET} is 53.6 kΩ and the recommended value for C_{SET} is 26 pF. The recommended value for R_{SET} is the minimum allowed value since it sets the display driver to the maximum allowed segment current. R_{SET} can be set to a higher value to set the segment current to a lower peak value whenever it is desirable. The effective value of C_{SET} includes not only the actual external capacitor used but also the stray capacitance from OSC to GND.

Example 10.9
▼

Daisy-chain two MAX6952s to drive eight of the cathode-row, monocolor, matrix LED displays GMC8975C made by Fairchild, assuming that the PM5 pin is used to drive the $\overline{CS}$ input of two MAX6952s. The connection of two MAX6952s with the HCS12 (24-MHz E-clock) is shown in Figure 10.30. Each MAX6952 is driving four matrix LED displays. Write a program to configure the SPI module to operate in master mode, and transfer data at 12 MHz, shift data in/out on the rising edge of the SCK clock. Display *MSU ECET* on the matrix LED displays without blinking.

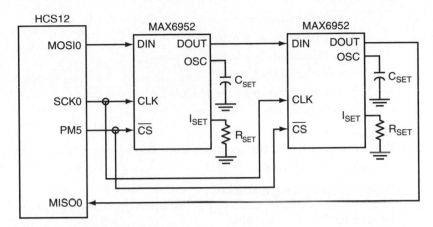

Figure 10.30 ■ HCS12 driving two MAX6952s

Solution: The SPI0 module should be configured with the following features:

- 12-MHz baud rate
- Master mode with interrupts disabled
- Shift data on the rising edge with clock idle low
- Shift data out most significant bit first
- Disable mode fault
- Stop SPI0 on wait mode

Two MAX6952 chips should be configured as follows:

Intensity10 register
We set the intensity of both MAX6952s to maximum by sending out the following value:

    0x01, 0xFF, 0x01, 0xFF

Sixteen bits need to be written to each MAX6952, of which the upper 8 bits are the address to select the Intensity10 register.

Intensity32 register
Set the second and third displays to maximum intensity by sending the same data as to Intensity10:

    0x02, 0xFF, 0x02, 0xFF

Scan-limit register
Configure the MAX6952 to drive four monocolor displays by writing the following value to the MAX6952:

    0x03, 0x01, 0x03, 0x01   ; drive four displays

Configuration register
Configure this register to

- Select P1 blink phase
- Not to clear digit data on both plane P0 and P1

- Reset blink timing counter on the rising edge of $\overline{CS}$
- Disable blink function
- Select slow blinking ("don't care")
- Select normal mode

Send the following values to the MAX6952s:

    0x04, 0x11, 0x04, 0x11

Display test register

Disable display test as the start by writing the following values to the MAX6952:

    0x07, 0x00, 0x07, 0x00

Digit 0 (rightmost digit) plane P0

Display space character and letter T, respectively, on display 0 of the first and second MAX6952 by sending the following data:

    0x20, 0x54, 0x20, 0x20

Digit 1 (second rightmost digit) plane P0

Display letter U and letter E, respectively, on display 1 of the first and second MAX6952 by sending the following data:

    0x21, 0x45, 0x21, 0x55

Digit 2 (second leftmost digit) plane P0

Display letter S and letter C, respectively, on display 2 of the first and second MAX6952 by sending the following data:

    0x22, 0x43, 0x22, 0x53

Digit 3 (leftmost digit) plane P0

Display letter M and letter E, respectively, on the display 3 of the first and second MAX6952 by sending the following data:

    0x23, 0x45, 0x23, 0x4D

The following C program configures the SPI and MAX6952 properly:

```c
#include "c:\cwHCS12\include\hcs12.h"
#include "c:\cwHCS12\include\spi0util.c"
void sendtomax(char x1, char x2, char x3, char x4);
void openspi0(void);
void main (void)
{
        openspi0();
        DDRM |= BIT5; // configure PM5 pin for output
        sendtomax(0x01, 0xFF, 0x01, 0xFF);      // set intensity for digits 0 and 1
        sendtomax(0x02, 0xFF, 0x02, 0xFF);      // set intensity for digits 2 and 3
        sendtomax(0x03, 0x01, 0x03, 0x01);      // set scan limit to drive four digits
        sendtomax(0x04, 0x11, 0x04, 0x11);      // set configuration register
        sendtomax(0x07, 0x00, 0x07, 0x00);      // disable test
        sendtomax(0x20, 0x54, 0x20, 0x20);      // value for digit 0
        sendtomax(0x21, 0x45, 0x21, 0x55);      // value for digit 1
        sendtomax(0x22, 0x43, 0x22, 0x53);      // value for digit 2
        sendtomax(0x23, 0x45, 0x23, 0x4D);      // value for digit 3
}
```

```
void sendtomax (char c1, char c2, char c3, char c4)
{
        char        temp;
        PTM &= ~BIT5;        /* enable SPI transfer to MAX6952 */
        putcspi0(c1);        /* send c1 to MAX6952 */
        putcspi0(c2);        /* send c2 to MAX6952 */
        putcspi0(c3);        /* send c3 to MAX6952 */
        putcspi0(c4);        /* send c4 to MAX6952 */
        PTM |= BIT5;         /* load data from shift register to latch */
}
void openspi0(void)
{
        SPI0BR   = 0x00;   /* set baud rate to 12 MHz */
        SPI0CR1  = 0x50;   /* disable Interrupt, set master mode, shift data on
                                rising edge, clock idle low */
        SPI0CR2  = 0x02;   /* disable mode fault, disable SPI in wait mode */
        WOMS     = 0;      /* enable Port S pull-up */
}
```

Example 10.10

Modify the previous example to blink the display at a slow rate.

Solution: We need to change the setting of the configuration register and also send the space character (0x20) to the four digits in plane P1. New data to be sent to the configuration registers are as follows.

0x04, 0x19, 0x04, 0x19

The main program should be modified as follows:

```
void main (void)
{
        openspi0();
        DDRM |= BIt5;                           // configure PM5 pin for output
        sendtomax(0x01, 0xFF, 0x01, 0xFF);
        sendtomax(0x02, 0xFF, 0x02, 0xFF);
        sendtomax(0x03, 0x01, 0x03, 0x01);
        sendtomax(0x04, 0x19, 0x04, 0x19);      // configuration register, blink at phase P1
        sendtomax(0x07, 0x00, 0x07, 0x00);      // disable test
        sendtomax(0x20, 0x54, 0x20, 0x20);      // value for digit 0 on plane P0
        sendtomax(0x21, 0x45, 0x21, 0x55);      // value for digit 1
        sendtomax(0x22, 0x43, 0x22, 0x53);      // value for digit 2
        sendtomax(0x23, 0x45, 0x23, 0x4D);      // value for digit 3
        sendtomax(0x40, 0x20, 0x40, 0x20);      // value for digit 0 on plane P1 (space)
        sendtomax(0x41, 0x20, 0x41, 0x20);      // value for digit 1    "
        sendtomax(0x42, 0x20, 0x42, 0x20);      // value for digit 2    "
        sendtomax(0x43, 0x20, 0x43, 0x20);      // value for digit 3    "

}
```

Example 10.11

For the circuit shown in Figure 10.30, write a program to display the following message and shift the information from right to left every second and enable blinking:

08:30:40 Wednesday, 72°F, humidity: 60%

Solution: We use plane P0 to shift the message once every half a second. The message in plane P0 is used to multiplex the display in half a second and the message sent to plane P1 is used to multiplex the display in the next half of a second. We display the space character in each matrix display using plane P1 and display the normal characters using plane P0. A delay function is invoked once every second to shift the message to plane P0. The program is as follows:

```
#include "c:\cwHCS12\include\hcs12.h"
#include "c:\cwHCS12\include\spiOutil.h"
#include "c:\cwHCS12\include\delay.h"
void    send2max (char x1, char x2, char x3, char x4);
void    openspi0 (void);
char    msgP0[41] = "08:30:40 Wednesday, 72°F, humidity: 60% ";

void main (void)
{
        char i1, i2, i3, i4;
        char j1, j2, j3, j4;
        char k;
        openspi0();
        DDRM |= BIT5; // configure PM5 pin for output
        send2max(0x01, 0xFF, 0x01, 0xFF);
        send2max(0x02, 0xFF, 0x02, 0xFF);
        send2max(0x03, 0x01, 0x03, 0x01);
        send2max(0x04, 0x1D, 0x04, 0x1D); // configuration register
        send2max(0x40, 0x20, 0x40, 0x20); // send space character to plane P1
        send2max(0x41, 0x20, 0x41, 0x20);
        send2max(0x42, 0x20, 0x42, 0x20);
        send2max(0x43, 0x20, 0x43, 0x20);
        k = 0;
        while (1) {
            i1 = k;
            i2 = (k+1)%40;
            i3 = (k+2)%40;
            i4 = (k+3)%40;
            j1 = (k+4)%40;
            j2 = (k+5)%40;
            j3 = (k+6)%40;
            j4 = (k+7)%40;
            sendtomax(0x20, msgP0[i1], 0x20, msgP0[j1]);
            sendtomax(0x21, msgP0[i2], 0x21, msgP0[j2]);
            sendtomax(0x22, msgP0[i3], 0x22, msgP0[j3]);
            sendtomax(0x23, msgP0[i4], 0x23, msgP0[j4]);
            delayby100ms(10);  /* wait for 1 s */
            k = (k+1)%40;
        }
}
```

```
void sendtomax (char c1, char c2, char c3, char c4)
{
     PTM &= ~BIT5;        // enable SPI transfer to MAX6952
     putcspi0(c1);        // send c1 to MAX6952
     putcspi0(c2);        // send c2 to MAX6952
     putcspi0(c3);        // send c3 to MAX6952
     putcspi0(c4);        // send c4 to MAX6952
     PTM |= BIT5;         // load data from shift register to latch
}
void openspi0(void)
{
     SPI0BR   = 0x00;   // set baud rate to 12 MHz
     SPI0CR1  = 0x50;   // disable interrupt, set master mode, shift data on
                        //     rising edge, clock idle low
     SPI0CR2  = 0x02;   // disable mode fault, disable SPI in wait mode
     WOMS     = 0;      // enable Port S pull-up
}
```

Example 10.12

Write a program to define fonts for three special characters as shown in Figure 10.31. Store the font of these three special characters at locations from 0x00 to 0x0E of the MAX6952.

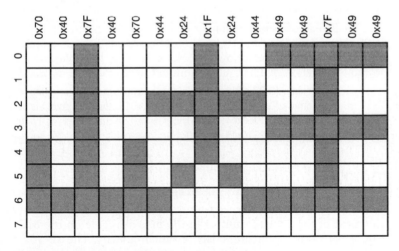

Figure 10.31 ■ User-definable font example

Solution: The program that sets fonts is very straightforward. The following program will set up the fonts as specified:

```
#include "c:\cwHCS12\include\hcs12.h"
#include "c:\cwHCS12\include\spi0util.h"
char fonts [15] = {0x70,0x40,0x7F,0x40,0x70,0x48,0x38,0x0F,0x38,0x48,0x49,
                   0x49,0x7F,0x49,0x49};
void send_font (char xc);
void openspi0 (void);
```

```
void main (void)
{
      char i;
      DDRM |= BIT5;          // configure PM5 pin for output
      openspi0();            // configure SPI module properly
      send_font(0x80);       // set font address pointer address to 0x00
      for (i = 0; i < 15; i++)
              send_font(fonts[i]);
}
void send_font(char xx)
{
      PTM &= ~BIT5;          // enable SPI transfer
      putcspi0(0x05);        // specify font address pointer
      putcspi0(xx);          // send a font value
      PTM |= BIT5;           // load data in shift register to destination
}
void openspi0(void)
{
      SPI0BR  = 0x00;  // set baud rate to 12 MHz
      SPI0CR1 = 0x50;  // disable interrupt, set master mode, shift data on
                       // rising edge, clock idle low
      SPI0CR2 = 0x02;  // disable mode fault, disable SPI in wait mode
      WOMS    = 0;     // enable Port S pull-up
}
```

▲

To display user-defined fonts, we need to send in the appropriate addresses in the range from 0x00 to 0x17. In Example 10.12, the addresses 0x00~0x02 should be used to access those three special characters. The following C statements will display those three Chinese characters followed by letters A, B, C, D, and E from left to right on the matrix displays shown in Figure 10.30:

```
sendtomax(0x20, 0x42, 0x20, 0x00); // 0x00 is the address of the first character font
sendtomax(0x21, 0x43, 0x21, 0x01); // 0x01 is the address of the second character font
sendtomax(0x22, 0x44, 0x22, 0x02); // 0x02 is the address of the third character font
sendtomax(0x23, 0x45, 0x23, 0x41);
```

10.13 Summary

When high-speed data transfer is not needed, using serial data transfer enables us to make the most of the limited number of I/O pins available on the MCU device. Serial data transfer can be performed asynchronously or synchronously. The SPI is a synchronous protocol created by Freescale for serial data exchange between peripheral chips and microcontrollers.

In the SPI format, a device (must be a microcontroller) is responsible for initiating the data transfer and generating the clock pulses for synchronizing data transfer. This device is referred to as the *SPI master*. All other devices in the same system are referred to as *SPI slaves*. The master device needs three signals to carry out the data transfer.

- SCK: a clock signal for synchronizing data transfer

- MOSI: serial data output from the master

- MISO: serial data input to the master

To transfer data to one or more SPI slaves, the MCU writes data into the SPI data register, and eight clock pulses are generated to shift out the data in the SPI data register from the MOSI pin. If the MISO pin of the MCU is also connected (to the slave), then 8 data bits are also shifted into the SPI data register. To read data from the slave, the MCU also needs to write data into the SPI data register to trigger clock pulses to be sent out from the SCK pin. However, the value written into the SPI data register is unimportant in this case.

When configured as a slave device, the HCS12 MCU also needs the fourth signal called Slave Select ($\overline{SS}$). The $\overline{SS}$ signal enables the HCS12 slave to respond to an SPI data transfer. Most slave peripheral devices have signals called $\overline{CE}$ (chip enable) or $\overline{CS}$ (chip select) to enable/disable the SPI data transfer.

Multiple peripheral devices with an SPI interface can be interfaced with a single MCU simultaneously. There are many different methods for interfacing multiple peripheral devices (with SPI interface) to the MCU. Two popular connection methods are

1. *Parallel connection.* In this method, the MISO, MOSI, and SCK signals of all the peripheral devices are connected to the same signals of the HCS12 device. The HCS12 MCU also needs to use certain unused I/O pins to control the $\overline{CS}$ (or $\overline{CE}$) inputs of each individual peripheral device. Using this method, the HCS12 MCU can exchange data with any selected peripheral device without affecting other peripheral devices.

2. *Serial connection.* In this method, the MOSI input of a peripheral device is connected to the MISO pin of its predecessor, and the MISO output of a peripheral device is connected to the MOSI input of its successor. The MOSI input of the peripheral device that is closest (in terms of connection) to the MCU is connected to the MOSI output of the MCU. The MISO output of the last peripheral device (in the loop) is connected to the MISO input of the MCU. The SCK inputs of all peripheral devices are tied to the SCK pin of the MCU. Using this method, the data to be sent to the last device in the loop will need to go through all other peripheral devices. The $\overline{CE}$ (or $\overline{CS}$) signals of all peripheral devices are controlled by the same signal.

In all of the peripheral devices with the SPI interface, shift registers such as 74HC595 and 74HC589 can be used to add parallel output and input ports to the MCU. The widely used digital temperature sensors are quite useful for displaying the ambient temperature. The TC72 and TC77 from Microchip and LM74 from National Semiconductor are digital temperature sensors with the SPI interface. The SPI interface is also often added to LED and LCD display drivers. The MAX7221 from Dallas-Maxim can drive up to eight seven-segment displays. The MC14489 from Freescale can drive up to five seven-segment displays. Both driver chips can be cascaded to drive more displays. Many A/D and D/A converters also have an SPI interface, for example, the 12-bit D/A converter MCP4922. Matrix displays have been widely used in recent years. They use the time-multiplexing method to display multiple digits. Due to their high demand on CPU time in performing time multiplexing, driver chips such as MAX6952 and MAX6953 have been designed to off-load the CPU from time-multiplexing matrix displays.

10.14 Exercises

E10.1 Configure the SPI0 module to operate with the following setting:
- Master mode with all interrupt enabled
- SCK0 idle high and data shifted on the falling edge

- Disable mode fault, disable bidirectional mode, stop SPI in wait mode
- Shift data least significant bit first
- Use $\overline{SS0}$ pin and enable $\overline{SS0}$ output
- Baud rate set to 4 MHz

E10.2 Configure the SPI0 module to operate with the following setting:

- Master mode with all interrupt disabled
- SCK0 idle high and data shifted on the rising edge
- Disable mode fault, disable bidirectional mode, SPI0 stops in wait mode
- Data shifted most significant bit first
- Use $\overline{SS0}$ pin and enable $\overline{SS0}$ output
- Baud rate set to 3 MHz

E10.3 Assume that there is an SPI-compatible peripheral output device that has the following characteristics:

- A CLK input pin that is used as the data-shifting clock signal. This signal is idle low and data is shifted in on the rising edge.
- An SI pin to shift in data on the falling edge of the CLK input.
- A $\overline{CE}$ pin, which enables the chip to shift in data when it is low.
- A highest data-shifting rate of 1 MHz.
- Most significant bit shifted in first.
- All interrupts disabled.
- The SPI0 module stopped in wait mode and freeze mode.
- The $\overline{SS0}$ pin used as a general I/O pin.

Describe how to connect the SPI0 pins for the HCS12 and this peripheral device and write an instruction sequence to configure the SPI subsystem properly for data transfer. Assume that the E-clock frequency is 24 MHz.

E10.4 The 74HC165 is another SPI-compatible shift register. This chip has both serial and parallel inputs and is often used to expand the number of parallel input ports. The block diagram of the 74HC165 is shown in Figure E10.4. The operation of this chip is illustrated in Table E10.4.

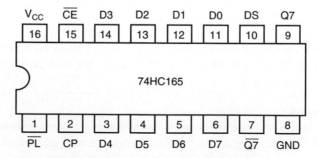

Figure E10.4 ■ The 74HC165 pin assignment

Operating Mode	Inputs					Qn Flip-Flops		Outputs	
	$\overline{PL}$	$\overline{CE}$	CP	D_S	D0–D7	Q0	Q1–Q6	Q7	$\overline{Q7}$
Parallel load	L	X	X	X	L	L	L-L	L	H
	L	X	X	X	H	H	H-H	H	L
Serial shift	H	L		1	X	L	q0–q5	q6	$\overline{q6}$
	H	L		h	X	H	q0–q5	q6	$\overline{q6}$
Hold	H	H	X	X	X	q0	q1–q6	q7	$\overline{q7}$

Table E10.4 ■ Function table of the 74HC165

The 74HC165 can be cascaded. To cascade, the Q7 output is connected to the DS input of its adjacent 74HC165. All 74HC165s should share the same shift clock SC. Suppose we want to use two 74HC165s to interface with two DIP switches so that four hex digits can be read in by the HCS12. Describe the circuit connection and write an instruction sequence to read the data from these two DIP switches.

E10.5 Suppose you are going to use four 74HC595s to drive four seven-segment displays using the circuit shown in Figure 10.10. Write an instruction sequence to display the number 1982 on these four seven-segment displays.

E10.6 Write a program to display the number 1, 2, . . . , 8 on the eight seven-segment displays in Figure 10.12. Display one digit at a time and turn off the other seven digits. Each digit is displayed for 1 s. Perform this operation continuously.

E10.7 Write the C language version of the program for Example 10.8.

E10.8 The Microchip TC77 is a small-footprint digital temperature sensor with a 13-bit resolution. This device uses a single pin for temperature I/O, which one would need to use the SPI bidirectional mode to read the temperature. The data sheet of this device is included in the complementary CD (it can also be downloaded from the Microchip website, http://www.microchip.com). Describe the circuit connection of this chip and the HCS12. Write a program to configure the TC77 to continuous conversion mode, read the temperature once every second, and display the temperature on the LCD. The block diagram and the pin assignment of the TC77 are shown in Figure E10.8.

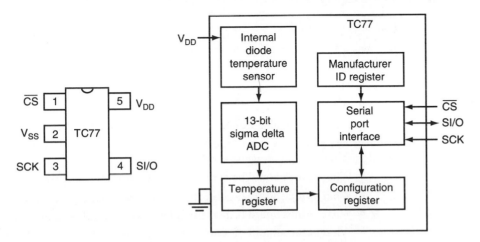

Figure E10.8 ■ Pin assignment and block diagram of the TC77

E10.9 Write a C program to generate a sine waveform using the circuit shown in Figure 10.20. Divide one period of the sine wave into 60 points and use these 60 points to represent the waveform. Every two adjacent points are separated by 3.0°.

E10.10 Use the OUT pin of the MCP4922 shown in Figure 10.20 to drive a speaker. Write a C program to generate a two-tone siren. The two frequencies of the siren are 1 kHz and 500 Hz.

E10.11 The Freescale MC14489 is a five-digit, seven-segment display driver chip. The data sheet is on the CD for this text (also available on Freescale's website). Describe how to interface one MC14489 with the HCS12 using the SPI subsystem. Write a program to display the value 12345 on the seven-segment displays driven by the MC14489.

E10.12 Define a new font for each of the characters shown in Figure E10.12 and write a program to display these four characters on the four matrix displays driven by one MAX6952. The PM5 signal drives the $\overline{CS}$ signal.

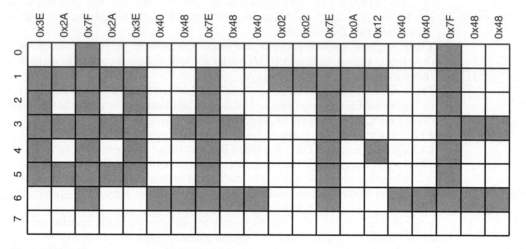

Figure E10.12 ■ User-definable font example

E10.13 Download the datasheet of the 10-bit, two-channel DAC LTC1661 from Linear Technology (www.linear-tech.com). Answer the following questions:

- How much time does it take to complete a DAC conversion?
- How many bytes need to be sent to LTC1661 in order to generate a voltage output?
- What is the range of the reference voltage?
- Describe a viable circuit connection for the LTC1661 and the HCS12.

E10.14 For the same LTC1661 interfacing with the HCS12, answer the following questions assuming the reference voltage is set to 5 V:

- What value should be sent in order to generate a 5-V output from V_{OUTA}?
- What value should be sent in order to generate a 0-V output from V_{OUTA}?
- What value should be sent in order to generate a 5-V output from V_{OUTB}?
- What value should be sent in order to generate a 0-V output from V_{OUTB}?

E10.15 Use the LTC1661 from Linear Technology to generate a two-frequency square wave from the V_{OUTA} pin. The frequency alternates between 1 and 4 kHz every 4 s.

10.15 Lab Exercises and Assignments

L10.1 Use the Freescale seven-segment display driver MC14489 (with SPI interface) to drive five seven-segment displays, and write a program to display the following pattern forever:

```
0
10
210
3210
43210
54321
65432
76543
87654
98765
09876
10987
21098
32109
```
go to the fifth pattern (43210) and repeat.

L10.2 Use the TC72 to measure the ambient temperature and the LCD on the demo board to display the current temperature:

1. Write a program to read the temperature once every 200 ms and display the new temperature reading on the LCD in two rows:

   Temperature is

   XXX.X°C

2. Change the ambient temperature using your hand, hot water in a plastic bag, ice in a plastic bag, and so on, to touch the TC72 and record the temperature reading.

L10.3 Use one MAX6952 and four matrix displays (e.g., common-cathode-row Fairchild GMC7175C) to perform the operation described in Exercise E10.12.

L10.4 Use the 12-bit DAC (MCP4922) to generate a sinusoidal waveform by dividing one cycle (360°) into 60 points. Take the output voltage from the V_{OUTA} pin and display the waveform using an oscilloscope.

L10.5 Use the 8-kBit EEPROM 25AA080B from Microchip to interface with the HCS12 using the SPI pins. Perform the following operations:

Step 1
Store the string "The 25AA080B" in the 25AA080B EEPROM starting the address 0x0000 to 0x000B.

Step 2
Read out the previous string from the EEPROM and display it on the first row of the LCD.

Step 3
Store the string "has SPI interface" in the 25AA080B starting from the address 0x0100 to 0x010F.

Step 4
Read out the previous string and display it on the second row of the LCD.

Step 5

Store the value 0 to 255 in the EEPROM memory locations from the address 0x0300 to 0x3FF.

Step 6

Read out the values stored in the 25AA080B in step 5 and display the values one at a time on the LEDs with each value displayed for 300 ms.

Step 7

Go to step 5.

The datasheet can be downloaded from www.microchip.com.

11

Inter-Integrated Circuit (I^{2}C) Interface

11.1 Objectives

After completing this chapter, you should be able to

- Understand the charactcristics of the I^{2}C protocol
- Describe the I^{2}C protocol in general
- Explain the I^{2}C signal components
- Explain the I^{2}C bus arbitration method
- Explain the I^{2}C data transfer format
- Interface with the real-time-clock chip DS1307
- Use the DS1631A to measure the ambient temperature
- Store and retrieve data in/from the serial EEPROM chip 24LC08B

11.2 The I²C Protocol

The inter-integrated circuit (I²C) serial interface protocol was developed by Philips in the late 1980s. Version 1.0 was published in 1992. This version supports

- Both the 100 kbps (*standard mode*) and the 400 kbps (*fast mode*) data rate
- 7-bit and 10-bit addressing
- Slope control to improve electromagnetic compatibility (EMC) behavior

After the publication of version 1.0, I²C was well received by the embedded application developers.

By 1998, the I²C protocol had become an industry standard; it has been licensed to more than 50 companies and implemented in over 1000 different integrated circuits. However, many applications require a higher data rate than that provided by the I²C protocol. Version 2.0 incorporates the *high-speed mode* (with a data rate of 3.4 Mbps) to address this requirement. Since the I²C module of the HCS12 does not support the high-speed mode, it will not be discussed in this chapter.

11.2.1 Characteristics of I²C Protocol

The I²C protocol has the following characteristics:

- *Synchronous in nature*. A data transfer is always initiated by a master device. A clock signal (SCL) synchronizes the data transfer. The clock rate can vary without disrupting the data. The data rate will simply change along with the changes in the clock rate.
- *Master/slave model*. The master device controls the clock line (SCL). This line dictates the timing of all data transfers on the I²C bus. Other devices can manipulate this line, but they can only force the line low. By forcing the line low, it is possible to clock more data into any device. This is known as clock stretching.
- *Bidirectional data transfer*. Data can flow in any direction on the I²C bus.
- *Serial interface method*. I²C uses only signals SCL and SDA. The SCL signal is the serial clock signal; the SDA signal is known as serial data. In reality, the SDA signal can carry both the address and data.

11.2.2 I²C Signal Levels

I²C can have only two possible electrical states: *float high* and *driven low*. A master or a slave device drives the I²C bus using an open-drain (or open-collector) driver. As shown in Figure 11.1, both the SDA and SCL lines are pulled up to V_{DD} via pull-up resistors. Because the driver circuit is open drain, it can pull the I²C bus only to low. When the clock or data output is low, the NMOS transistor is turned off. In this situation, no current flows from (to) the bus to (from) the NMOS transistor, and hence the bus line will be pulled to high by the pull-up resistor. Otherwise, the NMOS transistor is turned on and pulls the bus line to low.

The designer is free to use any resistor value for various speeds. But the calculation of what value to use depends on the capacitance of the driven line and the speed of the I²C communication. In general, the recommended values for the pull-up resistors are 2.2 and 1 kΩ for standard mode and fast mode, respectively. These values are found to work frequently. When the data rate is below 100 kbps, the pull-up resistor should be set to 4.7 kΩ.

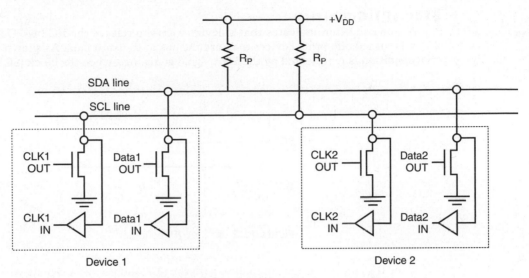

Figure 11.1 ■ Connecting standard- and fast-mode devices to the I²C bus

11.2.3 I²C Data Transfer Signal Components

An I²C data transfer consists of the following fundamental signal components:

- Start (**S**)
- Stop (**P**)
- Repeated start (**R**)
- Data
- Acknowledge (**A**)

START (S) CONDITION

A start condition indicates that a device would like to transfer data on the I²C bus. As shown in Figure 11.2, a start condition is represented by the SDA line going low when the clock (SCL) signal is high. The start condition will initialize the I²C bus. The timing details for the start condition will be taken care of by the microcontroller that implements the I²C bus. Whenever a data transfer using the I²C bus is to be initiated, the designer must tell the microcontroller that a start condition is wanted.

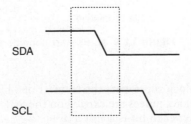

Figure 11.2 ■ I²C start condition

STOP (P) CONDITION

A stop condition indicates that a device wants to release the I²C bus. Once released (the driver is turned off), other devices may use the bus to transmit data. As shown in Figure 11.3, a stop condition is represented by the SDA signal going high when the clock (SCL) signal is high.

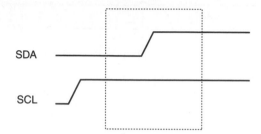

Figure 11.3 ■ Stop (P) condition

Once the stop condition completes, both the SCL and the SDA signals will be high. This is considered to be an *idle bus*. After the bus is idle, a start condition can be used to send more data.

REPEATED START (R) CONDITION

A repeated start signal is a start signal generated without first generating a stop signal to terminate the communication. This is used by the master to communicate with another slave or with the same slave in a different mode (transmit/receive mode) without releasing the bus. A repeated start condition indicates that a device would like to send more data instead of releasing the line. This is done when a start must be sent but a stop has not occurred. It prevents other devices from grabbing the bus between transfers. The timing diagram of a repeated start condition is shown in Figure 11.4. The repeated start condition is also called a *restart* condition. In the figure, there is no stop condition occurring between the start condition and the restart condition.

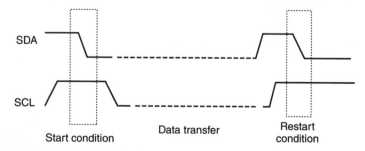

Figure 11.4 ■ Restart condition

DATA

The data block represents the transfer of 8 bits of information. The data is sent on the SDA line, whereas clock pulses are carried on the SCL line. The clock can be aligned with the data to indicate whether each bit is a 1 or a 0.

Data on the SDA line is considered valid only when the SCL signal is high. When SCL is not high, the data is permitted to change. This is how the timing of each bit works. Data bytes are used to transfer all kinds of information. When communicating with another I²C device, the 8 bits of data may be a control code, an address, or data. An example of 8-bit data is shown in Figure 11.5.

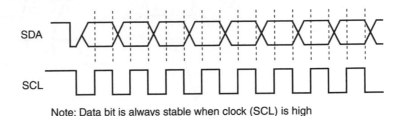

Note: Data bit is always stable when clock (SCL) is high

Figure 11.5 ■ I²C bus data elements

ACKNOWLEDGE (ACK) CONDITION

Data transfer in the I²C protocol needs to be acknowledged either positively (A) or negatively (NACK). As shown in Figure 11.6, a device can acknowledge (A) the transfer of each byte by bringing the SDA line low during the ninth clock pulse of SCL.

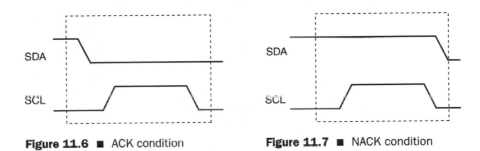

Figure 11.6 ■ ACK condition   **Figure 11.7** ■ NACK condition

If the device does not pull the SDA line to low and instead allows the SDA line to float high, it is transmitting a negative acknowledge (NACK). This situation is shown in Figure 11.7.

11.2.4 Synchronization

All masters generate their own clocks on the SCL line to transfer messages on the I²C bus. Data is valid only during the high period of the clock. A defined clock is therefore needed for the bit-by-bit arbitration procedure to take place. For most microcontrollers (including the HCS12 devices), the SCL clock is generated by counting down a programmable reload value using the instruction clock signal.

Clock synchronization is performed using the wired-AND connection of I²C interfaces to the SCL line. This means that a high-to-low transition on the SCL line will cause the devices concerned to start counting off their low period, and once a device clock has gone low, it will hold the SCL line in that state until the high state is reached (CLK1 in Figure 11.8). However, the transition from low to high of this clock may not change the state of the SCL line if another clock (CLK2) is still within its low period. The SCL line will therefore be held low by the device with the longest low period. Devices with shorter low periods enter a high *wait state* during this time.

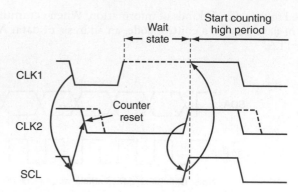

Figure 11.8 ■ Clock synchronization during the arbitration procedure

When all devices concerned have counted off their low period, the clock line will be released and go high. There will then be no difference between the device clocks and the state of the SCL line, and all the devices will start counting their high periods. In this way, a synchronized SCL clock is generated with its low period determined by the device with the longest clock low period and its high period determined by the one with the shortest clock high period.

Clock synchronization occurs when multiple masters attempt to drive the I²C bus and before the arbitration scheme can decide which master is the winner. The I²C bus arbitration process will be discussed in the next section.

HANDSHAKING

The clock synchronization mechanism can be used as a handshake in data transfer. Slave devices may hold the SCL low after completion of a 1-byte transfer (9 bits). In such a case, the slave halts the bus clock low and forces the master clock into wait states until it releases the SCL line.

CLOCK STRETCHING

The clock synchronization mechanism can be used by slaves to slow down the bit rate of a transfer. After the master has driven SCL low, the slave can drive SCL low for the required period and then release it. If the slave SCL low period is greater than the master SCL low period, then the resulting SCL bus signal low period is stretched.

11.2.5 Arbitration

I²C allows multiple master devices to coexist in the system. In the event that two or more master devices attempt to begin a transfer at the same time, an arbitration scheme is employed to force one or more masters to give up the bus. The master devices continue transmitting until one attempts a high while the other transmits a low. Since the bus driver has open drain, the bus will be pulled low. The master attempting to transfer a high signal will detect a low on the SDA line and give up the bus by switching off its data output stage. The winning master continues its transmission without interruption; the losing master becomes a slave and receives the rest of the transfer. This arbitration scheme is nondestructive: One device always wins, and no data is lost.

An example of the arbitration procedure is shown in Figure 11.9, where Data1 and Data2 are data driven by device 1 and device 2 and SDA is the resultant data on the SDA line. The moment there is a difference between the internal data level of the master generating Data1 and the actual level on the SDA line, its data output is switched off; this means that a high output level is then connected to the bus. This will not affect the data transfer initiated by the winning master.

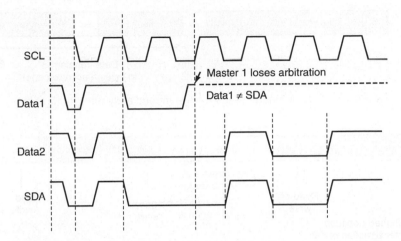

Figure 11.9 ■ Arbitration procedure of two masters

11.2.6 Data Transfer Format

I²C allows a master device to use either the 7-bit or the 10-bit address to specify a slave device for data transfer. The following are the possible I²C data transfer formats:

- *Master transmitter to slave receiver.* The transfer direction is not changed. An example of this format using the 7-bit addressing is shown in Figure 11.10.

- *Master reads slave immediately after the first byte (address byte).* At the moment of the first acknowledgement, the master transmitter becomes a master receiver and the slave receiver becomes a slave transmitter. The first acknowledgement is still generated by the slave. The stop condition is generated by the master, which has previously sent a negative acknowledgement ($\overline{A}$). An example of this format using the 7-bit addressing is shown in Figure 11.11.

- *Combined format.* During a change of direction within a transfer, both the start condition and the slave address are repeated, but with the R/$\overline{W}$ bit reversed. If a master receiver sends a repeated start condition, it has previously sent a negative acknowledgement. An example of this format in the 7-bit addressing is shown in Figure 11.12.

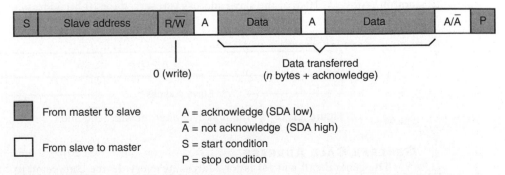

Figure 11.10 ■ A master transmitter addressing a slave receiver with a 7-bit address. The transfer direction is not changed

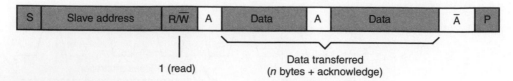

Figure 11.11 ■ A master reads a slave immediately after the first byte

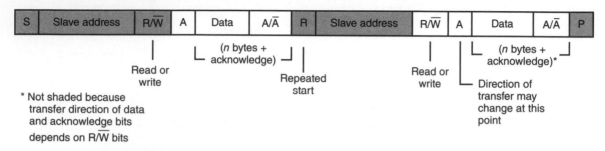

Figure 11.12 ■ Combined format

11.2.7 7-Bit Addressing

The addressing procedure for the I²C bus is such that the first byte after the start condition usually determines which slave will be selected by the master. The exception is the *general call address*, which can address all devices. When this address is used, all devices should, in theory, respond with an acknowledgement. However, devices can be made to ignore this address. The second byte of the general call address then defines the action to be taken.

The first byte after the start condition carries the 7-bit address and the direction of the message. The meaning of this byte is shown in Figure 11.13. When the least significant bit is 1, the master device will read information from the selected slave. Otherwise, the master will write information to a selected slave. When an address is sent, each device in a system compares the first 7 bits after the start condition with its address. If they match, the device considers itself addressed by the master as a slave receiver or slave transmitter, depending on the R/$\overline{\text{W}}$ bit. The I²C bus committee coordinates the allocation of I²C addresses. Two groups of eight addresses (0000xxx and 1111xxx) are reserved for the special purpose and shown in Table 11.1. The bit combination 11110xx of the slave address is reserved for 10-bit addressing.

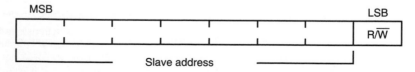

Figure 11.13 ■ The first byte after the start condition

GENERAL CALL ADDRESS

The general call address is for addressing every device connected to the I²C bus. However, if a device does not need any of the data supplied within the general call structure, it can ignore this address by not issuing an acknowledgement. If a device does require data from a general

Slave Address	R/$\overline{W}$ Bit	Description
0000 000	0	General call address
0000 000	1	Start byte [1]
0000 001	x	CBUS address [2]
0000 010	x	Reserved for different bus format [3]
0000 011	x	Reserved for future purpose
0000 1xx	x	Hs-mode master code
1111 1xx	x	Reserved for future purposes
1111 0xx	x	10-bit slave addressing

[1] No device is allowed to acknowledge at the reception of the start byte.
[2] The CBUS address has been reserved to enable the inter-mixing of CBUS compatible and I²C-bus compatible devices in the same system. I²C-bus compatible devices are not allowed to respond on reception of this address.
[3] The address reserved for a different bus format is included to enable I²C and other protocols to be mixed. Only I²C-bus compatible devices that can work with such formats and protocols are allowed to respond to this address.

Table 11.1 ■ Definition of bits in the first byte

call address, it will acknowledge this address and behave as a slave receiver. The second and following bytes will be acknowledged by every slave receiver capable of handling this data. A slave that cannot process one of these bytes must ignore it by not acknowledging. The meaning of the general call address is always specified in the second byte, as shown in Figure 11.14.

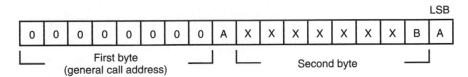

Figure 11.14 ■ General call address format

When bit B in Figure 11.14 is 0, the second byte has the following definitions:

- 00000110 (0x06)—*reset and write programmable part of slave address by hardware.* On receiving this 2-byte sequence, all devices designed to respond to the general call address will reset and take in the programmable part of their address. Precautions have to be taken to ensure that a device is not pulling down the SDA or SCL line after applying the supply voltage, since these low levels would block the bus.

- 00000100 (0x04)—*write programmable part of slave address by hardware.* All devices that define the programmable part of their addresses by hardware will latch this programmable part at the reception of this 2-byte sequence. The device will not reset.

- 00000000 (0x00). This code is not allowed to be used as the second byte.

Sequences of programming procedure are published in the appropriate device data sheets. The remaining codes have not been fixed, and devices must ignore them.

When bit B is 1, the 2-byte sequence is a *hardware general call.* This means that the sequence is transmitted by a hardware master device, such as a keyboard scanner, which cannot be programmed to transmit a desired slave address. Since a hardware master does not know in advance to which device the message has to be transferred, it can only generate this hardware general call and present its own address to identify itself to the system. The 7 bits remaining in the second byte contain the address of the hardware master. This address is recognized by an intelligent device (e.g., a microcontroller) connected to the bus, which will then direct the information from the hardware master. If the hardware master can also act as a slave, the slave address is identical to the master address. In some systems, an alternative could be that the hardware master transmitter is set in the slave receiver mode after the system reset. In this way, a system-configuring master can tell the hardware master transmitter (which is now in slave receiver mode) to which address data must be sent (see Figure 11.15). After this programming procedure, the hardware master remains in the master transmitter mode.

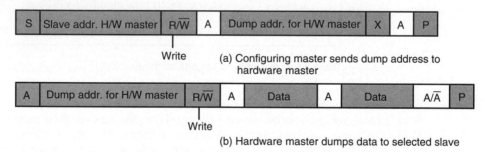

(a) Configuring master sends dump address to hardware master

(b) Hardware master dumps data to selected slave

Figure 11.15 ■ Data transfer by a hardware transmitter capable of dumping data directly to slave devices

11.2.8 10-Bit Addressing

Ten-bit addressing is compatible with, and can be combined with, 7-bit addressing. Using 10-bit addressing exploits the reserved combination 1111xxx for the first 7 bits of the first byte following a start (S) or repeated start (R) condition. The 10-bit addressing does not affect the existing 7-bit addressing. Devices with 7-bit and 10-bit addresses can be connected to the same I²C bus, and both 7-bit and 10-bit addressing can be used in standard- and fast-mode systems. Although there are eight possible combinations of the reserved address bits 1111xxx, only the four combinations 11110xx are used for 10-bit addressing. The remaining four combinations 11111xx are reserved for future I²C bus enhancements.

DEFINITIONS OF BITS IN THE FIRST 2 BYTES

The 10-bit slave address is formed from the first 2 bytes following a start (S) condition or repeated start (R) condition. The first 7 bits of the first byte are the combination 11110xx, of which the last 2 bits (xx) are the 2 most significant bits of the 10-bit address. The eighth bit of the first byte is the R/$\overline{\text{W}}$ bit that determines the direction of the message. A 0 in the least significant bit of the first byte means that the master will write information to the selected slave. A 1 in this position indicates that the master will read information from the selected slave.

FORMATS WITH 10-BIT ADDRESS

The following data formats are possible in 10-bit addressing:

- *Master transmitter transmits to slave receiver with a 10-bit slave address.* The transfer direction is not changed (see Figure 11.16). When a 10-bit address follows a start condition, each slave compares the first 7 bits of the first byte of the slave address (11110xx) with its own address and tests if the eighth bit is 0. It is possible that more than one device will find a match and generate an acknowledgement (A1). All slaves that find a match will compare the 8 bits of the second byte of the slave address with their own addresses, but only one slave will find a match and generate an acknowledgement (A2). The matching slave will remain addressed by the master until it receives a stop condition or a repeated start condition followed by a different slave address.

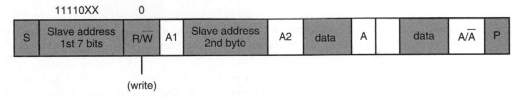

Figure 11.16 ■ A master transmitter addresses a slave receiver with a 10-bit address

- *Master receiver reads slave transmitter with a 10-bit address.* The transfer direction is changed after the second R/$\overline{\text{W}}$ bit (see Figure 11.17). Up to and including the acknowledgement bit A2, the procedure is the same as that described for a master transmitter addressing a slave receiver. After the repeated start condition, a matching slave remembers that it was addressed before. This slave then checks if the first 7 bits of the first byte of the slave address following repeated start (R) are the same as they were after the start condition (S), and tests if the eighth (R/$\overline{\text{W}}$) bit is 1. If there is a match, the slave considers that it has been addressed as a transmitter and generates acknowledgement bit A3. The slave transmitter remains addressed until it receives a stop condition (P) or until it receives another repeated start (R) condition followed by a different slave address. After a repeated start (R) condition, all other slave devices will also compare the first 7 bits of the first byte of the slave address (11110xx) with their own addresses and test the eighth bit.

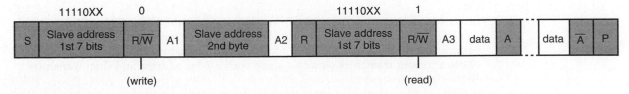

Figure 11.17 ■ A master receiver addresses a slave transmitter with a 10-bit address

However, none of them will be addressed because R/$\overline{W}$ = 1 (for the 10-bit address) or the 11110xx slave address (for the 7-bit address) does not match.

- *Combined format.* A master transmits data to a slave and then reads data from the same slave (see Figure 11.18). The same master occupies the bus all the time. The transfer direction is changed after the second R/$\overline{W}$ bit.

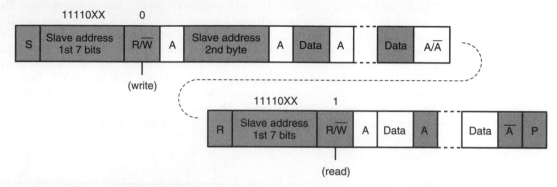

Figure 11.18 ■ **Combined format:** A master addresses a slave with a 10-bit address, then transmits data to this slave and reads data from this slave

- *Combined format.* A master transmits data to one slave and then transmits data to another slave (see Figure 11.19). The same master occupies the bus all the time.

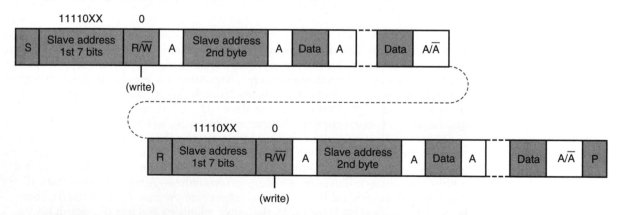

Figure 11.19 ■ **Combined format:** A master transmits data to two slaves, both with 10-bit addresses

- *Combined format.* Ten-bit and 7-bit addressing combined in one serial transfer (see Figure 11.20). After each start (S) condition or each repeated start (R) condition, a 10-bit or 7-bit slave address can be transmitted. Figure 11.20 shows how a master transmits data to a slave with a 7-bit address and then transmits

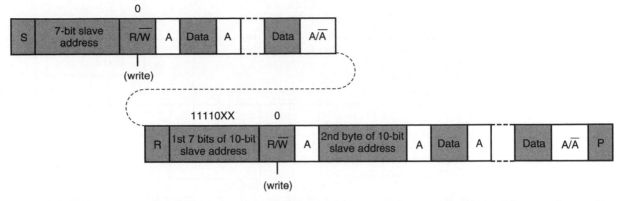

Figure 11.20 ■ **Combined format:** A master transmits data to two slaves, one with a 7-bit address and one with a 10-bit address

data to a second slave with a 10-bit address. The same master occupies the bus all the time.

11.3 An Overview of the HCS12 I²C Module

The I²C module of the HCS12 implements a subset of the I²C standard. It supports most of the master and slave functions and provides interrupts on start and stop bits in hardware to determine if a bus is free (multimaster function). Neither the 10-bit addressing nor the general call addressing is supported by the HCS12 I²C module. The I²C module of the HCS12 may operate at baud rates of up to 100 kbps with maximum capacitive bus loading. With a reduced bus slew rate, the device is capable of operating at a baud rate of up to 400 kbps, provided that the I²C bus slew rate is less than 100 ns. The maximum communications interconnect length and the number of devices that can be connected to the bus are limited by a maximum bus capacitance of 400 pF in all instances.

Two pins are used for data transfer.

- Serial clock (SCL)—PJ7/SCL
- Serial data (SDA)—PJ6/SDA

The I²C module works the same in normal, special, and emulation modes of the HCS12. It has two low-power modes: wait and stop modes. The I²C module has five registers to support its operation.

- I²C control register (IBCR)
- I²C status register (IBSR)
- I²C data I/O register (IBDR)
- I²C frequency divider register (IBFD)
- I²C address register (IBAD)

The block diagram of the I²C module is shown in Figure 11.21.

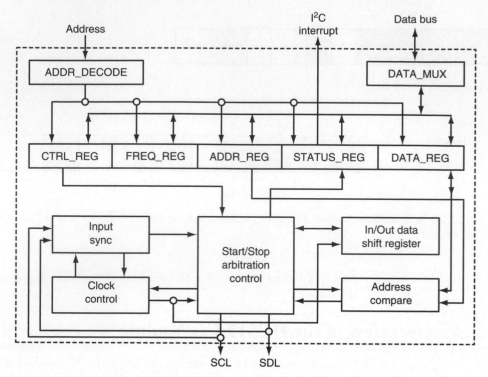

Figure 11.21 ■ I²C block diagram

11.4 Registers for I²C Operation

11.4.1 The I²C Address Register (IBAD)

The IBAD register contains the address to which the I²C module will respond when it is addressed as a slave. The contents of this register are shown in Figure 11.22.

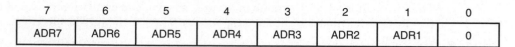

Figure 11.22 ■ I²C address register (IBAD)

11.4.2 The I²C Control Register (IBCR)

This register controls all the operation parameters except the baud rate of the I²C module. The contents of this register are shown in Figure 11.23.

When the MS/$\overline{\text{SL}}$ bit is changed from 0 to 1, a start signal is generated on the bus and the master mode is selected. When this bit is changed from 1 to 0, a stop signal is generated and the operation mode changes from master to slave. A stop signal should only be generated if the IBIF flag is set.

7	6	5	4	3	2	1	0
IBEN	IBIE	MS/$\overline{\text{SL}}$	Tx/$\overline{\text{Rx}}$	TxAK	RSTA	0	IBSWAI

Reset: 0   0   0   0   0   0   0   0

IBEN: I²C bus enable
  0 = I²C module is reset and disabled.
  1 = I²C module is enabled. This bit must be set before any other IBCR bits have
    any effect.
IBIE: I²C bus interrupt enable
  0 = interrupts from the I²C module are disabled.
  1 = interrupts from the I²C module enabled.
MS/$\overline{\text{SL}}$: master/slave mode select
  0 = slave mode.
  1 = master mode.
Tx/$\overline{\text{Rx}}$: transmit/receive mode select
  0 = receive.
  1 = transmit.
TxAK: transmit acknowledge
  0 = an acknowledge signal will be sent out to the I²C bus on the 9th clock bit
    after receiving 1 byte of data.
  1 = no acknowledge signal response is sent.
RSTA: repeat start
  0 = no action.
  1 = generate a repeat start cycle.
IBSWAI: I²C bus stop in wait mode
  0 = I²C module clock operates normally.
  1 = stop generating I²C module clock in wait mode.

Figure 11.23 ■ I²C control register (IBCR)

The Tx/$\overline{\text{Rx}}$ bit selects the direction of master and slave transfers. When addressed as a slave, this bit should be set by software according to the SRW bit in the status register. In the master mode, this bit should be set according to the type of transfer required. For address cycles, this bit will always be high.

The TxAK bit specifies the value driven onto SDA during data acknowledge cycles for both master and slave receivers. The I²C module will always acknowledge address matches, provided it is enabled, regardless of the value of TxAK. Values written to this bit are only used when the I²C module is a receiver, not a transmitter.

Writing a 1 to the RSTA bit will generate a repeated start condition on the I²C bus, provided it is the current master. This bit will always read as a low.

If the IBSWAI bit is set and the wai instruction is executed, all clock signals to the I²C module will be stopped and any transmission currently in progress will halt. If the CPU were woken up by a source other than the I²C module, then clocks would restart and the I²C would continue where it left off in the previous transmission. If the IBSWAI bit were cleared when the wai instruction is executed, the I²C internal clocks and interface would remain active, continuing the operation that is currently underway. It is also possible to configure the I²C such that it will wake up the CPU via an interrupt at the conclusion of the current operation.

11.4.3 The I²C Status Register (IBSR)

This register records the status of all I²C data transmission/reception activities. The contents of this register are shown in Figure 11.24.

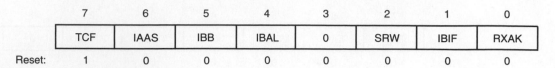

7	6	5	4	3	2	1	0
TCF	IAAS	IBB	IBAL	0	SRW	IBIF	RXAK

Reset: 1   0   0   0   0   0   0   0

TCF: data transferring bit
  0 = I²C transfer in progress.
  1 = I²C transfer complete.
IAAS: addressed as a slave
  0 = not addressed.
  1 = addressed as a slave.
IBB: bus busy bit
  0 = the bus enters idle state.
  1 = I²C bus is busy.
IBAL: arbitration lost
  0 = arbitration is not lost.
  1 = arbitration is lost.
SRW: slave read/write
  0 = slave receive, master writing to slave.
  1 = slave transmit, master reading from slave.
IBIF: I²C bus interrupt
  0 = no bus interrupt.
  1 = bus interrupt.
RXAK: receive acknowledge
  This bit reflects the value of SDA during the acknowledge bit of a cycle.
  0 = acknowledge received.
  1 = no acknowledge received.

Figure 11.24 ■ I²C status register (IBSR)

When a byte is being transferred, the TCF bit is cleared. It is set by the falling edge of the ninth clock of a byte transfer.

The IAAS bit is set (when the I²C module is configured as a slave) when its own address is matched with the calling address. The CPU may be interrupted if the IBIE bit is also set.

The IBB bit indicates the status of the I²C bus. When a start condition is detected, the IBB bit will be set. When a stop condition is detected, this bit is cleared.

The IBAL bit indicates the recent loss in I²C bus arbitration. The arbitration is lost in the following circumstances:

- The SDA signal is sampled low when the master drives a high during an address or data transmit cycle.
- The SDA signal is sampled low when the master drives a high during the acknowledge bit of a data receive cycle.
- A start cycle is attempted when the I²C bus is busy.
- A repeated start cycle is requested in slave mode.
- A stop condition is detected when the master did not request it.

When the IAAS bit is set, the SRW bit indicates the value of the R/$\overline{\text{W}}$ command bit of the calling address sent from the master. The SRW bit is valid when the I²C module is in the slave mode, a complete address transfer has occurred with an address match, and no other transfers have been initiated.

The IBIF bit is set when one of the following conditions occurs:

- Arbitration lost (IBAL bit is set).
- Byte transfer complete (TCF bit is set).
- Addressed as slave (IAAS bit is set).

This bit is cleared by writing a 1 to it.

11.4.4 I²C Data Register (IBDR)

In master transmit mode (the TxRx bit of the IBCR register set to 1), when data is written into the IBDR register a data transfer is initiated. The most significant bit is sent out first. In master receive mode, reading this register initiates the reception of the next byte. Nine clock pulses will be sent out on the SCL pin to shift in 8 data bits and send out the acknowledge bit. In slave mode, the same functions are available after an address match has occurred.

11.4.5 I²C Frequency Divider Register (IBFD)

The most important design consideration of the I²C module is to meet the timing requirements for the start and stop conditions so that data can be correctly transmitted over the bus line. As illustrated in Figure 11.25, there are four timing requirements to be met.

- SCL divider
- SDA hold time
- SCL hold time for start condition
- SCL hold time for stop condition

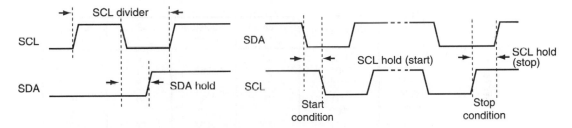

Figure 11.25 ■ SCL divider and SDA hold

The requirements of these four parameters are listed in Table 11.2. The SCL divider is equal to the bus frequency of the MCU divided by the SCL clock frequency.

Symbol	Parameter	Standard Mode		Fast Mode		Unit
		Min.	Max.	Min.	Max.	
f_{SCL}	SCL clock frequency	0	100	0	400	kHz
$t_{HD;STA}$	SCL hold (start)	4.0	–	0.6	–	µs
$t_{SU;STO}$	SCL hold (stop)	4.0	–	0.6	–	µs
$t_{HD;DAT}$	SDA hold	0	3.45	0	0.9	µs

Table 11.2 ■ I²C bus timing requirements

The I²C timing parameters are set by programming the I²C frequency divider register (IBFD). The contents of this register are shown in Figure 11.26. The contents of this register are used to prescale the bus clock for bit rate selection.

7	6	5	4	3	2	1	0
IBC7	IBC6	IBC5	IBC4	IBC3	IBC2	IBC1	IBC0

Figure 11.26 ■ I²C frequency divider register (IBFD)

The use of these 8 bits is as follows:
- IBC7~IBC6: multiply factor (shown in Table 11.3)
- IBC5~IBC3: prescaler divider (shown in Table 11.4)
- IBC2~IBC0: shift register tap points (shown in Table 11.5)

IBC7~IBC6	Multiply Factor
00	01
01	02
10	04
11	Reserved

Table 11.3 ■ Multiply factor

IBC5~IBC3	scl2start (clocks)	scl2stop (clocks)	scl2tap (clocks)	tap2tap (clocks)
000	2	7	4	1
001	2	7	4	2
010	2	9	6	4
011	6	9	6	8
100	14	17	14	16
101	30	33	30	32
110	62	65	62	64
111	126	129	126	128

Table 11.4 ■ Prescaler divider

IBC2~IBC0	SCL Tap (clocks)	SDA Tap (clocks)
000	5	1
001	6	1
010	7	2
011	8	2
100	9	3
101	10	3
110	12	4
111	15	4

Table 11.5 ■ I²C bus tap and prescale values

The circuit for controlling the I²C timing parameters is very complicated and Freescale does not disclose its details. Instead, Freescale provides the equations for computing these four timing parameters and a lookup table for them. A subset of the I²C divider and hold values is shown in Table 11.6; the complete lookup table can be found in the *I²C Block Guide* published by Freescale.

The equation for generating the divider value from the IBFD bits is

$$\text{SCL divider} = \text{MUL} \times 2 \times \{\text{scl2tap} + [(\text{SCL\_tap} - 1) \times \text{tap2tap}] + 2\} \tag{11.1}$$

The SDA hold delay is equal to the bus clock period multiplied by the SDA hold value shown in Table 11.6. The equation used to generate the SDA hold value from the IBFD bits is

$$\text{SDA hold} = \text{MUL} \times \{\text{scl2tap} + [(\text{SDA\_tap} - 1) \times \text{tap2tap}] + 3)\} \tag{11.2}$$

The equations for HCL hold values to generate the start and stop conditions from IBFD bits are as follows:

$$\text{SCL hold (start)} = \text{MUL} \times [\text{scl2start} + (\text{SCL\_tap} - 1) \times \text{tap2tap}] \tag{11.3}$$

$$\text{SCL hold (stop)} = \text{MUL} \times [\text{scl2stop} + (\text{SCL\_tap} - 1) \times \text{tap2tap}] \tag{11.4}$$

IBC[7:0] (hex)	SCL Divider (clocks)	SDA Hold (clocks)	SCL Hold (start)	SCL Hold (stop)
MUL = 1				
00	20	7	6	11
07	40	10	16	21
0B	40	9	16	21
14	80	17	34	41
18	80	9	38	41
1D	160	25	78	81
1F	240	33	118	121
20	160	17	78	81
25	320	49	158	161
27	480	65	238	241
28	320	33	158	161
MUL = 2				
40	40	14	12	22
45	60	18	22	32
47	80	20	32	42
4B	80	18	32	42
54	160	34	68	82
58	160	18	76	82
5D	320	50	156	162
5F	480	66	236	242
60	320	34	156	162
MUL = 4				
80	80	28	24	44
85	120	36	44	64
87	160	40	64	84
8B	160	36	64	84
94	320	68	136	164
98	320	36	152	164

Table 11.6 ■ I²C divider and hold values

Example 11.1

Assuming that the HCS12 is running with a 24-MHz bus clock, compute the values to be written into the IBFD register to set the baud rate to 100 and 400 kHz.

Solution

Case 1: Baud rate = 100 kHz
SCL divider = 24 MHz ÷ 100 kHz = 240

From Table 11.6, there is only one IBFD value ($1F) associated with this SCL divider value. Using this value to look up Tables 11.3, 11.4, and 11.5, we obtain the following parameter values:

- MUL   = 1
- scl2tap = 6
- tap2tap = 8
- scl2start = 6
- scl2stop = 9
- SDA_tap = 4
- SCL_tap = 15

SDA hold time	= MUL × {scl2tap + [(SDA_tap − 1) × tap2tap] + 3)}
	= 1 × {6 + [(4 − 1) × 8] + 3)}
	= 33 E cycles = 1.375 μs (< 3.45 μs)
SCL hold time (start)	= MUL × [scl2start + (SCL_tap − 1) × tap2tap]
	= 1 × [6 + (15 − 1) × 8] = 118 E cycles = 4.92 μs (> 4.0 μs)
SCL hold time (stop)	= MUL × [scl2stop + (SCL_tap − 1) × tap2tap]
	= 1 × [9 + (15 − 1) × 8] = 121 E cycles = 5.04 μs (> 4.0 μs)

The computed values satisfy the requirements.

Case 2: Baud rate = 400 kHz
SCL divider = 24 MHz ÷ 400 kHz = 60

From Table 11.6, there is only one corresponding IBFD value ($45) for this SCL divider value. Using this value to look up Tables 11.3, 11.4, and 11.5, we obtain the following parameter values:

- MUL   = 2
- scl2tap = 4
- tap2tap = 1
- scl2start = 2
- scl2stop = 7
- SDA_tap = 3
- SCL_tap = 10

SDA hold time	= 2 × {scl2tap + [(SDA_tap − 1) × tap2tap] + 3)}
	= 2 × {4 + [(3 − 1) × 1] + 3}
	= 18 E cycles = 0.75 μs (< 0.9 μs)
SCL hold time (start)	= MUL × [scl2start + (SCL_tap − 1) × tap2tap]
	= 2 × [2 + (10 − 1) × 1] = 22 E cycles = 0.917 μs (> 0.6 μs)
SCL hold time (stop)	= MUL × [scl2stop + (SCL_tap − 1) × tap2tap]
	= 2 × [7 + (10 − 1) × 1] = 32 E cycles = 1.33 μs (> 0.6 μs)

The computed values satisfy the requirements. A quicker way of timing verification is to use the numbers listed in Table 11.6 instead of using Equations 11.2 to 11.4.

▲

11.5 Programming the I²C Module

Before the I²C module can transmit and receive data correctly, it must be initialized properly. The initialization procedure is as follows:

- Compute the value that can obtain the SCL frequency from the E-clock and use it to update the IBFD register.
- Optionally load the IBAD register to define its slave address.
- Set the IBEN bit of the IBCR register to enable the I²C system.
- Modify the bits of the IBCR register to select master/slave mode, transmit/receive mode, and interrupt enable mode

The initialization of the I²C module can be performed using a macro or a subroutine. The subroutine that performs the I²C initialization is as follows:

```
; *************************** *************************** ******************
; The following function has two incoming parameters passed in accumulators A and B
; to set up baud rate and slave address, respectively.
; *************************** *************************** ******************
openI2C    bset    IBCR,IBEN      ; enable I²C module
           staa    IBFD           ; establish SCL frequency
           stab    IBAD           ; establish I²C module slave address
           bclr    IBCR,IBIE      ; disable I²C interrupt
           bset    IBCR,IBSWAI    ; disable I²C in wait mode
           rts
```

The C language version of the subroutine is as follows:

```
void openI2C(char ibc, char i2c_ID)
{
        IBCR |= IBEN;          /* enable I²C module */
        IBFD = ibc;            /* set up I2C baud rate */
        IBAD = i2c_ID;         /* set up slave address */
        IBCR &= ~IBIE;         /* disable I²C interrupt */
        IBCR |= IBSWAI;        /* disable I²C in wait mode */
}
```

A successful data exchange on the I²C bus requires the user to generate the I²C signal components in a proper order. The generation of these signal components is described in the following sections.

11.5.1 Generation of the Start Condition

All data transmissions in the I²C bus begin with the start condition. The start condition can only be sent when the bus is idle. If the MCU is connected to a multimaster bus system, the state of the IBB bit of the IBSR register must be tested to check whether the serial bus is busy.

If the bus is idle (IBB = 0), the start condition and the first byte can be sent. The first byte consists of the slave address and the R/$\overline{\text{W}}$ bit. The bus free time (i.e., the time between a *stop* condition and the following *start* condition) is built into the hardware that generates the start cycle.

Example 11.2

▼

Write a subroutine that generates a start condition and also sends out the address of the slave with which the MCU intends to exchange data. The address and the R/$\overline{\text{W}}$ bit value is passed in accumulator A.

Solution: The subroutine is as follows:

```
sendSlaveID   brset   IBSR,IBB,*          ; wait until I²C bus is free
              bset    IBCR,TXRX+MSSL      ; generate a start condition
              staa    IBDR               ; send out the slave address
              brclr   IBSR,IBIF,*         ; wait for address transmission to complete
              movb    #IBIF,IBSR          ; clear the IBIF flag
              rts
```

The C language version of the function is as follows:

```
void sendSlaveID (char cx)
{
    while (IBSR&IBB);           /* wait until I²C bus is idle */
    IBCR |= TXRX+MSSL;          /* generate a start condition */
    IBDR = cx;                  /* send out the slave address with R/W bit */
    while(!(IBSR & IBIF));      /* wait for address transmission to complete */
    IBSR = IBIF;                /* clear IBIF flag */
}
```

▲

11.5.2 I²C Data Transfer in Master Mode

When the HCS12 I²C module is configured as a master, it is responsible for initiating the data transmission and reception. By writing a byte into the IBDR register, nine clock pulses are generated to shift out 8 bits of data and shift in the acknowledgement bit. By reading from the IBDR register, nine clock pulses are triggered to shift in 8 bits of data and shift out the acknowledgement (positive or negative) bit.

After sending the slave ID and receiving the acknowledgement from the slave, the I²C module can send out a byte by performing the following three steps:

Step 1
Store the data byte in the IBDR register. This step will trigger nine clock pulses on the SCL pin to shift out 8 data bits and shift in the acknowledgement bit from the slave.

Step 2
Wait until the IBIF flag is set to 1, which indicates the byte has been shifted out.

Step 3
Clear the IBIF flag by writing a 1 to it.

If the data byte is stored in accumulator A, then the following instruction sequence can send a byte to a slave via the I²C bus:

```
staa    IBDR
brclr   IBSR,IBIF,*    ; wait until IBIF flag is set to 1
movb    #IBIF,IBSR     ; clear the IBIF flag
```

The following C statements send out the variable *cx* via the I²C bus:

```
IBDR = cx;                  /* send out the value cx */
while (!(IBSR & IBIF));      /* wait until the byte is shifted out */
IBSR − IBIF;                /* clear the IBIF flag */
```

To send multiple bytes to the slave, one can place the previous three instructions in a loop and use an index register to point to the data to be sent. The loop will be executed as many times as needed to send out all the data bytes.

After sending out the slave ID and receiving the acknowledgement from the slave, the I²C module can read a data byte by performing the following steps:

Step 1
Clear the Tx/Rx bit of the IBCR register to 0 for data reception. Clear the TxAK bit of the IBCR register to 0 if the user wants to acknowledge the data byte. This should be done when the user wants to receive multiple bytes from the same slave. If the user wants to read only 1 byte or when the byte to be read is the last byte of a read sequence, then the TxAK bit should be set to 1. Without receiving acknowledgement, the slave will stop driving the SDA line at the end of the eight clock pulses.

Step 2
Perform a dummy read. This action will trigger nine clock pulses to be sent out on the SCL pin to shift in 8 data bits and send out acknowledgement.

Step 3
Wait until the IBIF flag is set to 1.

Step 4
Clear the IBIF flag by writing a 1 to it.

The following instruction sequence reads a byte and acknowledges it:

```
bclr    IBCR,TXRX+TXAK    ; prepare to receive and acknowledge
ldaa    IBDR              ; a dummy read to trigger nine clock pulses
brclr   IBSR,IBIF,*       ; wait until the data byte is shifted in
movb    #IBIF,IBSR        ; clear the IBIF flag
ldaa    IBDR              ; place the received byte in A and also initiate the
                          ; next read sequence
```

The same read operation can be performed by the following C statements:

```
IBCR &= ~(TXRX + TXAK);   /* prepare to receive and acknowledge */
dummy = IBDR;             /* a dummy read */
while(!(IBSR & IBIF));    /* wait for the byte to shift in */
IBSR = IBIF;             /* clear the IBIF flag */
buf = IBDR;              /* place the received byte in buf and also initiate
                              the next read sequence */
```

The following instruction sequence reads a byte, sends out a negative acknowledgement, and also generates a stop condition:

```
bclr    IBCR,TXRX         ; prepare to receive
bset    IBCR,TXAK         ; to send negative acknowledgement
ldaa    IBDR              ; dummy read to trigger clock pulses
brclr   IBSR,IBIF,*       ; wait until the byte is shifted in
movb    #IBIF,IBSR        ; clear the IBIF flag
bclr    IBCR,MSSL         ; generate a stop condition
ldaa    IBDR              ; place the received byte in A
```

The same operation can be performed by the following C statements:

```
IBCR &= ~TXRX;           /* prepare to receive */
IBCR |= TXAK;            /* prepare not to acknowledge */
dummy = IBDR;            /* a dummy read to trigger nine clock pulses */
while(!(IBSR & IBIF));   /* wait for a byte to shift in */
IBSR = IBIF;             /* clear the IBIF flag */
IBCR &= ~MSSL;           /* generate a stop condition */
buf = IBDR;              /* place the received byte in buf */
```

11.5.3 I²C Data Transfer in Slave Mode

In slave mode, the I²C module cannot initiate any data transfer. After reset and a stop condition, the I²C module is in slave mode. Once the I²C module is enabled in slave mode, it waits for a start condition to occur. Following the start condition, 8 bits are shifted into the IBDR register. The value of the upper 7 bits of the received byte is compared with the IBAD register. If the address matches, the following events occur:

- The bit 0 of the address byte is copied into the SRW bit of the IBSR register.
- The IAAS bit is set to indicate address match.
- An ACK pulse is generated regardless of the value of the TxAK bit.
- The IBIF flag is set.

After the address match, the I²C module in slave mode should acknowledge every data byte received.

When operating in slave mode, the user program needs to make sure that the address match has occurred before it prepares to receive or transmit. The following instruction sequence can detect the address match and take appropriate actions:

```
            brset  IBSR,IAAS,addr_match    ; is address matched?
            ...
addr_match  brclr  IBSR,SRW,slave_rd
            bset   IBCR,TXRX               ; prepare to transmit data
            movb   tx_buf,IBDR             ; place data in IBDR to wait for SCL to shift it out
            brclr  IBSR,IBIF,*             ; wait for data to be shifted out
            ...
slave_rd    bclr   IBCR,TXAK+TXRX          ; prepare to receive and send ACK
            brclr  IBSR,IBIF,*             ; wait for data byte to shift in
            movb   #IBIF,IBSR              ; clear the IBIF flag
            movb   IBDR,rcv_buf            ; save the received data
            ...
```

Data transmission and reception in I²C protocol can also be made interrupt-driven. Using the interrupt mechanism to control data transmission and reception allows the MCU to perform other tasks during the waiting period.

11.6 The Serial Real-Time Clock DS1307

The DS1307 is a real-time clock (RTC) chip that uses BCD format to represent clock and calendar information. It has 56 bytes of nonvolatile SRAM for storing critical information. The address and data are transferred via the I²C bus. The clock and calendar provide seconds, minutes, hours, day, date, month, and year information. The end-of-the-month data is automatically adjusted for months with fewer than 31 days, including corrections for leap year. The clock operates in either the 24-hour or 12-hour format with an AM/FM indicator. The DS1307 has a built-in power-sense circuit that detects power failure and automatically switches to the battery supply. The pin assignment and the block diagram of the DS1307 are shown in Figure 11.27.

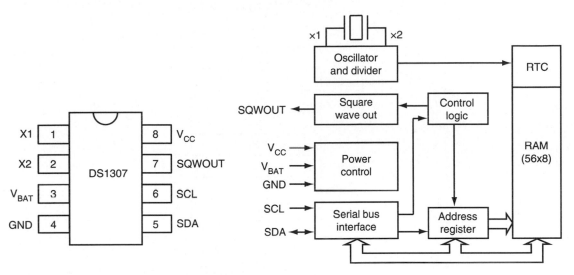

Figure 11.27 ■ DS1307 pin assignment and block diagram

11.6.1 Signal Functions

- V_{CC}, GND: *DC power input.* V_{CC} is the +5 V input. When a 5-V voltage is applied to this pin, the device is fully accessible and data can be written and read. When a 3-V battery is connected to the V_{BAT} pin and V_{CC} input is below $1.25 \times V_{BAT}$, reads and writes are inhibited. However, the timekeeping function continues unaffected by the lower input voltage. As V_{CC} falls below V_{BAT}, the RAM and timekeeper are switched over to the external power supply at V_{BAT}.

- V_{BAT}: *battery input* for any standard 3-V lithium cell or other energy source. Battery voltage must be held between 2.0 and 3.5 V for proper operation. The nominal write-protect trip-point voltage at which access to the RTC and user RAM is denied is set by the internal circuitry as $1.25 \times V_{BAT}$ nominal. A lithium battery with 48 mAh or greater will back up DS1307 for more than 10 years in the absence of power at 25°C.

- SCL: serial clock input.

- SDA: serial data input/output.

- SQWOUT: *square wave output driver.* When enabled (by setting the SQWE bit of the control byte to 1), the SQWOUT pin output may be programmed to one of the four square wave frequencies (1 Hz, 4 kHz, 8 kHz, and 32 kHz). The SQWOUT pin is open drain and requires an external pull-up resistor. SQWOUT will operate with either V_{CC} or V_{BAT} applied.

- X1, X2: *crystal connection.* These two pins are used for connections to a standard 32,768-Hz quartz crystal.

11.6.2 RTC and RAM Address Map

The address map for the RTC and RAM registers of the DS1307 is shown in Figure 11.28. The RTC registers are located at address locations $00 to $07. The RAM registers are located at address locations $08 to $3F. During a multibyte access, when the address pointer reaches $3F, it wraps around to location $00, the beginning of the clock space.

Address	
$00	Seconds
$01	Minutes
$02	Hours
$03	Day
$04	Date
$05	Month
$06	Year
$07	Control
$08 ⋮ $3F	RAM 56 × 8

Figure 11.28 ■ DS1307 address map

11.6.3 Clock and Calendar

The time and calendar information is obtained by reading the appropriate register bytes. The contents of the RTC registers are illustrated in Figure 11.29. The time and calendar are set or initialized by writing the appropriate register bytes, and the contents of the registers are in the BCD format. Bit 7 of register 0 is the clock halt (CH) bit. When this bit is set to 1, the oscillator is disabled. When this bit is a 0, the oscillator is enabled. This bit should be cleared to 0 after reset. The initial power-on states of all registers are not defined. It is important that the designer remembers to enable the oscillator during the initial configuration.

The DS1307 can be run in either 12-hour or 24-hour mode. Bit 6 of the hours register is defined as the 12- or 24-hour mode-select bit. When high, the 12-hour mode is selected. In the 12-hour mode, bit 5 is the AM/PM bit, with logic high being PM. In the 24-hour mode, bit 5 is the second 10-hour bit (20~23 hours).

11.6.4 The DS1307 Control Register

The contents of the control register are shown in the bottom row of Figure 11.29. Bit 7 controls the output level of the SQWOUT pin when the square output is disabled (i.e., the SQWE bit is 0). When bit 7 is 1, the SQWOUT output level is 1 when bit 4 (the SQWE bit) is set to 0. Otherwise, the SQWOUT output level is 0. The SQWE bit enables the SQWOUT pin (oscillator) output. The frequency of the SQWOUT output depends on the values of the RS1~RS0 bits. With the square wave output set to 1 Hz, the clock registers are updated on the falling edge of the square wave.

Bit 7						Bit 0
CH	10 seconds		Seconds			
0	10 minutes		Minutes			
0	12/24	10 HR / A/P	10 HR	Hours		
0	0	0	0	0	Day	
0	0	10 date		Date		
0	0	0	10 month	Month		
10 year				Year		
OUT	0	0	SQWE	0	0	RS1	RS0

Figure 11.29 ■ Contents of RTC registers

The frequency of the square wave output is set by programming the RS1 and RS0 bits with appropriate values. Table 11.7 lists the square wave frequencies that can be selected with the RS bits.

RS1	RS0	SQW Output Frequency
0	0	1 Hz
0	1	4.096 kHz
I	0	0.192 kHz
1	1	32.768 kHz

Table 11.7 ■ Square wave output frequency

11.6.5 Data Transfer

The DS1307 supports the standard mode (100 kbps baud rate) of the I²C bus. The device address of the DS1307 is %1101000. There are two types of data transfer between the DS1307 and an MCU: *slave receiver mode* and *slave transmitter mode*.

SLAVE RECEIVER MODE

In this mode, the MCU sends the device address of the DS1307, the address of the register to be accessed, and one or multiple data bytes to the DS1307. The following events occur:

- The MCU generates a start condition.
- The MCU sends a device address byte to the DS1307 with the direction bit ($R/\overline{W}$) set to 0.
- DS1307 acknowledges the address byte.
- The MCU sends the address of the register to be accessed to the DS1307. This value sets the register pointer. Only the address of the first register needs to be sent to the DS1307 during a multiple-byte transfer.

- The MCU sends one or multiple bytes of data to the DS1307 and the DS1307 acknowledges each byte received.
- The MCU generates the stop condition to terminate the data write.

SLAVE TRANSMITTER MODE

In this mode, the MCU sends the device address of the DS1307 and the address of the register to be read to the DS1307, and reads back one or multiple data bytes from the DS1307. The following events occur:

- The MCU generates a start condition.
- The MCU sends a device address byte to the DS1307 with the direction bit (R/$\overline{\text{W}}$) set to 0.
- DS1307 acknowledges the address byte.
- The MCU sends the address of the register to be accessed to the DS1307. This value sets the register pointer. The register pointer is incremented by 1 after each register transfer.
- The DS1307 acknowledges the register address byte.
- The MCU generates a restart condition.
- The MCU sends the device address to the DS1307 with the direction bit set to 1.
- The DS1307 acknowledges the device address byte.
- The DS1307 sends one or multiple bytes to the MCU, and the MCU acknowledges each byte received except the last byte.
- The MCU generates the stop condition to terminate the data read.

11.6.6 Circuit Connection

The circuit connection between the DS1307 and the HCS12 is shown in Figure 11.30. The HCS12 may also connect an LCD or LED circuit to display the current time and calendar.

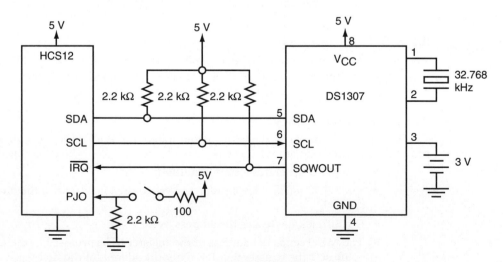

Figure 11.30 ■ Typical circuit connection between the HCS12 and the DS1307

After the HCS12 initializes the current time and calendar information, the DS1307 will interrupt the HCS12 periodically (most commonly, once per second) so that the HCS12 can update the time display periodically.

Example 11.3

▼

Write a function to configure the DS1307 to operate with the following setting:

- SQWOUT output enabled
- SQWOUT output set to 1 Hz
- SQWOUT idle high when it is disabled

Solution: The user needs to send the following bytes to the DS1307:

- Address byte $D0 (bit 0 is 0 to select write operation)
- Register address $07
- Control byte $90 (passed in accumulator B)

The following function will configure the DS1307 accordingly:

```
openDS1307    ldaa    #$D0                     ; place DS1307 ID in A
              jsr     sendSlaveID
              brclr   IBSR,RXAK,sndRegAdr      ; did DS1307 acknowledge?
              ldab    #$FF                     ; return −1 as error code
              rts
sndRegAdr     movb    #$07,IBDR                ; send out the control register address
              brclr   IBSR,IBIF,*              ; wait until the register address is shifted out
              movb    #IBIF,IBSR               ; clear the IBIF flag
              brclr   IBSR,RXAK,sndok          ; did DS1307 acknowledge?
              ldab    #$FF
              rts
sndok         stab    IBDR                     ; send out control byte
              brclr   IBSR,IBIF,*              ; wait until the control byte is shifted out
              movb    #IBIF,IBSR
              bclr    IBCR,MSSL                ; generate stop condition
              rts
```

The same configuration can be performed by the following C function:

```c
char openDS1307(char ctrl)
{
    sendSlaveID(0xD0);          /* send out DS1307's ID */
    if (IBSR & RXAK)            /* if DS1307 did not acknowledge, send error code */
            return −1;
    IBDR = 0x07;                /* send out control register address */
    while(!(IBSR & IBIF));
    IBSR = IBIF;                /* clear IBIF flag */
    if (IBSR & RXAK)            /* if DS1307 did not acknowledge, send error code */
            return −1;
    IBDR = ctrl;                /* send out control byte */
    while(!(IBSR & IBIF));
    IBSR = IBIF;
```

```
        if (IBSR & RXAK)          /* if DS1307 did not acknowledge, send error code */
            return −1;
        IBCR &= ~MSSL;            /* generate a stop condition */
            return 0;
    }
```

▲

Example 11.4

▼

Write a function to read the time and calendar information from the DIP switches and store them in a buffer to be sent to the DS1307. The DIP switches are driven by Port AD1.

Solution: The user must be informed to enter the time and calendar. Since there are 7 bytes of time and calendar information, the user will need to use the DIP switches to enter data seven times. The procedure for entering a byte of information is as follows:

1. The function outputs a message to ask the user to enter a value. The prompt message is
   Enter xxxx:

2. The user sets up a value using the DIP switches and presses the button connected to the PJ0 pin to remind the MCU to read the value.

3. The MCU reads the value of the DIP switches and sends it to the DS1307.

The function that reads the time and calendar information and stores them in a buffer is as follows:

```
#include   "c:\miniIDE\hcs12.inc"
           org     $1000
buf        ds.b    7
tready     ds.b    1                    ; flag to indicate a new time info. is ready
getTime    pshx
           pshy
           ldy     #buf                 ; Y is the pointer to the buffer
           movb    #$FF,ATD1DIEN        ; enable Port AD1 for digital inputs
           bclr    DDRJ,BIT0            ; enable PJ0 pin for input
           bset    PERJ,BIT0            ; enable pull-up or pull-down on PJ0 pin
           bclr    PPSJ,BIT0            ; enable pull-down, interrupt is rising edge triggered
           bset    PIEJ,BIT0            ; enable PJ0 interrupt
           cli                          ;        "
           ldaa    #$80                 ; set LCD cursor to the upper left corner
           jsr     cmd2LCD             ;        "
           ldx     #prompty            ; output the prompt "Enter year:"
           jsr     putsLCD             ;        "
waity      tst     tready              ; is new year info. ready?
           beq     waity               ;        "
           movb    PTAD1,1,y+          ; save year info. in buffer
           movb    #0,tready
           ldaa    #$80                 ; set LCD cursor to the upper left corner
           jsr     cmd2LCD             ;        "
           ldx     #promptm            ; output the prompt "Enter month:"
```

```
            jsr     putsLCD         ;        "
waitm       tst     tready          ; is new month info. ready?
            beq     waitm           ;        "
            movb    PTAD1,1,y+      ; save month info. in buffer
            movb    #0,tready        ; clear the ready flag
            ldaa    #$80            ; set LCD cursor to the upper left corner
            jsr     cmd2LCD         ;        "
            ldx     #prompte        ; output the prompt "Enter date:"
            jsr     putsLCD         ;        "
waite       lsl     tready          ; is new date info. ready?
            beq     waite           ;        "
            movb    PTAD1,1,y+      ; save date info. in buffer
            movb    #0,tready        ; clear the ready flag
            ldaa    #$80            ; set LCD cursor to the upper left corner
            jsr     cmd2LCD         ;        "
            ldx     #promptd        ; output the prompt "Enter day:"
            jsr     putsLCD         ;        "
waitd       tst     tready          ; is new day info. ready?
            beq     waitd           ;        "
            movb    PTAD1,1,y+      ; save day info. in buffer
            movb    #0,tready
            ldaa    #$80            ; set LCD cursor to the upper left corner
            jsr     cmd2LCD         ;        "
            ldx     #prompth        ; output the prompt "Enter hours:"
            jsr     putsLCD         ;        "
waith       tst     tready          ; is new hour info. ready?
            beq     waith           ;        "
            movb    PTAD1,1,y+      ; save hour info. in buffer
            movb    #0,tready
            ldaa    #$80            ; set LCD cursor to the upper left corner
            jsr     cmd2LCD         ;        "
            ldx     #promptmi       ; output the prompt "Enter minutes:"
            jsr     putsLCD         ;        "
waitmi      tst     tready          ; is new minute info. ready?
            beq     waitmi          ;        "
            movb    PTAD1,1,y+      ; save hour info. in buffer
            movb    #0,tready
            ldaa    #$80            ; set LCD cursor to the upper left corner
            jsr     cmd2LCD         ;        "
            ldx     #prompts        ; output the prompt "Enter seconds:"
            jsr     putsLCD         ;        "
waits       tst     tready          ; is new second info. ready?
            beq     waits           ;        "
            movb    PTAD1,1,y+      ; save second info. in buffer
            movb    #0,tready
            puly
            pulx
            rts
```

```
#include   "c:\miniIDE\lcd_util_dragon12.asm"
#include   "c:\miniIDE\delay.asm"
prompts    dc.b    "Enter seconds:",0
promptmi   dc.b    "Enter minutes:",0
prompth    dc.b    "Enter hours:",0
promptd    dc.b    "Enter day:",0
prompte    dc.b    "Enter date:",0
promptm    dc.b    "Enter month:",0
prompty    dc.b    "Enter year:",0
           end
```

▲

This function is interrupt-driven. It clears the flag *tready* to 0 after outputting the prompt message. When setting the DIP switches, the designer presses the button connected to the PJ0 pin. Pressing the PJ0 button causes an interrupt to be requested to the MCU. The interrupt service routine simply sets the flag tready to 1 and returns. Since the **getTime** function is polling this flag, it will detect the value change and read the value of the DIP switches. The PJ0 pin interrupt service routine is very simple.

```
PJ_ISR   movb   #1,tready
         movb   #1,PIFJ      ; clear the PIFJ0 flag
         rti
```

The C language version of the function will be left as an exercise problem.

Example 11.5

▼

Write a function to send the time and calendar information to the DS1307. The time and calendar information is pointed to by X. The device ID and the starting register address are passed in A and B, respectively. X points to the value of year and the second's value is located at [X] + 6.

Solution: The function that sends time and calendar information to the DS1307 is as follows:

```
sendTime     jsr     sendSlaveID                 ; send out device ID of the DS1307
             brclr   IBSR,RXAK,sndTimeOK1        ; did DS1307 acknowledge?
             ldab    #$FF                        ; return error code −1 if not acknowledged
             rts
sndTimeOK1   stab    IBDR                        ; send out register address for seconds
             brclr   IBSR,IBIF,*                 ; wait until seconds' address has been shifted out
             movb    #IBIF,IBSR                  ; clear the IBIF flag
             brclr   IBSR,RXAK,sndTimeOK2        ; did 1307 acknowledge?
             ldab    #$FF                        ; return error code −1 if not acknowledged
             rts
sndTimeOK2   ldy     #7                          ; byte count
             tfr     X,D                         ; set X to point to second's value
             addd    #6                          ;         "
             tfr     D,X                         ;         "
sndloop      movb    1,x−,IBDR                   ; send out 1 byte
             brclr   IBSR,IBIF,*
             movb    #IBIF,IBSR
```

```
                     brclr    IBSR,RXAK,sndTimeOK3    ; did DS1307 acknowledge?
                     ldab     #$FF                    ; return error code −1 if not acknowledged
                     rts
     sndTimeOK3      dbne     y,sndloop               ; continue until all bytes have been sent out
                     bclr     IBCR,MSSL               ; generate a stop condition.
                     ldab     #0                      ; return normal return code 0
                     rts
```

The same operation can be performed by the following C function:

```c
char sendTime(char *ptr, char ID)
{
    char i;
    sendSlaveID(0xD0);          /* send ID to DS1307 */
    if(IBSR & RXAK)             /* did DS1307 acknowledge? */
            return −1;
    IBDR = 0x00;                /* send out seconds register address */
    while(!(IBSR & IBIF));
    IBSR = IBIF;
    if(IBSR & RXAK)
            return −1;
    for(i = 6; i >= 0; i−−) {
            IBDR = *(ptr+i);
            while(!(IBSR&IBIF));
            IBSR = IBIF;
            If(IBSR & RXAK)
            return −1;
    }
    return 0;
}
```

Example 11.6

Write a C function to read the time of day from the DS1307 and save it in the array *cur_time*[0, . . . , 6].

Solution: The following C function will read the time of day from the DS1307:

```c
char readTime(char cx)
{
    char i, temp;
    sendSlaveID(0xD0);              /* generate a start condition and send DS1307's ID */
    if (IBSR & RXAK)
            return −1;              /* if DS1307 did not respond, return error code * 
    IBDR = cx;                      /* send address of seconds register */
    while(!(IBSR & IBIF));
    IBSR = IBIF;                    /* clear the IBIF flag */
    if (IBSR & RXAK)
            return −1;              /* if DS1307 did not respond, return error code */
```

```
IBCR |= RSTA;                    /* generate a restart condition */
IBDR = 0xD1;                     /* send ID and set R/W̄ flag to read */
while(!(IBSR & IBIF));
IBSR = IBIF;
if (IBSR & RXAK)
return −1;                       /* if DS1307 did not respond, return error code */
IBCR &= ~(TXRX + TXAK);          /* prepare to receive and acknowledge */
temp = IBDR;                     /* a dummy read to trigger nine clock pulses */
for (i = 0; i < 5; i++) {
while(!(IBSR & IBIF));           /* wait for a byte to shift in */
        IBSR = IBIF;             /* clear the IBIF flag */
cur_time[i] = IBDR;              /* save the current time in buffer */
}                                /* also initiate the next read */
while (!(IBSR & IBIF));          /* wait for the receipt of cur_time[5] */
        IBSR = IBIF;             /* clear IBIF flag */
IBCR |= TXAK;                    /* not to acknowledge cur_time[6] */
cur_time[5] = IBDR;              /* save cur_time[5] and initiate next read */
while (!(IBSR & IBIF));
IBSR = IBIF;
IBCR &= ~MSSL;                   /* generate stop condition */
cur_time[6] = IBDR;
return 0;
}
```

▲

Example 11.7
▼

Write a function to format the time information stored in the array *cur_time*[0, . . . , 6] so that it can be displayed on the LCD.

Solution: The time and calendar information are converted and stored in two character arrays: *hms*[0, . . . , 11] and *mdy*[0, . . . , 11]. The array *hms*[0, . . . , 11] holds the hours, minutes, and seconds, whereas the array *dmy*[0, . . . , 11] holds month, date, and year. The format of display is as follows:
In 24-hour mode,

hh:mm:ss:xx

mm:dd:yy

where xx stands for day of week (can be "SU," "MO," "TU," "WE," "TH," "FR," and "SA").
In 12-hour mode,

hh:mm:ss:ZM

xx:mm:dd:yy

where Z can be 'A' or 'P' to indicate AM or PM and xx stands for day of week.
The following C function will format the current time of day into two strings:

```
void formatTime(void)
{
    char  temp3;
    temp3 = cur_time[3] & 0x07;                         /* extract day-of-week */
```

```
        if (cur_time[2] & 0x40) {                              /* if 12-hour mode is used */
            hms[0] = 0x30 + ((cur_time[2] & 0x10) >> 4);       /* tens hour digit */
            hms[1] 0x30 + (cur_time[2] & 0x0F);                /* ones hour digit */
            hms[2] = ':';
            hms[3] = 0x30 + (cur_time[1] >> 4);                /* tens minute digit */
            hms[4] = 0x30 + (cur_time[1] & 0x0F);              /* ones minute digit */
            hms[5] = ':';
            hms[6] = 0x30 + ((cur_time[0] & 0x70) >> 4);       /* tens second digit */
            hms[7] = 0x30 + (cur_time[0] & 0x0F);              /* ones second digit */
            hms[8] = ':';
            if (cur_time[2] & 0x20)
                hms[9] = 'P';
            else
                hms[9] = 'A';
                hms[10] = 'M';
                hms[11] × 0;                                   /* terminate the string with a NULL */
            switch(temp3) {                                    /* convert to day of week */
                case 1:    mdy[0] = 'S';
                           mdy[1] = 'U';
                           break;
                case 2:    mdy[0] = 'M';
                           mdy[1] = 'O';
                           break;
                case 3:    mdy[0] = 'T';
                           mdy[1] = 'U';
                           break;
                case 4:    mdy[0] = 'W';
                           mdy[1] = 'E';
                           break;
                case 5:    mdy[0] = 'T';
                           mdy[1] = 'H';
                           break;
                case 6:    mdy[0] = 'F';
                           mdy[1] = 'R';
                           break;
                case 7:    mdy[0] = 'S';
                           mdy[1] = 'A';
                           break;
                default:   mdy[0] = 0x20;                       /* space */
                           mdy[1] = 0x20;
                           break;
            }
        mdy[2] = ':';
        mdy[3] = 0x30 + (cur_time[5] >> 4);                    /* month */
        mdy[4] = 0x30 + (cur_time[5] & 0x0F);
        mdy[5] = ':';
        mdy[6] = 0x30 + (cur_time[4] >> 4);                    /* date */
        mdy[7] = 0x30 + (cur_time[4] & 0x0F);
        mdy[8] = ':';
```

```
    mdy[9] = 0x30 + (cur_time[6] >> 4);                    /* year */
    mdy[10] = 0x30 + (cur_time[6] & 0x0F);
    mdy[11] = 0;                                           /* NULL character */
}
else {                                                     /* 24-hour mode */
    hms[0] = 0x30 + ((cur_time[2] & 0x30) >> 4);          /* hours */
    hms[1] = 0x30 + (cur_time[2] & 0x0F);
    hms[2] = ':';
    hms[3] = 0x30 + (cur_time[1] >> 4);                    /* minutes */
    hms[4] = 0x30 + (cur_time[1] & 0x0F);
    hms[5] = ':';
    hms[6] = 0x30 + ((cur_time[0] & 0x70) >> 4);          /* seconds */
    hms[7] = 0x30 + (cur_time[0] & 0x0F);
    hms[8] = ':';
    switch(temp3) {                                        /* convert to day of week */
        case 1:    hms[9]  = 'S';
                   hms[10] = 'U';
                   break;
        case 2:    hms[9]  = 'M';
                   hms[10] = 'O';
                   break;
        case 3:    hms[9]  = 'T';
                   hms[10] = 'U';
                   break;
        case 4:    hms[9]  = 'W';
                   hms[10] = 'E';
                   break;
        case 5:    hms[9]  = 'T';
                   hms[10] = 'H';
                   break;
        case 6:    hms[9]  = 'F';
                   hms[10] = 'R';
                   break;
        case 7:    hms[9]  = 'S';
                   hms[10] = 'A';
                   break;
        default:   hms[9]  = 0x20;                         /* space */
                   hms[10] = 0x20;
                   break;
    }
    hms[11] = 0; /* NULL character */
    mdy[0] = 0x30 + (cur_time[5] >> 4);                    /* month */
    mdy[1] = 0x30 + (cur_time[5] & 0x0F);
    mdy[2] = ':';
    mdy[3] = 0x30 + (cur_time[4] >> 4);                    /* date */
    mdy[4] = 0x30 + (cur_time[4] & 0x0F);
    mdy[5] = ':';
    mdy[6] = 0x30 + (cur_time[6] >> 4);                    /* year */
```

```
                mdy[7] = 0x30 + (cur_time[6] & 0x0F);
                mdy[8] = 0;                              /* NULL character */
        }
}
```

Example 11.8

▼

Write a function to display the current time and calendar information on the LCD.

Solution: This function is as follows:

```
void displayTime(void)
{
        cmd2lcd(0x83);       /* set cursor to row 1 column 3 */
        puts2lcd(hms);       /* output hours, minutes, and seconds */
        cmd2lcd(0xC3);       /* set cursor to row 2 column 3 */
        puts2lcd(mdy);       /* output month, date, and year */
}
```

▲

Example 11.9

▼

Write the $\overline{\text{IRQ}}$ interrupt service routine that reads the time from the DS1307, format the time and calendar information, and display them on the LCD.

Solution: The $\overline{\text{IRQ}}$ service routine simply calls the previous functions that have been written. The C language version of the service routine is as follows:

```
#pragma interrupt_handler irqISR          // CodeWarrior format
void  irqISR(void)
{
        readTime(0x00);       /* read all time registers starting from seconds */
        formatTime();         /* format time info into two strings */
        displayTime();        /* display the time on LCD */
}
```

▲

11.7 The Digital Thermometer and Thermostat DS1631A

Many embedded products, such as network routers and switches, are used in larger systems, and their failures due to overheating could severely damage the functioning or even cause the total failure of the larger system. Using a thermostat to warn of potential overheating is indispensable for the proper functioning of many embedded systems.

The digital thermostat device DS1631A from Dallas Semiconductor is one such product. The DS1631A will assert a signal (T_{OUT}) whenever the ambient temperature exceeds the *trip point* preestablished by the user. The DS1626 performs the same function but has an SPI interface.

11.7.1 Pin Assignment

As shown in Figure 11.31, the DS1631A is an eight-pin package. The SDA and SCL pins are used as the data and clock lines so that the DS1631A can be connected to an I²C bus. Pins A2~A0 are address inputs to the DS1631A. The T_{OUT} pin is the thermostat output, which is asserted whenever the ambient temperature is above the trip point set by the designer.

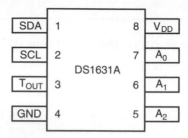

Figure 11.31 ■ Pin assignment of DS1631A

11.7.2 Functional Description

The block diagram of the DS1631A is shown in Figure 11.32. The DS1631A converts the ambient temperature into 9-, 10-, 11-, or 12-bit readings over a range of −55 to +125°C. The thermometer accuracy is ±0.5°C from 0 to +70°C with $3.0 \text{ V} \leq V_{DD} \leq 5.5 \text{ V}$.

The thermostat output T_{OUT} is asserted whenever the converted ambient temperature is equal to or higher than the value stored in the T_H register. The DS1631A automatically begins

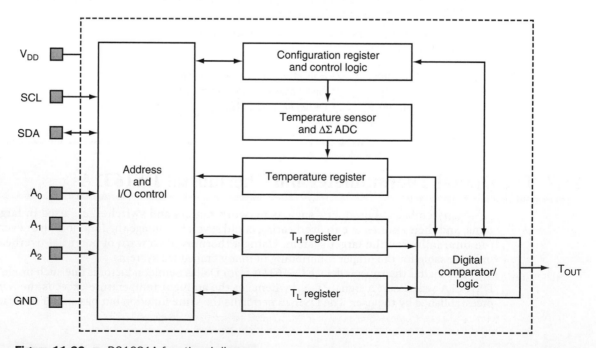

Figure 11.32 ■ DS1631A functional diagram

taking temperature measurements at power-up; this allows it to function as a stand-alone thermostat. The DS1631A conforms to the I²C bus specification.

11.7.3 DS1631A Registers

Four registers are involved in the operation of the DS1631A: Config, T_H, T_L, and Temperature registers. The SRAM-based Temperature register is a 2-byte register that holds the converted temperature value in two's complement format. The converted temperature is stored in the upper bits of the Temperature register with lower bits filled with 0s. When the most significant bit of this register is 1, the temperature is negative.

Both the T_H and T_L are EEPROM-based 2-byte registers. T_H holds the upper alarm temperature value in two's complement format. Whenever the converted temperature value is equal to or higher than the value in T_H, the T_{OUT} signal will be asserted. Once asserted, T_{OUT} can be de-asserted only when the converted temperature is lower than the value in the T_L register. The value in the T_L register is also represented in two's complement format.

The contents of the Config register are shown in Figure 11.33. The lower 2 bits of the configuration register are EEPROM-based, whereas the upper 6 bits are SRAM-based. The meaning of each bit is well defined. The Config register allows the designer to program various DS1631A

7	6	5	4	3	2	1	0	
DONE	THF	TLF	NVB	R1	R0	POL*	1SHOT*	power-up value=100011XX

*NV (EEPROM)

Done: temperature conversion done (read-only)
   0 = temperature conversion is in progress.
   1 = temperature conversion is complete. Will be cleared when the Temperature
       register is read.
THF: temperature high flag (read/write)
   0 = the measured temperature has not exceeded the value in T_H register.
   1 = the measured temperature has exceeded the value in T_H register. THF
       remains at 1 until it is overwritten with a 0 by the user, the power is
       recycled, or a software POR command is issued.
TLF: temperature low flag (read/write)
   0 = the measured temperature has not been lower than the value in T_L register.
   1 = at some point after power-up, the measured temperature is lower than the
       value stored in the T_L register. TLF remains at 1 until it is overwritten with a
       0 by the user, the power is recycled, or a software POR command is issued.
NVB: nonvolatile memory busy (read only)
   0 = NV memory is not busy.
   1 = A write to EEPROM memory is in progress.
R1:R0 : resolution bits (read/write)
   00 = 9-bit resolution (conversion time is 93.75 ms).
   01 = 10-bit resolution (conversion time is 187.5 ms).
   10 = 11-bit resolution (conversion time is 375 ms).
   11 = 12-bit resolution (conversion time is 750 ms).
POL: T_{OUT} polarity (read/write)
   0 = T_{OUT} active low.
   1 = T_{OUT} active high.
1SHOT: conversion mode (read/write)
   0 = continuous conversion mode. The **Start Convert** T command initiates
       continuous temperature conversions.
   1 = one-shot mode. The Start Convert T command initiates a single temperature
       conversion and then the device enters a low-power standby mode.

Figure 11.33 ■ DS1631A Config register

options such as conversion resolution, T_{OUT} polarity, and operation mode. It also provides information about conversion status, EEPROM activity, and thermostat activity. This register can be read from and written into using the **Access Config** command. When writing to the Config register, conversions should first be stopped using the **Stop Convert T** command if the device is in continuous conversion mode. Since the POL and 1SHOT bits are stored in EEPROM, they can be programmed prior to installation if desired. All other bits are in SRAM and are powered up in the state shown in Figure 11.33.

11.7.4 The DS1631A Operation

The DS1631A begins conversions automatically at power-up. It can be configured to perform continuous conversions (continuous conversion mode) or single conversions on command (one-shot mode).

The default resolution of the DS1631A is 12-bit. However, it can also be configured to 9-, 10-, or 11-bit. The resolution is changed via the R1:R0 bits of the Config register. A few samples of temperatures and their converted values are shown in Table 11.8. The lowest 4 bits are always zeros in the table because the resolution is 12-bit.

Both the T_H and T_L registers are in EEPROM. Their resolutions match the output temperature resolution and are determined by the R1:R0 bits. Writing to and reading from these two registers are achieved by using the **Access TH** and **Access TL** commands. When making changes to the T_H and T_L registers, conversions should first be stopped using the Stop Convert T command if the device is in continuous conversion mode.

Temperature (°C)	Digital Output (binary)	Digital Output (hex)
+125	0111 1101 0000 0000	0x7D00
+25.0625	0001 1001 0001 0000	0x1910
+10.125	0000 1010 0010 0000	0x0A20
+0.5	0000 0000 1000 0000	0x0080
0	0000 0000 0000 0000	0x0000
−0.5	1111 1111 1000 0000	0xFF80
−10.125	1111 0101 1110 0000	0xF5E0
−25.0625	1110 0110 1111 0000	0xE6F0
−55	1100 1001 0000 0000	0xC900

Table 11.8 ■ 12-bit resolution temperature/data relationship

Since the DS1631A automatically begins taking temperature measurements at power-up, it can function as a stand-alone thermostat. For stand-alone operation, the nonvolatile T_H and T_L registers and the POL and 1SHOT bits in the Config register should be programmed to the desired values prior to installation.

Table 11.8 shows that every 16 units (after ignoring the lowest 4 bits) of the conversion result correspond to 1°C. The conversion result cannot be higher than 0x7D00 or lower than 0xC900 because the range of temperature that can be handled by DS1631A cannot be higher than 125°C or lower than −55°C. An easy way to find the corresponding temperature of a certain conversion result is as follows:

Positive Conversion Result

Step 1
Truncate the lowest 4 bits.

Step 2
Divide the upper 12 bits by 16.

For example, the conversion result 0x7000 corresponds to 0x700/16 = 112°C. The conversion result 0x6040 corresponds to 0x604/16 = 96.25°C.

NEGATIVE CONVERSION RESULT

Step 1
Compute the two's complement of the conversion result.

Step 2
Truncate the lowest 4 bits.

Step 3
Divide the upper 12 bits of the two's complement of the conversion result by 16.

For example, the conversion result 0xE280 corresponds to −0x1D8/16 = −29.5°C.

11.7.5 DS1631A Command Set

The DS1631A supports the following commands:

- *Start Convert T [0x51].* This command initiates temperature conversions. If the part is in one-shot mode, only one conversion is performed. In continuous mode, continuous temperature conversions are performed until a **Stop Convert T** command is issued.

- *Stop Convert T [0x22].* This command stops temperature conversions when the device is in continuous conversion mode.

- *Read Temperature [0xAA].* This command reads the last converted temperature value from the 2-byte Temperature register.

- *Access TH [0xA1]* This command reads or writes the 2-byte T_H register.

- *Access TL [0xA2].* This command reads or writes the 2-byte T_L register.

- *Access Config [0xAC].* This command reads or writes the 1-byte Configuration register.

- *Software POR [0x54].* This command initiates a software power-on-reset operation, which stops temperature conversions and resets all registers and logic to their power-up states. The software POR allows the designer to simulate cycling the power without actually powering down the device.

11.7.6 I²C Communication with DS1631A

A typical circuit connection between the HCS12 microcontroller and a DS1631A is shown in Figure 11.34. The address input of the DS1631A is arbitrarily set to %001 in Figure 11.34. To initiate I²C communication, the HCS12 MCU asserts a start condition followed by a control byte containing the DS1631A device ID. The R/$\overline{\text{W}}$ bit of the control byte must be a 0 since the HCS12 MCU next will write a command byte to the DS1631A. The format for the control byte is shown in Figure 11.35. The DS1631A responds with an ACK after receiving the control byte. This must be followed by a command byte from the master, which indicates what type of operation is to be performed. The DS1631A again responds with an ACK after receiving the command byte. If the command byte is the **Start Convert T** or **Stop Convert T** command, the transaction is finished, and the master must issue a stop condition to signal the end of the communication sequence. If the command byte indicates a write or read operation, additional actions must occur.

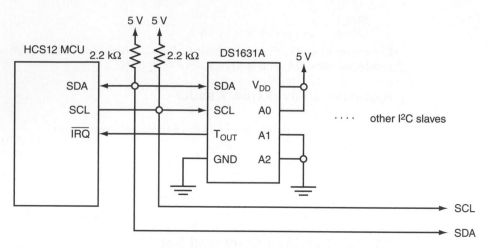

Figure 11.34 ■ Typical circuit connection between the HCS12 MCU and DS1631A

7	6	5	4	3	2	1	0
1	0	0	1	A_2	A_1	A_0	R/$\overline{W}$

Figure 11.35 ■ Control byte for DS1631A

WRITE DATA TO DS1631A

The master can write data to the DS1631A by issuing an **Access Config**, **Access TH**, or **Access TL** command following the control byte. Since the R/$\overline{W}$ bit in the control byte was a 0, the DS1631A is already prepared to receive data. Therefore, after receiving an ACK in response to the command byte, the master device can immediately begin transmitting data. When writing to the Config register, the master must send 1 byte of data, and when writing to the T_H or T_L register, the master must send 2 bytes of data. The upper byte is written first followed by the lower byte. After receiving each data byte, the DS1631A responds with an ACK, and the transaction is finished with a stop from the master.

READ DATA FROM DS1631A

The master can read data from the DS1631A by issuing an **Access Config**, **Access TH**, **Access TL**, or **Read Temperature** command following the control byte. After receiving an ACK in response to the command, the master must generate a Restart condition followed by a control byte with the same slave ID as the first control byte. However, this time the R/$\overline{W}$ bit must be set to 1 to inform the DS1631A that a read is being performed. After sending an ACK in response to this control byte, the DS1631A begins transmitting the requested data on the next clock cycle. One byte of data will be transmitted when reading from the configuration register, after which the master must respond with a NACK followed by a stop condition. For 2-byte reads (i.e., from the Temperature, T_H, or T_L register), the master must respond to the first data byte with an ACK and to the second byte with a NACK followed by a stop condition. The upper byte is read first followed by the lower byte. If only the most significant byte of data is needed, the master can issue a NACK followed by a stop condition after reading the first data byte.

Example 11.10

▼

Write a function to configure the DS1631A in Figure 11.34 to operate in continuous conversion mode and set the T_{OUT} polarity to active high. Assume that the I²C has only one master and there is no possibility in getting bus collision.

Solution: The procedure for configuring the DS1631A is as follows:

Step 1
Assert a start condition and send a control byte (0x92) to the DS1631A with $R/\overline{W} = 0$.

Step 2
Wait until the I²C transfer is completed and check to see if the DS1631A acknowledges. If not, exit.

Step 3
Send an Access-Config command to the DS1631A.

Step 4
Wait until the I²C bus is idle and check to see if the DS1631A acknowledges. If not, exit.

Step 5
Write a command byte to configure the DS1631A to operate in continuous conversion mode and set the T_{OUT} polarity to active high.

Step 6
Wait until the I²C bus is idle and check to see if the DS1631A acknowledges. If not, exit.

Step 7
Assert a stop condition to complete the whole process.

The assembly function that performs the desired configuration is as follows:

```
openDS1631   ldaa    #$92
             jsr     sendSlaveID
             brclr   IBSR,RXAK,openOK0    ; did DS1631A acknowledge?
             ldab    #$FF                 ; return error code 21
             rts
openOK0      movb    #$AC,IBDR            ; send the Access Config command
             brclr   IBSR,IBIF,*
             movb    #IBIF,IBSR           ; clear the IBIF flag
             brclr   IBSR,RXAK,openOK1    ; did DS1316A acknowledge?
             ldab    #$FF
             rts
openOK1      stab    IBDR                 ; sends configuration data
             brclr   IBSR,IBIF,*          ; wait until the byte has been shifted out
             movb    #IBIF,IBSR           ; clear the IBIF flag
             brclr   IBSR,RXAK,openOK2    ; did DS1316A acknowledge?
             ldab    #$FF
             rts
openOK2      bclr    IBCR,MSSL            ; generate a stop condition
             ldab    #0                   ; normal return code
             rts
```

To configure the DS1631A with the desired setting, use the following instruction sequence to call the previous function:

```
        ldab     #$0E
        jsr      openDS1631
```

The C language version of the function is as follows:

```
char openDS1631(char cy)
{
    sendSlaveID(0x92);          /* generate a start condition and send ID */
    if (IBSR & RXAK)
           return −1;           /* error code when DS1631A did not acknowledge */
    IBDR = 0xAC;                /* send command Access Config */
    while(!(IBSR & IBIF));
    IBSR = IBIF;                /* clear the IBIF flag */
    if (IBSR & RXAK)
           return −1;           /* error code when DS1631A did not acknowledge */
    IBDR = cy;                  /* send configuration byte */
    while(!(IBSR & IBIF));
    IBSR = IBIF;
    if (IBSR & RXAK)
           return −1;           /* error code when DS1631A did not acknowledge */
    IBCR  &= ~MSSL;            /* generate a stop condition */
    return 0;                   /* normal return code */
}
```

▲

Example 11.11

▼

Write a function to start temperature conversion.

Solution: The procedure for starting temperature conversion using the DS1631A is as follows:

Step 1
Assert a start condition and send a control byte (0x92) to the DS1631A with R/$\overline{\text{W}}$ = 0.

Step 2
Wait until the I²C bus is idle and check to see if the DS1631A acknowledges. If not, exit.

Step 3
Send a Start Convert T command to the DS1631A.

Step 4
Wait until the I²C transfer is completed and check to see if the DS1631A acknowledges. If not, exit.

Step 5
Assert a stop condition to complete the whole process.

The assembly program that implements this procedure is as follows:

```
startConv   ldaa    #$92
            jsr     sendSlaveID        ; generate a start condition and send DS1631A's ID
            brclr   IBSR,RXAK,startOK0 ; did DS1631A acknowledge?
            ldab    #$FF               ; return error code −1
            rts
```

```
startOK0    movb    #$51,IBDR          ; send Start Convert T command
            brclr   IBSR,IBIF,*        ; wait until the byte is shifted out
            movb    #IBIF,IBSR         ; clear the IBIF flag
            brclr   IBSR,RXAK,startOK1 ; did DS1631A acknowledge?
            ldab    #$FF
            rts
startOK1    bclr    IBCR,MSSL          ; generate a stop condition
            ldab    #0                 ; normal return code
            rts
```

The C language version of the function is as follows:

```
char startConv (void)
{
    sendSlaveID(0x92);                 /* generate a start condition and send slave ID */
    if (IBSR & RXAK)
            return −1;                 /* if not acknowledged, return error code −1 */
    IBDR = 0x51;                       /* send command to start a conversion */
    while(!(IBSR & IBIF));
    IBSR = IBIF;                       /* clear the IBIF flag */
    if (IBSR & RXAK)
            return −1;
    IBCR &= ~MSSL;                     /* generate a stop condition */
    return 0;
}
```

Example 11.12

Write a function to set the high thermostat temperature. The upper and lower bytes of the high thermostat temperatures are passed in stack.

Solution: The procedure for setting up the high thermostat temperature is as follows:

Step 1
Assert the start condition and send control byte 0x92 with R/$\overline{W}$ bit = 0.

Step 2
Wait until the I²C transfer is completed and check to see if the DS1631A acknowledges. If not, exit.

Step 3
Send an Access TH command to DS1631A.

Step 4
Wait until the I²C transfer is completed and check to see if the DS1631A acknowledges. If not, exit.

Step 5
Send the upper byte of the thermostat temperature to the DS1631A.

Step 6
Wait until the I²C transfer is completed and check to see if the DS1631A acknowledges. If not, exit.

Step 7

Send the lower byte of the thermostat temperature to the DS1631A.

Step 8

Wait until the I²C transfer is completed and check to see if the DS1631A acknowledges. If not, exit.

Step 9

Assert the stop condition.

The assembly program that implements this procedure is as follows:

```
THhi        equ     2                           ; offset of TH high byte from top of the stack
THlo        equ     3                           ; offset of TH low byte from top of the stack
setTH       ldaa    #$92
            jsr     sendSlaveID
            brclr   IBSR,RXAK,setTHok1          ; did DS1631A acknowledge?
            ldab    #$FF                        ; return error code −1
            rts
setTHok1    movb    #$A1,IBDR                   ; send out access TH command */
            brclr   IBSR,IBIF,*                 ; wait until command is shifted out
            movb    #IBIF,IBSR                  ; clear IBIF flag
            brclr   IBSR,RXAK,setTHok2          ; did DS1631A acknowledge?
            ldab    #$FF
            rts
setTHok2    ldaa    THhi,sp                     ; get the upper byte of TH from stack
            staa    IBDR                        ; send out TH high byte
            brclr   IBSR,IBIF,*
            movb    #IBIF,IBSR                  ; clear the IBIF flag
            brclr   IBSR,RXAK,setTHok3          ; did DS1631A acknowledge?
            ldab    #$FF
            rts
setTHok3    ldaa    THlo,sp                     ; get the lower byte of TH from stack
            staa    IBDR                        ; send out the lower byte of TH
            brclr   IBSR,IBIF,*                 ; "
            movb    #IBIF,IBSR                  ; clear the IBIF flag
            brclr   IBSR,RXAK,setTHok4          ; did DS1631A acknowledge?
            ldab    #$FF
            rts
setTHok4    bclr    IBCR,MSSL                   ; generate the stop condition
            ldab    #0                          ; normal return code
            rts
```

The caller of this function pushes the high- and then the low byte of the new TH value into the stack before calling this function.

The C language version of the function is as follows:

```c
char setTH(char hibyte, char lobyte)
{
        sendSlaveID(0x92);
        if(IBSR & RXAK)             /* did DS1631A acknowledge? */
            return −1;
        IBDR = 0xA1;                /* send command to access TH */
        while(!(IBSR & IBIF));
```

```
        IBSR = IBIF;                    /* clear the IBIF flag */
        if (IBSR & RXAK)
              return −1;
        IBDR = hibyte;                  /* send out the high byte of TH */
        while(!(IBSR & IBIF));
        IBSR = IBIF;
        if (IBSR & RXAK)
              return −1;
        IBDR = lobyte;                  /* send out the low byte of TH */
        while(!(IBSR & IBIF));
        IBSR = IBIF;
        if (IBSR & RXAK)
              return −1;
        IBCR &= ~MSSL;                  /* generate a stop condition */
        return 0;
}
```

Example 11.13

Write a function to read the Config register from the DS1631A and return its value in accumulator B.

Solution: The procedure for reading the Config register is as follows:

Step 1
Assert the start condition and send the control byte 0x92 with $R/\overline{W} = 0$.

Step 2
Wait until the control byte is shifted out and check to see if the DS1631A acknowledges. If not, exit.

Step 3
Send an Access Config command (0xAC) to the DS1631A.

Step 4
Wait until the command is shifted out and check to see if the DS1631A acknowledges. If not, exit.

Step 5
Assert a restart condition.

Step 6
Send the control byte 0x93 with $R/\overline{W} = 1$, which tells the DS1631A that a read is being performed.

Step 7
Wait until the control byte is shifted out and check to see if the DS1631A acknowledges. If not, exit.

Step 8
Read the Config register.

Step 9
Make sure that the bus is idle and send back a NACK to the DS1631A.

Step 10

Assert the stop condition.

The assembly program that implements this procedure is as follows:

```
readConf    ldaa    #$92                    ; generate a start condition and send out DS1631A
            jsr     sendSlaveID             ; ID
            brclr   IBSR,RXAK,rdConfok1      ; did DS1631A acknowledge?
            ldab    #$FF                    ; return error code −1
            rts
rdConfok1   movb    #$AC,IBDR               ; send out the Access Conf command
            brclr   IBSR,IBIF,*             ; wait until the command is shifted out
            movb    #IBIF,IBSR              ; clear the IBIF flag
            brclr   IBSR,RXAK,rdConfok2      ; did DS1631A acknowledge?
            ldab    #$FF                    ; return error code −1
            rts
rdConfok2   bset    IBCR,RSTA               ; generate a restart condition
            movb    #$93,IBDR               ; send out slave ID and set R/W̄ bit to 1
            brclr   IBSR,IBIF,*             ; wait until the command is shifted out
            movb    #IBIF,IBSR              ; clear the IBIF flag
            brclr   IBSR,RXAK,rdConfok3      ; did DS1631A acknowledge?
            ldab    #$FF                    ; return error code −1
            rts
rdConfok3   bclr    IBCR,TXRX              ; prepare to read
            bset    IBCR,TXAK              ; prepare to send NACK
            ldaa    IBDR                    ; perform a dummy read to trigger nine clock pulses
            brclr   IBSR,IBIF,*             ; wait until a byte has been shifted in
            movb    #IBIF,IBSR              ; clear the IBIF flag
            bclr    IBCR,MSSL              ; generate a stop condition
            ldab    IBDR                    ; place the Conf register value in B
            rts
```

The C language version of the function is as follows:

```c
char readConf(void)
{
    char temp;
    sendSlaveID(0x92);
    if(IBSR & RXAK)             /* did DS1631A acknowledge? */
           return −1;
    IBDR = 0xAC;               /* send command to access Conf register */
    while(!(IBSR & IBIF));
    IBSR = IBIF;
    if (IBSR & RXAK)
           return −1;
    IBCR |= RSTA;              /* generate a restart condition */
    IBDR = 0x93;              /* send out ID and set R/W̄ to 1 */
    while(!(IBSR & IBIF));
    IBSR = IBIF;
    if (IBSR & RXAK)
           return −1;
    IBCR &= ~TXRX;             /* prepare to receive */
```

```
        IBCR |= TXAK;               /* prepare to send NACK */
        temp = IBDR;                /* a dummy read to trigger nine clock pulses */
        while(!(IBSR & IBIF));
        IBSR = IBIF;
        IBCR &= ~MSSL;              /* generate a stop condition */
        return IBDR;
    }
```

Example 11.14

Write a subroutine to read the converted temperature and return the upper and lower bytes in double accumulator D. Assume that the temperature conversion has been started but this function needs to make sure that the converted temperature value is a result of the most recent Start Convert T command.

Solution: This subroutine calls the previous subroutine to make sure that the DONE bit is set to 1 and reads the converted temperature from the DS1631A. The assembly subroutine is as follows:

```
readTemp   ldx    #0                    ; initialize return error code
rdLoop     jsr    readConf              ; is temperature conversion done yet?
           cmpb   #-1                   ; is there any error?
           beq    rdErr                 ; "
           andb   #$80                  ; check DONE bit
           bpl    rdLoop                ; conversion not done yet?
           ldaa   #$92                  ; generate a start condition and send out
           jsr    sendSlaveID           ; the DS1631A ID
           brclr  IBSR,RXAK,rdTempok1   ; did DS1631A acknowledge?
           ldx    #-1
           rts
rdTempok1  movb   #$AA,IBDR             ; sends Read-Temperature command
           brclr  IBSR,IBIF,*          ;
           movb   #IBIF,IBSR
           brclr  IBSR,RXAK,rdTempok2   ; did DS1631A acknowledge?
           ldx    #-1
           rts
rdTempok2  bset   IBCR,RSTA             ; generate a restart condition
           movb   #$93,IBDR             ; send DS1631A's ID with R/W̄ set to 1
           brclr  IBSR,IBIF,*
           movb   #IBIF,IBSR
           brclr  IBSR,RXAK,rdTempok3   ; did DS1631A acknowledge?
           ldx    #-1
           rts
rdTempok3  bclr   IBCR,TXRX+TXAK        ; prepare to receive an ACK
           ldaa   IBDR                  ; perform a dummy read
           brclr  IBSR,IBIF,*          ; wait for high byte of temperature to shift in
           movb   #IBIF,IBSR
           bset   IBCR,TXAK             ; prepare to send NACK for the last read
```

ldaa	IBDR	; place the high byte of temperature in A
brclr	IBSR,IBIF,*	; wait for the low byte read to complete
movb	#IBIF,IBSR	; clear the IBIF flag
bclr	IBCR,MSSL	; generate a STOP condition
ldab	IBDR	; place the low byte of temperature in B
ldx	#0	; correct return code
rts		
rdErr	ldx	#−1
rts		

The C language version of the function is straightforward and is left as an exercise problem.

11.8 Interfacing the Serial EEPROM 24LC08B with I²C

Some applications require the use of a large amount of nonvolatile memory because these applications are powered by batteries and may be used in the field for an extended period of time. Many semiconductor manufacturers produce serial EEPROMs with a serial interface. Both serial EEPROMs with SPI and I²C interfaces are available.

The 24LC08B is a serial EEPROM from Microchip with the I²C interface. This device is an 8-kbit EEPROM organized as four blocks of 256 × 8-bit memory. Low-voltage design permits operation down to 2.5 V with standby and active currents of only 1 μA and 1 mA, respectively. The 24LC08B also has a pagewrite capability for up to 16 bytes of data.

11.8.1 Pin Assignment and Block Diagram

The pin assignment and the block diagram of the 24LC08B are shown in Figures 11.36 and 11.37, respectively. Pins A2, A1, and A0 are not used. The SCL and SDA pins are for I²C bus communications. The frequency of the SCL input can be as high as 400 kHz. The WP pin is used as the write protection input. When this pin is high, the 24LC08B cannot be written into. The pins A2~A0 are not used and can be left floating, grounded, or pulled to high.

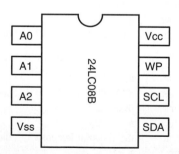

Figure 11.36 ■ 24LC08B PDIP package pin assignment

11.8.2 Device Addressing

Like any other I²C slave, the first byte sent to the 24LC08B after the start condition is the control byte. The contents of the control byte for the 24LC08B are shown in Figure 11.38.

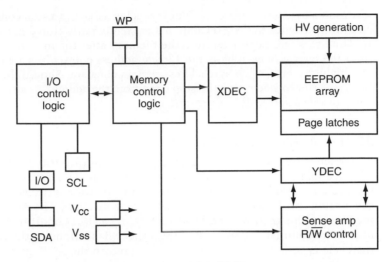

Figure 11.37 ■ Block diagram of 24LC08B

7	6	5	4	3	2	1	0
1	0	1	0	X	B1	B0	R/$\overline{W}$

Figure 11.38 ■ 24LC08B control byte contents

The upper 4 bits are the device ID of the 24LC08B, and the value represented by bits B1~B0 is the block address of the memory location to be accessed. For any access to the 24LC08B, the master must also send an 8-bit byte address after the control byte. There is an address pointer inside the 24LC08B. After the access of each byte, the address pointer is incremented by 1.

11.8.3 Write Operation

The 24LC08B supports byte write and pagewrite operations. In a *byte write* operation, the following operations need to be performed:

1. The master asserts the start condition and sends the control byte to the 24LC08B.
2. The 24LC08B acknowledges the data transmission.
3. The master sends the byte address to the 24LC08B.
4. The 24LC08B acknowledges the data transmission.
5. The master sends the data byte to the 24LC08B.
6. The 24LC08B acknowledges the data transmission.
7. The master asserts the stop condition.

In a *pagewrite* operation, the master can send up to 16 bytes of data to the 24LC08B. The write control byte, byte address, and the first data byte are transmitted to the 24LC08B in the same way as in a byte write operation. But instead of asserting a stop condition, the master transmits up to 16 data bytes to the 24LC08B that are temporarily stored in the on-chip page buffer. These 16 data bytes will be written into the memory after the master has asserted a stop condition. After the receipt of each byte, the 4 lower address pointer bits are internally incremented by 1. The highest 6 bits of the byte address remain constant. Should the master transmit more than 16 bytes prior to generating the stop condition, the address counter will roll over and the previously received data will be overwritten.

11.8.4 Acknowledge Polling

When the 24LC08B is writing the data held in the write buffer into the EEPROM array, it will not acknowledge any further write operation. This fact can be used to determine when the cycle is completed. Once the stop condition for a write command has been issued from the master, the device initiates the internal write cycle. The ACK polling can be initiated immediately. This involves the master's sending a start condition followed by the control byte for a write command ($R/\overline{W} = 0$). If the 24LC08B is still busy, then no ACK will be returned. If the cycle is complete, then the device will return the ACK, and the master can then proceed with the next read or write command. The polling process is illustrated in Figure 11.39.

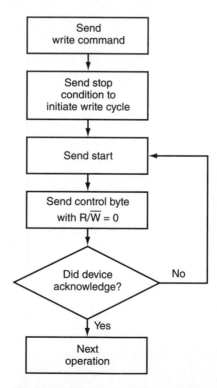

Figure 11.39 ■ Acknowledge polling flow

11.8.5 Read Operation

24LC08B supports three types of read operations: **current address read, random read,** and **sequential read.**

CURRENT ADDRESS READ

As explained earlier, the internal address counter is incremented by 1 after each access (read or write). The current address read allows the master to read the byte immediately following the location accessed by the previous read or write operation. On receipt of the slave address with the R/$\overline{W}$ bit set to 1, the 24LC08B issues an acknowledgement and transmits an 8-bit data byte. The master will not acknowledge the transfer but asserts a stop condition and the 24LC08B discontinues transmission.

RANDOM READ

Random read operations allow the master to access any memory location in a random manner. To perform this type of read operation, the address of the memory location to be read must be sent. The procedure for performing a random read is as follows:

Step 1
The master asserts a start condition and sends the control byte with the R/$\overline{W}$ bit set to 0 to the 24LC08B.

Step 2
The 24LC08B acknowledges the control byte.

Step 3
The master sends the address of the byte to be read to the 24LC08B.

Step 4
The 24LC08B acknowledges the byte address.

Step 5
The master asserts a repeated start condition.

Step 6
The master sends the control byte with R/$\overline{W}$ = 1.

Step 7
The 24LC08B acknowledges the control byte and sends data to the master.

Step 8
The master asserts NACK to the 24LC08B.

Step 9
The master asserts the stop condition.

SEQUENTIAL READ

If the master acknowledges the data byte returned by the random read operation, the 24LC08B transmits the next sequentially addressed byte as long as the master provides the clock signal on the SCL line. The master can read the whole chip using sequential read.

11.8.6 Circuit Connection Between the I²C Master and the 24LC08B

The circuit connection of a 24LC08B and the HCS12 MCU is similar to that in Figure 11.34. The diagram of a system with one DS1631A and one 24LC08B and the HCS12 MCU is shown in Figure 11.40.

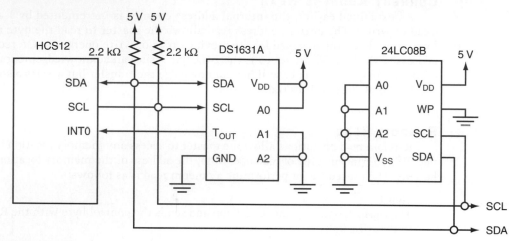

Figure 11.40 ■ Circuit connection of the HCS12MCU, 24LC08B, and DA1631A

Example 11.15

Write a function to read a byte from the 24LC08B. Pass the control byte and address in accumulators A and B to this subroutine. Return the data byte and error code in A and B, respectively. This function should check the error that the EEPROM did not acknowledge (implies that 24LC08B may have failed).

Solution: The following assembly subroutine implements the procedure for random read described earlier:

```
EErandomRead
                jsr     sendSlaveID         ; send out 24LC08B ID and block address
                brclr   IBSR,RXAK,ranRdok0  ; does EEPROM acknowledge?
                ldab    #$FF                ; return −1, if EEPROM does not acknowledge
                rts
ranRdok0        stab    IBDR                ; send out EEPROM memory address
                brclr   IBSR,IBIF,*         ; wait until the address is shifted out
                movb    #IBIF,IBSR          ; clear IBIF flag
                brclr   IBSR,RXAK,ranRdok1  ; does EEPROM acknowledge?
                ldab    #$FF                ; return −1, if EEPROM does not acknowledge
                rts
ranRdok1        bset    IBCR,RSTA           ; generate restart condition
                oraa    #$01                ; set R/W̄ bit for read
                staa    IBDR                ; resend the device ID
```

```
                    brclr   IBSR,IBIF,*              ; wait until the EEPROM ID is sent out
                    movb    #IBIF,IBSR              ; clear the IBIF flag
                    brclr   IBSR,RXAK,ranRdok2      ; does EEPROM acknowledge?
                    ldab    #$FF                    ; return −1, if EEPROM does not ackowledge
                    rts
        ranRdok2    bset    IBCR,TXAK              ; prepare to send NACK
                    bclr    IBCR,TXRX              ; perform reception
                    ldaa    IBDR                   ; dummy read to initiate reception
                    brclr   IBSR,IBIF,*            ; wait for a byte to shift in
                    movb    #IBIF,IBSR            ; clear the IBIF flag
                    bclr    IBCR,MSSL             ; generate a stop condition
                    ldaa    IBDR                   ; get the data byte
                    ldab    #0                     ; normal read status
                    rts
```

The C language version of the function is as follows:

```c
char EErandomRead(char ID, char addr)
{

    char dummy;
    SendSlaveID(ID);
    if (IBSR & RXAK)
        return −1;
    IBDR = addr;                /* send out EEPROM address */
    while(!(IBSR & IBIF));       /* wait until the address is shifted out */
    IBSR = IBIF;                 /* clear IBIF flag */
    if (IBSR & RXAK)
        return −1;
    IBCR |= RSTA;                /* generate restart condition */
    IBDR = ID | 0x01;            /* prepare to read */
    while (!(IBSR & IBIF));
    IBSR = IBIF;
    if (IBSR & RXAK)
        return −1;
    IBCR |= TXAK;                /* prepare to send NACK */
    IBCR &= ~TXRX;               /* perform reception */
    dummy = IBDR;                /* dummy read to trigger nine clock pulses */
    while(!(IBSR & IBIF));       /* wait for data to shift in */
    IBSR = IBIF;
    IBCR &= ~MSSL;               /* generate a stop condition */
    return IBDR;
}
```

▲

Sequential read is performed in a block read operation. When the MCU saves the contents of the IBDR register, it also initiates a sequential read if the MCU acknowledges the previously received byte.

Example 11.16

▼

Write a subroutine that writes a byte into the 24LC08B. The device ID, memory address, and data to be written are passed to this routine in the stack.

Solution: The assembly subroutine that writes a byte into the 24LC08B to the specified address is as follows:

```
EE_ID       equ    5                          ; stack offset for EE_ID
EE_addr     equ    4                          ; stack offset for EE_addr
EE_dat      equ    3                          ; stack offset for EE_data
EEbyteWrite psha
            ldaa   EE_ID,sp                    ; get the EEPROM ID from stack
            jsr    sendSlaveID                 ; generate start condition, send EEPROM ID
            brclr  IBSR,RXAK,bywriteok1        ; does EEPROM acknowledge?
            ldab   #$FF                        ; return −1 as the error code
            pula
            rts
bywriteok1  ldaa   EE_addr,sp                  ; get the address to be accessed from the stack
            staa   IBDR                        ; send address to I2C bus
            brclr  IBSR,IBIF,*                 ; wait until address is shifted out
            movb   #IBIF,IBSR                  ; clear the IBIF flag
            brclr  IBSR,RXAK,bywriteok2        ; does EEPROM acknowledge?
            ldab   #$FF                        ; return −1 as the error code
            pula
            rts
bywriteok2  ldaa   EE_dat,sp                   ; get the data byte to be written from stack
            staa   IBDR                        ; send out the data byte
            brclr  IBSR,IBIF,*                 ; wait until data byte is shifted out
            movb   #IBIF,IBSR                  ; clear the IBIF flag
            brclr  IBSR,RXAK,bywriteok3        ; does EEPROM acknowledge?
            ldab   #$FF                        ; return −1 as the error code
            pula
            rts
bywriteok3  bclr   IBCR,MSSL                   ; generate stop condition
            ldab   #0                          ; return error code 0
            pula
            rts
```

The C language version of the function is as follows:

```c
char EEbyteWrite(char ID, char addr, char data)
{
    SendSlaveID(ID);
    if (IBSR & RXAK)              /* error if EEPROM does not respond */
        return −1;
    IBDR = addr;                  /* send out address of the location to be written */
    while(!(IBSR & IBIF));
    IBSR = IBIF;                  /* clear the IBIF flag */
    if (IBSR & RXAK)              /* error if EEPROM does not respond */
        return −1;
    IBDR = data;                  /* send out the data byte */
```

```c
    while(!(IBSR&IBIF));
    IBSR = IBIF;                        /* clear the IBIF flag */
    if (IBSR & RXAK)                    /* error if EEPROM does not respond */
        return −1;
    IBCR &= ~MSSL;                      /* generate a stop condition */
        return 0;                       /* normal write code */
}
```

Example 11.17

Write a function that performs a *pagewrite* operation. The control byte, the starting address of the destination, and the pointer to the data in RAM to be written are passed to this function.

Solution: The following function writes a block of up to 16 bytes to the EEPROM. The block of data to be output is pointed to by X. The EEPROM ID and starting address to be output are passed in accumulators A and B, respectively. The number of bytes to be output is passed in Y. The error code −1 will be returned if the EEPROM does not respond to its ID.

```
EEpageWrite  jsr    sendSlaveID           ; generate start condition and send out slave ID
             brclr  IBSR,RXAK,pwriteok1   ; does the EEPROM acknowledge?
             ldab   #$FF                  ; return error code −1
             rts
pwriteok1    stab   IBDR                  ; send out the starting address to be written
             brclr  IBSR,IBIF,*           ; wait until the byte is shifted out
             movb   #IBIF,IBSR            ; clear the IBIF flag
             brclr  IBSR,RXAK,w_loop      ; does the EEPROM acknowledge?
             ldab   #$FF                  ; return error code −1
             rts
w_loop       cpy    #0
             beq    done_EEwrite          ; byte count is 0, done
             movb   1,x+,IBDR             ; send out 1 byte
             brclr  IBSR,IBIF,*           ; wait until the byte is shifted out
             movb   #IBIF,IBSR            ; clear the IBIF flag
             brclr  IBSR,RXAK,okNxt       ; receive ACK?
             ldab   #$FF
             rts
okNxt        dey                          ; decrement byte count
             bra    w_loop
done_EEwrite
             bclr   IBCR,MSSL             ; generate a stop condition
             ldab   #0                    ; return error code 0
             rts
```

The C function that performs the pagewrite operation is as follows:

```c
char EEpageWrite(char ID, char addr, char ByteCnt, char *ptr)
{
    SendSlaveID(ID);         /* send out EEPROM ID */
    if (IBSR & RXAK)
        return −1;           /* return −1 if EEPROM did not respond */
```

```
    IBDR = addr;                    /* send out starting address of pagewrite */
    while(!(IBSR & IBIF));          /* wait until the address is shifted out */
    IBSR = IBIF;                    /* clear IBIF flag */
    if (IBSR & RXAK)
        return −1;                  /* return −1 if EEPROM did not respond */
    while(ByteCnt) {
        IBDR = *ptr++;              /* send out 1 byte of data */
        while(!(IBSR & IBIF));
        IBSR = IBIF;
        if (IBSR & RXAK)
            return −1;              /* return −1 if EEPROM did not respond */
        ByteCnt−−;
    }
    IBCR &= ~MSSL;                  /* generate a stop condition */
    return 0;
}
```

In some applications, the program may need to write several blocks of data to different locations. An internal write cycle takes about 5 ms to complete. The designer can either call a delay function to wait for 5 ms or use acknowledge polling to find out whether the internal write has completed.

Example 11.18

Write a C function to implement the algorithm described in Figure 11.38.

Solution: The function **eeAckPoll()** will poll the EEPROM until it is not busy.

```
void eeAckPoll(char ID)
{
    SendSlaveID(ID);
    while(IBSR & RXAK){
        IBCR |= RSTA;               /* generate a restart condition */
        IBDR = ID;                  /* send out EEPROM ID */
        while(!(IBSR & IBIF));
        IBSR = IBIF;                /* clear the IBIF flag */
    };                              /* continue if EEPROM did not acknowledge */
    IBCR &= ~MSSL;                  /* generate a stop condition − indispensable */
}
```

11.9 Summary

The I²C protocol is an alternative to the SPI serial interface protocol. Compared with the SPI protocol, the I²C bus offers the following advantages:

- No chip-enable or chip-select signal for selecting slave devices
- Allows multiple master devices to coexist in a system because it provides easy bus arbitration

- Allows many more devices in the same I²C bus
- Allows resources to be shared by multiple master devices (microcontrollers)

However, the SPI has the following advantages over the I²C interface:

- Higher data rates (no longer true for I²C high-speed mode)
- Much lower software overhead to carry out data transmission

Data transfer over the I²C bus requires the user to generate the following signal components:

1. Start condition
2. Stop condition
3. ACK
4. Restart condition
5. Data

Whenever there are multiple master devices attempting to send data over the I²C bus, bus arbitration is carried out automatically. The loser is decided whenever it attempts to drive the data line to high, whereas another master device drives the same data line to low.

To select the slave device without using the chip-select (or chip-enable) signal, address information is used. Both 7-bit and 10-bit addresses are supported in the same I²C bus. Ten-bit addressing will be used in a system that consists of many slave devices. The I²C bus supports three speed rates:

- 100 kHz
- 400 kHz
- 3.4 MHz

Currently, the HCS12 devices support only 7-bit addressing and a 100-kHz clock rate.

Each data transfer starts with a start condition and ends with the stop condition. One or two control bytes will follow the start condition, which specifies the slave device to receive or send data. For each data byte, the receiver must assert either the ACK or NACK condition to acknowledge or unacknowledge, respectively, the data transfer.

The DS1307 is a real-time clock that can keep track of the current time and calendar information. After the current time and calendar information have been set up, the DS1307 will update it once per clock period. As long as the clock frequency input is accurate, the DS1307 can keep track of the time very accurately. The DS1307 can also optionally interrupt the MCU once per second so that the MCU can update the time display.

The DS1631A is a digital thermometer that allows the ambient temperature to be read using the I²C bus. It has an output signal (T_{OUT}) that will become active whenever the ambient temperature reaches the value preset by the designer. This feature is useful to provide early warning about potential overheating in a system. The DS1631A can be configured to operate in one-shot mode or continuous conversion mode. The default temperature conversion resolution is 12-bit. However, it can also be reduced to 9-, or 10-, or 11-bit if high resolution is not important compared to the required conversion time.

The 24LC08B is an EEPROM with an I²C interface. The capacity of the 24LC08B is 8K bits. This chip has an internal address pointer that will increment automatically after each access. This feature can increase the access efficiency when the access patterns are sequential.

Functions for performing basic read and write access to the I²C slave devices have been provided in this chapter. Designers can add their own processing functions to provide further processing or data formatting to make the data more user-friendly.

11.10 Exercises

E11.1 Does the I²C clock frequency need to be exactly equal to 100 or 400 kHz? Why?

E11.2 Suppose that the 7-bit address of an I²C slave is B'10101 A1 A0' with A1 tied to low and A0 pulled to high. What is the 8-bit hex write address for this device? What is the 8-bit hex read address?

E11.3 Assuming that the HCS12 is running with a 16-MHz E-clock, compute the values to be written into the IBFD register to set the baud rate to 100 and 200 kHz. Use Table 11.6.

E11.4 Assuming that the HCS12 is running with a 32-MHz E-clock, compute the values to be written into the IBFD register to set the baud rate to 200 and 400 kHz. Use Table 11.6.

E11.5 Write a function to set the TL value to the DS1631A in the assembly and C languages.

E11.6 Write a instruction sequence to call the openDS1631 function to configure the DS1631A to operate in one-shot mode, 10-bit resolution, and low active polarity for T_{OUT} output.

E11.7 Assuming that the DS1631A has been configured to operate in one-shot mode with 12-bit resolution, write a program to display the converted temperature in the format of three integer digits and one fractional digit in LCD.

E11.8 For the circuit shown in Figure 11.34, add an alarm speaker to the PWM1 output and set the high-temperature trip point to 50°C. Whenever temperature reaches 50°C or higher, turn on the alarm until the temperature drops down to 23°C.

E11.9 Write an assembly routine to implement the algorithm described in Figure 11.39.

E11.10 The MAX5812 is a DAC with 12-bit resolution and I²C interface. The MAX5821 data sheet can be downloaded from the website http://www.maxim-ic.com. What is the 7-bit address of this device? What is the highest operating frequency of this chip? How many commands are available to this chip? How is this device connected to the I²C bus?

E11.11 For the MAX5812 mentioned in E11.10, write a program to generate a sine wave from the OUT pin.

E11.12 The MCP23016 is an I/O expander from Microchip. This chip has an I²C interface and can add 16 I/O pins to the microcontroller. Download the data sheet of this device from Microchip's website. Show the circuit connection of this chip to the microcontroller, and write a sequence of instructions to configure the GP1.0, . . . , GP1.7 pins for input and configure the GP0.0, . . . , GP0.7 pins for output. Configure input polarity to active low.

E11.13 What are the corresponding temperatures for the conversion results 0x6800, 0x7200, 0x4800, 0xEE60, and 0xF280 output by the DS1631A? Assume that the 12-bit resolution is used.

E11.14 What will be the conversion results sent out by the DS1631A for the temperature values 40°C, 50°C, 80.5°C, −10.25°C, and −20.5°C? Assume that the 12-bit resolution is used.

E11.15 Write a function that performs a block read from the 24LC08B. The control byte, the starting address to be read, and the number of bytes to be read are passed in A, B, and Y, respectively. Return the error code in B and store the returned data in a buffer pointed to by X. This function should check the error of no acknowledgement.

E11.16 Write a function that performs a current address read to the 24LC08B. Pass the control byte in A to this subroutine. Return the data byte and error code in A and B, respectively. This function should check the error that the 24LC08B did not acknowledge.

11.11 Lab Exercises and Assignments

L11.1 Write a program to store 0 to 255 in the first block of the 24LC08B of your demo board and then stay in an infinite loop to read out one value from 24LC08B sequentially every half-second and display the value on eight LEDs of the demo board.

L11.2 Connect the DS1631A to the I²C bus of your HCS12 demo board. Write a program to set up the high- and low-temperature trip points to 40 and 20°C, respectively. Whenever the temperature goes above 40°C, turn on the alarm (speaker). Turn off the alarm when the temperature drops below 20°C. Display the temperature on the LCD display, and update the display once every second.

L11.3 Use the DS1307 (if any) on your demo board to set up a time-of-day display using the LCD. Enter the current time and calendar information using either the DIP switches or the keyboard. Update the current time-of-day display once a second. In addition, add alarm time to your program. Whenever the current time matches the alarm, generate the alarm for 1 min using the buzzer on the demo board.

12

Analog-to-Digital Converter

12.1 Objectives

After completing this chapter, you should be able to

- Explain the A/D conversion process

- Describe the resolution, the various channels, and the operation modes of the HCS12 A/D converter

- Interpret the A/D conversion results

- Describe the procedure for using the HCS12 A/D converter

- Configure the A/D converter for the application

- Use the temperature sensor TC1047A

- Use the humidity sensor IH-3605

- Use the barometric pressure sensor ASCX30AN from SenSym

12.2 Basics of A/D Conversion

Many embedded applications deal with nonelectric quantities, such as weight, humidity, pressure, massflow, airflow, temperature, light intensity, and speed. These quantities are *analog* in nature because they have a continuous set of values over a given range, in contrast to the discrete values of digital signals. To enable the microcontroller to process these quantities, they need to be represented in digital form; thus an *analog-to-digital* converter is required.

12.2.1 A Data Acquisition System

An A/D converter can deal only with electric voltage. A nonelectric quantity must be converted into a voltage before A/D conversion can be performed. The conversion of a nonelectric quantity to a voltage requires the use of a *transducer*. In general, a transducer is a device that converts the quantity from one form to another. For example, a temperature sensor is a transducer that can convert the temperature into a voltage. A *load cell* is the transducer that can convert a weight into a voltage.

A transducer may not generate an output voltage in the range suitable for A/D conversion. A voltage *scaler* (or *amplifier*) is often needed to amplify the transducer output voltage into a range that can be handled by the A/D converter. The circuit that performs the scaling and shifting of the transducer output is called a *signal-conditioning circuit*. The overall A/D process is illustrated in Figure 12.1.

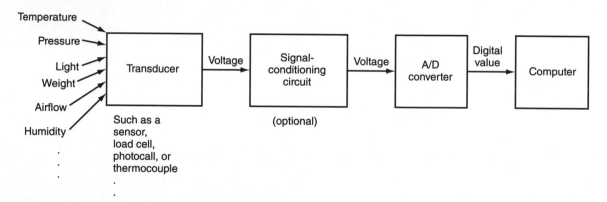

Figure 12.1 ■ The A/D conversion process

12.2.2 Analog Voltage and Digital Code Characteristic

An ideal A/D converter should demonstrate the linear input/output relationship shown in Figure 12.2. However, the output characteristic shown in Figure 12.2 is unrealistic because it requires the A/D converter to use an infinite number of bits to represent the conversion result. The output characteristic of an ideal A/D converter using n bits to represent the conversion result is shown in Figure 12.3. An n-bit A/D converter has 2^n possible output code values. The area between the dotted line and the staircase is called the *quantization error*. The value of $V_{DD}/2^n$ is the resolution of this A/D converter. Using n bits to represent the conversion result, the average *conversion error* is $V_{DD}/2^{n+1}$ if the converter is perfectly linear. For a real A/D converter, the output characteristic may have *nonlinearity*

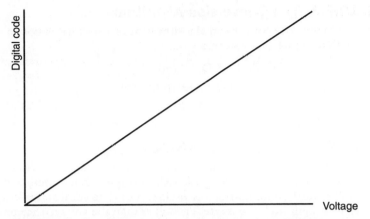

Figure 12.2 ■ An ideal A/D converter output characteristic

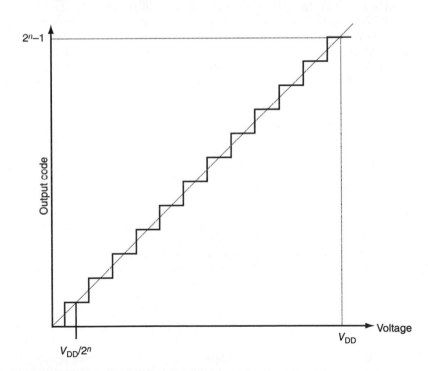

Figure 12.3 ■ Output characteristic of an ideal n-bit A/D converter

(the staircase may have unequal steps in some values) and *nonmonotonicity* (higher voltage may have smaller code).

Obviously, the more bits used in representing the A/D conversion result, the smaller the conversion error will be. Most microcontrollers use 8, 10, or 12 bits to represent the conversion result. Some microcontrollers (mainly 8051 variants from Silicon Laboratory, TI, and Analog Devices) use 16 bits or even 24 bits to represent conversion results. Whenever the on-chip A/D converter cannot provide the required accuracy, an external A/D converter should be considered.

12.2.3 A/D Conversion Algorithms

Many A/D conversion algorithms have been introduced in the past. These algorithms can be divided into four categories.

1. Parallel (Flash) A/D converter

2. Slope and double-slope A/D converter

3. Sigma-delta A/D converter

4. Successive-approximation A/D converter

PARALLEL (FLASH) A/D CONVERTERS

In this type of A/D converter, 2^n comparators are used. One of the inputs to each comparator is the input voltage to be converted; the other input corresponds to the voltage that represents one of the 2^n combinations of n-bit values. The comparator output will be high whenever the analog input (to be converted) is higher than the voltage that represents one of the 2^n combinations of the n-bit value. The largest n-bit value that causes the comparator output to become true is selected as the A/D conversion value through a priority encoder. It is obvious that this type of A/D converter will be very fast. However, they require a lot of hardware resources to implement and therefore are not suitable for implementing high-resolution A/D converters. This type of A/D converter is often used in applications that require high-speed but low resolution, such as a video signal. Over the years, several variations to this approach have been proposed to produce high-speed A/D converters. The most commonly used technique is to pipeline a flash A/D converter, which will reduce the amount of hardware required while still achieving high conversion speed.

SLOPE AND DOUBLE-SLOPE A/D CONVERTERS

This type of A/D converter is used in Microchip PIC14000 microcontrollers in which the charging and discharging of a capacitor is used to perform A/D conversion. It requires relatively simple hardware and is popular in low-speed applications, such as digital multimeters. In addition, high resolution (10- to 16-bit) can be achieved.

SIGMA-DELTA A/D CONVERTERS

This type of A/D converter uses the *oversampling* technique to perform A/D conversion. It has good noise immunity and can achieve high resolution. Sigma-delta A/D converters are becoming more and more popular in implementing high-resolution A/D converters. The only disadvantage is the slow conversion speed. However, this weakness is improving because of advancements in CMOS technology.

SUCCESSIVE-APPROXIMATION A/D CONVERTERS

The successive-approximation method approximates the analog signal to n-bit code in n steps. It may be used for low-frequency applications with large DC noise, such as an electrocardiograph. The block diagram of this method is shown in Figure 12.4.

It first initializes the successive-approximation register (SAR) to 0 and then performs a series of guessing, starting with the most-significant bit and proceeding toward the least-significant bit. The algorithm of the successive-approximation method is illustrated in Figure 12.5. For every bit of the SAR, the algorithm does the following operations:

- Guesses the bit to be a 1
- Converts the value of the SAR register to an analog voltage
- Compares the D/A output with the analog input and clears the bit to 0 if the D/A output is larger (which indicates that the guess is wrong)

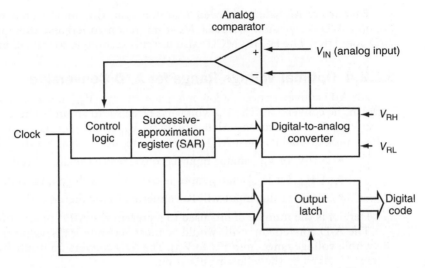

Figure 12.4 ■ Block diagram of a successive-approximation A/D converter

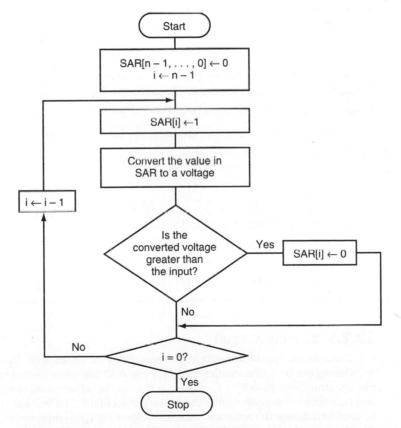

Figure 12.5 ■ Successive-approximation A/D conversion method

Because of its balanced speed and precision, this method has become one of the most popular A/D conversion methods. Most microcontrollers use this method to implement the A/D converter. The HCS12 MCUs also use this technique to implement A/D converters.

12.2.4 Optimal Voltage Range for A/D Conversion

An A/D converter needs a *low reference voltage* (V_{RL}) and a *high reference voltage* (V_{RH}) to perform the conversion. The V_{RL} voltage is often set to ground, whereas the V_{RH} voltage is often set to V_{DD}. Some microcontrollers simply tie V_{RL} to the ground voltage and leave only the V_{RH} voltage programmable. Most A/D converters are *ratiometric*, for the following reasons:

- A 0-V (or V_{RL}) analog input is converted to the digital code of n 0s.
- A V_{DD} (or V_{RH}) analog input is converted to the digital code of $2^n - 1$.
- A k-V analog input will be converted to the digital code of $k \times (2^n - 1) \div V_{DD}$.

Here, n is the number of bits used to represent the A/D conversion result.

The A/D conversion result would be most accurate if the value of the analog signal covers the whole voltage range from V_{RL} to V_{RH}. The A/D conversion result k corresponds to an analog voltage V_k given by the following equation:

$$V_k = V_{RL} + (\text{range} \times k) \div (2^n - 1) \tag{12.1}$$

where range $= V_{RH} - V_{RL}$.

Example 12.1

Suppose that there is a 10-bit A/D converter with $V_{RL} = 1$ V and $V_{RH} = 4$ V. Find the corresponding voltage values for the A/D conversion results of 25, 80, 240, 500, 720, 800, and 900.

Solution: Range $= V_{RH} - V_{RL} = 4$ V $- 1$ V $= 3$ V

The voltages corresponding to the A/D conversion results of 25, 80, 240, 500, 720, 800, and 900 are

$$1V + (3 \times 25) \div (2^{10} - 1) \ = 1.07 \text{ V}$$
$$1V + (3 \times 80) \div (2^{10} - 1) \ = 1.23 \text{ V}$$
$$1V + (3 \times 240) \div (2^{10} - 1) = 1.70 \text{ V}$$
$$1V + (3 \times 500) \div (2^{10} - 1) = 2.47 \text{ V}$$
$$1V + (3 \times 720) \div (2^{10} - 1) = 3.11 \text{ V}$$
$$1V + (3 \times 800) \div (2^{10} - 1) = 3.35 \text{ V}$$
$$1V + (3 \times 900) \div (2^{10} - 1) = 3.64 \text{ V}$$

12.2.5 Scaling Circuit

Some of the transducer output voltages are in the range of $0 \sim V_Z$, where $V_Z < V_{DD}$. Because V_Z sometimes can be much smaller than V_{DD}, the A/D converter cannot take advantage of the available full dynamic range (0 to $2^n - 1$, where n is the number of bits used to represent the conversion result), and therefore conversion results can be very inaccurate. The voltage scaling (amplifying) circuit can be used to improve the accuracy because it allows the A/D converter to utilize its full dynamic range. The diagram of a voltage scaling circuit is shown in Figure 12.6. Because the OP AMP has an infinite input impedance, the current that flows through the resistor R_2 will be the same as the current that flows through R_1. In addition, the voltage at the inverting input terminal (same as the voltage drop

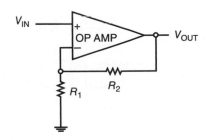

Figure 12.6 ■ A voltage scaler

across R_1) would be the same as that at the noninverting terminal (V_{IN}). Therefore, the voltage gain of this circuit is given by the following equation:

$$A_V = V_{OUT} \div V_{IN} = (R_1 + R_2) \div R_1 = 1 + R_2/R_1 \tag{12.2}$$

Example 12.2

Suppose the transducer output voltage ranges from 0 V to 200 mV. Choose the appropriate values for R_1 and R_2 to scale this range to 0~5 V.

Solution: 5 V ÷ 200 mV = 25; therefore R_2/R_1 = 24.

By choosing 240 kΩ for R_2 and 10 kΩ for R_1, we obtain a R_2/R_1 ratio of 24 and achieve the desired scaling goal.

12.2.6 Voltage Translation Circuit

Some transducers have output voltage in the range of $V_1 \sim V_2$ (V_1 can be negative and V_2 can be unequal to V_{DD}) instead of 0 V~V_{DD}. The accuracy of A/D conversion can be improved by using a circuit that shifts and scales the transducer output so that it falls in the full range of 0 V~V_{DD}.

An OP AMP circuit that can shift and scale the transducer output is shown in Figure 12.7c. This circuit consists of a summing circuit (Figure 12.7a) and an inverting circuit (Figure 12.7b). The voltage V_{IN} comes from the transducer output; V_1 is an adjusting voltage. By choosing appropriate values for V_1 and resistors R_0, R_1, R_2, and R_f, the desired voltage shifting and scaling can be achieved. Equation 12.5 shows that the resistance R_0 is an independent variable and can be set to a convenient value.

Example 12.3

Choose appropriate resistor values and the adjusting voltage so that the circuit shown in Figure 12.7c can shift the voltage from the range of −1.2 V~3.0 V to the range of 0 V~5 V.

Solution: Applying Equation 12.5,

$$0 = -1.2 \times (R_f/R_1) - (R_f/R_2) \times V_1$$
$$5 = 3.0 \times (R_f/R_1) - (R_f/R_2) \times V_1$$

By choosing $R_0 = R_1 = 10$ kΩ, $R_2 = 50$ kΩ, $R_f = 12$ kΩ, and $V_1 = -5$ V, we can translate and scale the voltage to the desired range. This example tells us that the selection of resistors and the voltage V_1 is a trial-and-error process at best.

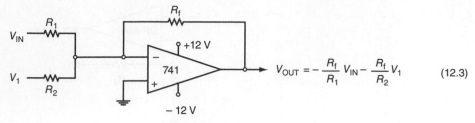

$$V_{OUT} = -\frac{R_f}{R_1} V_{IN} - \frac{R_f}{R_2} V_1 \qquad (12.3)$$

(a) Summing circuit

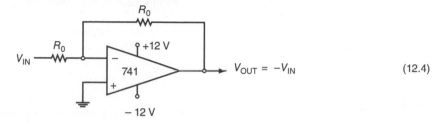

$$V_{OUT} = -V_{IN} \qquad (12.4)$$

(b) Inverting voltage follower

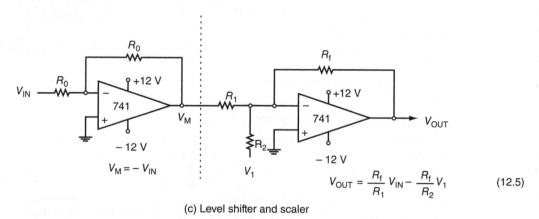

$$V_M = -V_{IN}$$

$$V_{OUT} = \frac{R_f}{R_1} V_{IN} - \frac{R_f}{R_2} V_1 \qquad (12.5)$$

(c) Level shifter and scaler

Figure 12.7 ■ Level shifting and scaling circuit

12.3 The HCS12 A/D Converter

The block diagram of the HCS12 A/D converter (ADC) is shown in Figure 12.8. An HCS12 microcontroller may have one or two 8-channel, 10-bit A/D converters. With a 2-MHz conversion clock, the ADC can perform an 8-bit single conversion in 6 μs or a 10-bit single conversion in 7 μs. An A/D conversion sequence can be started by writing into a register or by a valid signal on an external trigger input. The ADC uses the successive-approximation method to perform the conversion.

The HCS12 uses two 8-bit registers to hold a conversion result. The result can be stored either right- or left-justified. The A/D conversion is performed in a sequence from one to eight

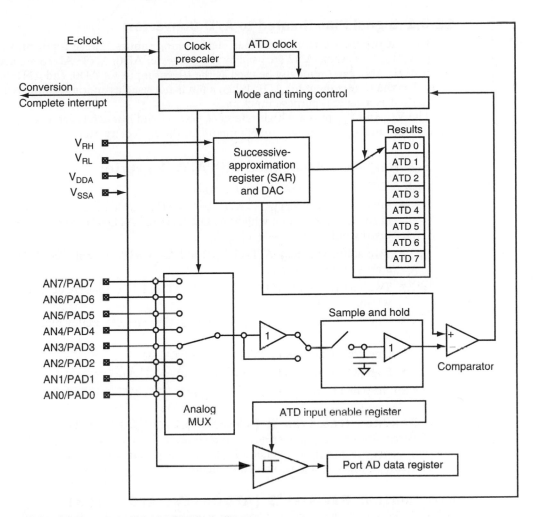

Figure 12.8 ■ The HCS12 ATD block diagram

samples. When a sequence consists of multiple samples, the samples may be taken from one channel or multiple channels. An interrupt may be generated on the completion of an A/D conversion.

The analog multiplexer in Figure 12.8 selects one of the three internal or one of the external signal sources for conversion. The sample selected by the multiplexer is amplified before charging the sample capacitor. The comparator indicates whether each successive output of the DAC is higher or lower than the sampled input. The control registers and associated logic select the conversion resolution, multiplexer input, and conversion sequencing mode, sample time, and ADC clock cycle. Each result is available in three formats: right-justified unsigned, left-justified signed, and left-justified unsigned.

12.3.1 Signal Pins Related to A/D Converter

Each of the two A/D modules has eight analog inputs. These input pins are labeled as AN0~AN15. Pins AN0~AN7 are used by the converter AD0; AN8~AN15 are used by the converter AD1. AN7 can optionally be used as the trigger input for AD0, and AN15 can optionally be used as the trigger input for AD1. When a pin is not used as an input to the A/D converter, it can be used as a general-purpose input pin.

The V_{RH} and V_{RL} pins are high-reference voltage and low-reference voltage inputs, respectively. V_{DDA} and V_{SSA} are power supply inputs for the A/D converters.

12.3.2 Registers Associated with A/D Converter

Each A/D converter has the following registers:

- Six control registers (ATDxCTL0~ATDxCTL5) that control the overall operation of the module, two of which (ATDxCTL0 and ATDxCTL1) are intended for factory testing only
- Two status registers (ATDxSTAT0 and ATDxSTAT1) that record the operation status of the converter
- Two test registers (ATDxTEST0 and ATDxTEST1) that are dedicated to factory testing purposes
- One input enable register (ATDxDIEN) that enables analog pins to be used as digital input
- One port data register (PORTADx or PTADx)
- Eight 16-bit result registers (ATDxDR0~ATDxDR7)

where x can be 0 or 1.

ATD CONTROL REGISTER 2 (ATD0CTL2, ATD1CTL2)

This register controls power-down, interrupt, and the external trigger. Writes to this register will abort the current conversion sequence but will not start a new conversion sequence. The contents of this register are shown in Figure 12.9.

ATD CONTROL REGISTER 3 (ATD0CTL3, ATD1CTL3)

This register sets the conversion sequence length, enables/disables the FIFO mode for results registers, and controls the ATD behavior in freeze mode. Writes to this register will abort the current conversion but will not start a new conversion. The contents of this register are shown in Figure 12.10.

If the FIFO bit is 0, the A/D conversion results map into result registers on the basis of the conversion sequence; the result of the first conversion appears in the first result register, the second result in the second result register, and so on. If the FIFO bit is 1, the result of a conversion will be stored in the result register specified by the conversion counter. The conversion counter does not reset at the start of a new conversion sequence. The conversion counter value is stored in the lowest 3 bits of the ATD status register 0.

ATD CONTROL REGISTER 4 (ATD0CTL4, ATD1CTL4)

This register sets the conversion clock frequency, the length of the second phase of the sample time, and the resolution of the A/D conversion. Writes to this register will abort the current conversion sequence but will not start a new sequence. The contents of this register are shown in Figure 12.11.

7	6	5	4	3	2	1	0
ADPU	AFFC	AWAI	ETRIGLE	ETRIGP	ETRIGE	ASCIE	ASCIF

Reset: 0   0   0   0   0   0   0   0

ADPU: ATD power-down bit
   0 = power-down ATD.
   1 = normal ATD operation.
AFFC: ATD fast flag clear all bit
   0 = ATD flag is cleared normally, i.e., read the status register before reading the result
      register.
   1 = any access to a result register will cause the associated CCF flag to clear
      automatically if it is set at the time.
AWAI: ATD power-down in wait-mode bit
   0 = ATD continues to run when the HCS12 is in wait mode.
   1 = halt conversion and power-down ATD during wait mode.
ETRIGLE: external trigger level/edge control
   This bit controls the sensitivity of the external trigger signal. Details are shown in Table
   12.1.
ETRIGP: external trigger polarity
   This bit controls the polarity of the external trigger signal. See Table 12.1 for details.
ETRIGE: external trigger mode enable
   0 = disable external trigger on ATD channel 7.
   1 = enable external trigger on ATD channel 7.
ASCIE: ATD sequence complete interrupt enable bit
   0 = disables ATD interrupt.
   1 = enables ATD interrupt on sequence complete (ASCIF = 1).
ASCIF: ATD sequence complete interrupt flag
   0 = no ATD interrupt occurred.
   1 = ATD sequence complete, interrupt pending.

Figure 12.9 ■ ATD control register 2 (ATDxCTL2, x = 0 or 1)

ETRIGLE	ETRIGP	External Trigger Sensitivity
0	0	Falling edge
0	1	Rising edge
1	0	Low level
1	1	High level

Table 12.1 ■ External trigger configurations

There are two stages in the analog signal sample time. The first sample stage is fixed at two conversion clock periods. The second stage is selected by SMP1 and SMP0 as shown in Table 12.2.

ATD CONTROL REGISTER 5 (ATD0CTL5, ATD1CTL5)

This register selects the type of conversion sequence and the analog input channels sampled. Writes to this register will abort the current conversion sequence and start a new conversion sequence. The contents of this register are shown in Figure 12.12.

7	6	5	4	3	2	1	0
0	S8C	S4C	S2C	S1C	FIFO	FRZ1	FRZ0

Reset: 0   0   0   0   0   0   0   0

S8C,S4C,S2C,S1C: conversion sequence limit
  0000 = 8 conversions.
  0001 = 1 conversion.
  0010 = 2 conversions.
  0011 = 3 conversions.
  0100 = 4 conversions.
  0101 = 5 conversions.
  0110 = 6 conversions.
  0111 = 7 conversions.
  1xxx = 8 conversions.
FIFO: result register FIFO mode
  0 = conversion results are placed in the corresponding result
    register up to the selected sequence length.
  1 = conversion results are placed in consecutive result registers
    (wrap around at end).
FRZ1 and FRZ0: background debug (freeze) enable bit
  00: continue conversions in active background mode.
  01: reserved.
  10: finish current conversion, then freeze.
  11: freeze immediately when background mode is active.

Figure 12.10 ■ ATD control register 3 (ATDxCTL3, x = 0 or 1)

7	6	5	4	3	2	1	0
SRES8	SMP1	SMP0	PRS4	PRS3	PRS2	PRS1	PRS0

Reset: 0   0   0   0   0   1   0   1

SRES8: ATD resolution select bit
  0 = 10-bit operation.
  1 = 8-bit operation.
SMP1 and SMP0: select sample time bits
  These bits are used to select the length of the second phase of the
  sample time in units of ATD conversion clock cycles. See Table
  12.2.
PRS4–PRS0: ATD clock prescaler bits
  These 5 bits are the binary value prescaler value PRS. The ATD
  conversion clock frequency is calculated as follows:

$$\text{ATDclock} = \frac{\text{E-clock}}{\text{PRS} + 1} \times 0.5$$

The ATD conversion frequency must be between 500 kHz and 2
MHz. The clock prescaler values are shown in Table 12.3.

Figure 12.11 ■ ATD control register 4 (ATDxCTL4, x = 0 or 1)

SMP1	SMP0	Length of 2nd Phase of Sample Time
0	0	2 A/D conversion clock periods
0	1	4 A/D conversion clock periods
1	0	8 A/D conversion clock periods
1	1	16 A/D conversion clock periods

Table 12.2 ■ Sample time select

Prescale Value	Total Divisor Value	Max. Bus Clock[1]	Min. Bus Clock[2]
00000	divide by 2	4 MHz	1 MHz
00001	divide by 4	8 MHz	2 MHz
00010	divide by 6	12 MHz	3 MHz
00011	divide by 8	16 MHz	4 MHz
00100	divide by 10	20 MHz	5 MHz
00101	divide by 12	24 MHz	6 MHz
00110	divide by 14	28 MHz	7 MHz
00111	divide by 16	32 MHz	8 MHz
01000	divide by 18	36 MHz	9 MHz
01001	divide by 20	40 MHz	10 MHz
01010	divide by 22	44 MHz	11 MHz
01011	divide by 24	48 MHz	12 MHz
01100	divide by 26	52 MHz	13 MHz
01101	divide by 28	56 MHz	14 MHz
01110	divide by 30	60 MHz	15 MHz
01111	divide by 32	64 MHz	16 MHz
10000	divide by 34	68 MHz	17 MHz
10001	divide by 36	72 MHz	18 MHz
10010	divide by 38	76 MHz	19 MHz
10011	divide by 40	80 MHz	20 MHz
10100	divide by 42	84 MHz	21 MHz
10101	divide by 44	88 MHz	22 MHz
10110	divide by 46	92 MHz	23 MHz
10111	divide by 48	96 MHz	24 MHz
11000	divide by 50	100 MHz	25 MHz
11001	divide by 52	104 MHz	26 MHz
11010	divide by 54	108 MHz	27 MHz
11011	divide by 56	112 MHz	28 MHz
11100	divide by 58	116 MHz	29 MHz
11101	divide by 60	120 MHz	30 MHz
11110	divide by 62	124 MHz	31 MHz
11111	divide by 64	128 MHz	32 MHz

[1] Maximum ATD conversion clock frequency is 2 MHz. The maximum allowed bus clock frequency is shown in this column.

[2] Minimum ATD conversion clock frequency is 500 kHz. The minimum allowed bus clock frequency is shown in this column.

Table 12.3 ■ Clock prescaler values

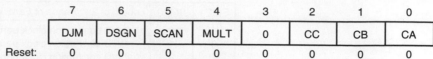

7	6	5	4	3	2	1	0
DJM	DSGN	SCAN	MULT	0	CC	CB	CA

Reset: 0   0   0   0   0   0   0   0

DJM: result register data justification
  0 = left-justified data in the result registers.
  1 = right-justified data in the result registers.
DSGN: result register data signed or unsigned representation
  0 = unsigned data representation in the result registers.
  1 = signed data representation in the result registers (not available
      in right justification).
SCAN: enable continuous channel scan bit
  0 = single conversion sequence.
  1 = continuous conversion sequences (scan mode).
MULT: enable multichannel conversion bit
  0 = sample only one channel.
  1 = sample across several channels.
CC, CB, and CA: channel select code
  The channel selection is shown in Table 12.4.

Figure 12.12 ■ ATD control register 5 (ATDxCTL5, x = 0 or 1)

CC	CB	CA	Analog Input Channel
0	0	0	AN0
0	0	1	AN1
0	1	0	AN2
0	1	1	AN3
1	0	0	AN4
1	0	1	AN5
1	1	0	AN6
1	1	1	AN7

Table 12.4 ■ Analog input channel select code

SRES8	DJM	DSGN	Result Data Formats Description and Bus Bit Mapping
1	0	0	8-bit/left-justified/unsigned—bits 8–15
1	0	1	8-bit/left-justified/signed—bits 8–15
1	1	x	8-bit/right-justified/unsigned—bits 0–7
0	0	0	10-bit/left-justified/unsigned—bits 6–15
0	0	1	10-bit/left-justified/signed—bits 6–15
0	1	x	10-bit/right-justified/unsigned—bits 0–9

Table 12.5 ■ Available result data formats

Table 12.5 summarizes the result data formats available and how they are set up using the control bits. Table 12.6 illustrates the difference between the signed and unsigned, left-justified and right-justified output codes for an input signal range between 0 and 5.12 V.

When single-channel mode is selected, the analog channel to be converted is specified by the channel selection code CC~CA. When multichannel mode is selected, the first

Input Signal $V_{RL} = 0\ V$ $V_{RH} = 5.12\ V$	Signed 8-bit Codes	Unsigned 8-bit Codes	Signed 10-bit Codes	Unsigned 10-bit Codes
5.120 V	7F	FF	7FC0	FFC0
5.100 V	7F	FF	7F00	FF00
5.080 V	7F	FE	7E00	FE00
2.580 V	01	81	0100	8100
2.560 V	00	80	0000	8000
2.540 V	FF	7F	FF00	7F00
0.020 V	81	01	8100	0100
0.000 V	80	00	8000	0000

Table 12.6 ■ Left-justified, signed and unsigned ATD output codes

analog channel examined is specified by the channel selection code (CC, CB, and CA); subsequent channels sampled in the sequence are determined by incrementing the channel selection code.

ATD Status Register 0 (ATD0STAT0, ATD1STAT0)

This read-only register contains the sequence complete flag, overrun flags for external trigger and FIFO mode, and the conversion counter. The contents of this register are shown in Figure 12.13.

If the AFFC flag is set to 1, the SCF flag can be cleared by reading a result register. Otherwise, it can be cleared by one of the following actions:

- Writing a 1 to SCF
- Writing to the ATDxCTL5 register (a new conversion is started)

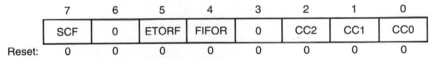

	7	6	5	4	3	2	1	0
	SCF	0	ETORF	FIFOR	0	CC2	CC1	CC0
Reset:	0	0	0	0	0	0	0	0

SCF: sequence complete flag
  0 = conversion sequence not completed.
  1 = conversion sequence has completed.
ETORF: external trigger overrun flag
  0 = no external trigger overrun has occurred.
  1 = external trigger overrun has occurred.
FIFOR: FIFO overrun flag
  0 = no overrun has occurred.
  1 = an overrun has occurred.
CC2, CC1, CC0: conversion counter
  The conversion counter points to the result register that will receive the result of the current conversion.
  In non-FIFO mode, this counter is reset to 0 at the begin and end of the conversion.
  In FIFO mode, this counter is not reset and will wrap around when its maximum value is reached.

Figure 12.13 ■ ATD status register 1 (ATDxSTAT0, $x = 0$ or 1)

While in edge trigger mode, if additional active edges are detected while a conversion sequence is in progress the ETORF flag is set. This flag is cleared when one of the following events occurs:

- Writing a 1 to ETORF
- Writing to ATDxCTL2, ATDxCTL3, or ATDxCTL4 (a conversion sequence is aborted)
- Writing to ATDxCTL5 (a new conversion sequence is started)

The setting of the FIFOR bit indicates that a result register has been written to before its associated conversion complete flag (CCF) has been cleared. This flag is cleared when one of the following events occurs:

- Writing a 1 to FIFOR
- Writing to ATDxCTL5

ATD TEST REGISTER 1 (ATD0TEST1, ATD1TEST1)

This register contains the SC bit used to enable special channel conversions. The contents of this register are shown in Figure 12.14.

	7	6	5	4	3	2	1	0
	0	0	0	0	0	0	0	SC
Reset:	0	0	0	0	0	0	0	0

SC: special channel conversion bit
If this bit is set, the special channel conversion can be selected using CC, CB, and CA of the ATDxCTL5 register. Table 12.7 shows the selection.
0 = special channel conversions disabled.
1 = special channel conversions enabled.

Figure 12.14 ■ ATD test register 1 (ATDxTEST1, $x = 0$ or 1)

SC	CC	CB	CA	Analog Input Channel
1	0	x	x	Reserved
1	1	0	0	V_{RH}
1	1	0	1	V_{RL}
1	1	1	0	$(V_{RH} + V_{RL})/2$
1	1	1	1	Reserved

Table 12.7 ■ Special channel select code

ATD STATUS REGISTER 1 (ATD0STAT1, ATD1STAT1)

This read-only register contains the conversion complete flags for all channels. The contents of this register are shown in Figure 12.15.

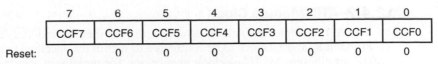

7	6	5	4	3	2	1	0
CCF7	CCF6	CCF5	CCF4	CCF3	CCF2	CCF1	CCF0

Reset: 0 0 0 0 0 0 0 0

CCFx: conversion complete flag x ($x = 7{\sim}0$)
   0 = conversion number x not completed.
   1 = conversion number x has completed, result in ATDyDRx ($y = 0$ or 1).

Figure 12.15 ■ ATD status register 1 (ATDxSTAT1, $x = 0$ or 1)

A CCFx flag can be cleared by one of the following events:
- Writing to ATDxCTL5 (a new conversion is started)
- If AFFC = 0 and read of ATDxSTAT1 followed by read of result register ATDxDRy
- If AFFC = 1 and read of result register ATDxDRy

ATD Input Enable Register (ATD0DIEN, ATD1DIEN)

This register allows the user to enable a Port ADx pin as a digital input. The contents of this register are shown in Figure 12.16.

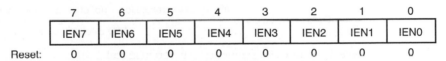

7	6	5	4	3	2	1	0
IEN7	IEN6	IEN5	IEN4	IEN3	IEN2	IEN1	IEN0

Reset: 0 0 0 0 0 0 0 0

IENx: ATD digital input enable on channel x ($x = 0{\sim}7$)
   0 = disable digital input buffer to PTADx.
   1 = enable digital input buffer to PTADx.

Figure 12.16 ■ ATD input enable register (ATDxDIEN, $x = 0$ or 1)

Port Data Register (PTAD0, PTAD1)

The data port associated with the ATD is input-only. The port pins are shared with the analog inputs AN0~AN7 or AN8~AN15.

ATD Conversion Result Registers (ATDxDRy, $x = 0{\sim}1$, $y = 0{\sim}7$)

Each result register is 16-bit and can be further divided into two 8-bit registers, ATDxDRHy and ATDxDRLy. Depending on the setting, the A/D conversion result can be stored right-justified or left-justified. The conversion result can be signed or unsigned. However, the conversion result is unsigned when it is stored right-justified.

12.4 The Functioning of the ATD Module

The operation of the ATD module is described in detail in this section.

12.4.1 Analog Input Multiplexer

The analog input multiplexer selects one of the external analog inputs to convert. The input analog signals are unipolar and must fall within the potential range of V_{SSA} and V_{DDA}. The input multiplexer includes protection circuitry to prevent crosstalk between channels when the applied input potentials are within specification.

12.4.2 ATD Module Clock

The ATD module needs clock signals to perform the conversion. A clock prescaler allows the ATD module's internal clock to be configured to within the specified frequency range (500 kHz to 2 MHz) for different MCU bus clock frequencies. For maximum accuracy, selecting the 500-kHz clock is recommended where conversion and sample times are not critical. The bus clock is divided by a programmable prescaler value (2, 4, 6, . . . , 64) to generate the ATD clock.

12.4.3 Sample-and-Hold Stage

As shown in Figure 12.8, a sample-and-hold stage accepts analog signals from the input multiplexer and stores them as a charge on the sample capacitor. The sampling process has two stages:

1. Initially, a sample amplifier (of unity gain) is used to buffer the input analog signal for two cycles to charge the sample capacitor almost to the input potential. This stage reduces charging and discharging the sample capacitor by the signal source.

2. The sample buffer is then disconnected and the input signal is directly connected to the storage node for programmable 2, 4, 8, or 16 cycles.

The conversion time of a sample is given by the following equation:

$$\text{Conversion time} = \frac{(\text{no. of bits in resolution} + \text{no. of programmed sample clocks} + 2)}{\text{ATD clock frequency}}$$

Conversion times for frequencies at 500 kHz and 2 MHz are shown in Table 12.8. For maximum accuracy, selecting 16 programmable sample cycles is recommended whenever the sample time is not critical. For fast-changing signals, sample time and source impedance should be kept low. The actual conversion time is the sum of the converter time and the sample time in one of the rightmost four columns.

ATD Clock Frequency	Resolution	Converter Time	2 + 2 Sample Clocks	2 + 4 Sample Clocks	2 + 8 Sample Clocks	2 + 16 Sample Clocks
2 MHz 2 MHz	8-bit[1] 10-bit[2]	4 µs 5 µs	2 µs	3 µs	5 µs	9 µs
500 kHz 500 kHz	8-bit 10-bit	16 µs 20 µs	8 µs	12 µs	20 µs	36 µs

[1] The fastest 8-bit resolution conversion time is 4 µs + 2 µs = 6 µs.
[2] The fastest 10-bit resolution conversion time is 5 µs + 2 µs = 7 µs.

Table 12.8 ■ ATD conversion timings

12.4.4 Input Channel Wraparound

In the case of a multiple-channel conversion sequence (MULT bit of the ATDxCTL5 register = 1), when the input selector goes past channel AN7, it wraps around to the channel AN0.

Example 12.4

Assuming that S8C~S1C (ATD0CTL3) are set to 0101 and CC~CA (ATD0CTL5) are set to 110, what is the conversion sequence for this setting?

Solution: The first channel to be converted is AN6 and there are five channels to be converted. The conversion sequence is as follows:

AN6, AN7, AN0, AN1, and AN2

12.4.5 FIFO Mode

All eight 16-bit result registers are organized into a circular ring. The conversion counter value in the ATDxSTAT0 register specifies the result register to hold the current conversion result. In the FIFO mode, the conversion count is not reset to 0 when a new conversion sequence is started. Since one can set the length of a conversion sequence to a value between 1 and 8, the first conversion result in a sequence may not be stored in the ATDxDR0 register.

Example 12.5

Assume that the following setting was programmed before a new conversion is started:
- The conversion counter value in the ATD0STAT0 register is 5.
- The channel select code of the ATD0CTL5 register is 6.
- The conversion sequence limit of the ATD0CTL3 register is set to 5.
- The MULT bit of the ATD0CTL5 register is set to 1.

How would the conversion results be stored when the FIFO mode is selected or not selected?

Solution: The conversion counter specifies the result register to hold the first conversion result. The channel-select code specifies the first analog input channel to be converted, and the conversion sequence limit specifies the number of channels to be converted. The conversion results will be stored as shown in Table 12.9.

Analog Channel	Result Stored in (FIFO mode)	Result Stored in (non-FIFO mode)
AN6	ATD0DR5	ATD0DR0
AN7	ATD0DR6	ATD0DR1
AN0	ATD0DR7	ATD0DR2
AN1	ATD0DR0	ATD0DR3
AN2	ATD0DR1	ATD0DR4

Table 12.9 ■ Conversion results storage

12.4.6 External Trigger Source

An external triggering function allows the user to synchronize the ATD conversion process with external events. The ATD module can be configured to trigger on edges or levels of different polarities. Several bits in the ATDxCTL2 register are used to configure this function. The external triggering capability allows the user to start an A/D conversion when it is desired. The external trigger source is AN7 for AD0 and AN15 for AD1. Connecting the trigger channels together on MCUs with two ATDs allows synchronized conversions on the two converters. Examples of the trigger source could be external logic, software driving an output pin, a timer output-compare output pin, or a PWM channel.

12.4.7 Signed and Unsigned Control

Conversion results can be signed or unsigned. Signed data is represented as two's complement. The ATDxCTL5 register controls this function. A signed conversion sequence treats the value $(V_{RH} - V_{RL})/2$ as zero reference. Input signals greater than this value produce positive results; input signals lower than this value result in negative results.

12.4.8 ATD Operation Modes

The ATD module has three operation modes: power-down mode, idle mode, and run mode.

POWER-DOWN MODE

The ATD module can be powered down to save power consumption in one of three ways.

1. By clearing the ADPU bit of the ATDxCTL2 register
2. By executing the stop instruction
3. By executing the wai instruction with the AWAI bit of the ATDxCTL2 register set to 1

The ATD module is in power-down mode out of reset. Once the command to power down has been received, the ATD module aborts any conversion sequence in progress and enters power-down mode. When the module is powered up again, the module requires a recovery time of about 20 μs to stabilize the bias settings in the analog electronics before conversions can be performed. Powering the module up and down does not change the contents of the register file, and in power-down mode, the control and result registers are still accessible.

IDLE MODE

Idle mode for the ATD module is defined as the state in which the ATD module is powered up and ready to perform A/D conversion, but is not actually performing a conversion at the present time. Access to all control, status, and result registers is available. The module is consuming near the maximum power.

RUN MODE

The run mode for the ATD module is defined as the state in which the ATD module is powered up and currently performing an A/D conversion. Complete access to all control, status, and result registers is available. The module consumes the maximum power.

12.5 Procedure for Performing A/D Conversion

Before the HCS12 can perform A/D conversion, the A/D module must be connected and configured properly. The procedure for performing an A/D conversion is as follows:

Step 1
Connect the hardware properly. The A/D-related pins must be connected as follows:

V_{DDA}: connect to 5 V

V_{SSA}: connect to 0 V

V_{RH}: 5 V or other positive value less than 5 V

V_{RL}: 0 V or other value less than V_{RH} but higher than 0 V

If the transducer output is not in the appropriate range, then a signal-conditioning circuit should be used to shift and scale it to between V_{RL} and V_{RH}.

Step 2
Configure ATD control registers 2 to 4 properly and wait for the ATD to stabilize (need to wait for 20 μs).

Step 3
Select the appropriate channel(s) and operation modes by programming ATD control register 5. Writing into ATD control register 5 starts an A/D conversion sequence.

Step 4
Wait until the SCF flag of the status register ATDxSTAT0 is set, then collect the A/D conversion results and store them in memory.

Example 12.6

▼

Write a subroutine to initialize the AD0 converter for the MC9S12DP256 and start the conversion with the following setting:

- Nonscan mode
- Select channel 7 (single-channel mode)
- Fast ATD flag clear all
- Stop ATD in wait mode
- Disable interrupt
- Perform four conversions in a sequence
- Disable FIFO mode
- Finish current conversion, then freeze when BDM becomes active
- 10-bit operation and two A/D clock periods of the second-stage sample time
- Choose 2 MHz as the conversion frequency for the 24-MHz E-clock
- Result is unsigned and right-justified

Solution: The settings of ATD control registers 2 to 5 are as follows:
The setting of *ATD control register 2* is as follows:

- Enable AD0 (set bits 7 to 1)
- Select fast flag clear all (set bits 6 to 1)

- Stop AD0 when in wait mode (set bits 5 to 1)
- Disable external trigger on channel 7 (set bits 4, 3, and 2 to 0)
- Disable AD0 interrupt (set bit 1 to 0)

Write the value $E0 into this control register.
The setting of the *ATD control register 3* is as follows:

- Perform four conversions
- Disable FIFO mode
- When BDM becomes active, complete the current conversion then freeze

Write the value $22 into ATD0CTL3.
The setting of the *ATD control register 4* is as follows:

- Select 10-bit operation (set bit 7 to 0)
- Set sample time to two A/D clock periods (set bits 6 and 5 to 00)
- Set the value of PRS4~PRS0 to 00101

Write the value $05 to this control register.
The setting of the *ATD control register 5* is as follows:

- Result register right-justified (set bit 7 to 1)
- Result is unsigned (set bit 6 to 0)
- Nonscan mode (set bit 5 to 0)
- Single-channel mode (set bit 4 to 0)
- Select channel 7 (set bits 2..0 to 111)

Write the value $87 to this control register
The following subroutine will perform the desired ATD configuration:

```
#include         "c:\miniide\hcs12.inc"

openAD0    movb    #$E0,ATD0CTL2
           jsr     wait20us          ; wait for 20 μs
           movb    #$22,ATD0CTL3
           movb    #$05,ATD0CTL4
           rts
wait20us   movb    #$90,TSCR1        ; enable TCNT and fast timer flag clear
           movb    #0,TSCR2          ; set TCNT prescaler to 1
           bset    TIOS,$01          ; enable OC0
           ldd     TCNT              ; start an OC0 operation
           addd    #480              ;              "
           std     TC0               ;              "
           brclr   TFLG1,C0F,*       ; wait for 20 μs
           rts
```

This routine does not write into the ATD0CTL5 register because that would start the conversion. We should write into the ATD0CTL5 register only when we want to perform the conversion.
The C language version of the subroutine is as follows:

```
void wait20us (void);
void openAD0 (void)
{
        ATD0CTL2 = 0xE0;
        wait20us();
```

```
        ATD0CTL3 = 0x22;
        ATD0CTL4 = 0x05;
}
void wait20us (void)
{
        TSCR1    = 0x90;
        TSCR2    = 0;
        TIOS     |= 0C0;
        TC0      = TCNT + 480;
        while(!(TFLG1 & C0F));
}
```

Example 12.7

Write a program to perform A/D conversion on the analog signal connected to the AN7 pin. Collect 20 A/D conversion results and store them at memory locations starting from $1000. Use the same configuration as in Example 12.6.

Solution: To collect 20 conversion results, we need to write into the ATD0CTL5 register five times. Each time we will wait for the SCF flag to be set and collect four results. The program is as follows:

```
#include "c:\miniide\hc12.inc"
         org    $1500
         lds    #$1500
         ldx    #$1000               ; use index register X as a pointer to the buffer
         jsr    openAD0              ; initialize the ATD0 converter
         ldy    #5
loop5    movb   #$87,ATD0CTL5        ; start an A/D conversion sequence
         brclr  ATD0STAT0,SCF,*
         movw   ATD0DR0,2,x+         ; collect and save the conversion results
         movw   ATD0DR1,2,x+         ; post-increment the pointer by 2
         movw   ATD0DR2,2,x+         ;       "
         movw   ATD0DR3,2,x+         ;       "
         dbne   y,loop5
         swi
         end
```

The C language version of the program is as follows:

```
#include "c:\cwHS12\include\hcs12.h"
void openAD0 (void);
int buf[20];
void main (void)
{
        int i;
        openAD0();
        for (i = 0; i < 5; i++) {
                ATD0CTL5 = 0x87;               /* start an A/D conversion */
                while (!(ATD0STAT0 & SCF));     /* wait for the A/D conversion to complete */
                buf[4*i + 0] = ATD0DR0;         /* save results right-justified */
                buf[4*i + 1] = ATD0DR1;
```

```
                              buf[4*i + 2] = ATDODR2;
                              buf[4*i + 3] = ATDODR3;
                      }
                 asm ("swi");
       }
```

Example 12.8

Write a C program to be run on the Dragon12 demo board to display the voltage (output of a potentiometer) connected to the AN7 pin. Configure the AD0 properly, perform five conversions per second, and display the voltage on the LCD.

Solution: The conversion result 1023 corresponds to 5 V. Therefore, we need to divide the conversion result by 204.6 to convert the result back to voltage. Since the HCS12 does not support floating-point operation directly, we multiply the conversion result by 10 and then divide the product by 2046 to obtain the voltage value. The C program is as follows:

```
#include "c:\cwHCS12\include\hcs12.h"
#include "c:\cwHCS12\include\delay.h"          // include delay.c in the project
#include "c:\cwHCS12\include\lcd_util.h"       // include lcd_util.c in the project
void openAD0(void);
main(void)
{
       char buffer[6];          /* used to hold the voltage value */
       int temp;
       char *msg1 = "Voltage = ";
       openlcd();
       buffer[1] = '.';         /* decimal point */
       buffer[3] = 0x20;        /* space character */
       buffer[4] = 'V';         /* volt character */
       buffer[5] = 0;           /* null character */

       openAD0();
       while(1) {
              ATDOCTL5 = 0x87;                          /* convert AN7, result right-justified */
              while(!(ATD0STAT0 & SCF));                /* wait for conversion to complete */
              buffer[0] = 0x30 + (ATD0DR0 * 10 )/2046;
              temp = (ATD0DR0 * 10) % 2046;             /* find the remainder */
              buffer[2] = 0x30 + (temp * 10)/2046;      /* compute the fractional digit */
              cmd2lcd(0x80);                            /* set LCD cursor to upper left corner*/
              puts2lcd(msg1);                           /* output the message "voltage =" */
              puts2lcd(&buffer[0]);                     /* output voltage string */
              delayby100ms(2);                          /* wait for 200 ms */
       }
       return 0;
}
void openAD0 (void)
{
       int i;
       ATDOCTL2 = 0xE0;   /* enable AD0, fast ATD flag clear, disable AD0 in wait mode */
```

```
    delayby10us(2);
    ATDOCTL3 = 0x0A;  /* perform one conversion */
    ATDOCTL4 = 0x25;  /* four cycles sample time, prescaler set to 12 */
}
```

When this program is running on the demo board, you can turn the potentiometer and see the voltage value changes on the LCD display.

12.6 Using the Temperature Sensor TC1047A

The TC1047A is a three-pin temperature sensor whose voltage output is directly proportional to the measured temperature. The TC1047A can accurately measure temperatures from −40°C to 125°C with a power supply from 2.7 to 5.5 V.

The output voltage range for these devices is typically 100 mV at −40°C, 500 mV at 0°C, 750 mV at +25°C, and +1.75 V at +125°C. As shown in Figure 12.17, the TC1047A has a 10 mV/°C voltage slope output response.

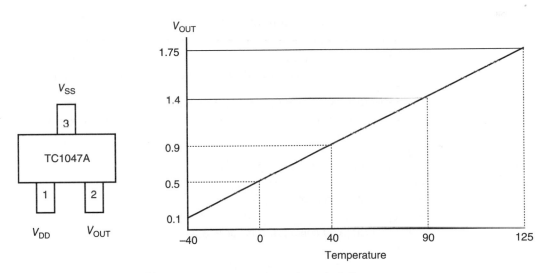

Figure 12.17 ■ TC1047A V_{OUT} versus temperature characteristic

Example 12.9

Describe a circuit connection and the required program to build a digital thermometer. Display the temperature in three integral and one fractional digits using the LCD. Measure and display the temperature over the whole range of TC1047A, that is, −40°C to +125°C. Update the display data five times per second and assume that the HCS12 operates with a 24-MHz E-clock.

Solution: Since the voltage output of the TC1047A is from 0.1 to 1.75 V for the temperature range of −40°C through 125°C, we need to use the circuit shown in Figure 12.7c to perform the

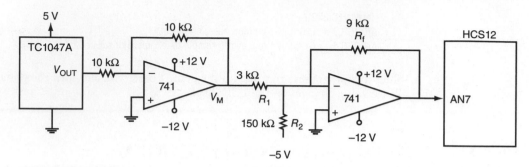

Figure 12.18 ■ Circuit connection between the TC1047A and the HCS12

translation and scaling for V_{OUT}. The circuit connection is shown in Figure 12.18. This circuit uses V_{DD} and V_{SS} as the reference voltages required in the process of A/D conversion. The range of the voltage connected to the AN7 pin is between 0 and 5 V for the given temperature range.

To convert the A/D conversion result to the corresponding temperature, we need to do the following:

1. Divide the conversion result by 6.2

2. Subtract 40 from the quotient

Since the HCS12 MCU cannot handle floating-point numbers directly, the operation of dividing by 6.2 would be implemented by multiplying the conversion result by 10 and then dividing the product by 62.

The procedure for starting an A/D conversion, reading the conversion result, calculating the corresponding temperature, and converting the temperature value into an ASCII string every 200 ms is as follows:

Step 1
Configure the A/D converter properly.

Step 2
Start an A/D conversion.

Step 3
Wait until the A/D conversion is complete.

Step 4
Multiply the conversion result by 10.

Step 5
Divide the product resulting from step 4 by 62 to obtain the temperature reading. Use the variables *quo* and *rem* to hold the quotient and remainder.

Step 6
Subtract 40 from *quo* to obtain the actual temperature.

Step 7
If $quo \geq 0$, go to step 9; otherwise, replace *quo* with its two's complement.

Step 8
If $rem \neq 0$, then

1. Decrement *quo* by 1

2. $rem \leftarrow 62 - rem$

Step 9

Compute the fractional digit by multiplying *rem* by 10 and then dividing the resulting product by 62.

Step 10

Compute the integral digits by performing repeated division by 10 to *quo*.

Step 11

Wait for 200 ms and then go to step 2.

The assembly program that implements this algorithm is as follows:

```
#include "c:\miniIDE\hcs12.inc"
period      equ         $2E                     ; ASCII code of period character
degree      equ         223                     ; ASCII code of degree character
            org         $1000
quo         ds.b        1
rem         ds.b        1
sign        ds.b        1
fract       ds.b        1
buf         ds.b        8                       ; to hold string to be output to LCD
            org         $1500
            lds         #$1500                  ; set up stack pointer
            ldy         #2                      ; wait for LCD to complete
            jsr         delayby100ms            ; internal configuration
            jsr         openlcd                 ; configure LCD
            ldaa        #$80                    ; set LCD cursor to upper
            jsr         cmd2lcd                 ; left corner
            ldx         #msg1                   ; output "Temperature = "
            jsr         puts2lcd                ;            "
            jsr         openAD0                 ; configure ATD0 module
forever     movb        #$20,buf                ; initialize the buffer contents to 0.0°C
            movb        #$20,buf+1              ;            "
            movb        #$30,buf+2              ;            "
            movb        #period,buf+3           ;            "
            movb        #$30,buf+4              ;            "
            movb        #degree,buf+5           ; degree character
            movb        #$43,buf+6              ; letter 'C'
            movb        #0,buf+7                ; null character
            movb        #$87,ATD0CTL5           ; start an ATD conversion sequence
            movb        #0,sign                 ; initialize sign to positive
            movb        #$30,fract              ; initialize fractional digit to 0
            brclr       ATD0STAT0,SCF,*         ; wait for the conversion to complete
            ldd         ATD0DR0                 ; read a conversion result
            ldy         #10                     ; compute result x 10/62
            emul                                ;            "
            ldx         #62                     ;            "
            ediv                                ;            "
            stab        rem                     ; save the remainder
            tfr         y,d                     ; transfer quotient to B
            subb        #40                     ; subtract temperature offset
            bhs         save_quo                ; if nonnegative, don't touch remainder
```

```
                 negb                              ; compute two's complement of quotient
                 stab      quo
                 movb      #1,sign                 ; temperature is negative
                 ldab      rem                     ; if remainder is 0, skip a few instructions
                 beq       convert                 ;
                 ldab      #62                     ; compute 62 − rem
                 subb      rem                     ;          "
                 stab      rem                     ;          "
                 bra       cal_fract
save_quo         stab      quo                     ; save updated quotient
cal_fract        ldab      rem
                 beq       convert                 ; come here when positive
                 ldaa      #10                     ; compute fractional digit
                 mul                               ;          "
                 ldx       #62                     ;          "
                 idiv                              ;          "
                 cmpb      #31                     ; round off fractional digit
                 blt       no_round                ;          "
                 inx                               ;          "
                 cpx       #10                     ;          "
                 bne       no_round                ;          "
                 inc       quo                     ;          "
                 bra       convert                 ; prepare to separate integer digits
no_round         tfr       x,d                     ; convert fractional digit to ASCII code
                 addb      #$30                    ;          "
                 stab      fract                   ;          "
convert          ldab      quo
                 clra                              ;          "
                 ldx       #10                     ; use repeated divide by 10 to separate
                 idiv                              ; integral digits
                 addb      #$30                    ;          "
                 stab      buf+2                   ; save the one's digit
                 tfr       x,d                     ; transfer quotient to D
                 tstb                              ; is quo zero?
                 beq       add_fra                 ; if integral part is 0, then add fraction digit
                 ldx       #10                     ; separate the ten's digit
                 idiv
                 addb      #$30                    ; convert and store the tens digit
                 stab      buf+1                   ;          "
                 tfr       x,d                     ; test hundreds digit
                 tstb                              ; is quotient 0?
                 beq       add_fra
                 movb      #$31,buf                ; hundreds digit, if any, is 1 only
add_fra          movb      fract,buf+4             ; insert fraction digit
                 ldaa      sign                    ; check the sign
                 beq       out_it
                 movb      #$2D,buf                ; when minus, add minus character
out_it           ldaa      #$C0                    ; set cursor to 2nd row
                 jsr       cmd2lcd                 ;          "
                 ldx       #spaces                 ; clear the 2nd row of the LCD
```

```
            jsr         puts2lcd                ;              "
            ldaa        #$C5                    ; set LCD cursor position
            jsr         cmd2lcd                 ;              "
            ldx         #buf                    ; output the temperature string
            jsr         puts2lcd                ;              "
            ldy         #2                      ; wait for 200 ms
            jsr         delayby100ms            ;              "
            jmp         forever                 ; continue
; ***************************************************************************
; The following function performs the AD0 configuration.
; ***************************************************************************
;
openAD0     movb        #$E0,ATD0CTL2           ; enable AD0, fast ATD flag clear, stop in wait mode
            ldy         #2
            jsr         delayby10us             ; wait until AD0 is stabilized
            movb        #$0A,ATD0CTL3           ; perform one A/D conversion
            movb        #$25,ATD0CTL4           ; set 4 cycles of sample time, set prescaler
                                                  divisor to 12
            rts
#include    "C:\miniIDE\lcd_util_dragon12.asm"
#include    "c:\miniIDE\delay.asm"
msg1        fcc         "Temperature = "
            dc.b        0
spaces      fcc         "              "
            dc.b        0
            end
```

The C language version of the program is as follows:

```
#include    "c:\egnu091\include\hcs12.h"
#include    "c:\cwHCS12\include\delay.h"         //include delay.c in the project
#include    "c:\cwHCS12\include\convert.h"       //include convert.c in the project
#include    "c:\cwHCS12\include\lcd_util.h"      //include lcd_util.c in the project
void    openAD0(void);
char    buf[8];
char    *msg1 = "temperature = ";
char    *blanks = "            ";
void    main (void)
{
        int     temp1,temp2;
        char    sign,fdigit,frem;
        char    *ptr;
        delayby100ms(2);            /* wait for LCD kit to initialize */
        openlcd();                  /* configure LCD kit */
        openAD0();                  /* configure AD0 module */
        cmd2lcd(0x80);              /* set cursor to upper left corner */
        puts2lcd(msg1);             /* output the message "temperature =" */
        while(1) {
                sign = 0;           /* initialize sign to be positive */
                ATD0CTL5 = 0x87;                    /* start a conversion with result right-justified */
                while(!(ATD0STAT0 & SCF));          /* wait until conversion is done */
                temp1 = (ATD0DR0 * 10)/62;          /* integer part of temperature */
```

```
                              temp2 = (ATD0DR0 * 10) % 62;        /* remainder part */
                              temp1 -= 40;        /* subtract the offset from the actual temperature */
                              if (temp1 < 0){        /* temperature is negative */
                                        sign = 1;
                                        temp1 = -temp1;                /* find the magnitude of temperature */
                                        if (temp2) { /* remainder not zero */
                                                  temp1--;
                                                  temp2 = 62 - temp2;
                                        }
                              }
                              fdigit = (temp2 * 10)/62;        /* compute the fractional digit */
                              frem = (temp2 * 10) % 62;
                              if (frem > 31) {
                                        fdigit++;
                                        if (fdigit == 10) {     /* round off the fractional digit */
                                                  fdigit = 0;
                                                  temp1++;
                                        }
                              }
                              if (sign) {
                                        ptr = &buf[1];        /* point to the first space to hold ASCII string */
                                        buf[0] = 0x2D;        /* store minus sign as the first character */
                              }
                              else
                                        ptr = &buf[0];
                              int2alpha(temp1,ptr);                /* convert the integer part to ASCII string */
                              ptr = &buf[0];
                              while(*ptr)                /* find the end of the integer string */
                                        ptr++;
                              *ptr++ = '.';
                              *ptr++ = fdigit + 0x30;
                              *ptr++ = 223;                /* add a degree character */
                              *ptr++ = 'C';                /* temperature in Celsius */
                              *ptr = 0;                /* terminate the temperature string */
                              cmd2lcd(0xC0);                /* move cursor to 2nd row */
                              puts2lcd(blanks);                /* clear the 2nd row */
                              cmd2lcd(0xC5);                /* set cursor to column 5 row 2 */
                              puts2lcd(&buf[0]);                /* output the temperature */
                              delayby100ms(2);                /* wait for 0.2 seconds */
                    }
          }
          void openAD0(void)
          {
                    ATD0CTL2 = 0xE0;                /* enable AD0, fast ATD flag clear, power-down on wait */
                    ATD0CTL3 = 0x0A;                /* perform one ATD conversion */
                    ATD0CTL4 = 0x25;                /* prescaler set to 12, select four cycles sample time */
                    delayby10us(2);
          }
```

This program can be tested on the Dragon12-Plus demo board by rotating the potentiometer. The circuit shown in Figure 12.18 contains a signal-conditioning circuit which consists of two Op-Amps (741), five resistors, and three different power supplies (-5 V, -12 V, and 12 V). This circuit allows the user to utilize the whole dynamic range of the HCS12 ADC. However, adding a signal-conditioning circuit adds cost to the resultant product. It should be used sparingly. By setting the V_{RH} input to 1.8 V, the designer can eliminate the use of the signal conditioning circuit shown in Figure 12.18. After changing the circuit, the designer will need to change the formula for translating the ADC result back to the temperature reading. This change is straightforward and hence is left as an exercise problem.

12.7 Using the IH-3605 Humidity Sensor

The IH-3605 is a humidity sensor made by Honeywell. It provides a linear output from 0.8 to 3.9 V in the full range of 0 to 100 percent *relative humidity* (RH) when powered by a 5-V power supply. It can operate in a range of 0 to 100 percent RH, $-40°$ to 185°F. The pins of the IH-3605 are shown in Figure 12.19. The specifications are listed in Table 12.10. The IH-3605 is light sensitive and should be shielded from bright light for best results.

Specification	Description
Total accuracy	±2% RH, 0%–100% TH @ 25°C
Interchangeability	±5% RH up to 60% RH, ±8% RH at 90% RH
Operating temperature	–40° to 85°C (–40° to 185°F)
Storage temperature	–51° to 110°C (–60° to 223°F)
Linearity	±0.5% RH typical
Repeatability	±0.5% RH
Humidity stability	±1% RH typical at 50% RH in 5 years
Temp. effect on 0% RH voltage	±0.007% RH/°C (negligible)
Temp. effect on 100% RH voltage	–0.22% RH/°C
Output voltage	$V_{OUT} = (V_S)(0.16$ to $0.78)$ nominal relative to supply voltage for 0%–100% RH; i.e., 1–4.9 V for 6.3-V supply; 0.8–3.9 V for 5-V supply; sink capability 50 A; drive capability 5 A typical; low-pass 1 kHz filter required. Turn on time < 0.1 s to full output.
VS supply requirement	4 to 9 V, regulated or use output/supply ratio; calibrated at 5 V
Current requirement	200 A typical @ 5 V, increased to 2 mA at 9 V

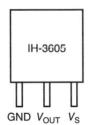

Figure 12.19 ■ Honeywell IH-3605 humidity sensor

Table 12.10 ■ Specifications of IH-3605

The IH-3605 can resist contaminant vapors, such as organic solvents, chlorine, and ammonia. It is unaffected by water condensate as well. Because of this capability, the IH-3605 has been used in refrigeration, drying, instrumentation, meteorology and many other applications.

Example 12.10

Construct a humidity measurement system that consists of an HCS12, an IH-3605 humidity sensor, and an LCD. The E-clock frequency of the HCS12 is 24 MHz.

Solution: Since the output of the IH-3605 is between 0.8 and 3.9 V with a 5-V power supply, it would be beneficial to use a circuit to translate and then scale it to between 0 and 5 V so that the best accuracy can be achieved. The circuit connection is shown in Figure 12.20. A set of resistor values and V_1 voltage are given in Figure 12.20. A low-pass filter that consists of a 1-kΩ resistor and a 0.16-μF capacitor is added to meet the requirement.

To translate from the conversion result to the relative humidity, divide the conversion result by 10.23. Since the HCS12 MCU does not support floating-point operation directly, we can multiply the conversion result by 100 (or 10) and then divide the product by 1023 (or 102). The C program that configures the A/D module, starts the A/D conversion, translates the conversion result to the relative humidity, converts the humidity into an ASCII string, and displays the string on the LCD is as follows:

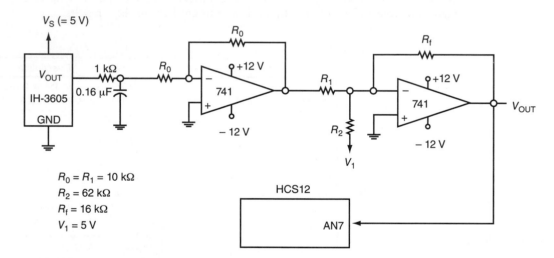

Figure 12.20 ■ Relative humidity measurement circuit

```
#include         "c:\cwHCS12\include\hcs12.h"
#include         "c:\cwHCS12\include\delay.c"
#include         "c:\cwHCS12\include\convert.c"
#include         "c:\cwHCS12\include\lcd_util_dragon12.c"
void openAD0(void);
char buf[10];
char *msg1 = "humidity = ";
char *blanks = "          ";
void main (void)
{
```

```c
          unsigned int quo,rem,frem;
          long temp;
          char *ptr,fdigit;
          delayby100ms(2);                  /* wait for LCD kit to self-initialize */
          openlcd();                        /* configure LCD kit */
          openAD0();                        /* configure AD0 module */
          cmd2lcd(0x80);                    /* set cursor to upper left corner */
          puts2lcd(msg1);                   /* output the message "humidity = " */
          while(1) {
                    ATD0CTL5 = 0x87;        /* start a conversion with result right-justified */
                    while(!(ATD0STAT0&SCF)); /* wait until conversion is done */
                    temp = (long)ATD0DR0 * 100; /* force compiler to use 32-bit to represent the product */
                    quo = temp/1023;        /* integer part of temperature */
                    rem = temp % 1023;      /* remainder part */
                    if (quo == 100)
                              rem = 0;      /* force highest humidity to 100% only */
                    fdigit = (rem * 10)/1023; /* compute the fractional digit */
                    frem = (rem * 10) % 1023;
                    if (frem > 511)
                              fdigit++;
                    if (fdigit == 10) {     /* round off the fractional digit */
                              fdigit = 0;
                              quo++;
                    }
                              ptr = &buf[0];
                    int2alpha(quo,ptr);     /* convert the integer part to ASCII string */
                    ptr = &buf[0];
                    while(*ptr)             /* find the end of the integer string */
                              ptr++;
                    *ptr++ = '.';           /* decimal point */
                    *ptr++ = fdigit + 0x30; /* fractional digit */
                    *ptr++ = '%';
                    *ptr = 0;               /* terminate the humidity string */
                    cmd2lcd(0xC0);          /* move cursor to 2nd row */
                    puts2lcd(blanks);       /* clear the 2nd row */
                    cmd2lcd(0xC5);          /* set cursor to column 5 row 2 */
                    puts2lcd(&buf[0]);      /* output the humidity */
                    delayby100ms(2);        /* wait for 0.2 s */
          }
}
void openAD0(void)
{
   ATD0CTL2 = 0xE0;          /* enable AD0, fast ATD flag clear, power-down on wait */
   ATD0CTL3 = 0x0A;          /* perform one ATD conversion */
   ATD0CTL4 = 0x25;          /* prescaler set to 12, select four cycles sample time */
   delayby10us(2);
}
```

This program can be tested on the Dragon12-Plus demo board by rotating the potentiometer. The potentiometer emulates the output from the signal conditioning circuit. Like Example 12.9, the user can eliminate the signal conditioning circuit by setting the V_{RH} input to a 3.9-V Zener diode. The user will also need to modify the translation formula from the ADC result back to the humidity. The modification of the program is straightforward and is left as an exercise.

12.8 Measuring Barometric Pressure

Barometric pressure refers to the air pressure existing at any point within the Earth's atmosphere. This pressure can be measured as an absolute pressure (with reference to absolute vacuum) or can be referenced to some other value or scale. The meteorology and avionics industries traditionally measure the absolute pressure and then reference it to a sea-level pressure value. This complicated process is used in generating maps of weather systems.

Mathematically, atmospheric pressure is exponentially related to altitude. Once the pressure at a particular location and altitude is measured, the pressure at any other altitude can be calculated. Several units have been used to measure the barometric pressure: **in−Hg**, **kPa**, **mbar**, or **psi**. A comparison of barometric pressure using four different units at sea level up to 15,000 ft is shown in Table 12.11.

Altitude (ft)	Pressure (in-Hg)	Pressure (mbar)	Pressure (kPa)	Pressure (psi)
0	29.92	1013.4	101.4	14.70
500	29.38	995.1	99.5	14.43
1000	28.85	977.2	97.7	14.17
6000	23.97	811.9	81.2	11.78
10,000	20.57	696.7	69.7	10.11
15,000	16.86	571.1	57.1	8.28

Table 12.11 ■ Altitude versus pressure data

There are three forms of pressure transducer: *gauge, differential,* and *absolute*. Both the gauge pressure (*psig*) and differential (*psid*) transducers measure pressure differentially. The abbreviation *psi* stands for "pounds per square inch"; the letters g and d stand for *gauge* and *differential*, respectively. A gauge pressure transducer measures pressure against ambient air, whereas the differential transducer measures against a reference pressure. An absolute pressure transducer measures the pressure against a vacuum (0 psia) and hence it measures the barometric pressure.

The SenSym ASCX30AN is a 0 to 30 psia (psi absolute) pressure transducer. The range of barometric pressure is between 28 to 32 inches of mercury (in-Hg) or 13.75 to 15.72 psia or 948 to 1083.8 mbar. The transducer output is about 0.15 V/psi, which would translate to an output voltage from 2.06 to 2.36 V. The complete specifications of the SenSym ASCX30AN are shown in Table 12.12. Since the range of V_{OUT} is very narrow, the designer will need to use a level shifting and scaling circuit in order to take advantage of the available dynamic range.

The pin assignment of the SenSym ASCX30AN is shown in Figure 12.21.

Characteristic	Min.	Typ.	Max.
Pressure range	0 psia	—	30 psia
Zero pressure offset	0.205	0.250	0.295
Full-scale span[2]	4.455	4.500	4.545
Output at FS pressure	4.660	4.750	4.840
Combined pressure non-linearity and pressure hysteresis[3]	—	±0.1	±0.5
Temperature effect on span[4]	—	±0.2	±1.0
Temperature effect on offset[4]	—	±0.2	±1.0
Response time (10%−90%)[5]	—	0.1	—
Repeatability	—	±0.05	—

[1] Reference conditions: TA = 25°C, supply voltage V_s = 5 V

[2] Full-scale span is the algebraic difference between the output voltage at full-scale pressure and the output at zero pressure. Full-scale span is ratiometric to the supply voltage.

[3] Pressure nonlinearity is based on the best-fit straight line. Pressure hysteresis is the maximum output difference at any point within the operating pressure range for increasing and decreasing pressure.

[4] Maximum error band of the offset voltage or span over the compensated temperature range, relative to the 25°C reading.

[5] Response time for 0 psi to full-scale pressure step response.

[6] If maximum pressure is exceeded, even momentarily, the package may leak or burst or the pressure-sensing die may burst.

Table 12.12 ■ ASCX30AN performance characteristics [1]

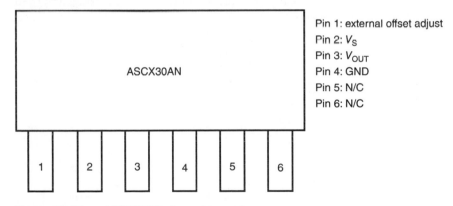

Pin 1: external offset adjust
Pin 2: V_S
Pin 3: V_{OUT}
Pin 4: GND
Pin 5: N/C
Pin 6: N/C

Figure 12.21 ■ ASCX30AN pin assignment

Example 12.11

Describe the circuit connection of the ASCX30AN and the voltage level shifting and scaling circuit, and write a program to measure and display the barometric pressure in units of mbar.

Solution: The circuit connection is shown in Figure 12.22. The circuit in Figure 12.22 will shift and scale V_{OUT} to the range of 0~5 V. According to Table 12.12, the typical value for the offset adjustment is around 0.25 V and can be achieved by using a potentiometer.

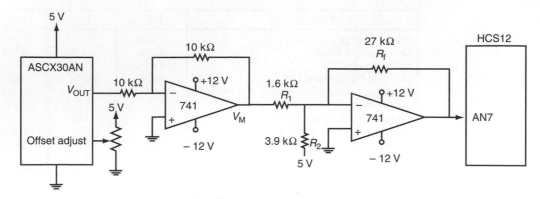

Figure 12.22 ■ Barometric pressure sensor output scaling and shifting circuit

The barometric pressure ranges from 948 to 1083.8 mbar. The A/D conversion result will be 0 and 1023 for these two values. Therefore, the A/D conversion result needs to be divided by 7.53 in order to find the corresponding barometric pressure. The following equation shows the conversion:

$$\text{Barometric pressure} = 948 + \text{A/D result}/7.53$$
$$\approx 948 + (\text{A/D result} \times 100)/753$$

The C program that performs A/D conversion, translates the result into barometric pressure, converts the barometric pressure into an ASCII string, and then displays the pressure on the LCD is as follows:

```
#include         "c:\cwHCS12\include\hcs12.h"
#include         "c:\cwHCS12\include\delay.h"        // include delay.c in project
#include         "c:\cwHCS12\include\convert.h"      // include convert.c in project
#include         "c:\cwHCS12\include\lcd_util.h"     // include lcd_util.c in project
void    openAD0(void);
char    buf[12];
char    *msg1 = "pressure = ";
char    *blanks = "            ";
void    main (void)
{
        unsigned int quo,rem,frem;
        long       temp;
        char       *ptr,fdigit;
        delayby100ms(2);                             /* wait for LCD kit to self-initialize */
        openlcd();                                   /* configure LCD kit */
        openAD0();                                   /* configure AD0 module */
        cmd2lcd(0x80);                               /* set cursor to upper left corner */
        puts2lcd(msg1);                              /* output the message "pressure = " */
        while(1) {
                ATD0CTL5 = 0x87;                     /* start a conversion with result right-justified */
                while(!(ATD0STAT0&SCF));             /* wait until conversion is done */
                temp = (long)ATD0DR0 * 100;
                quo = temp/753;                      /* integer part of pressure */
```

```
            rem = temp % 753;              /* remainder part */
            fdigit = (rem * 10)/753;       /* compute the fractional digit */
            frem = (rem * 10) % 753;
            if (frem > 377)
                    fdigit++;
            if (fdigit == 10) {            /* round off the fractional digit */
                    fdigit = 0;
                    quo++;
            }
            ptr = &buf[0];
            quo = quo + 948;
            int2alpha(quo,ptr);            /* convert the integer part to ASCII string */
            ptr = &buf[0];
            while(*ptr)                    /* find the end of the integer string */
                ptr++;
            *ptr++ = '.';
            *ptr++ = fdigit + 0x30;
            *ptr++ = 0x20;
            *ptr++ = 'm';
            *ptr++ = 'b';
            *ptr++ = 'a';
            *ptr++ = 'r';
            *ptr = 0;                      /* terminate the barometric pressure string */
            cmd2lcd(0xC0);                 /* move cursor to 2nd row */
            puts2lcd(blanks);              /* clear the 2nd row */
            cmd2lcd(0xC5);                 /* set cursor to column 5 row 2 */
            puts2lcd(&buf[0]);             /* output the pressure */
            delayby100ms(2);               /* wait for 0.2 s */
        }
}
void openAD0(void)
{
    ATD0CTL2 = 0xE0;         /* enable AD0, fast ATD flag clear, power-down on wait */
    ATD0CTL3 = 0x0A;         /* perform one ATD conversion */
    ATD0CTL4 = 0x25;         /* prescaler set to 12, select four cycles sample time */
    delayby10us(2);
}
```

▲

12.9 Summary

A data acquisition system consists of four major components: a transducer, a signal-conditioning circuit, an ADC, and a computer. The transducer converts a nonelectric quantity into a voltage. The transducer output may not be appropriate for processing by the ADC. The signal-conditioning circuit shifts and scales the output from a transducer to a range that can take advantage of the full capability of the ADC. The ADC converts an electric voltage into a digital value that will be further processed by the computer.

Due to the discrete nature of a digital system, the A/D conversion result has a quantization error. The accuracy of an ADC is dictated by the number of bits used to represent the analog quantity. The more bits used, the smaller the quantization error will be.

All of the A/D converters need a high-reference voltage (V_{RH}) and a low-reference voltage (V_{RL}) to perform the conversion. The best choice for the reference voltage is the Zener diode due to its stable breakpoint voltage and low price. By setting the high reference voltage to an appropriate value, the user may, in many cases, eliminate the need for the signal-conditioning circuit.

There are four major A/D conversion algorithms.

- Parallel (flash) A/D converter
- Slope and double-slope A/D converters
- Sigma-delta A/D converters
- Successive-approximation A/D converters

The HCS12 microcontroller uses the successive-approximation algorithm to perform the A/D conversion. All A/D conversion parameters are configured via four ATD control registers: ATDxCTL2, ATDxCTL3, ATDxCTL4, and ATDxCTL5.

The TC1047A temperature sensor, the IH-3605 humidity sensor, and the ASCX30AN pressure sensor are used as examples to illustrate the A/D conversion process. The TC1047A can measure a temperature in the range from −40 to 125°C. The IH-3605 can measure relative humidity from 0 to 100 percent. The ASCX30AN can measure a pressure in the range from 0 to 30 psi absolute. These three examples demonstrate the need for a good voltage shifting and scaling circuit.

There are applications that require A/D accuracy higher than that provided by the HCS12 microcontrollers. In this situation, the designer has the option of using an external A/D converter with higher precision or selecting a different microcontroller with higher A/D resolution. Many 8051 variants from Silicon Laboratory, TI, and Analog Devices have much higher A/D resolutions.

12.10 Exercises

E12.1 Design a circuit that can scale the voltage from the range of 0 mV~100 mV to the range of 0 V~5 V.

E12.2 Design a circuit that can shift and scale the voltage from the range of −80 mV~160 mV to the range of 0 V~5 V.

E12.3 Design a circuit that can shift and scale the voltage from the range of −50 mV~75 mV to the range of 0 V~5 V.

E12.4 Design a circuit that can shift and scale the voltage from the range of 2 V~2.5 V to the range of 0 V~5 V.

E12.5 Suppose that there is a 10-bit A/D converter with V_{RL} = 2 V and V_{RH} = 4 V. Find the corresponding voltage values for the A/D conversion results of 40, 100, 240, 500, 720, 800, and 1000.

E12.6 Suppose that there is a 12-bit A/D converter with V_{RL} = 1 V and V_{RH} = 4 V. Find the corresponding voltage values for the A/D conversion results of 80, 180, 480, 640, 960, 1600, 2048, 3200, and 4000.

E12.7 Write a few instructions to configure the HCS12 AD0 converter with the following parameters:

- E-clock frequency = 16 MHz
- Channel AN3
- Nonscan mode

- Unsigned result representation
- A/D result left-justified, single conversion sequence
- Perform eight conversions in a sequence
- Enable fast flag clear, non-FIFO mode
- Disable the ATD interrupt
- Power down ATD in wait mode
- In background debug mode, finish current conversion, then freeze
- 10-bit resolution, eight clock periods second-stage sample time, prescaler set to 8

E12.8 Write an instruction sequence to configure the HCS12DP256 A/D converter with the following characteristics:

- f_{OSC} = 8 MHz
- Channel AN0~AN3
- A/D result right-justified
- 8-bit resolution, unsigned result, continuous conversion
- Four conversions in a sequence
- Prescaler set to 2
- Four cycles second-stage sample time, non-FIFO mode
- Enable fast ATD flag clear, enable ATD interrupt
- External falling edge triggered (at ATD7 pin)
- Finish current conversion then freeze in background debug mode

E12.9 Assuming that S8C~S1C (ATD0CTL3) are set to 0111 and CC~CA (ATD0CTL5) are set to 101, what is the conversion sequence for this setting?

E12.10 Assume that the following setting was programmed before a new conversion is started:

- The conversion counter value in the ATD0STAT0 register is 4.
- The channel-select code of the ATD0CTL5 is 7.
- The conversion sequence limit of the ATD0CTL3 register is set to 6.
- The MULT bit of the ATD0CTL5 register is set to 1.

How would the conversion results be stored when the FIFO mode is selected or not selected?

E12.11 Assume that the following setting was programmed before a new conversion is started:

- The conversion counter value in the ATD0STAT0 register is 3.
- The channel-select code of the ATD0CTL5 is 5.
- The conversion sequence limit of the ATD0CTL3 register is set to 4.
- The MULT bit of the ATD0CTL5 register is set to 1.

How would the conversion results be stored when the FIFO mode is selected or not selected?

E12.12 At 8-bit resolution with the second-stage sampling time set to eight ATD clock cycles, how long does it take to complete the conversion of one sample at 500 kHz and at 2 MHz ATD clock frequencies, respectively?

E12.13 At 10-bit resolution with the second stage sampling time set to four ATD clock cycles, how long does it take to complete the conversion of one sample at 500 kHz and 2 MHz ATD clock frequencies, respectively?

E12.14 The LM35 from National Semiconductor is a Centigrade temperature sensor with three external connection pins. The pin assignment and circuit connection for converting temperature are shown in Figure E12.14. Use this device to construct a circuit to display the room temperature in Celsius. The temperature range to be converted is from $-27°C$ to $100°C$. Describe the circuit connection for a digital thermometer made up of the HCS12 and the LM35, and write a program to display temperature on the LCD.

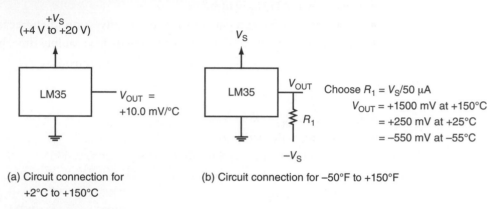

(a) Circuit connection for
    +2°C to +150°C

(b) Circuit connection for −50°F to +150°F

Figure E12.14 ■ Circuit connection for the LM35

E12.15 The LM34 from National Semiconductor is a Fahrenheit temperature sensor with three external connection pins. The pin assignment and circuit connection for converting the temperature are shown in Figure E12.15. Use this device to construct a circuit to display the room temperature in Fahrenheit. Assume that the temperature range of interest is from $-40°F$ to $215°F$. Use an LCD to display the temperature. Describe the circuit connection for a digital thermometer that is made up of the HCS12 and the LM34, and write a program to display temperature on the LCD.

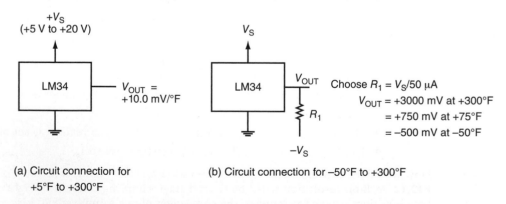

(a) Circuit connection for
    +5°F to +300°F

(b) Circuit connection for −50°F to +300°F

Figure E12.15 ■ Circuit connection for the LM34

E12.16 The Microbridge AWM3300V is a mass airflow sensor manufactured by Honeywell. The block diagram of the AWM3300V is shown in Figure E12.16. It is designed to measure

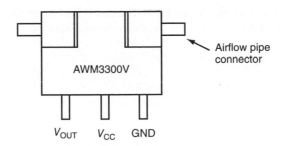

Figure E12.16 ■ Microbridge AWM3300V

the airflow. Its applications include air-conditioning, medical ventilation/anesthesia control, gas analyzers, gas metering, fume cabinets, and process control. The AWM3300V operates on a single 10-V ± 10-mV power supply. The sensor output (from V_{OUT}) corresponding to the airflow rate of 0~1.0 L/min is 1.0 V to 5.0 V. The AWM3300V can operate in the temperature range of −25 to 85°C. It takes 3 ms for the output voltage to settle after power-up. Design a circuit to measure and display the mass airflow using the AWM3300V. Write a program to configure the HCS12 A/D module, start the A/D conversion, and display the mass airflow in an LCD display. Update the display five times per second.

12.11 Lab Exercises and Assignments

L12.1 A/D converter testing. Perform the following steps.

Step 1
Set the function generator output (in the lab) to between 0 and 5 V.

Step 2
Connect the AN1 pin to the functional generator output. Set the frequency to about 10 kHz.

Step 3
Write a program to perform the following operations:

- Configure the A/D module properly.
- Start the A/D conversion.
- Take 64 samples, convert them, and store the conversion results at appropriate SRAM locations.
- Compute the average value and store it in D.

If your demo board has a square waveform output, then select an output with a frequency close to 16 kHz. Connect it to the AN1 pin.

L12.2 Digital thermometer. Use the LM34 temperature sensor, the 741 OP Amp, and the required resistors to construct a digital thermometer. The temperature should be displayed in three integral and one fractional digit. Use the LCD on your demo board to display the temperature. Update the temperature once every 200 ms.

L12.3 Barometric pressure measurement. The Motorola MPX4115A is a pressure sensor that can measure a pressure ranging from 15 to 115 kPa (2.2 to 16.7 psi) and has a corresponding voltage output from 0.2 to 4.8 V. The small outline package of this device has eight pins. Among these eight pins, only three pins carry useful signals:

Pin 2. V_S

Pin 3. GND

Pin 4. V_{OUT}

This pressure sensor is often used in aviation altimeters, industrial controls, engine control, and weather stations and weather reporting devices.

Connect an MPX4115AC6U (or MPX4115A6U, case 482) device to your HCS12 demo board to measure the current ambient barometric pressure once every second and display the pressure on the LCD display. The data sheet of the MPX4115A can be found on the complementary CD. Since the barometric pressure is in a narrow range (use the range mentioned in Example 12.11), you may need to construct a signal-conditioning circuit to translate and scale the voltage output of the pressure sensor to the range of 0 to 5 V.

13

Controller Area Network (CAN)

13.1 Objectives

After completing this chapter, you should be able to:

- Describe the layers of the CAN protocol
- Describe CAN's error detection capability
- Describe the formats of CAN messages
- Describe CAN message handling
- Explain CAN error handling
- Describe CAN fault confinement
- Describe CAN bit timing
- Explain CAN synchronization issue and methods
- Describe the CAN message structures
- Compute timing parameters to meet the requirements of your application
- Write programs to configure the HCS12 CAN module
- Write programs to transfer data over the CAN bus
- Design a CAN-based remote sensing system

13.2 Overview of Controller Area Network

The controller area network (CAN) was initially created by the German automotive system supplier Robert Bosch in the mid-1980s for automotive applications as a method for enabling robust serial communication. The goal was to make automobiles more reliable, safe, and fuel-efficient while at the same time decreasing wiring harness weight and complexity. Since its inception, the CAN protocol has gained widespread use in industrial automation and automotive/truck applications. The description of CAN in this chapter is based on the CAN Specification 2.0 published in September 1991 by Bosch.

13.2.1 Layered Approach in CAN

The CAN protocol specified the lowest two layers of the ISO seven-layer model: *data link* and *physical* layers. The data link layer is further divided into two sublayers: logical link control (LLC) layer and medium access control (MAC) layer.

- The LLC sublayer deals with message acceptance filtering, overload notification, and error recovery management.
- The MAC sublayer presents incoming messages to the LLC sublayer and accepts messages to be transmitted that are forwarded by the LLC sublayer. The MAC sublayer is responsible for message framing, arbitration, acknowledgement, error detection, and signaling. The MAC sublayer is supervised by a self-checking mechanism, called *fault confinement*, which distinguishes short disturbances from permanent failures.

The physical layer defines how signals are actually transmitted and deals with the description of bit timing, bit encoding, and synchronization. CAN bus driver/receiver characteristics and the wiring and connectors are not specified in the CAN protocol. These two aspects are not specified so that implementers can choose the most appropriate transmission medium and hence optimize signal-level implementations for their applications. The system designer can choose from multiple available media technologies including twisted pair, single wire, optical fiber, radio frequency (RF), infrared (IR), and so on. The layered CAN protocol is shown in Figure 13.1.

13.2.2 General Characteristics of CAN

The CAN protocol was optimized for systems that need to transmit and receive relatively small amounts of information (as compared to Ethernet or USB, which are designed to move much larger blocks of data). The CAN protocol has the following features:

CARRIER SENSE MULTIPLE ACCESS WITH COLLISION DETECTION (CSMA/CD)

The CAN protocol is a CSMA/CD protocol. Every node on the network must monitor the bus (carrier sense) for a period of no activity before trying to send a message on the bus. Once this period of no activity occurs, every node on the bus has an equal opportunity to transmit a message (multiple access). If two nodes happen to transmit at the same time, the nodes will detect the *collision* and take the appropriate action. In the CAN protocol, a nondestructive bitwise arbitration method is utilized. Messages remain intact after arbitration is completed even if collisions are detected. Message arbitration will not delay higher priority messages. To facilitate bus arbitration, the CAN protocol defines two bus states: *dominant* and *recessive*. The dominant state is represented by logic 0 (low voltage), whereas the recessive state is represented by logic 1 (high voltage). The dominant state will win over the recessive state.

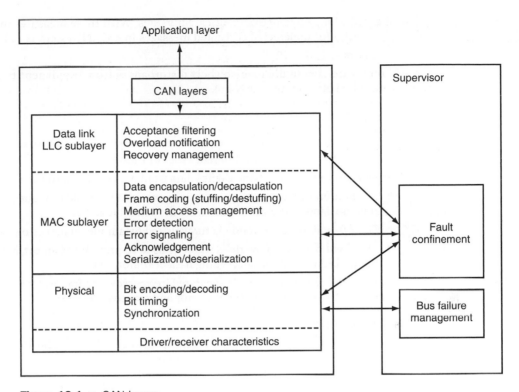

Figure 13.1 ■ CAN layers

MESSAGE-BASED COMMUNICATION

CAN is a message-based protocol rather than an address-based protocol. Embedded in each message is an *identifier*. This identifier allows messages to arbitrate the use of the CAN bus and also allows each node to decide whether to work on the message. The value of the identifier is used as the priority of the message. The lower the value, the higher the priority. Each node in the CAN system uses one or more filters to compare the identifier of the incoming message. Once the identifier passes the filter, the message will be worked on by the node. The CAN protocol also provides the mechanism for a node to request data transmission from another node. Since an address is not used in the CAN system, there is no need to reconfigure the system whenever a node is added to or deleted from a system. This capability allows the system to perform node-to-node or multicast communications.

ERROR DETECTION AND FAULT CONFINEMENT

The CAN protocol requires each sending node to monitor the CAN bus to find out if the bus value and the transmitted bit value are identical. For every message, cyclic redundancy check is calculated and the checksum is appended to the message. CAN is an asynchronous protocol and hence clock information is embedded in the message rather than transmitted as a separate signal. A message with long sequence of identical bits could cause a synchronization problem. To resolve this problem, the CAN protocol requires the physical layer to use bit stuffing to avoid a long sequence of identical bit values. With these measures implemented, the residual probability for undetected corrupted messages in a CAN system is as low as

Message error rate $\times$ 4.7 $\times$ 10^{-11}

CAN nodes are able to distinguish short disturbances from permanent failures. Defective nodes are switched off from the CAN bus.

13.3 CAN Messages

The CAN protocol defines four different types of messages.

- *Data frame.* A data frame carries data from a transmitter to the receivers.
- *Remote frame.* A remote frame is transmitted by a node to request the transmission of the data frame with the same identifier.
- *Error frame.* An error frame is transmitted by a node on detecting a bus error.
- *Overload frame.* An overload frame is used to provide for an extra delay between the preceding and the succeeding data or remote frames.

Data frames and remote frames are separated from preceding frames by an interframe space. Applications do not need to send or handle error and overload frames.

13.3.1 Data Frame

As shown in Figure 13.2, a data frame consists of seven different bit fields: start of frame, arbitration, control, data, CRC, ACK, and end of frame.

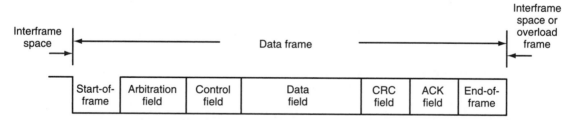

Figure 13.2 ■ CAN data frame

START-OF-FRAME FIELD

This field is a single dominant bit that marks the beginning of a data frame. A node is allowed to start transmission only when the bus is idle. All nodes have to synchronize to the leading edge of the field of the node that starts transmission first.

ARBITRATION FIELD

The format of the arbitration field is different for *standard format* and *extended format* frames, as illustrated in Figure 13.3. The identifier's length is 11 bits for the standard format and 29 bits for the extended format.

The identifier of the standard format corresponds to the base ID in the extended format. These bits are transmitted most significant bit first. The most significant 7 bits cannot all be recessive.

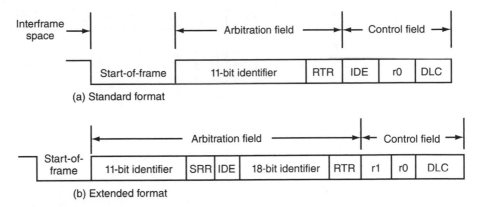

Figure 13.3 ■ Arbitration field

The identifier of the extended format comprises two sections: an 11-bit base ID and an 18-bit extended ID. Both the base ID and the extended ID are transmitted most significant bit first. The base ID defines the base priority of the extended frame.

The *remote transmission request* (RTR) bit in data frames must be dominant (0). Within a remote frame, the RTR bit has to be recessive (1).

The *substitute remote request* (SRR) bit is a recessive bit (1). The SRR bit of an extended frame is transmitted at the position of the RTR bit in the standard frame and therefore substitutes for the RTR bit in the standard frame. As a consequence, collisions between a standard frame and an extended frame, where the base IDs of both frames are identical, are resolved in such a way that the standard frame prevails over the extended frame.

The *identifier extension* (IDE) bit belongs to the arbitration field for the extended format and the control field for the standard format. The IDE bit in the standard format is transmitted dominant, whereas in the extended format the IDE bit is recessive.

CONTROL FIELD

The contents of this field are shown in Figure 13.4. The format of the control field is different for the standard format and the extended format. Frames in standard format include the data length code; the IDE bit, which is transmitted dominant; and the reserved bit **r0**. Frames in extended format include the data length code and two reserved bits, **r0** and **r1**. The reserved bits must be sent dominant, but the receivers accept dominant and recessive bits in all combinations. The data length code specifies the number of bytes contained in the data field. Data length can be 0 to 8, as encoded in Table 13.1.

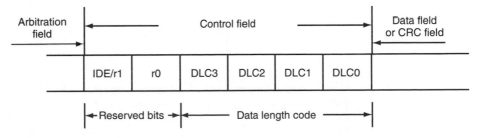

Figure 13.4 ■ Control field

DLC3	DLC2	DLC1	DLC0	Data Byte Count
d	d	d	d	0
d	d	d	r	1
d	d	r	d	2
d	d	r	r	3
d	r	d	d	4
d	r	d	r	5
d	r	r	d	6
d	r	r	r	7
r	d	d	d	8

Note: d = dominant, r = recessive.

Table 13.1 ■ CAN data length coding

DATA FIELD

The data field consists of the data to be transmitted within a data frame. It may contain from 0 to 8 bytes, each of which contains 8 bits that are transferred most significant bit first.

CRC FIELD

The CRC field contains the CRC sequence followed by a CRC delimiter, as shown in Figure 13.5. The frame-check sequence is derived from a cyclic redundancy code best suited to frames with bit counts less than 127. The CRC sequence is calculated by performing a polynomial division. The coefficients of the polynomial are given by the destuffed bit stream, consisting of the start-of-frame field, arbitration field, control field, data field (if present), and 15 0s. This polynomial is divided (the coefficients are calculated using modulo-2 arithmetic) by the generator polynomial

$$X^{15} + X^{14} + X^{10} + X^8 + X^7 + X^4 + X^3 + 1$$

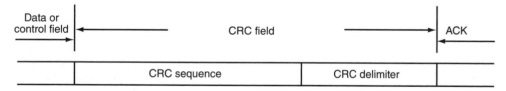

Figure 13.5 ■ CRC field

The remainder of this polynomial division is the CRC sequence. In order to implement this function, a 15-bit shift register **CRC_RG** (14:0) is used. If **nxtbit** denotes the next bit of the bit stream, given by the destuffed bit sequence from the start of frame until the end of the data field, the CRC sequence is calculated as follows:

```
CRC_RG = 0;                        /* initialize shift register */
do {
      crcnxt = nxtbit ^ CRC_RG(14);    /* exclusive OR */
      CRC_RG(14:1) = CRC_RG(13:0);    /* shift left by 1 bit */
      CRC_RG(0) = 0;
      if crcnxt
              CRC_RG(14:0) = CRC_RG(14:0) ^ 0x4599;
} while (!(CRC SEQUENCE starts or there is an error condition));
```

After the transmission/reception of the last bit of the data field, CRC_RG(14:0) contains the CRC sequence. The *CRC delimiter* is a single recessive bit.

ACK FIELD

As shown in Figure 13.6, the ACK field is 2 bits long and contains the ACK slot and the ACK delimiter. A transmitting node sends 2 recessive bits in the ACK field. A receiver that has received a valid message reports this to the transmitter by sending a dominant bit in the ACK slot (i.e., it sends ACK). A node that has received the matching CRC sequence overwrites the recessive bit in the ACK slot with a dominant bit. This bit will be received by the data frame transmitter and learn that the previously transmitted data frame has been correctly received. The *ACK delimiter* has to be a recessive bit. As a consequence, the ACK slot is surrounded by 2 recessive bits (the CRC delimiter and the ACK delimiter).

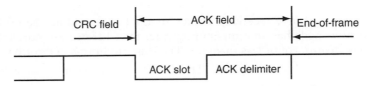

Figure 13.6 ■ ACK field

END-OF-FRAME FIELD

Each data frame and remote frame is delimited by a flag sequence consisting of 7 recessive bits. This 7-bit sequence is the *end-of-frame* sequence.

13.3.2 Remote Frame

A node that needs certain data can request the relevant source node to transmit the data by sending a remote frame. The format of a remote frame is shown in Figure 13.7. A remote frame

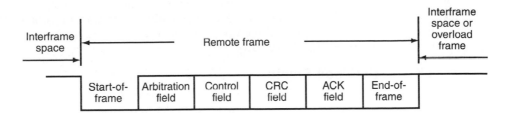

Figure 13.7 ■ Remote frame

consists of six fields: start of frame, arbitration, control, CRC, ACK, and end of frame. The polarity of the RTR bit in the arbitration field indicates whether a transmitted frame is a *data frame* (RTR bit dominant) or a *remote frame* (RTR bit recessive).

13.3.3 Error Frame

The error frame consists of two distinct fields. The first field is given by the superposition of error flags contributed from different nodes. The second field is the error delimiter. The format of the error frame is shown in Figure 13.8. In order to terminate an error frame correctly, an *error-passive node* may need the bus to be idle for at least 3 bit times (if there is a local error at

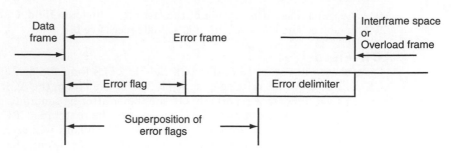

Figure 13.8 ■ Error frame

an error-passive receiver). Therefore, the bus should not be loaded to 100 percent. An error-passive node has an error count greater than 127 but no more than 255. An *error-active node* has an error count less than 127. There are two forms of error flags.

- *Active-error flag.* This flag consists of 6 consecutive dominant bits.
- *Passive-error flag.* This flag consists of 6 consecutive recessive bits unless it is overwritten by dominant bits from other nodes.

An error-active node signals an error condition by transmitting an *active-error* flag. The error flag's form violates the law of bit stuffing (to be discussed shortly) and applies to all fields from start of frame to CRC delimiter or destroys the fixed-form ACK field or end-of-frame field. As a consequence, all other nodes detect an error condition and each starts to transmit an error flag. Therefore, the sequence of dominant bits, which can be monitored on the bus, results from a superposition of different error flags transmitted by individual nodes. The total length of this sequence varies between a minimum of 6 and a maximum of 12 bits.

An error-passive node signals an error condition by transmitting a passive-error flag. The error-passive node waits for 6 consecutive bits of equal polarity, beginning at the start of the passive-error flag. The passive-error flag is complete when these equal bits have been detected.

The *error delimiter* consists of 8 recessive bits. After transmission of an error flag, each node sends recessive bits and monitors the bus until it detects a recessive bit. Afterward, it starts transmitting 7 more recessive bits.

13.3.4 Overload Frame

The *overload frame* contains two bit fields: *overload flag* and *overload delimiter*. There are three different overload conditions that lead to the transmission of an overload frame.

1. The internal conditions of a receiver require a delay of the next data frame or remote frame.
2. At least one node detects a dominant bit during intermission.
3. A CAN node samples a dominant bit at the eighth bit (i.e., the last bit) of an error delimiter or overload delimiter. The error counters will not be incremented.

The format of an overload frame is shown in Figure 13.9. An overload frame resulting from condition 1 is only allowed to start at the first bit time of an expected intermission, whereas an overload frame resulting from overload conditions 2 and 3 starts 1 bit after detecting the dominant bit.

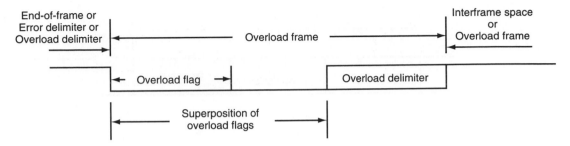

Figure 13.9 ■ Overload frame

No more than two overload frames may be generated to delay the next data frame or remote frame. The overload flag consists of 6 dominant bits. The format of an overload frame is similar to that of the active-error flag. The overload flag's form destroys the fixed form of the *intermission field*. As a consequence, all other nodes also detect an overload condition and each starts to transmit an overload flag. In the event that there is a dominant bit detected during the third bit of *intermission* locally at some node, it will interpret this bit as the start of frame.

The overload delimiter consists of 8 recessive bits. The overload delimiter has the same form as the error delimiter. After the transmission of an overload flag, the node monitors the bus until it detects a transition from a dominant to a recessive bit. At this point of time every bus node has finished sending its overload flag and all nodes start transmission of 7 more recessive bits in coincidence.

13.3.5 Interframe Space

Data frames and remote frames are separated from preceding frames by a field called *interframe space*. In contrast, overload frames and error frames are not preceded by an interframe space, and multiple overload frames are not separated by an interframe space.

For nodes that are not error-passive or have been receivers of the previous message, the interframe space contains the bit fields of *intermission* and *bus idle*, as shown in Figure 13.10. The interframe space of an error-passive node consists of three subfields: *intermission, suspend transmission*, and *bus idle*, as shown in Figure 13.11.

The intermission subfield consists of 3 recessive bits. During intermission, no node is allowed to start transmission of the data frame or remote frame. The only action permitted is

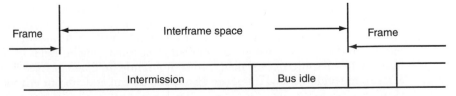

Figure 13.10 ■ Interframe space for non error-passive node or receiver of previous message

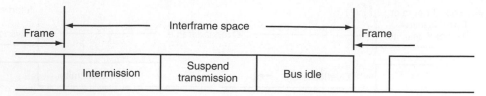

Figure 13.11 ■ Interframe space for error-passive nodes

signaling of an overload condition. The period of bus idle may be of arbitrary length. The bus is recognized to be free, and any node having something to transmit can access the bus. A message, pending during the transmission of another message, is started in the first bit following intermission. When the bus is idle, the detection of a dominant bit on the bus is interpreted as a start of frame. After an error-passive node has transmitted a frame, it sends 8 recessive bits following intermission, before starting to transmit a new message or recognizing the bus as idle. If, meanwhile, a transmission (caused by another node) starts, the node will become the receiver of this message.

13.3.6 Message Filtering

A node uses filter(s) to decide whether to work on a specific message. Message filtering is applied to the whole identifier. A node can optionally implement mask registers that specify which bits in the identifier are examined with the filter. If mask registers are implemented, every bit of the mask registers must be programmable; in other words, they can be enabled or disabled for message filtering. The length of the mask register can comprise the whole identifier or only part of it.

13.3.7 Message Validation

The point in time at which a message is taken to be valid is different for the transmitters and receivers of the message. The message is valid for the transmitter if there is no error until the end of frame. If a message is corrupted, retransmission will follow automatically and according to the rules of prioritization. In order to be able to compete for bus access with other messages, retransmission has to start as soon as the bus is idle. The message is valid for the receiver if there is no error until the last but 1 bit of the end of frame.

13.3.8 Bitstream Encoding

The frame segments including start-of-frame field, arbitration field, control field, data field, and CRC sequence are encoded by *bit stuffing*. Whenever a transmitter detects 5 consecutive bits of identical value in the bitstream to be transmitted, it automatically inserts a complementary bit in the actual transmitted bitstream. The remaining bit fields of the data frame or remote frame (CRC delimiter, ACK field, and end-of-frame field) are of fixed form and not stuffed. The error frame and overload frame are also of fixed form and are not encoded by the method of bit stuffing.

The bitstream in a message is encoded using the *non-return-to-zero* (NRZ) method. This means that during the total bit time the generated bit level is either dominant or recessive.

13.4 Error Handling

There are five types of errors. These errors are not mutually exclusive.

13.4.1 Bit Error

A node that is sending a bit on the bus also monitors the bus. When the bit value monitored is different from the bit value being sent, the node interprets the situation as an error. There are two exceptions to this rule.

- A node that sends a recessive bit during the stuffed bitstream of the arbitration field or during the ACK slot detects a dominant bit.
- A transmitter that sends a passive-error flag detects a dominant bit.

13.4.2 Stuff Error

A stuff error is detected whenever six consecutive dominant or six consecutive recessive levels occur in a message field.

13.4.3 CRC Error

The CRC sequence consists of the result of the CRC calculation by the transmitter. The receiver calculates the CRC in the same way as the transmitter. A CRC error is detected if the calculated result is not the same as that received in the CRC sequence.

13.4.4 Form Error

A form error is detected when a fixed-form bit field contains one or more illegal bits. For a receiver, a dominant bit during the last bit of the end-of-frame field is not treated as a form error.

13.4.5 Acknowledgement Error

An acknowledgement error is detected whenever the transmitter does not monitor a dominant bit in the ACK slot.

13.4.6 Error Signaling

A node that detects an error condition signals the error by transmitting an error flag. An error-active node will transmit an *active-error* flag; an error-passive node will transmit a *passive-error* flag. Whenever a node detects a bit error, a stuff error, a form error, or an acknowledgement error, it will start transmission of an error flag at the next bit time. Whenever a CRC error is detected, transmission of an error flag will start at the bit following the ACK delimiter, unless an error flag for another error condition has already been started.

13.5 Fault Confinement

13.5.1 CAN Node Status

A node in error may be in one of three states: error active, error passive, or bus off. An error-active node can normally take part in bus communication and sends an active-error flag when an error has been detected. An error-passive node must not send an active-error flag. It takes part in bus communication, but when an error has been detected, only a passive-error flag is sent.

After a transmission, an error-passive node will wait before initiating further transmission. A bus-off node is not allowed to have any influence on the bus.

13.5.2 Error Counts

The CAN protocol requires each node to implement *transmit error count* and *receive error count* to facilitate fault confinement. These two counts are updated according to 12 rules. These 12 rules can be found in the CAN specification. An error count value greater than roughly 96 indicates a heavily disturbed bus. It may be advantageous to provide the means to test for this condition. If during system start-up only one node is online and if this node transmits some message, it will get no acknowledgement, detect an error, and repeat the message. It can become error passive but not bus off for this reason.

13.6 CAN Message Bit Timing

The setting of a bit time in a CAN system must allow a bit sent out by the transmitter to reach the far end of the CAN bus and allow the receiver to send back an acknowledgement that reaches the transmitter. In a CAN environment, the *nominal bit rate* is defined to be the number of bits transmitted per second in the absence of resynchronization by an ideal transmitter.

13.6.1 Nominal Bit Time

The inverse of the nominal bit rate is the *nominal bit time*. A nominal bit time can be divided into four nonoverlapping time segments, as shown in Figure 13.12.

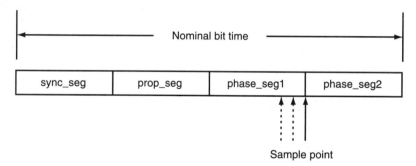

Figure 13.12 ■ Nominal bit time

The *sync_seg* segment is used to synchronize the various nodes on the bus. An edge is expected to lie within this segment. The *prop_seg* segment is used to compensate for the physical delay times within the network. It is twice the sum of the signal's propagation time on the bus line, the input comparator delay, and the output driver delay. The *phase_seg1* and *phase_seg2* segments are used to compensate for edge phase errors. These segments can be lengthened or shortened by synchronization. The *sample point* is the point in time at which the bus level is read and interpreted as the value of that respective bit. The sample point is located at the end of phase_seg1. A CAN controller may implement the three-samples-per-bit option in which the majority function is used to determine the bit value. Each sample is separated from the next sample by half a time quanta (CAN clock cycle). The *information processing time* is the time segment starting with the sample point reserved for calculation of the sample bit(s) level. The segments contained in a nominal bit time are represented in the unit of *time quantum*. The time quantum (t_Q) is a fixed unit of time that can be derived from the oscillator period (T_{OSC}). t_Q is expressed as a multiple of a *minimum time quantum*. This multiple is a programmable prescale factor. Thus, the time quantum can have the length of

Time quantum $= M \times$ minimum time quantum

where M is the value of the prescaler.

13.6.2 Length of Time Segments

The segments of a nominal bit time can be expressed in the unit of time quantum as follows:

- *sync_seg* is 1 time quantum long.
- *prop_seg* is programmable to be 1, 2, . . . , 8 time quanta long.
- *phase_seg1* is programmable to be 1, 2, . . . , 8 time quanta long.
- *phase_seg2* is the maximum of *phase_seg1* and information processing time and hence will be programmable from 2 to $8t_Q$.
- The information processing time is equal to or less than $2t_Q$ and is fixed at $2t_Q$ for the HCS12 CAN module.

The total number of time quanta in a bit time must be programmable over a range of at least 8 to 25.

13.7 Synchronization Issue

All CAN nodes must be synchronized while receiving a transmission; that is, the beginning of each received bit must occur during each node's *sync_seg* segment. This is achieved by synchronization. Synchronization is required because of phase errors between nodes, which may arise because of nodes having slightly different oscillator frequencies or because of changes in propagation delay when a different node starts transmitting.

Two types of synchronization are defined: *hard synchronization* and *resynchronization*. Hard synchronization is performed only at the beginning of a message frame, when each CAN node aligns the *sync_seg* of its current bit time to the recessive to dominant edge of the transmitted start-of-frame field. After a hard synchronization, the internal bit time is restarted with sync_seg. Resynchronization is subsequently performed during the remainder of the message frame whenever a change of bit value from recessive to dominant occurs outside of the expected *sync_seg* segment. Resynchronization is achieved by implementing a digital phase-lock loop (DPLL) function that compares the actual position of a recessive-to-dominant edge on the bus to the position of the expected edge.

13.7.1 Resynchronization Jump Width

There are three possibilities for the occurrence of the incoming recessive-to-dominant edge.

1. *After the sync_seg segment but before the sample point*. This situation is interpreted as a *late edge*. The node will attempt to resynchronize to the bitstream by increasing the duration of its phase_seg1 segment of the current bit by the number of time quanta by which the edge was late, up to the resynchronization jump width limit.

2. *After the sample point but before the sync_seg segment of the next bit*. This situation is interpreted as an *early bit*. The node will now attempt to resynchronize to the bitstream by decreasing the duration of its *phase_seg2* segment of the current bit by the number of time quanta by which the edge was early, up to the resynchronization jump width limit. Effectively, the *sync_seg* segment of the next bit begins immediately.

3. *Within the sync_seg segment of the current bit time*. This is interpreted as no synchronization error.

As a result of resynchronization, phase_seg1 may be lengthened or phase_seg2 may be shortened. The amount by which the phase buffer segments may be altered may not be greater than the *resynchronization jump width*, which is programmable to be between 1 and the smaller of 4 and the phase_seg1 time quanta.

Clocking information may be derived from transitions from 1 bit value to the other. The property that only a fixed maximum number of successive bits have the same value provides the possibility of resynchronizing a bus node to the bitstream during a frame.

The maximum length between two transitions that can be used for resynchronization is 29 bit times.

13.7.2 Phase Error of an Edge

The *phase error* of an edge is given by the position of the edge relative to sync_seg, measured in time quanta. The sign of phase error is defined as follows:

$e < 0$ if the edge lies after the sample point of the previous bit

$e = 0$ if the edge lies within sync_seg

$e > 0$ if the edge lies before the sample point

13.8 Overview of the HCS12 CAN Module

The Freescale implementation of the CAN 2.0A/B protocol for the HCS12 microcontroller family is referred to as MSCAN12. The MSCAN12 provides the following features:

- Full implementation of the CAN 2.0A/B protocol
- Five receive buffers with FIFO storage scheme
- Three transmit buffers with internal prioritization using a local priority concept
- Maskable identifier filter supporting two full-size extended identifier filters (two 32-bit) or four 16-bit filters or eight 8-bit filters
- Programmable wake-up functionality with integrated low-pass filter
- Programmable loopback mode supporting self-test operation
- Programmable listen-only mode for monitoring of CAN bus
- Separate signaling and interrupt capabilities for all CAN receiver and transmitter error states (warning, error passive, bus off)
- Clock source coming from either E-clock or oscillator clock
- Internal time for time stamping of received and transmitted message
- Three low-power modes: sleep, power-down, and MSCAN12 enable
- Global initialization of configuration registers

The block diagram of the MSCAN12 module is shown in Figure 13.13. As shown in the figure, each MSCAN module uses two signal pins: TxCAN and RxCAN.

All members of the automotive subfamily in the HCS12 family have implemented one or more CAN modules. The HCS12DP256 implements five CAN modules. For other HCS12 devices without an on-chip CAN module, a CAN controller chip such as the MCP2510 from Microchip or the PCA82C200 from Philips can be used to interface with the CAN bus. Most CAN controllers do not have enough driving capability and hence require a dedicated CAN bus

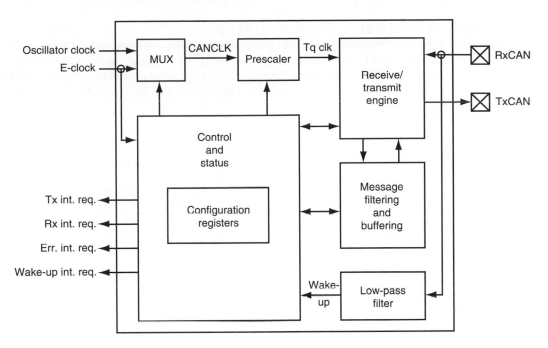

Figure 13.13 ■ MSCAN12 block diagram

transceiver chip such as the MCP2551 (from Microchip) or the PCA82C250 (from Philips) to interface with the CAN bus. In addition to providing large driving current, a CAN transceiver chip has current protection against defected CAN or defected nodes. A typical CAN system that includes one or more HCS12 devices is shown in Figure 13.14.

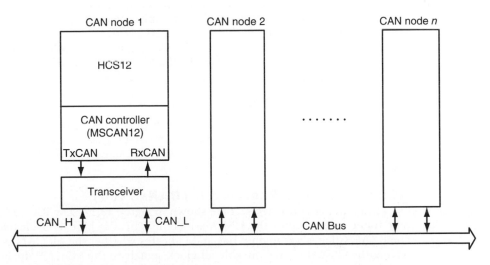

Figure 13.14 ■ A typical CAN system

13.9 MSCAN Module Memory Map

An HCS12 member may have up to five CAN modules and each CAN module occupies 64 bytes of memory space. The MSCAN register organization is shown in Figure 13.15. The detailed MSCAN register memory map is shown in Figure 13.16.

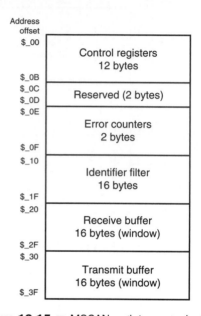

Figure 13.15 ■ MSCAN register organization

Each receive buffer and each transmit buffer occupies 16 bytes of space. Even though each MSCAN module has three transmit buffers, only one of them is accessible to the programmer at a time. Similarly, only one of the five receive buffers is accessible to the programmer at a time. Only the offset of each register relative to the start of MSCAN register block is shown in Figure 13.16. The address of each register is the sum of the offset and the base address of the MSCAN register block.

13.9.1 MSCAN Control Registers

The function of each MSCAN control register is explained in this section. Since the HCS12 may have up to five identical CAN modules having the same set of control registers, Freescale names each CAN register as CANxYYYY, where x indicates the CAN module and YYYY specifies the register.

MSCAN CONTROL REGISTER 0 (CANxCTL0)

The contents of this register are shown in Figure 13.17. The RXFRM bit is set whenever a valid frame is received by the MSCAN module. This flag bit can be cleared by writing a 1 to it. The RXACT flag is a read-only flag and indicates whether the MSCAN is receiving a message. Setting the CSWAI bit will disable all clock signals to the MSCAN during the wait mode and further reduce power consumption.

Address	Register Name	Access
$_00	MSCAN control register 0 (CANCTL0)	R/W
$_01	MSCAN control register 1 (CANCTL1)	R/W
$_02	MSCAN bus timing register 0 (CANBTR0)	R/W
$_03	MSCAN bus timing register 1 (CANBTR1)	R/W
$_04	MSCAN receiver flag register (CANRFLG)	R/W
$_05	MSCAN receiver interrupt enable register (CANRIER)	R/W
$_06	MSCAN transmitter flag register (CANTFLG)	R/W
$_07	MSCAN transmitter interrupt enable register (CANTIER)	R/W
$_08	MSCAN transmitter message abort request (CANTARQ)	R/W
$_09	MSCAN transmitter message abort acknowledge (CANTAAK)	R
$_0A	MSCAN transmit buffer selection (CANTBSEL)	R/W
$_0B	MSCAN identifier acceptance control register (CANIDAC)	R/W
$_0C	Reserved	
$_0D	Reserved	
$_0E	MSCAN receive error counter register (CANRXERR)	R
$_0F	MSCAN transmit error counter register (CANTXERR)	R
$_10	MSCAN identifier acceptance register 0 (CANIDAR0)	R/W
$_11	MSCAN Identifier acceptance register 1 (CANIDAR1)	R/W
$_12	MSCAN identifier acceptance register 2 (CANIDAR2)	R/W
$_13	MSCAN identifier acceptance register 3 (CANIDAR3)	R/W
$_14	MSCAN identifier mask register 0 (CANIDMR0)	R/W
$_15	MSCAN identifier mask register 1 (CANIDMR1)	R/W
$_16	MSCAN identifier mask register 2 (CANIDMR2)	R/W
$_17	MSCAN identifier mask register 3 (CANIDMR3)	R/W
$_18	MSCAN identifier acceptance register 4 (CANIDAR4)	R/W
$_19	MSCAN identifier acceptance register 5 (CANIDAR5)	R/W
$_1A	MSCAN identifier acceptance register 6 (CANIDAR6)	R/W
$_1B	MSCAN identifier acceptance register 7 (CANIDAR7)	R/W
$_1C	MSCAN identifier mask register 4 (CANIDMR4)	R/W
$_1D	MSCAN identifier mask register 5 (CANIDMR5)	R/W
$_1E	MSCAN identifier mask register 6 (CANIDMR6)	R/W
$_1F	MSCAN identifier mask register 7 (CANIDMR7)	R/W
$_20 $_2F	Foreground receive buffer (CANRXFG)	R
$_30 $_3F	Foreground transmit buffer (CANTXFG)	R/W

Figure 13.16 ■ CAN module memory map

The SYNCH bit indicates whether the MSCAN module is synchronized to the CAN bus. The MSCAN module has a 16-bit free-running timer. When the TIME bit is set to 1, the timer value will be assigned to each transmitted and received message within the transmit and receive buffers. The timestamp is assigned as soon as a message is acknowledged. The WUPE bit allows the MSCAN module to be woken up by the CAN bus activity. Setting the SLPRQ bit requests the MSCAN to enter the sleep mode. The sleep mode request is serviced only when the CAN bus is idle.

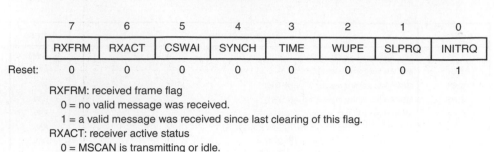

7	6	5	4	3	2	1	0
RXFRM	RXACT	CSWAI	SYNCH	TIME	WUPE	SLPRQ	INITRQ

Reset: 0   0   0   0   0   0   0   1

RXFRM: received frame flag
  0 = no valid message was received.
  1 = a valid message was received since last clearing of this flag.
RXACT: receiver active status
  0 = MSCAN is transmitting or idle.
  1 = MSCAN is receiving a message (including when arbitration is lost).
CSWAI: CAN stops in wait mode
  0 = the module is not affected during wait mode.
  1 = the module ceases to be clocked during wait mode.
SYNCH: synchronization status
  0 = MSCAN is not synchronized to the CAN bus.
  1 = MSCAN is synchronized to the CAN bus.
TIME: timer enable
  0 = disable internal MSCAN timer.
  1 = enable internal MSCAN timer and hence enable time stamp.
WUPE: wake-up enable
  0 = wake-up disabled (MSCAN ignores traffic on CAN bus).
  1 = wake-up enabled (MSCAN is able to restart).
SLPRQ: sleep mode request
  0 = running—the MSCAN functions normally.
  1 = sleep mode request—the MSCAN enters sleep mode when CAN is idle.
INITRQ: initialization mode request
  0 = normal operation.
  1 = MSCAN in initialization mode.

Figure 13.17 ■ MSCAN control register 0 (CANxCTL0, x = 0, 1, 2, 3, or 4)

To make sure that the CAN protocol is not violated, the MSCAN can only be configured when the MSCAN is in the initialization mode. The initialization mode is entered by setting the INITRQ bit. The registers CANxCTL1, CANxBTR0, CANxBTR1, CANxIDAC, CANxIDAR0~7, CANxIDMR0~7 can only be written by the CPU when the MSCAN is in the initialization mode. When the INITRQ bit is cleared by the CPU, the MSCAN restarts and then tries to synchronize to the CAN bus.

MSCAN CONTROL REGISTER 1 (CANxCTL1)

This register provides for various control and handshake status information of the MSCAN module. The contents of this register are shown in Figure 13.18. The MSCAN module must be enabled before it can operate. Setting the CANE bit enables the MSCAN module. The clock source of the CAN module can be either the E-clock or the oscillator clock. When the CAN bus is operating at a high data rate and the E-clock is generated by the PLL circuit, the user should select the oscillator clock as the clock source because the PLL circuit has jitter.

The loopback mode can be used to test the software. The loopback mode is entered by setting the LOOPB bit. The MSCAN module can be placed in the listen-only mode to monitor the CAN bus traffic without participating in the CAN data exchange. The listen-only mode

7	6	5	4	3	2	1	0
CANE	CLKSRC	LOOPB	LISTEN	0	WUPM	SLPAK	INITAK

Reset: 0   0   0   1   0   0   0   1

CANE: MSCAN enable
  0 = the MSCAN module is disabled.
  1 = the MSCAN module is enabled.
CLKSRC: MSCAN clock source
  0 = the MSCAN clock source is the oscillator clock.
  1 = the MSCAN clock source is the bus clock.
LOOPB: loopback self-test mode
  0 = loopback self-test disabled.
  1 = loopback self-test enabled.
LISTEN: listen-only mode
  0 = normal operation.
  1 = listen-only mode activated.
WUPM: wake-up mode
  0 = MSCAN wakes up the CPU after any recessive to dominant edge on the
    CAN bus and WUPE bit of the CANCTL0 register is set to 1.
  1 = MSCAN wakes up the CPU only in case of a dominant pulse on the CAN
    bus that has a length of T_{WUP} and the WUPE bit is set to 1.
SLPAK: sleep mode acknowledge
  0 = running (MSCAN functions normally).
  1 = sleep mode active (MSCAN has entered sleep mode).
INITAK: initialization mode acknowledge
  0 = normal operation (MSCAN operates normally).
  1 = initialization mode active (MSCAN is in initialization mode).

Figure 13.18 ■ MSCAN control register 1 (CANxCTL, x = 0, 1, 2, 3, or 4)

is entered by setting the LISTEN bit. When operating in a noisy environment, one can choose to enable the filter to filter out the noise on the CAN bus so that the CAN module won't be woken up by the noise.

There is a time delay from requesting to enter the sleep mode or initialization mode until the desired mode is entered. To make sure the requested mode is entered, one should check the SLPAK and INITAK bits.

MSCAN Bus Timing Register 0 (CANxBTR0)

The CANxBTR0 register allows the user to select the synchronization jump width and the divide factor to the selected clock source to derive the time quantum. The contents of the CANxBTR0 register are shown in Figure 13.19. The MSCAN uses the time quantum (t_Q) as the minimum unit for timing control. The synchronization jump width defines the maximum number of time quantum (t_Q) clock cycles a bit time can be shortened or lengthened to achieve resynchronization to data transitions on the bus. The time quantum (t_Q) is derived by dividing the prescale factor into the selected clock source. The value of the prescale factor is defined by bits BRP5~BRP0.

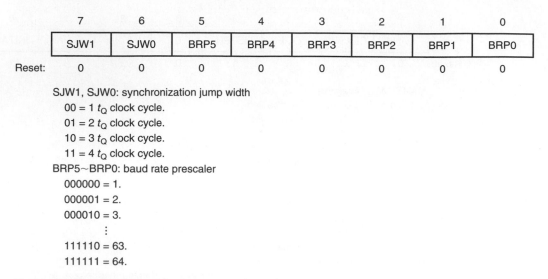

7	6	5	4	3	2	1	0
SJW1	SJW0	BRP5	BRP4	BRP3	BRP2	BRP1	BRP0

Reset: 0   0   0   0   0   0   0   0

SJW1, SJW0: synchronization jump width
  00 = 1 t_Q clock cycle.
  01 = 2 t_Q clock cycle.
  10 = 3 t_Q clock cycle.
  11 = 4 t_Q clock cycle.
BRP5~BRP0: baud rate prescaler
  000000 = 1.
  000001 = 2.
  000010 = 3.
    ⋮
  111110 = 63.
  111111 = 64.

Figure 13.19 ■ MSCAN control register 0 (CANxBTRO, x = 0, 1, 2, 3, or 4)

MSCAN BUS TIMING REGISTER 1 (CANxBTR1)

This register provides for control on phase_seg1 and phase_seg2 in Figure 13.12. The contents of the CANxBTR1 register are shown in Figure 13.20. When three samples are taken, the bit value is determined by the majority function of these samples. Two adjacent samples are separated by one time quantum.

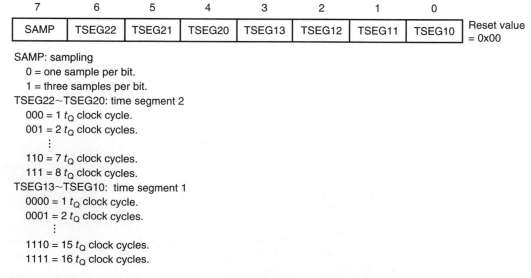

7	6	5	4	3	2	1	0	
SAMP	TSEG22	TSEG21	TSEG20	TSEG13	TSEG12	TSEG11	TSEG10	Reset value = 0x00

SAMP: sampling
  0 = one sample per bit.
  1 = three samples per bit.
TSEG22~TSEG20: time segment 2
  000 = 1 t_Q clock cycle.
  001 = 2 t_Q clock cycles.
    ⋮
  110 = 7 t_Q clock cycles.
  111 = 8 t_Q clock cycles.
TSEG13~TSEG10: time segment 1
  0000 = 1 t_Q clock cycle.
  0001 = 2 t_Q clock cycles.
    ⋮
  1110 = 15 t_Q clock cycles.
  1111 = 16 t_Q clock cycles.

Figure 13.20 ■ MSCAN control register 1 (CANxBTR1, x = 0, 1, 2, 3, or 4)

The bit time is determined by the oscillator frequency, the baud rate prescaler, and the number of time quanta (t_Q) clock cycles per bit. The bit time is given by the following expression:

$$\text{Bit time} = \frac{\text{prescaler value}}{f_{\text{CANCLK}}} \times (1 + \text{TimeSegment1} + \text{TimeSegment2})$$

TimeSegment1 consists of prop_seg and phase_seg1. TimeSegment2 is the same as phase_seg2.

MSCAN RECEIVER FLAG REGISTER (CANxRFLG)

This register contains status flags related to CAN reception. The contents of this register are shown in Figure 13.21. The WUPIF flag will be set to 1 if the MSCAN detects CAN bus activity and the WUPE bit of the CANxCTL0 register is set to 1. The CSCIF flag will be set to 1 when the MSCAN changes its current bus status due to the actual value of the transmit error counter and the receive error counter. If a valid message is received when all five receive buffers are full, then the OVRIF flag will be set to 1 to indicate this overflow condition. When the receive FIFO is not empty, the RXF flag is set to 1. The flag bits WUPIF, CSCIF, OVRIF, and RXF are cleared by writing a 1 to them.

7	6	5	4	3	2	1	0	
WUPIF	CSCIF	RSTAT1	RSTAT0	TSTAT1	TSTAT0	OVRIF	RXF	Reset value = 0x00

WUPIF: wake-up interrupt flag
    0 = no wake-up activity observed while in sleep mode.
    1 = MSCAN detected activity on the bus and requested wake-up.
CSCIF: CAN status change interrupt flag
    0 = no change in bus status occurred since last interrupt.
    1 = MSCAN changed current bus status.
RSTAT1-RSTAT0: receiver status bits
    00 = RxOK: 0 ≤ Receive error counter ≤ 96.
    01 = RxWRN: 96 < Receive error counter ≤ 127.
    10 = RxERR: 127 < Receive error counter.
    11 = Bus-off[1]: Transmit error counter > 255.
TSTAT1~TSTAT0: transmitter status bits
    00 = TxOK: 0 ≤ Transmit error counter ≤ 96.
    01 = TxWRN: 96 < Transmit error counter ≤ 127.
    10 = TxERR: 127 < Transmit error counter.
    11 = Bus-off: Transmit error counter > 255.
OVRIF: overrun interrupt flag
    0 = no data overrun occurred.
    1 = a data overrun detected.
RXF: receive buffer full flag
    0 = no new message available within the RxFG.
    1 = the receive FIFO is not empty. A new message is available in the RxFG.
Note 1. This information is redundant. As soon as the transmitter leaves its
    bus-off state, the receiver state skips to RxOK too.

Figure 13.21 ■ MSCAN receiver flag register (CANxRFLG, x = 0, 1, 2, 3, or 4)

MSCAN Receiver Interrupt Enable Register (CANxRIER)

This register enables and disables all interrupts related to the CAN receiver. The contents of this register are shown in Figure 13.22.

7	6	5	4	3	2	1	0	
WUPIE	CSCIE	RSTATE1	RSTATE0	TSTATE1	TSTATE0	OVRIE	RXFIE	Reset value = 0x00

WUPIE: wake-up interrupt enable
  0 = no interrupt request is generated from this event.
  1 = a wake-up event causes a wake-up interrupt request.
CSCIE: CAN status change interrupt enable
  0 = no interrupt request is generated from this event.
  1 = a CAN status change event causes an error interrupt request.
RSTATE1~RSTATE0: receiver status change interrupt enable
  00 = do not generate any CSCIF interrupt caused by receiver state changes.
  01 = generate CSCIF interrupt only if the receiver enters or leaves bus-off state.
  10 = generate CSCIF interrupt only if the receiver enters or leaves RxErr or bus-off state.
  11 = generate CSCIF interrupt on all state changes.
TSTATE1~TSTATE0: transmitter status change interrupt enable
  00 = do not generate any CSCIF interrupt caused by transmitter state changes.
  01 = generate CSCIF interrupt only if the transmitter enters or leaves bus-off state.
  10 = generate CSCIF interrupt only if the transmitter enters or leaves bus-off or TxErr state.
  11 = generate CSCIF interrupt on all state changes.
OVRIE: overrun interrupt enable
  0 = no interrupt request is generated from this event.
  1 = an overrun event causes an error interrupt request.
RXFIE: receive buffer interrupt enable
  0 = no interrupt request is generated from this event.
  1 = a receive buffer full event causes a receiver interrupt request.

Figure 13.22 ■ MSCAN receiver interrupt enable register (CANxRIER, x = 0, 1, 2, 3, or 4)

MSCAN Transmitter Flag Register (CANxTFLG)

The contents of this register are shown in Figure 13.23. Each of the three flags indicates whether its associated transmit buffer is empty, and thus not scheduled for transmission. These flags can be cleared by writing a 1 to them.

7	6	5	4	3	2	1	0	
0	0	0	0	0	TXE2	TXE1	TXE0	Reset value = 0x07

TXE2~TXE0: transmitter buffer x (x = 0,1, or 2) empty
  0 = the associated message buffer is full (loaded with a message due for transmission).
  1 = the associated message buffer is empty.

Figure 13.23 ■ MSCAN transmitter flag register (CANxTFLG, x = 0, 1, 2, 3, or 4)

MSCAN Transmitter Interrupt Enable Register (CANxTIER)

The contents of this register are shown in Figure 13.24.

7	6	5	4	3	2	1	0	
0	0	0	0	0	TXEIE2	TXEIE1	TXEIE0	Reset value = 0x00

TXEIE2~TXEIE0: transmitter k (k = 0,1, or 2) empty interrupt enable
  0 = disable interrupt from this buffer.
  1 = a transmitter empty event causes a transmitter empty interrupt request.

Figure 13.24 ■ MSCAN transmitter interrupt enable register (CANxTIER, x = 0, 1, 2, 3, or 4)

MSCAN Transmitter Message Abort Request Register (CANxTARQ)

When the application has a high-priority message to be sent but cannot find any empty transmit buffer to use, it can request to abort the previous messages that have been scheduled for transmission. The application can request to abort the message in any one of the transmit buffers. This is done via the CANxTARQ register. The contents of this register are shown in Figure 13.25.

7	6	5	4	3	2	1	0	
0	0	0	0	0	ABTRQ2	ABTRQ1	ABTRQ0	Reset value = 0x00

ABTRQ2~ABTRQ0: abort transmit request k (k = 0,1, or 2)
  0 = no abort request.
  1 = abort request pending.

Figure 13.25 ■ MSCAN transmitter message abort request register (CANxTARQ, x = 0, 1, 2, 3, or 4)

MSCAN Transmit Message Abort Acknowledge Register (CANxTAAK)

A message that is being transmitted cannot be aborted. Only those messages that have not been transmitted can be aborted. MSCAN answers the abort request by setting or clearing the associated bits in this register. The contents of this register are shown in Figure 13.26.

7	6	5	4	3	2	1	0	
0	0	0	0	0	ABTAK2	ABTAK1	ABTAK0	Reset value = 0x00

ABTAK2~ABTAK0: abort transmit k (k = 0,1, or 2) acknowledge
  0 = the message was not aborted.
  1 = the message was aborted.

Figure 13.26 ■ MSCAN transmitter message abort acknowledge register (CANxTAAK, x = 0, 1, 2, 3, or 4)

MSCAN Transmit Buffer Selection (CANxTBSEL)

This register allows the selection of the transmit message buffer, which will then be accessible in the CANxTXFG register space. The contents of this register are shown in Figure 13.27. The lowest-numbered bit that is set makes the respective transmit buffer accessible to the user. For example, the combination TX1 = 1 and TX0 = 1 selects transmit buffer 0, and the combination TX1 = 1 and TX0 = 0 selects transmit buffer TX1.

7	6	5	4	3	2	1	0	
0	0	0	0	0	TX2	TX1	TX0	Reset value = 0x00

TX2~TX0: transmit buffer k (*k* = 0,1, or 2) select bits
  0 = the associated message buffer is deselected.
  1 = the associated message buffer is selected, if it is the lowest-numbered bit.

Figure 13.27 ■ MSCAN transmitter buffer select register (CANxTBSEL, *x* = 0, 1, 2, 3, or 4)

When one wants to get the next available transmit buffer, the CANxTFLG register is read and this value is written into the CANxTBSEL register. Suppose that all three transmit buffers are available. The value read from CANxTFLG is therefore %00000111. When writing this value back to CANxTSEL, the transmit buffer 0 is selected as the foreground transmit buffer because the lowest-numbered bit set to 1 is at bit position 0. Reading back this value of CANxTSEL results in %00000001, because only the lowest-numbered bit position set to 1 is presented. This mechanism eases the application software selection of the next available transmit buffer. This situation is shown in the following instruction sequence:

```
ldaa   CAN1TFLG        ; value read is %00000111
staa   CAN1TBSEL       ; value written is %00000111
ldaa   CAN1TBSEL       ; value read is %00000001
```

MSCAN IDENTIFIER ACCEPTANCE CONTROL REGISTER (CANxIDAC)

This register provides for identifier acceptance control as shown in Figure 13.28. The IDHITx indicators are always related to the message in the foreground receive buffer (RxFG). When a message gets shifted into the foreground buffer of the receiver FIFO, the indicators are updated as well.

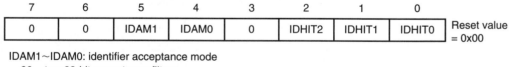

7	6	5	4	3	2	1	0	
0	0	IDAM1	IDAM0	0	IDHIT2	IDHIT1	IDHIT0	Reset value = 0x00

IDAM1~IDAM0: identifier acceptance mode
  00 = two 32-bit acceptance filters.
  01 = four 16-bit acceptance filters.
  10 = eight 8-bit acceptance filters.
  11 = filter closed
IDHIT2~IDHIT0: identifier acceptance hit indicator (read only)
  000 = filter 0 hit.
  001 = filter 1 hit.
  010 = filter 2 hit.
  011 = filter 3 hit.
  100 = filter 4 hit.
  101 = filter 5 hit.
  110 = filter 6 hit.
  111 = filter 7 hit.

Figure 13.28 ■ MSCAN identifier acceptance control register (CANxIDAC, *x* = 0, 1, 2, 3, or 4)

MSCAN RECEIVE ERROR COUNTER REGISTER (CANxRXERR)

This register reflects the status of the MSCAN receive error counter. This register can be read only in sleep or initialization mode. Reading this register in any other mode may return an incorrect value.

MSCAN TRANSMIT ERROR COUNTER REGISTER (CANxTXERR)

This register reflects the status of the MSCAN transmit error counter. Like CANxRXERR, this register can be read correctly only in sleep or initialization mode.

MSCAN IDENTIFIER ACCEPTANCE REGISTERS (CANxIDAR0~7)

On reception, each message is written into the background receive buffer. The CPU is only signaled to read the message if it passes the criteria in the identifier acceptance and identifier mask registers (accepted); otherwise, the message is overwritten by the next message (dropped).

The acceptance registers of the MSCAN are applied on the IDAR0 to IDAR3 registers of the incoming messages in a bit-by-bit manner. For extended identifiers, all four acceptance and mask registers are applied. For standard identifiers, only the first two are applied. The contents of the first bank and second bank acceptance registers are shown in Figures 13.29 and 13.30, respectively. The actual application of acceptance and mask registers is controlled by the CANxIDAC register.

AC7~AC0 comprise a user-defined sequence of bits with which the corresponding bits of the related identifier register (IDRn) of the receive message buffer are compared. The result of this comparison is then masked with the corresponding identifier mask register to determine if there is a hit.

MSCAN IDENTIFIER MASK REGISTERS (CANxIDMR0~7)

The identifier mask registers specify which of the corresponding bits in the identifier acceptance register are relevant for acceptance filtering. The contents of the first and second banks of the identifier mask registers are shown in Figures 13.31 and 13.32, respectively. If a mask bit is 1, its corresponding acceptance bit will be ignored.

	7	6	5	4	3	2	1	0	
	AC7	AC6	AC5	AC4	AC3	AC2	AC1	AC0	CANxIDAR0
Reset:	0	0	0	0	0	0	0	0	
	7	6	5	4	3	2	1	0	
	AC7	AC6	AC5	AC4	AC3	AC2	AC1	AC0	CANxIDAR1
Reset:	0	0	0	0	0	0	0	0	
	7	6	5	4	3	2	1	0	
	AC7	AC6	AC5	AC4	AC3	AC2	AC1	AC0	CANxIDAR2
Reset:	0	0	0	0	0	0	0	0	
	7	6	5	4	3	2	1	0	
	AC7	AC6	AC5	AC4	AC3	AC2	AC1	AC0	CANxIDAR3
Reset:	0	0	0	0	0	0	0	0	

Figure 13.29 ■ MSCAN identifier acceptance registers (first bank) (x = 0, 1, 2, 3, or 4)

7	6	5	4	3	2	1	0	
AC7	AC6	AC5	AC4	AC3	AC2	AC1	AC0	CANxIDAR4
Reset: 0	0	0	0	0	0	0	0	

7	6	5	4	3	2	1	0	
AC7	AC6	AC5	AC4	AC3	AC2	AC1	AC0	CANxIDAR5
Reset: 0	0	0	0	0	0	0	0	

7	6	5	4	3	2	1	0	
AC7	AC6	AC5	AC4	AC3	AC2	AC1	AC0	CANxIDAR6
Reset: 0	0	0	0	0	0	0	0	

7	6	5	4	3	2	1	0	
AC7	AC6	AC5	AC4	AC3	AC2	AC1	AC0	CANxIDAR7
Reset: 0	0	0	0	0	0	0	0	

Figure 13.30 ■ MSCAN identifier acceptance registers (second bank) (x = 0, 1, 2, 3, or 4)

7	6	5	4	3	2	1	0	
AM7	AM6	AM5	AM4	AM3	AM2	AM1	AM0	CANxIDMR0
Reset: 0	0	0	0	0	0	0	0	

7	6	5	4	3	2	1	0	
AM7	AM6	AM5	AM4	AM3	AM2	AM1	AM0	CANxIDMR1
Reset: 0	0	0	0	0	0	0	0	

7	6	5	4	3	2	1	0	
AM7	AM6	AM5	AM4	AM3	AM2	AM1	AM0	CANxIDMR2
Reset: 0	0	0	0	0	0	0	0	

7	6	5	4	3	2	1	0	
AM7	AM6	AM5	AM4	AM3	AM2	AM1	AM0	CANxIDMR3
Reset: 0	0	0	0	0	0	0	0	

Figure 13.31 ■ MSCAN identifier mask registers (first bank)

13.9.2 MSCAN Message Buffers

The receive message and transmit message buffers have the same outline. Each message buffer allocates 16 bytes in the memory map containing a 13-byte data structure (from identifier register 0 to data length register). The message buffer organization is illustrated in Figure 13.33.

7	6	5	4	3	2	1	0	
AM7	AM6	AM5	AM4	AM3	AM2	AM1	AM0	CANxIDMR4

Reset: 0   0   0   0   0   0   0   0

7	6	5	4	3	2	1	0	
AM7	AM6	AM5	AM4	AM3	AM2	AM1	AM0	CANxIDMR5

Reset: 0   0   0   0   0   0   0   0

7	6	5	4	3	2	1	0	
AM7	AM6	AM5	AM4	AM3	AM2	AM1	AM0	CANxIDMR6

Reset: 0   0   0   0   0   0   0   0

7	6	5	4	3	2	1	0	
AM7	AM6	AM5	AM4	AM3	AM2	AM1	AM0	CANxIDMR7

Reset: 0   0   0   0   0   0   0   0

Figure 13.32 ■ MSCAN identifier mask registers (second bank) (x = 0, 1, 2, 3, or 4)

Address	Register Name
$_x0	Identifier register 0
$_x1	Identifier register 1
$_x2	Identifier register 2
$_x3	Identifier register 3
$_x4	Data segment register 0
$_x5	Data segment register 1
$_x6	Data segment register 2
$_x7	Data segment register 3
$_x8	Data segment register 4
$_x9	Data segment register 5
$_xA	Data segment register 6
$_xB	Data segment register 7
$_xC	Data length register
$_xD	Transmit buffer priority register[1]
$_xE	Time stamp register high byte[2]
$_xF	Time stamp register low byte[2]

[1] Not applicable for receive buffer.
[2] Read-only for CPU.

Figure 13.33 ■ MSCAN message buffer organization

IDENTIFIER REGISTERS (IDR0~IDR3)

All four identifier registers are compared when a message with an extended identifier is received. The contents of these four identifier registers are shown in Figure 13.34. When a message with the standard identifier is received, only the first two identifier registers are compared. The meaning of the standard identifier is illustrated in Figure 13.35.

	7	6	5	4	3	2	1	0
IDR0	ID28	ID27	ID26	ID25	ID24	ID23	ID22	ID21
IDR1	ID20	ID19	ID18	SRR(=1)	IDE(=1)	ID17	ID16	ID15
IDR2	ID14	ID13	ID12	ID11	ID10	ID9	ID8	ID7
IDR3	ID6	ID5	ID4	ID3	ID2	ID1	ID0	RTR

Figure 13.34 ■ Receive/transmit message buffer extended identifier

	7	6	5	4	3	2	1	0
IDR0	ID10	ID9	ID8	ID7	ID6	ID5	ID4	ID3
IDR1	ID2	ID1	ID0	RTR	IDE(=0)			
IDR2								
IDR3								

Figure 13.35 ■ Receive/transmit message buffer standard identifier

The identifier consists of 29 bits (ID28~ID0) for the extended format. The ID28 bit is the most significant bit and is transmitted first. The CAN protocol uses the identifier to arbitrate simultaneous transmissions. The priority of an identifier is defined to be highest for the smallest binary number. The identifier consists of 11 bits (ID10~ID0) for the standard format. The ID10 bit is the most significant bit and is transmitted first. Similar to the extended identifier, the priority of a standard identifier is defined to be highest for the smallest binary number.

The SRR bit is used only in the extended format and must be set to 1. The IDE bit is used to indicate whether the identifier is an extended identifier. When set to 1, the identifier is used as an extended identifier.

The RTR bit is used to solicit transmission from other CAN nodes. In the case of a receive buffer, it indicates the status of the received frame and supports the transmission of an answering frame in software.

DATA SEGMENT REGISTERS (DSR0~DSR7)

The eight data segment registers, each with its DB7~DB0, contain the data to be transmitted or received. The number of bytes to be transmitted or received is determined by the data length code in the corresponding DLR.

DATA LENGTH REGISTER (DLR)

The data length register has only the lowest 4 bits implemented and is used to indicate the number of data bytes contained in the message.

TRANSMIT BUFFER PRIORITY REGISTER (TBPR)

This register defines the local priority of the associated message buffer. The local priority is used for the internal prioritization process of the MSCAN and is defined to be highest for the smallest binary number. The MSCAN implements the following internal prioritization mechanism:

- All transmission buffers with a cleared TXEx flag (bit 2, 1, or 0 of the CANxTFL register) participate in the prioritization immediately before the SOF (start of frame) is sent.

- The transmission buffer with the lowest local priority field wins the prioritization.

In case of more than one buffer having the same lowest priority, the message buffer with the lowest index number wins.

Time-stamp Register (TSRH, TSRL)

If the TIME bit of the CANxCTL0 register is set to 1, the MSCAN will write a special time-stamp to the respective registers in the active transmit or receive buffer as soon as a message has been acknowledged on the CAN bus. The timestamp is written on the bit sample point for the recessive bit of the ACK delimiter in the CAN frame. In the case of a transmission, the CPU can only read the timestamp after the respective transmit buffer has been flagged empty.

The timer value used for stamping is taken from a free-running internal CAN bit clock. A timer overrun is not indicated by the MSCAN. The contents of the timestamp registers are shown in Figure 13.36. The contents of timestamp registers are not sent out to the CAN bus.

	7	6	5	4	3	2	1	0	
	TSR15	TSR14	TSR13	TSR12	TSR11	TSR10	TSR9	TSR8	TSRH
Reset:	x	x	x	x	x	x	x	x	

	7	6	5	4	3	2	1	0	
	TSR7	TSR6	TSR5	TSR4	TSR3	TSR2	TSR1	TSR0	TSRL
Reset:	x	x	x	x	x	x	x	x	

Figure 13.36 ■ MSCAN timestamp registers

To facilitate access to the CAN buffers, the variable names for the transmit buffer and receive buffers are added to the *hcs12.inc* file for the assembly language and the *hcs12.h* file for the C language. These variable names are listed in Table 13.2a and b.

Name	Address	Description
CANxRIDR0	$_0	CAN foreground receive buffer x identifier register 0
CANxRIDR1	$_1	CAN foreground receive buffer x identifier register 1
CANxRIDR2	$_2	CAN foreground receive buffer x identifier register 2
CANxRIDR3	$_3	CAN foreground receive buffer x identifier register 3
CANxRDSR0	$_4	CAN foreground receive buffer x data segment register 0
CANxRDSR1	$_5	CAN foreground receive buffer x data segment register 1
CANxRDSR2	$_6	CAN foreground receive buffer x data segment register 2
CANxRDSR3	$_7	CAN foreground receive buffer x data segment register 3
CANxRDSR4	$_8	CAN foreground receive buffer x data segment register 4
CANxRDSR5	$_9	CAN foreground receive buffer x data segment register 5
CANxRDSR6	$_A	CAN foreground receive buffer x data segment register 6
CANxRDSR7	$_B	CAN foreground receive buffer x data segment register 7
CANxRDLR	$_C	CAN foreground receive buffer x data length register

1. x can be 0, 1, 2, 3, or 4,
2. The absolute address of each register is equal to the sum of the base address of the CAN foreground transmit buffer x and the address field of the corresponding register.

Table 13.2a ■ CAN foreground receive buffer x variable names

Name	Address	Description
CANxTIDR0	$_0	CAN foreground transmit buffer x identifier register 0
CANxTIDR1	$_1	CAN foreground transmit buffer x identifier register 1
CANxTIDR2	$_2	CAN foreground transmit buffer x identifier register 2
CANxTIDR3	$_3	CAN foreground transmit buffer x identifier register 3
CANxTDSR0	$_4	CAN foreground transmit buffer x data segment register 0
CANxTDSR1	$_5	CAN foreground transmit buffer x data segment register 1
CANxTDSR2	$_6	CAN foreground transmit buffer x data segment register 2
CANxTDSR3	$_7	CAN foreground transmit buffer x data segment register 3
CANxTDSR4	$_8	CAN foreground transmit buffer x data segment register 4
CANxTDSR5	$_9	CAN foreground transmit buffer x data segment register 5
CANxTDSR6	$_A	CAN foreground transmit buffer x data segment register 6
CANxTDSR7	$_B	CAN foreground transmit buffer x data segment register 7
CANxTDLR	$_C	CAN foreground transmit buffer x data length register
CANxTBPR	$_D	CAN foreground transmit buffer x priority register
CANxTSRH	$_E	CAN foreground transmit buffer x timestamp register high
CANxTSRL	$_F	CAN foreground transmit buffer x timestamp register low

1. x can be 0, 1, 2, 3, or 4.
2. The absolute address of each register is equal to the sum of the base address of the CAN foreground transmit buffer x and the address field of the corresponding register.

Table 13.2b ■ CAN foreground transmit buffer x variable names

13.9.3 Transmit Storage Structure

The design of the MSCAN transmit-structure achieves two goals.

- Providing the capability to send out a stream of scheduled messages without releasing the bus between the two messages
- Prioritizing messages so that the message with the highest priority is sent out first

As shown in Figure 13.37, the MSCAN has a triple transmit buffer scheme that allows multiple messages to be set up in advance and achieve a real-time performance. Only one of the three transmit buffers is accessible to the user at a time. A transmit buffer is made accessible to the user by writing an appropriate value into the CANxTBSEL register.

The procedure for transmitting a message includes the following steps:

1. Identifying an available transmit buffer by checking the TXEx flag associated with the transmit buffer
2. Setting a pointer to the empty transmit buffer by writing the CANxTFLG register to the CANxTBSEL register, making the transmit buffer accessible to the user
3. Storing the identifier, the control bits, and the data contents into one of the transmit buffers
4. Flagging the buffer as ready by clearing the associated TXE flag

After step 4, the MSCAN schedules the message for transmission and signals the successful transmission of the buffer by setting the associated TXE flag.

If there is more than one buffer scheduled for transmission when the CAN bus becomes available for arbitration, the MSCAN uses the local priority setting to choose the buffer with the highest priority and sends it out. The buffer having the smallest priority field has the

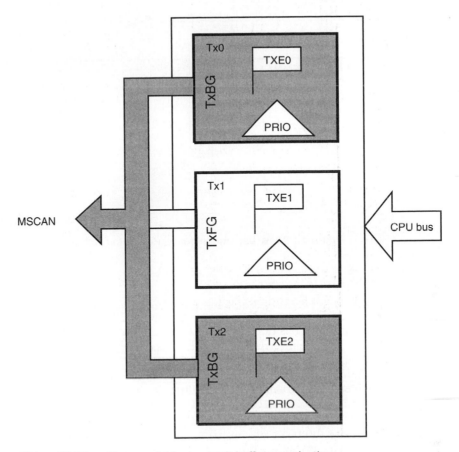

Figure 13.37 ■ User model for transmit buffer organization

highest priority and is scheduled for transmission first. The internal scheduling process takes place whenever the MSCAN arbitrates for the bus.

When a high-priority message is scheduled by the application, it may be necessary to abort a lower-priority message in one of the three transmit buffers. A message that is being transmitted cannot be aborted. One can make an abort request by setting an appropriate ABTRQ bit of the CANxTARQ register. The MSCAN then grants the request, if possible, by (1) setting the corresponding Abort Acknowledge flag (ABTAK) in the CANxTAAK register, (2) setting the associated TXE flag to release the buffer, and (3) generating a transmit interrupt. The transmit interrupt handler software can tell from the setting of the ABTAK flag whether the message was aborted (ABTAK = 1) or sent (ABTAK = 0).

13.9.4 Receive Storage Structure

As shown in Figure 13.38, the received messages are stored in a five-stage input FIFO data structure. The message buffers are alternately mapped into a single memory area, which is referred to as the *foreground receive buffer*. The application software reads the foreground receive buffer to access the received message. The background receive buffer is solely used to hold incoming CAN messages and is not accessible to the user.

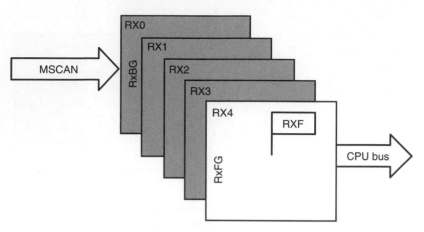

Figure 13.38 ■ User model for receive buffer organization

Whenever a valid message is received at the background receive buffer, it will be transferred to the foreground receive buffer and the RXF flag will be set to 1. The user's receive handler program has to read the received message from the RxFG and then reset the RXF flag to acknowledge the interrupt and to release the foreground buffer.

When the MSCAN module is transmitting, the MSCAN receives its own transmitted messages into the background receive buffer but does not shift it into the receiver FIFO or generate a receive interrupt. An overrun condition occurs when all receive message buffers in the FIFO are filled with correctly received messages with accepted identifiers and another message is correctly received from the bus with an accepted identifier. The latter message is discarded and an error interrupt with overrun indication is generated if enabled. The MSCAN is still able to transmit messages while the receiver FIFO is being filled, but all incoming messages are discarded. As soon as a receive buffer in the FIFO is available again, new valid messages will be accepted.

13.9.5 Identifier Acceptance Filter

The MSCAN identifier acceptance registers define the acceptance patterns of the standard or extended identifier. Any of these bits can be marked "don't care" in the MSCAN identifier mask registers.

A message is accepted only if its associated identifier matches one of the identifier filters. A filter hit is indicated to the application software by a RXF flag setting to 1 and the 3 hit bits in the CANxIDAC register. These hit bits identify the filter section that caused the acceptance. In case more than one hit occurs, the lower hit has priority. The identifier acceptance filter is programmable to operate in four different modes.

1. Two identifier acceptance filters with each filter applied to (a) the full 29 bits of the extended identifier, the SRR bit, the IDE bit, and the RTR bit or (b) the 11 bits of the standard identifier plus the RTR and IDE bits. This mode may cause up to two hits.

2. Four identifier acceptance filters with each filter applied to (a) the 14 most significant bits of the extended identifier plus the SRR and IDE bits or (b) the 11 bits of the standard identifier plus the RTR and IDE bits. This mode may cause up to four hits.

3. Eight identifier acceptance filters with each filter applied to the first 8 bits of the identifier. This mode implements eight independent filters for the first 8 bits of the extended or standard identifier. This may cause up to eight hits.

4. Closed filter. No CAN message is copied into the foreground buffer RxFG, and the RXF flag is never set.

13.9.6 MSCAN Clock System

The MSCAN clock generation circuitry is shown in Figure 13.39. This clock circuitry allows the MSCAN to handle CAN bus rates ranging from 10 kbps up to 1 Mbps. The CLKSRC bit in the CANxCTL1 register defines whether the internal CANCLK is connected to the output of a crystal oscillator or to the E-clock. The clock source has to be chosen such that the tight oscillator tolerance requirements (up to 0.4 percent) of the CAN protocol are met. Additionally, for a high CAN bus rate (1 Mbps), a 45 to 55 percent duty cycle of the clock is required.

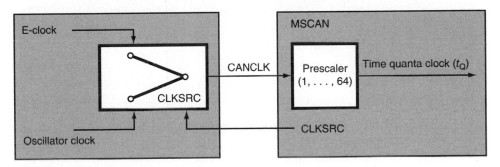

Figure 13.39 ■ MSCAN clocking scheme

If the E-clock is generated from a PLL, it is recommended to select the oscillator clock rather than the E-clock due to jitter considerations, especially at the faster CAN bus rate. For microcontrollers without a clock and reset generator (CRG), CANCLK is driven from the crystal oscillator.

A programmable prescaler generates the time quanta (t_Q) clock from CANCLK. A time quantum is the atomic unit of time handled by the MSCAN. The frequency of the time quantum is derived from the CANCLK using the following expression:

$$f_{t_Q} = f_{CANCLK} \div \text{prescaler value}$$

Slightly deviated from Figure 13.12, the MSCAN divides a bit time into three segments (shown in Figure 13.40) rather than four segments.

- Sync_seg. This segment is fixed at 1 time quantum. Signal edges are expected to happen within this segment.
- Time segment 1. This segment includes the **prop_seg** and the **phase_seg1** of the CAN standard. It can be programmed by setting the TSEG1 parameter of the CANxBTR1 register to consist of 4 to 16 time quanta.
- Time segment 2. This segment represents the **phase_seg2** of the CAN standard. It can be programmed by setting the TSEG2 parameter of the CANxBTR1 register to be 2 to 8 time quanta long.

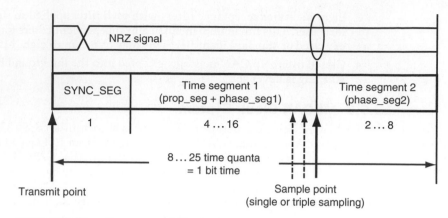

Figure 13.40 ■ Segments within the bit time

The bit rate of the CAN bus can be derived from the frequency of the time quantum using the following expression:

Bit rate = f_{t_Q} ÷ (number of time quanta)

By computing the number of time quanta contained in a bit time, we can figure out the appropriate values for time segment 1 and time segment 2. This will be elaborated in Section 13.12.

13.9.7 MSCAN Interrupt Operation

The MSCAN supports four interrupt vectors, any of which can be individually masked.

TRANSMIT INTERRUPT

At least one of the three transmit buffers is empty and can be loaded to schedule a message for transmission. The TXEx flag of the empty message buffer is set.

RECEIVE INTERRUPT

A receive interrupt may be requested when a message is successfully received and shifted into the foreground buffer of the receive FIFO. This interrupt is generated immediately after receiving the EOF symbol. The RXF flag is set. If there are multiple messages in the receiver FIFO, the RXF flag is set as soon as the next message is shifted to the foreground buffer.

WAKE-UP INTERRUPT

Activity on the CAN bus occurring during the MSCAN internal sleep mode may generate the wake-up interrupt.

ERROR INTERRUPT

An overrun of the receiver FIFO, error, warning, or bus-off condition may generate an error interrupt.

INTERRUPT ACKNOWLEDGE

Interrupts are directly associated with one or more status flags in either the CANxRFLG or the CANxTFLG register. Interrupts are pending as long as one of the corresponding flags is set. The flags in these registers must be cleared within the interrupt handler to avoid repeated interrupts. The flags are cleared by writing a 1 to the corresponding bit position. A flag cannot be cleared if the respective condition still prevails.

13.9.8 MSCAN Initialization

The MSCAN needs to be configured before it can operate properly. The procedures for configuring the MSCAN out of reset and in normal mode are different.

MSCAN INITIALIZATION OUT OF RESET

The procedure to initialize the MSCAN after reset is as follows:

- Enable the CAN module by setting the CANE bit of the CANxCTL1 register to 1.
- Request to enter the initialization mode by setting the INITRQ bit of the CANxCTL0 register to 1.
- Make sure that the CAN initialization mode is entered by waiting until the INITAK bit of the CANxCTL1 register has been set to 1.
- Write to the configuration registers (CANxCTL1, CANxBTR0, CANxBTR1, CANxIDAC, CANxIDAR0~7, CANxIDMR0~7) in initialization mode (both the INITRQ and INITAK bits are set).
- Clear the INITRQ bit to leave initialization mode and enter normal mode.

MSCAN INITIALIZATION IN NORMAL MODE

- Make sure that the MSCAN transmission queue is empty and bring the module into sleep mode by asserting the SLPRQ bit and waiting for the SLPAK bit to be set.
- Enter the initialization mode.
- Write to the configuration registers in initialization mode.
- Clear the INITRQ bit to leave the initialization mode and continue in normal mode.

13.10 Physical CAN Bus Connection

The CAN protocol is designed for data communication over a short distance. It does not specify what medium to use for data transmission. The user can choose optical fiber, shielded cable, or unshielded cable as the transmission medium. Using a shielded or unshielded cable is recommended for a short-distance communication.

A typical CAN bus system setup using a cable is illustrated in Figure 13.41. The resistor R_T is the terminating resistor. Each node uses a transceiver to connect to the CAN bus. The CAN bus transceiver is connected to the bus via two bus terminals, CAN_H and CAN_L, which

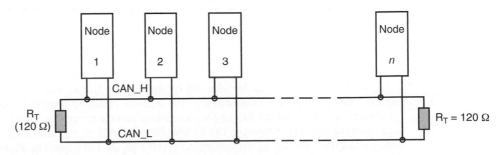

Figure 13.41 ■ A typical CAN bus setup using cable

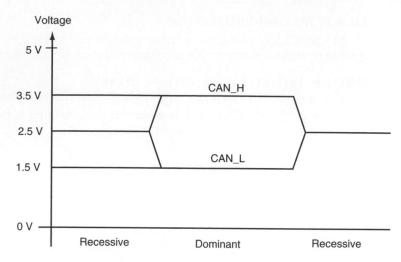

Figure 13.42 ■ Nominal CAN bus levels

provide differential receive and transmit capabilities. The nominal CAN bus levels for the 5-V power supply are shown in Figure 13.42.

Many semiconductor companies produce CAN bus transceivers. The Microchip MCP2551, Philips PCA82C250/251, Texas Instruments SN65HVD251, and MAXIM MAX3050/3057 are among the most popular. All of these chips are compatible. TI SN65HVD251, Philips PCA82C250, and Microchip MCP2551 are drop-in replaceable with each other.

13.10.1 The MCP2551 CAN Transceiver

The MCP2551 provides differential transmit and receive capability for the CAN protocol controller (or module). It operates at speeds up to 1 Mbps. The MCP2551 converts the digital signals generated by a CAN controller to signals suitable for transmission over the CAN bus cabling. It also provides a buffer between the CAN controller and the high-voltage spikes that can be generated on the CAN bus by outside sources, such as EMI, ESD, electrical transients, and so on. The block diagram of the MCP2551 is shown in Figure 13.43.

The MCP2551 transceiver outputs will drive a minimum load of 45 Ω, allowing a maximum of 112 nodes to be connected to the same CAN bus. The RxD pin reflects the differential bus voltage between CAN_H and CAN_L. The low and high states of the RxD output pin correspond to the dominant and recessive states of the CAN bus, respectively. The R_S input allows the user to select one of the three operation modes.

- High speed
- Slope control
- Standby

The high-speed mode is selected by grounding the R_S pin. In this mode, the transmission output drivers have fast output rise and fall times to support the high-speed CAN bus. The slope-control mode further reduces the electromagnetic interference (EMI) by limiting the rise and fall times of the CAN_H and CAN_L signals. The slope, or *slew rate*, is controlled by connecting an external resistor (R_{EXT}) between the R_S pin and the ground pin. Figure 13.44 illustrates typical slew rate values as a function of the slope-control resistance value.

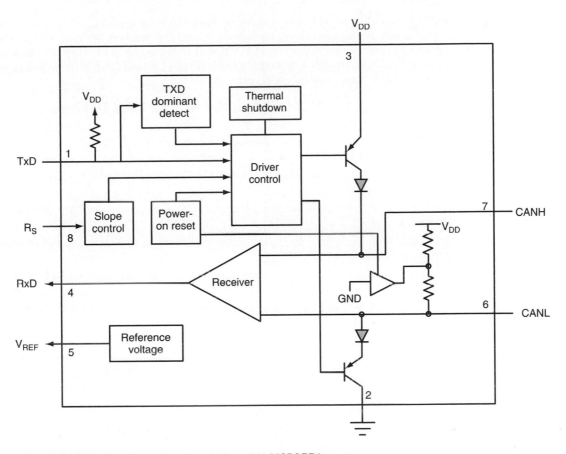

Figure 13.43 ■ The block diagram of Microchip MCP2551

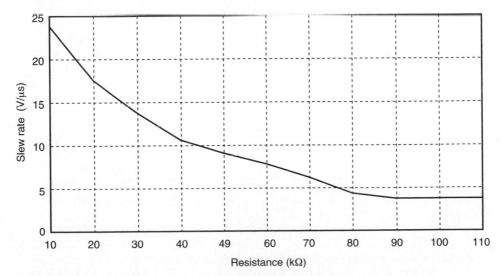

Figure 13.44 ■ Slew rate versus slope control resistance value

The MCP2551 can be placed in the standby or sleep mode by applying a high voltage at the R_S pin. In sleep mode, the transmitter is switched off, and the receiver operates at a lower current. The receive pin on the controller side (RxD) is still functioning but will operate at a slower rate. The attached microcontroller can monitor the RxD pin for CAN bus activity and place the transceiver into normal operation via the R_S pin.

13.10.2 Interfacing the MCP2551 to the HCS12 CAN Devices

A typical method of interfacing the MCP2551 transceiver to the HCS12 CAN module is shown in Figure 13.45. The maximum achievable bus length in a CAN bus network is determined by the following physical factors:

1. The loop delay of the connected bus nodes (CAN controller and transceiver) and the delay of the bus line.

2. The differences in bit time quantum length due to the relative oscillator tolerance between nodes.

3. The signal amplitude drop due to the series resistance of the bus cable and the input resistance of bus node.

The resultant equation after taking these three factors into account would be very complicated. The bus length that can be achieved as a function of the bit rate in the high-speed mode and with CAN bit timing parameters being optimized for maximum propagation delay is

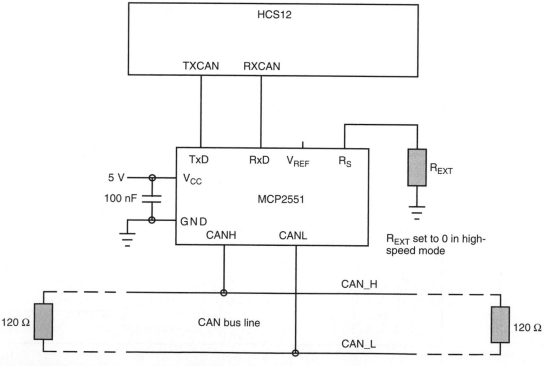

Figure 13.45 ■ Interfacing the MCP2551 with the HCS12

Bit Rate (kbps)	Bus Length
1000	40
500	100
250	250
125	500
62.5	1000

Table 13.3 ■ CAN bus bit rate/bus length relation

Bus Length/Number of Nodes	32	64	100
100 m	0.25 mm² or AWG 24	0.25 mm² or AWG 24	0.25 mm² or AWG 24
250 m	0.34 mm² or AWG 22	0.50 mm² or AWG 20	0.50 mm² or AWG 20
500 m	0.75 mm² or AWG 18	0.75 mm² or AWG 18	0.75 mm² or AWG 18

Table 13.4 ■ Minimum recommended bus wire cross section for the trunk cable

shown in Table 13.3. The types and the cross sections suitable for the CAN bus trunk cable that has more than 32 nodes connected and spans more than 100 meters are listed in Table 13.4.

13.11 Setting the CAN Timing Parameters

The *nominal bit rate* of a CAN network is uniform throughout the network and is given by the equation

$$f_{NBT} = 1/t_{NBT} \tag{13.1}$$

where t_{NBT} is the *nominal bit time*. As described earlier, a bit time is divided into four separate nonoverlapping time segments: *sync_seg*, *prop_seg*, *phase_seg1*, and *phase_seg2*.

The purpose of the *sync_seg* segment is to inform all nodes that a bit time has just started. The existence of the propagation delay segment, *prop_seg,* is due to the fact that the CAN protocol allows for nondestructive arbitration between nodes contending for access to the bus and the requirement for *in-frame acknowledgement*. In the case of nondestructive arbitration, more than one node may be transmitting during the arbitration field. Each transmitting node samples data from the bus in order to determine whether it has won or lost the arbitration and also to receive the arbitration field in case it loses the arbitration. When each node samples a bit, the value sampled must be the logical superposition of the bit values transmitted by each of the nodes arbitrating for bus access. In the case of the acknowledge field, the transmitting node transmits a recessive bit but expects to receive a dominant bit; in other words, a dominant value must be sampled at the sample time. The length of the *prop_seg* segment must be selected for the earliest possible sample of the bit by a node until the transmitted bit values from all the transmitting nodes have reached all of the nodes. When multiple nodes are arbitrating for the control of the CAN bus, the node that transmitted the earliest must wait until the node that transmitted the latest in order to find out if it won or lost. That is, the worst-case value for t_{PROP_SEG} is given by

$$t_{PROP\_SEG} = t_{PROP(A, B)} + t_{PROP(D, A)} \tag{13.2}$$

where $t_{\mathrm{PROP(A,\,B)}}$ and $t_{\mathrm{PROP(B,\,A)}}$ are the propagation delays from node A to B and node B to node A, respectively. In the worst case, node A and node B are at the two ends of the CAN bus. The propagation delay from node A to node B is given by

$$t_{\mathrm{PROP(A,\,B)}} = t_{\mathrm{BUS}} + t_{\mathrm{Tx}} + t_{\mathrm{Rx}} \tag{13.3}$$

where t_{BUS}, t_{Tx}, and t_{Rx} are data traveling time on the bus, transmitter propagation delay, and receiver propagation delay, respectively.

Let node A and node B be two nodes at opposite ends of the CAN bus, then the worst-case value for $t_{\mathrm{PROP\_SEG}}$ is

$$t_{\mathrm{PROP\_SEG}} = 2 \times (t_{\mathrm{BUS}} + t_{\mathrm{Tx}} + t_{\mathrm{Rx}}) \tag{13.4}$$

The minimum number of *time quanta* (t_{Q}) that must be allocated to the *prop_seg* segment is therefore

$$\mathrm{prop\_seg} = \mathrm{round\_up}\,(t_{\mathrm{PROP\_SEG}} \div t_{\mathrm{Q}}) \tag{13.5}$$

where the **round_up ()** function returns a value that equals the argument rounded up to the next integer value.

In the absence of bus errors, bit stuffing guarantees a maximum of 10 bit periods between resynchronization edges (5 dominant bits followed by 5 recessive bits and then followed by a dominant bit). This represents the worst-case condition for the accumulation of phase error during normal communication. The accumulated phase error must be compensated for by resynchronization and therefore must be less than the programmed resynchronization jump width (t_{RJW}). The accumulated phase error is due to the tolerance in the CAN system clock, and this requirement can be expressed as

$$(2 \times \Delta f) \times 10 \times t_{\mathrm{NBT}} < t_{\mathrm{RJW}} \tag{13.6}$$

where Δf is the largest crystal oscillator frequency variation (in percentage) of all CAN nodes in the network.

Real systems must operate in the presence of electrical noise, which may induce errors on the CAN bus. A node transmits an error flag after it detects an error. In the case of a local error, only the node that detects the error will transmit the error flag. All other nodes receive the error flag and transmit their own error flags as an echo. If the error is global, all nodes will detect it within the same bit time and will therefore transmit error flags simultaneously. A node can therefore differentiate between a local error and a global error by detecting whether there is an echo after its error flag. This requires that a node can correctly sample the first bit after transmitting its error flag.

An error flag from an error-active node consists of 6 dominant bits, and there could be up to 6 dominant bits before the error flag, if, for example, the error was a stuff error. A node must therefore correctly sample the 13th bit after the last resynchronization. This can be expressed as

$$(2 \times \Delta f) \times (13 \times t_{\mathrm{NBT}} - t_{\mathrm{PHASE\_SEG2}}) < \mathrm{MIN}\,(t_{\mathrm{PHASE\_SEG1}}, t_{\mathrm{PHASE\_SEG2}}) \tag{13.7}$$

where the function MIN(arg1,arg2) returns the smaller of the two arguments.

Thus there are two clock tolerance requirements that must be satisfied. The selection of bit timing values involves consideration of various fundamental system parameters. The requirement of the *prop_seg* value imposes a trade-off between the maximum achievable bit rate and the maximum propagation delay, due to the bus length and the characteristics of the bus driver circuit. The highest bit rate can only be achieved with a short bus length, a fast bus driver circuit, and a high-frequency CAN clock source with high tolerance.

The procedure for determining the optimum bit timing parameters that satisfy the requirements for proper bit sampling is as follows:

Step 1

Determine the minimum permissible time $(t_{\mathrm{PROP\_SEG}})$ for the *prop_seg* segment using Equation 13.4.

Step 2
Choose the CAN system-clock frequency (and hence set the CAN bus time quantum t_Q).
The CAN system clock will be the CPU oscillator output or the E-clock divided by a prescale
factor. The CAN system clock is chosen so that the desired CAN bus *nominal bit time* (t_{NBT})
is an integer multiple of time quanta (CAN system-clock period) from 8 to 25.

Step 3
Calculate *prop_seg* duration. The number of time quanta required for the *prop_seg* can
be calculated by using Equation 13.5. If the result is greater than 8, go back to step 2 and
choose a lower CAN system-clock frequency.

Step 4
Determine *phase_seg1* and *phase_seg2*. Subtract the *prop_seg* value and 1 (for *sync_seg*) from
the time quanta contained in a bit time. If the difference is less than 3, then go back to step
2 and select a higher CAN system-clock frequency. If the difference is an odd number greater
than 3, then add 1 to the *prop_seg* value and recalculate. If the difference is equal to 3, then
phase_seg1 = 1 and *phase_seg2* = 2 and only one sample per bit may be chosen. Otherwise,
divide the remaining number by 2 and assign the result to *phase_seg1* and *phase_seg2*.

Step 5
Determine the resynchronization jump width (RJW). RJW is the smaller of 4 and *phase_seg1*.

Step 6
Calculate the required oscillator tolerance from Equations 13.6 and 13.7. If *phase_seg1* > 4,
it is recommended to repeat steps 2 to 6 with a larger value for the prescaler. Conversely,
if *phase_seg1* < 4, it is recommended to repeat step 2 to 6 with a smaller value for the
prescaler, as long as *prop_seg* < 8, as this may result in a reduced oscillator tolerance
requirement. If the prescaler is already equal to 1 and a reduced oscillator tolerance is still
required, the only option is to consider using a clock source with higher frequency.

Example 13.1

▼

Calculate the CAN bit segments for the following system constraints:

Bit rate = 100 kbps

Bus length = 25 m

Bus propagation delay = 5×10^{-9} s/m

MCP2551 transceiver plus receiver propagation delay = 150 ns at 85°C

CPU oscillator frequency = 8 MHz

Solution: Let's follow the procedure described earlier.

Step 1
Physical delay of bus = 125 ns

t_{PROP_SEG} = 2 × (125 ns + 150 ns) = 550 ns

Step 2
Try the prescaler of 4 for the CAN system clock of 8 MHz, which gives a time quantum of
500 ns. One bit time is 10 µs. This gives 10,000/500 = 20 time quanta per bit.

Step 3
Prop_seg = round_up (550 ns ÷ 500 ns) = round_up (1.1) = 2. Phase_seg1 + phase_seg2
= 20 − 1 − prop_seg = 17 > 16.

Set prescaler to 8. Then one time quantum is 1 µs and one bit time has 10 time quanta.
The new prop_seg = round_up (550 ns ÷ 1000 ns) = 1.

Step 4
Subtracting 1 for prop_seg and 1 for sync_seg from 10 time quanta per bit gives 8. Since the result is equal to 8 and is even, divide it by 2 (quotient is 4) and assign it to phase_seg1 and phase_seg2.

Step 5
RJW is the smaller one of 4 and phase_seg1 and is 4.

Step 6
From Equation 13.6,

$\Delta f <$ RJW $\div$ (20 $\times$ NBT) $= 4 \div (20 \times 10) = 2.0$ percent

From Equation 13.7,

$\Delta f <$ MIN (phase_seg1, phase_seg2) $\div$ 2(13 $\times$ NBT $-$ phase_seg2) $= 4 \div 252 = 1.59$ percent

The desired oscillator tolerance is the smaller of these values, 1.59 percent.

In summary,

Prescaler	= 8
Nominal bit time	= 10
prop_seg	= 1
sync_seg	= 1
phase_seg1	= 4
phase_seg2	= 4
RJW	= 4
Oscillator tolerance	= 1.59 percent

Example 13.2

Calculate the bit segments for the following system constraints:

Bit rate = 500 kbps

Bus length = 50 m

Bus propagation delay = 5×10^{-9} s/m

MCP2551 transmitter plus receiver propagation delay = 150 ns at 85°C

CPU oscillator frequency = E-clock frequency = 16 MHz

Solution: Follow the standard procedure to perform the calculation.

Step 1
Physical delay of bus = 250 ns

$t_{PROP\_SEG} = 2 \times (250 \text{ ns} + 150 \text{ ns}) = 800 \text{ ns}$

Step 2
A prescaler of 2 for a CAN system clock of 16 MHz gives a time quantum of 125 ns. The nominal bit time is 2 µs. This gives 2000/125 = 16 time quanta per bit.

Step 3
Prop_seg = round_up (800 ns $\div$ 125 ns) = round_up (6.4) = 7.

Step 4

Subtracting 7 for prop_seg and 1 for sync_seg from 16 time quanta per bit gives 8.
Since the result is greater than 4 and is even, divide it by 2 (quotient is 4) and assign it to phase_seg1 and phase_seg2.

Step 5

RJW (= 4) is the smaller of 4 and phase_seg1.

Step 6

From Equation 13.6,

$$\Delta f < RJW \div (20 \times NBT) = 4 \div (20 \times 16) = 1.25 \text{ percent}$$

From Equation 13.7,

$$\Delta f < MIN (phase\_seg1, phase\_seg2) \div 2(13 \times NBT - phase\_seg2) = 4 \div 408 = 0.98 \text{ percent}.$$

The required oscillator tolerance is the smaller of these values, 0.98 percent.

Since phase_seg2 = 4, there is no need to try other prescale values.

In summary,

Prescaler	= 2
Nominal bit time	= 16
prop_seg	= 7
sync_seg	= 1
phase_seg1	= 4
phase_seg2	= 4
RJW	= 4
Oscillator tolerance	= 0.98 percent

13.12 MSCAN Configuration

The timing parameters for the MSCAN module need be set only once and should be done immediately after MCU reset. Other parameters such as acceptance filters may be changed after reset configuration.

Example 13.3

▼

Write a program to configure the MSCAN module 1 after reset to operate with the timing parameters computed in Example 13.1 and the following setting:

- Enable wake-up
- Disable time stamping
- Select oscillator as the clock source to the MSCAN
- Disable loopback mode, disable listen-only mode
- One sample per bit
- Accept messages with extended identifiers that start with T1 and P1 (use two 32-bit filters)

Solution: The first step is to enter the initialization mode and make sure the initialization is entered. This can be done as follows:

```
bset      CAN1CTL0,INITRQ
brclr     CAN1CTL1,INITAK,*
```

The assembly program that performs the desired configuration is as follows:

```
openCan1  bset    CAN1CTL1,CANE       ; required after reset
          bset    CAN1CTL0,INITRQ     ; request to enter initialization mode
w1        brclr   CAN1CTL1,INITAK,w1  ; make sure initialization mode is entered
          movb    #$84,CAN1CTL1       ; enable CAN1, select oscillator as clock source
                                      ; enable wake-up filter
          movb    #$C7,CAN1BTR0       ; set jump width to 4 tQ, prescaler set to 8
          movb    #$B4,CAN1BTR1       ; set phase_seg2 to 4 tQ, phase_seg1 to 4 tQ,
                                      ; set prop_seg to 1 tQ, 3 samples per bit
          movb    #$54,CAN1IDAR0      ; acceptance identifier T1
          movb    #$3C,CAN1IDAR1      ;   "
          movb    #$40,CAN1IDAR2      ;   "
          movb    #$00,CAN1IDAR3      ;   "
          movb    #$00,CAN1IDMR0      ; acceptance mask for extended identifier T1
          movb    #$00,CAN1IDMR1      ;   "
          movb    #$3F,CAN1IDMR2      ;   "
          movb    #$FF,CAN1IDMR3      ;   "
          movb    #$50,CAN1IDAR4      ; acceptance identifier P1
          movb    #$3C,CAN1IDAR5      ;   "
          movb    #$40,CAN1IDAR6      ;   "
          movb    #$00,CAN1IDAR7      ;   "
          movb    #$00,CAN1IDMR4      ; acceptance mask for extended identifier P1
          movb    #$00,CAN1IDMR5      ;   "
          movb    #$3F,CAN1IDMR6      ;   "
          movb    #$FF,CAN1IDMR7      ;   "
          clr     CAN1IDAC            ; set two 32-bit filter mode
          movb    #$25,CAN1CTL0       ; stop clock on wait mode, enable wake-up
          bclr    CAN1CTL0,INITRQ     ; leave initialization mode
          rts
```

The C language version of the function is as follows:

```
void openCan1(void)
{
   CAN1CTL1   |= CANE;              /* enable CAN, required after reset */
   CAN1CTL0   |= INITRQ;            /* request to enter initialization mode */
   while(!(CAN1CTL1&INITAK));       /* wait until initialization mode is entered */
   CAN1CTL1   = 0x84;               /* enable CAN1, select oscillator as MSCAN clock
                                        source, enable wake-up filter */
   CAN1BTR0   = 0xC7;               /* set SJW to 4, set prescaler to 8 */
   CAN1BTR1   = 0xB4;               /* set phase_seg2 to 4 tQ, phase_seg1 to 4 tQ,
                                        prop_seg to 1 tQ */
   CAN1IDAR0  = 0x54;               /* set acceptance identifier T1 */
   CAN1IDAR1  = 0x3C;               /*" " */
   CAN1IDAR2  = 0x40;               /*" " */
   CAN1IDAR3  = 0x00;               /*" " */
   CAN1IDMR0  = 0x00;               /* acceptance mask for T1 */
```

```
        CAN1IDMR1  = 0x00;          /* " */
        CAN1IDMR2  = 0x3F;          /* " */
        CAN1IDMR3  = 0xFF;          /* " */
        CAN1IDAR4  = 0x50;          /* set acceptance identifier P1 */
        CAN1IDAR5  = 0x3C;          /* " */
        CAN1IDAR6  = 0x40;          /* " */
        CAN1IDAR7  = 0x00;          /* " */
        CAN1IDMR4  = 0x00;          /* acceptance mask for P1 */
        CAN1IDMR5  = 0x00;          /* " */
        CAN1IDMR6  = 0x3F;          /* " */
        CAN1IDMR7  = 0xFF;          /* " */
        CAN1IDAC   = 0x00;          /* select two 32-bit filter mode */
        CAN1CTL0   = 0x25;          /* stop clock on wait mode, enable wake-up */
        CAN1CTL0  &= ~INITRQ;       /* exit initialization mode *
}
```

▲

The MSCAN application can be very complicated; the initial configuration of the MSCAN may need to be changed to satisfy the changing environment. The following example illustrates the change of acceptance filters:

Example 13.4

▼

Write an instruction sequence to change the configuration of the CAN1 module so that it will accept messages with a standard identifier starting with letter T or P.

Solution: The procedure described in Section 13.9.8 should be followed to reconfigure the CAN1 module. In this example, we set up two sets of identical acceptance filters to accept standard identifiers that start with letters T and P. The instruction sequence that performs the reconfiguration is as follows:

```
ct1        brset  CAN1TFLG,$07,tb_empty   ; wait until all transmit buffers are empty
           bra    ct1
tb_empty   bset   CAN1CTL0,SLPRQ          ; request to enter sleep mode
ct2        brclr  CAN1CTL1,SLPAK,ct2      ; wait until sleep mode is entered
           bset   CAN1CTL0,INITRQ         ; request to enter initialization mode
ct3        brclr  CAN1CTL1,INITAK,ct3     ; wait until initialization mode is entered
           movb   #$10,CAN1IDAC           ; select four 16-bit acceptance mode
           movb   #$54,CAN1IDAR0          ; set up filter for letter T for standard
           movb   #0,CAN1IDAR1            ; identifier
           movb   #$50,CAN1IDAR2          ; set up filter for letter P for standard
           clr    CAN1IDAR3              ; identifier
           clr    CAN1IDMR0              ; acceptance mask for T
           movb   #$F7,CAN1IDMR1          ; check IDE bit only (0 for standard identifier)
           clr    CAN1IDMR2             ; acceptance mask for P
           movb   #$F7,CAN1IDMR3          ; check IDE bit only (must be 0)
           movb   #$54,CAN1IDAR4          ; set up filter for letter T for standard
           movb   #0,CAN1IDAR5            ; identifier
           movb   #$50,CAN1IDAR6          ; set up filter for letter P for standard
           clr    CAN1IDAR7             ; identifier
           clr    CAN1IDMR4             ; acceptance mask for T
```

```
        movb    #$F7,CAN1IDMR5          ; check IDE bit only (0 for standard identifier)
        clr     CAN1IDMR6              ; acceptance mask for P
        movb    #$F7,CAN1IDMR7          ; check IDE bit only (must be 0)
        bclr    CAN1CTL0,INITRQ        ; exit initialization mode
        bclr    CAN1CTL0,SLPRQ         ; exit sleep mode
```

▲

13.13 Data Transmission and Reception in MSCAN

When the application has only a small amount of data to be transmitted every now and then, it would be easy to find empty transmit buffers to send the message. There is no need to use interrupt in this case. However, if the application has many messages to be sent out at one time, then it may not be able to find an empty transmit buffer to hold the message to be transmitted at some point. In this situation, the designer should consider using the interrupt-driven approach to control the transmission of messages to allow the CPU to perform other operations.

For a received message, the application would take action on the basis of the identifier of the message. If the message arrival time is unpredictable, then it is more convenient to use the interrupt-driven approach. If the arrival times of messages are more predictable, then the designer can consider using either the polling or the interrupt-driven approach.

13.13.1 MSCAN Data Transmission Programming

To prepare data for transmission, there are two options.

1. Identify an empty transmit buffer and place data directly on the available transmit buffer.

2. Store data in a buffer and copy the buffer to the first available transmit buffer that is identified.

The first method can be used in the polling method. The second method involves more overhead. However, it is suitable for both the polling and interrupt-driven approaches.

Example 13.5

▼

Write a function to send out the message stored at a buffer pointed to by index register X from the CAN1 module. The function should find an available buffer to hold the message to be sent out.

Solution:

```
tbuf        equ     0                       ; tbuf offset from top of stack
snd2can1    pshy
            pshb
            leas    −1,sp                   ; allocate 1 byte for local variable
sloop1      brset   CAN1TFLG,$01,tb0        ; is transmit buffer 0 empty?
            brset   CAN1TFLG,$02,tb1        ; is transmit buffer 1 empty?
            brset   CAN1TFLG,$04,tb2        ; is transmit buffer 2 empty?
            bra     sloop1                  ; if necessary wait until one buffer is empty
```

```
tb0         movb    #0,tbuf,sp              ; mark transmit buffer 0 empty
            bra     tcopy
tb1         movb    #1,tbuf,sp              ; mark transmit buffer 1 empty
            bra     tcopy
tb2         movb    #2,tbuf,sp              ; mark transmit buffer 2 empty
tcopy       movb    CAN1TFLG,CAN1TBSEL      ; make the empty transmit buffer accessible
            ldy     #CAN1TIDR0             ; set y to point to the start of the transmit buffer
            ldab    #7                     ; always copy seven words (place word count in B)
cploop      movw    2,x+,2,y+
            dbne    b,cploop
            ldab    tbuf,sp
            cmpb    #0
            beq     istb0
            cmpb    #1
            beq     istb1
            movb    #$04,CAN1TFLG          ; mark buffer 2 ready for transmission
            bra     dcopy
istb0       movb    #$01,CAN1TFLG          ; mark buffer 0 ready for transmission
            bra     dcopy
istb1       movb    #$02,CAN1TFLG          ; mark buffer 1 ready for transmission
dcopy       leas    1,sp                   ; deallocate local variables
            pulb
            puly
            rts
```

The C language version of the function is as follows:

```
void snd2can1(char *ptr)

{
    int    tb,i,*pt1,*pt2;
    pt1  = (int *)ptr;              /* convert to integer pointer */
    while(1) {                      /* find an empty transmit buffer */
        if(CAN1TFLG & 0x01){
            tb = 0;
            break;
        }
        if(CAN1TFLG & 0x02){
            tb = 1;
            break;
        }
        if(CAN1TFLG & 0x04){
            tb = 2;
            break;
        }
    }
    CAN1TBSEL = CAN1TFLG;           /* make empty transmit buffer accessible */
    pt2 = (int *)&CAN1TIDR0;        /* pt2 points to the IDR0 of TXFG */
    for (i = 0; i < 7; i++)         /* copy the whole transmit buffer */
        *pt2++ = *pt1++;
```

```
        if (tb == 0)
             CAN1TFLG = 0x01;         /* mark buffer 0 ready for transmission */
        else if (tb == 1)
             CAN1TFLG = 0x02;         /* mark buffer 1 ready for transmission */
        else
             CAN1TFLG = 0x04;         /* mark buffer 2 ready for transmission */

    }
```

Example 13.6

Write a program to send out the string "3.5 V" from CAN1 and use V1 as its identifier. Set transmit buffer priority to the highest.

Solution: The assembly program that performs the desired function is as follows:

```
        org     $1000
tbuf0   ds      16
        org     $2000
        movb    #$56,tbuf0          ; identifier V1
        movb    #$3C,tbuf0+1        ;           "
        movb    #$40,tbuf0+2        ;           "
        movb    #0,tbuf0+3          ;           "
        movb    #$33,tbuf0+4        ; data "3"
        movb    #$2E,tbuf0+5        ; data "."
        movb    #$35,tbuf0+6        ; data "5"
        movb    #$20,tbuf0+7        ; data " "
        movb    #$56,tbuf0+8        ; data "V"
        movb    #5,tbuf0+12         ; data length (= 5)
        movb    #0,tbuf0+13         ; set transmit buffer priority to highest
        ldx     #tbuf0
        jsr     snd2can1            ; call subroutine to perform the actual transmission
        swi
```

The sequence of C statements that performs the same function is as follows:

```
void snd2can1(char *ptr);
        char tbuf0[16];
        tbuf0[0]  = 'V';          /* identifier V1 */
        tbuf0[1]  = 0x3C;         /* " */
        tbuf0[2]  = 0x40;         /* " */
        tbuf0[3]  = 0;            /* " */
        tbuf0[4]  = '3';          /* letter 3 */
        tbuf0[5]  = '.';          /* character.*/
        tbuf0[6]  = '5';          /* letter 5 */
        tbuf0[7]  = 0x20;         /* space */
        tbuf0[8]  = 'V';          /* letter V */
        tbuf0[12] = 5;            /* data length */
        tbuf0[13] = 0;            /* tbuf0 priority */
        snd2can1(tbuf0);
```

13.13.2 MSCAN Data Reception Programming

Incoming CAN bus messages are sent by other nodes and their arrival times are unpredictable. Therefore, it would be more efficient to use the interrupt-driven approach to handle data reception on the CAN bus. After setting up acceptance filters, one can enable the reception interrupt and wait for the incoming messages to arrive while performing other operations.

Example 13.7

▼

Assuming that the CAN1 receiver has been set up to accept messages with extended identifiers T1 and V1, filter 0 is set up to accept the identifier started with T1, whereas filter 1 is set up to accept the identifier started with V1. Write the interrupt handling routine for the RXF interrupt. If the acceptance is caused by filter 0, the RXF service routine would display the following message on a 20 × 2 LCD:

Temperature is
    xxx.y°F

If the acceptance of the message is caused by filter 1, the RXF interrupt service routine would display the following message:

Voltage is
    x.y V

Solution: The service routine will check the RXF bit of the CAN1RFLG register to make sure that the interrupt is caused by the RXF flag. If not, it returns. This routine also ignores the remote transfer request. The assembly language version of the service routine is as follows:

```
; ******************************************************************************
; The following function is the service routine for the RXF interrupt. It only services hit 0 and
; hit 1. It ignores the RTR.
; ******************************************************************************
can1Rx_ISR  brset    CAN1RFLG,RXF,RxfSet      ; is the RXF flag set to 1?
            rti                               ; if not, do nothing
RxfSet      ldaa     #$80                     ; set cursor to row 0 column 0
            jsr      cmd2LCD
            ldab     CAN1IDAC                 ; check into IDHIT bits
            andb     #$07                     ; mask out higher 5 bits
            beq      hit0                     ; filter 0 hit?
            cmpb     #1                       ; filter 1 hit?
            beq      hit1
            rti                               ; not hit 0 or hit 1, do thing
hit0        ldab     CAN1RDLR                 ; get the byte count of incoming data
            beq      rxfDone                  ; byte count 0, return
            ldx      #t1_line1                ; output "Temperature is"
            jsr      puts2lcd                 ;      "
            ldaa     #$C4                     ; set cursor to row 2 column 4
            jsr      cmd2LCD                  ;      "
            ldx      #CAN1RDSR0
outLoop1    ldaa     1,x+                     ; output 1 byte at a time
            jsr      putc2lcd                 ;      "
            dbne     b,outLoop1               ;      "
            rti
```

```
hit1          ldab     CAN1RDLR                    ; get the byte count of incoming data
              beq      rxfDone                     ; byte count 0, return
              ldx      #v1_line1                   ; output "Voltage is"
              jsr      puts2lcd                    ;         "
              ldaa     #$C4                        ; set cursor to row 2 column 4
              jsr      cmd2LCD                     ;         "
              ldx      #CAN1RDSR0                  ; x points to data segment register 0
outLoop2      ldaa     1,x+
              jsr      putc2lcd
              dbne     b,outLoop2
rxfDone       rti
t1_line1      fcc      "Temperature is"
              dc.b     0
v1_line1      fcc      "Voltage is"
              dc.b     0
```

The C language version (compiled by ICC12 compiler) of the service routine is as follows:

```
#include       "c:\iccv712\include\hcs12.h"
#include       "c:\iccv712\include\delay.h"
#include       "c:\iccv712\include\lcd_util.h"
void           RxISR(void);
void           open_can1(void);
char           *t1Msg = "Temperature is";
char           *v1Msg = "Voltage is";
int main (void)
{
    asm("ldd #_RxISR");                 //pass the interrupt vector of CAN1Rx interrupt
    asm("pshd");                        //                "
    asm("ldd #21");                     //place the vector number of CAN1Rx in D
    asm("ldx $EEA4");
    asm("jsr 0,X");
    open_can1();
    openLCD();
    CAN1RIER = 0x01;                    /* enable CAN1 RXF interrupt only */
    asm("cli");
    while(1);                           /* wait for RXF interrupt */
    return 0;
}
;*********************************************************************
// The following function handles the RXF interrupt. It checks if the RXF flag
// is set. If not, return. It also ignores the RTR request.
;*********************************************************************
#pragma interrupt_handler RxISR
void RxISR (void)
{
    char tmp,i,*ptr;
    if (!(CAN1RFLG & RXF))              /* interrupt not caused by RXF, return */
        return;
    CAN1RFLG = RXF;                     /* clear RXF flag */
    tmp = CAN1IDAC & 0x07;              /* extract filter hit info */
    if (tmp == 0) {                     /* filter 0 hit */
```

```
            if (CAN1RDLR == 0)                /* data length 0, do nothing */
                return;
            cmd2lcd(0x80);                     /* set LCD cursor to first row */
            puts2lcd(t1Msg);                   /* output "Temperature is" on LCD */
            cmd2lcd(0xC4);                     /* set LCD cursor to second row */
            ptr = (char *)&CAN1RDSR0;          /* ptr points to the first data byte */
            for (i = 0; i < CAN1RDLR; i++)
                putc2lcd(*ptr++);              /* output temperature value on the LCD second row */
        }
        else if (tmp == 1) {                   /* filter 1 hit */
            if(CAN1RDLR == 0)                  /* data length 0, do nothing */
                return;
            cmd2lcd(0x80);                     /* set LCD cursor to first row */
            puts2lcd(v1Msg);                   /* output "Voltage is" on the first row of LCD */
            cmd2lcd(0xC4);                     /* set LCD cursor to second row */
            ptr = (char *)&CAN1RDSR0;          /* ptr points to the first data byte */
            for(i = 0; i < CAN1RDLR; i++)
                putc2lcd(*ptr++);              /* output voltage value on the second row of LCD */
        }
        else asm("nop");                       /* other hit, do nothing */
}
```

This service routine can be modified to handle the remote transmission request (when the RTR bit of the received message is set to 1). The message to be transmitted in response to this request should be stored in a buffer and kept up to date. The modification to this service routine is straightforward, and hence is left as an exercise.

13.13.3 Putting It All Together

The program pieces needed for CAN data communications are already in place. This section uses one example to combine them.

Example 13.8

Write a C program to be run in a CAN environment using the same timing parameters as computed in Example 13.1. Each CAN node measures the voltage (in the range from 0 to 5 V) and sends it out from the CAN bus and also receives the voltage message sent over the CAN bus by other nodes. Configure the CAN1 module to receive messages having an extended identifier starting with V1. The transmission and reception are to proceed as follows:

■ The program measures the voltage connected at the AN7 pin every 200 ms and sends out the value with identifier V1. The voltage is represented in the format x.y V. After sending out a message, the program outputs the following message on the first row of the LCD:

    Sent: x.y V

■ Message reception is interrupt-driven. Whenever a new message is accepted, the program outputs the following message on the second row of the LCD:

    Received x.y V

Solution: The C program that performs the desired operation is as follows:

```c
#include "c:\iccv712\include\hcs12.h"
#include "c:\iccv712\include\delay.h"
#include "c:\iccv712\include\lcd_util.h"
char *msg1 = "Sent: ";
char *msg2 = "Received: ";
void RxISR(void);
void openAD0(void);
void OpenCan1(void);
void MakeBuf(char *pt1, char *pt2);
void snd2can1(char *ptr);
int main(void)
{
    char buffer[6];                            /* to hold measured voltage */
    char buf[16];                              /* transmit data buffer */
    int temp;
    asm("ldd #_RxISR");                        // pass the interrupt vector of CAN1Rx interrupt
    asm("pshd");                               //           "
    asm("ldd #21");                            // place the vector number of CAN1Rx in D
    asm("ldx $EEA4");
    asm("jsr 0,X");
    openLCD();                                 /* configure LCD kit */
    OpenCan1();                                /* configure CAN1 module */
    buffer[1] = '.';                           /* decimal point */
    buffer[3] = 0x20;                          /* space character */
    buffer[4] = 'V';                           /* volt character */
    buffer[5] = 0;                             /* null character */
    openAD0();                                 /* configure AD0 module */
    CAN1RIER = 0x01;                           /* enable RXF interrupt only */
    asm("cli");                                /* enable interrupt globally */

    while(1) {
        ATD0CTL5 = 0x87;                       /* convert AN7, result right-justified */
        while(!(ATD0STAT0 & SCF));             /* wait for conversion to complete */
        buffer[0] = 0x30 + (ATD0DR0*10)/2046;  /* integral digit of voltage */
        temp = (ATD0DR0 * 10) % 2046;          /* find the remainder */
        buffer[2] = 0x30 + (temp * 10)/2046;   /* compute the fractional digit */
        MakeBuf(&buf[0],&buffer[0]);           /* format data for transmission */
        snd2can1(&buf[0]);                     /* send out voltage on CAN bus */
        cmd2LCD(0x80);                         /* set LCD cursor to first row */
        putsLCD(msg1);                         /* output the message "sent: x.y V" */
        putsLCD(&buffer[0]);                   /*          "         */
        delayby100ms(2);                       /* wait for messages to arrive for 0.2 s */
    }
    return 0;
}
/**************************************************************************/
/* The following function formats a buffer into the structure of a CAN transmit buffer so */
/* that it can be copied into any empty transmit buffer for transmission.          */
/**************************************************************************/
```

```
void MakeBuf(char *pt1, char *pt2)
{
    char i;
    *pt1 = 'V';                /* set V1 as the transmit identifier */
    *(pt1+1) = 0x3C;           /* " */
    *(pt1+2) = 0x40;           /* " */
    *(pt1+3) = 0;              /* " */
    for(i = 4; i < 9; i++)  /* copy voltage data */
        *(pt1 + i) = *(pt2 + i − 4);
    *(pt1+12) = 5;                     /* set data length to 5 */
}
/*************************************************************************/
/* The following function handles the RXF interrupt. It checks if the RXF    */
/* is set. If not, return. It also ignores the RTR request.            */
/*************************************************************************/
#pragma interrupt_handler RxISR
void  RxISR (void)
{
    char tmp,i,*ptr;
    if (!(CAN1RFLG & RXF))                /* interrupt not caused by RXF, return */
            asm("rti")
    CAN1RFLG = RXF;                       /* clear RXF flag */
    tmp = CAN1IDAC & 0x07;                /* extract filter hit info */
    if (tmp == 0){                        /* filter 0 hit */
        if (CAN1RDLR == 0)                /* if data length is 0, do nothing */
            asm("rti")
        cmd2LCD(0xC0);                    /* set LCD cursor to second row */
        putsLCD(msg2);                    /* output "received: " */
        ptr = (char *)&CAN1RDSR0;         /* ptr points to the first data byte */
        for (i = 0; i < CAN1RDLR; i++)
            putcLCD(*ptr++);              /* output "x.y V" */
        asm("rti")
    }
    else asm("rti")                       /* other hit, do nothing */
}
void openAD0 (void)
{
    int i;
    ATD0CTL2 − 0xE0;
    delayby10us(2);
    ATD0CTL3 = 0x0A;                      // perform one conversion
    ATD0CTL4 = 0x25;                      // 4 cycles sample time, prescaler set to 12
}
void OpenCan1(void)
{
    CAN1CTL1 |= CANE;                     // enable CAN, required after reset
    CAN1CTL0   |= INITRQ;                 // request to enter initialization mode
    while(!(CAN1CTL1&INITAK));            // wait until initialization mode is entered
    CAN1CTL1  = 0x84;                     // enable CAN1, select oscillator as MSCAN clock
                                          // source, enable wake-up filter
```

```
          CAN1BTR0  = 0xC7;              // set SJW to 4, set prescaler to 8
          CAN1BTR1  = 0xB4;              // set phase_seg2 to 4 t_Q, phase_seg1 to 4 t_Q,
                                         // prop_seg to 1 t_Q
          CAN1IDAR0 = 0x56;             // set acceptance identifier V1
          CAN1IDAR1 = 0x3C;             // "
          CAN1IDAR2 = 0x40;             // "
          CAN1IDAR3 = 0x00;             // "
          CAN1IDMR0 = 0x00;             // acceptance mask for V1
          CAN1IDMR1 = 0x00;             // "
          CAN1IDMR2 = 0x3F;             // "
          CAN1IDMR3 = 0xFF;             // "
          CAN1IDAR4 = 0x56;             // set acceptance identifier V1
          CAN1IDAR5 = 0x3C;             // "
          CAN1IDAR6 = 0x40;             // "
          CAN1IDAR7 = 0x00;             // "
          CAN1IDMR4 = 0x00;             // acceptance mask for V1
          CAN1IDMR5 = 0x00;             // "
          CAN1IDMR6 = 0x3F;             // "
          CAN1IDMR7 = 0xFF;             // "
          CAN1IDAC  = 0x00;             // select two 32-bit filter mode
          CAN1CTL0  = 0x25;             // stop clock on wait mode, enable wake-up
          CAN1CTL0  &= ~INITRQ;         /* exit initialization mode */
}
void snd2can1(char *ptr)
{
     int tb,i,*pt1,*pt2;
     pt1 = (int *)ptr;                  /* convert to integer pointer */
     while(1) {
          if(CAN1TFLG & TXE0){
               tb = 0;
               break;
          }
          if(CAN1TFLG & TXE1){
               tb = 1;
               break;
          }
          if(CAN1TFLG & TXE2){
               tb = 2;
               break;
          }
     }
     CAN1TBSEL = CAN1TFLG;              // make empty transmit buffer accessible
     pt2 = (int *)&CAN1TIDR0;           // pt2 points to the IDR0 of TXFG
     for (i = 0; i < 7; i++)            // copy the whole transmit buffer
          *pt2++ = *pt1++;
     if (tb == 0)
          CAN1TFLG = TXE0;             // mark buffer 0 ready for transmission
```

```
        else if (tb == 1)
              CAN1TFLG = TXE1;              // mark buffer 1 ready for transmission
        else
              CAN1TFLG = TXE2;              // mark buffer 2 ready for transmission
   }
```

▲

This example can be extended to construct a distributed sensor system. There are applications that need to monitor temperature, pressure, humidity, and any other parameter in a small area that can be spanned by a CAN network. The purpose might be to monitor these environmental parameters and take appropriate actions in case abnormal values appear. In this type of application, each CAN node sends out the environmental parameters periodically to the CAN bus. A controlling CAN node keeps track of values sent from each node and does some processing. If the controlling CAN node discovers any abnormal situation, it may remind the operator to take appropriate action. To make sure each CAN node works normally, the controlling CAN node may set a counter for each distributed node. If any CAN node did not send in environmental parameters within the time-out period, it will remind the operator to take some corrective actions.

To make the CAN network more fault-tolerant, the designer may consider adding the second CAN bus to each node. The multiple CAN modules of many HCS12 devices serve this purpose well.

13.14 Summary

The CAN bus specification was initially proposed as a data communication protocol for automotive applications. However, it can also fulfill the data communication needs of a wide range of applications, from high-speed networks to low-cost multiplexed wiring.

The CAN protocol has gone through several revisions. The latest revision is 2.0A/B. The CAN 2.0A uses standard identifiers whereas the CAN 2.0B specification accepts extended identifiers. The HCS12 MSCAN module supports the CAN 2.0A and 2.0B specifications.

The CAN protocol supports four types of messages: *data frame, remote frame, error frame,* and *overload frame*. Users need to transfer only data frames and remote frames. The other two types of frames are only used by the CAN controller to control data transmission and reception. Data frames are used to carry normal data; remote frames are used to request other nodes to send messages with the same identifier as in the remote frame.

The CAN protocol allows all nodes on the bus to transmit simultaneously. When there are multiple transmitters on the bus, they arbitrate, and the CAN node transmitting a message with the highest priority wins. The simultaneous transmissions of multiple nodes will not cause any damage to the CAN bus. CAN data frames are acknowledged in-frame; that is, a receiving node sets a bit only in the acknowledge field of the incoming frame. There is no need to send a separate acknowledge frame.

The CAN bus has two states: *dominant* and *recessive*. The dominant state represents logic 0, and the recessive state represents logic 1 for most CAN implementations. When one node drives the dominant voltage to the bus while other nodes drive the recessive level to the bus, the resultant CAN bus state will be the dominant state.

Synchronization is a critical issue in the CAN bus. Each bit time is divided into four segments: *sync_seg, prop_seg, phase_seg1,* and *phase_seg2*. The sync_seg segment signals the start of a bit time. The sample point is between phase_seg1 and phase_seg2. At the beginning of each frame, every node performs a hard synchronization to align its sync_seg segment of its current bit time to the recessive-to-dominant edge of the transmitted start of frame. Resynchronization is performed during the remainder of the frame, whenever a change of bit value from recessive to dominant occurs outside of the expected *sync_seg* segment.

A CAN module (also called CAN controller) requires a CAN bus transceiver, such as the Philips PCA82C250 or the Microchip MCP2551, to interface with the CAN bus. Most CAN controllers allow more than 100 nodes to connect to the CAN bus. The CAN trunk cable could be a shielded cable, unshielded twisted pair, or simply a pair of insulated wires. It is recommended that shielded cable be used for high-speed transfer when there is a radio frequency interference problem. Up to a 1-Mbps data rate is achievable over a distance of 40 m.

The CAN module of the HCS12 MCU has six major modes of operation.

- Initialization mode
- Disable mode
- Normal operation mode
- Listen-only mode
- Loopback mode
- Sleep mode
- Stop mode

All configuration operations can only be performed in the initialization mode. Designers must make sure that the initialization mode is entered before performing the configuration. Most CAN parameters are configured immediately after reset. However, applications may need to change the configuration during the normal operation. Under this situation, the designer will need to put the CAN into sleep mode before performing reconfiguration. The first step in CAN configuration is to initialize the timing parameters. The procedure for doing this is described in Section 13.12. All data transfer and reception with other nodes are performed in the normal operation mode.

The CAN module has three transmit buffers and five receive buffers. The first 13 bytes of a transmit buffer and a receive buffer are identical. In addition, the transmit buffer also has a transmit buffer priority register and two transmit timestamp registers. One of the three transmit buffers is mapped to the foreground transmit buffer space and is made accessible to the designer. Five receive buffers are organized as a FIFO, and only the head of the FIFO is mapped to the receive foreground space and becomes accessible to the designer.

Each transmit and receive buffer has eight data registers for holding data, four identifier registers for identification and arbitration purposes, and one length-count register for indicating the number of data segment registers that contain valid data. Data transmission involves copying data into identifier registers, data segment registers, and the data length register, setting the priority of the transmit buffer, and clearing the transmit buffer empty bit to mark the buffer ready for transmission. Data reception in the CAN bus is often interrupt-driven due to the unpredictability of message arrival times.

13.15 Exercises

E13.1 Calculate the bit segments for the following system constraints assuming that the Microchip MCP2551 transceiver is used:

Bit rate = 400 kbps

Bus length = 50 m

Bus propagation delay = 5×10^{-9} s/m

Microchip MCP2551 transceiver and receiver propagation delay = 150 ns at 85°C

Oscillator frequency = bus frequency = 24 MHz

E13.2 Calculate the bit segments for the following system constraints assuming that the Microchip MCP2551 transceiver is used:

Bit rate = 200 kbps

Bus length = 100 m

Bus propagation delay = 5×10^{-9} s/m

Microchip MCP2551 transceiver and receiver propagation delay = 150 ns at 85°C

Oscillator frequency = bus frequency = 24 MHz

E13.3 Calculate the bit segments for the following system constraints assuming that the Microchip MCP2551 transceiver is used:

Bit rate = 1 Mbps

Bus length = 20 m

Bus propagation delay = 5×10^{-9} s/m

MCP2551 transceiver and receiver propagation delay = 150 ns at 85°C

Oscillator frequency = bus frequency = 32 MHz

E13.4 Calculate the bit segments for the following system constraints assuming that the Microchip MCP2551 transceiver is used:

Bit rate = 800 kbps

Bus length = 20 m

Bus propagation delay = 5×10^{-9} s/m

MCP2551 transceiver and receiver propagation delay = 150 ns at 85°C

Oscillator frequency = 16 MHz (PLL is used to convert to 24-MHz bus frequency)

E13.5 Write a subroutine to configure the CAN1 module of the HCS12DP256 with the bit timing parameters computed in E13.2. Enable receive interrupt but disable transmit interrupt. Configure the CAN so that it accepts only messages with the extended or standard identifiers starting with H, P, or T. Use an 8-bit filter to control the acceptance.

E13.6 Write a C function to perform the same setting as in E13.4.

E13.7 Provide three different ways of accepting messages that have the letter K as the first letter of the identifier.

E13.8 Write an assembly program to send out the string "Monday" using WD as the first two characters of the extended identifier. The remaining bits of the identifier are cleared to 0.

E13.9 Write a C program to send out the string "too high" using ans as the first three letters of the extended identifier. The remaining bits of the extended identifier are cleared to 0.

13.16 Lab Exercises and Assignments

L13.1 Practice data transfer over the CAN bus using the following procedures:

Step 1
Use a pair of insulated wires about 20 m long. Connect the CAN_H pins of two demo boards (e.g., the SSE256 or Dragon12 demo boards) with one wire and connect the CAN_L pins of

both demo boards with another wire. Connect both ends of these two wires together with
120-Ω resistors (SSE256 has terminating resistors on board). The circuit connection is shown
in Figure L13.1.

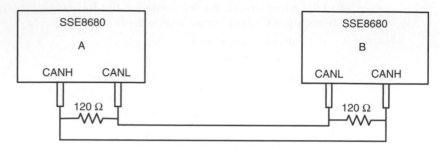

Figure L13.1 ■ CAN circuit connection for L13.1

Step 2
Write a program to be downloaded onto one of the demo boards (called board A) that
performs A/D conversion 10 times every second. Use the on-board potentiometer to
generate a voltage to be converted. Send out the A/D conversion result over the CAN
bus every 100 ms. Use the letter W as the identifier of the data frame. (Use the timing
parameters computed in E13.4 for the SSE256 demo board.)

Step 3
Write a program to be downloaded onto another demo board (called board B). This program
will send out the number of data frames received so far over the CAN bus. This program
will use the letter R as the identifier. After the number reaches 99, the program will reset
the number to 0 and start over again. Each digit of the decimal number is encoded in
ASCII code.

Step 4
Board A will display the number received over the CAN bus on eight LEDS in BCD
format.

Step 5
Board B will display the received A/D result in an LCD display.

L13.2 This assignment requires three students to wire three HCS12 demo boards together using
the CAN bus. Assign each of these three demo boards with a number 0, 1, and 2, respectively.
Write a program to be run on each demo board and perform the following procedure:

Step 1
Use two potentiometers (one is already on board) to emulate the temperature sensor and
barometric sensor outputs (0 to 5 V). Connect the demo boards as shown in Figure L13.2.
The temperature sensor represents the temperature range from −40 to 125°C. The
barometric sensor represents the pressure range from 948 to 1083.8 mbar.

Step 2
Each demo board performs temperature and barometric pressure measurements once per
second and stores the result in separate buffers formatted to have the same structure as a
transmit buffer. When storing the measurement data, use Ti (for node i) as the identifier
for temperature value and use Pi (for node i) as the identifier for pressure value.

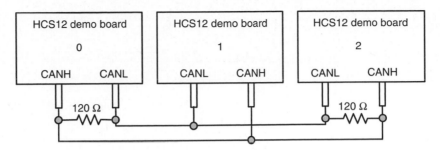

Figure L13.2 ■ CAN circuit connection for L13.2

Step 3

After performing temperature and pressure measurements and storing data in buffers, each demo board uses a random number generator to generate a number k in the range of 0 to 2 and waits for that amount of time (seconds) to send out a remote transmission request to request node k to send in temperature or pressure data and display them on the LCD. The received data should be displayed as follows:

      received Tk = xxx.y

or

      received Pk = xxxx.y

The demo board that transmits the data should display the following message:

      send Tk = xxx.y

or

      send Pk = xxxx.y

14

Internal Memory Configuration and External Expansion

14.1 Objectives

After completing this chapter, you should be able to

- Understand the overall HCS12 memory system organization and mapping

- Erase and program the on-chip flash memory

- Control the operation, programming, and protection of the on-chip flash memory

- Erase and program the on-chip EEPROM

- Understand the external memory expansion issue

- Make memory space assignment

- Design address decoders and the memory control circuit

- Perform timing analysis for the memory system

14.2 Introduction

Each of the HCS12 members has a certain amount of on-chip SRAM, EEPROM, and flash memory. I/O-related registers are implemented in SRAM. These memories share the same memory space and are assigned to a certain default memory space after reset. To meet the requirements of different applications, these memories can be remapped to other locations. This can be achieved by programming the appropriate initialization register associated with each of these memories. Both the EEPROM and flash memory must be erased before they can be programmed correctly. The erasure and programming methods for both the flash memory and EEPROM are similar.

Most of the HCS12-based embedded systems operate in the single-chip mode and do not need external memory. An HCS12 member may have from a minimum of 32 kB of internal memory up to a maximum of 512 kB of on-chip memory. The memory requirements of a wide range of applications can be satisfied by different members of the HCS12 microcontroller family. In case an application needs more than 512 kB of memory, the designer has the option to add external memory to the HCS12 device. After adding external memory, an HCS-based embedded system may have up to 1 MB of system memory. The designer should also consider using a 32-bit microcontroller, because a 32-bit microcontroller may prove to be more cost-effective and has several additional advantages.

- No paging overhead. The HCS12 has a 16-bit program counter and hence can only address up to 64 kB of program memory directly. Accessing more than 64 kB of memory would require a certain type of paging or banking scheme, which has been proved to be very inefficient. A 32-bit microcontroller does not have such a problem.

- Much larger memory space. A 32-bit microcontroller supports 4 GB of memory space, which can satisfy the memory requirements of many demanding applications.

- Much higher performance for the same clock frequency. A 32-bit microcontroller can process 32-bit data in one operation, whereas the HCS12 can handle only 16 bits of data at a time.

Many microcontrollers multiplex the address and data signals on the same set of signal pins in order to economize on the use of signal pins. The HCS12 is no exception. In the HCS12, the address and data signals are time multiplexed on Port A and Port B pins. However, external memory devices require address signals to be stable throughout the whole access cycle. This can be achieved by using latches to make a copy of the address signals so that the address signals remain valid for the whole access cycle.

When adding external memory to the HCS12, we must make sure that the timing requirements are satisfied. Each bus cycle takes exactly one E-clock cycle to complete. However, the bus cycle can be stretched to support slower memory devices. A bus cycle can be stretched by one to three E-clock cycles.

14.3 Internal Resource Remapping

The internal register block, SRAM, EEPROM, and flash memory have default locations within the 64-kB standard address space but may be reassigned to other locations during program execution by setting bits in mapping registers INITRG, INITRM, and INITEE. These registers can be written only once in normal operation modes. It is advisable to explicitly establish these resource locations during the initialization phase of program execution, even if default values are chosen, in order to protect the registers from inadvertent modification later. Writes to the mapping registers require one cycle to take effect. To ensure that there are no unintended operations, a write to one of these registers should be followed by a NOP instruction.

If conflict occurs when mapping resources, the register block will take precedence over other resources; RAM or EEPROM addresses occupied by the register block will not be available for storage. When active, BDM ROM takes precedence over other resources, although a conflict between BDM ROM and register space is not possible. Table 14.1 shows mapping precedence. Only one module will be selected at a time. In the case of more than one module sharing a space, only the module with the highest precedence will be selected. Mapping more than one module to the same location won't damage the system. However, it is not wise to map two or more modules to the same space because a significant amount of memory could become unusable.

In expanded modes, all address space not used by internal resources is by default external memory.

Precedence	Resource
1	BDM firmware or register space
2	Internal register space
3	SRAM block
4	EEPROM
5	Flash memory
6	Remaining external memory

Table 14.1 ■ Mapping precedence

14.3.1 Register Block Mapping

After reset, the register block (1 or 2 kB) can be remapped to any 2 kB space within the first 32 kB of the system address space. Mapping of internal registers is controlled by 4 bits in the INITRG register. The contents of this register are shown in Figure 14.1. Only bit 6 to bit 3 are implemented. The INITRG register can be written into once in normal mode but many times in special modes.

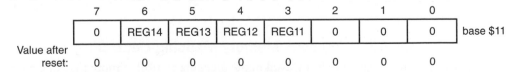

7	6	5	4	3	2	1	0	
0	REG14	REG13	REG12	REG11	0	0	0	base $11

Value after
reset: 0   0   0   0   0   0   0   0

REG14~REG11: internal register map position
These 4 bits along with bit 7 specify the upper 5 bits of the 16-bit register address. These 4 bits can be written only once in normal modes and can be written many times in special modes. There is no restriction on the reading of this register.

Figure 14.1 ■ Contents of the INITRG register

Example 14.1

Write an instruction sequence to remap the register block to $4000.

Solution: The upper 5 bits of the INITRG register should be 01000. The following instruction sequence will achieve the desired assignment:

```
movb    #$40,INITRG
nop                           ; wait for the remap to take effect
```

14.3.2 SRAM Mapping

SRAM can be remapped to any 2-kB boundary within the 64-kB memory space. The remapping is controlled by the INITRM register. The contents of the INITRM are shown in Figure 14.2. Bits RAM15 to RAM11 determine the upper 5 bits of the base address for the system's internal SRAM array. The INITRM register can be written into once in normal modes but many times in special modes.

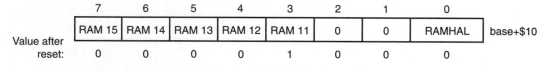

	7	6	5	4	3	2	1	0	
	RAM 15	RAM 14	RAM 13	RAM 12	RAM 11	0	0	RAMHAL	base+$10
Value after reset:	0	0	0	0	1	0	0	0	

RAM15~RAM11: internal RAM map position
    These bits determine the upper 5 bits of the base address for
    the system's internal RAM array.
RAMHAL: RAM high-align
    0 = aligns the RAM to the lowest address ($0000) of the
        mappable space.
    1 = aligns the RAM to the highest address ($FFFF) of the
        mapping space.

Figure 14.2 ■ RAM initialization register (INITRM)

14.3.3 EEPROM Mapping

The EEPROM block can be remapped to any 2-kB boundary. The remapping of EEPROM is controlled by the INITEE register. The contents of this register are shown in Figure 14.3. The EEPROM is activated by bit 0 of this register. Bits EE15 to EE11 determine the upper 5 bits of the base address for the system's internal EEPROM array. The INITEE register can be written any time.

14.3.4 Miscellaneous Memory Mapping Control

A few registers are involved in the mapping control of more than one memory module. The functions of these registers are described in this section.

MISCELLANEOUS SYSTEM CONTROL REGISTER (MISC)

This register enables/disables the on-chip ROM (including flash memory and EEPROM) and allows us to stretch the length of the external bus cycle. The contents of this register are shown in Figure 14.4. The value of the ROMONE pin is latched to the ROMON bit of the MISC register on the rising edge of the $\overline{\text{RESET}}$ signal. If the ROMONE signal is high when it is

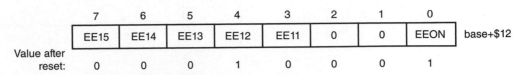

	7	6	5	4	3	2	1	0	
	EE15	EE14	EE13	EE12	EE11	0	0	EEON	base+$12
Value after reset:	0	0	0	1	0	0	0	1	

EE15~EE11: internal EEPROM map position
These bits specify the upper 5 bits of the 16-bit registers address.
These 5 bits can be written only once in normal modes and can be
written many times in special modes. There is no restriction on the
reading of this register.
EEON: internal EEPROM on (enabled reading).
 0 = removes EEPROM from the map.
 1 = places the on-chip EEPROM in the memory map.

Figure 14.3 ■ Contents of the INITEE register

	7	6	5	4	3	2	1	0	
Reset value:	0	0	0	0	EXSTR1	EXSTR0	ROMHM	ROMON	base+$13
Expanded or emulation	0	0	0	0	1	1	0	1̲	
Peripheral or single chip	0	0	0	0	1	1	0	1	
Special test	0	0	0	0	1	1	0	0	

EXSTR1~EXSTR0: external access stretch bits 1 and 0
 00 = no stretch to external bus cycle.
 01 = stretch the external bus cycle by one E cycle.
 10 = stretch the external bus cycle by two E cycles.
 11 = stretch the external bus cycle by three E cycles.
ROMHM: flash EEPROM or ROM only in second half of memory map
 0 = The fixed page(s) of flash EEPROM or ROM in the lower half of
  the memory map can be accessed.
 1 = Disable direct access to the flash EEPROM or ROM in the lower
  half of the memory map. These physical locations of flash memory
  can still be accessed through the program page window.
ROMON: enable flash memory or ROM
 0 = disable the flash memory or ROM in the memory map.
 1 = enable the flash memory or ROM in the memory map.
Note: 1. The reset state of this bit is determined at the chip integration level.

Figure 14.4 ■ Contents of the MISC register

latched, then the on-chip flash memory is enabled and the reset startup routine will be executed
from the on-chip flash memory rather than from the external ROM. After reset, this pin is used
as the $\overline{ECS}$ signal. This register can only be written once in normal and emulation modes but
can be written any time in special modes.

MEMORY SIZE REGISTER ZERO (MEMSIZ0)

This register is read only. It reflects the size of the on-chip I/O register, EEPROM, and
SRAM. The contents of this register are shown in Figure 14.5. The allocated system register
space, EEPROM space, RAM memory space, and the RAM reset base address are summarized
in Table 14.2.

7	6	5	4	3	2	1	0	
REG_SW0	0	EEP_SW1	EEP_SW0	0	RAM_SW2	RAM_SW1	RAM_SW0	base+$1C

Reset: -- -- -- -- -- -- -- --

REG_SW0: allocated system register space
  0 = allocated system register space size is 1 kB.
  1 = allocated system register space size is 2 kB.
EEP_SW1~EEP_SW0: allocated system EEPROM memory space
  00 = 0 kB
  01 = 2 kB
  10 = 4 kB
  11 = 8 kB
RAM_SW2~RAM_SW0: Allocated system RAM memory space
  The allocated system RAM space size is as given in Table 14.2.

Figure 14.5 ■ Memory size register zero (MEMSIZ0)

RAM_SW2: RAM_SW0	Allocated RAM Space	RAM Mappable Region	INITRM Bits Used	RAM Reset Base Address[1]
000	2 kB	2 kB	RAM15~RAM11	$0800
001	4 kB	4 kB	RAM15~RAM12	$0000
010	6 kB	8 kB[2]	RAM15~RAM13	$0800
011	8 kB	8 kB	RAM15~RAM14	$0000
100	10 kB	16 kB[2]	RAM15~RAM14	$1800
101	12 kB	16 kB[2]	RAM15~RAM14	$1000
110	14 kB	16 kB[2]	RAM15~RAM14	$0800
111	16 kB	16 kB	RAM15~RAM14	$0000

[1] The RAM reset base address is based on the reset value of the INITRM register, $09.
[2] Alignment of the allocated RAM space within the RAM mappable region is dependent on the value of RAMHAL.

Table 14.2 ■ Allocated RAM memory space

MEMORY SIZE REGISTER ONE (MEMSIZ1)

This register is read-only and reflects the state of the flash or ROM physical memory and paging switches at the core boundary, which are configured at system integration. This register allows read visibility to the state of these switches. The contents of this register are shown in Figure 14.6. Table 14.3 shows that the HCS12 supports 1 MB of memory space.

7	6	5	4	3	2	1	0	
ROM_SW1	ROM_SW0	0	0	0	0	PAG_SW1	PAG_SW0	base+$1D

Reset: -- -- -- -- -- -- -- --

ROM_SW1~ROM_SW0: allocated system flash or ROM physical memory space
  00 = 0 kB
  01 = 16 kB
  10 = 48 kB
  11 = 64 kB
PAG_SW1~PAG_SW0: allocated off-chip flash or ROM memory space
  The allocated off-chip flash or ROM memory space size is as given in Table 14.3.

Figure 14.6 ■ Memory size register one (MEMSIZ1)

PAG_SW1~PAG_SW0	Off-Chip Space	On-Chip Space
00	876 kB	128 kB
01	768 kB	256 kB
10	512 kB	512 kB
11	0 kB	1 MB

Table 14.3 ■ Allocated off-chip memory options

PROGRAM PAGE INDEX REGISTER (PPAGE)

This register allows for integrating up to 1 MB of flash or ROM into the system by using the 6 page index bits to page 16-kB blocks into the program page window located from $8000 to $BFFF. CALL and RTC instructions have special access to read and write this register without using the address bus. The contents of this register are shown in Figure 14.7.

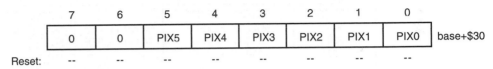

PIX5~PIX0: Program page index bits 5~0

Figure 14.7 ■ Program page index register (PPAGE)

14.4 Expanded Memory Mapping

The HCS12 uses the PPAGE register to determine which of the 64 possible 16-kB pages is active in the program window. Only one page at a time can occupy the window, and the value in the associated register must be changed to access a different page of memory.

Since all memory locations above 64 kB must be accessed via this page window, the direct concatenation of the page number and the page offset does not result in linear addresses. This also causes a problem in encoding machine code using S records because S records use linear addresses in addressing machine code. Freescale uses the following scheme to solve this problem:

1. Use $00 to $3F as the page numbers for the 64 16-kB pages.

2. Use higher page numbers to select on-chip flash memory. For example, for the HCS12 members with 256-kB on-chip flash memory, assign $30~$3F as the page numbers of the 16 on-chip 16-kB pages. Use $00 to $2F as page numbers for external memory. The mapping from PPAGE to S records is illustrated in Table 14.4.

3. Use linear addresses to address external memory and also use the linear method to address machine codes for S records.

4. Compute the page number and page address within the $8000~$BFFF page window using the following equations:

$$PageNum = SRecAddr/PPAGEWinSize \qquad (14.1)$$

$$PageWinAddr = (SRecAddr \% PPAGEWinSize) + PPAGEWinStart \qquad (14.2)$$

where

PageNum is from $00 to $3F.

SRecAddr is the linear address used in S records.

PPAGEWinSize = 16 kB (or 16384)

PageWinAddr is a number between $8000 and $BFFF.

PPAGEWinStart = $8000

5. Compute SRecAddr from PageNum and PageWinAddr as follows:

$$SRecAddr = PageNum \times PPAGEWinSize + PageWinAddr - PPAGEWinStart \qquad (14.3)$$

6. When addressing external memory, the highest 6 address bits appear on XADDR19~XADDR14 and the lowest 14 address bits appear on A13~A0.

PPAGE Value	S-Record Address Range	A15~A0	Page + A13~A0	Memory Type
$00~$2F	$00000~$BFFFF	1	$00000~$BFFFF	Off-chip memory
$30	$C0000~$C3FFF	$0000~$3FFF	$C0000~$C3FFF	On-chip flash
$31	$C4000~$C7FFF	$4000~$7FFF	$C4000~$C7FFF	On-chip flash
$32	$C8000~$CBFFF	$8000~$BFFF	$C8000~$CBFFF	On-chip flash
$33	$CC000~$CFFFF	$C000~$FFFF	$CC000~$CFFFF	On-chip flash
$34	$D0000~$D3FFF	$0000~$3FFF	$D0000~$D3FFF	On-chip flash
$35	$D4000~$D7FFF	$4000~$7FFF	$D4000~$D7FFF	On-chip flash
$36	$D8000~$DBFFF	$8000~$BFFF	$D8000~$DBFFF	On-chip flash
$37	$DC000~$DFFFF	$C000~$FFFF	$DC000~$DFFFF	On-chip flash
$38	$E0000~$E3FFF	$0000~$3FFF	$E0000~$E3FFF	On-chip flash
$39	$E4000~$E7FFF	$4000~$7FFF	$E4000~$E7FFF	On-chip flash
$3A	$E8000~$EBFFF	$8000~$BFFF	$E8000~$EBFFF	On-chip flash
$3B	$EC000~$EFFFF	$C000~$FFFF	$EC000~$EFFFF	On-chip flash
$3C	$F0000~$F3FFF	$0000~$3FFF	$F0000~$F3FFF	On-chip flash
$3D	$F4000~$F7FFF	$4000~$7FFF	$F4000~$F7FFF	On-chip flash
$3E	$F8000~$FBFFF	$8000~$BFFF	$F8000~$FBFFF	On-chip flash
$3F	$FC000~$FFFFF	$C000~$FFFF	$FC000~$FFFFF	On-chip flash

Note: 1. Repetition of $0000~$3FFF, $4000~$3FFF, $8000~$BFFF, and $C000~$FFFF for 12 times.

Table 14.4 ■ MC9S12DP256 PPAGE to S-record address mapping

Example 14.2

What are the PageNum and PageWinAddr for the SRecAddr of $E1003?

Solution: Apply Equations 14.1 and 14.2 as follows:

PageNum   = $E1003/$4000 = $38
PageWinAddr = ($E1003 % $4000) + $8000 = $1003 + $8000 = $9003

Example 14.3

▼

What is the corresponding SRecAddr for the pair of (PageNum, PageWinAddr) equal to ($20, $A003)?

Solution: Apply Equation 14.3 as follows:

SRecAddr = $20 × $4000 (16K) + $A003 − $8000 = $82003

▲

14.5 On-Chip Flash Memory

All HCS12 members have a certain amount of on-chip flash memory. The amount of flash memory may be 32 kB, 64 kB, 128 kB, 256 kB, or 512 kB. A flash memory larger than 64 kB is divided into two or more 64-kB blocks. These 64-kB flash memory blocks can be read, programmed, or erased concurrently.

The algorithms for erasing and programming flash memory require very little user involvement. Programming and erasure are performed by sending commands to the command register. Interrupts may be requested when a command is completed or when the command buffer is empty. A two-stage command pipeline is implemented to accelerate the command execution. The flash memory includes a flexible protection scheme against accidental programming and erasure. The flash memory also implements security measures to prevent the application code from being pirated.

The discussion of flash memory will be based on the 256-kB flash memory of several HCS12 members.

14.5.1 Flash Memory Map

The memory map of the 256-kB flash memory is shown in Figure 14.8.

The entire flash memory is divided into pages of 16 kB in size. One page is always accessible at $4000 to $7FFF and another fixed page is always accessible at $C000 to $FFFF. The memory space from $0000 to $3FFF has been occupied by registers, EEPROM, and SRAM and hence is not available for flash memory. As shown in Figure 14.8, the first 64 kB of the flash memory is referred to as block 0 and is assigned page numbers $3C to $3F. The pages with addresses from $4000 to $7FFF, $8000 to $BFFF, and $C000 to $FFFF are assigned the page numbers $3E, $3D, and $3F, respectively. Even though it is unnecessary, one can access the pages in address ranges of $4000 to $7FFF and $C000 to $FFFF via the page window.

14.5.2 Flash Memory Protection

Each flash block may be protected against accidental erasure or programming. Flash protection is controlled by a flash protection register (FPROT). For microcontrollers that have multiple flash blocks, there is a separate flash protection register for each flash block. These flash protection registers share a common address, with the active register selected by the bank select bits of the flash configuration register (FCNFG). During the HCS12 reset sequence (execution of reset startup routine), the flash protection registers for each flash block are loaded from the programmed bytes within a flash block. For example, for the MC9S12DP256, the locations $FF0A, $FF0B, $FF0C, and $FF0D store the protection information of blocks three, two, one, and zero, respectively.

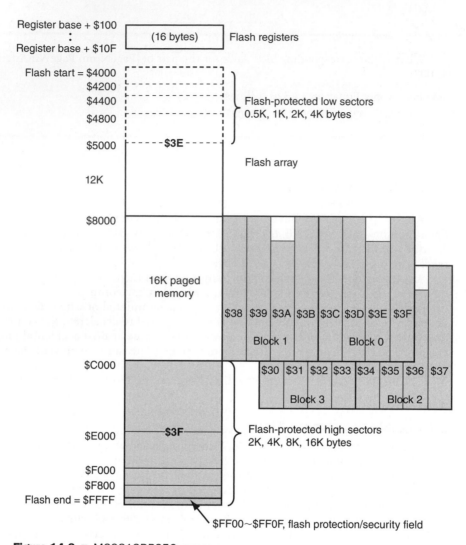

Figure 14.8 ■ MC9S12DP256 memory map

The contents of each FPROT register determine whether the entire block or just subsections are protected from being accidentally erased or programmed. Each flash block can be entirely protected or can have one or two separate protected areas. One of these areas, known as the *lower protected block*, starts at a point 32 kB (the page with the second-largest page number in the same block) below the maximum flash block address and is extendable toward higher addresses. The other area, known as the *upper protected block*, ends at the top of the flash block and is extended toward lower addresses. The upper and lower protected blocks do not meet up. In general, the upper protected area of flash block 0 is used to hold bootloader code since it contains the reset and interrupt vectors. The lower protected area of block 0 and the protected areas of the other flash blocks can be used for critical parameters that would not change when the application program is updated.

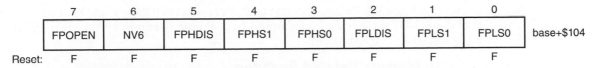

7	6	5	4	3	2	1	0	
FPOPEN	NV6	FPHDIS	FPHS1	FPHS0	FPLDIS	FPLS1	FPLS0	base+$104

Reset: F F F F F F F F

FPOPEN: opens the flash for program or erase
  0 = the whole flash block is protected. In this case, bits 5 to 0 have no effect.
  1 = the flash sectors not protected are enabled for program or erase.
NV6: not volatile flag bit
  This bit is available for nonvolatile flag usage.
FPHDIS: flash protection higher address range disable
  0 = higher address range protection enabled.
  1 = higher address range protection disabled.
FPHS1~FPHS0: flash protection higher address size
  00 = 2 kB.
  01 = 4 kB.
  10 = 8 kB.
  11 = 16 kB.
FPLDIS: flash protection lower address range disable
  0 = lower address range protection enabled.
  1 = lower address range protection disabled.
FPLS1~FPLS0: flash protection lower address size
  00 = 512 bytes.
  01 = 1 kB.
  10 = 2 kB.
  11 = 4 kB.

Figure 14.9 ■ FPROT register

The contents of the FPROT register are shown in Figure 14.9. The FPROT register is readable in normal and special modes. The FPOPEN, FPHDIS, and FPLDIS bits in the FPROT register can only be written into the protected state (i.e., 0). The FPLS1~FPLS0 bits can be written anytime until the FPLDIS bit is cleared. The FPHS1~FPHS0 bits can be written anytime until the FPHDIS bit is cleared.

To change the flash protection that will be loaded on reset, the upper sector of the flash memory must be unprotected, then the flash protect/security byte located as described in Table 14.5 must be written into.

Trying to program or erase any of the protected areas will result in a protection violation error and the bit PVIOL of the FSTAT register will be set to 1. A bulk erase is possible only if the

Address	Size (bytes)	Description
$FF00~$FF07	8	Backdoor comparison keys
$FF08~$FF09	5	Reserved
$FF0A	1	Block 3 flash protection byte
$FF0B	1	Block 2 flash protection byte
$FF0C	1	Block 1 flash protection byte
$FF0D	1	Block 0 flash protection byte
$FF0E	1	Reserved
$FF0F	1	Flash options/security byte

Table 14.5 ■ Flash protection/options field

protection for that block is fully disabled; that is, FOPEN = 1, FPHDIS = 1, and FPLDIS = 1. The contents of the FPROT register are loaded from 1 byte of four locations, $FF0A~$FF0D. This is indicated by the letter F in Figure 14.9.

14.5.3 Flash Memory Related Registers

In addition to the FPROT registers, the following registers are related to the configuration and operation control of the flash memory.

FLASH CLOCK DIVIDER REGISTER (FLCKDIV)

This register is not banked and is used to control the timed events in programming and erasure algorithms. The contents of this register are shown in Figure 14.10. Bit 7 is read only; other bits are readable and writable.

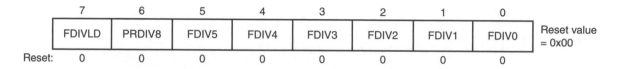

	7	6	5	4	3	2	1	0	
	FDIVLD	PRDIV8	FDIV5	FDIV4	FDIV3	FDIV2	FDIV1	FDIV0	Reset value = 0x00
Reset:	0	0	0	0	0	0	0	0	

FDIVLD: clock divider loaded
  0 = register has not been written.
  1 = register has been written to since the last reset.
PRDIV8: enable prescaler by 8
  0 = the input oscillator clock is directly fed into the FCLKDIV divider.
  1 = enables a divide-by-8 prescaler, to divide the flash module input oscillator
      clock before feeding into the CLKDIV divider.
FDIV[5:0]: clock divider bits
  The combination of PRDIV8 and FDIV[5:0] effectively divides the flash module
  input oscillator clock down to a frequency of 150 kHz~200 kHz. The maximum
  divide ratio is 512.

Figure 14.10 ■ Flash clock divider register (FCLKDIV)

In order to guarantee that the on-chip flash memory and EEPROM can be programmed and erased properly, the clock signal that controls the timing of programming and erasure should be configured properly. The only clock configuration to be done is to set the prescaler of the clock by programming this register. The startup routine of the HCS12 needs to check the FDIVLD bit to make sure that the prescaler has been configured. Depending on the oscillator frequency, the designer may choose to insert or bypass a divide-by-8 prescaler to the clock signal for controlling the programming and erasure of the flash memory and EEPROM. The procedure for configuring the clock prescaler is discussed in Section 14.5.6.

FLASH SECURITY REGISTER (FSEC)

This register holds all the bits associated with the device security. The contents of this register are shown in Figure 14.11. All the bits of this register are readable but not writable. This register is loaded from the flash protection/options field byte at $FF0F during the reset sequence, indicated by F in Figure 14.11. This register is not banked. The SEC1 and SEC0 bits define the security state of the device. If the flash is unsecured using the Backdoor Key Access, the SEC bits are forced to 10. The HCS12 security function is discussed in Section 14.5.4.

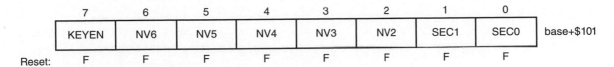

7	6	5	4	3	2	1	0	
KEYEN	NV6	NV5	NV4	NV3	NV2	SEC1	SEC0	base+$101

Reset: F F F F F F F F

KEYEN: enable backdoor key to security
  0 = backdoor to flash is disabled.
  1 = backdoor to flash is enabled.
NV6~NV2: nonvolatile flag bits
  These 5 bits are available to the user as nonvolatile flags.
SEC[1:0]: memory security bits
  00 = secured.
  01 = secured.
  10 = unsecured.
  11 = secured.

Figure 14.11 ■ Flash security register (FSEC)

FLASH TEST MODE REGISTER (FTSTMOD)

This register is not banked and is used primarily to control the nonvolatile memory test modes. The contents of this register are shown in Figure 14.12. The WRALL bit is writable only in special modes. The purpose of this bit is to launch a command on all blocks in parallel. This can be useful for bulk erase and blank check operations.

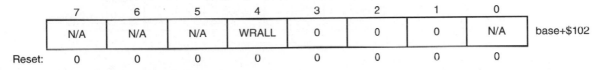

7	6	5	4	3	2	1	0	
N/A	N/A	N/A	WRALL	0	0	0	N/A	base+$102

Reset: 0 0 0 0 0 0 0 0

WRALL: write to all register banks
  0 = write only to the bank selected via BKSEL.
  1 = write to all register banks.

Figure 14.12 ■ Flash test mode register (FTSTMOD)

FLASH CONFIGURATION REGISTER (FCNFG)

This register enables the flash interrupts, gates the security backdoor writes, and selects the register bank to be operated on. The contents of this register are shown in Figure 14.13. This register is not banked.

For an HCS12 device with 256 kB of flash memory, there are four 64-kB blocks. Each block has a set of control registers. Only one set of them is available at a time. The BKSEL1 and BKSEL0 bits select one bank to be accessed.

FLASH STATUS REGISTER (FSTAT)

The programming and erase of flash memory is controlled by a finite state machine. The FSTAT register defines the flash state machine command status and flash array access, protection, and bank verify status. This register is banked. The contents of this register are shown in Figure 14.14.

The bits CBEIF, PVIOL, and ACCERR are readable and writable, whereas bits CCIF and BLANK are readable but not writable.

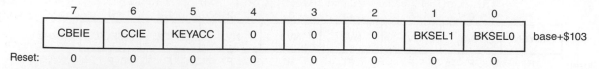

7	6	5	4	3	2	1	0	
CBEIE	CCIE	KEYACC	0	0	0	BKSEL1	BKSEL0	base+$103

Reset: 0 0 0 0 0 0 0 0

CBEIE: command buffer empty interrupt enable
    0 = command buffer empty interrupts disabled.
    1 = an interrupt will be requested whenever the CBEIF flag is set.
CCIE: command completion interrupt enable
    0 = command complete interrupts disabled.
    1 = an interrupt will be requested whenever the CCIF flag is set.
KEYACC: enable security key writing
    0 = flash writes are interpreted as the start of a program or erase sequence.
    1 = writes to flash array are interpreted as keys to open the backdoor. Reads
       of the flash array return invalid data.
BKSEL[1:0]: register bank select
    00 = flash 0.
    01 = flash 1.
    10 = flash 2.
    11 = flash 3.

Figure 14.13 ■ Flash configuration register (FCNFG)

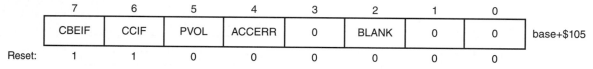

7	6	5	4	3	2	1	0	
CBEIF	CCIF	PVOL	ACCERR	0	BLANK	0	0	base+$105

Reset: 1 1 0 0 0 0 0 0

CBEIF: command buffer empty interrupt flag
    0 = command buffers are full.
    1 = command buffers are ready to accept a new command.
CCIF: command completion interrupt flag
    0 = command in progress.
    1 = all commands are completed.
PVOL: protection violation
    0 = no protection violation has occurred.
    1 = a protection violation has occurred.
ACCERR: flash access error
    0 = no failure.
    1 = access error has occurred.
BLANK: array has been verified as erased.
    0 = if an erase and verify command has been requested, and the CCIF flag is set,
       then a zero in BLANK indicates that the block is not erased.
    1 = flash block verifies as erased.

Figure 14.14 ■ Flash status register (FSTAT)

       The CBEIF flag indicates that the address, data, and command buffers are empty so that a new command sequence can be started. The CBEIF flag is cleared by writing a 1 to it. Writing a 0 to the CBEIF flag has no effect on CBEIF but sets ACCERR, which can be used to abort a command sequence.

The CCIF flag indicates that there are no more commands pending. The CCIF flag is cleared when CBEIF is cleared and sets automatically on completion of all active and pending commands. Writes to the CCIF flag have no effect.

The PVIOL flag indicates an attempt was made to program or erase an address in a protected flash memory area. The PVIOL flag is cleared by writing a 1 to it. Writing a 0 has no effect.

The ACCERR flag indicates an illegal access to the selected flash array. This can be a violation of the command sequence, issuing an illegal command, or the execution of a CPU STOP instruction while a command is executing. This flag is cleared by writing a 1 to it. Writing a 0 to the ACCERR flag has no effect.

FLASH COMMAND REGISTER (FCMD)

This register defines the flash commands. This register is banked; its contents are shown in Figure 14.15.

7	6	5	4	3	2	1	0	
0	CMDB6	CMDB5	0	0	CMDB2	0	CMDB0	base+$106

Reset:

1	1	0	0	0	0	0	0

CMDB6, CMDB5, CMDB2, and CMDB0: command bits
Valid commands include the following:
$05 = erase and verify.
$20 = program a word (2 bytes).
$40 = sector erase.
$41 = bulk erase.

Figure 14.15 ■ Flash command register (FCMD)

To perform these commands, we must make sure that the following prerequisites are satisfied:

1. The FCLKDIV register has been configured correctly.
2. The sector to be erased is not be protected.
3. The flash word to be programmed has been erased and not be protected.
4. The first flash address is word aligned (A0 = 0).
5. If the flash address is in the range of $8000 to $BFFF, then the PPAGE register has been written to select the desired page.
6. If the MCU has multiple flash blocks, the ACCERR and PVIOL flags in all blocks have been cleared.
7. The BSEL bits in the FCNFG register have been written to select the desired block to be programmed or erased.

FLASH 16-BIT ADDRESS REGISTER (FADDRHI AND FADDRLO)

Bit 15 of the address register is tied to 0. In normal modes, the FADDR register reads zero and is not writable. The FADDRHI and FADDRLO registers can be written in special modes by writing to the address base+$108 and base+$109 in the register space.

FLASH 16-BIT DATA BUFFER AND REGISTER (FDATA)

In normal modes, all FDATA bits read zero and are not writable. In special modes, all FDATA bits are readable and writable when writing to an address within the flash address range.

14.5.4 Securing the Flash Memory

The flash security feature is designed to prevent unauthorized access to the nonvolatile memory. The memory contents are secured by programming the security bits within the flash options/security byte at the address $FF0F. On devices that have a memory page window, the flash options/security byte is also available at the address $BF0F by setting the value of the PPAGE register to $3F. The contents of this byte are copied into the flash security register (FSEC) during the reset sequence.

The flash sector $FE00 to $FFFF must be erased before the flash options/security byte is programmed. The flash is programmed by aligned word only, so the address $FF0E must be written as the word address to be programmed to program the flash options/security byte. The flash options/security byte can be erased or programmed only when this sector is not protected. Secured operation takes effect on the next reset after programming the security bits of the FSEC register to a secure value. The effects that the secured operation has on the HCS12 microcontroller are listed in Table 14.6.

Operation Mode	Effects
Normal single-chip mode	1. Background debug module operation is completely disabled. 2. Flash and EEPROM commands PROG, bulk erase, sector erase, erase and verify, and sector modify remain enabled.
Special single-chip mode	1. BDM firmware commands are disabled. 2. BDM hardware commands are restricted to register space. 3. Flash and EEPROM commands are limited to bulk erase only.
Expanded modes	1. BDM operation is completely disabled. 2. External access to internal flash and EEPROM is disabled. 3. Internal visibility (IVIS) and CPU pipe (IPIPE) information is disabled. 4. Flash and EEPROM commands cannot be executed from external memory in normal expanded mode.

Table 14.6 ■ Effects of secured operations on the HCS12 operation modes

14.5.5 Unsecuring the Microcontroller

When in secure mode, the HCS12 microcontroller can be unsecured by one of the following methods:

BACKDOOR KEY ACCESS

In normal modes (single chip and expanded), security can be temporarily disabled by the backdoor key access method. This method requires the following actions to be taken:

- Program the backdoor key at $FF00~$FF07 to a valid value.
- Set the KEYEN1 and KEYEN0 bits of the flash option/security byte to 10 (enabled).
- In single-chip mode, design applications to have the capability to write to the backdoor key locations.

The backdoor key access method is useful because it allows debugging of a secured microcontroller without having to erase the flash memory. This is useful for failure analysis. The backdoor key is not allowed to have the value of $0000 or $FFFF.

The backdoor key access sequence includes

1. Setting the KEYACC bit in the flash configuration register (FCNFG)
2. Writing the first 16-bit word of the backdoor key to $FF00
3. Writing the second 16-bit word of the backdoor key to $FF02
4. Writing the third 16-bit word of the backdoor key to $FF04
5. Writing the fourth 16-bit word of the backdoor key to $FF06
6. Clearing the KEYACC bit in the flash configuration register FCNFG

Since the flash memory cannot be read when the KEYACC bit is set, the code for the backdoor key access sequence must be executed from SRAM.

If all four 16-bit words match the flash contents at $FF00 to $FF07, the microcontroller will be unsecured and the security bits SEC1 and SEC0 of the FSEC register will be forced to unsecured state 10. The contents of the flash options/security byte are not changed by this procedure, and so the HCS12 will revert to the secure state after the next reset, unless further action is taken. If any one of the four 16-bit words does not match the flash contents at $FF00 to $FF07, the HCS12 microcontroller will remain secured.

REPROGRAMMING THE SECURITY BITS

In normal single-chip mode, security can also be disabled by means of erasing and reprogramming the security bits within the flash options/security byte to the unsecured value. Since the erase operation will erase the entire sector from $FE00 to $FFFF, the backdoor key and the interrupt vectors will also be erased. Therefore, this method is not recommended for normal single-chip mode. The application software can erase and program the flash options/security byte only if the flash sector containing the flash options/security byte is not protected. Thus, flash protection is a useful way to prevent this from occurring. The microcontroller will enter the unsecured state after the next reset following the programming of the security bits to the unsecured value. One of the following conditions must be satisfied in order for this method to work:

■ The application software previously programmed into the MCU has been designed to have the capability to erase and program the flash options/security byte, and the flash section containing the flash options/security byte is not protected.

■ Security is first disabled using the backdoor key method allowing the BDM to be used to issue commands to erase and program the flash options/security byte, and the flash section containing the flash options/security byte is not protected.

COMPLETE MEMORY ERASE

The microcontroller can be unsecured in special single-chip modes by erasing the entire EEPROM and flash contents. When a secure microcontroller is reset into special single-chip mode, the BDM firmware verifies whether the EEPROM and flash are erased. If any EEPROM or flash location is not erased, only BDM hardware commands are enabled. BDM hardware commands can then be used to write to the EEPROM and flash registers and also to bulk erase the EEPROM and all flash blocks. When next reset to the special single-chip mode occurs, the BDM firmware will again verify whether all EEPROM and flash are erased and, if this is the case, will enable all BDM commands, allowing the flash options/security byte to be programmed to the unsecured value. The security bits SEC1 and SEC0 in the FSEC register will indicate the unsecure state following the next reset.

14.5.6 Configuring the FCLKDIV Register

It is critical to configure the FCLKDIV register properly because this register controls the timing of the programming and erasure operations for the flash memory. It is necessary to divide the oscillator down to within the 150- to 200-kHz range.

Let

> FCLK be the clock of the flash timing control block
>
> T_{bus} be the period of the E-clock
>
> INT(x) take the integer part of x [e.g., INT(3.5) = 3]

Example 14.4

▼

Assume that f_{bus} = 24 MHz and f_{osc} = 16 MHz, respectively. Determine an appropriate value to be written into the FCLKDIV register to set the timing of programming and erasure properly for the flash memory and EEPROM.

Solution: Follow the logic flow illustrated in Figure 14.16.

1. T_{bus} = 41.7 ns (< 1 μs)
2. Initialize the PRDIV8 bit to 0.
3. f_{osc} = 16 MHz (> 12.8 MHz)
4. Set PRDIV8 to 1 and set PRDCLK to $f_{osc}/8$ = 2 MHz.
5. PRDCLK × (5 + 0.0417 μs) = 10.08, not an integer.
6. Set FDIV[5:0] to INT(PRDCLK[MHz] × (5 + T_{bus}[μs])) = 10.
7. FCLK = PRDCLK/(1 + FDIV[5:0]) = 2 MHz ÷ 11 = 181.81 kHz
8. 1/FCLK[MHz] + T_{bus}[μs] = 5.5 (> 5) and FCLK > 150 kHz, so stop.
9. Write the value of $4A into the FCLKDIV register.

▲

14.5.7 Flash Memory Programming and Erasure Algorithms

The HCS12 uses a command state machine to supervise the write sequencing for the flash memory programming and erasing. Before starting a command sequence, it is necessary to verify that there is no pending access error or protection violation in any flash blocks. It is then required to set the PPAGE register and the FCNFG register. The procedure for this initial setup is as follows:

1. Verify that all ACCERR and PVIOL flags in the FSTAT register are cleared in all banks. This requires checking the contents of the FSTAT register for all combinations of the BKSEL bits in the FCNFG register.

2. Write to bits BKSEL in the FCNFG register to select the bank of registers corresponding to the flash block to be programmed or erased.

3. Write to the PPAGE register to select one of the pages to be programmed if programming is to be done in the $8000~$BFFF address range. There is no need to set PPAGE when programming in the $4000~$7FFF or $C000~$FFFF address ranges.

After this initialization procedure, the CBEIF flag should be tested to ensure that the address, data, and command buffers are empty. If the CBEIF flag is set to 1, the program/erase

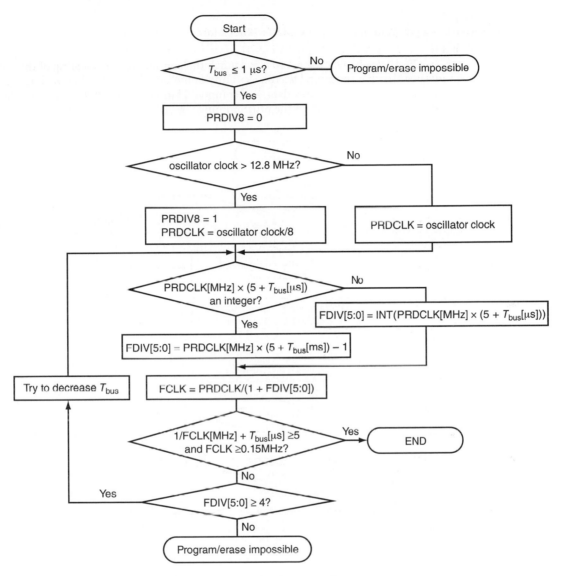

Figure 14.16 ■ PRDIV8 and FDIV bits determination procedure

command write sequence can be started. The following three-step procedure should be strictly adhered to when programming or erasing the flash memory:

Step 1

Write the aligned data word to be programmed to the valid flash address space. The address and data will be stored in internal buffers. For programming, all address bits are valid. For flash memory erasure, the value of the data bytes is "don't care." For bulk erasure, the address can be anywhere in the available address space of the block to be erased. For sector erasure, the address bits 7 to 0 are ignored for the flash. A sector has 256 bytes.

Step 2

Write the program or erase command to the command buffer.

Step 3

Clear the CBEIF flag by writing a 1 to it to launch the command. The clearing of the CBEIF flag indicates that the command was successfully launched. The CBEIF flag will be set again, indicating that the address, data, and command buffers are ready for a new command sequence to begin. A summary of the programming algorithm is shown in Figure 14.17.

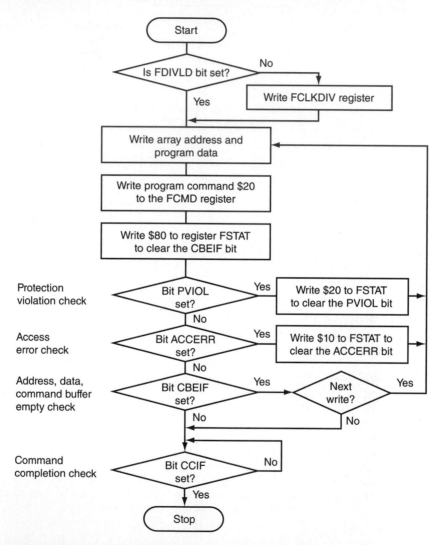

Figure 14.17 ■ HCS12 flash memory program algorithm

If an HCS12 device has multiple flash memory blocks, we need to clear the ACCERR and PVIOL flags in all blocks and also set the BKSEL bits in the FCNFG register properly to select the desired block for erasure or programming.

Example 14.5

▼

Write a function to clear the ACCERR and PVIOL flags in all four blocks in the HCS12 devices with 256 kB of on-chip flash memory.

Solution: The following function clears the ACCERR and PVIOL flags in all four blocks in the HCS12 devices with 256 kB of on-chip flash memory:

```
clearflags   bclr   FCNFG,$03              ; select bank 0
             movb   #ACCERR+PVIOL,FSTAT    ; clear the ACCERR and PVIOL flags
             bset   FCNFG,$01              ; select bank 1
             movb   #ACCERR+PVIOL,FSTAT    ; clear the ACCERR and PVIOL flags
             bset   FCNFG,$03              ; select bank 3
             movb   #ACCERR+PVIOL,FSTAT    ; clear the ACCERR and PVIOL flags
             bclr   FCNFG,$01              ; select bank 2
             movb   #ACCERR+PVIOL,FSTAT    ; clear the ACCERR and PVIOL flags
             rts
```

The C language version of the function is as follows:

```
void clearflags (void)
{
    FCNFG   &= ~0x03;          // select bank 0
    FSTAT   = ACCERR+PVIOL;    // clear the ACCERR and PVIOL flags
    FCNFG   |= 0x01;           // select bank 1
    FSTAT   = ACCERR+PVIOL;    // clear the ACCERR and PVIOL flags
    FCNFG   |= 0x03;           // select bank 3
    FSTAT   = ACCERR+PVIOL;    // clear the ACCERR and PVIOL flags
    FCNFG   &= 0xFE;           // select bank 2
    FSTAT   = ACCERR+PVIOL;    // clear the ACCERR and PVIOL flags
}
```

▲

Example 14.6

▼

Write a function that erases a sector of flash memory. Index register X contains a word-aligned address within the sector to be erased. Return a 1 in B if the command buffer is not empty.

Solution: The assembly routine that erases the sector pointed to by index register X is as follows:

```
EraseFSector brclr  FSTAT,CBEIF,err2ES      ; erase prohibited if command buffer not empty
             std    0,X                     ; write any data to sector address
             movb   #SectorErase,FCMD       ; write sector erase command
             movb   #CBEIF,FSTAT            ; launch the erase command
             brclr  FSTAT,ACCERR+PVIOL,OK2ER ; no error flag?
err2ES       ldab   #1                      ; return error code 1
             rts
OK2ER        brclr  FSTAT,CCIF,OK2ER        ; wait until command is completed
             ldab   #0                      ; erase successfully, return code 0
             rts
```

The C language version of the function is as follows:

```
int EraseFSector(int *pt)
{
        if (!(FSTAT & CBEIF))
                return 1;                       /* command buffer not empty, erase prohibited */
        *pt = 0x00;                             /* write any data to the sector */
        FCMD = SectorErase;                     /* write sector erase command */
        FSTAT = CBEIF;                          /* launch the erase command */
        if (FSTAT & (ACCERR+PVIOL))
                return 1;                       /* return error code 1 */
        while(!(FSTAT & CCIF));                 /* wait until erase command is completed */
        return 0;                               /* return normal code */
}
```

Example 14.7

Write a function that performs a bulk erasure operation to the flash memory. The index register contains a word-aligned address of a word inside the sector to be bulk erased. Return a 1 in accumulator B if bulk erasure is not allowed.

Solution: The assembly language function that performs the bulk erasure operation to the flash memory is as follows:

```
BulkEraseF  brset FPROT,FPOPEN+FPHDIS+FPLDIS,doBL    ; Is bulk erasure allowed?
            ldab  #1                                 ; return error code 1
            rts
doBL        brclr FSTAT,CBEIF,errBL                  ; bulk erase prohibited if CBEIF == 0
            std   0,X                                ; write any data to sector address
            movb #BulkErase,FCMD                     ; write bulk erase command
            movb #CBEIF,FSTAT                        ; launch the erase command
            brclr FSTAT,ACCERR+PVIOL,OK2BR           ; no error flag?
errBL       ldab  #1                                 ; return error code 1
            rts
OK2BR       brclr FSTAT,CCIF,*                       ; wait until command is completed
            ldab  #0                                 ; erase successfully, return code 0
            rts
```

The C language version of the function is as follows:

```
int bulkeraseF(int *ptr)
{
    if(FPROT&(FPOPEN | FPHDIS | FPLDIS)!= 0xA4)
            return 1;                   /* can't bulk erase */
    if(!(FSTAT & CBEIF))
            return 1;                   /* command buffer isn't empty, bulk erase not allowed */
    else {
            *ptr = 0x00;                /* write anything to flash block location */
            FCMD = BulkErase;           /* write bulk erase command */
            FSTAT = CBEIF;              /* launch bulk erase command */
```

```
        if (FSTAT & (ACCERR | PVIOL))
            return 1;                /* error flag is set, command failed */
        while(!(FSTAT & CCIF));      /* wait until command completion */
        return 0;
    }
}
```

Example 14.8

Write a function that programs a block of words to the flash memory. The number of words to be programmed, the starting address of the flash memory to be programmed, and the starting address of data are passed to this function in B, X, and Y, respectively.

Solution: The assembly function that performs the flash programming is as follows:

```
feProgBlok  tstb                               ; check word count
            bne     doFLprog                   ; word count is valid
            rts                                ; return if word count is zero
doFLprog    pshb                               ; save the word count in stack
fepwait1    brclr   FSTAT,CBEIF,fepwait1       ; wait until command buffer is empty
            movw    2,y+,2,x+                  ; write data word to flash address
            movb    #Program,FCMD              ; write program command
            movb    #CBEIF,FSTAT               ; launch the command
            brclr   FSTAT,ACCERR+PVIOL,progK   ; is there any error?
            pulb
            ldab    #1                         ; return error code 1
            rts
progOK      dec     0,SP                       ; one word less to be programmed
            bne     fepwait1                   ; more words to be programmed?
            pulb
            clrb                               ; return error code 0
            rts
```

The C language version of the function is as follows:

```c
int feProgBlok(char cnt, int *destptr, int *srcptr)
{
    if(cnt == 0)
        return 0;                       /* if word count is 0, do nothing */
    while(cnt){
        if(FSTAT & CBEIF){              // if command buffer is not empty, do nothing
            *destptr++ = *srcptr++;     // write data word to flash location
            FCMD = Program;             // write program command
            FSTAT = CBEIF;              // launch program command
            if(FSTAT & (ACCERR+PVIOL))
                return 1;               // program error?
            cnt--;
        }
    }
}
```

```
        while(!(FSTAT&CCIF));      // wait for the last command to complete
        return 0;
}
```

Example 14.9

Write a function that performs the erase-and-verify command to the flash memory. The index register contains a word-aligned address to the flash block to be erased and verified.

Solution: The assembly routine that performs the erase-and-verify command is as follows:

```
feraseverify    brclr   FSTAT,CBEIF,cantE            ; command buffer not empty
                std     0,x                         ; write any data to flash sector address
                movb    #EraseVerify,FCMD           ; write the command
                movb    #CBEIF,FSTAT                ; launch the erase and verify command
                brclr   FSTAT,ACCERR+PVIOL,EVNoErr
                ldab    #1                          ; return error code 1
                rts
EVNoErr         brclr   FSTAT,CCIF,EVNoErr          ; wait until command is done
                brset   FSTAT,BLANK,EVFOK           ; successful erase and verify?
cantE           ldab    #1                          ; flash is not blank
                rts
EVFOK           clrb                                ; erase and verify OK
                rts
```

The C function that performs the erase-and-verify operation is as follows:

```
int feraseverify(int *ptr)
{
    if(!(FSTAT & CBEIF))
        return 1;               // command buffer not empty, returns
    *ptr = 0x00;                // write data to flash sector address
    FCMD = EraseVerify;         // write erase verify command
    FSTAT = CBEIF;              // launch the command
    if(FSTAT&(ACCERR | PVIOL))
        return 1;               // errors have occurred
    while(!(FSTAT & CCIF));     // wait until command is completed
    if(FSTAT & BLANK)
        return 0;               // command completed successfully
    else
        return 1;
}
```

14.6 The On-Chip EEPROM Memory

An HCS12 device may have 1, 2, or 4 kB of on-chip EEPROM to store nonvolatile data or programs to be executed out of reset. The EEPROM is organized as an array of 2-byte words. For example, the 4-kB EEPROM is organized as 2048 rows of 2 bytes (1 word). The EEPROM block's erase sector size is two rows or two words.

The programming and erasure of the EEPROM are similar to those of the flash memory. Both the sector and bulk erasures are supported. An erased bit reads 1 and a programmed bit reads 0. The high voltage required for programming and erasure is generated internally by an on-chip charge pump.

The discussion of the HCS12 on-chip EEPROM will be based on the 4-kB EEPROM available in the HCS12DP256.

14.6.1 EEPROM Memory Map

The EEPROM can be mapped to any 2-kB boundary, as explained in Section 14.3.3. The memory map of the 4-kB EEPROM is illustrated in Figure 14.18. The whole 4 kB of EEPROM can be protected by setting the EPOPEN bit of the EPROT register. A 16-byte field is reserved inside the EEPROM module from the address $_FF0 to $_FFF. The byte at $_FFD stores the EEPROM protection information. The EEPROM has hardware interlocks that protect data from accidental corruption. A protected sector is located at the higher address end of the EEPROM block, just below $_FFF. The protected sector can be sized from 64 bytes to 512 bytes.

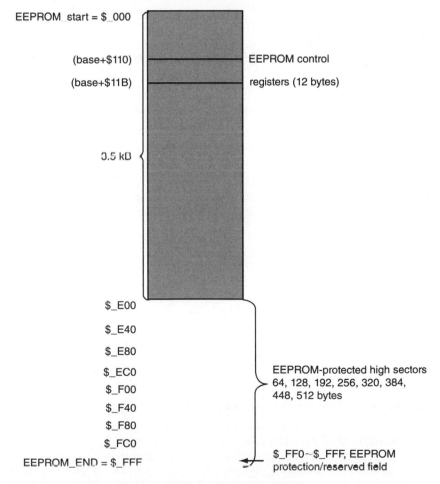

Figure 14.18 ■ HCS12 4-kB EEPROM memory map

14.6.2 EEPROM Associated Registers

The following registers control the operation of the EEPROM block:

EEPROM Clock Divider Register (ECLKDIV)

The contents of this register are shown in Figure 14.19. The EDIVLD bit is read-only. This register controls the timing of the EEPROM programming and erasure operation. The contents of this register are identical to those of the FCLKDIV register. Bits 0~6 can be written once after reset. The method for selecting an appropriate dividing value for the EEPROM is the same as that for the flash memory.

7	6	5	4	3	2	1	0	
EDIVLD	PRDIV8	EDIV5	EDIV4	EDIV3	EDIV2	EDIV1	EDIV0	base+$100

Reset: 0 0 0 0 0 0 0 0

EDIVLD: clock divider loaded
    0 = register has not been written.
    1 = register has been written to since the last reset.
PRDIV8: enable prescaler by 8
    0 = the input oscillator clock is directly fed into the FCLKDIV divider.
    1 = enables a divide-by-8 prescaler, to divide the flash module input oscillator
       clock before feeding into the CLKDIV divider.
EDIV[5:0]: clock divider bits
    The combination of PRDIV8 and FDIV[5:0] effectively divides the flash module
    input oscillator clock down to a frequency of 150 kHz~200 kHz. The maximum
    divide ratio is 512.

Figure 14.19 ■ EEPROM clock divider register (ECLKDIV)

EEPROM Configuration Register (ECNFG)

This register enables the EEPROM interrupts. The contents of this register are shown in Figure 14.20. This register is readable and writable any time.

7	6	5	4	3	2	1	0	
CBEIE	CCIE	0	0	0	0	0	0	base+$113

Reset: 0 0 0 0 0 0 0 0

CBEIE: command buffer empty interrupt enable
    0 = command buffer empty interrupts disabled.
    1 = an interrupt will be requested whenever the CBEIF flag is set.
CCIE: command completion interrupt enable
    0 = command complete interrupts disabled.
    1 = an interrupt will be requested whenever the CCIF flag is set.

Figure 14.20 ■ EEPROM configuration register (ECNFG)

EEPROM Protection Register (EPROT)

This register defines which EEPROM sectors are protected against programming or erasure. The contents of this register are shown in Figure 14.21. This register is loaded from EEPROM array at $_FFD in the reset sequence as indicated by the F letter. All bits are readable; bits NV[6:4]

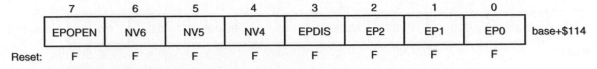

7	6	5	4	3	2	1	0	
EPOPEN	NV6	NV5	NV4	EPDIS	EP2	EP1	EP0	base+$114
Reset: F	F	F	F	F	F	F	F	

EPOPEN: opens the EEPROM for programming or erasure
   0 = the whole EEPROM array is protected. In this case, EPDIS and EP bits within
      the protection register have no effect.
   1 = the EEPROM sectors not protected are enabled for program or erase.
NV6~NV4: not volatile flag bit
   These bits are available for nonvolatile flag usage.
EPDIS: EEPROM protection address range disable
   0 = protection enabled.
   1 = protection disabled.
EP2~EP0: EEPROM protection address size
   000 = 64 bytes ($_FC0~$_FFF).
   001 = 128 bytes ($_F80~$_FFF).
   010 = 192 bytes ($_F40~$_FFF).
   011 = 256 bytes ($_F00~$_FFF).
   100 = 320 bytes ($_EC0~$_FFF).
   101 = 384 bytes ($_E80~$_FFF).
   110 = 448 bytes ($_E40~$_FFF).
   111 = 512 bytes ($_E00~$_FFF).

Figure 14.21 ■ EPROT register

are not writable. The EPOPEN and EPDIS bits in the EPROT register can only be written to the protected state (i.c., 0). The EP[2:0] bits can be written any time until the EPDIS bit is cleared. If the EPOPEN bit is cleared, then the state of the EPDIS and EP[2:0] bits is irrelevant. Trying to alter any of the protected areas will result in a protection violation error and the PVIOL bit of the ESTAT register will be set. A bulk erasure is only possible when protection is fully disabled.

EEPROM Status Register (ESTAT)
The contents of this register are shown in Figure 14.22.
The bits CBEIF, PVIOL, and ACCERR are readable and writable; the bits CCIF and BLANK are readable but not writable.

EEPROM Command Register (ECMD)
The ECMD register defines the EEPROM commands. The contents of this register are shown in Figure 14.23. The definitions of EEPROM commands have been included in the *hcs12.inc* file and are ready for use.

14.6.3 EEPROM Protection
The EEPROM can be protected against accidental erasure and programming. EEPROM protection is controlled by the EPROT register. During the HCS12 reset sequence, the EPROT register is loaded from the EEPROM protection byte located within the EEPROM. The EPROT register is located within the smallest EEPROM protected area, so protecting EEPROM always protects the EEPROM protection byte, thus guaranteeing the reset state of EEPROM protection. The value of the EPROT register determines whether the entire EEPROM or just subsections are protected from being accidentally erased or programmed. Trying to program or erase any of the protected areas will result in a protection violation error and the PVIOL bit in the ESTAT register will be set. A bulk erasure is possible only if protection is fully disabled.

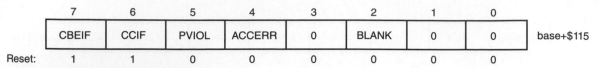

CBEIF: command buffer empty interrupt flag
  0 = command buffers are full.
  1 = command buffers are ready to accept a new command.
CCIF: command completion interrupt flag
  0 = command in progress.
  1 = all commands are completed.
PVIOL: protection violation
  0 = no failure.
  1 = a protection violation has occurred.
ACCERR: flash access error
  0 = no failure.
  1 = access error has occurred.
BLANK: array has been verified as erased
  0 = if an erase and verify command has been requested, and the CCIF flag is set,
    then a zero in BLANK indicates that the block is not erased.
  1 = flash block verifies as erased.

Figure 14.22 ■ EEPROM status register (ESTAT)

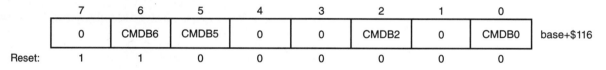

CMDB6, CMDB5, CMDB2, and CMDB0: command bits
  Valid commands include the following:
  $05 = erase and verify.
  $20 = program a word (2 bytes).
  $40 = sector erase.
  $41 = bulk erase.
  $60 = sector modify.

Figure 14.23 ■ EEPROM command buffer and register (ECMD)

14.6.4 Configuring the ECLKDIV Register

The timing for EEPROM programming and erasure is controlled by the ECLKDIV register. The bit definitions and configuration are identical to those of the FCLKDIV register for the flash memory. We need to write a value to divide the oscillator frequency down to a range from 150 to 200 kHz.

14.6.5 Programming and Erasure of EEPROM

The algorithm for programming the EEPROM is illustrated in Figure 14.24. It is almost identical to the algorithm for programming the flash memory.

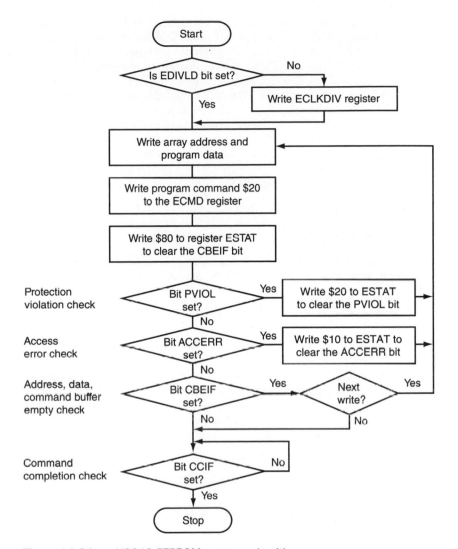

Figure 14.24 ■ HCS12 EEPROM program algorithm

The execution of the first four EEPROM commands as listed at the end of Section 14.6.2 follows the same three-step sequence as their counterparts in flash memory:

1. Write the aligned data word to be programmed to the valid EEPROM address space. The address and data will be stored in internal buffers. For programming, all address bits are valid. For erasure, the value of the data bytes is "don't care." For bulk erasure, the address can be anywhere in the available address space of the block to be erased. For sector erasure, the address bits 1 and 0 are ignored for the EEPROM.

2. Write the program or erase command to the command register.

3. Clear the CBEIF flag to launch the command. The clearing of the CBEIF bit indicates that the command was successfully launched.

The completion of the command is indicated by the setting of the CCIF flag. An erroneous command write sequence will abort and set the appropriate flag. If set, the designer must clear the ACCERR or PVIOL flags before commencing another command write sequence. By writing a 0 to the CBEIF flag, the command sequence can be aborted after writing a word to the EEPROM address space or after writing a command to the ECMD register and before the command is launched. Writing a 0 to the CBEIF flag in this way will set the ACCERR flag.

The sector-modify command ($60) is a two-step command, which first erases a sector (two words) of EEPROM and then reprograms one of the words in that sector. The EEPROM sector that is erased by the command is the sector containing the address of the aligned data write that starts the valid command sequence. That same address is reprogrammed with the data that is written. By launching a sector-modify command and then pipelining a program command, it is possible to completely replace the contents of an EEPROM sector.

Example 14.10

▼

Write a function that erases a sector (4 bytes) of EEPROM. Index register X contains a word-aligned EEPROM address within the sector to be erased.

Solution: The following assembly function erases the specified EEPROM sector by following the three-step procedure:

```
eraseEEsector
                movb    #ACCERR+PVIOL,ESTAT          ; clear error flags
                brclr   ESTAT,CBEIF,EERErr          ; command buffer not empty, return
                std     0,X                         ; write any data to EEPROM sector
                movb    #SectorErase,ECMD           ; write sector-erase command
                movb    #CBEIF,ESTAT                ; launch erase command
                brclr   ESTAT,ACCERR+PVIOL,EEROK    ; no error?
EERErr          ldab    #1                          ; error code set to 1
                rts
EEROK           brclr   ESTAT,CCIF,EEROK            ; wait until command completion
                clrb
                rts
```

The C language version of the function is as follows:

```
int eraseEEsector (int *ptr)
{
    ESTAT = ACCERR | PVIOL;     // clear error flags
    if(!(ESTAT & CBEIF))
            return 1;           // command buffer not empty, can't issue new command
    *ptr = 0x00;                // write any data to EEPROM sector location
    ECMD = SectorErase;         // write sector-erase command
    ESTAT = CBEIF;              // launch the command
    if(ESTAT & (ACCERR | PVIOL))
            return 1;           // error occurred
    while(!(ESTAT&CCIF));       // wait for command completion
    return 0;                   // command completed correctly
}
```

▲

Example 14.11

▼

Write a function that performs bulk erasure to the EEPROM. Index register X contains a word-aligned EEPROM address.

Solution: The assembly function that performs EEPROM bulk erasure is as follows:

```
bulkeraseEE  movb   #ACCERR+PVIOL,ESTAT          ; clear error flags
             brclr  ESTAT,CBEIF,EEBEErr          ; command buffer not empty, return
             std    0,X                          ; write any data to EEPROM
             movb   #BulkErase,ECMD              ; write bulk-erase command
             movb   #CBEIF,ESTAT                 ; launch bulk-erase command
             brclr  ESTAT,ACCERR+PVIOL,EEBROK    ; check error ; no error?
EEBEErr      ldab   #1                           ; error code set to 1
             rts
EEBROK       brclr  ESTAT,CCIF,EEBROK            ; wait until command completion
             clrb
             rts
```

The C language version of the function is straightforward and hence is left as an exercise.

Example 14.12

▼

Write a function that bulk-erases the EEPROM and verifies if the whole EEPROM is blank. The index register X contains a valid word-aligned address to the EEPROM.

Solution: The assembly function that performs the specified operation is as follows:

```
eeraseverify  movb   #ACCERR+PVIOL,ESTAT        ; clear error flags
              brclr  ESTAT,CBEIF,cantEE          ; command buffer not empty
              std    0,x                         ; write any data to EEPROM address
              movb   #EraseVerify,ECMD           ; write the command
              movb   #CBEIF,ESTAT                ; launch the erase and check command
              brclr  ESTAT,ACCERR+PVIOL,EEEVOK
              ldab   #1                          ; return error code 1
              rts
EEEVOK        brclr  ESTAT,CCIF,EEEVOK           ; wait until command is done
              brset  ESTAT,BLANK,EVEOK           ; successful erase and verify?
cantEE        ldab   #1                          ; EEPROM is not blank
              rts
EVEOK         clrb
              rts
```

The C language version of the function is straightforward and is left as an exercise.

Example 14.13

▼

Write a function that programs a word to the EEPROM. X contains the address of the EEPROM location to be programmed, and double accumulator D holds the data to be programmed. A 0 is returned in accumulator B if the command is performed correctly; otherwise, a 1 is returned.

Solution: The function that performs the specified function is as follows:

```
eeprogram    movb    #ACCERR+PVIOL,ESTAT    ; clear error flags
             brclr   ESTAT,CBEIF,cantPRE       ; command buffer not empty
             std     0,x                       ; write data to EEPROM address
             movb    #Program,ECMD             ; write the command
             movb    #CBEIF,ESTAT              ; launch the erase and check command
             brclr   ESTAT,ACCERR+PVIOL,EEPROK
cantPRE      ldab    #1                        ; return error code 1
             rts
EEPROK       brclr   ESTAT,CCIF,EEPROK         ; wait until command is done
             clrb                              ; successful program code is 0
             rts
```

The C language version of the function is as follows:

```
int eeprogram (unsigned int data, unsigned int *ptr)
{
      ESTAT = ACCERR+PVIOL;     // clear error flag
      if(!(ESTAT&CBEIF))
            return 1;
      *ptr = data;              // write data to word-aligned address
      ECMD = Program;           // write program command
      ESTAT = CBEIF;            // launch command
      if(ESTAT &(ACCERR|PVIOL))
            return 1;
      while(!(ESTAT&CCIF));     // wait for command to complete
      return 0;
}
```

Example 14.14

Write a function that executes the sector-modify command to change the contents of an EEPROM sector. The index register X holds the word-aligned address of the first EEPROM word to be modified. The index register Y holds the word-aligned address of the first word of data to be programmed.

Solution: The assembly function that performs the desired function is as follows:

```
eesectormodify
             movb    #ACCERR+PVIOL,ESTAT    ; clear error flags
             brclr   ESTAT,CBEIF,cantmod       ; command buffer not empty
             movw    0,Y,0,X                   ; write data to EEPROM address
             movb    #SectorModify,ECMD        ; write sector-modify command
             movb    #CBEIF,ESTAT              ; launch the erase and check command
             brclr   ESTAT,ACCERR+PVIOL,EEModOK
cantmod      ldab    #1                        ; return error code 1
             rts
```

```
EEModOK     brclr    ESTAT,CBEIF,*                    ; wait for command buffer to empty
            movw     2,Y,2,X                          ; write second data word to second word of
                                                      ; EEPROM
            movb     #Program,ECMD                    ; write program command
            movb     #CBEIF,ESTAT                     ; launch the program command
            brclr    ESTAT,ACCERR+PVIOL,EEPR2OK
            ldab     #1
            rts
EEPR2Ok     brclr    ESTAT,CCIF,EEPR2OK               ; wait for program command completion
            clrb                                      ; successful program code is 0
            rts
```

The C language version of the function is as follows:

```
int EESectorModify(unsigned int *src, unsigned int *dest)
{
     ESTAT = ACCERR | PVIOL;      // clear error flags
     if(!(ESTAT&CBEIF))
             return 1;            // command buffer not empty is error
     *dest = *src;                // write first data word
     ECMD = SectorModify;         // write sector-modify command
     ESTAT = CBEIF;               // launch the sector-modify command
     if(ESTAT&(ACCERR | PVIOL))
             return 1;            // command failed
     while(!(ESTAT&CBEIF));       // wait for command buffer to become empty
     *(dest+1) = *(src+1);        // write second data word
     ECMD = Program;              // write the program command
     ESTAT = CBEIF;               // launch the program command
     if(ESTAT&(ACCERR | PVIOL))
             return 1;            // command failed
     while(!(ESTAT&CCIF));        // wait for command buffer to become empty
             return 0;
}
```

14.7 HCS12 External Memory Interface

The HCS12 can access external memory when it is reset into one of the expanded modes. Address and data signals are multiplexed onto the Port A and Port B pins; the Port E and Port K pins supply control signals that are required for accessing external memories.

14.7.1 HCS12 Pins for External Memory Interfacing

The HCS12 external memory interface uses 27 pins and is implemented across four I/O ports (A, B, E, and K). The signal pins used in external memory interface are listed in Table 14.7. The signals ADDR0, . . . , ADDR15 and XADDR14, . . . , XADDR19 are used to select a memory location to access and are referred to as the *address bus*. The signals DATA0, . . . , DATA15 are used to carry data and are referred to as the *data bus*. The remaining signals are referred to as the *control bus*. When the external memory is not paged, only the lower 16 address/data pins are used. However, if expanded memory is enabled, then ADDR14 and ADDR15 are not used. Instead, signals

Signal Name	Function
ADDR0/DATA0	EMI address bit 0 or data bit 0
ADDR1/DATA1	EMI address bit 1 or data bit 1
ADDR2/DATA2	EMI address bit 2 or data bit 2
ADDT3/DATA3	EMI address bit 3 or data bit 3
ADDT4/DATA4	EMI address bit 4 or data bit 4
ADDT5/DATA5	EMI address bit 5 or data bit 5
ADDT6/DATA6	EMI address bit 6 or data bit 6
ADDT7/DATA7	EMI address bit 7 or data bit 7
ADDR8/DATA8	EMI address bit 8 or data bit 8
ADDR9/DATA9	EMI address bit 9 or data bit 9
ADDR10/DATA10	EMI address bit 10 or data bit 10
ADDT11/DATA11	EMI address bit 11 or data bit 11
ADDT12/DATA12	EMI address bit 12 or data bit 12
ADDT13/DATA13	EMI address bit 13 or data bit 13
ADDT14/DATA14	EMI address bit 14 or data bit 14
ADDT15/DATA15	EMI address bit 15 or data bit 15
XADDR14	EMI extended address bit 14
XADDR15	EMI extended address bit 15
XADDR16	EMI extended address bit 16
XADDR17	EMI extended address bit 17
XADDR18	EMI extended address bit 18
XADDR19	EMI extended address bit 19
R/$\overline{\text{W}}$	Read/write
$\overline{\text{LSTRB}}$	Lower byte strobe
ECLK	E-clock
$\overline{\text{ECS}}$/ROMONE	Emulated chip select/on-chip ROM enable
$\overline{\text{XCS}}$	External data chip select

Note: EMI stands for external memory interface

Table 14.7 ■ HCS12 external memory interface signal pins

XADDR19, . . . , XADDR14 are used as expanded address signals. In the following discussion, we will use A0, . . . , A15, D0, . . . , D15, and XA14, . . . , XA19 to refer to ADDR0, . . . , ADDR15, DATA0, . . . , DATA15, and XADDR14, . . . , XADDR19, respectively. The R/$\overline{\text{W}}$ signal is used to indicate the direction of data transfer. When this signal is high, the MCU reads data from external memory chips. When this signal is low, the MCU writes data to the external memory. The $\overline{\text{LSTRB}}$ signal is used to indicate whether the lower data bus (DA7, . . . , DA0) carries valid data. There are two situations in which the lower data bus does not carry valid data.

1. In expanded narrow mode, external memory data pins are connected to the D15, . . . , D8 pins. D7, . . . , D0 are not used to carry data.

2. In expanded wide mode, the MCU may execute instructions to write byte data to memory locations at even addresses.

The signals $\overline{\text{LSTRB}}$, R/$\overline{\text{W}}$, and A0 indicate the type of bus access that is taking place. Accesses to the internal RAM module are the only type of access that would produce $\overline{\text{LSTRB}}$ = A0 = 1, because the internal RAM is specifically designed to allow misaligned 16-bit accesses in a single cycle. In these cases, the data at the given address is on the lower half of the data bus, and the data at *address + 1* is on the upper half of the data bus. These are summarized in Table 14.8.

LSTRB	A0	R/W̄	Type of Access
1	0	1	8-bit read of an even address
0	1	1	8-bit read of an odd address
1	0	0	8-bit write of an even address
0	1	0	8-bit write of an odd address
0	0	1	16-bit read of an even address
1	1	1	16-bit read of an odd address (low/high bytes swapped)
0	0	0	16-bit write of an even address
1	1	0	16-bit write of an odd address (low/high bytes swapped)

Table 14.8 ■ Access type versus bus control signals

The ECLK signal is provided as a timing reference for external memory accesses. This signal is controlled by 3 bits.

- The NECLK bit of the PEAR register
- The IVIS bit of the MODE register
- The ESTR bit of the EBICTL register

The NECLK bit enables/disables the E-clock signal. The IVIS bit allows the E-clock to be visible when the MCU is performing an internal access cycle. The ESTR bit allows the E-clock to be stretched.

The $\overline{\text{ECS}}$ signal is used as the chip-select signal for external memory chips. Because it is asserted immediately after the address signals become valid, the $\overline{\text{ECS}}$ signal is often used to latch address signals so that they can remain valid throughout a complete bus cycle. The rising edge of the E-clock can also be used to latch address signals. The $\overline{\text{XCS}}$ signal, available in some HCS12 devices only, is also used as the external chip-select signal for external chips. It can also be used to latch address signals because of its timing.

14.7.2 Waveforms of Bus Signals

The waveform of a typical digital signal is shown in Figure 14.25. A bus signal cannot rise from low to high or drop from high to low instantaneously. The time needed for a signal to rise from 10 to 90 percent of the power supply voltage is referred to as the *rise time* (t_R). The time needed for a signal to drop from 90 to 10 percent of the power supply voltage is referred to as the *fall time* (t_F).

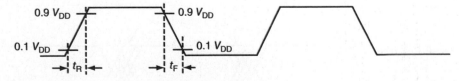

Figure 14.25 ■ A typical digital waveform

A single bus signal is often represented as a set of line segments (see Figure 14.26). The horizontal axis and vertical axis represent the time and the magnitude (in volts), respectively, of the signal. Multiple signals of the same nature, such as address and data, are often grouped together and represented as parallel lines with crossovers, as illustrated in Figure 14.27. A crossover in the waveform represents the point at which one or multiple signals change values. The HCS12 literature shows that Freescale uses 0.65 V_{CC} and 0.35 V_{CC} as high input voltage (V_{IH}) and low input voltage (V_{IL}). All timing indications use these two values as the reference points.

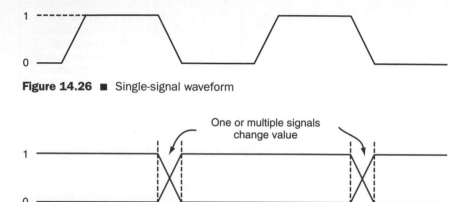

Figure 14.26 ■ Single-signal waveform

Figure 14.27 ■ Multiple-signal waveform

Sometimes a signal value is unknown because the signal is changing. Hatched areas in the timing diagram, shown in Figure 14.28, represent single and multiple unknown signals. Sometimes one or multiple signals are not driven (because their drivers are not enabled) and hence cannot be received. An undriven signal is said to be *floating* or in a *high-impedance* state. Single and multiple floating signals are represented by a value between the high and low levels, as shown in Figure 14.29.

In a microcontroller system, a bus signal falls into one of the three categories: *address*, *data*, or *control*.

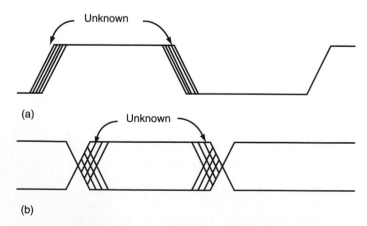

Figure 14.28 ■ Unknown signals: (a) single signal, (b) multiple signal

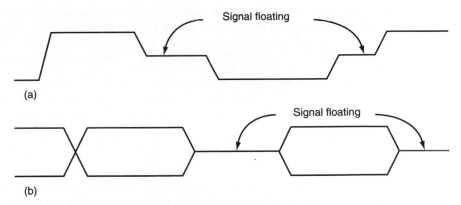

Figure 14.29 ■ Floating signals: (a) single signal, (b) multiple signal

14.7.3 Bus Transactions

A bus transaction includes sending the address and receiving or sending the data. A *read* transaction (also called a *read bus cycle*) transfers data from memory to either the CPU or the I/O device, and a *write* transaction (also called a *write bus cycle*) writes data to the memory. In a read transaction, the address is first sent down the bus to the memory, together with the appropriate control signals indicating a read. In Figure 14.30, this means pulling the read signal to high. The memory responds by placing the data on the bus and driving the $\overline{\text{Ready}}$ signal to low. The $\overline{\text{Ready}}$ signal (asserted low) indicates that the data on the data bus is valid.

In Figure 14.30, a read bus cycle takes one clock cycle to complete. For some microcontrollers, the Ready signal is used to extend the bus cycle to more than one clock cycle to

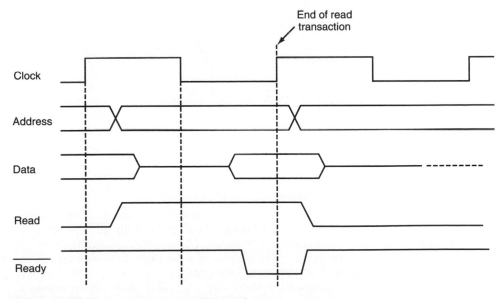

Figure 14.30 ■ A typical bus read transaction

accommodate slower memory components. When data from external memory is not ready, the control circuitry does not pull the $\overline{\text{Ready}}$ signal to low. When the $\overline{\text{Ready}}$ signal is sampled to be low, the microcontroller copies the data into the CPU. The HCS12 microcontroller does not use the $\overline{\text{Ready}}$ signal to accommodate slower memory. In a write bus cycle, the CPU sends both the address and the data and requires no return of data.

In a bus transaction, there must be a device that can initiate a read or write transaction. The device that can initiate a bus transaction is called a *bus master*. A microcontroller is always a bus master. A device such as a memory chip that cannot initiate a bus transaction is called a *bus slave*.

In a bus transaction, there must be a signal to synchronize the data transfer. The signal that is used most often is a clock signal. The bus is *synchronous* when a clock signal is used to synchronize the data transfer. In a synchronous bus, the timing parameters of all signals use the clock signal as the reference. As long as all timing requirements are satisfied, the bus transaction will be successful. An *asynchronous bus*, on the other hand, is not clocked. Instead, self-timed, handshaking protocols are used between the bus sender and receiver. Figure 14.31 shows the steps of a master performing a read on an asynchronous bus. A synchronous bus is often used between the CPU and the memory system, whereas an asynchronous bus is often used to handle different types of I/O devices. A synchronous bus is usually faster because it avoids the overhead of synchronizing the bus for each transaction.

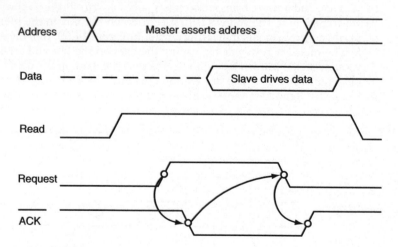

Figure 14.31 ■ Asynchronous read bus transaction

14.7.4 Bus Multiplexing

Designers of microcontrollers prefer to minimize the number of signal pins because it will make the chip less expensive to manufacture and use. By multiplexing the address bus and the data bus, many signal pins can be saved. The drawback of multiplexing bus signals is that the achievable bus transaction performance is compromised. Most 8-bit and many 16-bit microcontrollers multiplex their address and data buses.

For any bus transaction, the address signal input to the memory chips must be stable throughout the whole bus transaction. The memory system will need to use a circuit to latch the address signals so that they stay valid throughout the bus cycle. In a microcontroller that multiplexes the address and data buses, the address signals are placed on the multiplexed bus first, along with certain control signals to indicate that the address signals are valid. After the address signals are on the bus long enough so that the external logic has time to latch them, the microcontroller stops driving the address signals and either waits for the memory devices to place data on the multiplexed bus (in a read bus cycle) or places data on the multiplexed bus (in a write bus cycle).

14.7.5 The HCS12 Bus Cycles

When adding external memory chips to the HCS12, it is important to make sure that all the timing requirements of the MCU and memory chips are satisfied. The timing requirements of the HCS12 are specified using a diagram as shown in Figure 14.32. The value for each timing requirement is listed in Table 14.9. The interval when the ECLK signal is high can be stretched by one to three minimum ECLK cycles (t_{cyc}) to accommodate slower memory devices.

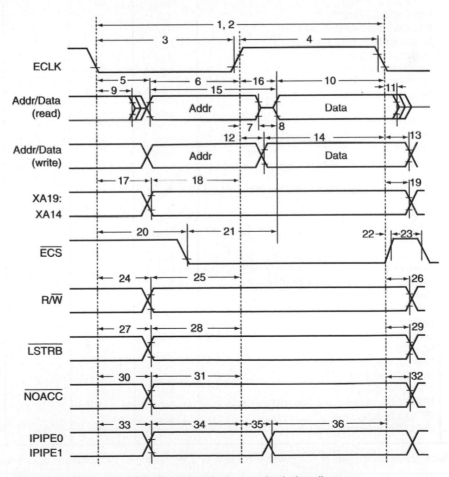

Figure 14.32 ■ The HCS12 read/write bus cycle timing diagram

The timing diagram in Figure 14.32 does not clearly describe how a read bus cycle and a write bus cycle proceed. A read bus cycle starts from the HCS12 driving the address signals onto the address bus to select a location to read. The sequence of events that occurs when the HCS12 family MCU performs a read from an external memory is illustrated in Figure 14.33. In a read bus cycle, the R/$\overline{\text{W}}$ signal stays high and the external memory drives the data onto the data bus.

A write cycle also starts with the HCS12 driving the address signals onto the address bus to select a memory location to receive the data. The sequence of events that occurs when the HCS12 family MCU performs a write access to the external memory is shown in Figure 14.34. In a write bus cycle, the R/$\overline{\text{W}}$ signal goes low and the MCU drives the data onto the data bus.

Num	Parameter Name	Symbol	Min	Typ	Max	Unit
1	Frequency of operation (E-clock)	f_o	0	–	25.0	MHz
2	Cycle time	t_{cyc}	40	–	–	ns
3	Pulse width, E low	PW_{EL}	19	–	–	ns
4	Pulse width, E high	PW_{EH}	19	–	–	ns
5	Address delay time	t_{AD}	–	–	8	ns
6	Address valid time to E rise ($PW_{EL} - t_{AD}$)	t_{AV}	11	–	–	ns
7	Muxed address hold time	t_{MAH}	2	–	–	ns
8	Address hold to data valid	t_{AHDS}	7	–	–	ns
9	Data hold to address	t_{DHA}	2	–	–	ns
10	Read data setup time	t_{DSR}	13	–	–	ns
11	Read data hold time	t_{DHR}	0	–	–	ns
12	Write data delay time	t_{DDW}	–	–	7	ns
13	Write data hold time	t_{DHW}	2	–	–	ns
14	Write data setup time[1] ($PW_{EH} - t_{DDW}$)	t_{DSW}	12	–	–	ns
15	Address access time[1] ($t_{cyc} - t_{AD} - t_{DSR}$)	t_{ACCA}	19	–	–	ns
16	E high access time[1] ($PW_{EH} - t_{DSR}$)	t_{ACCE}	6	–	–	ns
17	Nonmultiplexed address delay time	t_{NAD}	–	–	6	ns
18	Nonmuxed address valid to E rise ($PW_{EL} - t_{NAD}$)	t_{NAV}	15	–	–	ns
19	Nonmultiplexed address hold time	t_{NAH}	2	–	–	ns
20	Chip-select delay time	t_{CSD}	–	–	16	ns
21	Chip-select access time[1] ($t_{cyc} - t_{CSD} - t_{DSR}$)	t_{ACCS}	11	–	–	ns
22	Chip-select hold time	t_{CSH}	2	–	–	ns
23	Chip-select negated time	t_{CSN}	8	–	–	ns
24	Read/write delay time	t_{RWD}	–	–	7	ns
25	Read/write valid time to E rise ($PW_{EL} - t_{RWD}$)	t_{RWV}	14	–	–	ns
26	Read/write hold time	t_{RWH}	2	–	–	ns
27	Low strobe delay time	t_{LSD}	–	–	7	ns
28	Low strobe valid time to E rise ($PW_{EL} - t_{LSD}$)	t_{LSV}	14	–	–	ns
29	Low strobe hold time	t_{LSH}	2	–	–	ns
30	NOACC strobe delay time	t_{NOD}	–	–	7	ns
31	NOACC valid time to E rise ($P_{WEL} - t_{NOD}$)	t_{NOV}	14	–	–	ns
32	NOACC hold time	t_{NOH}	2	–	–	ns
33	IPIPO[1:0] delay time	t_{POD}	2	–	7	ns
34	IPIPO[1:0] valid time to E rise ($PW_{EL} - t_{POD}$)	t_{POV}	11	–	–	ns
35	IPIPO[1:0] delay time[1] ($PW_{EL} - t_{P1V}$)	t_{P1D}	2	–	25	ns
36	IPIPO[1:0] valid time to E fall	t_{P1V}	11	–	–	ns

[1]Affected by clock stretch: add $N \times t_{cyc}$, where $N = 0$, 2, or 3, depending on the number of clock stretches.

Table 14.9 ■ HCS12 expanded bus timing characteristics

Microcontroller   External memory

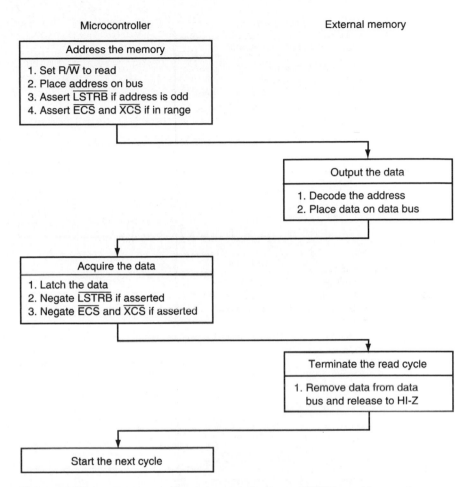

Figure 14.33 ■ Sequence of events occurring in an HCS12 read bus cycle

14.8 Issues Related to Adding External Memory

Adding external memory becomes necessary when the application gets larger than the on-chip memory. When external memory is added, the designer should also consider treating external peripheral chips (when possible) as memory devices because it makes the programming of the peripheral chips easier. When adding external memory, there are three issues that need to be considered.

- Memory space assignment
- Address decoder and control circuitry design
- Timing verification

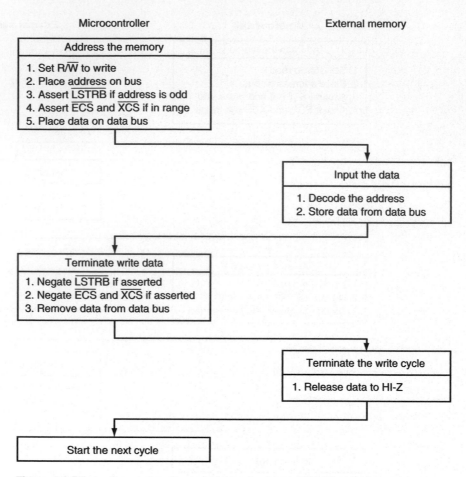

Figure 14.34 ■ Sequence of events occurring in an HCS12 write bus cycle

14.8.1 Memory Space Assignment

Any space unoccupied by the on-chip flash memory can be assigned to external memory devices. When making a memory space assignment, the designer has two options to choose from.

- *Equal size assignment.* In this method, the available memory space is divided into blocks of equal size and then each block is assigned to a memory device without regard for the actual size of each memory-mapped device. A memory-mapped device could be a memory chip or a peripheral device. Memory space tends to be wasted using this approach because most memory-mapped peripheral chips need only a few bytes to be assigned to their internal registers.

- *Demand assignment.* In this approach, the designer assigns the memory space according to the size of the memory devices.

Example 14.15

▼

Suppose that a designer is to design an HCS12DG256-based embedded product that requires 256 kB of external 16-bit SRAM, 256 kB of 16-bit EEPROM, and a parallel peripheral interface (PPI) that requires only 4 bytes of address space. The only SRAM available to this designer is the 128K × 8 SRAM chips (this chip has 128K locations with each location containing 8 bits). The only available EEPROM is the 128K × 8 EEPROM chips. Suggest a workable memory space assignment.

Solution: The designer is to design a 16-bit-wide memory system using the 8-bit-wide SRAM and EEPROM chips. Two 8-bit-wide memory chips are needed to construct a 16-bit memory module. One 16-bit-wide SRAM module can provide the 256-kB capacity. One 16-bit-wide EEPROM module is needed to offer the 256-kB capacity.

The on-chip FLASH memory occupies the space from \$C0000 to \$FFFFF. The address space from 0x00000 to 0xBFFFF is available for assignment. The following memory space assignment will be appropriate for this project:

SRAM: \$00000~\$3FFFF   ; 256 kB

EEPROM: \$40000~\$7FFFF   ; 256 kB

PPI: \$BFFFC~\$BFFFF   ; 4 bytes

▲

14.8.2 Address Decoder Design

The function of an address decoder is to make sure that no more than one memory device is enabled to drive the data bus at a time. If there are two or more memory devices driving the same bus lines, *bus contention* will occur and could cause severe damage to the system. All memory devices or peripheral devices have control signals such as *chip enable* (CE), *chip select* (CS), or *output enable* (OE) to control their read and write operations. These signals are often asserted low. The address decoder outputs will be used as the chip-select or chip-enable signals of external memory devices.

Two address-decoding schemes have been used: *full decoding* and *partial decoding*. A memory device is said to be *fully decoded* when each of its addressable locations responds to only a single address on the system bus. A memory component is said to be *partially decoded* when each of its addressable locations responds to more than one address on the system bus. Memory components such as DRAM, SRAM, EPROM, EEPROM, and flash memory chips use the full address-decoding scheme more often, whereas peripheral chips or devices use the partial address-decoding scheme more often.

Address decoder design is closely related to memory space assignment. For the address space assignment made in Example 14.15, the higher address signals are used as inputs to the decoder, and the lower address signals are applied to the address inputs of memory devices. Before *programmable logic devices* (PLDs) became popular and inexpensive, designers used transistor-transistor logic (TTL) chips such as the 74LS138 as address decoders. However, the off-the-shelf TTL decoders force designers to use equal-size memory space assignment. When PLDs became popular and inexpensive, designers started to use them to implement address decoders. PLDs allow the designer to implement demand assignment.

One of the methods for implementing the address decoder is to use one of the hardware description languages (HDL), such as ABEL, CUPL, VHDL, or VERILOG. Low-density PLDs, such as GAL18V10, GAL20V8, GAL20V8, SPLD16V8, SPLD20V8, and SPLD20V8, are often used to implement address decoders for their ability to implement product terms of many variables. The Generic Array Logic (GAL) devices are produced by Lattice Semiconductor. The Simple Programmable Logic Devices (SPLDs) are produced by Atmel. Both ABEL and CUPL are simple hardware description languages that are very suitable for describing circuit behaviors for address decoders. ABEL is supported by Lattice Semiconductor, and CUPL is supported by Atmel.

14.8.3 Timing Verification

When designing a memory system, the designer needs to make sure that timing requirements for both the microcontroller and the memory system are satisfied. In a read cycle, the most critical timing requirements are the *data setup time* (t_{DSR}, parameter 10) and *data hold time* (t_{DHR}, parameter 11) required by the HCS12 microcontroller. In addition, the designer must make sure that the address setup time and hold time requirements for the memory devices are met. The control signals needed by memory devices during a read cycle must be asserted at the appropriate times.

In a write cycle, the most critical timing requirements are the *write data setup time* and *write data hold time* required by the memory devices. As in a read cycle, the address setup time and address hold time must also be satisfied. Control signals required during a write cycle must also be generated at proper times.

14.9 Memory Devices

The control circuit designs for interfacing the SRAM, the EPROM, the EEPROM, and the flash memory to the HCS12 MCU are quite similar. The following sections illustrate how to add SRAM and EEPROM chips with the 128K × 8 organization to the HCS12 microcontroller.

14.9.1 The K6R1008C1D

The K6R1008C1D is a 128K × 8 bit asynchronous SRAM from Samsung that operates with a 5-V power supply. The K6R1008C1D has a short access time (10 ns) and three-state outputs. The pin assignment of the K6R1008C1D is shown in Figure 14.35. The address signals A16, . . . , A0 select one of the 128K locations within the chip to be read or written. Pins I/O8, . . . , I/O1 carry the data to be transferred between the chip and the microcontroller. The chip-select ($\overline{CS}$) input allows/disallows the read/write access request to the K6R1008C1D. The $\overline{OE}$ signal is the output-enable signal. When the $\overline{OE}$ signal is high, all eight I/O pins will be in the high-impedance state.

Depending on the assertion times of control signals, there are two timing diagrams for the read cycle and three timing diagrams for the write cycle (shown in Figures 14.36 and 14.37). The values of the related timing parameters for the read and write cycles are listed in Table 14.10.

In Figure 14.36a and b, the $\overline{CS}$ signal must be asserted for at least t_{RC} ns during a read cycle. The signal that is asserted the latest determines the time that data will become available. For example, in Figure 14.36b, the $\overline{OE}$ signal is asserted the latest; therefore, data becomes valid t_{OE} ns later. Data pins will go to the high-impedance state t_{OHZ} ns after the $\overline{OE}$ signal goes to high or t_{HZ} ns after the $\overline{CS}$ signal goes to high.

In Figure 14.37a, the $\overline{OE}$ is controlled by a signal. Whenever the MCU wants to write data into the SRAM, the $\overline{OE}$ signal is pulled to high. In Figure 14.37b and c, the $\overline{OE}$ signal is tied to

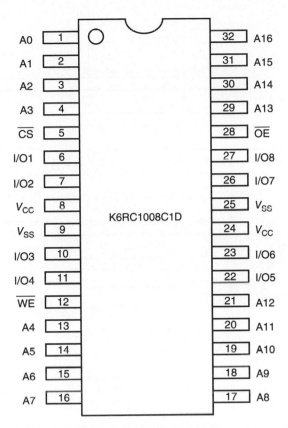

Figure 14.35 ■ The K6R1008C1D pin assignment

low. We see that the data out gets out of the high-impedance state t_{OW} ns after the $\overline{WE}$ and $\overline{CS}$ signals go to high. The write cycles in Figure 14.37a and b are $\overline{WE}$-controlled because the $\overline{WE}$ signal is asserted later than the $\overline{CS}$ signal. The write cycle in Figure 14.37c is $\overline{CS}$-controlled because the $\overline{CS}$ signal is asserted later than the $\overline{WE}$ signal.

14.9.2 The AT28C010 EEPROM

The AT28C010 is a 128K × 8 electrically erasable, programmable, read-only memory. It needs only a 5-V power supply to operate and achieves access times ranging from 120 to 200 ns.

The AT28C010 supports a page-write operation that can write from 1 to 128 bytes. The device contains a 128-byte page register to allow writing up to 128 bytes simultaneously. During a write cycle, the address and 1 to 128 bytes of data are internally latched, freeing the address and data bus for other operations. Following the initiation of a write cycle, the device will automatically write the latched data using an internal control timer. The end of an internal write cycle can be detected by polling the I/O7 pin or checking whether the I/O6 pin has stopped toggling.

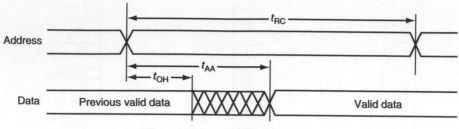

(a) Read cycle 1 ($\overline{OE}$ and $\overline{CE}$ are asserted in the whole cycle)

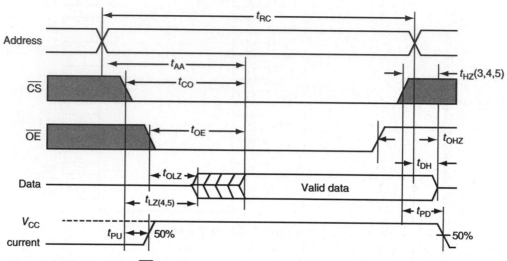

(b) Read cycle 2 ($\overline{OE}$ controlled)

Notes:

1. $\overline{WE}$ is high for read cycle.
2. All read cycle timing is referenced from the last valid address to the first transition address.
3. t_{HZ} and t_{OHZ} are defined as the time at which the output achieves the open circuit condition and are not referenced to the V_{OH} or V_{OL} level.
4. At any given temperature and voltage condition, $t_{HZ}(max)$ is less than $t_{LZ}(min)$ both for a given device and from device to device.
5. Transition is measured ±200 mV from steady-state voltage with load. This parameter is sampled and not 100% tested.
6. Device is continuously selected with $\overline{CS} = V_{IL}$.
7. For common I/O applications, minimization or elimination of bus contention conditions is necessary during the read and write cycle.

Figure 14.36 ■ K6R1008C1D read cycle timing diagram

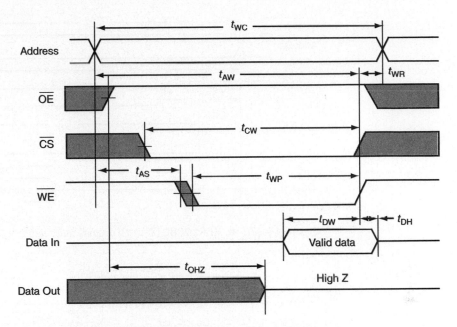

Figure 14.37a ■ K6R1008C1D write cycle timing diagram

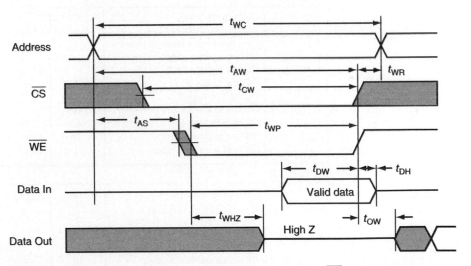

Figure 14.37b ■ K6R1008C1D write cycle timing diagram ($\overline{\text{OE}}$ tied to low)

The device utilizes internal error correction for extended endurance and improved data retention. To prevent an unintended write operation, an optional software data protection mechanism is available. The device also includes an additional 128 bytes of EEPROM for device identification or tracking. The pin assignment for the AT28C010 is shown in Figure 14.38.

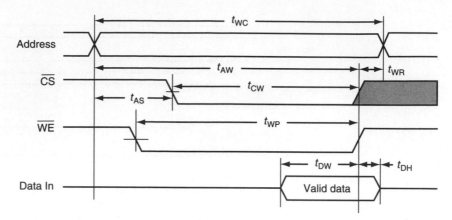

Figure 14.37c ■ K6R1008C1D write cycle timing diagram ($\overline{CS}$ controlled)

Parameter	Description	K6R1008C1D-10		Unit
		Min	Max	
Read Cycle				
t_{RC}	Read cycle time	10	–	ns
t_{AA}	Address to data valid	–	10	ns
t_{OH}	Data hold from address change	3	–	ns
t_{CO}	$\overline{CS}$ low to data valid	–	10	ns
t_{OE}	$\overline{OE}$ low to data valid	–	5	ns
t_{OLZ}	$\overline{OE}$ low to low Z	0	–	ns
t_{OHZ}	$\overline{OE}$ high to high Z	–	5	ns
t_{LZ}	$\overline{CS}$ low to low Z	3	–	ns
t_{HZ}	$\overline{CS}$ high to high Z	–	5	ns
t_{PU}	$\overline{CS}$ low to power-up	0	–	ns
t_{PD}	$\overline{CS}$ high to power-down	–	10	ns
Write Cycle				
t_{WC}	Write cycle time	10	–	ns
t_{CW}	$\overline{CS}$ low to write end	7	–	ns
t_{AW}	Address valid to end of write	7	–	ns
t_{AS}	Address setup to write start	0	–	ns
t_{WP}	$\overline{WE}$ pulse width ($\overline{OE}$ high)	7	–	ns
t_{WP1}	$\overline{WE}$ pulse width ($\overline{OE}$ low)	10	–	ns
t_{DW}	Data setup to write end	5	–	ns
t_{DH}	Data hold from write end	0	–	ns
t_{OW}	$\overline{WE}$ high to low Z	3	–	ns
t_{WHZ}	$\overline{WE}$ low to high Z	0	5	ns
t_{WR}	Write recovery time	0	–	ns

Table 14.10 ■ K6R1008C1D read and write timing parameters

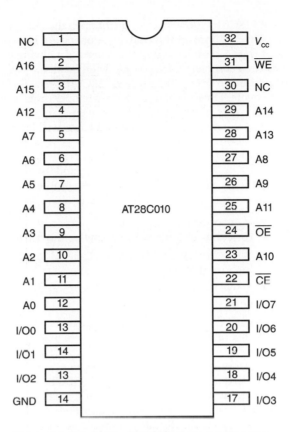

Figure 14.38 ■ The AT28C010 pin assignment

DEVICE OPERATION

The AT28C010 is accessed like a SRAM. When the $\overline{CE}$ and $\overline{OE}$ signals are low and the $\overline{WE}$ signal is high, the data stored at the location determined by the address pins is driven on the I/O pins. The I/O pins are put in the high-impedance state when either the $\overline{CE}$ or $\overline{OE}$ signal is high.

A byte-write operation is started by a low pulse on the $\overline{WE}$ or the $\overline{CE}$ input with the $\overline{CE}$ or the $\overline{WE}$ input low (respectively) and the $\overline{OE}$ pin high. The address inputs are latched on the falling edge of the $\overline{CE}$ or the $\overline{WE}$ signal, whichever occurs last. The data is latched by the first rising edge of the $\overline{CE}$ or the $\overline{WE}$ signal.

A page-write operation is initiated in the same manner as a byte-write operation; after the first byte is written, it can then be followed by 1 to 127 additional bytes. Each successive byte must be loaded within 150 μs (t_{BLC}) of the previous byte. If the t_{BLC} limit is exceeded, the AT28C010 will cease accepting data and begin the internal programming operation. All bytes involved in a page-write operation must reside on the same page as defined by the state of the A16, . . . , A7 inputs. For each high-to-low transition of the $\overline{WE}$ signal during the page-write operation, the address signals A16–A7 must be the same. The inputs A6–A0 are used to specify which bytes within the page are to be written. The bytes may be loaded in any order and may be altered within the same load period. Only bytes that are specified for writing will be written.

The AT28C010 allows the designer to poll the I/O7 pin to find out if an internal write operation has completed. Before an internal write operation has completed, a read of the last byte written will result in the complement of the written data being presented on the I/O7 pin. Once the write cycle has been completed, true data is valid on all outputs, and the next write cycle can begin.

In addition to data polling, the AT28C010 provides another method for determining the end of a write cycle. During the internal write operation, successive attempts to read data from the device will result in I/O6 toggling between 1 and 0. Once the write operation has completed, the I/O6 pin will stop toggling, and valid data will be read.

DATA PROTECTION

Atmel has incorporated both hardware and software features to protect the memory against inadvertent write operations. The hardware protection method works as follows:

- V_{CC} *sense*. If V_{CC} is below 3.8 V, the write function is inhibited.
- V_{CC} *power-on delay*. Once V_{CC} has reached 3.8 V, the device will automatically time out for 5 ms before allowing a write operation.
- *Write inhibit*. Holding the $\overline{OE}$ signal low or the $\overline{CE}$ signal high or the $\overline{WE}$ signal high inhibits write cycles.
- *Noise filter*. Pulses shorter than 15 ns on the $\overline{WE}$ or the $\overline{CE}$ input will not initiate a write cycle.

A software data protection (SDP) feature is included that can be enabled to prevent inadvertent write operations. The SDP is enabled by the host system issuing a series of three write commands; 3 specific bytes of data are written to three specific addresses. After writing the 3-byte command sequence and after the t_{WC} delay, the entire AT28C010 will be protected against inadvertent write operations. It should be noted that once protected, the host may still perform a byte write or page write to the AT28C010. This is done by preceding the data to be written by the same 3-byte command sequence used to enable the SDP. Once set, the SDP will remain active unless the disable command sequence is issued. Power transitions do not disable the SDP, and the SDP will protect the AT28C010 during the power-up and power-down conditions. After setting the SDP, any attempt to write to the device without the 3-byte command sequence will start the internal write timer. No data will be written into the device.

The algorithm for enabling software data protection is as follows:

Step 1
Write the value of 0xAA to the memory location at 0x5555.

Step 2
Write the value of 0x55 to the memory location at 0x2AAA.

Step 3
Write the value of 0xA0 to the memory location at 0x5555. At the end of write, the write-protect state will be activated. After this step, write operation is also enabled.

Step 4
Write any value to any location (1–128 bytes of data are written).

Step 5
Write last byte to last address.

Software data protection can be disabled any time when it is undesirable. The algorithm for disabling software data protection is as follows:

Step 1
Write the value of 0xAA to the memory location at 0x5555.

Step 2
Write the value of 0x55 to the memory location at 0x2AAA.

Step 3
Write the value of 0x80 to the memory location at 0x5555.

Step 4
Write the value of 0xAA to the memory location at 0x5555.

Step 5
Write the value of 0x55 to the memory location at 0x2AAA.

Step 6
Write the value of 0x20 to the memory location at 0x5555. After this step, software data protection is exited.

Step 7
Write any value(s) to any location(s).

Step 8
Write the last byte to the last address.

DEVICE IDENTIFICATION

An extra 128 bytes of EEPROM memory are available to the designer for device identification. By raising the voltage at the A9 pin to 12 V ± 0.5 V and using the address locations 0x1FF80 to 0x1FFFF, the bytes may be written to or read from in the same manner as the regular memory array.

READ AND WRITE TIMING

The read cycle timing diagram is shown in Figure 14.39. The AT28C010 has three read access times.

- Address access time t_{ACC}
- CE access time t_{CE}
- $\overline{OE}$ access time t_{OE}

Each of the read access times is measured by assuming that the other control signals and/or the address have been valid. For example, the address access time is the time from the moment that the address inputs to the AT28C010 become valid until data is driven out of data pins

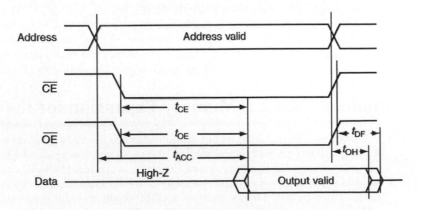

Figure 14.39 ■ AT28C010 read timing diagram

assuming that the $\overline{CE}$ and $\overline{OE}$ signals have been valid (low) before the required moment (t_{CE} or t_{OE} ns before data become valid). After address change, the data value will change immediately (data hold time is 0). The values of read timing parameters are listed in Table 14.11.

Symbol	Parameter	AT28C010-12		AT28C010-15		AT28C010-20		Unit
		Min	Max	Min	Max	Min	Max	
t_{ACC}	Address to output delay		120		150		200	ns
$t_{CE(1)}$	$\overline{CE}$ to output delay		120		150		200	ns
$t_{OE(2)}$	$\overline{OE}$ to output delay	0	50	0	55	0	55	ns
$t_{DF(3),(4)}$	$\overline{OE}$ to output float	0	50	0	55	0	55	ns
t_{DH}	Output hold from $\overline{OE}$, $\overline{CE}$ address, which occurred first	0		0		0		ns

1. $\overline{CE}$ may be delayed up to $t_{ACC} - t_{CE}$ after the address transition without impact on t_{ACC}.
2. $\overline{OE}$ may be delayed up to $t_{CE} - t_{OE}$ after the falling edge of $\overline{CE}$ without impact on t_{CE} or by $t_{ACC} - t_{OE}$ after an address change without impact on t_{ACC}.
3. t_{DF} is specified from $\overline{OE}$ or $\overline{CE}$, whichever occurs first (C_L = 5 pF).
4. This parameter is characterized and is not 100 percent tested.

Table 14.11 ■ AT28C010 read characteristics

The write cycle timing diagram is shown in Figure 14.40. There are two write timing diagrams: $\overline{WE}$-controlled and $\overline{CE}$-controlled diagrams, depending on which signal is asserted the latest. In a $\overline{WE}$-controlled write timing waveform, the $\overline{CE}$ signal is asserted (goes low) earlier than the $\overline{WE}$ signal and becomes inactive (goes high) after the $\overline{WE}$ signal goes high. In a $\overline{CE}$-controlled write cycle, the $\overline{WE}$ signal goes low earlier and returns to high later than the $\overline{CE}$ signal. In Figure 14.40, the control signal that is asserted the latest must have a minimal pulse width of 100 ns. The address input must not be valid after the assertion (going low) of the latest control signal and must remain valid for at least 50 ns after the assertion of the same signal. The values of write cycle timing parameters are shown in Table 14.12. The write data must be valid at least 50 ns before the latest control signal starts to rise. The write data need not be stable after the latest control signal (data hold time) rises.

The page-mode write cycle timing diagram is shown in Figure 14.41, and the values of the timing parameters are shown in Table 14.13.

The write timing waveform illustrates only how the CPU writes data into the EEPROM. The EEPROM still needs to initiate an internal programming process to write data into the specified location. The CPU can find out whether the internal programming process has been completed by using the data polling or the toggle bit polling method.

14.10 Example of External Memory Expansion for the HCS12

One can choose to use or not to use expanded memory (paging) when adding external memory chips to the HCS12, but being limited to adding less than 64 kB of external memory without using expanded memory doesn't make much sense. When expanded memory is used, we can add up to 1 MB of external memory if the on-chip flash memory is disabled. The external memory can be 8 or 16 bits wide. In an 8-bit-wide external memory, the HCS12 performs two read bus cycles when handling 16-bit data. This reduces the external memory performance to

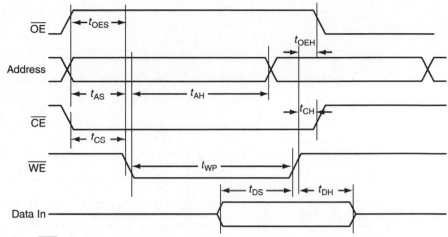

(a) $\overline{\text{WE}}$-controlled write timing waveform

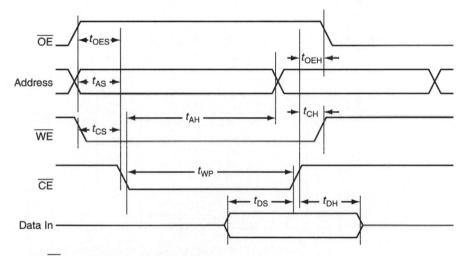

(b) $\overline{\text{CE}}$-controlled write timing waveform

Figure 14.40 ■ AT28C010 write-cycle timing waveform

Symbol	Parameter	Min	Max	Unit
t_{AS}, t_{OES}	Address, $\overline{\text{OE}}$ setup time	0		ns
t_{AH}	Address hold time	50		ns
t_{CS}	Chip-select setup time	0		ns
t_{CH}	Chip-select hold time	0		ns
t_{WP}	Write pulse width ($\overline{\text{WE}}$ or $\overline{\text{CE}}$)	100		ns
t_{DS}	Data setup time	50		ns
t_{DH}, t_{OEH}	Data, $\overline{\text{OE}}$ hold time	0		ns

Table 14.12 ■ AT28C010 write characteristics

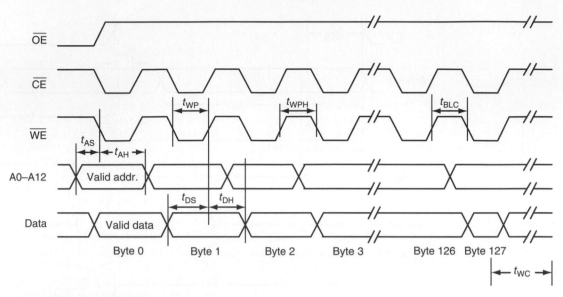

Figure 14.41 ■ AT28C010 page-mode write waveform

Symbol	Parameter	Min	Max	Unit
t_{WC}	Write cycle time		10	ms
t_{AS}	Address setup time	0		ns
t_{AH}	Address hold time	50		ns
t_{DS}	Data setup time	50		ns
t_{DH}	Data hold time	0		ns
t_{WP}	Write pulse width	100		ns
t_{BLC}	Byte load cycle time		150	μs
t_{WPH}	Write pulse width high	50		ns

Table 14.13 ■ AT28CO10 page-write characteristics

half but may lower the system cost slightly. When adding 8-bit-wide external memory, the data output of memory is connected to Port A.

This section illustrates how to add 256 kB of SRAM and 256 kB of EEPROM to the HCS12 by using the paging technique. The application program needs to go through the paging window located from $8000 to $BFFF in order to access these two memory modules.

14.10.1 Memory Space Assignment

Because this design will be using two 128-kB SRAM chips (K6R1008C1D) to construct 256 kB of 16-bit-wide memory, the lowest 13 address signals (A13, . . . , A1) and the expanded address signals XA14, . . . , XA17 will be used as address inputs for both SRAM chips. The A0 signal will not be needed. Because these two chips must be selected by the same chip-select signal, the address space must be divided into 256-kB blocks when making memory space assignment to the SRAM module. The following assignment will be appropriate:

- SRAM module: $00000–$3FFFF
- EEPROM module: $40000–$7FFFF

14.10.2 Address Latch

Since the address signals A15, . . . , A0 and data signals D15, . . . , D0 are time multiplexed, they need to be latched into a memory device such as the dual octal latch 74ABT16373B (made by Philips) and held valid throughout the whole bus cycle. This device uses the falling edge of its enable inputs (1E and 2E) to latch the data inputs. The $\overline{ECS}$ signal (or $\overline{XCS}$ for some devices) of the HCS12 can be used to drive this input and latch the address signals A15, . . . , A0 into the 74ABT16373B. The pin assignment of the 74ABT16373B is shown in Figure 14.42.

The values of 1D0, . . . , 1D7 and 2D0, . . . , 2D7 are latched into 1Q0, . . . , 1Q7 and 2Q0, . . . , 2Q7 on the falling edges of the 1E and 2E signals, respectively. The propagation delay from the falling edge of 1E (or 2E) to 1Q0, . . . , 1Q7 (or 2Q0, . . . , 2Q7) is 4.4 ns. The propagation delay from 1Dx ($x = 0, . . . ,7$) to 1Qx is also 4.4 ns in the worst case. When the 1E (or 2E) signal is high, the latch is transparent; that is, the value on 1Dx (2Dx) will appear on 1Qx (2Qx) after the propagation delay. The required data setup and hold times for 1Dx (and 2Dx) are 1 ns and 0.5 ns, respectively. The outputs 1Q0, . . . , 1Q7 (2Q0, . . . , 2Q7) will be in high impedance if the $\overline{IOE}$ ($\overline{2OE}$) signal is high.

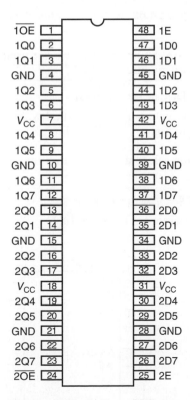

Figure 14.42 ■ Pin assignment of 74ABT16373B

At 25 MHz, address signals A15, ..., A0 (driven by the HCS12) become valid 8 ns after the falling edge of the E-clock or up to 8 ns before the falling edge of the $\overline{ECS}$ signal. Since A15, ..., A0 stay valid until 2 ns after the rising edge of the E-clock, A15, ..., A0 have a hold time of at least 6 ns. Therefore, the input-setup and hold-time requirements of the 74ABT16373B are satisfied. Since the propagation delay relative to the 1E and 2E signals is 4.4 ns, A15, ..., A0 will become valid (to memory chips) 0.4 ns after the rising edge of the E-clock or 20.4 ns after the start of a bus cycle. The address latching circuit is shown in Figure 14.43.

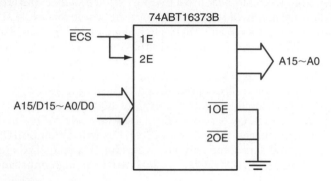

Figure 14.43 ■ Address latch circuit for the HCS12

The latched outputs A0, A14, and A15 are not used in a paged memory module. A0 is not used because the memory module is 16 bit.

14.10.3 Address Decoder Design

In this simple memory design example, the 1-MB memory space is divided into four 256-kB modules. Therefore, the highest 2 address bits, XA19 and XA18, should be used as decoder inputs. A 3-to-8 decoder (e.g., 74F138) or a dual 2-to-4 decoder (e.g., 74F139) can be used as the address decoder. The chip-select signal equations for the SRAM and EEPROM are

$$\overline{SRAM\_CS} \quad = \ !(!XA19 * !XA18 * ECS)$$

$$\overline{EEPROM\_CS} = \ !(!XA19 * XA18 * ECS)$$

The decoder circuit is shown in Figure 14.44.

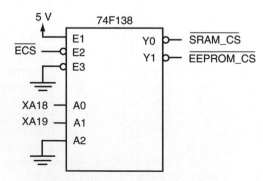

Figure 14.44 ■ HCS12 external memory decoder circuit

14.10.4 Write-Control Signals

Because the HCS12 may sometimes write to a byte rather than a word of the external memory, the $\overline{WE}$ signal for the upper byte (even address) and lower byte (odd address) must be generated separately. Let $\overline{WE1}$ and $\overline{WE0}$ be the write-enable signals of the high and low bytes; then

$$\overline{WE1} = !(LSTRB * !A0 * !R/\overline{W} + \overline{LSTRB}* !A0 * !R/\overline{W}) = !(!A0 * !R/\overline{W})$$

$$\overline{WE0} = !(\overline{LSTRB} * A0 * !R/\overline{W} + LSTRB * A0 * !R/\overline{W}) = !(A0 * !R/\overline{W})$$

These two control signals can be generated by using an inverter chip and a NAND gate chip. A simpler solution is to generate the chip-select signals and write-enable signals using a single GAL device, such as the GAL16V8. The GAL16V8 has a propagation delay of 3.5 ns. By using a GAL16V8 to generate the chip-select and write-enable signals as shown in Figure 14.45, the decoder circuit shown in Figure 14.44 can be eliminated. In Figure 14.45, the output-enable signal $(\overline{OE})$ required to control the output of SRAM and EEPROM can be generated without additional cost. The $\overline{OE}$ signal is simply the complement of the R/$\overline{W}$ signal output of the HCS12.

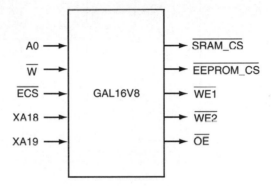

Figure 14.45 ■ Address decoder and write-enable signal-generating circuit

The design of the circuit shown in Figure 14.45 can be done by using the ispLEVER software from Lattice Semiconductor.

14.10.5 Example HCS12 External Memory Design

As shown in Figure 14.46, the circuit required to interface with external SRAMs and EEPROMs is minimal. A 16-bit latch is needed to keep address signals A15~A0 valid throughout the whole bus cycle. A GAL16V8 chip is used to generate the chip-select signals for selecting the SRAM and EEPROM modules. The $\overline{ECS}$ signal is asserted (goes low) only when the address outputs are valid. It must be included as one of the inputs to the address decoder circuit to make sure that the decoder asserts one of its outputs to low when its address inputs XA18 and XA19 are valid.

Example 14.16

▼

For the circuit shown in Figure 14.46, can the SRAM be accessed without stretching the E-clock? Assume that the E-clock is 25 MHz.

Solution: The timing verification consists of the read cycle and the write cycle.

ADDRESS TIMING

The address inputs A12, . . . , A0 to the memory chips become valid 20.5 ns after the start of a bus cycle; the address inputs A16, . . . , A13 become valid 6 ns after the start of a bus cycle.

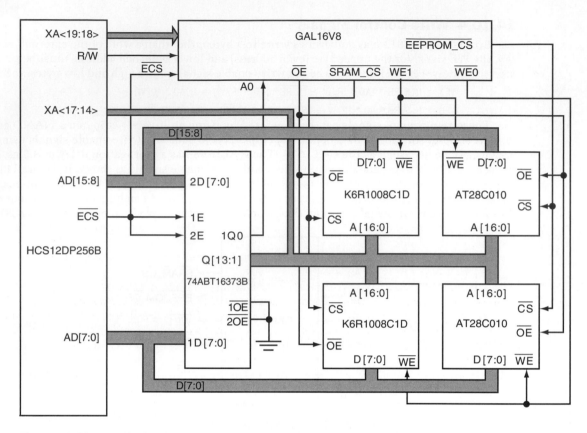

Figure 14.46 ■ HCS12DP256B paged external memory example

Therefore, the complete address inputs A16, . . . , A0 to the memory chips become valid 20.5 ns after the start of a bus cycle.

CHIP-SELECT SIGNALS TIMING

The $\overline{\text{ECS}}$ signal becomes valid 16 ns after the start of a bus cycle. XA19, . . . , XA18 are valid 6 ns after the start of a bus cycle. The GAL16V8 has a propagation delay of 3.5 ns. Therefore, the chip-select signals $\overline{\text{SRAM\_CS}}$ and $\overline{\text{EEPROM\_CS}}$ become valid 19.5 ns after the start of a bus cycle. Because the HCS12 stops driving the XA19, XA18, and $\overline{\text{ECS}}$ signals 2 ns after the end of a bus cycle, these two chip-select signals will become invalid 5.5 ns after the end of a bus cycle.

WRITE-ENABLE SIGNALS TIMING

Since $\overline{\text{WE1}}$ and $\overline{\text{WE0}}$ are a function of A0 and R/$\overline{\text{W}}$ and A0 is valid later than R/$\overline{\text{W}}$, the valid times of $\overline{\text{WE1}}$ and $\overline{\text{WE0}}$ are determined by A0. Since A0 is valid 20.5 ns after the start of a bus cycle and the propagation delay of the GAL16V8 is 3.5 ns, $\overline{\text{WE1}}$ and $\overline{\text{WE0}}$ will become valid 24 ns after the start of a bus cycle. The HCS12 deasserts the R/$\overline{\text{W}}$ signal 2 ns after the end of a bus cycle; both the $\overline{\text{WE1}}$ and $\overline{\text{WE0}}$ signals become invalid 5.5 ns after the end of a write cycle.

OUTPUT-ENABLE SIGNAL TIMING

The $\overline{OE}$ signal is the complement of the R/$\overline{W}$ signal and is generated by the GAL16V8. The R/$\overline{W}$ signal is valid 7 ns after the start of a bus cycle. Therefore, the $\overline{OE}$ signal will become valid 10.5 ns after the start of a bus cycle and become invalid 5.5 ns after the end of a bus read cycle.

READ CYCLE TIMING VERIFICATION

For the K6R1008C1D SRAM, address inputs A16~A0 become valid later than $\overline{OE}$ and $\overline{CS}$. Therefore, read data is valid 30.5 ns after the start of a read cycle or 9.5 ns before the end of a read cycle. The HCS12DP256 requires a read data setup time of 13 ns. Therefore, the read data does not provide enough setup time for the HCS12DP256 to correctly read it. One needs to stretch the E-clock by one period.

WRITE CYCLE TIMING VERIFICATION

The address inputs to the K6R1008C1D will stay valid until the new address is latched in the next bus cycle. Therefore, A16~A0 stay valid for a whole E-clock period (40 ns at 25 MHz) and satisfy the minimum requirement 10 ns (t_{WC}).

The write-enable signals $\overline{WE1}$ and $\overline{WE0}$ have a low pulse width of 21.5 ns (40 ns − 24 ns + 5.5 ns) and satisfy the requirement of the SRAM.

The HCS12 drives the write data 28 ns after the start of a write cycle (or 12 ns before the end of a write cycle). Both the $\overline{WE}$ and $\overline{CS}$ signal inputs to the K5R1008C1D are deasserted 5.5 ns after the end of a write cycle. This provides 17.5 ns of write data setup time for the K5R1008C1D and exceeds the minimum requirement of 5 ns. The K5R1008C1D requires a write data hold time of 0 ns. Since the $\overline{CS}$ and $\overline{WE}$ signals become invalid 5.5 ns after the end of a write cycle. The write data must remain valid for at least 5.5 ns after a write cycle. The HCS12 stops driving the write data 2 ns after the end of a write bus cycle. This is shorter than the requirement. However, the HCS12 does not drive the multiplexed address/data bus for 8 ns after the start of the next bus cycle. During this period, the capacitance of the printed circuit board can keep the write data for a little while longer and satisfies the write data hold-time requirement. Because both the write data setup- and hold-time requirements are satisfied, the E-clock needs not be stretched for the write cycle.

Because the E-clock stretching applies to all the bus cycles, both the read and write cycles need to be stretched for one E-clock period for the K5R1008C1D.

▲

Example 14.17

▼

Provide an analytical proof that the capacitance of the printed circuit board can hold the data voltage for enough time to satisfy the data hold-time requirement of the SRAM.

Solution: The voltage of the data input to the K5R1008C1D is degraded by the following leakage currents:

1. Input current into the HCS12 data pin (typically on the order of 2.5 μA for Freescale microcontroller products)
2. Input current into the EEPROM chips (on the order of 10 μA)
3. Other leakage paths (assumed to be as large as 10 μA)

The capacitance of the printed circuit board is estimated to be 20 pF per foot. Let C, I, ΔV, and Δt be the printed circuit board capacitance of one data line, total leakage current, voltage change due to leakage current, and the time it takes for the voltage to degrade by ΔV, respectively.

The elapsed time before the data bus signal degrades to an invalid level can be estimated by the following equation:

$$\Delta t \approx C\Delta V \div I$$

The voltage degradation of 2.5 V (for a 5-V V_{CC}) is considered enough to cause data input to the SRAM to change the logic value from 1 to 0. The data bus line on the printed circuit board is normally not longer than 1 ft for a single-board computer, so 1 ft will be used as its length. The elapsed time before the data bus signal degrades to an invalid level is

$$\Delta t \approx 20 \text{ pF} \times 2.5 \text{ V} \div 22.5 \text{ µA} \approx 2.2 \text{ µs}$$

Although this equation is oversimplified, it does give some idea about the order of the time over which the charge across the data bus capacitor will hold after the HCS12 stops driving the data bus. Even if the leakage current is 10 times larger, the Δt value will be 225 ns and is still longer than the minimum hold-time requirement of the K5R1008C1D.

Example 14.18

For the circuit shown in Figure 14.46, can the EEPROM be accessed without stretching the E-clock, assuming that the 120-ns access time version of the AT28C010 is used? If not, how many cycles should the E-clock be stretched?

Solution: The timing analysis of this problem differs from that of the SRAM mainly in the read/write data setup times. According to the analysis done in Example 14.16, A16, . . . , A0 become valid the latest (30.5 ns after the start of a read cycle). The data output from the AT28C010 becomes valid 150.5 ns after the start of a read cycle. For a 25-MHz E-clock, this is 110.5 ns (150.5 ns – 40 ns) too late from the end of a read cycle. Therefore, the read cycle for the AT28C010 needs to be stretched for at least three E-clock periods at this frequency. After stretching the E-clock by three cycles, the read data is valid 9.5 ns before the end of the read cycle. This doesn't provide enough data setup time for the HCS12. The only solution is to lower the frequency of the E-clock. By slowing the E-clock frequency to 24 MHz, the read data setup time is increased to 16.17 ns and satisfies the requirement. The AT28C010 provides 0 ns of data hold time, which is acceptable for the HCS12.

According to Example 14.16, the HCS12 provides 17.5 ns of write data setup time, which is shorter than the requirement (50 ns minimum) of the AT28C010. Stretching the E-clock by one E-clock period can satisfy this requirement. However, the write pulse width (both the $\overline{\text{EEPROM\_CS}}$ and $\overline{\text{WE0}}$) must be at least 100 ns. This would require the write bus cycle to be stretched by at least two E-clock periods. Since the read cycle needs to be stretched by three E-clock periods, the EBICTL and MISC registers need to be programmed to provide this requirement.

Example 14.19

Write an instruction sequence to configure the EBICTL and MISC registers to stretch the external bus cycles by three E-clock periods.

Solution: The value to be written into the EBICTL is $01. This value enables the E-clock stretch. The value to be written into the MISC register is $0D, which will stretch the bus cycles by three E-clock periods. The following instruction sequence will achieve the required stretching:

```
movb #$01,EBICTL    ; enable E stretch
movb #$0D,MISC      ; stretch E by 3 periods
```

The control signals that are multiplexed on Port E pins must be programmed properly in order for external bus cycles to proceed properly. The designer needs to program the PEAR register to enable the required signals, including R/$\overline{W}$, E-clock, $\overline{LSTRB}$, and others. The signals NOACC, IPIPE1, and IPIPE0 need not be enabled if the HCS12 is not designed to be debugged by the in-circuit emulator.

Example 14.20

▼

Give a value to be written into the PEAR register to achieve the following setting:
- Enable E-clock output
- Enable the R/$\overline{W}$ output
- Enable the $\overline{LSTRB}$ output
- Disable the NOACC, IPIPE1, and IPEP0 outputs

Solution: The value to be written into the PEAR register to achieve this setting is $0C.

▲

14.11 Summary

The HCS12 has on-chip memory resources, including I/O registers, SRAM, EEPROM, and flash memory. The I/O registers, SRAM, and EEPROM can be remapped to other memory spaces when the application requires it. It is possible to have two or more memory resources overlapped. The HCS12 assigns precedence to each memory resource. The memory resource with the highest precedence is selected when two or more memory resources overlap in their spaces. The remapping of memory resources is performed by programming their associated initialization registers.

Limited by its 16-bit architecture, the HCS12 uses the paging technique to allow programs to access memory space larger than the 64-kB limit. The paging window is located at $8000 to $BFFF. An additional 6-bit PPAGE register is used to facilitate the mapping from the paging window to the actual memory block to be accessed. The HCS12 uses this technique to allow the application program to access up to 1 MB of memory space. Whenever the program wants to access expanded memory, it sets the address to be in the range of this window and also sets the appropriate value to the PPAGE register. After this, the hardware maps to the corresponding location to perform the access. Writing programs to take advantage of the expanded memory incurs quite a bit of overhead in assembly language. Neither the asmIDE nor the MiniIDE supports expanded memory. Many C compilers (including the ICC12 from Imagecraft and the CodeWarrior from Metrowerks) support the expanded memory. It is recommended that C language be used to write application programs whenever expanded memory is needed.

Both the flash memory and the EEPROM need to be erased before they can be programmed with new values. The algorithms of erasure and programming for these two memory technologies are quite similar. Before erasing and programming the flash memory and EEPROM, the designer needs to make sure that they are not protected. The erasure and programming operations are controlled by a clock signal that is derived by dividing the E-clock by a prescaler. The following four operations are common in flash memory and EEPROM erasure and programming:
- Clear the error flags in ESTAT (or FSTAT).
- Write a word to a location in the sector to be erased or programmed.
- Write a command to the command register (ECMD or FCMD).
- Clear the CBEIF flag to launch the command.

Both the flash memory and EEPROM can be protected from inadvertent erasure. The software stored in the flash memory can be secured from being pirated.

Certain applications require more memory than the amount available in any of the HCS12 members. Adding external memory is one solution to this problem. External memory can be added only in the expanded modes. There are two expanded modes: narrow and wide. The expanded narrow mode allows the designer to add 8-bit-wide external memory; the expanded wide mode allows the user to add 16-bit-wide memory to the HCS12.

Most 8- and 16-bit microcontrollers multiplex their address and data buses onto the same set of pins in order to lower the pin count and device cost. All of the read and write cycles take exactly one E-clock cycle to complete. However, the HCS12 allows the user to stretch the E-clock so that the slower but less expensive memory chips can be used in the product. A read or write cycle starts with the E-clock going low. When the E-clock is low, the HCS12 drives the address onto the multiplexed address/data bus. The control circuitry outside the MCU latches the address to keep it valid throughout the whole bus cycle. When the E-clock goes high, the HCS12 stops driving the address onto the bus and either waits for the read data to be returned from the memory devices or drives data to be written into external memory on the multiplexed bus.

There are three major issues in the external memory expansion:

- Memory space assignment
- Address decoder design and certain additional signals generation
- Timing verification

14.12 Exercises

E14.1 Write a program to erase the first 100 words of the on-chip EEPROM and then to write the values 1, 2, . . . , 100 to these 100 words, read them back, and store them in the first 100 words of the on-chip SRAM.

E14.2 Write an instruction sequence to remap the register block so that it starts at $1000.

E14.3 Write an instruction sequence to remap the SRAM block so that it starts at $2000.

E14.4 Write an instruction sequence to disable E-clock stretch and disable on-chip ROM for expanded mode.

E14.5 Write an instruction sequence to map the EEPROM block to start from $0000.

E14.6 Give an instruction to protect the highest 64 bytes of the EEPROM.

E14.7 Write an instruction sequence to protect the 1 kB in the lower address range and the 4 kB in the higher address range of the flash 3 of the HCS12DG256.

E14.8 Write a C function that erases the EEPROM in bulk.

E14.9 Write a C function that erases the EEPROM in bulk and verifies it.

E14.10 What is the slowest SRAM read access time that does not require the designer to stretch the E-clock of the HCS12? Assume the E-clock frequency is 25 MHz.

14.13 Lab Exercise and Assignment

L14.1 Write a program to store the first 100 prime numbers (starting from 2 and upward) in the on-chip EEPROM starting from address $400. Read out the prime numbers and display them to the terminal monitor with each line displaying five numbers.

Appendix A: Instruction Set Reference

Freescale Semiconductor, Inc.

Appendix A. Instruction Set Reference

A.1 Contents

A.2 Introduction

This appendix provides quick references for the instruction set, opcode map, and encoding.

Freescale Semiconductor, Inc.

**For More Information On This Product,
Go to: www.freescale.com**

Freescale Semiconductor, Inc.

Instruction Set Reference

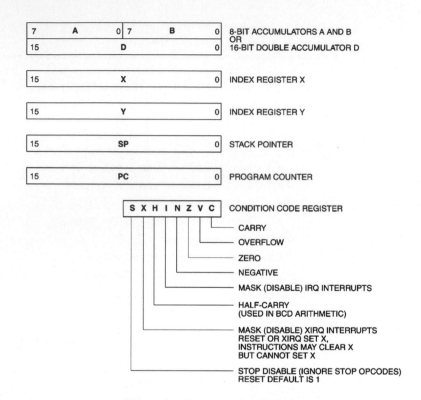

Figure A-1. Programming Model

Freescale Semiconductor, Inc.

A.3 Stack and Memory Layout

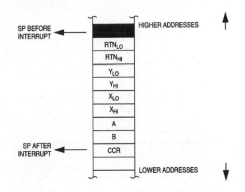

A.4 Interrupt Vector Locations

$FFFE, $FFFF	Power-On (POR) or External Reset
$FFFC, $FFFD	Clock Monitor Reset
$FFFA, $FFFB	Computer Operating Properly (COP Watchdog Reset
$FFF8, $FFF9	Unimplemented Opcode Trap
$FFF6, $FFF7	Software Interrupt Instruction (SWI)
$FFF4, $FFF5	XIRQ
$FFF2, $FFF3	IRQ
$FFC0–$FFF1 (M68HC12)	Device-Specific Interrupt Sources
$FF00–$FFF1 (HCS12)	Device-Specific Interrupt Sources

For More Information On This Product,
Go to: www.freescale.com

Freescale Semiconductor, Inc.

Instruction Set Reference

A.5 Notation Used in Instruction Set Summary

CPU Register Notation

Accumulator A — A or a	Index Register Y — Y or y
Accumulator B — B or b	Stack Pointer — SP, sp, or s
Accumulator D — D or d	Program Counter — PC, pc, or p
Index Register X — X or x	Condition Code Register — CCR or c

Explanation of Italic Expressions in Source Form Column

abc — A or B or CCR

abcdxys — A or B or CCR or D or X or Y or SP. Some assemblers also allow T2 or T3.

abd — A or B or D

abdxys — A or B or D or X or Y or SP

dxys — D or X or Y or SP

msk8 — 8-bit mask, some assemblers require # symbol before value

opr8i — 8-bit immediate value

opr16i — 16-bit immediate value

opr8a — 8-bit address used with direct address mode

opr16a — 16-bit address value

oprx0_xysp — Indexed addressing postbyte code:

    oprx3,–xys Predecrement X or Y or SP by 1 . . . 8

    oprx3,+xys Preincrement X or Y or SP by 1 . . . 8

    oprx3,xys– Postdecrement X or Y or SP by 1 . . . 8

    oprx3,xys+ Postincrement X or Y or SP by 1 . . . 8

    oprx5,xysp 5-bit constant offset from X or Y or SP or PC

    abd,xysp Accumulator A or B or D offset from X or Y or SP or PC

oprx3 — Any positive integer 1 . . . 8 for pre/post increment/decrement

oprx5 — Any integer in the range –16 . . . +15

oprx9 — Any integer in the range –256 . . . +255

oprx16 — Any integer in the range –32,768 . . . 65,535

page — 8-bit value for PPAGE, some assemblers require # symbol before this value

rel8 — Label of branch destination within –256 to +255 locations

rel9 — Label of branch destination within –512 to +511 locations

rel16 — Any label within 64K memory space

trapnum — Any 8-bit integer in the range $30-$39 or $40-$FF

xys — X or Y or SP

xysp — X or Y or SP or PC

Operators

+ — Addition

– — Subtraction

• — Logical AND

+ — Logical OR (inclusive)

Continued on next page

For More Information On This Product,
Go to: www.freescale.com

Freescale Semiconductor, Inc.

Freescale Semiconductor, Inc.

Operators (continued)

⊕ — Logical exclusive OR

× — Multiplication

÷ — Division

$\overline{M}$ — Negation. One's complement (invert each bit of M)

: — Concatenate
Example: A : B means the 16-bit value formed by concatenating 8-bit accumulator A with 8-bit accumulator B.
A is in the high-order position.

⇒ — Transfer
Example: (A) ⇒ M means the content of accumulator A is transferred to memory location M.

⇔ — Exchange
Example: D ⇔ X means exchange the contents of D with those of X.

Address Mode Notation

INH — Inherent; no operands in object code

IMM — Immediate; operand in object code

DIR — Direct; operand is the lower byte of an address from $0000 to $00FF

EXT — Operand is a 16-bit address

REL — Two's complement relative offset; for branch instructions

IDX — Indexed (no extension bytes); includes:
5-bit constant offset from X, Y, SP, or PC
Pre/post increment/decrement by 1 . . . 8
Accumulator A, B, or D offset

IDX1 — 9-bit signed offset from X, Y, SP, or PC; 1 extension byte

IDX2 — 16-bit signed offset from X, Y, SP, or PC; 2 extension bytes

[IDX2] — Indexed-indirect; 16-bit offset from X, Y, SP, or PC

[D, IDX] — Indexed-Indirect; accumulator D offset from X, Y, SP, or PC

Machine Coding

dd — 8-bit direct address $0000 to $00FF. (High byte assumed to be $00).

ee — High-order byte of a 16-bit constant offset for indexed addressing.

eb — Exchange/Transfer post-byte. See **Table A-5** on page 436.

ff — Low-order eight bits of a 9-bit signed constant offset for indexed addressing, or low-order byte of a 16-bit constant offset for indexed addressing.

hh — High-order byte of a 16-bit extended address.

ii — 8-bit immediate data value.

jj — High-order byte of a 16-bit immediate data value.

kk — Low-order byte of a 16-bit immediate data value.

lb — Loop primitive (DBNE) post-byte. See **Table A-6** on page 437.

ll — Low-order byte of a 16-bit extended address.

mm — 8-bit immediate mask value for bit manipulation instructions.
Set bits indicate bits to be affected.

Freescale Semiconductor, Inc.

Instruction Set Reference

pg — Program page (bank) number used in CALL instruction.

qq — High-order byte of a 16-bit relative offset for long branches.

tn — Trap number $30–$39 or $40–$FF.

rr — Signed relative offset $80 (–128) to $7F (+127).
Offset relative to the byte following the relative offset byte, or
. low-order byte of a 16-bit relative offset for long branches.

xb — Indexed addressing post-byte. See **Table A-3** on page 434
and **Table A-4** on page 435.

Access Detail
Each code letter except (,), and comma equals one CPU cycle. Uppercase = 16-bit
operation and lowercase = 8-bit operation. For complex sequences see the *CPU12
Reference Manual* (CPU12RM/AD) for more detailed information.

f — Free cycle, CPU doesn't use bus

g — Read PPAGE internally

I — Read indirect pointer (indexed indirect)

i — Read indirect PPAGE value (CALL indirect only)

n — Write PPAGE internally

O — Optional program word fetch (P) if instruction is misaligned and has
an odd number of bytes of object code — otherwise, appears as
a free cycle (f); Page 2 prebyte treated as a separate 1-byte instruction

P — Program word fetch (always an aligned-word read)

r — 8-bit data read

R — 16-bit data read

s — 8-bit stack write

S — 16-bit stack write

w — 8-bit data write

W — 16-bit data write

u — 8-bit stack read

U — 16-bit stack read

V — 16-bit vector fetch (always an aligned-word read)

t — 8-bit conditional read (or free cycle)

T — 16-bit conditional read (or free cycle)

x — 8-bit conditional write (or free cycle)

( ) — Indicate a microcode loop

, — Indicates where an interrupt could be honored

Special Cases

PPP/P — Short branch, PPP if branch taken, P if not

OPPP/OPO — Long branch, OPPP if branch taken, OPO if not

Reference Manual   CPU12 — Rev. 3.0

416   Instruction Reference   Freescale
                    For More Information On This Product,
                          Go to: www.freescale.com

Freescale Semiconductor, Inc.

Instruction Set Reference

Condition Codes Columns

- — Status bit not affected by operation.
- 0 — Status bit cleared by operation.
- 1 — Status bit set by operation.
- Δ — Status bit affected by operation.
- ⇓ — Status bit may be cleared or remain set, but is not set by operation.
- ⇑ — Status bit may be set or remain cleared, but is not cleared by operation.
- ? — Status bit may be changed by operation but the final state is not defined.
- ! — Status bit used for a special purpose.

Freescale Semiconductor, Inc.

For More Information On This Product,
Go to: www.freescale.com

Freescale Semiconductor, Inc.

Instruction Set Reference

Table A-1. Instruction Set Summary (Sheet 1 of 14)

Source Form	Operation	Addr. Mode	Machine Coding (hex)	Access Detail HCS12	Access Detail M68HC12	S X H I	N Z V C
ABA	(A) + (B) ⇒ A Add Accumulators A and B	INH	18 06	OO	OO	– – Δ –	Δ Δ Δ Δ
ABX	(B) + (X) ⇒ X *Translates to LEAX B,X*	IDX	1A E5	Pf	PP[1]	– – – –	– – – –
ABY	(B) + (Y) ⇒ Y *Translates to LEAY B,Y*	IDX	19 ED	Pf	PP[1]	– – – –	– – – –
ADCA #opr8i ADCA opr8a ADCA opr16a ADCA oprx0_xysp ADCA oprx9,xysp ADCA oprx16,xysp ADCA [D,xysp] ADCA [oprx16,xysp]	(A) + (M) + C ⇒ A Add with Carry to A	IMM DIR EXT IDX IDX1 IDX2 [D,IDX] [IDX2]	89 ii 99 dd B9 hh ll A9 xb A9 xb ff A9 xb ee ff A9 xb A9 xb ee ff	P rPf rPO rPf rPO frPP fIfrPf fIPrPf	P rfP rOP rfP rPO frPP flPrfP flPrfP	– – Δ –	Δ Δ Δ Δ
ADCB #opr8i ADCB opr8a ADCB opr16a ADCB oprx0_xysp ADCB oprx9,xysp ADCB oprx16,xysp ADCB [D,xysp] ADCB [oprx16,xysp]	(B) + (M) + C ⇒ B Add with Carry to B	IMM DIR EXT IDX IDX1 IDX2 [D,IDX] [IDX2]	C9 ii D9 dd F9 hh ll E9 xb E9 xb ff E9 xb ee ff E9 xb E9 xb ee ff	P rPf rPO rPf rPO frPP fIfrPf fIPrPf	P rfP rOP rfP rPO frPP flfrfP flPrfP	– – Δ –	Δ Δ Δ Δ

ADDA #opr8i
ADDA opr8a
ADDA opr16a
ADDA oprx0_xysp
ADDA oprx9,xysp
ADDA oprx16,xysp
ADDA [D,xysp]
ADDA [oprx16,xysp]
	(A) + (M) ⇒ A
	Add without Carry to A
		IMM
		DIR
		EXT
		IDX
		IDX1
		IDX2
		[D,IDX]
		[IDX2]
			8B ii
			9B dd
			BB hh ll
			AB xb
			AB xb ff
			AB xb ee ff
			AB xb
			AB xb ee ff
				P
				rPf
				rPO
				rPf
				rPO
				frPP
				fIfrPf
				fIPrPf
					P
					rfP
					rOP
					rfP
					rPO
					frPP
					flfrfP
					flPrfP
						– – Δ –	Δ Δ Δ Δ

ADDB #opr8i
ADDB opr8a
ADDB opr16a
ADDB oprx0_xysp
ADDB oprx9,xysp
ADDB oprx16,xysp
ADDB [D,xysp]
ADDB [oprx16,xysp]
	(B) + (M) ⇒ B
	Add without Carry to B
		IMM
		DIR
		EXT
		IDX
		IDX1
		IDX2
		[D,IDX]
		[IDX2]
			CB ii
			DB dd
			FB hh ll
			EB xb
			EB xb ff
			EB xb ee ff
			EB xb
			EB xb ee ff
				P
				rPf
				rPO
				rPf
				rPO
				frPP
				fIfrPf
				fIPrPf
					P
					rfP
					rOP
					rfP
					rPO
					frPP
					flfrfP
					flPrfP
						– – Δ –	Δ Δ Δ Δ

ADDD #opr16i
ADDD opr8a
ADDD opr16a
ADDD oprx0_xysp
ADDD oprx9,xysp
ADDD oprx16,xysp
ADDD [D,xysp]
ADDD [oprx16,xysp]
	(A:B) + (M:M+1) ⇒ A:B
	Add 16-Bit to D (A:B)
		IMM
		DIR
		EXT
		IDX
		IDX1
		IDX2
		[D,IDX]
		[IDX2]
			C3 jj kk
			D3 dd
			F3 hh ll
			E3 xb
			E3 xb ff
			E3 xb ee ff
			E3 xb
			E3 xb ee ff
				PO
				RPf
				RPO
				RPf
				RPO
				fRPP
				fIfRPf
				fIPRPf
					OP
					RfP
					ROP
					RfP
					RPO
					fRPP
					flfRfP
					flPRfP
						– – – –	Δ Δ Δ Δ

ANDA #opr8i ANDA opr8a ANDA opr16a ANDA oprx0_xysp ANDA oprx9,xysp ANDA oprx16,xysp ANDA [D,xysp] ANDA [oprx16,xysp]	(A) • (M) ⇒ A Logical AND A with Memory	IMM DIR EXT IDX IDX1 IDX2 [D,IDX] [IDX2]	84 ii 94 dd B4 hh ll A4 xb A4 xb ff A4 xb ee ff A4 xb A4 xb ee ff	P rPf rPO rPf rPO frPP fIfrPf fIPrPf	P rfP rOP rfP rPO frPP flfrfP flPrfP	– – – –	Δ Δ 0 –
ANDB #opr8i ANDB opr8a ANDB opr16a ANDB oprx0_xysp ANDB oprx9,xysp ANDB oprx16,xysp ANDB [D,xysp] ANDB [oprx16,xysp]	(B) • (M) ⇒ B Logical AND B with Memory	IMM DIR EXT IDX IDX1 IDX2 [D,IDX] [IDX2]	C4 ii D4 dd F4 hh ll E4 xb E4 xb ff E4 xb ee ff E4 xb E4 xb ee ff	P rPf rPO rPf rPO frPP fIfrPf fIPrPf	P rfP rOP rfP rPO frPP flfrfP flPrfP	– – – –	Δ Δ 0 –
ANDCC #opr8i	(CCR) • (M) ⇒ CCR Logical AND CCR with Memory	IMM	10 ii	P	P	⇓ ⇓ ⇓ ⇓	⇓ ⇓ ⇓ ⇓

Note 1. Due to internal CPU requirements, the program word fetch is performed twice to the same address during this instruction.

Reprinted with permission of Freescale Semiconductor, Inc.

Freescale Semiconductor, Inc.

Table A-1. Instruction Set Summary (Sheet 2 of 14)

Source Form	Operation	Addr. Mode	Machine Coding (hex)	Access Detail HCS12	Access Detail M68HC12	S X H I	N Z V C
ASL opr16a ASL oprx0_xysp ASL oprx9,xysp ASL oprx16,xysp ASL [D,xysp] ASL [oprx16,xysp] ASLA ASLB	 Arithmetic Shift Left Arithmetic Shift Left Accumulator A Arithmetic Shift Left Accumulator B	EXT IDX IDX1 IDX2 [D,IDX] [IDX2] INH INH	78 hh ll 68 xb 68 xb ff 68 xb ee ff 68 xb 68 xb ee ff 48 58	rPwO rPw rPwO frPwP fIfrPw fIPrPw O O	rOPw rPw rPOw frPPw fIfrPw fIPrPw O O	– – – –	Δ Δ Δ Δ
ASLD	 Arithmetic Shift Left Double	INH	59	O	O	– – – –	Δ Δ Δ Δ
ASR opr16a ASR oprx0_xysp ASR oprx9,xysp ASR oprx16,xysp ASR [D,xysp] ASR [oprx16,xysp] ASRA ASRB	 Arithmetic Shift Right Arithmetic Shift Right Accumulator A Arithmetic Shift Right Accumulator B	EXT IDX IDX1 IDX2 [D,IDX] [IDX2] INH INH	77 hh ll 67 xb 67 xb ff 67 xb ee ff 67 xb 67 xb ee ff 47 57	rPwO rPw rPwO frPwP fIfrPw fIPrPw O O	rOPw rPw rPOw frPPw fIfrPw fIPrPw O O	– – – –	Δ Δ Δ Δ
BCC rel8	Branch if Carry Clear (if C = 0)	REL	24 rr	PPP/P[1]	PPP/P[1]	– – – –	– – – –
BCLR opr8a, msk8 BCLR opr16a, msk8 BCLR oprx0_xysp, msk8 BCLR oprx9,xysp, msk8 BCLR oprx16,xysp, msk8	(M) • (mm) ⇒ M Clear Bit(s) in Memory	DIR EXT IDX IDX1 IDX2	4D dd mm 1D hh ll mm 0D xb mm 0D xb ff mm 0D xb ee ff mm	rPwO rPwP rPwO rPwP frPwPO	rPOw rPPw rPOw rPwP frPwOP	– – – –	Δ Δ 0 –
BCS rel8	Branch if Carry Set (if C = 1)	REL	25 rr	PPP/P[1]	PPP/P[1]	– – – –	– – – –
BEQ rel8	Branch if Equal (if Z = 1)	REL	27 rr	PPP/P[1]	PPP/P[1]	– – – –	– – – –
BGE rel8	Branch if Greater Than or Equal (if N ⊕ V = 0) (signed)	REL	2C rr	PPP/P[1]	PPP/P[1]	– – – –	– – – –
BGND	Place CPU in Background Mode see CPU12 Reference Manual	INH	00	VfPPP	VfPPP	– – – –	– – – –
BGT rel8	Branch if Greater Than (if Z + (N ⊕ V) = 0) (signed)	REL	2E rr	PPP/P[1]	PPP/P[1]	– – – –	– – – –
BHI rel8	Branch if Higher (if C + Z = 0) (unsigned)	REL	22 rr	PPP/P[1]	PPP/P[1]	– – – –	– – – –
BHS rel8	Branch if Higher or Same (if C = 0) (unsigned) same function as BCC	REL	24 rr	PPP/P[1]	PPP/P[1]	– – – –	– – – –
BITA #opr8i BITA opr8a BITA opr16a BITA oprx0_xysp BITA oprx9,xysp BITA oprx16,xysp BITA [D,xysp] BITA [oprx16,xysp]	(A) • (M) Logical AND A with Memory Does not change Accumulator or Memory	IMM DIR EXT IDX IDX1 IDX2 [D,IDX] [IDX2]	85 ii 95 dd B5 hh ll A5 xb A5 xb ff A5 xb ee ff A5 xb A5 xb ee ff	P rPf rPO rPf rPO frPP fIfrPf fIPrPf	P rfP rOP rfP rPO frPP fIfrfP fIPrfP	– – – –	Δ Δ 0 –
BITB #opr8i BITB opr8a BITB opr16a BITB oprx0_xysp BITB oprx9,xysp BITB oprx16,xysp BITB [D,xysp] BITB [oprx16,xysp]	(B) • (M) Logical AND B with Memory Does not change Accumulator or Memory	IMM DIR EXT IDX IDX1 IDX2 [D,IDX] [IDX2]	C5 ii D5 dd F5 hh ll E5 xb E5 xb ff E5 xb ee ff E5 xb E5 xb ee ff	P rPf rPO rPf rPO frPP fIfrPf fIPrPf	P rfP rOP rfP rPO frPP fIfrfP fIPrfP	– – – –	Δ Δ 0 –
BLE rel8	Branch if Less Than or Equal (if Z + (N ⊕ V) = 1) (signed)	REL	2F rr	PPP/P[1]	PPP/P[1]	– – – –	– – – –
BLO rel8	Branch if Lower (if C = 1) (unsigned) same function as BCS	REL	25 rr	PPP/P[1]	PPP/P[1]	– – – –	– – – –

Note 1. PPP/P indicates this instruction takes three cycles to refill the instruction queue if the branch is taken and one program fetch cycle if the branch is not taken.

Freescale Semiconductor, Inc.

Instruction Set Reference

Table A-1. Instruction Set Summary (Sheet 3 of 14)

Source Form	Operation	Addr. Mode	Machine Coding (hex)	Access Detail HCS12	Access Detail M68HC12	S X H I	N Z V C
BLS rel8	Branch if Lower or Same (if C + Z = 1) (unsigned)	REL	23 rr	PPP/P[1]	PPP/P[1]	– – – –	– – – –
BLT rel8	Branch if Less Than (if N ⊕ V = 1) (signed)	REL	2D rr	PPP/P[1]	PPP/P[1]	– – – –	– – – –
BMI rel8	Branch if Minus (if N = 1)	REL	2B rr	PPP/P[1]	PPP/P[1]	– – – –	– – – –
BNE rel8	Branch if Not Equal (if Z = 0)	REL	26 rr	PPP/P[1]	PPP/P[1]	– – – –	– – – –
BPL rel8	Branch if Plus (if N = 0)	REL	2A rr	PPP/P[1]	PPP/P[1]	– – – –	– – – –
BRA rel8	Branch Always (if 1 = 1)	REL	20 rr	PPP	PPP	– – – –	– – – –
BRCLR opr8a, msk8, rel8 BRCLR opr16a, msk8, rel8 BRCLR oprx0_xysp, msk8, rel8 BRCLR oprx9,xysp, msk8, rel8 BRCLR oprx16,xysp, msk8, rel8	Branch if (M) • (mm) = 0 (if All Selected Bit(s) Clear)	DIR EXT IDX IDX1 IDX2	4F dd mm rr 1F hh ll mm rr 0F xb mm rr 0F xb ff mm rr 0F xb ee ff mm rr	rPPP rfPPP rPPP rfPPP PrfPPP	rPPP rfPPP rPPP rffPPP frPffPPP	– – – –	– – – –
BRN rel8	Branch Never (if 1 = 0)	REL	21 rr	P	P	– – – –	– – – –
BRSET opr8, msk8, rel8 BRSET opr16a, msk8, rel8 BRSET oprx0_xysp, msk8, rel8 BRSET oprx9,xysp, msk8, rel8 BRSET oprx16,xysp, msk8, rel8	Branch if ($\overline{M}$) • (mm) = 0 (if All Selected Bit(s) Set)	DIR EXT IDX IDX1 IDX2	4E dd mm rr 1E hh ll mm rr 0E xb mm rr 0E xb ff mm rr 0E xb ee ff mm rr	rPPP rfPPP rPPP rfPPP PrfPPP	rPPP rfPPP rPPP rffPPP frPffPPP	– – – –	– – – –
BSET opr8, msk8 BSET opr16a, msk8 BSET oprx0_xysp, msk8 BSET oprx9,xysp, msk8 BSET oprx16,xysp, msk8	(M) + (mm) ⇒ M Set Bit(s) in Memory	DIR EXT IDX IDX1 IDX2	4C dd mm 1C hh ll mm 0C xb mm 0C xb ff mm 0C xb ee ff mm	rPwO rPwP rPwO rPwP frPwPO	rPwO rPPw rPOw rPwP frPwOP	– – – –	Δ Δ 0 –
BSR rel8	(SP) – 2 ⇒ SP; RTN$_H$:RTN$_L$ ⇒ M$_{(SP)}$:M$_{(SP+1)}$ Subroutine address ⇒ PC Branch to Subroutine	REL	07 rr	SPPP	PPPS	– – – –	– – – –
BVC rel8	Branch if Overflow Bit Clear (if V = 0)	REL	28 rr	PPP/P[1]	PPP/P[1]	– – – –	– – – –
BVS rel8	Branch if Overflow Bit Set (if V = 1)	REL	29 rr	PPP/P[1]	PPP/P[1]	– – – –	– – – –
CALL opr16a, page CALL oprx0_xysp, page CALL oprx9,xysp, page CALL oprx16,xysp, page CALL [D,xysp] CALL [oprx16, xysp]	(SP) – 2 ⇒ SP; RTN$_H$:RTN$_L$ ⇒ M$_{(SP)}$:M$_{(SP+1)}$ (SP) – 1 ⇒ SP; (PPG) ⇒ M$_{(SP)}$; pg ⇒ PPAGE register; Program address ⇒ PC Call subroutine in extended memory (Program may be located on another expansion memory page.) Indirect modes get program address and new pg value based on pointer.	EXT IDX IDX1 IDX2 [D,IDX] [IDX2]	4A hh ll pg 4B xb pg 4B xb ff pg 4B xb ee ff pg 4B xb 4B xb ee ff	gnSsPPP gnSsPPP gnSsPPP fgnSsPPP fIignSsPPP fIignSsPPP	gnfSsPPP gnfSsPPP gnfSsPPP fgnfSsPPP fIignSsPPP fIignSsPPP	– – – –	– – – –
CBA	(A) – (B) Compare 8-Bit Accumulators	INH	18 17	OO	OO	– – – –	Δ Δ Δ Δ
CLC	0 ⇒ C *Translates to ANDCC #$FE*	IMM	10 FE	P	P	– – – –	– – – 0
CLI	0 ⇒ I *Translates to ANDCC #$EF* (enables I-bit interrupts)	IMM	10 EF	P	P	– – – 0	– – – –
CLR opr16a CLR oprx0_xysp CLR oprx9,xysp CLR oprx16,xysp CLR [D,xysp] CLR [oprx16,xysp] CLRA CLRB	0 ⇒ M Clear Memory Location 0 ⇒ A Clear Accumulator A 0 ⇒ B Clear Accumulator B	EXT IDX IDX1 IDX2 [D,IDX] [IDX2] INH INH	79 hh ll 69 xb 69 xb ff 69 xb ee ff 69 xb 69 xb ee ff 87 C7	PwO Pw PwO PwP PIfw PIPw O O	wOP Pw PwO PwP PIfPw PIPPw O O	– – – –	0 1 0 0
CLV	0 ⇒ V *Translates to ANDCC #$FD*	IMM	10 FD	P	P	– – – –	– – 0 –

Note 1. PPP/P indicates this instruction takes three cycles to refill the instruction queue if the branch is taken and one program fetch cycle if the branch is not taken.

Reprinted with permission of Freescale Semiconductor, Inc.

Freescale Semiconductor, Inc.

Table A-1. Instruction Set Summary (Sheet 4 of 14)

Source Form	Operation	Addr. Mode	Machine Coding (hex)	Access Detail		S X H I	N Z V C
				HCS12	M68HC12		

CMPA #opr8i
CMPA opr8a
CMPA opr16a
CMPA oprx0_xysp
CMPA oprx9,xysp
CMPA oprx16,xysp
CMPA [D,xysp]
CMPA [oprx16,xysp]
	(A) − (M)
	Compare Accumulator A with Memory
		IMM
		DIR
		EXT
		IDX
		IDX1
		IDX2
		[D,IDX]
		[IDX2]
			81 ii
			91 dd
			B1 hh ll
			A1 xb
			A1 xb ff
			A1 xb ee ff
			A1 xb
			A1 xb ee ff
				P
				rPf
				rPO
				rPf
				rPO
				frPP
				fIfrPf
				fIPrPf
					P
					rfP
					rOP
					rfP
					rPO
					frPP
					fIfrfP
					fIPrfP
						– – – –	Δ Δ Δ Δ

CMPB #opr8i
CMPB opr8a
CMPB opr16a
CMPB oprx0_xysp
CMPB oprx9,xysp
CMPB oprx16,xysp
CMPB [D,xysp]
CMPB [oprx16,xysp]
	(B) − (M)
	Compare Accumulator B with Memory
		IMM
		DIR
		EXT
		IDX
		IDX1
		IDX2
		[D,IDX]
		[IDX2]
			C1 ii
			D1 dd
			F1 hh ll
			E1 xb
			E1 xb ff
			E1 xb ee ff
			E1 xb
			E1 xb ee ff
				P
				rPf
				rPO
				rPf
				rPO
				frPP
				fIfrPf
				fIPrPf
					P
					rfP
					rOP
					rfP
					rPO
					frPP
					fIfrfP
					fIPrfP
						– – – –	Δ Δ Δ Δ

COM opr16a COM oprx0_xysp COM oprx9,xysp COM oprx16,xysp COM [D,xysp] COM [oprx16,xysp] COMA COMB	(M̄) ⇒ M equivalent to $FF − (M) ⇒ M$ 1's Complement Memory Location (Ā) ⇒ A Complement Accumulator A (B̄) ⇒ B Complement Accumulator B	EXT IDX IDX1 IDX2 [D,IDX] [IDX2] INH INH	71 hh ll 61 xb 61 xb ff 61 xb ee ff 61 xb 61 xb ee ff 41 51	rOPwO rPw rPwO frPwP fIfrPw fIPrPw O O	rOPwO rPw rPOw frPPw fIfrPw fIPrPw O O	– – – –	Δ Δ 0 1
CPD #opr16i CPD opr8a CPD opr16a CPD oprx0_xysp CPD oprx9,xysp CPD oprx16,xysp CPD [D,xysp] CPD [oprx16,xysp]	(A:B) − (M:M+1) Compare D to Memory (16-Bit)	IMM DIR EXT IDX IDX1 IDX2 [D,IDX] [IDX2]	8C jj kk 9C dd BC hh ll AC xb AC xb ff AC xb ee ff AC xb AC xb ee ff	PO RPf RPO RPf RPO fRPP fIfRPf fIPRPf	OP RfP ROP RfP RPO fRPP fIfRfP fIPRfP	– – – –	Δ Δ Δ Δ
CPS #opr16i CPS opr8a CPS opr16a CPS oprx0_xysp CPS oprx9,xysp CPS oprx16,xysp CPS [D,xysp] CPS [oprx16,xysp]	(SP) − (M:M+1) Compare SP to Memory (16-Bit)	IMM DIR EXT IDX IDX1 IDX2 [D,IDX] [IDX2]	8F jj kk 9F dd BF hh ll AF xb AF xb ff AF xb ee ff AF xb AF xb ee ff	PO RPf RPO RPf RPO fRPP fIfRPf fIPRPf	OP RfP ROP RfP RPO fRPP fIfRfP fIPRfP	– – – –	Δ Δ Δ Δ
CPX #opr16i CPX opr8a CPX opr16a CPX oprx0_xysp CPX oprx9,xysp CPX oprx16,xysp CPX [D,xyap] CPX [oprx16,xysp]	(X) − (M:M+1) Compare X to Memory (16-Bit)	IMM DIR EXT IDX IDX1 IDX2 [D,IDX] [IDX2]	8E jj kk 9E dd BE hh ll AE xb AE xb ff AE xb ee ff AE xb AE xb ee ff	PO RPf RPO RPf RPO fRPP fIfRPf fIPRPf	OP RfP ROP RfP RPO fRPP fIfRfP fIPRfP	– – – –	Δ Δ Δ Δ
CPY #opr16i CPY opr8a CPY opr16a CPY oprx0_xysp CPY oprx9,xysp CPY oprx16,xysp CPY [D,xysp] CPY [oprx16,xysp]	(Y) − (M:M+1) Compare Y to Memory (16-Bit)	IMM DIR EXT IDX IDX1 IDX2 [D,IDX] [IDX2]	8D jj kk 9D dd BD hh ll AD xb AD xb ff AD xb ee ff AD xb AD xb ee ff	PO RPf RPO RPf RPO fRPP fIfRPf fIPRPf	OP RfP ROP RfP RPO fRPP fIfRfP fIPRfP	– – – –	Δ Δ Δ Δ
DAA	Adjust Sum to BCD Decimal Adjust Accumulator A	INH	18 07	OfO	OfO	– – – –	Δ Δ ? Δ
DBEQ abdxys, rel9	(cntr) − 1 ⇒ cntr if (cntr) = 0, then Branch else Continue to next instruction Decrement Counter and Branch if = 0 (cntr = A, B, D, X, Y, or SP)	REL (9-bit)	04 lb rr	PPP (branch) PPO (no branch)	PPP	– – – –	– – – –

Freescale Semiconductor, Inc.

Instruction Set Reference

Table A-1. Instruction Set Summary (Sheet 5 of 14)

Source Form	Operation	Addr. Mode	Machine Coding (hex)	Access Detail HCS12	Access Detail M68HC12	S X H I	N Z V C
DBNE abdxys, rel9	(cntr) − 1 ⇒ cntr If (cntr) not = 0, then Branch; else Continue to next instruction Decrement Counter and Branch if ≠ 0 (cntr = A, B, D, X, Y, or SP)	REL (9-bit)	04 lb rr	PPP (branch) PPO (no branch)	PPP	– – – –	– – – –
DEC opr16a DEC oprx0_xysp DEC oprx9,xysp DEC oprx16,xysp DEC [D,xysp] DEC [oprx16,xysp] DECA DECB	(M) − $01 ⇒ M Decrement Memory Location (A) − $01 ⇒ A Decrement A (B) − $01 ⇒ B Decrement B	EXT IDX IDX1 IDX2 [D,IDX] [IDX2] INH INH	73 hh ll 63 xb 63 xb ff 63 xb ee ff 63 xb 63 xb ee ff 43 53	rPwO rPw rPwO frPwP fIfrPw fIPrPw O O	rOPw rPw rPOw frPPw fIfrPw fIPrPw O O	– – – –	Δ Δ Δ –
DES	(SP) − $0001 ⇒ SP *Translates to* LEAS −1,SP	IDX	1B 9F	Pf	PP[1]	– – – –	– – – –
DEX	(X) − $0001 ⇒ X Decrement Index Register X	INH	09	O	O	– – – –	– Δ – –
DEY	(Y) − $0001 ⇒ Y Decrement Index Register Y	INH	03	O	O	– – – –	– Δ – –
EDIV	(Y:D) ÷ (X) ⇒ Y Remainder ⇒ D 32 by 16 Bit ⇒ 16 Bit Divide (unsigned)	INH	11	ffffffffffO	ffffffffffO	– – – –	Δ Δ Δ Δ
EDIVS	(Y:D) ÷ (X) ⇒ Y Remainder ⇒ D 32 by 16 Bit ⇒ 16 Bit Divide (signed)	INH	18 14	OffffffffffO	OffffffffffO	– – – –	Δ Δ Δ Δ
EMACS opr16a [2]	(M(X):M(X+1)) × (M(Y):M(Y+1)) + (M~M+3) ⇒ M~M+3 16 by 16 Bit ⇒ 32 Bit Multiply and Accumulate (signed)	Special	18 12 hh ll	ORROfffRRfWWP	ORROfffRRfWWP	– – – –	Δ Δ Δ Δ
EMAXD oprx0_xysp EMAXD oprx9,xysp EMAXD oprx16,xysp EMAXD [D,xysp] EMAXD [oprx16,xysp]	MAX((D), (M:M+1)) ⇒ D MAX of 2 Unsigned 16-Bit Values N, Z, V and C status bits reflect result of internal compare ((D) − (M:M+1))	IDX IDX1 IDX2 [D,IDX] [IDX2]	18 1A xb 18 1A xb ff 18 1A xb ee ff 18 1A xb 18 1A xb ee ff	ORPf ORPO OfRPP OfIfRPf OfIPRPf	ORfP ORPO OfRPP OfIfRPf OfIPRfP	– – – –	Δ Δ Δ Δ
EMAXM oprx0_xysp EMAXM oprx9,xysp EMAXM oprx16,xysp EMAXM [D,xysp] EMAXM [oprx16,xysp]	MAX((D), (M:M+1)) ⇒ M:M+1 MAX of 2 Unsigned 16-Bit Values N, Z, V and C status bits reflect result of internal compare ((D) − (M:M+1))	IDX IDX1 IDX2 [D,IDX] [IDX2]	18 1E xb 18 1E xb ff 18 1E xb ee ff 18 1E xb 18 1E xb ee ff	ORPW ORPWO OfRPWP OfIfRPW OfIPRPW	ORPW ORPWO OfRPWP OfIfRPW OfIPRPW	– – – –	Δ Δ Δ Δ
EMIND oprx0_xysp EMIND oprx9,xysp EMIND oprx16,xysp EMIND [D,xysp] EMIND [oprx16,xysp]	MIN((D), (M:M+1)) ⇒ D MIN of 2 Unsigned 16-Bit Values N, Z, V and C status bits reflect result of internal compare ((D) − (M:M+1))	IDX IDX1 IDX2 [D,IDX] [IDX2]	18 1B xb 18 1B xb ff 18 1B xb ee ff 18 1B xb 18 1B xb ee ff	ORPf ORPO OfRPP OfIfRPf OfIPRPf	ORfP ORPO OfRPP OfIfRfP OfIPRfP	– – – –	Δ Δ Δ Δ
EMINM oprx0_xysp EMINM oprx9,xysp EMINM oprx16,xysp EMINM [D,xysp] EMINM [oprx16,xysp]	MIN((D), (M:M+1)) ⇒ M:M+1 MIN of 2 Unsigned 16-Bit Values N, Z, V and C status bits reflect result of internal compare ((D) − (M:M+1))	IDX IDX1 IDX2 [D,IDX] [IDX2]	18 1F xb 18 1F xb ff 18 1F xb ee ff 18 1F xb 18 1F xb ee ff	ORPW ORPWO OfRPWP OfIfRPW OfIPRPW	ORPW ORPWO OfRPWP OfIfRPW OfIPRPW	– – – –	Δ Δ Δ Δ
EMUL	(D) × (Y) ⇒ Y:D 16 by 16 Bit Multiply (unsigned)	INH	13	ffO	ffO	– – – –	Δ Δ – Δ
EMULS	(D) × (Y) ⇒ Y:D 16 by 16 Bit Multiply (signed)	INH	18 13	OfO (if followed by page 2 instruction) OffO	OfO OfO	– – – –	Δ Δ – Δ
EORA #opr8i EORA opr8a EORA opr16a EORA oprx0_xysp EORA oprx9,xysp EORA oprx16,xysp EORA [D,xysp] EORA [oprx16,xysp]	(A) ⊕ (M) ⇒ A Exclusive-OR A with Memory	IMM DIR EXT IDX IDX1 IDX2 [D,IDX] [IDX2]	88 ii 98 dd B8 hh ll A8 xb A8 xb ff A8 xb ee ff A8 xb A8 xb ee ff	P rPf rPO rPf rPO frPP fIfrPf fIPrPf	P rfP rOP rfP rPO frPP fIfrfP fIPrfP	– – – –	Δ Δ 0 –

Notes:
1. Due to internal CPU requirements, the program word fetch is performed twice to the same address during this instruction.
2. opr16a is an extended address specification. Both X and Y point to source operands.

For More Information On This Product,
Go to: www.freescale.com

Freescale Semiconductor, Inc.

Table A-1. Instruction Set Summary (Sheet 6 of 14)

Source Form	Operation	Addr. Mode	Machine Coding (hex)	Access Detail HCS12	M68HC12	S X H I	N Z V C
EORB #opr8i EORB opr8a EORB opr16a EORB oprx0_xysp EORB oprx9,xysp EORB oprx16,xyep EORB [D,xysp] EORB [oprx16,xysp]	(B) ⊕ (M) ⇒ B Exclusive-OR B with Memory	IMM DIR EXT IDX IDX1 IDX2 [D,IDX] [IDX2]	C8 ii D8 dd F8 hh ll E8 xb E8 xb ff E8 xb ee ff E8 xb E8 xb ee ff	P rPf rPO rPf rPO frPP fIfrPf fIPrPf	P rfP rOP rfP rPO frPP fIfrfP fIPrfP	– – – –	Δ Δ 0 –
ETBL oprx0_xysp	(M:M+1)+ [(B)×((M+2:M+3) – (M:M+1))] ⇒ D 16-Bit Table Lookup and Interpolate Initialize B, and index before ETBL. <ea> points at first table entry (M:M+1) and B is fractional part of lookup value (no indirect addr. modes or extensions allowed)	IDX	18 3F xb	ORRfffffffP	ORRfffffffP	– – – – C Bit is undefined in HC12	Δ Δ – Δ ?
EXG abcdxys,abcdxys	(r1) ⇔ (r2) (if r1 and r2 same size) or $00:(r1) ⇒ r2 (if r1=8-bit; r2=16-bit) or (r1$_{low}$) ⇔ (r2) (if r1=16-bit; r2=8-bit) r1 and r2 may be A, B, CCR, D, X, Y, or SP	INH	B7 eb	P	P	– – – –	– – – –
FDIV	(D) ÷ (X) ⇒ X: Remainder ⇒ D 16 by 16 Bit Fractional Divide	INH	18 11	OfffffffffO	OfffffffffO	– – – –	– Δ Δ Δ
IBEQ abdxys, rel9	(cntr) + 1 ⇒ cntr If (cntr) = 0, then Branch else Continue to next instruction Increment Counter and Branch if = 0 (cntr = A, B, D, X, Y, or SP)	REL (9-bit)	04 1b rr	PPP (branch) PPO (no branch)	PPP	– – – –	– – – –
IBNE abdxys, rel9	(cntr) + 1 ⇒ cntr if (cntr) not = 0, then Branch; else Continue to next instruction Increment Counter and Branch if ≠ 0 (cntr = A, B, D, X, Y, or SP)	REL (9-bit)	04 1b rr	PPP (branch) PPO (no branch)	PPP	– – – –	– – – –
IDIV	(D) ÷ (X) ⇒ X; Remainder ⇒ D 16 by 16 Bit Integer Divide (unsigned)	INH	18 10	OfffffffffO	OfffffffffO	– – – –	– Δ 0 Δ
IDIVS	(D) ÷ (X) ⇒ X; Remainder ⇒ D 16 by 16 Bit Integer Divide (signed)	INH	18 15	OfffffffffO	OfffffffffO	– – – –	Δ Δ Δ Δ
INC opr16a INC oprx0_xysp INC oprx9,xysp INC oprx16,xysp INC [D,xysp] INC [oprx16,xysp] INCA INCB	(M) + $01 ⇒ M Increment Memory Byte (A) + $01 ⇒ A Increment Acc. A (B) + $01 ⇒ B Increment Acc. B	EXT IDX IDX1 IDX2 [D,IDX] [IDX2] INH INH	72 hh ll 62 xb 62 xb ff 62 xb ee ff 62 xb 62 xb ee ff 42 52	rPwO rPw rPwO frPwP fIfrPw fIPrPw O O	rOPw rPw rPOw frPwP fIfrPw fIPrPw O O	– – – –	Δ Δ Δ –
INS	(SP) + $0001 ⇒ SP Translates to LEAS 1,SP	IDX	1B 81	Pf	PP[1]	– – – –	– – – –
INX	(X) + $0001 ⇒ X Increment Index Register X	INH	08	O	O	– – – –	– Δ – –
INY	(Y) + $0001 ⇒ Y Increment Index Register Y	INH	02	O	O	– – – –	– Δ – –
JMP opr16a JMP oprx0_xysp JMP oprx9,xysp JMP oprx16,xysp JMP [D,xysp] JMP [oprx16,xysp]	Routine address ⇒ PC Jump	EXT IDX IDX1 IDX2 [D,IDX] [IDX2]	06 hh ll 05 xb 05 xb ff 05 xb ee ff 05 xb 05 xb ee ff	PPP PPP PPP fPPP fIfPPP fIfPPP	PPP PPP PPP fPPP fIfPPP fIfPPP	– – – –	– – – –

Note 1. Due to internal CPU requirements, the program word fetch is performed twice to the same address during this instruction.

Freescale Semiconductor, Inc.

Instruction Set Reference

Table A-1. Instruction Set Summary (Sheet 7 of 14)

Source Form	Operation	Addr. Mode	Machine Coding (hex)	Access Detail HCS12	Access Detail M68HC12	S X H I	N Z V C
JSR opr8a JSR opr16a JSR oprx0_xysp JSR oprx9,xysp JSR oprx16,xysp JSR [D,xysp] JSR [oprx16,xysp]	(SP) − 2 ⇒ SP; RTN$_H$:RTN$_L$ ⇒ M$_{(SP)}$:M$_{(SP+1)}$; Subroutine address ⇒ PC Jump to Subroutine	DIR EXT IDX IDX1 IDX2 [D,IDX] [IDX2]	17 dd 16 hh ll 15 xb 15 xb ff 15 xb ee ff 15 xb 15 xb ee ff	SPPP SPPP PPPS PPPS fPPPS fIfPPPS fIfPPPS	PPPS PPPS PPPS PPPS fPPPS fIfPPPS fIfPPPS	- - - -	- - - -
LBCC rel16	Long Branch if Carry Clear (if C = 0)	REL	18 24 qq rr	OPPP/OPO[1]	OPPP/OPO[1]	- - - -	- - - -
LBCS rel16	Long Branch if Carry Set (if C = 1)	REL	18 25 qq rr	OPPP/OPO[1]	OPPP/OPO[1]	- - - -	- - - -
LBEQ rel16	Long Branch if Equal (if Z = 1)	REL	18 27 qq rr	OPPP/OPO[1]	OPPP/OPO[1]	- - - -	- - - -
LBGE rel16	Long Branch Greater Than or Equal (if N ⊕ V = 0) (signed)	REL	18 2C qq rr	OPPP/OPO[1]	OPPP/OPO[1]	- - - -	- - - -
LBGT rel16	Long Branch if Greater Than (if Z + (N ⊕ V) = 0) (signed)	REL	18 2E qq rr	OPPP/OPO[1]	OPPP/OPO[1]	- - - -	- - - -
LBHI rel16	Long Branch if Higher (if C + Z = 0) (unsigned)	REL	18 22 qq rr	OPPP/OPO[1]	OPPP/OPO[1]	- - - -	- - - -
LBHS rel16	Long Branch if Higher or Same (if C = 0) (unsigned) same function as LBCC	REL	18 24 qq rr	OPPP/OPO[1]	OPPP/OPO[1]	- - - -	- - - -
LBLE rel16	Long Branch if Less Than or Equal (if Z + (N ⊕ V) = 1) (signed)	REL	18 2F qq rr	OPPP/OPO[1]	OPPP/OPO[1]	- - - -	- - - -
LBLO rel16	Long Branch if Lower (if C = 1) (unsigned) same function as LBCS	REL	18 25 qq rr	OPPP/OPO[1]	OPPP/OPO[1]	- - - -	- - - -
LBLS rel16	Long Branch if Lower or Same (if C + Z = 1) (unsigned)	REL	18 23 qq rr	OPPP/OPO[1]	OPPP/OPO[1]	- - - -	- - - -
LBLT rel16	Long Branch if Less Than (if N ⊕ V = 1) (signed)	REL	18 2D qq rr	OPPP/OPO[1]	OPPP/OPO[1]	- - - -	- - - -
LBMI rel16	Long Branch if Minus (if N = 1)	REL	18 2B qq rr	OPPP/OPO[1]	OPPP/OPO[1]	- - - -	- - - -
LBNE rel16	Long Branch if Not Equal (if Z = 0)	REL	18 26 qq rr	OPPP/OPO[1]	OPPP/OPO[1]	- - - -	- - - -
LBPL rel16	Long Branch if Plus (if N = 0)	REL	18 2A qq rr	OPPP/OPO[1]	OPPP/OPO[1]	- - - -	- - - -
LBRA rel16	Long Branch Always (if 1=1)	REL	18 20 qq rr	OPPP	OPPP	- - - -	- - - -
LBRN rel16	Long Branch Never (if 1 = 0)	REL	18 21 qq rr	OPO	OPO	- - - -	- - - -
LBVC rel16	Long Branch if Overflow Bit Clear (if V=0)	REL	18 28 qq rr	OPPP/OPO[1]	OPPP/OPO[1]	- - - -	- - - -
LBVS rel16	Long Branch if Overflow Bit Set (if V = 1)	REL	18 29 qq rr	OPPP/OPO[1]	OPPP/OPO[1]	- - - -	- - - -

LDAA #opr8i
LDAA opr8a
LDAA opr16a
LDAA oprx0_xysp
LDAA oprx9,xysp
LDAA oprx16,xysp
LDAA [D,xysp]
LDAA [oprx16,xysp]
	(M) ⇒ A
	Load Accumulator A
		IMM
		DIR
		EXT
		IDX
		IDX1
		IDX2
		[D,IDX]
		[IDX2]
			86 ii
			96 dd
			B6 hh ll
			A6 xb
			A6 xb ff
			A6 xb ee ff
			A6 xb
			A6 xb ee ff
				P
				rPf
				rPO
				rPf
				rPO
				frPP
				fIfrPf
				fIPrPf
					P
					rPf
					rOP
					rfP
					rPO
					frPP
					fIfrfP
					fIPrfP
						- - - -	Δ Δ 0 -

LDAB #opr8i
LDAB opr8a
LDAB opr16a
LDAB oprx0_xysp
LDAB oprx9,xysp
LDAB oprx16,xysp
LDAB [D,xysp]
LDAB [oprx16,xysp]
	(M) ⇒ B
	Load Accumulator B
		IMM
		DIR
		EXT
		IDX
		IDX1
		IDX2
		[D,IDX]
		[IDX2]
			C6 ii
			D6 dd
			F6 hh ll
			E6 xb
			E6 xb ff
			E6 xb ee ff
			E6 xb
			E6 xb ee ff
				P
				rPf
				rPO
				rPf
				rPO
				frPP
				fIfrPf
				fIPrPf
					P
					rfP
					rOP
					rfP
					rPO
					frPP
					fIfrfP
					fIPrfP
						- - - -	Δ Δ 0 -

LDD #opr16i
LDD opr8a
LDD opr16a
LDD oprx0_xysp
LDD oprx9,xysp
LDD oprx16,xysp
LDD [D,xysp]
LDD [oprx16,xysp]
	(M:M+1) ⇒ A:B
	Load Double Accumulator D (A:B)
		IMM
		DIR
		EXT
		IDX
		IDX1
		IDX2
		[D,IDX]
		[IDX2]
			CC jj kk
			DC dd
			FC hh ll
			EC xb
			EC xb ff
			EC xb ee ff
			EC xb
			EC xb ee ff
				PO
				RPf
				RPO
				RPf
				RPO
				fRPP
				fIfRPf
				fIPRPf
					OP
					RfP
					ROP
					RfP
					RPO
					fRPP
					fIfRfP
					fIPRfP
						- - - -	Δ Δ 0 -

Note 1. OPPP/OPO indicates this instruction takes four cycles to refill the instruction queue if the branch is taken and three cycles if the branch is not taken.

Freescale Semiconductor, Inc.

Table A-1. Instruction Set Summary (Sheet 8 of 14)

Source Form	Operation	Addr. Mode	Machine Coding (hex)	Access Detail HCS12	M68HC12	SXHI	NZVC

LDS #opr16i
LDS opr8a
LDS opr16a
LDS oprx0_xysp
LDS oprx9_xysp
LDS oprx16,xysp
LDS [D,xysp]
LDS [oprx16,xysp]
	(M:M+1) ⇒ SP
	Load Stack Pointer
		IMM
		DIR
		EXT
		IDX
		IDX1
		IDX2
		[D,IDX]
		[IDX2]
			CF jj kk
			DF dd
			FF hh ll
			EF xb
			EF xb ff
			EF xb ee ff
			EF xb
			EF xb ee ff
				PO
				RPf
				RPO
				RPf
				RPO
				fRPP
				fIfRPf
				fIPRPf
					OP
					RfP
					ROP
					RfP
					RPO
					fRPP
					fIfRfP
					fIPRfP
						– – – –	Δ Δ 0 –

LDX #opr16i
LDX opr8a
LDX opr16a
LDX oprx0_xysp
LDX oprx9_xysp
LDX oprx16,xysp
LDX [D,IDX]
LDX [oprx16,xysp]
	(M:M+1) ⇒ X
	Load Index Register X
		IMM
		DIR
		EXT
		IDX
		IDX1
		IDX2
		[D,IDX]
		[IDX2]
			CE jj kk
			DE dd
			FE hh ll
			EE xb
			EE xb ff
			EE xb ee ff
			EE xb
			EE xb ee ff
				PO
				RPf
				RPO
				RPf
				RPO
				fRPP
				fIfRPf
				fIPRPf
					OP
					RfP
					ROP
					RfP
					RPO
					fRPP
					fIfRfP
					fIPRfP
						– – – –	Δ Δ 0 –

LDY #opr16i
LDY opr8a
LDY opr16a
LDY oprx0_xysp
LDY oprx9_xysp
LDY oprx16,xysp
LDY [D,xysp]
LDY [oprx16,xysp]
	(M:M+1) ⇒ Y
	Load Index Register Y
		IMM
		DIR
		EXT
		IDX
		IDX1
		IDX2
		[D,IDX]
		[IDX2]
			CD jj kk
			DD dd
			FD hh ll
			ED xb
			ED xb ff
			ED xb ee ff
			ED xb
			ED xb ee ff
				PO
				RPf
				RPO
				RPf
				RPO
				fRPP
				fIfRPf
				fIPRPf
					OP
					RfP
					ROP
					RfP
					RPO
					fRPP
					fIfRfP
					fIPRfP
						– – – –	Δ Δ 0 –

LEAS oprx0_xysp LEAS oprx9_xysp LEAS oprx16,xysp	Effective Address ⇒ SP Load Effective Address into SP	IDX IDX1 IDX2	1B xb 1B xb ff 1B xb ee ff	Pf PO PP	PP[1] PO PP	– – – –	– – – –
LEAX oprx0_xysp LEAX oprx9_xysp LEAX oprx16,xysp	Effective Address ⇒ X Load Effective Address into X	IDX IDX1 IDX2	1A xb 1A xb ff 1A xb ee ff	Pf PO PP	PP[1] PO PP	– – – –	– – – –
LEAY oprx0_xysp LEAY oprx9_xysp LEAY oprx16,xysp	Effective Address ⇒ Y Load Effective Address into Y	IDX IDX1 IDX2	19 xb 19 xb ff 19 xb ee ff	Pf PO PP	PP[1] PO PP	– – – –	– – – –
LSL opr16a LSL oprx0_xysp LSL oprx9_xysp LSL oprx16,xysp LSL [D,xysp] LSL [oprx16,xysp] LSLA LSLB	[diagram] C ⟵ b7...b0 ⟵ 0 Logical Shift Left same function as ASL Logical Shift Accumulator A to Left Logical Shift Accumulator B to Left	EXT IDX IDX1 IDX2 [D,IDX] [IDX2] INH INH	78 hh ll 68 xb 68 xb ff 68 xb ee ff 68 xb 68 xb ee ff 48 58	rPwO rPw rPwO frPPw fIfrPw fIPrPw O O	rOPw rPw rPOw frPPw fIfrPw fIPrPw O O	– – – –	Δ Δ Δ Δ
LSLD	[diagram] C ⟵ b7 A b0 ⟵ b7 B b0 ⟵ 0 Logical Shift Left D Accumulator same function as ASLD	INH	59	O	O	– – – –	Δ Δ Δ Δ
LSR opr16a LSR oprx0_xysp LSR oprx9_xysp LSR oprx16,xysp LSR [D,xysp] LSR [oprx16,xysp] LSRA LSRB	[diagram] 0 ⟶ b7...b0 ⟶ C Logical Shift Right Logical Shift Accumulator A to Right Logical Shift Accumulator B to Right	EXT IDX IDX1 IDX2 [D,IDX] [IDX2] INH INH	74 hh ll 64 xb 64 xb ff 64 xb ee ff 64 xb 64 xb ee ff 44 54	rPwO rPw rPwO frPwP fIfrPw fIPrPw O O	rOPw rPw rPOw frPPw fIfrPw fIPrPw O O	– – – –	0 Δ Δ Δ
LSRD	[diagram] 0 ⟶ b7 A b0 ⟶ b7 B b0 ⟶ C Logical Shift Right D Accumulator	INH	49	O	O	– – – –	0 Δ Δ Δ
MAXA oprx0_xysp MAXA oprx9_xysp MAXA oprx16,xysp MAXA [D,IDX] MAXA [oprx16,xysp]	MAX((A), (M)) ⇒ A MAX of 2 Unsigned 8-Bit Values N, Z, V and C status bits reflect result of internal compare ((A) – (M)).	IDX IDX1 IDX2 [D,IDX] [IDX2]	18 18 xb 18 18 xb ff 18 18 xb ee ff 18 18 xb 18 18 xb ee ff	OrPf OrPO OfrPP OfIfrPf OfIPrPf	OrfP OrPO OfrPP OfIfrfP OfIPrfP	– – – –	Δ Δ Δ Δ

Note 1. Due to internal CPU requirements, the program word fetch is performed twice to the same address during this instruction.

For More Information On This Product,
Go to: www.freescale.com

Reprinted with permission of Freescale Semiconductor, Inc.

Freescale Semiconductor, Inc.

Table A-1. Instruction Set Summary (Sheet 9 of 14)

Source Form	Operation	Addr. Mode	Machine Coding (hex)	Access Detail HCS12	M68HC12	S X H I	N Z V C
MAXM oprx0_xysp MAXM oprx9_xysp MAXM oprx16,xysp MAXM [D,xysp] MAXM [oprx16,xysp]	MAX((A), (M)) ⇒ M MAX of 2 Unsigned 8-Bit Values N, Z, V and C status bits reflect result of internal compare ((A) – (M)).	IDX IDX1 IDX2 [D,IDX] [IDX2]	18 1C xb 18 1C xb ff 18 1C xb ee ff 18 1C xb 18 1C xb ee ff	OrPw OrPwO OfrPwP OfIfrPw OfIPrPw	OrPw OrPwO OfrPwP OfIfrPw OfIPrPw	– – – –	Δ Δ Δ Δ
MEM	μ (grade) ⇒ M_Yγ; (X) + 4 ⇒ X; (Y) + 1 ⇒ Y; A unchanged if (A) < P1 or (A) > P2 then μ = 0, else μ = MIN[((A – P1)×S1, (P2 – (A))×S2, \$FF] where: A = current crisp input value; X points at 4-byte data structure that describes a trapezoidal membership function (P1, P2, S1, S2); Y points at fuzzy input (RAM location). See CPU12 Reference Manual for special cases.	Special	01	RRfOw	RRfOw	– – ? –	? ? ? ?
MINA oprx0_xysp MINA oprx9_xysp MINA oprx16,xysp MINA [D,xysp] MINA [oprx16,xysp]	MIN((A), (M)) ⇒ A MIN of 2 Unsigned 8-Bit Values N, Z, V and C status bits reflect result of internal compare ((A) – (M)).	IDX IDX1 IDX2 [D,xysp] [IDX2]	18 19 xb 18 19 xb ff 18 19 xb ee ff 18 19 xb 18 19 xb ee ff	OrPf OrPO OfrPP OfIfrPf OfIPrPf	OrPf OrPO OfrPP OfIfrPf OfIPrPf	– – – –	Δ Δ Δ Δ
MINM oprx0_xysp MINM oprx9_xysp MINM oprx16,xysp MINM [D,xysp] MINM [oprx16,xysp]	MIN((A), (M)) ⇒ M MIN of 2 Unsigned 8-Bit Values N, Z, V and C status bits reflect result of internal compare ((A) – (M)).	IDX IDX1 IDX2 [D,IDX] [IDX2]	18 1D xb 18 1D xb ff 18 1D xb ee ff 18 1D xb 18 1D xb ee ff	OrPw OrPwO OfrPwP OfIfrPw OfIPrPw	OrPw OrPwO OfrPwP OfIfrPw OfIPrPw	– – – –	Δ Δ Δ Δ

MOVB #opr8, opr16a[1]
MOVB #opr8i, oprx0_xysp[1]
MOVB opr16a, oprx0_xysp[1]
MOVB opr16a, oprx0_xysp[1]
MOVB oprx0_xysp, opr16a[1]
MOVB oprx0_xysp, oprx0_xysp[1]
	(M₁) ⇒ M₂
	Memory to Memory Byte-Move (8-Bit)
		IMM-EXT
		IMM-IDX
		EXT-EXT
		EXT-IDX
		IDX-EXT
		IDX-IDX
			18 0B ii hh 11
			18 08 xb ii
			18 0C hh 11 hh 11
			18 09 xb hh 11
			18 0D xb hh 11
			18 0A xb xb
				OPwP
				OPwO
				OrPwPO
				OrPw
				OrPwP
				OrPwO
					OPwP
					OPwO
					OrPwPO
					OPrPw
					OrPwP
					OrPwO
						– – – –	– – – –

MOVW #oprx16, opr16a[1]
MOVW #opr16i, oprx0_xysp[1]
MOVW opr16a, opr16a[1]
MOVW opr16a, oprx0_xysp[1]
MOVW oprx0_xysp, opr16a[1]
MOVW oprx0_xysp, oprx0_xysp[1]
	(M:M+1₁) ⇒ M:M+1₂
	Memory to Memory Word-Move (16-Bit)
		IMM-EXT
		IMM-IDX
		EXT-EXT
		EXT-IDX
		IDX-EXT
		IDX-IDX
			18 03 jj kk hh 11
			18 00 xb jj kk
			18 04 hh 11 hh 11
			18 01 xb hh 11
			18 05 xb hh 11
			18 02 xb xb
				OPWPO
				OPPW
				ORPWPO
				ORPW
				ORPWP
				ORPWO
					OPWPO
					OPPW
					ORPWPO
					OPRPW
					ORPWP
					ORPWO
						– – – –	– – – –

MUL	(A) × (B) ⇒ A:B 8 by 8 Unsigned Multiply	INH	12	O	ffO	– – – –	– – – Δ
NEG opr16a NEG oprx0_xysp NEG oprx9,xysp NEG oprx16,xysp NEG [D,xysp] NEG [oprx16,xysp]	0 – (M) ⇒ M equivalent to ($\overline{M}$) + 1 ⇒ M Two's Complement Negate	EXT IDX IDX1 IDX2 [D,IDX] [IDX2]	70 hh 11 60 xb 60 xb ff 60 xb ee ff 60 xb 60 xb ee ff	rPwO rPw rPwO frPwP fIfrPw fIPrPw	rOPw rPw rPOw frPPw fIfrPw fIPrPw	– – – –	Δ Δ Δ Δ
NEGA	0 – (A) ⇒ A equivalent to ($\overline{A}$) + 1 ⇒ A Negate Accumulator A	INH	40	O	O		
NEGB	0 – (B) ⇒ B equivalent to ($\overline{B}$) + 1 ⇒ B Negate Accumulator B	INH	50	O	O		
NOP	No Operation	INH	A7	O	O	– – – –	– – – –
ORAA #opr8i ORAA opr8a ORAA opr16a ORAA oprx0_xysp ORAA oprx9,xysp ORAA oprx16,xysp ORAA [D,xysp] ORAA [oprx16,xysp]	(A) + (M) ⇒ A Logical OR A with Memory	IMM DIR EXT IDX IDX1 IDX2 [D,IDX] [IDX2]	8A ii 9A dd BA hh 11 AA xb AA xb ff AA xb ee ff AA xb AA xb ee ff	P rPf rPO rPf rPO frPP fIfrPf fIPrPf	P rfP rOP rfP rPO frPP fIfrfP fIPrfP	– – – –	Δ Δ 0 –

Note 1. The first operand in the source code statement specifies the source for the move.

Freescale Semiconductor, Inc.

Freescale Semiconductor, Inc.

Table A-1. Instruction Set Summary (Sheet 10 of 14)

Source Form	Operation	Addr. Mode	Machine Coding (hex)	Access Detail HCS12	Access Detail M68HC12	S X H I	N Z V C
ORAB #opr8i ORAB opr8a ORAB opr16a ORAB oprx0,xysp ORAB oprx9,xysp ORAB oprx16,xysp ORAB [D,xysp] ORAB [oprx16,xysp]	(B) + (M) ⇒ B Logical OR B with Memory	IMM DIR EXT IDX IDX1 IDX2 [D,IDX] [IDX2]	CA ii DA dd FA hh ll EA xb EA xb ff EA xb ee ff EA xb EA xb ee ff	P rPf rPO rPf rPO frPP fIfrPf fIPrPf	P rfP rOP rfP rPO frPP fIfrfP fIPrfP	– – – –	Δ Δ 0 –
ORCC #opr8i	(CCR) + M ⇒ CCR Logical OR CCR with Memory	IMM	14 ii	P	P	⇑ – ⇑ ⇑	⇑ ⇑ ⇑ ⇑
PSHA	(SP) – 1 ⇒ SP; (A) ⇒ M(SP) Push Accumulator A onto Stack	INH	36	Os	Os	– – – –	– – – –
PSHB	(SP) – 1 ⇒ SP; (B) ⇒ M(SP) Push Accumulator B onto Stack	INH	37	Os	Os	– – – –	– – – –
PSHC	(SP) – 1 ⇒ SP; (CCR) ⇒ M(SP) Push CCR onto Stack	INH	39	Os	Os	– – – –	– – – –
PSHD	(SP) – 2 ⇒ SP; (A:B) ⇒ M(SP):M(SP+1) Push D Accumulator onto Stack	INH	3B	OS	OS	– – – –	– – – –
PSHX	(SP) – 2 ⇒ SP; (XH:XL) ⇒ M(SP):M(SP+1) Push Index Register X onto Stack	INH	34	OS	OS	– – – –	– – – –
PSHY	(SP) – 2 ⇒ SP; (YH:YL) ⇒ M(SP):M(SP+1) Push Index Register Y onto Stack	INH	35	OS	OS	– – – –	– – – –
PULA	(M(SP)) ⇒ A; (SP) + 1 ⇒ SP Pull Accumulator A from Stack	INH	32	ufO	ufO	– – – –	– – – –
PULB	(M(SP)) ⇒ B; (SP) + 1 ⇒ SP Pull Accumulator B from Stack	INH	33	ufO	ufO	– – – –	– – – –
PULC	(M(SP)) ⇒ CCR; (SP) + 1 ⇒ SP Pull CCR from Stack	INH	38	ufO	ufO	Δ ⇓ Δ Δ	Δ Δ Δ Δ
PULD	(M(SP):M(SP+1)) ⇒ A:B; (SP) + 2 ⇒ SP Pull D from Stack	INH	3A	UfO	UfO	– – – –	– – – –
PULX	(M(SP):M(SP+1)) ⇒ XH:XL; (SP) + 2 ⇒ SP Pull Index Register X from Stack	INH	30	UfO	UfO	– – – –	– – – –
PULY	(M(SP):M(SP+1)) ⇒ YH:YL; (SP) + 2 ⇒ SP Pull Index Register Y from Stack	INH	31	UfO	UfO	– – – –	– – – –
REV	MIN-MAX rule evaluation Find smallest rule input (MIN). Store to rule outputs unless fuzzy output is already larger (MAX). For rule weights see REVW. Each rule input is an 8-bit offset from the base address in Y. Each rule output is an 8-bit offset from the base address in Y. $FE separates rule inputs from rule outputs. $FF terminates the rule list. REV may be interrupted.	Special	18 3A	Orf(t,tx)O (exit + re-entry replaces comma above if interrupted) ff + Orf(t,	Orf(t,tx)O ff + Orf(t,	– – ? –	? ? Δ ?
REVW	MIN-MAX rule evaluation, Find smallest rule input (MIN), Store to rule outputs unless fuzzy output is already larger (MAX). Rule weights supported, optional. Each rule input is the 16-bit address of a fuzzy input. Each rule output is the 16-bit address of a fuzzy output. The value $FFFE separates rule inputs from rule outputs. $FFFF terminates the rule list. REVW may be interrupted.	Special	18 3B	ORf(t,Tx)O (loop to read weight if enabled) (r,RfRf) (exit + re-entry replaces comma above if interrupted) ffff + ORf(t,	ORf(t,Tx)O (r,RfRf) fff + ORf(t,	– – ? –	? ? Δ !

Freescale Semiconductor, Inc.

Instruction Set Reference

Table A-1. Instruction Set Summary (Sheet 11 of 14)

Source Form	Operation	Addr. Mode	Machine Coding (hex)	Access Detail HCS12	M68HC12	S X H I	N Z V C
ROL opr16a ROL oprx0_xysp ROL oprx9,xysp ROL oprx16,xysp ROL [D,xysp] ROL [oprx16,xysp] ROLA ROLB	 Rotate Memory Left through Carry Rotate A Left through Carry Rotate B Left through Carry	EXT IDX IDX1 IDX2 [D,IDX] [IDX2] INH INH	75 hh ll 65 xb 65 xb ff 65 xb ee ff 65 xb 65 xb ee ff 45 55	rPwO rPw rPwO frPwP fIfrPw fIPrPw O O	rOPw rPw rPOw frPPw fIfrPw fIPrPw O O	– – – –	Δ Δ Δ Δ
ROR opr16a ROR oprx0_xysp ROR oprx9,xysp ROR oprx16,xysp ROR [D,xysp] ROR [oprx16,xysp] RORA RORB	 Rotate Memory Right through Carry Rotate A Right through Carry Rotate B Right through Carry	EXT IDX IDX1 IDX2 [D,IDX] [IDX2] INH INH	76 hh ll 66 xb 66 xb ff 66 xb ee ff 66 xb 66 xb ee ff 46 56	rPwO rPw rPwO frPwP fIPrPw fIPrPw O O	rOPw rPw rPOw frPPw fIfrPw fIPrPw O O	– – – –	Δ Δ Δ Δ
RTC	$(M_{(SP)}) \Rightarrow PPAGE; (SP) + 1 \Rightarrow SP;$ $(M_{(SP)}:M_{(SP+1)}) \Rightarrow PC_H:PC_L;$ $(SP) + 2 \Rightarrow SP$ Return from Call	INH	0A	uUnfPPP	uUnPPP	– – – –	– – – –
RTI	$(M_{(SP)}) \Rightarrow CCR; (SP) + 1 \Rightarrow SP$ $(M_{(SP)}:M_{(SP+1)}) \Rightarrow B:A; (SP) + 2 \Rightarrow SP$ $(M_{(SP)}:M_{(SP+1)}) \Rightarrow X_H:X_L; (SP) + 4 \Rightarrow SP$ $(M_{(SP)}:M_{(SP+1)}) \Rightarrow PC_H:PC_L; (SP) - 2 \Rightarrow SP$ $(M_{(SP)}:M_{(SP+1)}) \Rightarrow Y_H:Y_L; (SP) + 4 \Rightarrow SP$ Return from Interrupt	INH	0B	uUUUUPPP (with interrupt pending) uUUUVfPPP	uUUUUPPP uUUUUfVfPPP	Δ ⇓ Δ Δ	Δ Δ Δ Δ
RTS	$(M_{(SP)}:M_{(SP+1)}) \Rightarrow PC_H:PC_L;$ $(SP) + 2 \Rightarrow SP$ Return from Subroutine	INH	3D	UfPPP	UfPPP	– – – –	– – – –
SBA	$(A) - (B) \Rightarrow A$ Subtract B from A	INH	18 16	OO	OO	– – – –	Δ Δ Δ Δ
SBCA #opr8i SBCA opr8a SBCA opr16a SBCA oprx0_xysp SBCA oprx9,xysp SBCA oprx16,xysp SBCA [D,xysp] SBCA [oprx16,xysp]	$(A) - (M) - C \Rightarrow A$ Subtract with Borrow from A	IMM DIR EXT IDX IDX1 IDX2 [D,IDX] [IDX2]	82 ii 92 dd B2 hh ll A2 xb A2 xb ff A2 xb ee ff A2 xb A2 xb ee ff	P rPf rPO rPf rPO frPP fIfrPf fIPrPf	P rfP rOP rfP rPO frPP fIfrfP fIPrfP	– – – –	Δ Δ Δ Δ
SBCB #opr8i SBCB opr8a SBCB opr16a SBCB oprx0_xysp SBCB oprx9,xysp SBCB oprx16,xysp SBCB [D,xysp] SBCB [oprx16,xysp]	$(B) - (M) - C \Rightarrow B$ Subtract with Borrow from B	IMM DIR EXT IDX IDX1 IDX2 [D,IDX] [IDX2]	C2 ii D2 dd F2 hh ll E2 xb E2 xb ff E2 xb ee ff E2 xb E2 xb ee ff	P rPf rPO rPf rPO frPP fIfrPf fIPrPf	P rfP rOP rfP rPO frPP fIfrfP fIPrfP	– – – –	Δ Δ Δ Δ
SEC	$1 \Rightarrow C$ *Translates to ORCC #$01*	IMM	14 01	P	P	– – – –	– – – 1
SEI	$1 \Rightarrow I$; (inhibit I interrupts) *Translates to ORCC #$10*	IMM	14 10	P	P	– – – 1	– – – –
SEV	$1 \Rightarrow V$ *Translates to ORCC #$02*	IMM	14 02	P	P	– – – –	– – 1 –
SEX abc,dxys	$\$00:(r1) \Rightarrow r2$ if r1, bit 7 is 0 or $\$FF:(r1) \Rightarrow r2$ if r1, bit 7 is 1 Sign Extend 8-bit r1 to 16-bit r2 r1 may be A, B, or CCR r2 may be D, X, Y, or SP *Alternate mnemonic for TFR r1, r2*	INH	B7 eb	P	P	– – – –	– – – –

**For More Information On This Product,
Go to: www.freescale.com**

Reprinted with permission of Freescale Semiconductor, Inc.

Freescale Semiconductor, Inc.

Instruction Set Reference

Table A-1. Instruction Set Summary (Sheet 12 of 14)

Source Form	Operation	Addr. Mode	Machine Coding (hex)	Access Detail HCS12	M68HC12	S X H I	N Z V C
STAA opr8a STAA opr16a STAA oprx0_xysp STAA oprx9,xysp STAA oprx16,xysp STAA [D,xysp] STAA [oprx16,xysp]	(A) ⇒ M Store Accumulator A to Memory	DIR EXT IDX IDX1 IDX2 [D,IDX] [IDX2]	5A dd 7A hh ll 6A xb 6A xb ff 6A xb ee ff 6A xb 6A xb ee ff	Pw PwO Pw PwO PwP PIfw PIPw	Pw wOP Pw PwO PwP PIfPw PIPPw	– – – –	Δ Δ 0 –
STAB opr8a STAB opr16a STAB oprx0_xysp STAB oprx9,xysp STAB oprx16,xysp STAB [D,xysp] STAB [oprx16,xysp]	(B) ⇒ M Store Accumulator B to Memory	DIR EXT IDX IDX1 IDX2 [D,IDX] [IDX2]	5B dd 7B hh ll 6B xb 6B xb ff 6B xb ee ff 6B xb 6B xb ee ff	Pw PwO Pw PwO PwP PIfw PIPw	Pw wOP Pw PwO PwP PIfPw PIPPw	– – – –	Δ Δ 0 – .
STD opr8a STD opr16a STD oprx0_xysp STD oprx9,xysp STD oprx16,xysp STD [D,xysp] STD [oprx16,xysp]	(A) ⇒ M, (B) ⇒ M+1 Store Double Accumulator	DIR EXT IDX IDX1 IDX2 [D,IDX] [IDX2]	5C dd 7C hh ll 6C xb 6C xb ff 6C xb ee ff 6C xb 6C xb ee ff	PW PWO PW PWO PWP PIfW PIPW	PW WOP PW PWO PWP PIfPW PIPPW	– – – –	Δ Δ 0 –
STOP	(SP) – 2 ⇒ SP; RTN$_H$:RTN$_L$ ⇒ M$_{(SP)}$:M$_{(SP+1)}$; (SP) – 2 ⇒ SP; (Y$_H$:Y$_L$) ⇒ M$_{(SP)}$:M$_{(SP+1)}$; (SP) – 2 ⇒ SP; (X$_H$:X$_L$) ⇒ M$_{(SP)}$:M$_{(SP+1)}$; (SP) – 2 ⇒ SP; (B:A) ⇒ M$_{(SP)}$:M$_{(SP+1)}$; (SP) – 1 ⇒ SP; (CCR) ⇒ M$_{(SP)}$; STOP All Clocks Registers stacked to allow quicker recovery by interrupt. If S control bit = 1, the STOP instruction is disabled and acts like a two-cycle NOP.	INH	18 3E	(entering STOP) OOSSSSSf (exiting STOP) fVfPPP (continue) ff (if STOP disabled) OO	OOSSSSfSs fVfPPP fO OO	– – – –	– – – –
STS opr8a STS opr16a STS oprx0_xysp STS oprx9,xysp STS oprx16,xysp STS [D,xysp] STS [oprx16,xysp]	(SP$_H$:SP$_L$) ⇒ M:M+1 Store Stack Pointer	DIR EXT IDX IDX1 IDX2 [D,IDX] [IDX2]	5F dd 7F hh ll 6F xb 6F xb ff 6F xb ee ff 6F xb 6F xb ee ff	PW PWO PW PWO PWP PIfW PIPW	PW WOP PW PWO PWP PIfPW PIPPW	– – – –	Δ Δ 0 –
STX opr8a STX opr16a STX oprx0_xysp STX oprx9,xysp STX oprx16,xysp STX [D,xysp] STX [oprx16,xysp]	(X$_H$:X$_L$) ⇒ M:M+1 Store Index Register X	DIR EXT IDX IDX1 IDX2 [D,IDX] [IDX2]	5E dd 7E hh ll 6E xb 6E xb ff 6E xb ee ff 6E xb 6E xb ee ff	PW PWO PW PWO PWP PIfW PIPW	PW WOP PW PWO PWP PIfPW PIPPW	– – – –	Δ Δ 0 –
STY opr8a STY opr16a STY oprx0_xysp STY oprx9,xysp STY oprx16,xysp STY [D,xysp] STY [oprx16,xysp]	(Y$_H$:Y$_L$) ⇒ M:M+1 Store Index Register Y	DIR EXT IDX IDX1 IDX2 [D,IDX] [IDX2]	5D dd 7D hh ll 6D xb 6D xb ff 6D xb ee ff 6D xb 6D xb ee ff	PW PWO PW PWO PWP PIfW PIPW	PW WOP PW PWO PWP PIfPW PIPPW	– – – –	Δ Δ 0 –

SUBA #opr8i
SUBA opr8a
SUBA opr16a
SUBA oprx0_xysp
SUBA oprx9,xysp
SUBA oprx16,xysp
SUBA [D,xysp]
SUBA [oprx16,xysp]
	(A) – (M) ⇒ A
	Subtract Memory from Accumulator A
		IMM
		DIR
		EXT
		IDX
		IDX1
		IDX2
		[D,IDX]
		[IDX2]
			80 ii
			90 dd
			B0 hh ll
			A0 xb
			A0 xb ff
			A0 xb ee ff
			A0 xb
			A0 xb ee ff
				P
				rPf
				rPO
				rPf
				rPO
				frPP
				fIfrPf
				fIPrPf
					P
					rfP
					rOP
					rfP
					rPO
					frPP
					fIfrfP
					fIPrfP
						– – – –	Δ Δ Δ Δ

For More Information On This Product,
Go to: www.freescale.com

Freescale Semiconductor, Inc.

Instruction Set Reference

Table A-1. Instruction Set Summary (Sheet 13 of 14)

Source Form	Operation	Addr. Mode	Machine Coding (hex)	Access Detail HCS12	Access Detail M68HC12	S X H I	N Z V C

SUBB #opr8i
SUBB opr8a
SUBB opr16a
SUBB oprx0_xysp
SUBB oprx9,xysp
SUBB oprx16,xysp
SUBB [D,xysp]
SUBB [oprx16,xysp]
	(B) − (M) ⇒ B
	Subtract Memory from Accumulator B
		IMM
		DIR
		EXT
		IDX
		IDX1
		IDX2
		[D,IDX]
		[IDX2]
			C0 ii
			D0 dd
			F0 hh ll
			E0 xb
			E0 xb ff
			E0 xb ee ff
			E0 xb
			E0 xb ee ff
				P
				rPf
				rPO
				rPf
				rPO
				frPP
				fIfrPf
				fIPrPf
					P
					rfP
					rOP
					rfP
					rPO
					frPP
					fIfRfP
					fIPrfP
						– – – –	Δ Δ Δ Δ

SUBD #opr16i
SUBD opr8a
SUBD opr16a
SUBD oprx0_xysp
SUBD oprx9,xysp
SUBD oprx16,xysp
SUBD [D,xysp]
SUBD [oprx16,xysp]
	(D) − (M:M+1) ⇒ D
	Subtract Memory from D (A:B)
		IMM
		DIR
		EXT
		IDX
		IDX1
		IDX2
		[D,IDX]
		[IDX2]
			83 jj kk
			93 dd
			B3 hh ll
			A3 xb
			A3 xb ff
			A3 xb ee ff
			A3 xb
			A3 xb ee ff
				PO
				RPf
				RPO
				RPf
				RPO
				fRPP
				fIfRPf
				fIPRPf
					OP
					RfP
					ROP
					RfP
					RPO
					fRPP
					fIfRfP
					fIPRfP
						– – – –	Δ Δ Δ Δ

SWI	(SP) − 2 ⇒ SP; RTN$_H$:RTN$_L$ ⇒ M$_{(SP)}$:M$_{(SP+1)}$; (SP) − 2 ⇒ SP; (Y$_H$:Y$_L$) ⇒ M$_{(SP)}$:M$_{(SP+1)}$; (SP) − 2 ⇒ SP; (X$_H$:X$_L$) ⇒ M$_{(SP)}$:M$_{(SP+1)}$; (SP) − 2 ⇒ SP; (B:A) ⇒ M$_{(SP)}$:M$_{(SP+1)}$; (SP) − 1 ⇒ SP; (CCR) ⇒ M$_{(SP)}$ 1 ⇒ I; (SWI Vector) ⇒ PC Software Interrupt	INH	3F	VSPSSPSsP* (for Reset) VfPPP	VSPSSPSsP* VfPPP	– – – 1 1 1 – 1	– – – – – – – –
*The CPU also uses the SWI microcode sequence for hardware interrupts and unimplemented opcode traps. Reset uses the VfPPP variation of this sequence.							
TAB	(A) ⇒ B Transfer A to B	INH	18 0E	OO	OO	– – – –	Δ Δ 0 –
TAP	(A) ⇒ CCR Translates to TFR A , CCR	INH	B7 02	P	P	Δ ⇓ Δ Δ	Δ Δ Δ Δ
TBA	(B) ⇒ A Transfer B to A	INH	18 0F	OO	OO	– – – –	Δ Δ 0 –
TBEQ abdxys,rel9	If (cntr) = 0, then Branch; else Continue to next instruction Test Counter and Branch if Zero (cntr = A, B, D, X,Y, or SP)	REL (9-bit)	04 lb rr	PPP (branch) PPO (no branch)	PPP	– – – –	– – – –
TBL oprx0_xysp	(M) + [(B) × ((M+1) − (M))] ⇒ A 8-Bit Table Lookup and Interpolate Initialize B, and index before TBL. <oa> points at first 8-bit table entry (M) and B is fractional part of lookup value. (no indirect addressing modes or extensions allowed)	IDX	18 3D xb	ORfffP	OrrfffP	– – – – C Bit is undefined in HC12	Δ Δ – Δ ?
TBNE abdxys,rel9	If (cntr) not = 0, then Branch; else Continue to next instruction Test Counter and Branch if Not Zero (cntr = A, B, D, X,Y, or SP)	REL (9-bit)	04 lb rr	PPP (branch) PPO (no branch)	PPP	– – – –	– – – –
TFR abcdxys,abcdxys	(r1) ⇒ r2 or $00:(r1) ⇒ r2 or (r1[7:0]) ⇒ r2 Transfer Register to Register r1 and r2 may be A, B, CCR, D, X, Y, or SP	INH	B7 eb	P	P	– – – – or Δ ⇓ Δ Δ	– – – – Δ Δ Δ Δ
TPA	(CCR) ⇒ A Translates to TFR CCR ,A	INH	B7 20	P	P	– – – –	– – – –

**For More Information On This Product,
Go to: www.freescale.com**

Reprinted with permission of Freescale Semiconductor, Inc.

Freescale Semiconductor, Inc.

Table A-1. Instruction Set Summary (Sheet 14 of 14)

Source Form	Operation	Addr. Mode	Machine Coding (hex)	Access Detail HCS12	M68HC12	S X H I	N Z V C
TRAP *trapnum*	(SP) − 2 ⇒ SP; RTN$_H$:RTN$_L$ ⇒ M$_{(SP)}$:M$_{(SP+1)}$; (SP) − 2 ⇒ SP; (Y$_H$:Y$_L$) ⇒ M$_{(SP)}$:M$_{(SP+1)}$; (SP) − 2 ⇒ SP; (X$_H$:X$_L$) ⇒ M$_{(SP)}$:M$_{(SP+1)}$; (SP) − 2 ⇒ SP; (B:A) ⇒ M$_{(SP)}$:M$_{(SP+1)}$; (SP) − 1 ⇒ SP; (CCR) ⇒ M$_{(SP)}$; 1 ⇒ I; (TRAP Vector) ⇒ PC Unimplemented opcode trap	INH	18 tn tn = $30–$39 or $40–$FF	OVSPSSPSsP	OfVSPSSPSsP	– – – 1	– – – –
TST *opr16a* TST *oprx0_xysp* TST *oprx9,xysp* TST *oprx16,xysp* TST [D,xysp] TST [oprx16,xysp] TSTA TSTB	(M) − 0 Test Memory for Zero or Minus (A) − 0 Test A for Zero or Minus (D) − 0 Test B for Zero or Minus	EXT IDX IDX1 IDX2 [D,IDX] [IDX2] INH INH	F7 hh ll E7 xb E7 xb ff E7 xb ee ff E7 xb E7 xb ee ff 97 D7	rPO rPf rPO frPP fIfrPf fIPrPf O O	rOP rfP rPO fₗPP fIfrfP fIPrfP O O	– – – –	Δ Δ 0 0
TSX	(SP) ⇒ X *Translates to TFR SP,X*	INH	B7 75	P	P	– – – –	– – – –
TSY	(SP) ⇒ Y *Translates to TFR SP,Y*	INH	B7 76	P	P	– – – –	– – – –
TXS	(X) ⇒ SP *Translates to TFR X,SP*	INH	B7 57	P	P	– – – –	– – – –
TYS	(Y) ⇒ SP *Translates to TFR Y,SP*	INH	B7 67	P	P	– – – –	– – – –
WAI	(SP) − 2 ⇒ SP; RTN$_H$:RTN$_L$ ⇒ M$_{(SP)}$:M$_{(SP+1)}$; (SP) − 2 ⇒ SP; (Y$_H$:Y$_L$) ⇒ M$_{(SP)}$:M$_{(SP+1)}$; (SP) − 2 ⇒ SP; (X$_H$:X$_L$) ⇒ M$_{(SP)}$:M$_{(SP+1)}$; (SP) − 2 ⇒ SP; (B:A) ⇒ M$_{(SP)}$:M$_{(SP+1)}$; (SP) − 1 ⇒ SP; (CCR) ⇒ M$_{(SP)}$; WAIT for interrupt	INH	3E	OSSSSsf (after interrupt) fVfPPP	OSSSfSsf VfPPP	– – – – or – – – 1 or – 1 – 1	– – – – – – – – – – – –
WAV	$\displaystyle\sum_{i=1}^{B} S_i F_i \Rightarrow Y{:}D$ and $\displaystyle\sum_{i=1}^{B} F_i \Rightarrow X$ Calculate Sum of Products and Sum of Weights for Weighted Average Calculation Initialize B, X, and Y before WAV. B specifies number of elements. X points at first element in S$_i$ list. Y points at first element in F$_i$ list. All S$_i$ and F$_i$ elements are 8-bits. If interrupted, six extra bytes of stack used for intermediate values	Special	18 3C	Of(frr,ffff)O (add if interrupt) SSS + UUUrr,	Off(frr,fffff)O SSSf + UUUrr	– – ? –	? Δ ? ?
wavr pseudo-instruction	*see WAV* Resume executing an interrupted WAV instruction (recover intermediate results from stack rather than initializing them to zero)	Special	3C	UUUrr,ffff (frr,ffff)O (exit + re-entry replaces comma above if interrupted) SSS + UUUrr,	UUUrrfffff (frr,ffff)O SSSf + UUUrr	– – ? –	? Δ ? ?
XGDX	(D) ⇔ (X) *Translates to EXG D, X*	INH	B7 C5	P	P	– – – –	
XGDY	(D) ⇔ (Y) *Translates to EXG D, Y*	INH	B7 C6	P	P	– – – –	– – – –

Appendix B: Number System Issue

B.1 Introduction

Computers were initially designed to deal with numerical data. Due to the on-and-off nature of electricity, numbers were represented in binary base from the beginning of the electronic computer age.

However, the binary number system is not natural to us because human beings have been mainly using the decimal number system for thousands of years. A binary number needs to be converted to a decimal number before it can be quickly interpreted by a human being.

For identification, a subscript 2, 8, or 16 is added to a number to indicate the base of the given number. For example, 101_2 is a binary number; 234_8 is an octal number, and 2479_{16} is a hexadecimal number. No subscript is used for a decimal number.

B.2 Converting from Binary to Decimal

A binary number is represented by two symbols: 0 and 1. Here, we refer to 0 and 1 as *binary digits*. A binary digit is also referred to as a *bit*. In the computer, 8 bits are referred to as a *byte*. Depending on the computer, either 16 bits or 32 bits are referred to as a *word*. The values of 0 and 1 in the binary number system are identical to their counterparts in the decimal number system. To convert a binary number to a decimal, we compute a weighted sum of every binary digit contained in the binary number. If a specific bit is k bits from the rightmost bit, then its weight is 2^k. For example,

$$
\begin{aligned}
10100100_2 &= 2^7 + 2^5 + 2^2 &&= 128 + 32 + 4 &&= 164 \\
11011001_2 &= 2^7 + 2^6 + 2^4 + 2^3 + 2^0 &&= 128 + 64 + 16 + 8 + 1 &&= 217 \\
10010010.101_2 &= 2^7 + 2^4 + 2^1 + 2^{-1} + 2^{-3} &&= 146.625
\end{aligned}
$$

B.3 Converting from Decimal to Binary

A decimal integer can be converted to a binary number by performing the repeated division-by-2 operation until the quotient becomes 0. The remainder resulting from the first division is the least significant binary digit whereas the remainder resulting from the last division is the most significant digit.

Example B.1

Convert the decimal number 73 to binary.

Solution: The procedure of conversion is shown in Figure B.1.

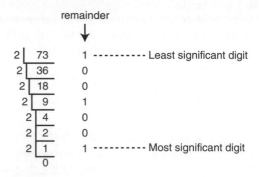

Figure B.1 ■ An example of decimal-to-binary conversion

$$73 = 1001001_2$$

Example B.2

Convert the decimal number 95 to binary.

Solution: The repeated division-by-2 process is shown in Figure B.2.

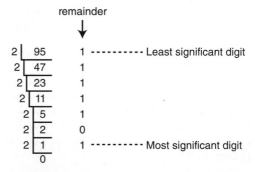

Figure B.2 ■ Another example of decimal-to-binary conversion

$$95 = 1011111_2$$

If a decimal number has a *fractional part*, then it needs to be converted using a different method. The fractional part can be converted to a decimal by performing the repeated-multiplication-by-2 operation to the fraction until either the fraction part becomes 0 or the *required accuracy* is achieved. The resulting integer binary digit of the first multiplication is the most significant binary digit of the fractional part whereas the resulting integer binary digit of the last multiplication is the least significant binary digit of the fractional part.

Let m and k be the number of digits of the decimal fraction and the binary fraction, respectively; then the desired accuracy has been achieved if and only if the following expression is true:

$2^{-k} < 10^{-m}$

A few pairs of k and m values that satisfy the previous relationship are shown in Table B.1. To be more accurate, if the resultant fractional part of the last multiplication is 5 or larger, then we should round it up by adding 1 to the least significant binary digit.

k	m
>=4	1
>=7	2
>=10	3
>=14	4

Table B.1 ■ k and m that satisfies $2^{-k} < 10^{-m}$

Example B.3

Convert the decimal fraction 0.6 to binary.

Solution: According to Table B.1, we need to perform four repeated-multiplication-by-2 operations as shown in Figure B.3.

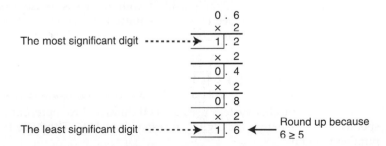

Figure B.3 ■ Convert a decimal fraction to a binary fraction

$0.6_{10} = 0.1001_2 + 0.0001_2 = 0.1010_2$

Example B.4

Convert the decimal number 59.75 to binary.

Solution: Since the given number has both the integral and fractional parts, they need to be converted separately and then combined together. The conversion process is shown in Figure B.4.

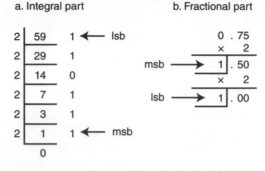

Converted number = 111011.11^2

Figure B.4 ■ Convert a decimal number with both integral and fractional parts to binary

B.4 Why Octal and Hexadecimal Numbers?

After the computation, the computer needs to output the result so that the user can see and determine whether the result is correct. The result can be displayed in either binary or decimal format. To display a number in binary format is not convenient because it takes many 0s and 1s to represent a large number. To display a number in decimal format will require the computer to perform some complicated conversion operations. The method to convert a binary number to decimal is illustrated in Example 4.5. A compromise is to use octal or hexadecimal format to represent a number. Hexadecimal representation is used more often than octal representation. The shorthand of *hexadecimal* is *hex*.

B.5 Convert from Binary to Octal

The octal number system uses digits 0 through 7 to represent a number. The digit 0 corresponds to 000_2 whereas the digit 7 corresponds to 111_2. There are two steps to convert a binary number to octal.

Step 1
Partition the given number (a string of 0s and 1s) from right to the left into groups of 3 bits. Add leading 0s if the leftmost group has less than 3 bits.

Step 2
Convert each 3-bit group to the corresponding octal digit (0 to 7).

Example B.5

▼

Convert the following two binary numbers into octal representation:

a. 101011001_2

b. 1010101010_2

Solution

a. $10101001_2 = 10,101,001 = 010,101,001_2 = 252_8$

b. $1010101010_2 = 1,010,101,010 = 001,010,101,010_2 = 1252_8$

▲

B.6 Convert from Octal to Binary

To convert from octal to binary, simply convert each octal digit to its 3-bit binary equivalent and delete the leading 0s from the resultant binary number.

Example B.6

▼

Convert the following octal numbers into binary:

a. 532_8

b. 246_8

Solution

a. $532_8 = 101,011,010_2$

b. $246_8 = 010,100,110_2 = 10,100,110_2$

▲

B.7 Convert from Binary to Hexadecimal

The hex number system uses digits 0 through 9 and letters A through F to represent a number. The digit 0 corresponds to 0000_2, and the letter A corresponds to 1010, whereas the letter F corresponds to 1111_2. There are two steps to convert a binary number to hex.

Step 1
Partition the given number (a string of 0s and 1s) from the right to the left into groups of 4 bits. Add leading 0s if the leftmost group has fewer than 4 bits.

Step 2
Convert each 4-bit group to the corresponding hex digit (0 to F).

Example B.7

▼

Convert the following two binary numbers into hex representation:

a. 100101011001_2

b. 1011010101110_2

Solution

    a. $100101011001_2 = 1001,0101,1001_2 = 959_{16}$

    b. $1011010101010_2 = 1,0110,1010,1010_2 = 0001,0110,1010,1110_2 = 16AE_{16}$

▲

B.8 Convert from Hexadecimal to Binary

To convert from hex to binary, simply convert each hex digit into its 4-bit binary equivalent and delete the leading 0s from the resultant binary number.

Example B.8

▼

Convert the following hex numbers into binary:

    a. $5CB_{16}$

    b. $2A6_{16}$

Solution

    a. $5CB_{16} = 0101,1100,1011_2 = 101,1100,1011_2$

    b. $2A6_{16} = 0010,1010,0110_2 = 10,1010,0110_2$

▲

B.9 Specifying the Number Base

To facilitate the specification of the number base used in a number representation, we use a notation that adds a suffix to a number to indicate the base used in the number representation. The suffixes for hex, binary, and octal are H (or h), B (or b), and O (or o, Q, q), respectively. Decimal numbers do not use a suffix. For example,

      10101011B   or   10101011b

specifies the binary number 10101011_2.

      123O,   123o,   123q,   or   123Q

specifies the octal number 123_8.

      A097H   or   A097h

refers to the hexadecimal number $A097_{16}$.

      3467

is a decimal number.

Hex numbers are also represented by adding the prefix 0x to the number. For example, 0x1000 stands for the hex number 1000_{16}. In this text, we will mix the use of these two methods for hexadecimal numbers.

B.10 Binary Addition and Subtraction

Addition in binary representation follows the familiar rules of decimal addition. When adding two numbers, add the successive bits and any carry. You will need the following addition rules:

$$0 + 0 = 0$$
$$0 + 1 = 1$$
$$1 + 1 = 0 \text{ carry} = 1$$
$$1 + 1 + 1 = 1 \text{ carry} = 1$$

A carry generated in any bit position is added to the next higher bit.

Binary numbers are subtracted using the following rules:

$$0 - 0 = 0$$
$$0 - 1 = 1 \text{ with a borrow 1}$$
$$1 - 0 = 1$$
$$1 - 1 = 0$$

Example B.9

Add the following pairs of binary positive numbers:

a. 1110110_2 and 1100100_2

b. 01101_2 and 10001_2

Solution: The addition process is shown in Figure B.5.

```
a. carry                    b. carry
   1     1                           1
   1 1 1 0 1 1 0              0 1 1 0 1
 + 1 1 0 0 1 0 0            + 1 0 0 0 1
 ───────────────           ───────────
   1 1 0 1 1 0 1 0            1 1 1 1 0
```

Figure B.5 ■ Examples of binary addition

Example B.10

Perform the following binary subtractions:

a. $11001_2 - 110_2$

b. $110011_2 - 11001_2$

Solution: The subtraction process is shown in Figure B.6.

```
a.                             b.
    1 1     borrow               1 1        borrow
    1 1 0 0 1  minuend           1 1 0 0 1 1  minuend
 +      1 1 0  subtrahend      −   1 1 0 0 1  subtrahend
  ─────────────                ──────────────
    1 0 0 1 1  difference        0 1 1 0 1 0  difference
```

Figure B.6 ■ Examples of binary subtraction

B.11 Two's Complement Numbers

In a computer, the number of bits that can be used to represent a number is fixed. As a result, computers always manipulate numbers that are fixed in length.

Another restriction on number representation in a computer is that both positive and negative numbers must be represented. However, a computer does not include the plus and minus signs in a number. Instead, all modern computers use the two's complement number system to represent positive and negative numbers. In the two's complement system, all numbers that have a most significant bit (MSB) set to 0 are *positive*, and all numbers with an MSB set to 1 are *negative*. Positive two's complement numbers are identical to binary numbers, except that the MSB must be a 0. If N is a positive number, then its two's complement N_C is given by

$$N_C = 2^n - N \tag{B.1}$$

where n is the number of bits available for representing N.

The two's complement of N is used to represent $-N$. Machines that use the two's complement number system can represent integers in the range

$$-2^{n-1} \le N \le 2^{n-1} - 1 \tag{B.2}$$

Example B.11
▼

Find the range of integers that can be represented by an 8-bit two's complement number system.

Solution: The range of integers that can be represented by the 8-bit two's complement number system is

$$-2^7 \le N \le 2^7 - 1 \quad \text{or} \quad -128_{10} \le N \le 127_{10}$$

▲

Example B.12
▼

Represent the negative binary number -11001_2 in 8-bit two's complement format.

Solution: Use Equation B.1 and the subtraction method in Section B.11.

$$N_C = 2^8 - 11001 = 11100111_2$$

An easy way to find the two's complement of a binary number is to flip every bit from 0 to 1 or 1 to 0 and then add 1 to it.

▲

B.12 Two's Complement Subtraction

Subtraction can be performed by adding the two's complement of the subtrahend. This allows the same piece of hardware to perform addition and subtraction. All of today's computers use this method to perform subtraction.

After the addition, throw away the carry generated from the most significant bit and the result is the desired difference. If the most significant bit is 0, then the resultant difference is positive. Otherwise, it is negative.

Example B.13

Subtract 13 from 97 using two's complement arithmetic.

Solution: The 8-bit binary representations of 13 and 97 are 00001101_2 and 01100001_2, respectively. The two's complement of 13 is 11110011_2. The value of $97 - 13$ can be computed as shown in Figure B.7.

```
    0 1 1 0 0 0 0 1
 +  1 1 1 1 0 0 1 1
  ------------------
  1 0 1 0 1 0 1 0 0   = 84₁₀
```

Throw away
carry

Figure B.7 ■ Subtraction by adding two's complement of subtrahend

Example B.14

Subtract 98 from 65 using two's complement arithmetic.

Solution: The 8-bit binary representations of 98 and 65 are 01100010_2 and 01000001_2, respectively. The two's complement of 98 is 10011110_2. The value of $65 - 98$ is computed in Figure B.8.

```
    0 1 0 0 0 0 0 1
 +  1 0 0 1 1 1 1 0
  ------------------
  1 1 0 1 1 1 1 1   = –33₁₀
```

Figure B.8 ■ Subtraction by adding two's complement of subtrahend

The binary number 11011111_2 is the two's complement of 00100001_2 (33). Therefore, the resultant difference is -33 and is correct.

B.13 Overflow

Overflow can occur with either addition or subtraction in two's complement representation. During addition, overflow occurs when the sign of the sum of two numbers with like signs differs from the sign of two numbers. Overflow never occurs when adding two numbers with unlike signs. In subtraction, overflow can occur when subtracting two numbers with unlike signs. If

Negative − positive = positive
or   Positive − negative = negative

then overflow has occurred.

Overflow never occurs when subtracting two numbers with like signs.

Example B.15

Does overflow occur in the following 8-bit operations?

a. $01111111_2 - 00000111_2$
b. $01100101_2 + 01100000_2$
c. $10010001_2 - 01110000_2$

Solution: These subtraction operations are illustrated in Figure B.9 to B.11.

a. Negation of 00000111_2 is 11111001_2. $01111111_2 - 00000111_2$ is performed as follows:

```
     0 1 1 1 1 1 1 1
   + 1 1 1 1 1 0 0 1
   1 0 1 1 1 1 0 0 0
```
Carry out to be discarded

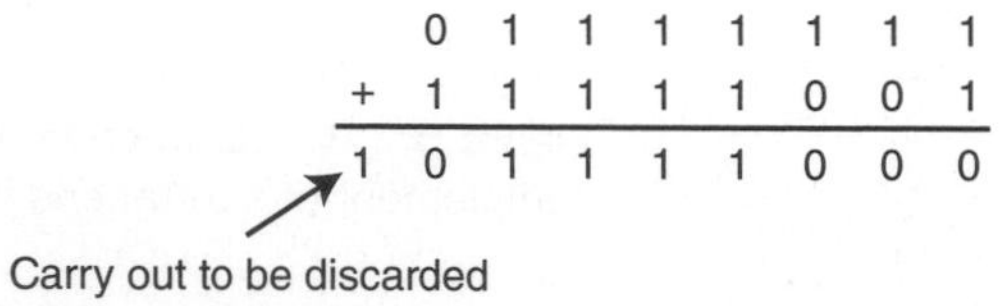

Figure B.9 ■ Example of addition that causes no overflow

The difference is $01111000_2 = 120_{10}$. There is no overflow.

b. The sum of these two numbers is as follows:

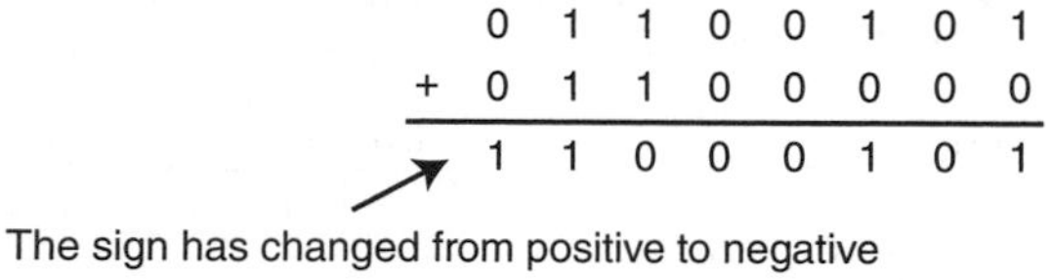

```
     0 1 1 0 0 1 0 1
   + 0 1 1 0 0 0 0 0
     1 1 0 0 0 1 0 1
```
The sign has changed from positive to negative

Figure B.10 ■ Example of addition that causes overflow

Overflow has occurred.

c. The two's complement of 01110000_2 is 10010000_2. $10010001_2 - 01110000_2$ is performed as follows:

```
                              1 0 0 1 0 0 0 1
                            + 1 0 0 1 0 0 0 0
Carry out to be discarded → 1 0 0 1 0 0 0 0 1
```
The sign has changed from negative to positive

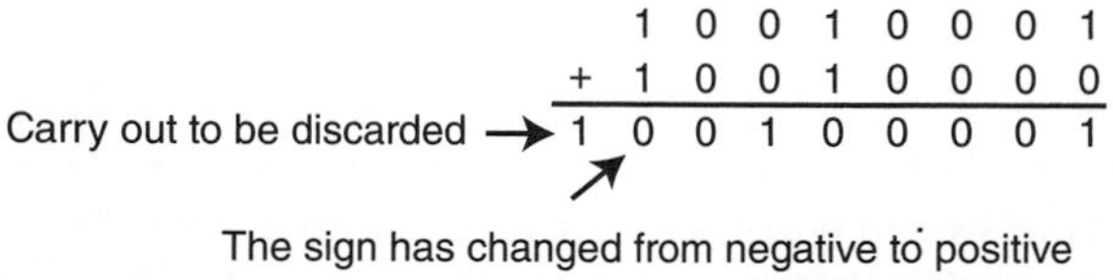

Figure B.11 ■ Another example of addition that causes overflow

Overflow has occurred.

An important benefit of using two's complement arithmetic is that the same computer hardware can perform signed and unsigned (all numbers are nonnegative) addition and subtraction without modification. It is up to the user to interpret the number to be negative.

B.14 Representing Nonnumeric Data

Computers are also used to process nonnumeric data. Nonnumeric data is often in the form of character strings. A character can be a letter, a digit, or a special character symbol. A unique number is assigned to each character of the character set so that it can be differentiated by the computer. To facilitate the processing of characters by the computer and the exchange of information, all users must use the same number to represent (or encode) each character.

Two character code sets are in widespread use today: ASCII (American Standard Code for Information Interchange) and EBCDIC (Extended Binary Coded Decimal Interchange Code). Several types of mainframe computers use EBCDIC for internal storage and processing of characters. In EBCDIC, each character is represented by a unique 8-bit number, and a total of 256 different characters can be represented. Most microcomputer and minicomputer systems use the ASCII character set. A computer does not depend on any particular set, but most I/O devices that display characters require the use of ASCII codes.

ASCII characters have a 7-bit code. They are usually stored in a fixed-length 8-bit number. The $2^7 = 128$ different codes are partitioned into 95 printable characters and 33 control characters. The control characters define communication protocols and special operations on peripheral devices. The printable characters consist of

26	uppercase letters (A–Z)
26	lowercase letters (a–z)
10	digits (0–9)
1	blank space
32	special-character symbols, including ! @ # $ % ^ & * ( ) − _ = + ' [ ] ; : ''' , < . > / ? { }

The complete ASCII code set is shown in Table B.2.

Seven-bit Hexadecimal Code	Character	Seven-bit Hexadecimal Code	Character	Seven-bit Hexadecimal Code	Character	Seven-bit Hexadecimal Code	Character
00	NUL	20	SP	40	@	60	`
01	SOH	21	!	41	A	61	a
02	STX	22	"	42	B	62	b
03	ETX	23	#	43	C	63	c
04	EOT	24	$	44	D	64	d
05	ENQ	25	%	45	E	65	e
06	ACK	26	&	46	F	66	f
07	BEL	27	'	47	G	67	g
08	BS	28	(	48	H	68	h
09	HT	29	)	49	I	69	i
0A	LF	2A	*	4A	J	6A	j
0B	VT	2B	+	4B	K	6B	k
0C	FF	2C	,	4C	L	6C	l
0D	CR	2D	−	4D	M	6D	m
0E	SO	2E	.	4E	N	6E	n
0F	SI	2F	/	4F	O	6F	o
10	DLE	30	0	50	P	70	p
11	DC1	31	1	51	Q	71	q
12	DC2	32	2	52	R	72	r
13	DC3	33	3	53	S	73	s
14	DC4	34	4	54	T	74	t
15	NAK	35	5	55	U	75	u
16	SYN	36	6	56	V	76	v
17	ETB	37	7	57	W	77	w
18	CAN	38	8	58	X	78	x
19	EM	39	9	59	Y	79	y
1A	SUB	3A	:	5A	Z	7A	z
1B	ESC	3B	;	5B	[	7B	{
1C	FS	3C	<	5C	\	7C	|
1D	GS	3D	=	5D	]	7D	}
1E	RS	3E	>	5E	^	7E	~
1F	US	3F	?	5F	−	7F	DEL

Table B.2 ■ Complete ASCII code table

Appendix C: Summary of Features of HCS12 Devices

Device	RAM (bytes)	Flash (bytes)	EEPROM (bytes)	Timer	I/O	Serial	MUX	A/D	PWM	Operating frequency (MHz)	Operating voltage (V)	Packaging
MC9S12A128	4K	128K	2K	8-ch 16-bit IC, OC, PA	Up to 91	Up to 2 SCI 2 SPI IIC	–	Up to 2 × 8 ch 10-bit	Up to 8-ch 8-bit or 4-ch 16-bit	25.0	5.0	80-pin QFP (FU) 112-pin LQFP (PV)
MC9S12A32	2K	32K	1K	8-ch 16-bit ECT	Up to 59	2 SCI 1 SPI	–	8-ch 10-bit	7-ch 8-bit or 3-ch 16-bit	25.0	3.0, 5.0	80-pin QFP (FU)
MC9S12A512	14K	512K	4K	8-ch 16-bit ECT	Up to 91	2 SCI 3 SPI I²C	–	2 × 8 ch 10-bit	8-ch 8-bit or 4-ch 16-bit	25, 33	5.0	112-pin LQFP(PV)
MC9S12A64	4K	64K	1K	8-ch 16-bit IC, OC, PA	Up to 91	Up to 2 SCI 1 SPI IIC	–	Up to 2 × 8 ch 10-bit	Up to 8-ch 8-bit or 4-ch 16-bit	25.0	5.0	80-pin QFP (FU) 112-pin LQFP (PV)
MC9S12C32	2K	32K	0K	8-ch 16-bit IC, OC, or PWM	Up to 60	SCI SPI	CAN	8-ch 10-bit	See timer	16, 25	3.15–5.5	48-pin QFP(FA) 52-pin QFP(PB) 80-pin QFP(FU)
MC9S12D32	2K	32K	1K	8-ch 16-bit ECT	Up to 59	2 SCI 1 SPI	CAN	8-ch 10-bit	7-ch 8-bit or 3-ch 16-bit	25.0	3.0, 5.0	80-pin QFP (FU)
MC9S12D64	4K	64K	1K	8-ch 16-bit IC, OC, PA	Up to 91	Up to 2 SCI 1 SPI IIC	CAN	Up to 2 × 8 ch 10-bit	Up to 8-ch 8-bit or 4-ch 16-bit	25.0	5.0	80-pin QFP (FU) 112-pin LQFP (PV)
MC9S12DJ64	4K	64K	1K	8-ch 16-bit IC, OC, PA	Up to 91	Up to 2 SCI 1 SPI IIC	1 CAN 1 J1850	Up to 2 × 8 ch 10-bit	Up to 8-ch 8-bit or 4-ch 16-bit	25.0	5.0	80-pin QFP (FU) 112-pin LQFP (PV)
MC9S12DB128	8K	128K	2K	8-ch 16-bit IC, OC, PA	Up to 91	Up to 2 SCI 2 SPI	1 CAN byte flight	Up to 2 × 8 ch 10-bit	Up to 8-ch 8-bit or 4-ch 16-bit	25.0	5.0	80-pin QFP (FU) 112-pin LQFP (PV)

Table C.1 ■ Features of the HCS12 family devices (1 of 3)

Device	RAM (bytes)	Flash (bytes)	EEPROM (bytes)	Timer	I/O	Serial	MUX	A/D	PWM	Operating frequency (MHz)	Operating voltage (V)	Packaging
MC9S12DG128	8K	128K	2K	8-ch 16-bit IC, OC, PA	Up to 91	Up to 2 SCI 2 SPI IIC	2 CAN	Up to 2 x 8 ch 10-bit	Up to 8-ch 8-bit or 4-ch 16-bit	25.0	5.0	80-pin QFP (FU) 112-pin LQFP (PV)
MC9S12DJ128	8K	128K	2K	8-ch 16-bit IC, OC, PA	Up to 91	Up to 2 SCI 2 SPI IIC	2 CAN 1 J1850	Up to 2 x 8 ch 10-bit	Up to 8-ch 8-bit or 4-ch 16-bit	25.0	5.0	80-pin QFP (FU) 112-pin LQFP (PV)
MC9S12DP512	14K	512K	4K	8-ch 16-bit ECT	Up to 91	2 SCI 3 SPI I2C	5 CAN	2 x 8 ch 10-bit	8-ch 8-bit or 4-ch 16-bit	25, 33	5.0	112-pin LQFP (PV)
MC9S12DT128	8K	128K	2K	8-ch 16-bit IC, OC, PA	Up to 91	2 SCI 3 SPI I2C	3 CAN	2 x 8 ch 10-bit	8-ch 8-bit or 4-ch 16-bit	25.0	5.0	112-pin LQFP (PV)
MC9S12A256	12K	256K	4K	8-ch 16-bit IC, OC, PA	Up to 91	Up to 2 SCI 3 SPI I2C	–	Up to 2 x 8 ch 10-bit	Up to 8-ch 8-bit or 4-ch 16-bit	25	5.0	80-pin QFP (FU) 112-pin LQFP (PV)
MC9S12DG256	12K	256K	4K	8-ch 16-bit IC, OC, PA	Up to 91	2 SCI 3 SPI I2C	2 CAN	2 x 8 ch 10-bit	8-ch 8-bit or 4-ch 16-bit	25.0	5.0	112-pin LQFP(PV)
MC9S12DJ256	12K	256K	4K	8 ch 16-bit IC, OC, PA	Up to 91	2 SCI 3 SPI I2C	2 CAN 1 J1850	Up to 2 x 8 ch 10-bit	8-ch 8-bit or 4-ch 16-bit	25	5.0	80-pin QFP (FU) 112-pin LQFP (PV)
MC9S12DP256	12K	256K	4K	8-ch 16-bit IC,OC,PA	Up to 91	2 SCI 3 SPI I2C	5 CAN	2 x 8 ch 10-bit	8-ch 8-bit or 4-ch 16-bit	25.0	5.0	112-pin LQFP (PV)
MC9S12DT256	12K	256K	4K	8-ch 16-bit IC, OC, PA	Up to 91	2 SCI 3 SPI I2C	3 CAN	2 x 8 ch 10-bit	8-ch 8-bit or 4-ch 16-bit	25.0	5.0	112-pin LQFP (PV)
MC9S12E64	4K	64K	0	3 4-ch 16-bit IC, OC, or PWM	Up to 90	1 SCI 1 SPI 1 I2C	–	16-ch 10-bit	See timer	16.0, 25.0	3.3–5.0	80-pin QFP (FU) 112-pin LQFP (PV)

Table C.1 ■ Features of the HCS12 family devices (2 of 3)

Device	RAM (bytes)	Flash (bytes)	EEPROM (bytes)	Timer	I/O	Serial	MUX	A/D	PWM	Operating frequency (MHz)	Operating voltage (V)	Packaging
MC9S12E128	8K	128K	0	3 4-ch 16-bit IC, OC, or PWM	Up to 90	1 SCI 1 SPI 1 I²C	–	16-ch 10-bit	See timer	16.0, 25.0	3.3–5.0	80-pin QFP (FU) 112-pin LQFP (PV)
MC9S12H128B	12K	128K	4K	8-ch 16-bit IC, OC, PA	99 plus 18 in	1 SCI 1 SPI 1 I²C	2 CAN	16-ch 10-bit	6-ch 8-bit or 3-ch 16-bit	16.0	5.0	112-pin LQFP (PV)
MC9S12H256B	12K	256K	4K	8-ch 16-bit IC, OC, PA	99 plus 18 in	1 SCI 1 SPI 1 I²C	2 CAN	16-ch 10-bit	6-ch 8-bit or 3-ch 16-bit	16.0	5.0	112-pin LQFP (PV) 144-pin LQFP (FV)
MC9S12NE64	8K	64K	0	4-ch 16-bit IC,OC,or PWM	Up to 70	2 SCI SPI	–	8-ch 10-bit	See timer	16, 25	3.3–5.0	80-pin TQFP-EP(TU) 112-pin LQFP (PV)
MC9S12UF32	3.5K	32K	0	8-ch 16-bit IC, OC, or PWM	Up to 75	SCI USB 2.0	–	–	See timer	30	5.0	100-pin LQFP (PU) 64-pin LQFP
MC9S12T64	2K + 2K CALRAM	64K	–	8-ch 16-bit IC, OC, PA	25	2 SCI 1 SPI	–	8-ch 10-bit	8-ch 8-bit or 4-ch 16-bit	16.0	5.0	80-pin QFP (PK)

Note: 1. The MC9S12NE64 has an on-chip integrated media access controller (EMAC), 10/100 Ethernet PHY (EPHY), and was designed for interfacing with Ethernet.
2. The MC9S12UF32 was designed for interfacing with USB 2.0 interface.

Table C.1 ■ Features of the HCS12 family devices (3 of 3)

Appendix D: Tutorial for Using the AsmIDE

D.1 Introduction

AsmIDE, a freeware IDE developed by Eric Engler, combines a simple text editor, project manager, terminal program, and cross assemblers for developing assembly programs for the Freescale HC11, HC12, and HCS12 microcontrollers. The latest version of AsmIDE can be downloaded from www.geocities.com/englere_geo.

D.2 Starting the AsmIDE

The AsmIDE can be started by clicking on its icon. The screen should look like Figure D.1 after the AsmIDE is started. The upper half of the window in Figure D.1 is used to enter the assembly program. The lower half of the AsmIDE window is used to display messages after an assembly process or as the terminal window to communicate with the D-Bug12 monitor.

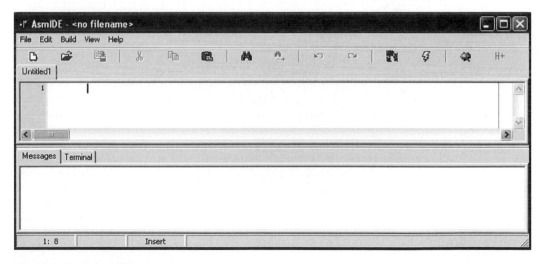

Figure D.1 ■ AsmIDE startup screen

D.3 Communicating with the Demo Board

To communicate with the demo board with its MCU programmed with the D-Bug12 monitor, click on **Terminal** in the lower half of the AsmIDE window. After this, the lower half becomes the terminal window. Press the reset button of the demo board and the screen will change to Figure D.2.

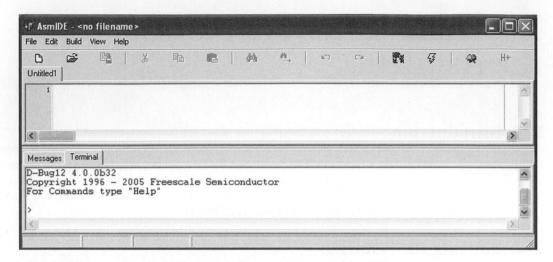

Figure D.2 ■ AsmIDE terminal window after pressing the reset button

You can then enter D-Bug12 commands in the terminal window to display registers and memory contents, modify memory and register contents, set breakpoints, trace instruction execution, and download the program onto the demo board for execution.

If the D-Bug12 command prompt does not appear on the screen, the most likely three causes are as follows:

1. The demo board is not powered up.
2. The RS232 cable connection is wrong.
3. The baud rate of the terminal program does not match that of the demo board.

One should make sure that the demo board has been powered up and also make sure that the RS232 cable is connected to the right connector (if the demo board has two RS232 connectors). If the D-Bug12 prompt still does not appear, then check the baud rate setting. Select **View** from the IDE menu and click on **Options.** Then click on **COM** port and the screen should change to that shown in Figure D.3. To set COM port options, click on **Set COM Options** and a dialog box will appear. Make sure the settings of the terminal program are identical to those in Figure D.4. Another option that the user needs to set is *assembler*. Click on **Assembler** in the dialog box and make sure the settings are the same as those in Figure D.5.

The user can learn the D-Bug12 monitor commands by walking through the Section 3.6.3. After getting familiar with the D-Bug12 monitor commands, the user can continue to walk through the following sections.

D.4 Entering an Assembly Program

When a program is first created, we need to open a new file to hold it. To do that, we press the **File** menu from the AsmIDE window and select **New,** as shown in Figure D.6. After the **New** command is selected from the menu, a popup confirm box as shown in Figure D.7 appears and asks the user to confirm the request. The user simply clicks on **Yes** to get rid of the popup confirm box and start to enter a new program.

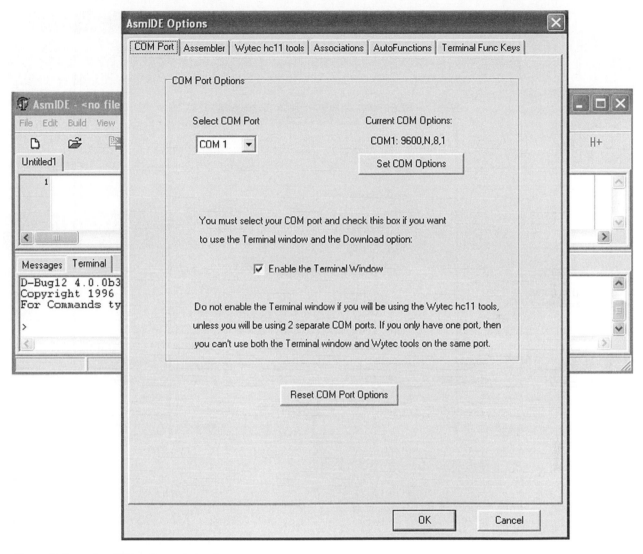

Figure D.3 ■ AsmIDE options dialog box

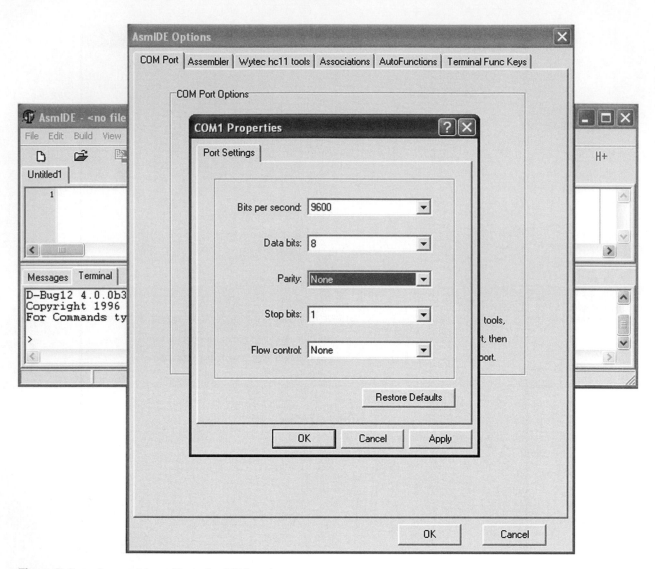

Figure D.4 ■ Appropriate settings for COM port

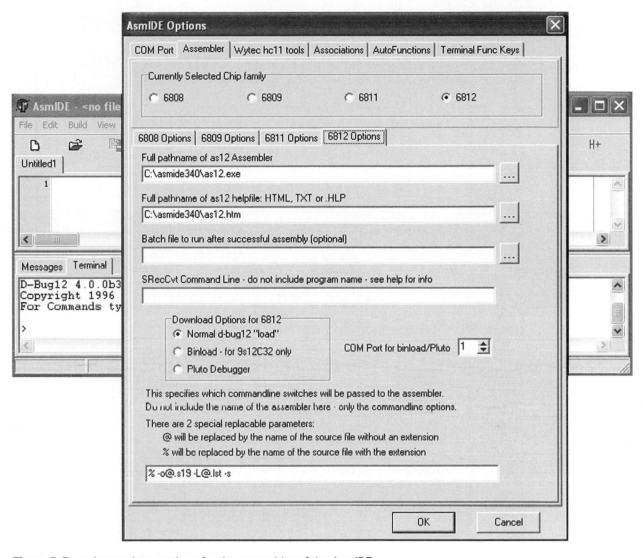

Figure D.5 ■ Appropriate settings for the assembler of the AsmIDE

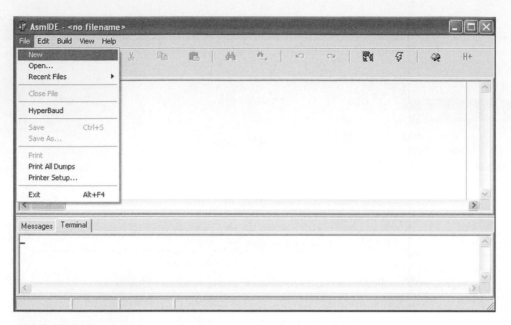

Figure D.6 ■ Select New from the file menu to create a new file

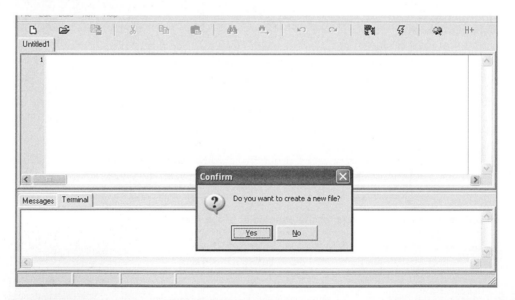

Figure D.7 ■ Confirmation dialog for new file creation

The editing window (upper half of the IDE window) may be too small and should be resized. After adjusting the window size, the user enters the program that converts a hex number into BCD digits in Example 2.13. This program performs the repeated divide-by-10 operation to the given number and adds hex number $30 to each remainder to convert to its corresponding ASCII code. The result is shown in Figure D.8.

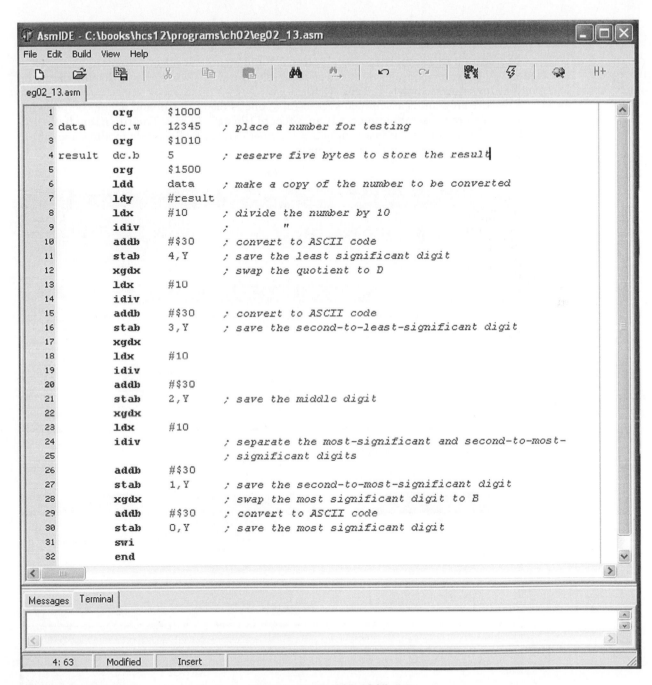

```
 1          org     $1000
 2 data     dc.w    12345       ; place a number for testing
 3          org     $1010
 4 result   dc.b    5           ; reserve five bytes to store the result
 5          org     $1500
 6          ldd     data        ; make a copy of the number to be converted
 7          ldy     #result
 8          ldx     #10         ; divide the number by 10
 9          idiv                ;           "
10          addb    #$30        ; convert to ASCII code
11          stab    4,Y         ; save the least significant digit
12          xgdx                ; swap the quotient to D
13          ldx     #10
14          idiv
15          addb    #$30        ; convert to ASCII code
16          stab    3,Y         ; save the second-to-least-significant digit
17          xgdx
18          ldx     #10
19          idiv
20          addb    #$30
21          stab    2,Y         ; save the middle digit
22          xgdx
23          ldx     #10
24          idiv                ; separate the most-significant and second-to-most-
25                              ; significant digits
26          addb    #$30
27          stab    1,Y         ; save the second-to-most-significant digit
28          xgdx                ; swap the most significant digit to B
29          addb    #$30        ; convert to ASCII code
30          stab    0,Y         ; save the most significant digit
31          swi
32          end
```

Figure D.8 ■ Enter the program to convert hex number to BCD ASCII digits

D.5 Assembling the Program

To assemble an assembly program, press the **Build** menu and select **Assemble,** as shown in Figure D.9. If the program is assembled successfully, the status window will display the corresponding message, as shown in Figure D.10.

After the program is assembled successfully, one is ready to run and debug the program. The output of the assembly process is an S-record file *eg2_13.s19*. The file name has a suffix **s19.** The S-record format is a common file format defined by Freescale to allow tools from different vendors to work on the same project.

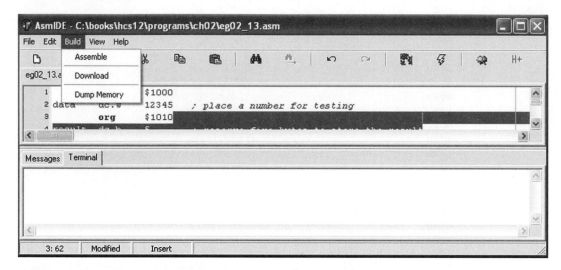

Figure D.9 ■ Prepare to assemble the program eg02_13.asm

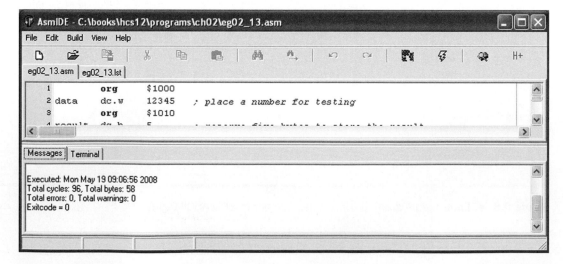

Figure D.10 ■ Message window shows that previous assembling is successful

D.6 Downloading the S-Record File onto the Demo Board for Execution

To download the S19 file onto the demo board, one needs to select the Terminal window by pressing **Terminal.** The D-Bug12 monitor prompt will appear. If not, then press the reset button on the demo board. Type the **load** command at the command prompt and the **enter** key. After that, press the **Build** menu and select **Download** as shown in Figure D.11.

After the **Download** command is selected, a popup dialog box, as shown in Figure D.12, will appear. This dialog box allows us to specify the file to be downloaded. Click on the

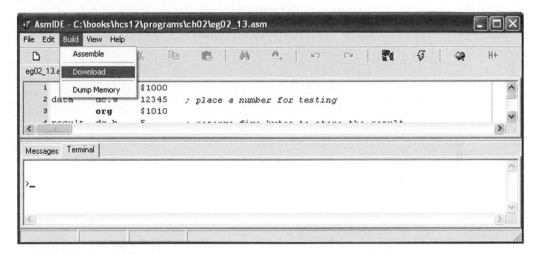

Figure D.11 ■ Select the Download command

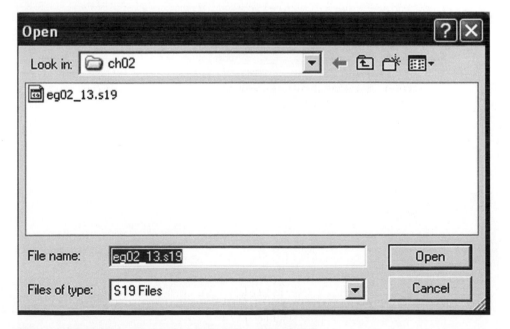

Figure D.12 ■ Select an S-record file to be downloaded

Open button on the popup window and the file will be transferred to the demo board. AsmIDE has an auto-download option which can be enabled by selecting the **AutoFunctions** menu and clicking on appropriate options in Figure D.5. However, too much automation is not always a good choice.

D.6 Running and Debugging the Program

The procedure for running and debugging the user program after downloading the S19 file onto the demo board is identical to that described in Section 3.6.7.

Appendix E: Tutorial for Using EGNU to Develop C Programs

E.1 Introduction

The GNU-m68hc11 C compiler is a freeware C compiler for the Freescale 68HC11, 68HC12, and HCS12 microcontrollers. It can be downloaded from the website at http://www .gnu-m68hc11.org. The EGNU IDE is a freeware integrated development environment for the GNU C (GCC) compiler to support the Freescale 68HC11, 68HC12, and HCS12 micro-controllers running on Windows platforms. The EGNU IDE, developed by Eric Engler, can be downloaded from the website at http://www.geocities.com/englere_geo. Both need to be down-loaded and installed. Assume that GCC is installed at the directory *c:\gnu* whereas the EGNU IDE is installed at the directory *c:\egnu*. To facilitate the use of the EGNU IDE, one should create a shortcut (or icon) for it and place it on the Windows screen. EGNU IDE can only work with demo boards programmed with the D-Bug12 monitor.

E.2 The GNU C Compiler

The GNU C (GCC) compiler supports three versions of the C standard. The corrected ver-sion of the 1990 ISO C standard is fully supported. The support for the latest 1999 C standard is incomplete. GCC also has limited support for the traditional (pre-ISO) C language. However, the use of the pre-ISO C language is not encouraged.

GCC does not provide library support for the HCS12 at the moment. Users will need to deal with the peripheral registers directly to perform the desired function. The EGNU IDE invokes GCC to compile the user programs.

E.3 The Procedure for Using the EGNU IDE

The procedure for using the EGNU IDE to enter, compile, and download a C program into an HCS12 demo board with the D-Bug12 monitor for execution is as follows:

Step 1
Invoke the EGNU IDE by double clicking its icon. The IDE window will appear as shown in Figure E.1. In Figure E.1, the block in gray color is used as the *workspace* for program entering and editing. The *project manager* block displays the name of the current active project and the files contained in the project. The pane at the bottom of the window is the *status window*. By default, it is used to display the output of the project build process (referred to as **Make log** in Figure E.1). The status pane can be used to display compiler output by clicking the keyword **Compiler** above the status pane. The result is shown in Figure E.2.

The status pane can also be used as the *terminal window* that displays the commu-nications between the PC and the monitor on the demo board. The screen would change

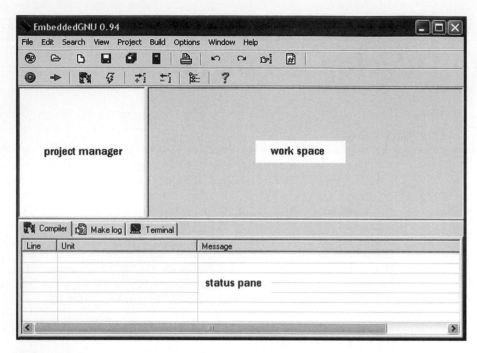

Figure E.1 ■ Embedded GNU IDE startup window

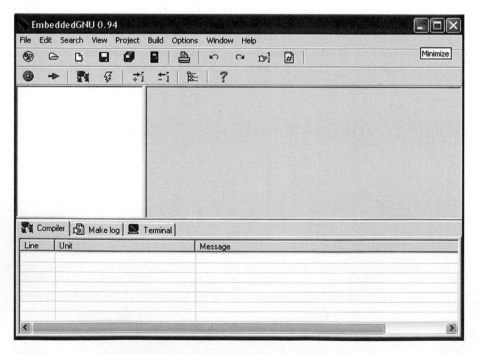

Figure E.2 ■ EGNU window with the status pane used for compiler output

to that in Figure E.3 if the user clicks on the keyword **Terminal** above the status pane and then presses the Enter key once or twice when the demo board is powered on.

Figure E.3 shows the power-on status for a demo board with the D-Bug12 monitor. When the status pane is used as the terminal window, we can enter any D-Bug12 monitor command.

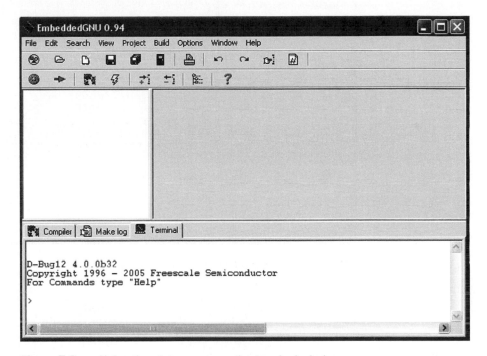

Figure E.3 ■ Using the status pane as the terminal window

Step 2
Click on the **File** menu and select **New Source File** to create a new file for holding a program. The resultant screen is shown in Figure E.4. This step must be repeated as many times as the number of files required in the project. As shown in Figure E.4, one can also press the **Ctrl** + **U** keys to open a new file. To open an existing file, we select **Reopen File ->** instead of **New Source File.**

Step 3
Type in the C program and save it in an appropriate directory (name it *hcs12_timer.c*). An example is shown in Figure E.4. The example program performs the following functions:

1. Defines the *cnt* variable and initializes it to 0.

2. Configures port B for output and outputs the value of *cnt* to it.

3. Calls the **delayby100ms()** function to wait for 0.5 s and then increments the variable *cnt*. The argument to this function specifies the number of 100-ms time delays to be created.

4. Output the value of *cnt* to port B and go to 3.

Port B is driving eight light-emitting diodes (LEDs). When this program is running, you can see the change of the LED pattern.

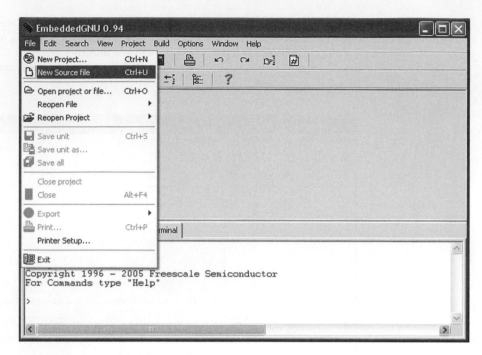

Figure E.4 ■ Creating a new file for entering a program

Step 4

Create a new project by pressing the **File** menu, moving the mouse over **New Project . . .** and clicking on it. The resultant screen is shown in Figure E.5. The user will be asked to enter the project name as shown in Figure E.6. In this tutorial, the name **count** is entered. After the project name is entered, one should also click on OK. As shown in Figure E.7, the screen has changed to ask the user to select the directory to save the project information. One should select an appropriate directory and click on the **save** button on the screen. After this, another popup dialog box appears on the screen (shown in Figure E.8) that will allow the user to select the Hardware Profile and turn on/off compiler switches. The default compiler switches are acceptable. This tutorial selects the Dragon12 demo board as its hardware profile. This tutorial does not use GNU Embedded Library. Therefore, the user should select No for this choice (in the middle of the dialog box). After the project options are set properly, click on OK. The screen is changed to that in Figure E.9.

Step 5

Add files to the newly created project. Press the **Project** menu (shown in Figure E.10) and select **Add to project. . . .** A popup dialog box (shown in Figure E.11) appears to allow the user to enter the name of the file to be added to the project. Since the project and the files in the project are often stored in the same directory, one can simply click on the file name (*hcs12_timer*) and then click on **Open** in Figure E.11 to complete the add file operation. The screen changes to that in Figure E.12 after adding *hcs12_timer.c* to the project. Please remember that EGNU IDE requires all the C files belong to the same project to be stored in the same directory.

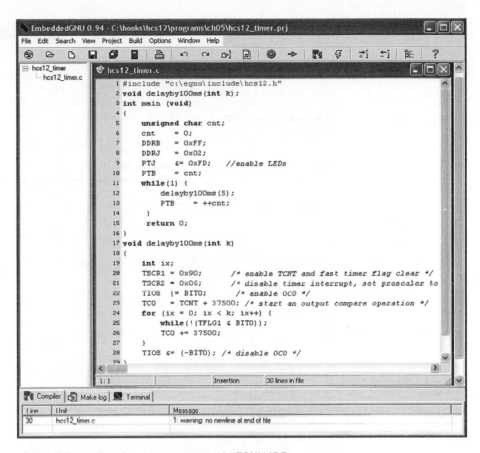

Figure E.5 ■ Entering a new program in EGNU IDE

Figure E.6 ■ Create a new project and name it count

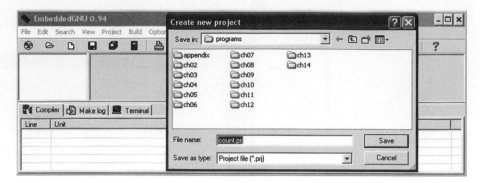

Figure E.7 ■ EGNU IDE screen after entering project name and clicking OK

Project options

Make Options

Hardware Profile

CML12S-DP256
Dragon12
evbplus2
hc11e20
hc11e9
hcs12c32

Make with GEL (GNU Embedded Library)
⦿ No   ◯ Yes

Compiler switches:

Note: Don't include processor choice here. That comes from the hardware profile.

-Os -fno-ident -fno-common -fomit-frame-pointer -mshort -fsigned-char

✓ OK   ✗ Cancel

Figure E.8 ■ Dialog for project options

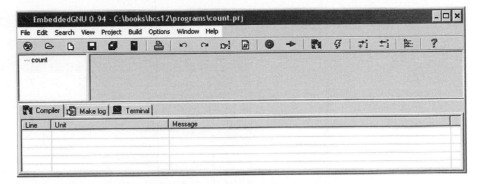

Figure E.9 ■ Screen appearance after the project name has been set

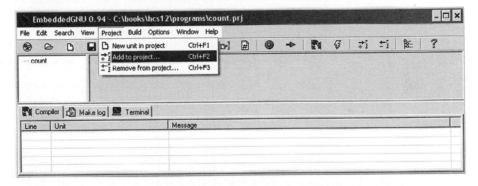

Figure E.10 ■ Screen for adding files to a project

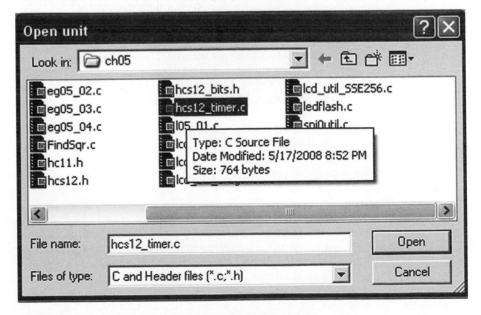

Figure E.11 ■ Dialog box for adding a file to the project

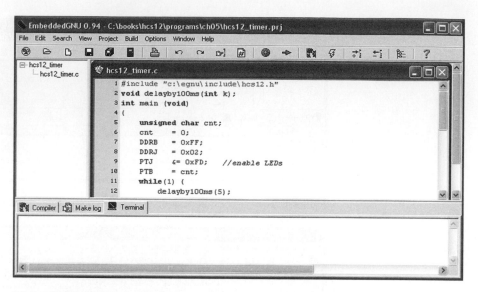

Figure E.12 ■ Screen after adding *hcs_timer.c* to the project

Step 6

Set up options for the project and environment. Pressing the **Options** menu will reveal that there are two options to be set. The processor options have been set and can be ignored. Select **Environment options** and a popup dialog box will appear, as shown in Figure E.13, which shows the **Directories** options. Users can enter the search paths for several types of files in this dialog box. If users have created their own library and included files, they may enter them in this dialog box. The setting in Figure E.13 is acceptable and need not be changed for this tutorial.

After setting the options for **Directories**, make sure that the options for **Preferences** are set properly. The dialog box (shown in Figure E.14) for setting them will appear after the user clicks on the **Preferences** menu. The **Default** directory option is the default place for the EGNU IDE to look for files to be added to the project. Users should change it to match their own setting. The directory **c:\books\hcs12\programs** is where this author stores all the programs in his HCS12 book.

The third option is related to the text editor. Click on **Editor** and the dialog box for setting editor options appears as shown in Figure E.14. Users can set the available options according to their preference. After the options for the editor have been set properly, click on <u>O</u>K.

The fourth option is related to syntax colors. Click on **Syntax colors** and a dialog box appears as shown in Figure E.15. EGNU IDE allows the user to set the color of nine different components in a program, including the comment, identifier, keyword, number, space, string, symbol, white space, and directive. The color of any component is indicated by the word **FG** on the color map. The color for comments is shown in Figure E.16.

The fifth option to be set is related to associations. The default settings are acceptable. The sixth option to be set is related to the COM port. Click on **COM Port** and its

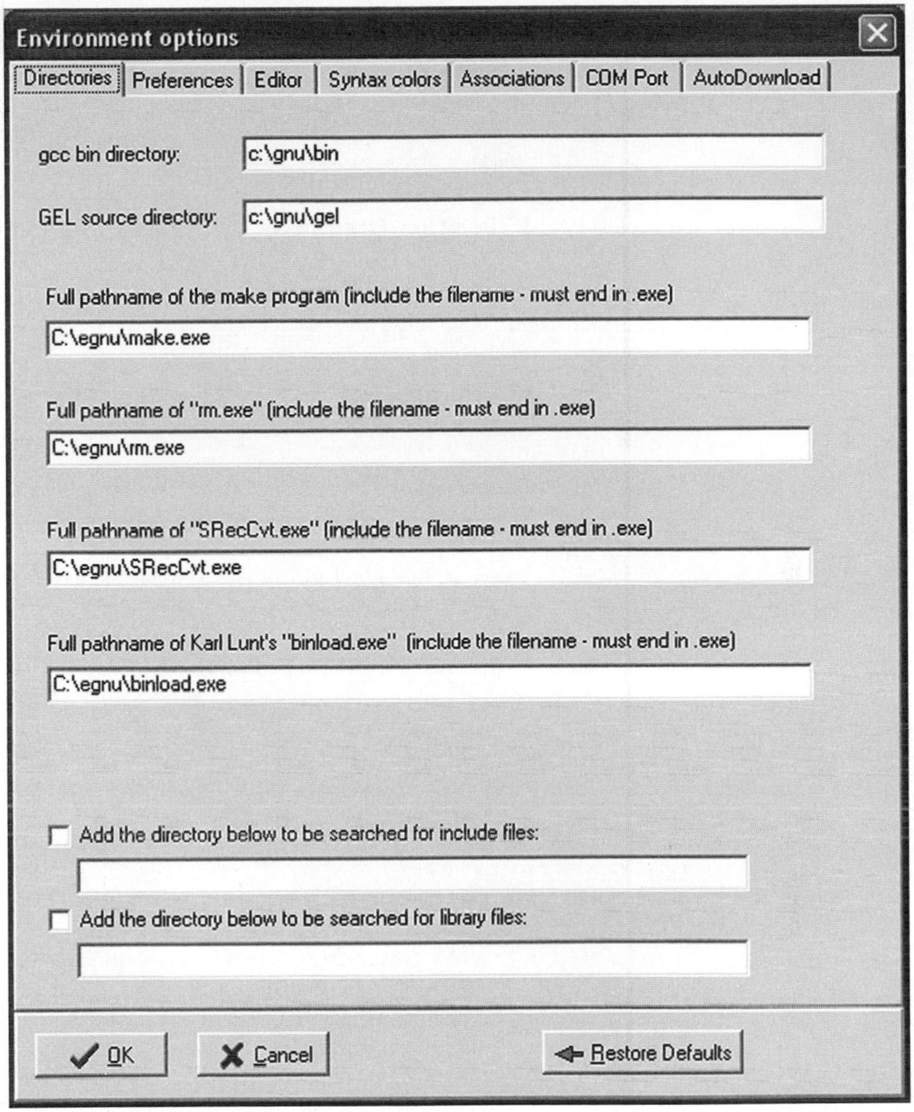

Figure E.13 ■ Dialog for setting environment options

associated dialog box will appear as shown in Figure E.17. The user should select the appropriate COM port (default is COM 1) and 8 data bits, 1 stop bit, and no parity.

Environment options need only be set once. EGNU IDE saves the environment options after they have been set. Therefore, step 6 can be skipped for the next project unless the user wants to change some of them.

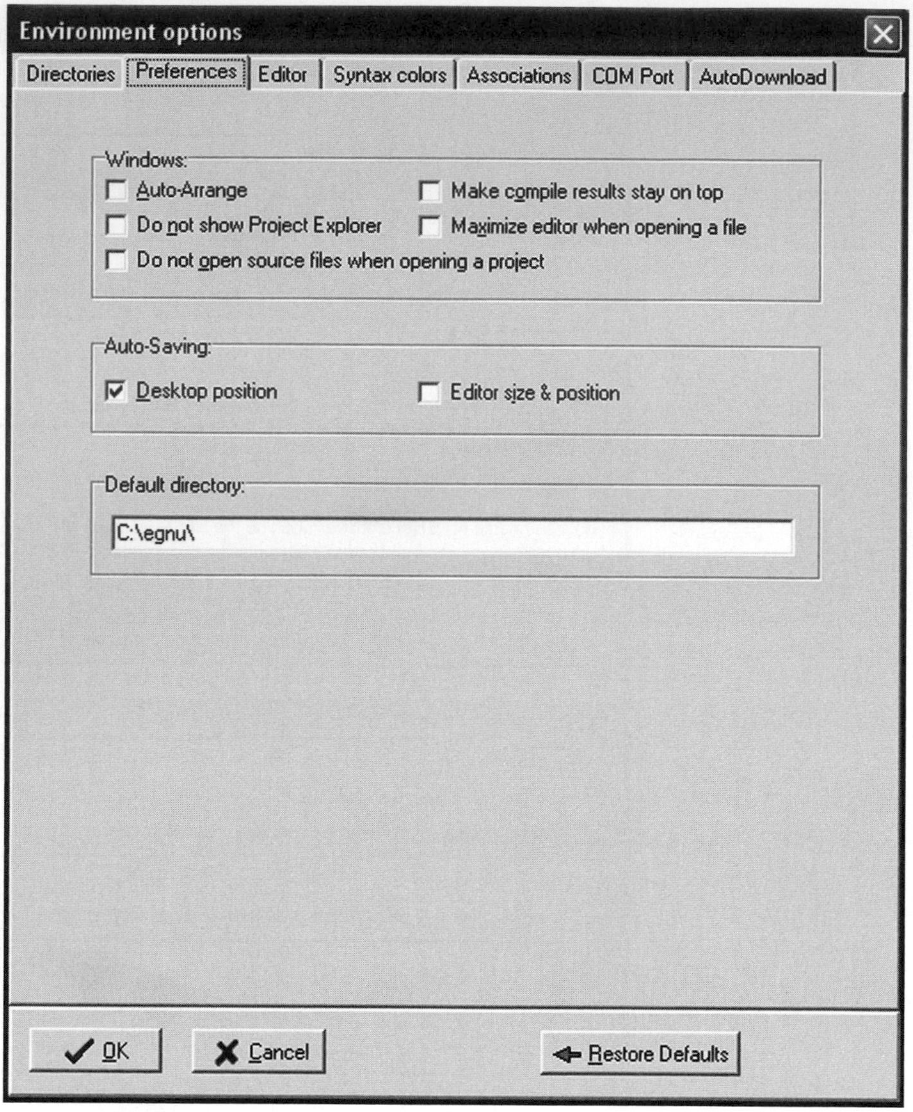

Figure E.14 ■ Dialog for setting preferences

Step 7

Build the project. Press the **Build** menu and select **Make**. The EGNU IDE will display the make result as shown in Figure E.18. This example project has no errors and the resultant machine code has 526 bytes. The machine code is stored in a file with *s19* as its file name extension. Users who are interested in the *s19* file format should refer to the appropriate document from Freescale.

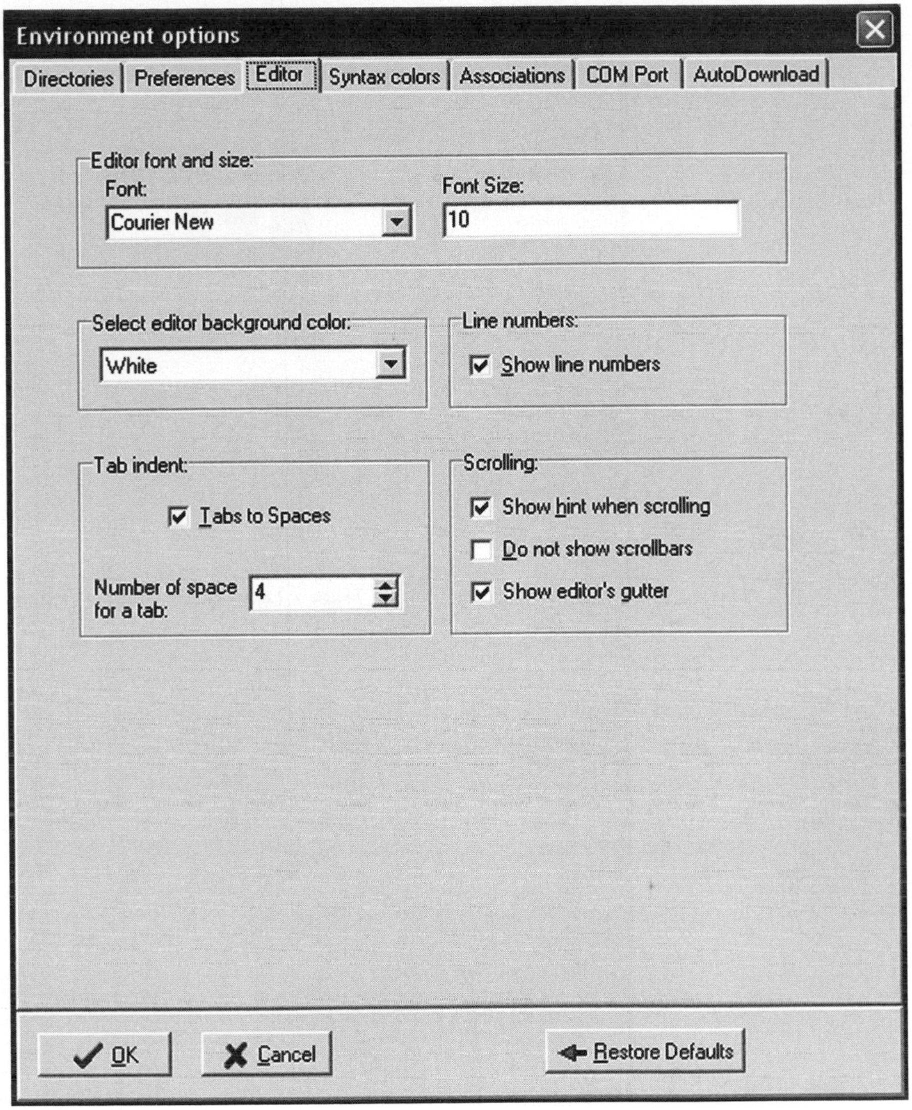

Figure E.15 ■ Setting editor options

Step 8
Download the program onto the demo board for execution. Make the Terminal window active by clicking on **Terminal** above the status pane of the EGNU IDE window. Type **load** and press the **enter** key on the terminal window. Press the **Build** menu and select **Download** (or press the function key **F10**). The *S19* file

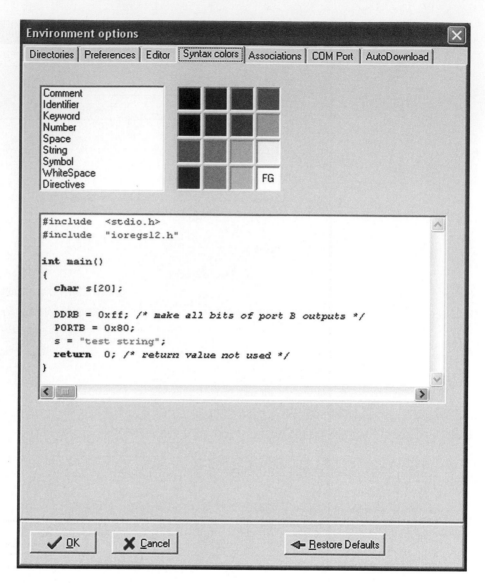

Figure E.16 ■ Dialog for setting syntax colors

of the current project will be downloaded onto the demo board and the D-Bug12 command prompt will reappear on the screen as follows:

```
>load
done
>
```

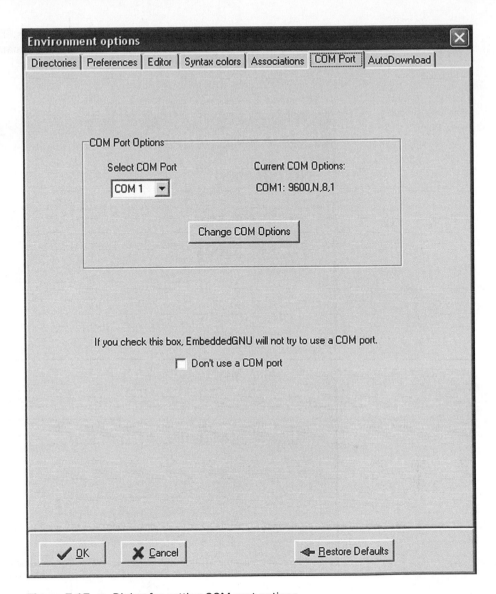

Figure E.17 ■ Dialog for setting COM port options

Step 9
Execute and debug the program. The user can use all the D-Bug12 commands to debug and execute the downloaded program on the demo board as illustrated in Chapter 3. The EGNU IDE sets the program starting address to $2000. To run the program that was just downloaded, enter **g 2000.** After this, when the example program is running,

the user should be able to see the change in the LED patterns. To redisplay the D-Bug12 monitor prompt, press the reset button on the demo board and then hit the enter key of the keyboard.

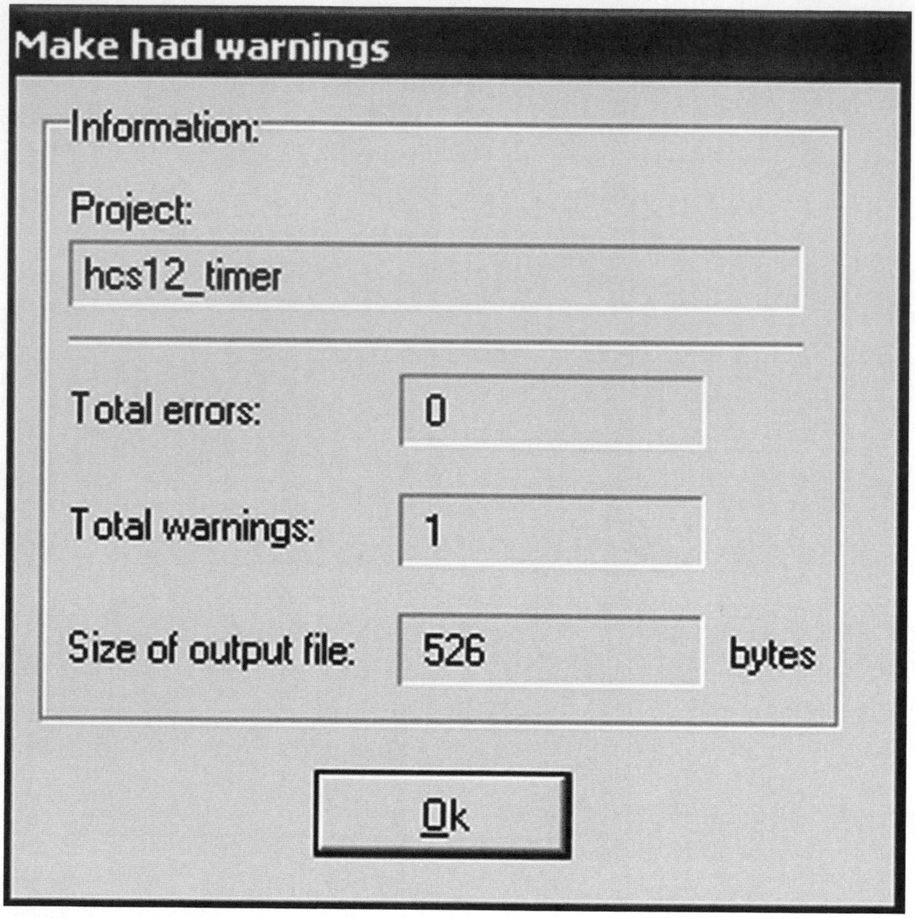

Figure E.18 ■ Project make results

E.4 Library C Functions for the GNU C Compiler

There are functions common to many applications. These functions should be categorized and placed in several files so that they can be reused by many programs. Shown in Table E.1 are three groups of functions written for the GNU C compiler. However, they are also usable by CodeWarrior and ImageCraft ICC.

Prototype	Function
Category: time delay	
void delay by 10 us (int k);	Creates a time delay that is equal to 10 k us
void delay by 50 us (int k);	Creates a time delay that is equal to 50 k us
void delay by 1 ms (int k);	Creates a time delay that is equal to k ms
void delay by 10 ms (int k);	Creates a time delay that is equal to 10 k ms
void delay by 100 ms (int k);	Creates a time delay that is equal to 100 k ms
void delay by 1 s(int k);	Creates a time delay that is equal to k sec
Category: terminal I/O	
int putch(char cx);	Outputs the character cx to the sci0 port
int puts(const char *cptr);	Outputs the string pointed to by cptr to the sci0 port
int newline(void);	Outputs a carriage return followed by a linefeed
int putsr(char *cptr);	Outputs a string stored in SRAM that is pointed to by cptr
char getch (void);	Reads a character from the sci0 port using polling method
int gets(char *ptr);	Reads a string terminated by a carriage return and saves it in a buffer pointed to by ptr
Category: Data conversion	
int int2alpha(int xx, char arr[]);	Converts the 16-bit integer xx into a decimal string and stores it in an array arr[]
int long2alpha(long xx, char arr[]);	Converts the 32-bit integer xx into a decimal string and stores it in an array arr[]
int alpha2int(char *ptr);	Converts the ASCII string pointed to by ptr into a 16-bit binary number
long alpha2long(char *ptr);	Converts the ASCII string pointed to by ptr into a 32-bit binary number
int reverse(char arr[], int k, char sign);	Reverses a string that is stored in the character array with k characters and its sign is indicated by sign

Table E.1 ■ Library functions for EGNU IDE

Example E.1

Write a program to convert a 32-bit integer to a string and then output it to the SCI0 port.

Solution: The program is as follows:

```
#include "c:\egnu\include\hcs12.h"
#include "c:\egnu\include\convert.c"
#include "c:\egnu\include\stdio.c"
const char *msg + "Prepare to convert!\n";
int main(void)
{
    long temp;
    char buf[20];          /* used to hold the converted ASCII string */
    temp = 993450123;      /* number to be converted */
    puts(msg);             /* output a message */
    newline();             /* move to the beginning of the next line */
```

```
        long2alpha(temp,buf);      /* convert the 32-bit number to an ASCII string */
        putsr(&buf[0]);            /* output the converted string */
        newline();
            return 0;
    }
```

Example E.2

Modify the C program in Example 6.5 so that it can be compiled using the GNU C compiler and run in a demo board programmed with the D-Bug12 monitor.

Solution: EGNU does not support flash programming and hence it relies on the D-Bug12 monitor to set the interrupt vector in SRAM. This program uses inline assembly instructions to set up the interrupt vector for the RTI. The user also needs to include the *SetClk.c* file into the project when compiling the program. The following program will run without modification on the Dragon12-Plus demo board:

```
; ********************************************************************************
// This program is written for the EGNU C compiler.
; ********************************************************************************
#include "c:\egnu\include\hcs12.h"
#include "c:\egnu\include\vectors12.h"
#include "c:\egnu\include\SetClk.h"
#define INTERRUPT_attribute_((interrupt))
void INTERRUPT rtiISR(void);
int seq,ix,count;
char segPat[13] = {0x06,0x5B,0x4F,0x66,0x6D,0x7D,0x07,0x7F,0x67,0x3F,0x06,0x5B,0x4F};
char digit[4] = {0xFE,0xFD,0xFB,0xF7};
void main (void) {
        asm("ldd #rtiISR");
        asm("pshd");
        asm("ldd #56");
        asm("ldx 0xEEA4");
        asm("jsr 0,x");
        seq      = 0;
        ix       = 0;
        count    = 400;
        SetClk8();
        RTICTL   = 0x40;              // RTI interrupt interval set to 2**10 OSCCLK cycles
        DDRB     = 0xFF;              // configure Port B for output
        DDRP     = 0xFF;              // configure Port P for output
        CRGINT   |= RTIE;            // enable RTI interrupt
        asm("CLI");                   //enable interrupt globally
        while(1);
}
void INTERRUPT rtiISR(void) {
        CRGFLG = 0x80;               // clear RTIF bit
```

```
PTB = segPat[seq+ix];
PTP = digit[ix];
ix++;
if (ix == 4)
ix = 0;
count--;
if(count == 0){
    seq++;
    count = 400;
}
if(seq == 10)
seq = 0;
}
```

▲

Appendix F: Music Note Frequencies

Note	Frequency (Hz)	Wavelength (cm)	Note	Frequency (Hz)	Wavelength (cm)
C_0	16.35	2100	$G^\#_1/A^b_1$	51.91	665
$C^\#_0/D^b_0$	17.32	1990	A_1	55.00	627
D_0	18.35	1870	$A^\#_1/B^b_1$	58.27	592
$D^\#_0/E^b_0$	19.45	1770	B_1	61.74	559
E_0	20.60	1670	C_2	65.41	527
F_0	21.83	1580	$C^\#_2/D^b_2$	69.30	498
$F^\#_0/G^b_0$	23.12	1490	D_2	73.42	470
G_0	24.50	1400	$D^\#_2/E^b_2$	77.78	444
$G^\#_0/A^b_0$	25.96	1320	E_2	82.41	419
A_0	27.50	1250	F_2	87.31	395
$A^\#_0/B^b_0$	29.14	1180	$F^\#_2/G^b_2$	92.50	373
B_0	30.87	1110	G_2	98.00	352
C_1	32.70	1050	$G^\#_2/A^b_2$	103.83	332
$C^\#_1/D^b_1$	34.65	996	A_2	110.00	314
D_1	36.71	940	$A^\#_2/B^b_2$	116.54	296
$D^\#_1/E^b_1$	38.89	887	B_2	123.47	279
E_1	41.20	837	C_3	130.81	264
F_1	43.65	790	$C^\#_3/D^b_3$	138.59	249
$F^\#_1/G^b_1$	46.25	746	D_3	146.83	235
G_1	49.00	704	$D^\#_3/E^b_3$	155.56	222

Table F.1 ■ Complete list of frequencies of music notes (page 1)

$A_4 = 440$ Hz
Speed of sound = 345 m/s
(Middle C is C_4)

Note	Frequency (Hz)	Wavelength (cm)	Note	Frequency (Hz)	Wavelength (cm)
E_3	164.81	209	C_5	523.25	65.9
F_3	174.61	198	$C^{\#}_5/D^{b}_5$	554.37	62.2
$F^{\#}_3/G^{b}_3$	185.00	186	D_5	587.33	58.7
G_3	196.00	176	$D^{\#}_5/E^{b}_5$	622.25	55.4
$G^{\#}_3/A^{b}_3$	207.65	166	E_5	659.26	52.3
A_3	220.00	157	F_5	698.46	49.4
$A^{\#}_3/B^{b}_3$	233.08	148	$F^{\#}_5/G^{b}_5$	739.99	46.6
B_3	246.94	140	G_5	783.99	44.0
C_4	261.63	132	$G^{\#}_5/A^{b}_5$	830.61	41.5
$C^{\#}_4/D^{b}_4$	277.18	124	A_5	880.00	39.2
D_4	293.66	117	$A^{\#}_5/B^{b}_5$	932.33	37.0
$D^{\#}_4/E^{b}_4$	311.13	111	B_5	987.77	34.9
E_4	329.63	105	C_6	1046.50	33.0
F_4	349.23	98.8	$C^{\#}_6/D^{b}_6$	1108.73	31.1
$F^{\#}_4/G^{b}_4$	369.99	93.2	D_6	1174.66	29.4
$G4$	392.00	88.0	$D^{\#}_6/E^{b}_6$	1244.51	27.7
$G^{\#}_4/A^{b}_4$	415.30	83.1	E_6	1318.51	26.2
A_4	440.00	78.4	F_6	1396.91	24.7
$A^{\#}_4/B^{b}_4$	466.16	74.0	$F^{\#}_6/G^{b}_6$	1479.98	23.3
B_4	493.88	69.9	G_6	1567.98	22.0

Table F.1 ■ Complete list of frequencies of music notes (page 2)

Note	Frequency (Hz)	Wavelength (cm)
$G^\#_6/A^b_6$	1661.22	20.8
A_6	1760.00	19.6
$A^\#_6/B^b_6$	1864.66	18.5
B_6	1975.53	17.5
C_7	2093.00	16.5
$C^\#_7/D^b_7$	2217.46	15.6
D_7	2349.32	14.7
$D^\#_7/E^b_7$	2489.02	13.9
E_7	2637.02	13.1
F_7	2793.83	12.3
$F^\#_7/G^b_7$	2959.96	11.7
G_7	3135.96	11.0
$G^\#_7/A^b_7$	3322.44	10.4
A_7	3520.00	9.8
$A^\#_7/B^b_7$	3729.31	9.3
B_7	3951.07	8.7
C_8	4186.01	8.2
$C^\#_8/D^b_8$	4434.92	7.8
D_8	4698.64	7.3
$D^\#_8/E^b_8$	4978.03	6.9

Table F.1 ■ Complete list of frequencies of music notes (page 3)

Appendix G: Vector Table Template in C for the HCS12 (Used in CodeWarrior)

```c
extern void near irqISR(void);          // irqISR () is in a different file
extern void near rtiISR(void);          // rtiISR() is in a different file
// put the function name of your interrupt handler into this constant array
// at the ISR location that you want.
#pragma CODE_SEG_NEAR_SEG NON_BANKED /* interrupt section for this module. placement will be in NON_
BANKED area. */
_interrupt void UnImplementedISR(void)
{
        for(;;);                        // do nothing, simply return
}
// added redirected ISR vectors when BootLoader is enabled.
// the application cannot have a reset vector (resides in BootLoader).
#pragma CODE_SEG DEFAULT                /* change code section to default. */
typedef void (*near tlsrFunc)(void);
const tlsrFunc _vect[] @0xFF80 = {
/* Interrupt table */
        UnimplementedISR,              // reserved
        UnimplementedISR,              // reserved
        UnimplementedISR,              // reserved
        UnimplementedISR,              // reserved
        UnimplementedISR,              // reserved
        UnimplementedISR,              // reserved
        UnimplementedISR,              // PWM emergency shutdown
        UnimplementedISR,              // port P
        UnimplementedISR,              // reserved
        UnimplementedISR,              // reserved
        UnimplementedISR,              // reserved
        UnimplementedISR,              // reserved
        UnimplementedISR,              // reserved
        UnimplementedISR,              // reserved
        UnimplementedISR,              // reserved
        UnimplementedISR,              // reserved
        UnimplementedISR,              // reserved
        UnimplementedISR,              // reserved
        UnimplementedISR,              // reserved
        UnimplementedISR,              // reserved
        UnimplementedISR,              // CAN1 transmit
        UnimplementedISR,              // CAN1 receive
        UnimplementedISR,              // CAN1 errors
        UnimplementedISR,              // CAN1 wake-up
        UnimplementedISR,              // CAN0 transmit
```

```
        UnimplementedISR,                        // CAN0 receive
        UnimplementedISR,                        // CAN0 errors
        UnimplementedISR,                        // CAN0 wake-up
        UnimplementedISR,                        // flash
        UnimplementedISR,                        // EEPROM
        UnimplementedISR,                        // reserved
        UnimplementedISR,                        // reserved
        UnimplementedISR,                        // IIC bus
        UnimplementedISR,                        // BDLC
        UnimplementedISR,                        // CRG self clock mode
        UnimplementedISR,                        // CRG PLL clock
        UnimplementedISR,                        // pulse accumulator B overflow
        UnimplementedISR,                        // modulus down counter underflow
        UnimplementedISR,                        // portH
        UnimplementedISR,                        // portJ
        UnimplementedISR,                        // ATD1
        UnimplementedISR,                        // ATD0
        UnimplementedISR,                        // SCI1
        UnimplementedISR,                        // SCI0
        UnimplementedISR,                        // SPI0
        UnimplementedISR,                        // pulse accumulator input edge
        UnimplementedISR,                        // pulse accumulator A overflow
        UnimplementedISR,                        // enhanced capture timer overflow
        UnimplementedISR,                        // enhanced capture timer Ch7
        UnimplementedISR,                        // enhanced capture timer Ch6
        UnimplementedISR,                        // enhanced capture timer Ch5
        UnimplementedISR,                        // enhanced capture timer Ch4
        UnimplementedISR,                        // enhanced capture timer Ch3
        UnimplementedISR,                        // enhanced capture timer Ch2
        UnimplementedISR,                        // enhanced capture timer Ch1
        UnimplementedISR,                        // enhanced capture timer Ch0
        rtiISR,                                  // real time interrupt
        IrqISR,                                  // IRQ
        UnimplementedISR,                        // XIRQ
        UnimplementedISR,                        // swi
        UnimplementedISR,                        // unimplemented instruction trap
        UnimplementedISR,                        // COP failure reset
        UnimplementedISR                         // clock monitor fail reset
        /*_startup, by default in library*/      // reset vector
};
```

References

1. Analog Devices. *AD7302 Datasheet*. Norwood, MA: Analog Devices, 1997.
2. Bosch. *CAN Specification, Version 2.0*. Stuggart Germany Bosch 1991.
3. CANopen. *Cabling and Connector Pin Assignment, Version 1.1*. Stuttgart, Germany: Bosch, April 2001.
4. Dallas Semiconductor. *DS1307 Datasheet.*, Dallas, TX: Dallas Semiconductor, 2004.
5. Dallas Semiconductor. *DS1631A Datasheet*. Dallas, TX: Dallas Semiconductor, 2004.
6. Honeywell. *IH-3605 Datasheet*. Morristown, NJ: Honeywell, 1999.
7. Honeywell. *ASCX30AN Datasheet*. Morristown, NJ: Honeywell, 2004.
8. Han-Way Huang. *MC68HC11: An Introduction*. 2d ed. Clifton Park, NY: Delmar Thomson Learning, 2001.
9. Han-Way Huang. *Using the MCS51 Microcontroller*. New York: Oxford University Press, 1998.
10. Han-Way Huang. *MC68HC12: An Introduction*. Clifton Park, NY: Delmar Thomson Learning, 2002.
11. Han-Way Huang. *PIC Microcontroller*. Clifton Park, NY: Delmar Thomson Learning, 2004.
12. Linear Technology. *LTC1661 Datasheet*. Milpitas, CA: Linear Technology, 1999.
13. Maxim Semiconductor. *MAX6952 Datasheet*. Sunnyvale, CA: Maxim Semiconductor, 2004.
14. Microchip. *MCP2551 Data Sheet DS21667D*. Chandler, AZ: Microchip Technology, 2003.
15. Microchip. *24LC08B Data Sheet DS21710A*. Chandler, AZ: Microchip Technology, 2002.
16. Microchip. *24LC16 Data Sheet DS21703D*. Chandler, AZ: Microchip Technology, 2003.
17. Microchip. *MCP23016 Data Sheet DS21710A*. Chandler, AZ: Microchip Technology, 2002.
18. Microchip. *TC1047A Data Sheet DS21498B*. Chandler, AZ: Microchip Technology, 2002.
19. Microchip. *TC72 Data Sheet DS21743A*. Chandler, AZ: Microchip Technology, 2002.
20. Freescale. *MC9S12DP256B Device User Guide, V02.14, 9S12DP256BDGV2/D*. Austin, TX: Freescale, 2003.
21. Freescale. *S12ATD10B8CV2/D Block User Guide*. Austin, TX: Freescale, 2002.
22. Freescale. *S12CPUV2 Reference Manual*. Austin, TX: Freescale, 2003.
23. Freescale. *Background Debug Module V4, S12BDMV3*. Austin, TX: Freescale, 2003.
24. Freescale. *Breakpoint Module Block User Guide, S12BKPV1/D*. Austin, TX: Freescale, 2003.
25. Freescale. *Clock and Reset Generator Block User Guide, S12CRGV2/D*. Austin, TX: Freescale, 2002.
26. Freescale. *MC9S12DP256 Port Integration Module Block User Guide, V02.07, S12DP256PIMV2*. Austin, TX: Freescale, 2002.
27. Freescale. *Enhanced Captured Timer Block User Guide, S12ECT16B8CV1/D*. Austin, TX: Freescale, 2004.
28. Freescale. *EETS4K Block User Guide V02.03, HCS12EETS4KUG/D*. Austin, TX: Freescale, 2001.
29. Freescale. *FTS256K Block User Guide V02.02, HCS12FTS256KUG/D*. Austin, TX: Freescale, 2001.
30. Freescale. *HCS12 Inter-Integrated Circuit Block Guide, V02.08, S12IICV2/D*. Austin, TX. Freescale, 2004.

31. Freescale. *HCS12 Interrupt Module V1, S12INTV1/D*. Austin, TX: Freescale, 2003.

32. Freescale. *Multiplexed External Bus Interface (MEBI) Module V3 Block User Guide, S12MEBIV3/D*. Rev. 3.00. Austin, TX: Freescale, 2003.

33. Freescale. *Module Mapping Control V4, S12MMCV4/D*. Rev. 4.00. Austin, TX: Freescale, 2003.

34. Freescale. *MSCAN Block Guide,V02.15, S12MSCANV2/D*. Austin, TX: Freescale, 2004.

35. Freescale. *PWM_8B8C Block User Guide, V01.17, S12PWM8B8CV1/D*. Austin, TX: Freescale, 2004.

36. Freescale. *HCS12 Serial Communications Interface Block Guide, V02.05, S12SCIV2/D*. Austin, TX: Freescale, 2001.

37. Freescale. *SPI Block User Guide, V02.07, S12SPIV2/D*, Austin, TX: Freescale, 2000.

38. Freescale. *Reference Guide for D-Bug12, Version 4.x.x*. Austin, TX: Freescale, 2004.

39. National Semiconductor. *LM34 Precision Fahrenheit Temperature Sensor Datasheet*. Santa Clara, CA: National Semiconductor, 2000.

40. National Semiconductor. *LM35 Precision Centigrade Temperature Sensor Datasheet*. Santa Clara, CA: National Semiconductor, 2000.

41. Texas Instruments. *TLV5616 Datasheet*. Dallas, TX: Texas Instruments, 2004.

Glossary

Accumulator   A register in a computer that contains an operand to be used in an arithmetic operation.

Acknowledgement error   An error detected whenever the transmitter does not monitor a dominant bit in the ACK slot in a CAN bus protocol.

Activation record   Another term for stack frame.

Address access time   The amount of time it takes for a memory component to send out valid data to the external data pins after address signals have been applied (assuming that all other control signals have been asserted).

Addressing   Applying a unique combination of high and low logic levels to select a corresponding unique memory location.

Address multiplexing   A technique that allows the same address pin to carry different signals at different times; used mainly by DRAM technology. Address multiplexing can dramatically reduce the number of address pins required by DRAM chips and the size of the memory chip package.

Algorithm   A set of procedure steps represented in pseudo-code that is designed to solve certain computation issues.

Arithmetic logic unit (ALU)   The part of the processor in which all arithmetic and logical operations are performed.

Array   An ordered set of elements of the same type. The elements of the array are arranged so that there is a zeroth, first, second, third, and so forth. An array may be one-, two-, or multidimensional.

American Standard Code for Information Interchange (ASCII)   A code that uses 7 bits to encode all printable and control characters.

Assembler   A program that converts a program in assembly language into machine instructions so that it can be executed by a computer.

Assembler directive   A command to the assembler for defining data and symbols, setting assembling conditions, and specifying output format. Assembler directives do not produce machine code.

Assembly instruction   A mnemonic representation of a machine instruction.

Assembly program   A program written in assembly language.

Automatic variable   A variable defined inside a function that comes into existence when the function is entered and disappears when the function returns.

Background receive buffer   The receive buffer where the incoming frame is placed before it is forwarded to the foreground receive buffer.

Barometric pressure   The air pressure existing at any point within the Earth's atmosphere.

BDM mode   A mode provided by the HCS12 and some other Freescale microcontrollers to facilitate the design of debugging tools. This mode has been utilized to implement source-level debuggers.

Binary Coded Decimal (BCD)   Four binary digits representing a decimal digit. The binary codes 0000_2–1001_2 correspond to the decimal digits 0 to 9.

Bit error   An error interpreted in the CAN bus protocol by a node sending a bit on the bus when it monitors the bus and detects a bit value that is different from the bit value being sent.

Branch instruction   An instruction that causes the program flow to change.

Break   The transmission or reception of a low for at least one complete character time.

Breakpoint   A memory location in a program where the user program execution will be stopped and the monitor program will take over the CPU control and display the contents of CPU registers.

Bubble sort   A simple sorting method in which an array or a file to be sorted is gone through sequentially several times. Each iteration consists of comparing each element in the array or file with its successor ($x[i]$ with $x[i + 1]$) and interchanging the two elements if they are not in proper order (either ascending or descending)

Bus   A set of signal lines through which the processor of a computer communicates with memory and I/O devices.

Bus cycle timing diagram   A diagram that describes the transitions of all the involved signals during a read or write operation.

Bus multiplexing   A technique that allows more than one set of signals to share the same group of bus lines.

Bus off   The situation in which a CAN node has a transmit error count above 256.

CAN transceiver   A chip used to interface a CAN controller to the CAN bus.

Central processing unit (CPU)   The combination of the register file, the ALU, and the control unit.

Charge pump   A circuit technique that can raise a low voltage to a level above the power supply. A charge pump is often used in an A/D converter, in EEPROM and EPROM programming, and so on.

Clock monitor reset   A mechanism that detects whether the frequency of the system clock signal inside a CPU is lower than a certain value.

Clock stretching   In the I^2C protocol, the SCL bus signal driven low for an interval longer than the master's SCL low period by the slave device when it needs more time to make data ready.

Column address strobe (CAS)   The signal used by DRAM chips to indicate that column address logic levels are applied to the address input pins.

Comment   A statement that explains the function of a single instruction or directive or a group of instructions or directives. Comments make a program more readable.

Communication program   A program that allows a PC to communicate with another computer.

Computer   An electronic device consisting of hardware, that includes four major parts: the central processing unit, the memory unit, the input unit, and the output unit; and software that is a sequence of instructions that controls the operations of the hardware.

Computer operate properly (COP) watchdog timer   A special timer circuit designed to detect software processing errors. If software is written correctly, then it should complete all operations within a certain amount of time. Software problems can be detected by enabling a watchdog timer so that the software resets the watchdog timer before it times out.

Contact bounce   A phenomenon in which a mechanical key switch will go up and down several times before it settles down when it is pressed.

Control unit   The part of the processor that decodes and monitors the execution of instructions. It arbitrates the use of computer resources and makes sure that all computer operations are performed in proper order.

Controller Area Network (CAN)   A serial communication protocol initially proposed to be used in automotive applications. In this protocol, data is transferred frame by frame. Each frame can carry up to 8 bytes of data. Each data frame is acknowledged in frame.

CRC error (Cyclic redundancy check error) In data communications, the CRC sequence consists of the result of the CRC calculation by the transmitter. The receiver calculates the CRC in the same way as the transmitter. A CRC error is detected if the calculated result is not the same as that received in the CRC sequence.

Cross assembler An assembler that runs on one computer but generates machine instructions that will be executed by another computer that has a different instruction set.

Cross compiler A compiler that runs on one computer but generates machine instructions that will be executed by another computer that has a different instruction set.

D/A converter A circuit that can convert a digital value into an analog voltage.

Data hold time The length of time over which the data must remain stable after the edge of the control signal that latches the data.

Datapath The part of the processor that consists of a register file and the ALU.

Data setup time The amount of time over which the data must become valid before the edge of the control signal that latches the data.

D-Bug12 monitor A monitor program designed for the HCS12 microcontroller.

Data communication equipment (DCE) Equipment such as a modem, concentrator, router, and so on.

Demo board A single-board computer that contains the target microcontroller as the CPU and a monitor to help the user perform embedded product development.

Direct mode An addressing mode that uses an 8-bit value to represent the address of a memory location.

Dominant level A voltage level in a CAN bus that will prevail when a voltage level at this state and a different level (*recessive level*) are applied to the CAN bus at the same time.

Data terminal equipment (DTE) A computer or terminal.

Dynamic memories Memory devices that require periodic refreshing of the stored information, even when power is on.

EIA Electronic Industry Association.

Electrically erasable programmable read-only memory (EEPROM) A type of read-only memory that can be erased and reprogrammed using electric signals. EEPROM allows each individual location inside the chip to be erased and reprogrammed.

Embedded system A product that uses a microcontroller as the controller to provide the features. End users are interested in these features rather than the power of the microcontroller. A cell phone, a charge card, a weather station, and a home security system are examples of embedded systems.

Erasable programmable read-only memory (EPROM) A type of read-only memory that can be erased by subjecting it to strong ultraviolet light. It can be reprogrammed using an EPROM programmer. A quartz window on top of the EPROM chip allows light to be shone directly on the silicon chip inside.

Error active A CAN node that has both the transmit error count and receive error count lower than 127.

Error passive A CAN node that has either the transmit error count or the receive error count between 128 and 256.

Exception Software interrupts such as an illegal opcode, an overflow, division by zero, or an underflow.

Expanded mode An operation mode in which the HCS12 can access external memory components by sending out address signals. A 1-MB memory space is available in this mode.

Extended mode   An addressing mode that uses a 16-bit value to represent the address of a memory location.

Fall time   The amount of time a digital signal takes to go from logic high to logic low.

Floating signal   An undriven signal.

Foreground receive buffer   The CAN receive buffer that is accessible to the programmer.

Form error   An error detected when a fixed-form bit field contains one or more illegal bits in the CAN bus protocol.

Frame pointer   A pointer used to facilitate access to parameters in a stack frame.

Framing error   A data communication error in which a received character is not properly framed by the start and stop bits.

Full-duplex link   A communication link that allows both transmission and reception to proceed simultaneously.

General call address   A special value (all 0s) used by the master to address every device connected to the I²C bus.

Global memory   Memory that is available to all programs in a computer system.

Half duplex link   A communication link that can be used for either transmission or reception, but only in one direction at a time.

Hard synchronization   The synchronization performed by all CAN nodes at the beginning of a frame.

Hardware breakpoint   A hardware circuit that compares address and data values to predetermined data in setup registers. A successful comparison places the CPU in background debug mode or initiates a software interrupt (SWI).

Identifier acceptance filter   A group of registers that can be programmed to be compared with the identifier bits of the incoming frames to make an acceptance decision.

Idle   A continuous logic high on the RxD line for one complete character time.

Illegal opcode   A binary bit pattern of the opcode byte for which an operation is not defined.

Immediate mode   An addressing mode that will be used as the operand of the instruction.

Indexable data structure   A data structure in which each element is associated with an integer that can be used to access it. Arrays and matrices are examples of indexable data structures.

Indexed addressing mode   An addressing mode that uses the sum of the contents of an index register and a value contained in the instruction to specify the address of a memory location. The value to be added to the index register can be a 5-, 9-, or 16-bit signed value. The contents of an accumulator (A, B, or D) can also be used as the value to be added to the index register to compute the address.

Inline assembly instruction   Assembly instructions that are embedded in a high-level language program.

Input capture   The HCS12 function that captures the value of the 16-bit free-running main timer into a latch when the falling or rising edge of the signal connected to the input-capture pin arrives.

Input handshake   A protocol that uses two handshake signals to make sure that the peripheral chip receives data correctly from the input device.

Input port   The part of the microcontroller that consists of input pins, the input data register, and other control circuitry to perform the input function.

Instruction queue   A circuit in the CPU to hold the instruction bytes prefetched by the CPU. The HCS12 and some other microprocessors utilize the bus idle time to perform instruction prefetch in the hope of enhancing the processor throughput.

Instruction tagging   Putting a tag on an instruction so that when detected by the HCS12 CPU, the CPU enters background debug mode, does not execute the tagged instruction, and stops executing the application program.

Integrated development environment   A piece of software that combines a text editor, a terminal program, a cross compiler and/or cross assembler, and/or simulator to allow the user to perform program development activities without quitting any one of the programs.

Interframe space   A field in the CAN bus that is used to separate data frames or remote frames from the previous frames.

Inter-integrated circuit   A serial communication protocol proposed by Philips in which two wires (SDA and SCL) are used to transmit data. One wire carries the clock signal, and the other wire carries the data or address information. This protocol allows multiple master devices to coexist in the same bus system.

Interrupt   An unusual event that requires the CPU to stop normal program execution and perform some service to the event.

Interrupt overhead   The time spent handling an interrupt, consisting of saving and restoring the registers and executing the instructions contained in the service routine.

Interrupt priority   The order in which the CPU will service interrupts when all of them occur at the same time.

Interrupt service   The service provided to a pending interrupt by the CPU's execution of a program called a service routine.

Interrupt vector   The starting address of an interrupt service routine.

Interrupt-vector table   A table that lists all interrupt vectors.

I/O synchronization   A mechanism that can make sure that CPU and I/O devices exchange data correctly.

ISO   International Standard Organization.

Keyboard debouncing   A process that can eliminate the keyswitch bouncing problem so that the computer can detect correctly whether a key has indeed been pressed.

Keyboard scanning   A process that is performed to detect whether any key has been pressed by the user.

Key wake-up   A mechanism, associated with I/O ports, that can generate interrupt requests to wake up a sleeping CPU.

Label field   The field in an assembly program statement that represents a memory location.

Linked list   A data structure that consists of linked nodes. Each node consists of two fields, an information field and a next address field. The information field holds the element on the list, and the next address field contains the address of the next node in the list.

Load cell   A transducer that can convert a weight into a voltage.

Local variable   Temporary variables that exist only when a subroutine is called. They are used as loop indices, working buffers, and so on. Local variables are often allocated in the system stack.

Low-power mode   An operation mode in which less power is consumed. In CMOS technology, the low-power mode is implemented by either slowing down the clock frequency or turning off some circuit modules within a chip.

Machine instruction   A set of binary digits that tells the computer what operation to perform.

Mark   A term used to indicate a binary 1.

Maskable interrupts   Interrupts that can be ignored by the CPU. This type of interrupt can be disabled by setting a mask bit or by clearing an enable bit.

Masked ROM (MROM)   A type of ROM that is programmed when the chip is manufactured.

Matrix   A two-dimensional data structure that is organized into rows and columns. The elements of a matrix are of the same length and are accessed using their row and column numbers (i, j), where i is the row number and j is the column number.

Memory   Storage for software and information.

Memory capacity   The total amount of information that a memory device can store; also called memory density.

Memory organization   A description of the number of bits that can be read from or written into a memory chip during a read or write operation.

Microcontroller   A computer system implemented on a single, very-large-scale integrated circuit. A microcontroller contains everything that is in a microprocessor and may contain memories, an I/O device interface, a timer circuit, an A/D converter, and so on.

Microprocessor   A CPU packaged in a single integrated circuit.

Mode fault   An SPI error that indicates that there may have been a multimaster conflict for system control. Mode fault is detected when the master SPI device has its $\overline{SS}$ pin pulled low.

Modem   A device that can accept digital bits and change them into a form suitable for analog transmission (modulation) and can also receive a modulated signal and transform it back to its original digital representation (demodulation).

Multidrop   A data communication scheme in which more than two stations share the same data link. One station is designated as the master, and the other stations are designated as slaves. Each station has its own unique address, with the primary station controlling all data transfers over the link.

Multiprecision arithmetic   Arithmetic operations (add, subtract, multiply, or divide) performed by a computer that deals with operands longer than the computer's word length.

Multitasking   A computing technique in which CPU time is divided into slots that are usually 10 to 20 ms in length. When multiple programs are resident in the main memory waiting for execution, the operating system assigns a program to be executed to one time slot. At the end of a time slot or when a program is waiting for completion of I/O, the operating system takes over and assigns another program to be executed.

Nibble   A group of 4-bit information.

Nonmaskable interrupts   Interrupts that the CPU cannot ignore.

Nonvolatile memory   Memory that retains stored information even when power to the memory is removed.

Null modem   A circuit connection between two DTEs in which the leads are interconnected in such a way as to fool both DTEs into thinking that they are connected to modems. A null modem is only used for short-distance interconnections.

Object code   The sequence of machine instructions that results from the process of assembling and/or compiling a source program.

Output-compare   An HCS12 timer function that allows the user to make a copy of the value of the 16-bit main timer, add a delay to the copy, and then store the sum in a register. The output-compare function compares the sum with the main timer in each of the following E-clock cycles. When these two values are equal, the circuit can trigger a signal change on an output-compare pin and may also generate an interrupt request to the HCS12.

Output handshake   A protocol that uses two handshake signals to make sure that an output device correctly receives the data driven by the peripheral chip (sent by the CPU).

Output port   The part of the circuit in a microcontroller that consists of output pins, data register, and control circuitry to send data to the output device.

Overflow   A condition that occurs when the result of an arithmetic operation cannot be accommodated by the preset number of bits (say, 8 or 16 bits); it occurs fairly often when numbers are represented by fixed numbers of bits.

Parameter passing   The process and mechanism of sending parameters from a caller to a subroutine, where they are used in computations; parameters can be sent to a subroutine using CPU registers, the stack, or global memory.

Parity error   An error in which an odd number of bits change value; it can be detected by a parity checking circuit.

Phase_seg1 and Phase_seg2   Segments that are used to compensate for edge phase errors. These segments can be lengthened or shortened by synchronization.

Physical layer   The lowest layer in the layered network architecture. This layer deals with how signals are transmitted, the descriptions of bit timing, bit encoding, and synchronization.

Physical time   In the HCS12 timer system, the time represented by the count in the 16-bit main timer counter.

Point-to-point   A data communication scheme in which two stations communicate as peers.

Precedence of operators   The order in which operators are processed.

Program   A set of instructions that the computer hardware can execute.

Program counter (PC)   A register that keeps track of the address of the next instruction to be executed.

Program loops   A group of instructions or statements that are executed by the processor more than once.

PROM (programmable read-only memory)   A type of ROM that allows the end user to program it once and only once using a device called a PROM programmer.

Prop_seg   The segment within a bit time used to compensate for the physical delay times within the CAN network.

Pseudo-code   An expressive method that combines the use of plain English and statements similar to certain programming languages to represent an algorithm.

Pull   The operation that removes the top element from a stack data structure.

Pulse accumulator   A timer function that uses a counter to count the number of events that occur or measure the duration of a single pulse.

Pulse-width modulation   A timer function that allows the user to specify the frequency and duty cycle of the digital waveform to be generated.

Push   The operation that adds a new element to the top of a stack data structure.

Queue   A data structure to which elements can be added at only one end and removed from only the other end. The end to which new elements can be added is called the *tail* of the queue, and the end from which elements can be removed is called the *head* of the queue.

Random-access memory (RAM)   RAM allows read and write access to every location inside the memory chip. Furthermore, read access and write access take the same amount of time for any location within the RAM chip.

Receiver overrun   A data communication error in which a character or a number of characters were received but not read from the buffer before subsequent characters are received.

Refresh An operation performed on dynamic memories in order to retain the stored information during normal operation.

Refresh period The time interval within which each location of a DRAM chip must be refreshed at least once in order to retain its stored information.

Register A storage location in the CPU. It is used to hold data and/or a memory address during the execution of an instruction.

Relative mode An addressing mode that uses an 8- or 16-bit value to specify the branch distance for branch instructions. If the sign of the value is negative, then the branch is a backward branch. Otherwise, the branch is a forward branch.

Remote frame A frame sent out by a CAN node to request another node to send data frames.

Repeated start condition This condition is used in the I²C protocol. A repeated start signal is a start signal generated without first generating a stop signal to terminate the current communication. This condition is used to change the direction of data transfer or change the partner of data communication.

Reset A signal or operation that sets the flip-flops and registers of a chip or microprocessor to some predefined values or states so that the circuit or microprocessor can start from a known state.

Reset handling routine The routine that will be executed when the microcontroller or microprocessor gets out of the reset state.

Reset state The state in which the voltage level of $\overline{\text{RESET}}$ the pin of the HCS12 is low. In this state, a default value is established for most on-chip registers, including the program counter. The operation mode is established when the HCS12 exits the reset state.

Resynchronization All CAN nodes perform resynchronization within a frame whenever a change of bit value from recessive to dominant occurs outside of the expected *sync_seg* segment after the hard synchronization.

Resynchronization jump width The amount of lengthening in *phase_seg1* or shortening in *phase_seg2* in order to achieve resynchronization in every bit within a frame. The resynchronization jump width is programmable to between 1 and 4.

Return address The address of the instruction that immediately follows the subroutine call instruction (either JSR or BSR).

Rise time The amount of time a digital signal takes to go from logic low to logic high.

ROM (read-only memory) A type of memory that is nonvolatile in the sense that when power is removed from ROM and then reapplied, the original data are still there. ROM data can only be read—not written—during normal computer operation.

Row address strobe (RAS) The signal used by DRAM chips to indicate that row address logic levels are applied to the address input pins.

RS232 An interface standard recommended for interfacing between a computer and a modem. This standard was established by EIA in 1960 and has since been revised several times.

Serial Peripheral Interface (SPI) A protocol proposed by Freescale that uses three wires to perform data communication between a master device and a slave device.

Seven-bit addressing In the I²C protocol, the master device may use a 7-bit value to specify the device for communication.

Signal conditioning circuit A circuit added to the output of a transducer to scale and shift the voltage output from the transducer to a range that can take advantage of the whole dynamic range of the A/D converter being used.

Simplex link   A line that is dedicated either for transmission or reception, but not both.

Simulator   A program that allows the user to execute microcontroller programs without having the hardware.

Single-chip mode   The operation mode in which the HCS12 functions without external address and data buses.

Source code   A program written in either an assembly language or a high-level language; also called a source program.

Source-level debugger   A program that allows the user to find problems in user code at the high-level language (such as C) or assembly-language level.

Space   A term used to indicate a binary 0.

Special test mode   The HCS12 operation mode used primarily during Freescale's internal production testing.

Stack   A last-in-first-out data structure whose elements can be accessed only from one end. A stack structure has a top and a bottom. A new item can be added only to the top, and the stack elements can be removed only from the top.

Stack frame   A region in the stack that holds incoming parameters, the subroutine return address, local variables, saved registers, and so on.

Standard timer module   The timer module implemented in some HCS12 members. This module consists of a 16-bit main timer, 8 channels of output-compare/input-capture function, and one 16-bit pulse accumulator.

Start condition   A signal condition that the I^2C protocol requires all data transmission to start with. A start condition is a condition that the SDA signal goes from high to low when the SCL signal is high.

Static memories   Memory devices that do not require periodic refreshing in order to retain the stored information as long as power is applied.

Status register   A register located in the CPU that keeps track of the status of instruction execution by noting the presence of carries, zeros, negatives, overflows, and so on.

Stepper motor   A digital motor that rotates certain degrees clockwise or counterclockwise whenever a certain sequence of values is applied to the motor.

Stop condition   In the I^2C protocol, all data transfer must be terminated by a stop condition. A stop condition is generated by bringing the SDA line from low to high when the SCL signal is high.

String   A sequence of characters terminated by a NULL character or other special character.

Subroutine   A sequence of instructions that can be called from various places in the program and will return to the caller after its execution. When a subroutine is called, the return address will be saved on the stack.

Subroutine call   The process of invoking the subroutine to perform the desired operations. The HCS12 has BSR, JSR, and CALL instructions for making subroutine calls.

Successive approximation method   A method for performing an A/D conversion that works from the most significant bit toward the least significant bit. For every bit, the algorithm guesses the bit to be 1, converts the resultant value into the analog voltage, and then compares it with the input voltage. If the converted voltage is smaller than the input voltage, the guess is right. Otherwise, the guess is wrong and the bit is cleared to 0.

***Switch* statement**   A multiway decision based on the value of a control expression.

Sync_seg   In the CAN format, the sync_seg segment is the segment within a bit time used to synchronize all CAN nodes.

Ten-bit addressing   In the I²C protocol, a master device can use either a 7- or 10-bit value to address a slave device. When a 10-bit value is used, it is called 10-bit addressing.

Temperature sensor   A transducer that can convert temperature into a voltage.

Text editor   A program that allows the end user to enter and edit text and program files.

Thermocouple   A transducer that converts a high temperature into a voltage.

Transducer   A device that can convert a nonelectric quantity into a voltage.

Trap   A software interrupt; an exception.

UART (Universal Asynchronous Receiver and Transmitter)   An interface chip that allows the microprocessor to perform asynchronous serial data communication.

Union   A variable that may hold (at different times) objects of different types and sizes, with the compiler keeping track of size and alignment requirements.

Vector   A vector is a one-dimensional data structure in which each element is associated with an index *i*. The elements of a vector are of the same length.

Volatile memory   Semiconductor memory that loses its stored information when power is removed.

Volatile variable   A variable that has a value that can be changed by something other than user code. A typical example is an input port or a timer register. These variables must be declared as *volatile* so that the compiler makes no assumptions about their values while performing optimizations.

Write collision   The SPI error that occurs when an attempt is made to write to the SPDR register while data transfer is taking place.

Index